WATER TREATMENT MEMBRANE PROCESSES

Other Books of Interest from McGraw-Hill

AWWA · WATER QUALITY AND TREATMENT

Biswas · WATER RESOURCES: ENVIRONMENTAL PLANNING, MANAGEMENT, AND DEVELOPMENT

Brater, King, Lindell, Wei · HANDBOOK OF HYDRAULICS

Dodson · STORM WATER POLLUTION CONTROL: INDUSTRY AND CONSTRUCTION NPDES COMPLIANCE

Hoggan · COMPUTER-ASSISTED FLOODPLAIN HYDROLOGY AND HYDRAULICS

Maidment · HANDBOOK OF HYDROLOGY

Mays · WATER RESOURCES HANDBOOK

Morris, Jiahua · RESERVOIR SEDIMENTATION HANDBOOK

Parmley · HYDRAULICS FIELD MANUAL

Zipparo, Hazen · DAVIS' HANDBOOK OF APPLIED HYDRAULICS

WATER TREATMENT MEMBRANE PROCESSES

American Water Works Association
Research Foundation

Lyonnaise des Eaux

Water Research Commission
of South Africa

Editorial Group

Joël Mallevialle
Peter E. Odendaal
Mark R. Wiesner

McGraw-Hill

New York   San Francisco   Washington, D.C.   Auckland   Bogotá
Caracas   Lisbon   London   Madrid   Mexico City   Milan
Montreal   New Delhi   San Juan   Singapore
Sydney   Tokyo   Toronto

Library of Congress Cataloging-in-Publication Data

Water treatment membrane processes / American Water Works Association
   Research Foundation, Lyonnaise des Eaux, Water Research
   Commission of South Africa.
      p.   cm.
   Includes index.
   ISBN 0-07-001559-7 (hardcover)
    1. Water—Purification—Membrane filtration.   I. AWWA Research
   Foundation. II. Lyonnaise des Eaux (Firm)   III. South Africa,
   Water Research Commission.
   TD442.5.W38   1996
   628.1'64—dc20   96-17666
                                               CIP

McGraw-Hill

A Division of The McGraw-Hill Companies

ISBN 0-07-001559-7

*The sponsoring editor for this book was Larry S. Hager, the editing
supervisor was Virginia Carroll, and the production supervisor was
Pamela A. Pelton. It was set in Times Roman by North Market Street
Graphics.*

Printed and bound by R. R. Donnelley & Sons Company.

This book is printed on acid-free paper.

Disclaimer

This report was produced by the American Water Works Association Research Foundation (AWWARF), Lyonnaise des Eaux (LdE), and Water Research Commission (WRC). AWWARF, LdE, and WRC assume no responsibility for the opinions or statements of fact expressed in the report. **The mention of trade names for commercial products does not represent or imply the approval or endorsement of AWWARF, LdE, or WRC.** This report is presented solely for informational purposes.

CONTENTS

Index follows Chapter 18

CONTRIBUTORS

This book would not have been possible without the dedicated efforts of the authors who have volunteered many long hours of their time to successfully produce it. The AWWA Research Foundation, Lyonnaise des Eaux, and the Water Research Commission of South Africa thank all authors for their significant contributions to this effort. To produce this book, a select group of membrane experts from Europe, South Africa, and North America was assembled to assimilate the vast body of knowledge on the subject into a comprehensive state-of-the-science manual on the use of membrane processes for water treatment applications. Initially, one primary author and co-author(s) were assigned for each chapter. A mix of North American, European, and South African experts were involved with each chapter to maximize the cross-fertilization of their respective membrane experiences. Because of the significant contributions to the work by both authors and co-authors, we will not distinguish between the two. All participants played a critical role in the development and quality of this document.

Christophe Anselme *Lyonnaise des Eaux, Le Pecq, France.* (CHAPS. 10, 15)

Philippe Aptel *Université Paul Sabatier, Toulouse, France* (CHAPS. 2, 4, 8, 13)

Isabelle Baudin *Lyonnaise des Eaux, Le Pecq, France* (CHAP. 15)

Jean-Luc Bersillon *Vandoeuvre Cedex, France* (CHAP. 14)

Christopher A. Buckley *University of Natal, Durban, South Africa* (CHAPS. 2, 3, 5, 11, 18)

Mark M. Clark *University of Illinois, Urbana, Illinois* (CHAP. 15)

Hans-Curt Flemming *University of Duisburg, Duisburg, Germany* (CHAP. 6)

Quentin E. Hurt *University of Natal, Durban, South Africa* (CHAP. 3)

Joseph G. Jacangelo *Montgomery Watson, Herndon, Virginia* (CHAP. 11)

Ed P. Jacobs *University of Stellenbosch, Stellenbosch, South Africa* (CHAPS. 9, 10)

Jean-Michel Laîné *Lyonnaise des Eaux, Le Pecq, France* (CHAP. 16)

Joël Mallevialle *Lyonnaise des Eaux, Aguas Argentina, Buenos Aires, Argentina* (CHAP. 1)

Jacques Manem *Lyonnaise des Eaux, Le Pecq, France* (CHAP. 17)

Peter E. Odendaal *Water Research Commission, Pretoria, South Africa* (CHAP. 1)

Harry F. Ridgway *Orange County Water District, Fountain Valley, California* (CHAP. 6)

Ron Sanderson *University of Stellenbosch, Stellenbosch, South Africa* (CHAP. 17)

Japie J. Schoeman *Environmentek, CSIR, Pretoria, South Africa* (CHAPS. 7, 12)

Michael J. Semmens *University of Minnesota, Minneapolis, Minnesota* (CHAPS. 8, 13)

James S. Taylor *University of Central Florida, Orlando, Florida* (CHAP. 9)

Mark A. Thompson *Malcolm Pirnie, Inc., Newport News, Virginia* (CHAPS. 12, 14)

Susan Wadley *University of Natal, Durban, South Africa* (CHAP. 18)

Mark R. Wiesner *Rice University, Houston, Texas* (CHAPS. 1, 4, 5, 16)

FOREWORD

The AWWA Research Foundation is a nonprofit corporation that is dedicated to the implementation of research efforts to help local utilities respond to regulatory requirements and traditional high-priority concerns of the industry. The research agenda is developed through a process of grassroots consultation with members, utility subscribers, and working professionals. Under the umbrella of a five-year plan, the AWWARF Research Advisory Council prioritizes the suggested projects based upon current and future needs, applicability, and past work; the recommendations are forwarded to the AWWARF board of trustees for final selection.

This publication is a result of one of those sponsored studies, and it is hoped that its findings will be applied in communities throughout the world. The following report serves as a means of not only communicating the results of the water industry's centralized research program but also as a tool to enlist the further support of nonmember utilities and individuals.

Projects are managed closely from their inception to the final report by the foundation's staff and a large cadre of volunteers who willingly contribute their time and expertise. The foundation serves a planning and management function and awards contracts to other institutions, such as water utilities, universities, and engineering firms. The funding for this research effort comes primarily from the subscription program. The program, through which water utilities subscribe to the research program and make an annual payment proportionate to the volume of water they deliver, offers a cost-effective and fair method for funding research in the public interest.

A broad spectrum of water supply issues is addressed by the foundation's research agenda: resources, treatment and operations, distribution and storage, water quality and analysis, toxicology, economics, and management. The ultimate purpose of the coordinated effort is to assist local water suppliers in providing the highest possible quality of water economically and reliably. The true benefits are realized when the results are implemented at the utility level. The foundation's trustees are pleased to offer this publication as a contribution toward that end.

This book is the culmination of a joint cooperative effort between the AWWA Research Foundation, Lyonnaise des Eaux, and the Water Research Commission. These three organizations, with the assistance of expert authors, have collaborated to produce a state-of-the-science manual on the application of membrane processes for drinking water—a precious resource. This book is intended for all individuals interested in membrane technology. Topics covered include principles of membrane processes, membrane performance, practical applications, design, operation, and economics.

GEORGE W. JOHNSTONE
Chairman, Board of Trustees
AWWA Research Foundation

JAMES F. MANWARING, P.E.
Executive Director
AWWA Research Foundation

SYMBOLS

a	activity, mol/L^3
A	area, L^2
b_s	shear coefficient, $1/t$
b_d	decay coefficient, $1/t$
B_t	solute permeation coefficient, L/t
c	concentration, M/L^3 or mol/L^3
c_e	equilibrium concentration, M/L^3
C	heat capacity, L^2/t^2T or E/MT
d	differential operator
d	diameter (with subscript), L
d_e	characteristic diameter, L
d_h	hydraulic diameter, L
d_i	molecular particle diameter of species i, L
d_p	particle diameter, L
d_{pore}	pore diameter, L
d_t	tube diameter, L
D	diffusion coefficient (with subscript), L^2/t
e	base of natural logarithms
e_o	charge of an electron (1.602×10^{-19} coulombs)
E	applied electrical potential, V
E_{cell}	cell potential, V
E	energy, ML^2/t^2 or E
E_a	activation energy, ML^2/t^2mol or E/mol
$\dot{E}$	rate of energy consumption, ML^2/t^3 or E/t
f_μ	friction factor
F	thermodynamic force (with subscript), E/mol, also viscoelectric correction factor
F	Faraday's constant, At/mol (96,490 coulombs/mol)
F_{ij}	thermodynamic force (with subscripts)
f	fraction of open pore space, dimensionless
g	gravitational acceleration (981 cm/s^2)
g_{ij}	Flory-Huggins interaction parameter of species i and j, ML^2/t^2mol or E/mol
G	Gibbs free energy, ML^2/t^2 or E, also ratio of mean electrostatic potential to zeta potential
h	channel thickness, L
H	enthalpy, ML^2/t^2 or E
H_u	Henry's constant (unitless)
H_v	latent heat of vaporization, ML^2/t^2 or E
i_o	exchange current density, A/L^2
i	electric current density, A/L^2
i_{lim}	limiting current density, A/L^2
I	ionic strength, mol/L

I	electric current, A
J_i	flux of species i, M/L^2t or mol/L^2t
J	solvent (water) flux, L/t (electrolyte flux in electrodialysis)
k	mass transfer coefficient (with subscripts), L/t
$\boldsymbol{k}$	Boltzmann constant (1.3805×10^{-23} J/°K)
k	rate constant, L/T; also conductivity, Ω^{-1}/m^2
k_f	forward reaction rate constant $(\text{mol}/L^3)^{1/n}\, t^{-1}$, n = order of reaction
k_i	mass transfer coefficient for species i, L/t
k_r	reverse reaction rate constant $(\text{mol}/L^3)^{1/n}\, t^{-1}$, n = order of reaction
k_s	saturation coefficient, L/t
K_a	apparent Freundlich constant
K_f	Freundlich coefficient
K_{eq}	equilibrium coefficient, various units
K_d	distribution coefficient, dimensionless
K_m	half-saturation coefficient, M/L^3
K_{sp}	solubility product, various units
L	length, L
L_p	hydraulic permeability, hydraulic conductivity or filtration coefficient, L^2t/M or L/tp, L^3t/M, or L^2
L_u	phenomenological conductance coefficient, various units
m	molal concentration, mol/M
$\overline{M}$	molecular weight, M/mol
n	number of moles
n_e	number of electrons
n_{pr}	number of membrane pairs
n	Freundlich exponent
$\overline{N}$	Avogadro constant ($6.022 \times 10^{23}\text{mol}^{-1}$)
Δp	pressure drop or loss (transmembrane pressure), M/Lt^2 or p
Δp_{mod}	pressure drop across membrane module or pressure vessel, M/Lt^2 or p
p	pressure or partial pressure (with subscript), M/Lt^2 or p
P	power consumption
P	permeability coefficient, various units
Pe	Peclet number vd/D, dimensionless
q_e	mass of adsorbate per mass of adsorbent, M/M
q	charge, At
q	heat flux, M/t^3 or E/L^2t
Q	heat, ML^2/t^2 or E
$\dot{Q}$	heat consumption rate, ML^2/t^2, or E/t
Q	volumetric flow rate, L^3/t
r	recovery (dimensionless)
r	radial coordinate, radial position, or radius (with subscript), L
r_p	particle radius, L
r_{pore}	pore radius, L
r_t	tube radius, L
r_u	phenomenological resistance coefficient, various units
R	rejection (observed or global), dimensionless
$\boldsymbol{R}$	gas constant, $ML^2/t^2\text{mol}$
R	resistance coefficient (with subscript), various units
R_c	cake resistance, $1/L$
R_e	electrical resistance, V/A
R_m	membrane resistance, $1/L$
Re	Reynolds number, vd/ν, dimensionless

S	entropy, ML^2/t^2T or E/T
S	permselectivity
S_i	solubility coefficient of gaseous species i, various units
Sc	Schmidt number, v/D, dimensionless
Sh	Sherwood number, kL/D, dimensionless
t_b	time to backflush, t
$\bar{t}$	time to backflush, t
t	mean detention time
t	time, t
t	transference number (with subscript), dimensionless
$\bar{t}$	transference number in the membrane
$t_{1/2}$	half-life, t
T	temperature, T
T_b	normal boiling point, T
T_c	critical temperature, T
T_g	glass transition temperature, T
u_i	absolute mobility of species, i, t mol/M or L^2mol/Et
u	velocity (usually axial), L/t
U	absolute mobility
v	velocity (usually lateral or radial direction), L/t
V	volume, L^3
$\bar{V_i}$	partial molar volume of species i, L^3/mol
w	channel width, L
w_i	mass fraction of species i, dimensionless
W	weight, also used as a generic constant
x, y, z	rectilinear coordinates, L
x_i	mole fraction of species i, dimensionless
X	fixed-charge density of membrane
Y	yield, dimensionless
z_i	valency of species i (with sign)
α	transfer coefficient, dimensionless
α_{ij}	separation factor
β	electroosmotic coefficient
β_i	enrichment factor, y'', y', $y = c, p, w$, etc., dimensionless
γ	surface tension, M/t^2 or E/L^2
γ_i	activity coefficient of species i, a_1/c_1, dimensionless
γ	mean activity coefficient, $\sqrt{\gamma \cdot \gamma}$ dimensionless
Γ_y	selectivity coefficient K_1K_1/D_1K_1 or D_1s_1/D_1s_1, dimensionless
δ_{cp}	thickness of concentration-polarization layer, L
δ	solubility parameter (with subscript), $(E/L^3)^{1/2}$
δ_f	biofilm thickness
δ_m	membrane thickness, L
δ_c	cake thickness
∂	partial differential operator
Δ	difference operator
ϵ	porosity, dimensionless
ϵ_o	permittivity of a vacuum (8.854×10^{-12} c^{2}m^{-1}J^{-1})
ϵ_w	dielectric constant of water ($= 80$)
η	efficiency (with subscript), dimensionless
$\eta(\phi)$	relative viscosity, dimensionless
η_c	current efficiency
κ	reciprocal Debye length, L^{-1}

θ	tortuosity factor, dimensionless
λ	nondimensional particle size
μ	absolute viscosity $(ML^{-1}t^{-1})$
μ_i	chemical potential, also growth rate
μ_{max}	maximum growth rate
ν	kinematic viscosity, L^2/t, also vibrational frequency, t^{-1}
ν_i	number of ions per molecule of electrolyte i
x	current utilization, dimensionless
Π	osmotic pressure, M/Lt^2 or p
ρ	density or mass concentration (with subscript), M/L^3
σ	reflection coefficient, dimensionless
$\sum$	summation operator
τ	shear stress, also diffusion time scale
ϕ	volume fraction, dimensionless, also free energy dissipation, E
ψ	electric potential, V
ω	solute permeability coefficient, $t$mol$/LM$, $t$mol$/M$
ζ	zeta potential

Diacritical Marks

~	per mole, also dimensionless
^	per unit mass, also per length
-	average value, also may indicate value in membrane
.	time rate of change

Superscripts

'	value in feed stream or on high-pressure side of membrane: value in phase external to the membrane
"	value in extract, permeate, product, or on low-pressure side of membrane
°	standard reference state
*	equilibrium value

Subscripts

A	anode
ave	average
b	brine
A, B, etc.	particular components
bulk	interior of stream outside of boundary layer
C	cathode
d	diluate
ef	effective
f	feed
g	gas phase
i	general species index or solute species i
int	solution interface
in	inlet
j	species j
l	liquid
m	membrane or confined to membrane
obs	observed or apparent
out	outlet
p	product, permeate, or particle
pac	powdered activated carbon

pore	pore
r	retentate or reject
red	reduced
s	solution
t	tube
w	water or solvent
v	vapor, also volume
o	initial value
ol	overall liquid phase
og	overall gas phase
ox	oxidized

CHAPTER 1
THE EMERGENCE OF MEMBRANES IN WATER AND WASTEWATER TREATMENT

Joël Mallevialle
Lyonnaise des Eaux
Auguas Argentinas
Buenos Aires, Argentina

Peter E. Odendaal
Water Research Commission
Pretoria, South Africa

Mark R. Wiesner
Department of Environmental Science and Engineering
Rice University
Houston, Texas

A *membrane* or, more properly, a *semipermeable membrane,* is a thin layer of material that is capable of separating materials as a function of their physical and chemical properties when a driving force is applied across the membrane. Membranes may be classified by the range of materials separated and the driving forces employed. For example, *microfiltration* (MF) and *reverse osmosis* (RO) are two membrane processes that use pressure to transport water across the membrane. MF membranes are capable of removing only particulate matter, while RO membranes retain many solutes as water permeates through the membrane. Electrodialysis is also capable of separating ionic solutes from water, but, in this case, ions are transported across the membrane and the driving force is an electrical potential.

A detailed discussion of these and other membrane technologies is provided in subsequent chapters of this book with the goal of introducing principles and applications of membranes in environmental engineering. In this chapter, we explore factors responsible for the emergence of membrane technologies in water and wastewater treatment.

Since the development of synthetic asymmetric membranes in 1960, interest in membrane processes for water and wastewater treatment has grown steadily, and these technologies are now the object of substantial international research, development, commercial activity, and full-scale application. This relatively recent global increase in the use of membranes in environmental engineering applications can be attributed to at least three categories of factors: (1) increased regulatory pressure to provide better treatment for both potable and waste waters; (2) increased demand for water requiring exploitation of water resources of lower quality than those relied upon previously; and (3) market forces surrounding the development and commercialization of the membrane technologies as well as the water and wastewater industries themselves. An appreciation of the reasons membrane technologies have emerged in environmental engineering practice is of more than historical interest—these driving forces for change give some indication of where the current application of membrane technologies may be most appropriate and where future applications may be anticipated.

1.1 REGULATORY PRESSURE

The potential range of applications for membrane processes in water and wastewater treatment is wide. In the field of drinking water treatment, new regulations on filtration, disinfection, and disinfection by-products have generated considerable interest in the use of membrane processes for particle removal, for the removal of organic materials that may be precursors to disinfection by-products,[1, 2] and for membrane disinfection.[3–10] Similarly, more stringent requirements in wastewater treatment for nutrient removal and dechlorination have created interest in membrane disinfection and solid-liquid separation.[11, 12]

A key regulatory development in the United States was the passage of the 1986 Amendments to the Safe Drinking Water Act (SDWA). The SDWA Amendments have a particularly strong impact on requirements for treatment to reduce turbidity and chemically disinfect water. The Surface Water Treatment Rule (SWTR), developed in response to the SDWA Amendments, established treatment requirements to ensure the removal or inactivation of specific protozoa, viruses, and bacteria. These requirements might be met by increasing the dosage of chemical disinfectants such as chlorine. However, higher doses of chlorine may result in higher concentrations of oxidation by-products such as chloroform and other trihalomethanes (THMs). Moreover, the SDWA Amendments require that a new set of maximum contaminant levels (MCLs) be established for THMs and a host of other currently unregulated oxidation by-products. The SDWA Amendments also mandated new drinking water regulations on synthetic organic chemicals, lead, and copper in drinking water. Potable water treatment has traditionally focused on processes for liquid-solid separation rather than on processes for removing dissolved contaminants from water. Thus, the effect of the SDWA Amendments has been to force water treatment professionals to consider the nonconventional treatment processes such as membrane technologies that alone, or in conjunction with liquid-solid separation, will be capable of meeting the anticipated standards.

In Europe, similar developments surrounding the need to improve disinfection while avoiding the formation of oxidation by-products have stimulated interest in membrane processes. Regulations on synthetic organic chemicals, especially pesticides, have also been important factors. For instance, in the European Union the existing standard for the herbicide atrazine is 0.1 µg/L (it is 3 µg/L in the United States). The achievement of such standards will, in many instances, require new treatment technologies, and membranes are likely to play an important role.

1.2   WATER SCARCITY

Water scarcity in this context refers to limited water supplies for meeting prevailing or projected water demand at a specific location. Water supplies may be limited in quantity or quality. The importance of qualitative limitations is illustrated by the fact that, while approximately 97 percent of the earth's water is contained in the oceans, the high salt content of 35,000 mg/L makes this vast resource virtually useless for beneficial application without treatment.

An additional 2 percent of the earth's water is present as ice at the polar caps and as glaciers: 0.3 percent is present in the atmosphere and only 0.1 percent in rivers and lakes. Groundwater aquifers account for the remaining 0.6 percent of the earth's water. About half of the groundwater occurs at depths greater than 800 m. There are approximately 5×10^{15} m^3 of fresh waters in rivers, lakes, and shallow aquifers that the world's 6000 million people typically depend on for their needs. These waters, if not seriously polluted, can be treated by conventional means to yield safe potable water. If today's world population consumed water in quantities comparable to the industrialized nations, there would be only 100 times more readily available fresh water than daily human needs dictate to serve as storage in the hydrologic cycle and take care of environmental needs and future population growth. Thus, it is clear from a global perspective that, as the world's population grows and industrialization continues, there will indeed be a scarcity of fresh water resources. However, this analysis assumes that all of the world's water resources are available to everyone.

In fact, fresh water and rainfall are unevenly distributed over the land masses. As a result, many areas of the world today are subject to serious and recurring droughts. In many arid regions, groundwater resources contain high salt concentrations as a result of natural processes. In addition, humankind—through lack of planning and irresponsible practices—has seriously polluted, and continues to pollute, available fresh water supplies, thereby creating additional water quality–based scarcities.

Membrane processes can play a key role in reducing water scarcity. They may be used to treat wastewaters before discharge to surface water, to recover materials used in industry before they enter waste streams, and, of course, to treat waters for potable use. In this last regard, membranes may enable us to utilize water resources such as the oceans that were previously inaccessible due to technical or economic considerations. These capabilities of membranes have been significant in driving their use in water and wastewater treatment, particularly in areas with scarce water supplies.

1.2.1   Desalination of Seawater and Brackish Water

Desalinated seawater and saline groundwater has become a major source of water in arid regions of the Middle East, which boast about two-thirds of the world's desalting plant capacity. Distillation technology dominated the desalination scene until about 1970. Since then, improvements in RO and *electrodialysis* (ED) technologies have resulted in substantial increases in their application. In 1988, there were 1742 RO plants—that is 49.4 percent of the total 3527 desalination plants in the world. ED now accounts for 564 plants or 16 percent of the total. On the basis of installed capacity, RO and ED account for approximately 23 percent and 5 percent, respectively, of the world's desalination capacity. By far the largest RO plants have been installed in Bahrain: a 45,420 m^3/d plant in Ras Abu-Jarjur, which desalinates highly brackish water, and a 56,000 m^3/d plant at Al Dur. *Ultrafiltration* (UF) and *microfiltration* (MF) have also been demonstrated as effective pretreatment technologies for desalination. These figures clearly reflect the increased competitiveness of membrane technology. The reason for this trend is that the real cost of membrane technology (RO in particular) has steadily decreased over the last two decades.

In arid regions which are remote from water supply services, communities rely primarily on groundwater supplies. Unfortunately, groundwaters in arid regions are often highly mineralized and unsuitable for sustained human consumption. High total dissolved solids (TDS) concentrations can impart unacceptable tastes to the water, affect the digestive system of sensitive consumers, and have negative economic impacts due to corrosion or scaling. The World Health Organization's recommended upper limit for the concentration of TDS in drinking water is 1000 mg/L.

Water supply problems in such areas can be alleviated by desalination of the available brackish water, not only for human consumption but also for stock watering. The desalination facility should preferably be robust and easy to operate and maintain. Solar distillation has not found wide acceptance due to relatively high capital costs, large areas required to produce relatively small volumes of water, and material failure as a result of prolonged exposure to the sun. RO and ED have proven to be viable options. Although the cost of desalination of brackish water by RO or ED is still relatively high compared with the cost of treating fresh water by conventional means, it is certainly an economically feasible alternative to transporting water over long distances.

An interesting combination of regulatory push and the pull of water quality–based scarcity has propelled the use of membranes for potable water treatment in Florida. The shallow groundwaters used by many municipalities throughout Florida as a source of potable water contain high concentrations of hardness (calcium and magnesium) as well as high concentrations of *natural organic matter* (NOM), which may react with chlorine to form unacceptably high levels of THMs. These factors combined to make Florida a leading state in the United States for the application of nanofiltration or "softening" membranes. These membranes are capable of removing large percentages of NOM and divalent ions such as calcium at pressures substantially lower than those required for RO.

1.2.2 Wastewater Reuse and Recycling

In areas of water scarcity, the upgrading of treated municipal wastewaters for indirect potable and direct industrial reuse, as well as internal industrial recycling, have become attractive means of extending existing water supplies. In many of these applications, opportunities exist for the incorporation of membrane technologies. In one full-scale application, RO is used at the Water Factory 21 in Orange County, California, as part of a system that indirectly reclaims treated municipal effluent via groundwater infiltration.

Reclamation for direct potable reuse is practiced in Windhoek, the capital of Namibia. Technology for producing high-quality potable water from secondary municipal wastewaters has also been demonstrated in various pilot- and demonstration-scale facilities. The most notable of these was conducted at the Denver, Colorado, demonstration water reclamation facility, which included RO in the treatment train. In this instance, RO is used to reduce concentrations of both TDS and organic pollutants. In both the Denver and Orange County facilities, RO is used only as a polishing step for highly treated wastewater. However, RO has been demonstrated to be an effective treatment process for reclamation when pretreatment is designed only to provide suitable water for the RO unit rather than accommodating reclamation needs.

Internal industrial recycling is not only effective in extending water supplies but may also limit the discharge of pollutants and enable the recovery of useful materials. Such applications of membranes for resource recovery and pollution prevention are receiving greater attention as industrialized nations shift their regulatory efforts from end-of-pipe treatment to source reduction.

1.3   MARKET FORCES

A complex and sometimes more controversial force driving the introduction of membranes in environmental engineering applications is the change in the market, the size and features of which are intimately related to changes in the available technologies. Municipal water treatment is not generally regarded as an open market. The vast majority of water treatment plants are held by municipalities that do not act according to common market incentives or business rules.

The evaluation of cost-effectiveness for a project is usually poorly made, the amortization period is abnormally long, and the willingness to take any risk in the design is limited. The restricted competition between both plant designers and manufacturers presents little incentive for changes. However, as the industry is becoming privatized, this situation seems to be changing. Municipalities are contracting private companies, not only to operate and maintain plants but also to design and construct them. This new situation leads to more innovative designs. Another factor is the opening up of world markets due to a tremendous growth of the internal market of Japan, particularly in industrial and wastewater treatment. The 1984 sales of water treatment equipment in Japan were twice those in the United States. This growth is coupled with ambitious research programs that increase the likelihood that Japan, as it has done in so many other areas of technological competition, could seize a major share of the world treatment market. One can speculate, however, that this will stimulate competition, forcing other firms around the world to become more innovative in developing new treatment technologies.

Besides the water market, it is important to consider the membrane market itself. In recent years, membranes and membrane processes have become industrial products of substantial technical and commercial importance. The worldwide sales of synthetic membranes in 1990 were in excess of US $2000 million. Taking into consideration that in most industrial applications membranes account for about 40 percent (sometimes only 20 percent) of the total investment costs for a complete membrane plant, the total annual sales for the membrane-based industry is close to US $5000 million.[13, 14]

Membranes and membrane processes have found a very broad range of applications, but, in spite of impressive sales and a growth rate of the industry of about 12 to 15 percent each year, the use of membranes in industrial-scale separation processes is not without technical and economic problems. Technical problems are related to insufficient membrane selectivities, relatively poor transmembrane fluxes (fouling problems), general process operating problems, and lack of application know-how. Economic problems originate from the multitude of different membrane products and processes with very different price structures in a wide range of applications which are distributed by a great number of sales companies, very often as individual products. This has led to relatively large production volumes for some products such as hemodialyzers, disposable items used only once for a few hours and sold in relatively large and uniform market segments. Other membranes used in special applications in the food or pharmaceutical industry are expected to last for several years in operation and can be sold only in relatively small quantities to small market segments. Consequently, production volumes are low and prices are high.[14]

The rapid development of new membrane products and processes makes it difficult to predict the growth rate of the market with reasonable accuracy. In addition, the comparison between different figures, such as total sales, as published in the literature, is often very difficult to make, due to the many ways these figures are calculated (membranes, systems, financial costs, bids, etc.).

The distribution of the world membrane market according to the different technologies was published in 1992 by H. Strathman[13] (Fig. 1.1). These data must be compared with the ones published in *Membrane & Separation Technology News* in February 1993 (Fig. 1.2). The main differences can be explained by the fact that one takes into account entire membrane systems and the other accounts for only the membranes and modules. A common "accounting" mistake is to assume that total sales for membrane companies such as Pall, Millipore, and Gelman, are all for membranes. One of the largest markets for membranes today is that for dialysers used in artificial kidneys (US \$900 million). The water treatment market is estimated to be in excess of US \$500 million, half of which corresponds to the use of microfiltration.[13] As far as the municipal drinking water market is concerned, at the beginning of 1994, 31,000 m^3/day of water were produced by microfiltration; 63,000 m^3/day by ultrafiltration; 500,000 m^3/day by nanofiltration, mainly in Florida; and 3,000,000 m^3/day by reverse osmosis (desalination). At the end of 1996, approximately 100,000 m^3/day and 130,000 m^3/day will be produced by MF and UF, respectively.

The overall membrane market is rather unevenly distributed with about 75 percent of the market located in the United States, Europe, and Japan (Fig. 1.3a). Worldwide, more than 100 companies are involved one way or another in membrane technology,[15] but only about 60 companies are both membranes and modules manufacturers. The other companies are involved mainly in process design and plant engineering using membranes as components. The regional sales distribution is shown in Fig. 1.3b.

1.3.1 Case History

As one of the world's leading groups in water distribution and wastewater treatment services, the Lyonnaise des Eaux provides drinking water to more than 45 million

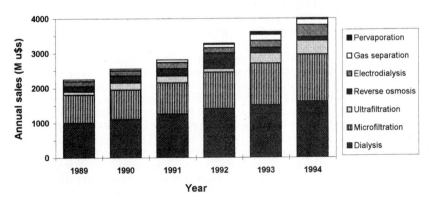

FIGURE 1.1 Annual sales of membranes according to the different processes (1991–1994 correspond to expected sales). *(After H. Strathman, 1992.)*

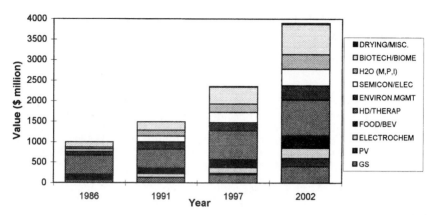

FIGURE 1.2   World membranes and modules by application. *(After* Membrane & Separation Technology News, *1993.)*

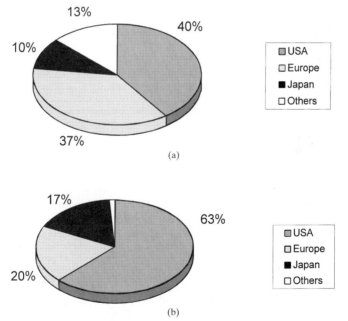

FIGURE 1.3   (*a*) Regional distribution of the membrane market; (*b*) regional sales distribution of the membrane industry. *(After H. Strathman, 1992.)*

clients the world over, including 14 million in France. Its subsidiary Degrémont is a leading water treatment engineering company in the world. From 1984 to 1985, taking into account the different forces for potential changes in the drinking water field, the R&D managers at the Lyonnaise des Eaux launched an effort to develop and apply membrane technologies for water and wastewater treatment.

To comply with regulations that are increasingly stringent, the treatment trains were increasingly complex. A typical treatment train to produce drinking water from the river Seine in the Paris area included preozonation-coagulation-flocculation-settling-sand filtration-ozonation-granular activated carbon and post-disinfection (Fig. 1.4). In terms of innovation, the returns on R&D expenditures on conventional technologies were flattening out from 1984 to 1986. To reverse this tendency and move the curve further up the graph, a breakthrough in technology was required (Fig. 1.5).

By looking at the new technologies developed in food, pharmaceutical, and nuclear industries, membrane separation seemed to be a promising technique. A few Japanese companies were already using membranes for small water recycling systems in tall buildings, and reverse osmosis was significantly developed for water desalination in the Middle East. It was decided that one goal of the membrane efforts would be to replace conventional clarification and disinfection processes with ultrafiltration. The differential advantages of such a technology were better and more reliable water quality, no use of chemical reagents, automation, and compactness. However, the use of membranes then was restricted to high-value products processing such as drugs or food. The market price of the membranes was therefore not compatible with the considerably lower price of the potable water. The Lyonnaise des Eaux then decided to develop its own UF membranes suitable for water treatment in an ambitious research program, needing partnerships from membrane manufacturers, as well as a large amount of work on the membrane process itself, in order to make it applicable to water treatment. The cooperation framework offered by the European Economic Community (EEC) through the BRITE program was chosen to undertake the basic research effort. On the other hand, the EUREKA framework was ideal to undertake the applied research program, as well as the development effort needed to bring the scientific findings of the basic R&D to the industrial application level.

As of today in 1996, 30 filtration plants are installed and supply water to consumers. The plants' capacities range from 50 up to 5500 m^3/day. A new company (AQUASOURCE), affiliate of Lyonnaise des Eaux and Degrémont, was created in 1991 in order to manufacture and market the products of this research.

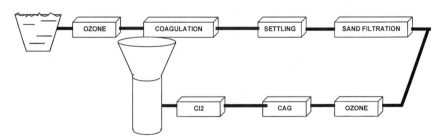

FIGURE 1.4   Typical treatment train to produce drinking water.

Innovation

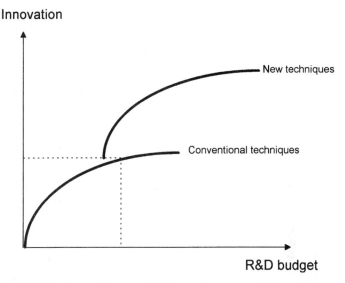

FIGURE 1.5   New techniques enabled improved returns on R&D expenditures.

REFERENCES

1. M. L. Arora, and B. M. Michalczyk, 1986, "Pilot Scale Evaluation of Ultrafiltration to Remove Humic Substances from Groundwater," *Proceedings of the Annual Conference of the American Water Works Association,* Denver, Colo.

2. J. S. Taylor, D. M. Thompson, and J. K. Carswell, 1987, "Applying Membrane Processes to Groundwater Sources for Trihalomethane Precursor Control," *Journal American Water Works Association,* 79(8):72–82.

3. V. P. Olivieri, et al., 1991, "Continuous Microfiltration of Surface Water," *Proceedings of the Membrane Technologies in the Water Industry,* Orlando, Fla.

4. J. G. Jacangelo, et al., 1991, "Low-Pressure Membrane Filtration for Removing Giardia and Microbial Indicators," *Journal American Water Works Association* 83(9):97–106.

5. F. Bourdon, M. M. Bourbigot, and M. Faivre, 1988, "Microfiltration tangentielle des eaux souterraines d'origine karstique," *L'Eau, L'Industrie, Les Nuisances* 121(9):35–41.

6. F. Duclert, C. Moulin, and M. Rumeau, 1990, "Results of Six Years of Practical Experience in the Potabilization of Water by Ultrafiltration," *Proceedings of the 5th World Filtration Congress,* Nice, France.

7. A. T. Pain, et al., 1990, "Iron Removal in Groundwater by Crossflow Micro- and Ultrafiltration," *Proceedings of the 5th World Filtration Congress,* Nice, France.

8. J. L. Bersillon, et al., 1989, "L'ultrafiltration appliquée au traitement de l'eau potable: le cas d'un petit système," *L'Eau, L'Industrie, Les Nuisances* 130(9):61–64.

9. J. L. Bersillon, C. Anselme, and J. Mallevialle, 1991, "Ultrafiltration in Drinking Water Treatment: Long Term Estimation of Operating Conditions and Water Quality for 3 Water Production Plants," *Proceedings of the Membrane Technologies in the Water Industry,* Orlando, Fla.

10. V. Mandra, et al., 1992, "Utilsation de l'Ultrafiltration pour la Desinfection d'Eau Residuaire Urbaine," *Proceedings of the Colloque la Ville et L'Eau,* Paris, France.

11. V. P. Olivieri, et al., 1991, "Continuous microfiltration of secondary wastewater effluent," *Proceedings of the Membrane Technologies in the Water Industry,* Orlando, Fla.

12. M. Kolega, et al., 1990, "Disinfection and Clarification of Treated Sewage by Advanced Microfiltration," *Proceedings of the 25th Annual Conference of the International Association on Water Pollution Research and Control,* Kyoto, Japan.

13. H. Strathman, May 1992, "Economic Assessment of Membrane Processes," *Proceedings of the EEC-Brazil Workshop on Membrane Separation Processes,* Rio de Janeiro, Brazil.

14. R. W. Baker, et al., 1990, *Membrane Separation Systems: A Research and Development Needs Assessment,* National Technical Information Service, U.S. Department of Commerce, NTIS-PR 360.

15. H. Strathman, 1989, "Economic Evaluation of Membrane Technology," in L. Cecill and J. C. Toussaint (eds.), *Future Industrial Prospects of Membrane Processes,* Elsevier Science Publishers Ltd., Amsterdam.

CHAPTER 2
CATEGORIES OF MEMBRANE OPERATIONS

Philippe Aptel
Laboratoire de Génie Chimique, CNRS
Université Paul Sabatier
Toulouse, France

Chris A. Buckley
Pollution Research Group
Department of Chemical Engineering
University of Natal
Durban, South Africa

2.1 INTRODUCTION

The present chapter introduces the different categories of membrane operations. The purpose is to define the most usual terms, introducing the concept of a membrane and a membrane operation, as well as the applications in water treatment. The preparation procedures of the main membrane types are explained. Module configuration and system design are briefly discussed.

2.2 MEMBRANE OPERATIONS AND WATER TREATMENT

2.2.1 Drinking Water

The primary driving force behind the industrial development of membranes has been desalination for municipal drinking water supplies. *Electrodialysis* (ED) and *reverse osmosis* (RO) have been used since the early '60s in competition with distillation processes. The latter include *multistage-flash evaporation, multiple-effect evaporation,* and *vapor compression.* Today, there are approximately 8900 desalting units comprising over 15.6 million m³/day of installed capacity. Distillation technologies account for approximately 60 percent of the world's desalting capacity, whereas RO and ED account for about 35 and 6 percent, respectively. However, over the past years, the percentage of worldwide desalting capacity by membrane processes has increased, while the reverse has been observed for distillation: RO treatment capacity has increased from 20 to more than 30 percent, while distillation has decreased from 68 to 56 percent. Although there is greater desalting capacity by distillation, there is a greater number of treatment plants employing membrane processes. Membrane plants account for almost 70 percent of the total number of desalting plants constructed. In general, they are several times smaller than distillation facilities. The average capacities of RO and ED plants are 990 and 660 m³/day, respectively, whereas MSF plants average 7000 m³/day. The largest RO plants are in the United States (Yuma: 250,000 m³/day) and in Saudi Arabia (Riyadh-Salbukh: 50,000 m³/day).

In the late '80s, *nanofiltration* (NF) gained considerable interest for water softening and, more recently, for removal of disinfection by-product precursors. Nanofiltration is now the second-largest application of membrane processes, even if it is regarded as conventional technology only in Florida. There are approximately 150 plants with a total capacity in excess of 600,000 m³/day.

More recently, there has been increasing interest in employing membrane technologies for removing particulates, microorganisms, and colloidal material from potable water supplies by *ultrafiltration* (UF). The object of UF is not to solve problems such as desalting, softening, or micropollutant removal, but to replace conventional physicochemical clarification and disinfection by a physical unit operation using membranes far less open than sand filters. There are today approximately 30 plants in operation or under construction with a capacity of 60,000 m³/day.

Cross-flow microfiltration (CFMF) plants have also been installed (12,000 m³/day). However, due to severe fouling problems and nonabsolute removal of the smallest microorganisms (viruses), CFMF is not as attractive as UF.

The latest developments in membrane technologies for potable water treatment are still at the demonstration plant or pilot research level. They include combined UF/powdered activated carbon adsorption treatment to ensure organic micropollutant removal, combined UF/oxidation for iron and manganese removal, and ED or combined UF/bioreactor for nitrate removal.

2.2.2 Industrial Water

As for drinking water, demineralization by RO has been used since the '60s in various industries:

- Electronics for ultrapure water
- Pharmaceuticals for dialysis bath makeup
- Food industry for carbonated beverage preparation
- Power station for boiler feed water

Ultrapure water is a typical example illustrating the contribution of membranes in the growth of the electronics industry. The semiconductor industry has long required water that is chemically, physically, and biologically very pure; however, the extremely advanced miniaturization of the components and integrated circuits has brought about ever more demanding standards for the quality of ultrapure water. The chemical part of these standards has evolved only slightly (as 18 MΩ·cm is quite near the theoretical limit). The physical and biological characteristic requirements have become much stricter:

- The *total organic carbon* (TOC) must be reduced to 50 µg/L and even to 20 µg/L, instead of 0.5 to 1 mg/L as previously.
- The number of bacteria must be less than 10 per liter.
- The sizes of particles taken into consideration have gone from 0.5 to 0.1, and even 0.05 µm.

In typical treatment routes, membranes are combined with other more standard treatments to give the system exceptional reliability, even with feedwaters whose composition is changing with time and is not fully known.

Dissolved inorganic pollution is removed to a level of 90 to 95 percent of the initial content by RO. It must be pointed out that an absolutely saltfree permeate cannot be produced by RO. However RO relieves the ion exchangers located downstream. This is very important since the regeneration of the ion exchangers introduces impurities.

Particle, organic, and bacteriological pollution are treated by combining two types of processes:

- Disinfection/oxidation processes (ozone and UV) that destroy the bacteria and oxidize the organics
- Membrane processes that remove particle and organic pollution

Microfiltration (disposable cartridges) are used at three points of the flow sheet:

- As a safety precaution upstream of the RO treatment of the makeup water
- Downstream from the ion exchangers to retain any possible resin fines
- At the point of use to retain bacteria and particles

Ultrafiltration is used on the production loop at the end of the treatment so as to stop viruses, macromolecules (including pyrogens), and particles. Most new plants have adopted UF instead of MF at the point of use.

Reverse osmosis is also more and more commonly used on the production loop, instead of UF or MF, to reduce dissolved organic pollution (DOC) as much as possible.

2.2.3   Industrial Effluents and Domestic Wastewater

Membrane operations are applied to a number of environmental problems as the result of more stringent regulations. For economical reasons, applications are still generally limited to the cases where contaminants and/or water can be recovered for recycle or reuse. Two typical examples are electropaint recovery and cheese whey treatment. Ultrafiltration has been employed in the electrodeposition paint industry for over 20 years. During this time, ultrafiltration has evolved from a simple means of controlling excess bath conductivity to a highly sophisticated technology which also controls the paint concentration, produces filtrate for countercurrent rinsing of painted components and therefore plays an equally important role in water recycling, paint recovery, and pollution abatement. It is an excellent early example of cleaner production methods.

Cheese whey is the supernatant liquid produced in cheese making or in casein processing. The world production per year is around 3 million tons of dry matter containing 10 percent protein. Using ultrafiltration to eliminate 75 percent of the volume, one can produce a retentate with a protein content similar to that of milk, which can be used as a milk substitute in cattle feed. Further purification by diafiltration leads to protein concentrates used for their functional properties (high water solubility, foamability, emulsification, gelation) in the food industry for human nutrition (beverages, cakes, ice cream).

Although the concept of coupling ultrafiltration and activated sludge was first commercialized in the '60s, the application has only recently started to attract serious attention. These hybrid processes have several advantages:

- The process can be significantly more compact than conventional processes.
- Higher biomass concentrations can be achieved, which results in reduced quantities of excess sludge.
- The effluent can be particulate free and totally disinfected.

More than 100 installations are in operation in Japanese buildings recovering around 30,000 m^3/day which is reused for flushing toilets.

A number of other examples will be discussed in other chapters showing that membranes already play a major role in environmental management.

2.3 CLASSIFICATION OF MEMBRANES AND MEMBRANE OPERATIONS

2.3.1 Definition of a Membrane

A membrane can be defined as a thin film separating two phases and acting as a selective barrier to the transport of matter (Fig. 2.1). This definition includes the definition of a permselective membrane and implies that a chemical potential difference exists between the two phases. It is very important to point out here that a membrane is not defined as a *passive* material but better as a *functional* material. In other words, even if (permselective) membranes may be characterized by their structure, their performances in terms of fluxes and selectivities are mainly dependent on the nature of the elements contained in the two phases and on the driving force which is applied. This is why we choose to classify membranes according to the type of separation they are able to perform rather than according to their structure, and only then discuss the structure best adapted to improve the performances of the separations.

2.3.2 Definition of a Membrane Operation

The term *membrane operation* is recommended rather than the term *membrane process*. In general, a process is supposed to consist of two or more operations. A membrane operation can be defined as an operation where a feed stream is divided into two streams: a permeate containing material which has passed through the membrane and a retentate containing the nonpermeating species (Fig. 2.2). Membrane operations can be used to concentrate or to purify a solution or a suspension (solvent-solute or particle separation) and to fractionate a mixture (solute-solute separation).

Among the separation operations, membrane offers basic advantages:

- Separation takes place at ambient temperature without phase change, which offers an energetic advantage compared to distillation. This explains, for example, the success of reverse osmosis and electrodialysis for water desalination.

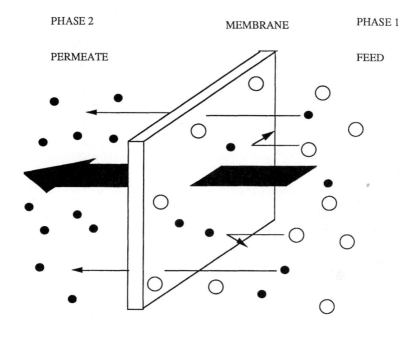

DRIVING FORCE ($\Delta C, \Delta P, \Delta \Psi$)

FIGURE 2.1 Definition of a permselective membrane.

- Separation takes place without accumulation of products inside the membrane. Membranes are then well adapted to be run continuously without a regeneration cycle as in ion-exchange resin operations or without an elution cycle as in chromatography.

- Separation does not need the addition of chemical additives, as is the case with azeotropic distillation or in water clarification by settlement or conventional filtration. This gives advantages for the quality of the product and leads to less pollutant wastes and explains the success of pervaporation for the fractionation of azeotropic mixtures and ultrafiltration for water clarification.

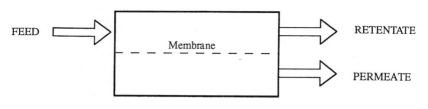

Three - end modules

FIGURE 2.2 Principle of a membrane operation.

2.3.3 Membrane Operations

As for all transport phenomena, the transmembrane flux for each element can be written by the following simple expression:

$$\text{Flux} = \text{force} \times \text{concentration} \times \text{mobility} \tag{2.1}$$

In most cases, the concentration will vary with distance through the membrane and through the boundary layers at the two interfaces of membrane fluid; it is appropriate therefore to treat Eq. (2.1) as a local equation where the local forces are the gradients of chemical potential $d\mu/dx$ of every component that can be transported.

The variation of the chemical potential of component i can be expressed as a sum of terms:

$$d\mu_i = \text{RT}\, d \ln a_i + V_i\, dP + z_i F\, d\Psi \tag{2.2}$$

Whereas the activity a_i (product of concentration by activity coefficient) is not under the arbitrary control of the operator, the pressure P and electric potential Ψ can be varied in order to improve the separation between the mobile components. Note that an applied pressure acts on every component in proportion to its molar volume V_i, while an electric field acts on every ionic species according to its valency z_i and does not affect nonionic species. Equation (2.2) can help to choose qualitatively the functionality of the membrane.

A general classification of membrane operation (Table 2.1) can be obtained by considering the following parameters:

- Driving force
- Mechanism of separation
- Membrane structure
- Phases in contact

TABLE 2.1 Technically Relevant Main Membrane Operations in Water Treatment

Membrane operation	Driving force	Mechanism of separation	Membrane structure	Phase 1*	Phase 2
Microfiltration	Pressure	Sieve	Macropores	L	L
Ultrafiltration	Pressure	Sieve	Mesopores	L	L
Nanofiltration	Pressure	Sieve + (solution/ diffusion, + exclusion)	Micropores	L	L
Reverse osmosis	Pressure	Solution/diffusion + exclusion	Dense	L	L
Pervaporation	Activity (partial pressure)	Solution-diffusion	Dense	L	G
Membrane stripping	Activity (partial pressure)	Evaporation	Macropores (gaz membrane)	L	G
Membrane distillation	Activity (temperature)	Evaporation	Macropores (gaz membrane)	L	L
Dialysis	Activity (concentration)	Diffusion	Mesopores	L	L
Electrodialysis	Electrical potential	Ion exchange	Ion exchange	L	L

* Phase 1 is the feed.

2.3.3.1 Pressure-Driven Membrane Operations. These are membrane operations in which the driving force is a pressure difference across the membrane.

Reverse Osmosis (RO). Reverse osmosis is a pressure-driven membrane operation in which the solvent of the solution is transferred through a dense membrane tailored to retain salts and low-molecular-weight solutes. If a concentrated saline solution is separated from pure water by such a membrane, the difference in chemical potential tends to promote the diffusion of water from the diluted compartment to the concentrated compartment in order to equalize the concentrations. At equilibrium, the difference in the levels between the two compartments corresponds to the osmotic pressure of the saline solution.

To produce "pure" water from a saline solution, the osmotic pressure of the solution must be exceeded in the brine. In the same way, it may be said that, in order to obtain economically viable flows, at least twice the osmotic pressure must be exerted; for instance, for seawater, pressures of 5 to 8 MPa are used.

Nanofiltration (NF). Nanofiltration, also called *low-pressure reverse osmosis* or *membrane softening,* lies between RO and ultrafiltration in terms of selectivity of the membrane which is designed for removal of multivalent ions (calcium and magnesium) in softening operations. More recently, nanofiltration has been employed for organics control. In nanofiltration, the monovalent ions are poorly rejected by the membrane. This explains why nanofiltration leads to an osmotic backpressure which is much lower than that experienced with RO. As a consequence, the operating pressure used in NF is much lower than in RO (typically 0.5 to 1.5 MPa).

Ultrafiltration (UF). In water treatment, ultrafiltration can be defined as a clarification and disinfection membrane operation. UF membranes are porous and allow only the coarsest solutes (macromolecules) to be rejected, and, a fortiori, all types of microorganisms as viruses and bacteria, or all types of particles. Since the low-molecular-weight solutes are not retained by UF, the osmotic backpressure can be neglected, and the operating pressure is kept low (50 to 500 kPa).

Microfiltration (MF). A major difference between MF and UF is membrane pore size; those of MF are 0.1 μm or greater. The primary application for this operation is particulate removal (clarification). Pressures are similar to UF.

2.3.3.2 Permeation Operations. These are membrane operations where the driving force is activity difference across the membrane. They are applied to solutions or mixtures. When applied to solutions, it is the solvent which is transferred through the membrane.

Gas Permeation (GP). Gas permeation is a gas/gas membrane separation process in which the activity difference is maintained through a pressure difference across a dense membrane. Examples of applications are hydrogen recovery from the blown-off gases from an ammonia synthesis operation and enriching air with nitrogen or oxygen.

Gas Diffusion. This is the same operation as the aforementioned, but the membrane is porous. This process has been developed for isotope enrichment in the nuclear industries.

Pervaporation (PV). Pervaporation is a liquid/vapor separation operation in which a liquid is partially vaporized through a dense membrane. The activity difference is generally maintained by creating a partial vacuum on the permeate side in such a way that the pressure is kept below the vapor pressure of at least one component of the liquid in contact with the upstream face of a dense membrane. Pervaporation is mainly used for the dehydration of alcoholic azeotropes. More recent studies concern volatile organic compound (VOC) removal from wastewater or even drinking water.

Membrane Stripping (MS). This process differs from PV in that the membrane is porous and air sweeping is generally used to maintain a partial pressure below the vapor pressure of the volatiles to be removed.

Membrane Distillation (MD). This process differs from MS in that an air gap is maintained between the downstream side of the membrane and a cold wall. The temperature difference between the feed and the cold wall leads to a partial pressure difference. Most studies on this nonconventional process concern desalination of water.

2.3.3.3 Dialysis Operations.

These are membrane operations applied to solutions in which it is the solute which is transferred through the membrane. The driving force is an activity or an electrical potential difference in the absence of any transmembrane pressure difference.

Dialysis (DIA). This term refers to operations in which the driving force is a transmembrane concentration difference. Selective passage of ions and low-molecular-weight solutes occurs with the rejection of larger colloidal and high-molecular-weight solutes. The main application is hemodialysis, the largest market for membranes (54 million hemodialyzers of roughly one m^2 each are sold annually). Other applications include purification of bioproducts or alcohol reduction in beer.

Donnan Dialysis. This operation is a special variant of dialysis which combines the Donnan exclusion attainable with ion-exchange membranes with a transmembrane concentration gradient driving force as used in conventional dialysis. This process could also be defined as a continuous ion-exchange process. It is used for recovering valuable ions from dilute solutions by exchanging ions of the same sign from a concentrated solution that is cheaper than the material to be recovered.

Electrodialysis (ED). Electrodialysis is an operation by which ions are driven through ion-selective membranes under the influence of an electrical potential. By alternating cation- and anion-selective membranes in a stacked arrangement with thin channels between them, it is possible to produce alternating channels of fluid that are enriched and depleted, respectively, in ions. The most important large-scale application of electrodialysis is the production of potable water from brackish water. More recently, a first ED unit has been installed in Italy for nitrate removal from an underground water resource. Other applications are the removal of salts and acids from pharmaceutical solutions and in food processing, and for the production of salts from seawater. Recent developments have used bipolar membranes between the cationic and anionic membranes leading to three distinct channels and allowing acids and alkalis to be produced from a salt feed.

2.3.4 Classification of Membranes

Permselective membranes can be classified according to different criteria as mechanisms of separation, physical morphology, and chemical nature.

2.3.4.1 Classification According to Separation Mechanism.

There are essentially three mechanisms of separation which depend on one specific property of the components to be selectively removed or retained by the membrane:

- Separation based on large differences in the size (sieve effect). Main operations are MF, UF, DIA.
- Separation based on the differences in solubility and diffusivity of materials in the membrane (solution-diffusion mechanism). This is typically the case with GP, PV, and RO.

- Separation based on differences in the charges of the species to be separated (electrochemical effect) as in ED and in Donnan dialysis.

The classification of membranes based on separation mechanisms leads to three main classes: porous membranes (sieve effect); nonporous, or dense, membranes (solution-diffusion mechanism); and electrically charged membranes, also called ion-exchange membranes (electrochemical effect).

Porous Membranes. In porous membranes (Fig. 2.3), fixed pores are present. Using the definition of pore sizes as adopted by the IUPAC (1985).

- Macropores are larger than 50 nm.
- Mesopores are in the range of 2 to 50 nm.
- Micropores are smaller than 2 nm.

This means that MF, UF, NF, and DIA membranes are porous membranes. These definitions can lead to some confusion since microfiltration membranes contain macropores while nanofiltration membranes contain micropores. In fact, NF could be classified in an intermediate class between porous and nonporous membranes since solution-diffusion and even electrochemical effects have to be introduced in the equations of mass transfer.

Nonporous Membranes. These membranes can be considered as dense media. Diffusion of species takes place in the free volume which is present between the macromolecular chains of the membrane material. GP, PV, and RO membranes are of this type.

Ion-Exchange Membranes. Ion-exchange membranes are a specific type of nonporous membranes. They consist of highly swollen gels carrying fixed positive or negative charges. A membrane with fixed positive charges (for example $-NR_3^+$) is called an anion-exchange membrane, whereas a cation-exchange membrane has fixed negative charges (for example, $-SO^{3-}$).

2.3.4.2 Classification According to Morphology.
For pressure-driven and permeation membrane operations, the flux of permeate is inversely proportional to the thickness of the membrane. In fact, it was the development of anisotropic membranes which led to the breakthrough in industrial applications. These membranes (Figs. 2.4 to 2.6) consist of a very thin top layer, called the *skin,* supported by a thicker and more porous supporting sublayer. The skin has the main functions of the membrane, since the overall flux and selectivity depend only on the structure of the skin. Its thickness is in the range of 0.1 to 0.5 μm, roughly 1 percent of the thickness of the porous sublayer. The supporting layer possesses negligible resistance to mass transfer and is present for mechanical support only. Membranes of these designs are normally produced on a porous substrate material (frequently a spun-bonded nonwoven polyester). The carrier material forms an integral part of the membrane, since it imparts mechanical strength to the membrane. It is usual to distinguish two types of anisotropic membranes: asymmetric and composite.

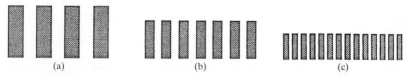

FIGURE 2.3 Schematic representation of isotropic porous membranes: (*a*) macropores > 50 nm; (*b*) 2 < mesopores > 50 nm; (*c*) micropores < 2 nm.

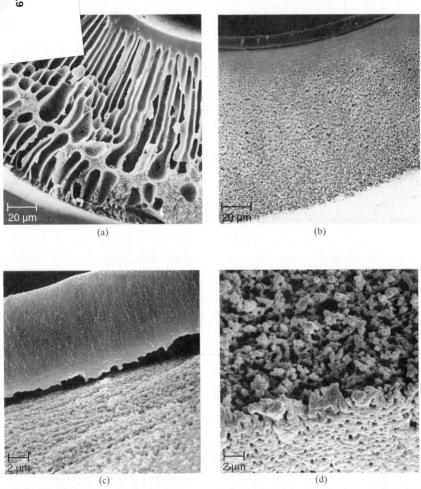

FIGURE 2.4 Scanning electron micrographs showing cross sections of asymmetric polysulfone hollow fibers with an internal skin: (a) finger-pore substructure; (b) graded pore sponge substructure; (c) internal skin of fiber b (during preparation of the sample, the skin was drawn before being fractured; thus the micrograph shows the skin surface which has been folded back on the porous substructure); (d) external surface of fiber b.

Asymmetric Membranes.   These are anisotropic membranes prepared from the same material (Figs. 2.4 and 2.5).

Composite Membranes.   These are anisotropic membranes where the top layer and sublayer originate from different materials (Fig. 2.6). Each layer can be optimized independently. Generally, the porous layer is already an asymmetric membrane.

2.3.4.3   *Classification According to Geometry.* Membranes can be prepared in two geometries: flat and cylindrical. On the basis of differences in dimensions, the following types of cylindrical membranes may be distinguished:

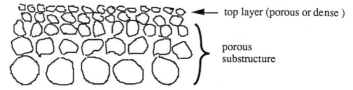

FIGURE 2.5 Schematic drawing of an asymmetric membrane.

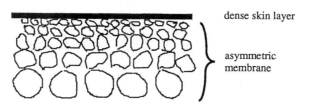

FIGURE 2.6 Schematic drawing of a composite membrane.

- Tubular membranes with internal diameter larger than 3 mm
- Hollow-fiber membranes with diameter smaller than 3 mm

Hollow fibers are geometrically the smallest tubelike membrane available, with outside diameters that range from 80 to 500 μm. They are used in RO, GP, and hemodialysis. With larger diameter, they are used in UF and MF, in which case they are also called *capillary membranes*.

2.3.4.4 *Classification According to Chemical Nature.* Synthetic membranes can be made from a large number of different materials: organic (polymers) or inorganic (metals, ceramics, glasses, etc.)

Organic Membranes. See Fig. 2.7. Basically, all polymers can be used, but for processing requirements and the lifetime of the membrane, only a limited number are used in practice. The most widely used are cellulose and its derivatives. These hydrophilic polymers are low cost, have low tendencies for adsorption, and are used not only in all pressure-driven processes but also in hemodialysis and gas permeation. In water treatment, cellulose esters (mainly di and triacetate) membranes have the advantage of being relatively resistant to chlorine and, despite their sensitivity to acid or alkaline hydrolysis, to temperature, and to biological degradation, they are widely used for desalination, softening, disinfection, and clarification.

Another important class of hydrophilic membrane polymers is the polyamides. Aromatic polyamides were the second type of polymer, after cellulose diacetate, to be used in desalination because of its permselective property and better thermal, chemical, and hydrolytic stability than cellulose esters. The amide group ($-CO-NH-$), however, has a great sensitivity to oxydative degradation and cannot tolerate exposure to even traces of chlorine.

Polyacrylonitrile (PAN) is also commonly used for ultrafiltration and hemodialysis membranes. Less hydrophilic than the two previous polymers, it does not have a permselective property, and it is not used in RO.

Another widely used class of polymers is the polysulphone (PSf) and polyethersulphone (PES). These polymers are not hydrophilic and have a relatively high

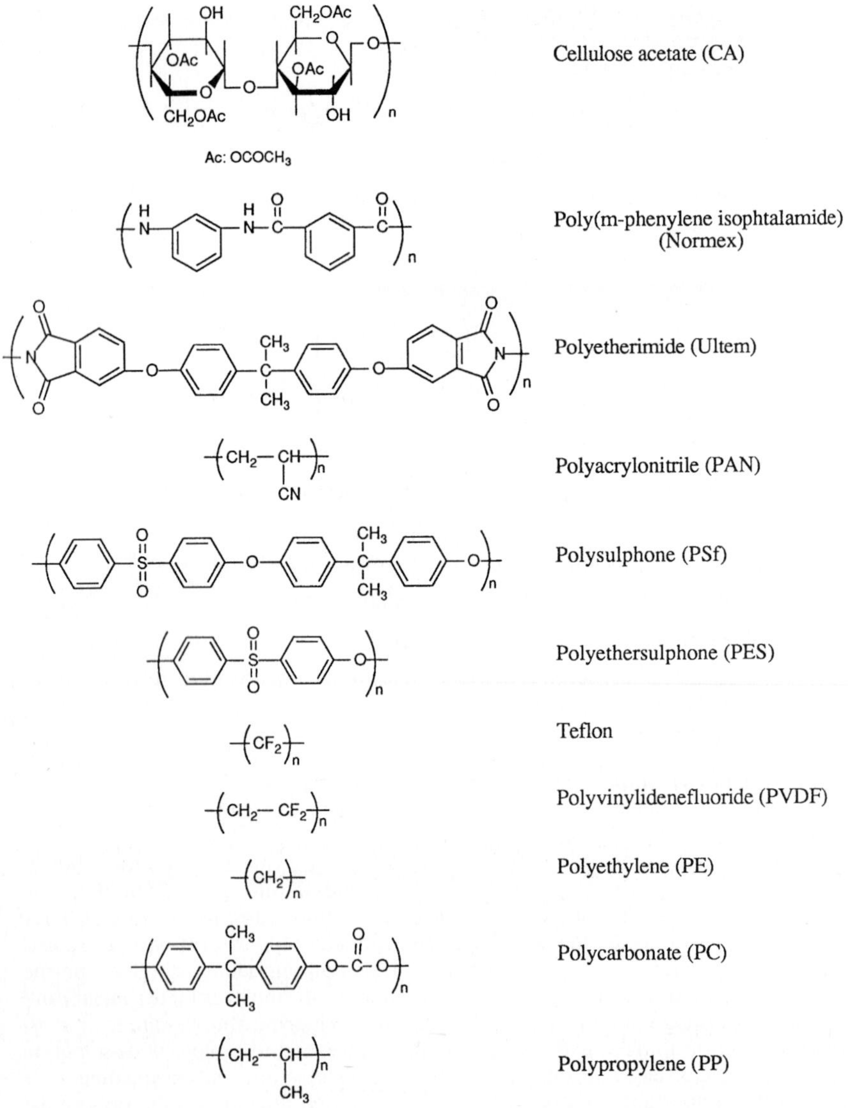

FIGURE 2.7   Molecular structures of main organic membrane materials.

adsorption tendency, but they have very good chemical, mechanical, and thermal stability. They are commonly used as UF membranes, as support for composite membranes, or as hemodialysis membranes. Most of the PES and PSf membranes are modified by blending with hydrophilic polymers to provide the membranes with better antifouling properties.

Because of their excellent chemical and thermal stability, the following hydrophobic polymers are often used as macroporous membranes: polytetrafluoroethylene

(PTFE), polyvinylidene fluoride (PVDF), polyethylene (PE), polycarbonate (PC), or isotactic polypropylene (PP). In water treatment, PP is commonly used as microfiltration membranes despite its sensitivity to chlorine.

Inorganic Membranes.   Inorganic materials generally possess superior chemical, mechanical, and thermal stability relative to polymeric materials. However, these materials have the disadvantages of being very brittle and more expensive than organic membranes. This explains why their main fields of application are limited to the chemical industries for aggressive/high-temperature fluids treatment and to pharmaceutical and dairy industries when heat sterilization is needed.

Ceramic membranes represent the main class of inorganic membranes. Ceramics are oxides, nitrides, or carbides of metals such as aluminium, zirconium, or titanium.

2.4   PREPARATION OF MEMBRANES

There are a number of different techniques used to prepare organic membranes: sintering, stretching, track etching, coating, and phase inversion. Coating techniques are used to make composite dense membranes; sintering, stretching, and track etching can make only microfiltration membranes. These techniques will not be discussed further. Phase inversion is a far more general technique.

2.4.1   Asymmetric Membranes Prepared by Phase Inversion

The most important commercial membranes are asymmetric or asymmetric-based composite membranes. They are generally prepared by the so-called phase inversion process, in which a polymer is dissolved in an appropriate solvent and cast as a 0.1- to 1-mm-thick film. A nonsolvent is then added to this liquid film, causing phase separation and precipitation. At the interface between the polymer solution and the nonsolvent (Fig. 2.8), the solvent and the nonsolvent exchange by diffusion. The solvent diffuses into the coagulation bath with a flux J_s whereas the nonsolvent will diffuse into the case film (J_{ns}). By an appropriate choice of the solvent and the nonsolvent in order to have $J_s > J_{ns}$, the polymer composition in the cast film will increase, while the nonsolvent/solvent ratio increases, as shown by line AB on the schematic ternary diagram (Fig. 2.9). At point B the composition shifts into the two-phase region where demixing then precipitation occur (point C). During these steps the exchange between solvent and nonsolvent still goes on, and the final composition is represented by point D on the ternary diagram. Due to the concentration profile which exists in the cast film, the first layer which precipitates is on the top of the cast film, and its composition is richer in polymer than the deeper layers. As a consequence, the successive layers which precipitate are less and less concentrated in polymer (or more and more porous). In summary, it is useful to remember that the anisotropic structure of the membrane depends on thermodynamic and kinetic factors which can be estimated knowing the following data:

- Nature of polymer
- Nature of solvent and nonsolvent
- Composition of casting solution
- Composition of coagulating bath
- Gelation and crystallization behavior of the polymer

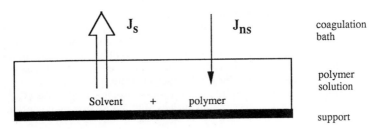

FIGURE 2.8   Principle of the formation of a membrane by phase inversion.

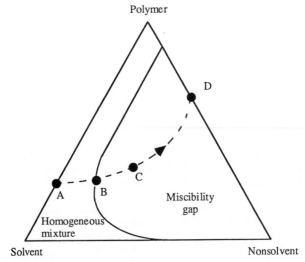

FIGURE 2.9   Phase diagram showing the formation of a membrane by addition of a nonsolvent to a homogeneous polymer solution: A—composition of the casting solution; B—composition of the ternary mixture where demixing occurs; C—point of solidification; D—composition of the membrane after complete exchange between solvent and nonsolvent.

- Location of the liquid-liquid demixing gap
- Temperature of the casting solution and the coagulation bath
- Evaporation time

Most of these parameters are not independent of each other and, usually, additives are also added in the casting solution. This makes the problem of anticipating the final structure quite difficult. On the other hand, the choice is open to tailor a large range of membranes from the same polymer: from very dense skinned anisotropic membranes for GP or RO to isotropic porous membranes for MF.

The principle of manufacturing flat-sheet membranes is schematically shown in Fig. 2.10. After filtration and degassing, the solution (A) is pumped (B) through a casting knife (C) and cast as a thin fluid film onto a nonwoven fabric or directly on a metallic casting belt. After a short residence in the air, the cast film enters into a coagulation bath (D). Following gelation, the membrane is washed free of solvent (E). Before collecting the membrane on a take-up roll (F), other treatments can also be applied, such as heat treatment, conditioning, and drying.

2.4.2 Composite Membranes Prepared by Interfacial Polymerization

This technique, which is widely used to prepare RO membranes, consists of polymerizing two reactive monomers or prepolymers on the skin of a UF membrane (Fig. 2.11). The membrane is immersed in an aqueous solution containing a reactive monomer 1, or a prepolymer. The film is then immersed in a second bath containing a water-immiscible solvent with the other monomer (2). The reaction takes place at the interface to form a dense top layer. One of the advantages of this technique is that the first polymerized layers offer great resistance to the diffusion of the reactants, resulting in an extremely thin film of thickness within the 50-nm range.

2.4.3 Preparation of Hollow Fibers by Phase Inversion

Hollow fibers can be prepared from the same materials used to cast flat-sheet membranes. The fibers can be spun directly as a membrane or as a substrate which is post-treated to get a composite hollow fiber. The technology employed in the fabrication of synthetic fiber applies also to the spinning of hollow-fiber membranes.

In melt spinning, a polymer melt is extruded into a cooler atmosphere, which induces phase transition: the controlled solidification of the nascent filament determines its characteristics. The resulting fiber is usually a dense, isotropic membrane; addition of a removable additive to the dope yields a porous membrane.

In the dry process, the dope consists of the polymer dissolved in a volatile solvent. Evaporation of solvent induces phase transition and produces a porous, is-, or anisotropic membrane.

In the wet process, the extruded mixture is coagulated in a nonsolvent in liquid or vapor phase.

The dry-wet spinning technique is a combination of the last two methods; the spinneret is positioned above a coagulation bath allowing evaporation or cooling to take place in the air gap (Fig. 2.12).

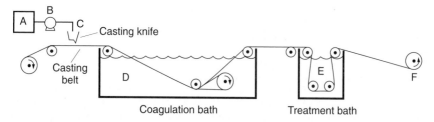

FIGURE 2.10 Schematic representation of a flat-sheet membrane casting machine.

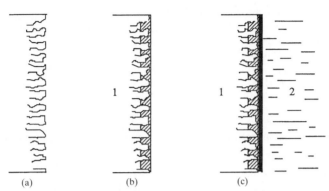

FIGURE 2.11   Schematic drawing of the formation of a composite membrane via interfacial polymerization: (*a*) support layer (UF asymmetric membrane); (*b*) immersion of the support in an aqueous solution of monomer 1; (*c*) immersion in a water-immiscible solution of monomer 2 and formation of very thin film at the surface of the support.

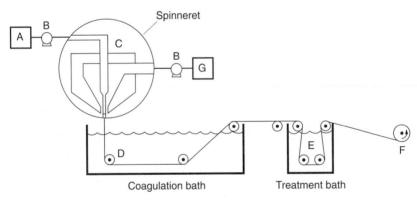

FIGURE 2.12   Schematic representation of a hollow-fiber spinning machine.

The possibility to induce phase inversion from the bore side or lumen, from the shell or outside, or from both sides of the hollow filament is the first main difference between flat-sheet and hollow-fiber membranes. A second important difference leads to the fact that, while flat-sheet membranes are cast on a nonwoven film to reinforce the mechanical properties, the nascent filament is self-supporting. As a consequence, drawing can be applied before complete solidification.

2.4.4   Preparation of Inorganic Membranes

Ceramic pastes derived from powders as alumina (Al_2O_3) and zirconia (ZrO_2) are extruded and then sintered at high temperature to give macroporous supports with pore diameters larger than 1 μm. Flat, tubular, or multichannel supports can be obtained. Suspensions of submicronic powders are then laid on the support in successive layers to get microfiltration membranes with lower pore diameters. Sol-gel processes starting from suspensions of colloidal particles are used to form ultrafiltration layers exhibiting pore diameters down to 3 nm.

2.5 MODULE CONFIGURATIONS

The single operational unit into which membranes are engineered for use is referred to as a *module*. This operational unit consists of the membranes, pressure support structures, feed inlet and concentrate outlet ports, and permeate draw-off points. Modules are designed to achieve three essential objectives:

- To ensure, at membrane level, a sufficient circulation of the fluid to be treated in order to limit the phenomena of concentration polarization and of particle deposits
- To produce a compact module, i.e., one providing a maximum exchange surface per unit volume (maximum packing density)
- To avoid any leak between the feed and the permeate compartments

The first two objectives tend to reduce the cost of the module for producing a given volume of treated fluid, but they also tend to increase the energy cost of separation: high circulation velocity and small passage cross section will give a great head loss. The third objective could appear as trivial, but in practice most of the problems that occur come from a leak due to a defective assembly and not from faulty membranes.
The module must also satisfy other requirements such as:

- Ease of cleaning (hydraulic, chemical, sterilization)
- Ease of assembly and disassembly
- Low hold-up volume

Four major types of modules are found on the market: plate and frame, spiral wound, tubular, and hollow fiber.

2.5.1 Plate and Frame

These modules are made up of stacked flat-sheet membranes and support plates. Their design is derived from that of filter presses (Fig. 2.13a). The feed circulates between the membranes of two adjacent plates. The thickness of the liquid sheet is in the range of 0.5 to 3.0 mm. The packing density of plate-and-frame units is about 100 to 400 m^2/m^3. The plates ensure the mechanical support of the membrane and, at the same time, the drainage of the permeate. The plates may be corrugated on the feed side to improved mass transfer. Their arrangement makes it possible to bring about, in parallel and/or in series, circulation. Large unitary assemblies with a surface of up to 100 m^2 can thus be formed. Units are easily disassembled to gain access for manual cleaning or replacement of the membranes. In some of the designs, permeate is collected from individual support plates, which makes the location of faulty membranes a simple matter.

2.5.2 Spiral Wound

An envelope of two flat-sheet membranes enclosing a flexible porous sheet (permeate collector) is sealed on three of its edges. The open edge is connected and rolled up onto a perforated tube which carries the permeate (Fig. 2.13b). Several "sandwiches" are thus fastened and separated from one another by a feed-side spacer. This spacer not only maintains an open flow channel for feed flow, but also fulfills the very important function of inducing turbulence, thus reducing concentration

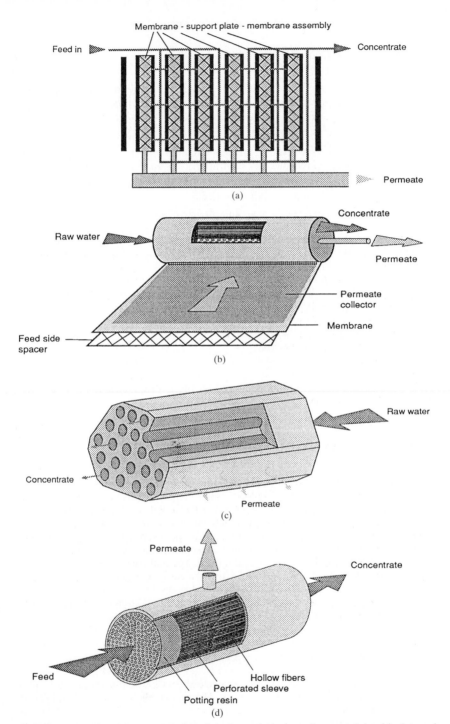

FIGURE 2.13 Schematic representation of the four principal membrane modules: (*a*) plate and frame; (*b*) spiral module; (*c*) tubular module; (*d*) hollow fiber.

2.18

polarization. The spacer may be a mesh or it may be a corrugated spacer. The feed flows parallel to the permeate tube axis.

The diameter of an element can be as much as 300 mm, and its length can be up to 1.5 m. Several elements (two to six) can be inserted into a single cylindrical pressure vessel. These are much more compact (700 to 1000 m^2/m^3) and cause a lower head loss than the plate-and-frame module. The spiral-wound module is, however, more sensitive to clogging than open-channel flat-sheet systems due to the spacer, and they cannot be used directly without pretreatment on turbid water.

2.5.3 Tubular

The tubular module in one form is the simplest configuration in which the membrane is cast on the inside wall of a porous support tube. These tubes have internal diameters ranging from 6 to 40 mm. Individual tubes may be placed inside stainless steel or PVC sleeves for smaller-scale units or bunched together in bundles of 3 to 151 tubes in a cylindrical housing with appropriate end plates.

Inorganic membranes may be formed on multichannel ceramic supports containing up to 19 parallel flow channels (Fig. 2.13c). Each multichannel membrane element is housed individually or in parallel sets (up to 99 elements), thus providing membrane modules with various total membrane surface areas (0.2 to 7.4 m^2).

The hydrodynamics of the flow is perfectly defined and circulation velocities up to 6 $m \cdot s^{-1}$ are possible if a highly turbulent flow is necessary. These modules do not need fine prefiltration of the feed and are easy to clean. They are particularly well adapted to the treatment of very viscous fluids. Their main disadvantage is that they have a low packing density, thus increasing the capital cost.

2.5.4 Hollow Fiber

The fibers are gathered in a bundle of several thousand, even several million. Flow of the feed takes place either inside the fibers (inside-out configuration) or outside the fibers (outside-in configuration). In the first case, the watertightness between the feed and the permeate flows is provided by a potting resin which forms a tube plate at each end of the bundle. After hardening of the resin, the bundle is cut in such a way that the open ends of all the fibers appear. In many designs (Fig. 2.13d), the pressure housing is sealed in the same operation, avoiding the need of rings that are the main sources of leaks in modules. In the outside-in configuration, the bundle is often arranged in a U shape; the fibers are sealed at only one end.

As the packing density is inversely proportional to the diameter, these units are very compact (from 1000 m^2/m^3 in UF modules, up to 10,000 m^2/m^3 in RO modules). Several bundles can be potted in a single housing, providing units with large membrane surface area able to produce up to 220 m^3/d in UF (70 m^2 module from Aquasource) and 140 m^3/d in RO (B-9 Twin permeator from Du Pont).

Operating velocities in hollow-fiber modules are normally low and modules can even be operated without recirculation (dead-end mode). Hollow fibers thus operate in the laminar flow region. But, even in this regime, shear rates can be high due to the very small flow channels.

Another advantage that has led to the success of UF and MF hollow fibers in water treatment is the backflushing capability resulting from the fibers being self-supporting. In UF, backflush is carried out by placing the permeate under a pressure greater than the feed pressure. The change in direction of the flow through the wall of the fiber makes it possible to detach the cake of particles deposited on the surface (Fig. 2.14). This cake is then transported out of the module by circulating flow through the module. During this operation it is also possible to flush the particles, which eventually block the entry of the hollow fibers. In MF, due to the larger pore dimensions, air backflush may be used.

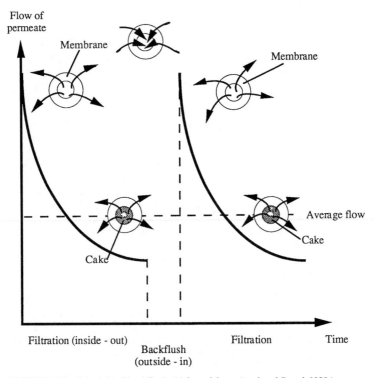

FIGURE 2.14 Principle of backflush. *(Adapted from Aptel and Rovel, 1989.)*

2.5.5 Rotating Disc and Cylinder Modules

Different devices based on a rotating disc or cylinder have been developed (Belfort et al., 1994). These modules have the advantage that they promote secondary flows to help depolarize the solute and particle buildup at the membrane-solution interface. This results in significantly improved performance for pressure-driven filtration in terms of flux. The limitations are high energy consumption to rotate the equipment, and difficulties in maintenance and scaling up the capacity of the modules. Table 2.2 summarizes the advantages of the different types of modules according to various criteria.

TABLE 2.2 Comparison of Different Types of Module

Criteria	Plate and frame	Spiral-wound	Tubular	Hollow fiber Fine (RO)	Capillary (UF/MF)	Rotating disc or cylinder
Packing density	+	++	–	+++	+++	–
Ease of cleaning						
–in situ	+	–	++	–	–	+
–by backflush	–	–	–(1)	–	+++	–
Cost of module	+	+++	–	+++	+++	–
Pressure drop	–	++	+++	++	++	+++
Hold-up volume	+	+	–	+++	++	–
Quality of pretreatment required	+	–	+++	–	++	+++

– Clear disadvantage
+++ Clear advantage
(1): With the exception of certain ceramic modules where the layer forming the membrane is chemically bound to its support

2.6 SYSTEM DESIGN

2.6.1 Principle

The principle of using a module is simple. A pump ensures pressurization of the feed and circulation along the membrane. A valve is placed on the retentate (or concentrate) line to maintain the pressure inside the module. The permeate is drawn off at a pressure P_P, generally close to atmospheric pressure (Pa).

The choice of the pump as well as the adjustment of the valve allow independent setting of the mean transmembrane pressure P_{tm} and the conversion (or yield) Y. The transmembrane pressure is defined as:

$$P_{tm} = \frac{P_{in} - P_{out}}{P_P}$$ (2.3)

where P_{in} = pressure at the inlet of the module
  P_{out} = pressure at the outlet

The conversion is the ratio between the flow of permeate and the feed flow at the inlet of the module ($Y = 0.5/10.5 = 0.0476$, or 4.8 percent, for the example shown in Fig. 2.15a).

For a given module, it is advantageous to work at a high conversion ratio; this indeed limits the capital cost for the pump and the pipes and also the energy consumed by the circulation of the retentate. Nevertheless, if the conversion ratio is very high, the concentration factor in the module can reach values such that:

- The solubility product of the various compounds is exceeded.
- The viscosity becomes excessive.

Scaling will occur in RO and NF and a protein gel may appear in UF and MF, together with a progressive clogging of the circulation channels.

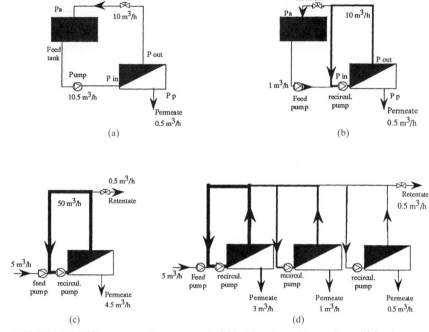

FIGURE 2.15 Different types of arrangements: (*a*) batch system with open loop; (*b*) batch system with closed loop; (*c*) single-stage continuous system; (*d*) reject-staged arrangement with recirculation. *(Adapted from Aptel and Rovel, 1989.)*

In the classical mode of operation of plate-and-frame or tubular modules, pressure-driven membranes are operated under conditions of high cross-flow velocities to limit the polarization phenomena and/or thickness of the filtration cake. This leads to the choice of a low conversion in a single element and the necessity to combine several elements in series to obtain reasonable conversions at relatively low energy consumption.

Conversion can be generally higher in spiral-wound and hollow-fiber modules. In the latter case when sequential backflushes are used, the dead-end mode of operation with 100 percent conversion has proven to be as effective as cross-flow in several cases (see Chap. 11). Table 2.3 summarizes the conversion figures generally maintained for each module.

TABLE 2.3 Typical Conversion (*Y* in Percent) for Each Module Element in Pressure-Driven Membrane Processes

	Plates (per plate)	Hollow fiber	Spiral wound (per element)	Tubular (per tube)
RO/NF	5–15	30–60	10–25	0.2–2
UF/MF	1–5	5–15*, 100†	2–10	0.5–5

* Tangential filtration
† Dead-end (filtration cycle)

2.6.2 Full-Scale Applications

Full-scale membrane facilities comprise series/parallel modules and operate according to various modes. Thus, these are systems which range from the intermittent single-stage system to the continuous multistage system.

2.6.2.1 Batch System with Open Loop. The system includes a feed tank, a recirculation pump, and an assembly of modules (Fig. 2.15a). A valve placed on the retentate line sets the pressure across the membranes. As the permeate is drawn off, the concentration in the tank increases, but the flow of permeate Q_P decreases (or the pressure increases if the system is operated at constant Q_P). One of the disadvantages of this system is the high energy requirements, since only a small part of the liquid pressurized and circulated actually goes through the membrane. Nevertheless, it is generally retained for applications in which the volumes to be treated are low, since the investment cost is lower than for the other operating modes; a typical example is the treatment of soluble oils (a few hundred liters per day) with UF.

2.6.2.2 Batch System with Closed Loop. The preceding system is modified by placing a force-feed pump between the tank and the recirculation pump (Fig. 2.15b). The energy gain compared to the open-loop system is all the greater as the outlet pressure P_{out} is high. Taking into account the flows indicated in Fig. 2.15a and b, the different energies involved for the circulation are, as a first approximation:

$$E_o \text{ (open loop)} \propto 10\,(P_{out} - P_P) + 10\,(P_{in} - P_{out}) \tag{2.4}$$

$$E_c \text{ (closed loop)} \propto 0.5\,(P_{out} - P_P) + 10\,(P_{in} - P_{out}) \tag{2.5}$$

$$E_o - E_c \propto 9.5\,(P_{out} - P_P) \tag{2.6}$$

whereas the energy necessary for permeation is the same in both cases:

$$E_P = 0.5\,(P_{in} - P_{out}) - P_P \tag{2.7}$$

This explains why, for the same application as in Sec. 2.6.2.1, this system is chosen when the flows to be treated are greater than 200 to 300 L/d.

2.6.2.3 Single-Stage Continuous System. When the liquid to be treated is produced continuously, the retentate is drawn off continuously, and the feed tank is no longer useful. The valve placed on the retentate line controls the conversion, which in the case of Fig. 2.15c is 90 percent. The disadvantage in this case is that the system operates with a high concentration in the loop and thus a low flow of permeate. However, this system, combined with a backflush system, is chosen for water clarification units using UF or MF hollow-fiber modules. For other applications where there is a considerable decrease in permeate flow with the concentration factor, a multistage installation should be considered.

2.6.2.4 Multistaged Arrangement with Recirculation. This system is mainly used in large-scale UF or MF installations. Each stage is fed by the retentate of the preceding one. The number of modules installed in parallel in each stage gives the desired stage conversion. Only the last stage operates with high conversion; the average flow per module is thus greater than that of an installation operating with equal conversion in a single stage. Practically, three stages are used (Fig. 2.15d).

2.6.2.5 Multistaged Arrangement Without Recirculation. In RO or NF, the average operating pressure is much greater than the pressure drop through the modules. Recirculation pumps are, then, not required.

2.7 CONCLUSION

This chapter has attempted to provide the reader with some of the basics of membrane science and technology. Most of the concepts introduced in this chapter are detailed in subsequent specialized chapters. Other, more complete, information can also be found in the books and reviews listed at the end of the chapter.

REFERENCES

Aimar, P., and P. Aptel (eds.), 1992, "Membrane Processes," *Proceedings of Euromembranes 92,* Lavoisier, Paris.

Aptel, P., and J. M. Rovel, 1991, "Separation by Membranes," in Degrémont (ed.), *Water Treatment Handbook,* 6th ed., Lavoisier, Paris.

Belfort, G., R. H. Davis, and A. L. Zydney, 1994, "The Behavior of Suspensions and Macromolecular Solutions in Crossflow Microfiltration," *J. Membrane Sci.,* **96**:1–58.

Bhave, R. R. (ed.), 1991, *Inorganic Membranes. Synthesis, Characteristics and Applications,* Van Nostrand Reinhold, New York.

Brun, J. P., *Les Techniques à Membranes,* Masson, Paris, 1989.

Bungay, P. M., H. K. Londsdale, and M. N. Pinho (eds.), 1983, *Synthetic Membranes: Science and Engineering and Applications,* NATO ASI Series, Series C: Mathematical and Physical Sciences, vol. 181, D. Reidel Pub. Co., Dordrecht.

Crespo, J. G., and K. W. Böddeker (eds.), 1994, *Membrane Processes in Separation and Purification,* NATO ASI Series E: Applied Sciences, vol. 272, Kluwer Academic Publishers, Dordrecht.

Howell, J. A., V. Sanchez, and R. W. Field (eds.), 1993, *Membranes in Bioprocessing: Theory and Applications,* Chapman & Hall, London.

Huang, R. Y. M. (ed.), 1991, *Pervaporation Membrane Separation Processes,* Elsevier, Amsterdam.

IUPAC, 1985, "Reporting Physisorption Data," *Pure Appl. Chem.,* **57**:603.

Kesting, R. E., 1985, *Synthetic Polymeric Membranes,* John Wiley, New York.

Lloyd, D. R. (ed.), 1985, *Material Science of Synthetic Membranes,* ACS Symp. Ser., Series 269, American Chemical Society, Washington, D.C.

Mulder, M., 1991, *Basic Principle of Membrane Technology,* Kluwer Academic Publishers, Dordrecht.

Rautenbach, R., and R. Albrecht, 1989, *Membrane Processes,* John Wiley, New York.

CHAPTER 3
MEMBRANE APPLICATIONS: A CONTAMINANT-BASED PERSPECTIVE

Chris A. Buckley and Quentin E. Hurt
Pollution Research
University of Natal
Durban, South Africa

While membranes were initially applied in specialized situations such as seawater desalination, membrane technology has grown into a multibillion-dollar international business expanding at around 15 percent per year (Cross, 1992). Water purification is a major segment of the membrane market, but other areas in which membranes are employed include effluent treatment, bioreactors, metal recovery, solvent dewatering, and paint recovery. The dynamic growth in membrane technology applications has been driven by both commercial and environmental forces. Membrane processes do not generally require the addition of aggressive chemicals, can be operated at ambient temperatures, form an absolute barrier to the flow of contaminants, and are space efficient—features that make them both economically and environmentally attractive.

The nature of the feed to be treated by the membrane process will determine the best membrane for each application. Figure 3.1 illustrates the range of membrane types used and their areas of application in terms of the size of the contaminant of concern. The term *contaminant* is used loosely to describe the molecule or particle of concern within the feed. The successful application of membrane technology will depend on the concentration, isolation, or removal of the contaminant. Reverse osmosis is applicable to the separation of ionic solutions such as salts, ultrafiltration to macromolecules and colloids, and microfiltration to a range of particles. The membrane processes may be used in series to remove progressively smaller contaminants. To some extent, this categorization has become blurred as membranes find application for removal of particles smaller than their pore size should theoretically permit, for instance, the removal of organisms by microfiltration. This phenomenon arises because the interactions between some membrane-solute systems result in the apparent *tightening* of the membrane.

This chapter reviews examples of the application of membrane processes in the treatment of contaminated aqueous streams. The chapter is divided among the three

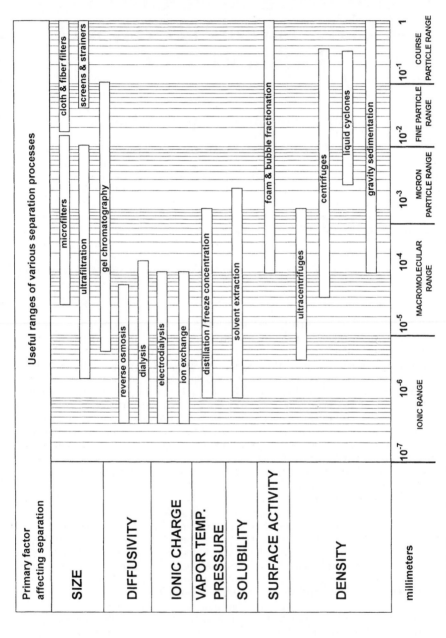

FIGURE 3.1 Ranges of separation processes.

3.2

major membrane processes: microfiltration (solid-liquid separation), ultrafiltration (membrane disinfection), and reverse osmosis (organics and inorganics), although these distinctions become rather blurred with advances in membrane materials and production mechanisms. The range of membrane applications for water treatment is reviewed, paying particular attention to the potential for membrane processes in conventional water treatment. Niche processes are discussed briefly. The potential for membranes in cases of nonadditive water treatment and for allowing contaminant recycle and reuse is examined.

3.1 SOLID-LIQUID SEPARATION

The foremost membrane technologies for removal of solids from aqueous streams are the similar processes of *cross-flow microfiltration* (CFMF), *dead-end microfiltration* (DEMF), and *tubular filter press* (TFP). The microfiltration membranes have porous structures and are commonly synthetic polymers woven as textile fibers or sintered materials presented as polymer, ceramic, or metallic, channeled or tubular blocks. Microfiltration can often be applied in processes where traditional filtration processes have been used, although microfiltration can remove particles an order of magnitude smaller than can conventional filtration units (down to 0.1 µm). CFMF has been recommended for metals and solid removal in a range of processes, usually in conjunction with sulphide precipitation (Bradbury Ltd., 1993). This section will concentrate on the potential applications of CFMF and DEMF and explain the differences in the processes.

Both CFMF and DEMF may be used to remove solids from suspensions. CFMF has typically been used as a clarification and thickening solution. DEMF has been used for dewatering of concentrated suspensions. Both processes make use of tubular supports as the filtration medium. The processes are theoretically similar in that TFP represents an extreme operating condition of CFMF (Swart, 1994).

During CFMF, as the name suggests, the feed stream flows tangentially across the surface of the filtration membrane. A slurry is pumped into an array of horizontal collapsible fabric tubes. Solid particles accumulate against the filter wall as the permeate is removed, forming a filter cake. The thickness of the cake increases initially as the solids are deposited but is limited by the tangential shearing action of the feed slurry. A clarified liquid permeate is produced, and the retentate is a concentrated slurry. A number of published applications for CFMF exist, and these are discussed in this section.

TFP was developed as a cake recovery process. TFP is operated in dead-end mode, with all liquid passing through the filter cake and cloth filter media. Cake deposition and cleaning cycles are alternately used. TFP may be viewed as a limiting case of CFMF where the cross-flow velocity is zero. Published examples of applications of TFP have been limited to work undertaken by the University of Natal, Durban, in South Africa and EXXPRESS Technologies Ltd. in the United Kingdom. Documented examples of applications are dewatering of power generation plant sludges and water treatment works sludges (Rencken, 1992, and Treffry-Goatley et al., 1987).

Flexible-flow geometry allows the slurry to be fed from the outside or the inside of the tubes. The flow is characterized by a two-stage flow regime: an initial rapid flux decline followed by a gradual reduction. Hence, microfiltration techniques incorporate a cake removal or cleaning stage. Cleaning can be accomplished by air or filtrate backflushing, flow reversal (with or without the use of sponge balls), brushing or spraying the outside of the tubes to dislodge the cake on the internal surface of the tubes, or the use of roller cleaners.

3.1.1 Replacement or Improvement of Conventional Treatment

3.1.1.1 Chemical Flocculation. In instances where chemical flocculants are used to concentrate contaminants, membrane processes may provide an attractive alternative to additive treatment. Preliminary tests on a skim latex solution (containing approximately 5% dry rubber content) with micro- and ultrafiltration membranes produced excellent separation with the permeate containing undesirable proteinaceous matter while the required latex emulsion was retained by the membrane (Cartwright, 1992). Tests were continuing at higher system pressures to achieve higher recoveries and more practical flux rates.

3.1.1.2 Clarification. Ultrafiltration has been successfully applied to aid clarification in the traditional wastewater treatment process chain. Vial et al. (1992) optimized an ultrafiltration process to treat the effluent from a secondary clarifier at a municipal wastewater treatment plant. The results of their work were used to design a plant producing 480 m^3/d. A microstrainer, equalization tank, and 200-μm prefilter were used for feed pretreatment. A heat exchanger heated the feed and recycled concentrate. Trials were performed using PVDF membranes, type FS 61. Results obtained showed that suspended solids could be removed to less than 1 mg/L and average chemical oxygen demand (COD) reduced by 60 percent to 30 mg/L. The authors concluded that the plate-and-frame arrangement they had used was not economically viable but felt that membranes that produced higher filtration rates (such as hollow-fine fibers) should significantly reduce the investment cost.

Similar results have been achieved using cross-flow microfiltration units where costs would be expected to be lower. A number of examples of microfiltration units are discussed in this chapter in Sec. 3.2. As a corollary, cross-flow microfiltration of secondary sewage effluents polishes the water, often reducing the suspended solids and turbidity to less than 1 mg/L and 1 NTU, respectively.

Cross-flow microfiltration has also been used effectively as a clarification and disinfection unit for primary sewage treatment (Kolega et al., 1991). The existing treatment plant provided screening, grit removal, scum removal, and sedimentation. A Memtec microfiltration unit was installed to process the effluent from this plant. The unit reduced the suspended solids from between 103 and 389 to less than 1 mg/L. Similarly, turbidity was reduced from between 70 and 250 to less than 1 NTU. Biological oxygen demand (BOD), oils and grease, and phosphorus were also reduced. Furthermore, all indicator microorganisms were eliminated in the filtration stage. The flux of permeate from the unit was 120 m^3/h but was dependent on the suspended solids loading.

Membrane processes are capable of performing clarification operations. It is difficult to assess the applicability in terms of economic or throughput considerations, as units typically perform more than one function (such as clarification and disinfection). Full-scale units have not been reported, although the success of recent pilot-scale studies seems to indicate that the use of membrane units, particularly cross-flow microfiltration, will become more common in the near future. Some such applications are discussed in the relevant chapters of this book.

3.1.1.3 Centrifugation. Separation of cells from growth media is a common primary downstream processing stage following biological reactions. Centrifugation has been the favored option for this type of separation but cross-flow microfiltration has recently become a viable alternative (Mackay and Salusbury, 1988). This application can be extended to a range of similar applications for water treatment where cell debris and colloids are components of the feed stream.

While centrifuges are capable of up to 95 percent retention of yeast, bacteria, and cell debris, they are unsuitable for highly viscous fungal broths and shear-sensitive suspensions. Cross-flow microfiltration has the advantage in these systems of potential 100 percent retention of solids, good biological retention, and flexibility of module design and operation. However, the perceived disadvantages are those typically associated with cross-flow filtration: time-dependent permeate flux, high membrane replacement and pumping costs, and the relatively poor economics of large-scale implementation.

The choice of the most appropriate process depends on the application but some general guidelines can be given. Centrifugation's major costs are depreciation and maintenance, while microfiltration's are membrane replacement and pumping. Where the biological components are large (>6 mm) cells, it is likely that centrifugation will be the most appropriate choice. For processing mixtures containing small (1 to 3 mm) cells, the difficulties of separation increase significantly, although they are not as acute as those faced when dewatering cell debris (<1 mm). If the cells can be strongly aggregated, centrifugation will be the best option. Where the feed contains discrete cells, only laboratory and pilot-scale work will indicate the best option.

The advantages of centrifugation will increasingly outweigh those of microfiltration as the scale of operation increases. At small (<20,000 L) scale, cross-flow microfiltration may be more attractive, while at large (>20,000 L), centrifugation may be the better choice. Unless aggregation is possible though, centrifugation becomes difficult and processing rates become low or high, G-force machines are required for any operation exceeding 5000 L.

Even at small scale, the use of microfiltration for dewatering cell debris poses a problem due to acute membrane fouling. It is generally the rheological behavior of the fermentation broth which dictates the microfiltration downtime and membrane replacement costs. Viscous and fouling feeds are not ideal candidates for microfiltration. For centrifugation, it is the cell or aggregate size and density and the scale of operation which will most likely influence the process choice. At present, centrifugation is the preferred process for large-scale operation, but the flexibility of microfiltration for a range of feeds might make it more appropriate outside the biotechnology sectors.

3.1.1.4 Bioreactors.   A modification of conventional biological treatment processes is the replacement of secondary sedimentation tanks with membrane units. The membrane process has the advantages of allowing a higher biomass concentration in the reactor (Smith et al., 1969), it should prevent sludge bulking, and it may preclude the necessity for a separate disinfection process. Unit arrangements typically consist of a bioreactor with an adjacent membrane filtration unit. The activated sludge is recycled through the system, and the permeate is withdrawn through the membrane. Alternatively, the membrane may be placed, possibly within a separation tank, inside the reactor. Aeration of the reactor may reduce fouling of such an in situ membrane.

Chiemchaisri et al. (1992) investigated the use of a domestic hollow-fine fiber membrane in a bioreactor and separation tank configuration. A high degree of organic stabilization was obtained without sludge wastage in the unit. By manipulating the aeration rate, nitrogen removal was improved to over 90 percent. Rejection of 4 to 6 log viruses by the gel layer formed at the membrane surface was observed.

Magara et al. (1992) devised a bioreactor and ultrafiltration unit to treat sewage sludge. The full-scale unit was anticipated to produce effluent with levels of BOD, SS, COD, total nitrogen, and total phosphorus of about 5, 0, 35, 20, and less than 0.3 mg/L, respectively. The stated benefits of the system were that it would require less than or equivalent investment to a conventional biological denitrification plant, and it would require less land and fewer operators. The authors stated that the process would be the *major system for collective night-soil treatment plants in Japan.*

A laboratory pilot-plant bioreactor using membrane ultrafiltration showed promising results in wastewater treatment (Chaize and Huyard, 1991). The bioreactor had a working volume of 4.5 L and was equipped with a draft tube to prevent foaming difficulties. The temperature was maintained at 20°C, the air flow rate at 80 L/h, and the mixing speed between 800 and 1200 r/min. The ultrafiltration unit consisted of a plate-and-frame module manufactured by De Danske Sukker-fabrieker. Both polysulphone (GR51PP) and cellulose (ETNA) membranes were investigated. A linear velocity across the membranes of 1.5 m/s and a pressure of 1 to 2 bars were used in the investigation. Indications were that, after 150 days of operation, the cellulose membranes returned a higher flux but both membrane sets produced good-quality effluent. At the optimum sludge and hydraulic retention times of 100 days and 2 hours, respectively, all indicator organisms were removed and suspended solids were reduced to below the measurable limit. Soluble ammonia and nitrates in the process effluent were less than 10 and less than 0.5 mg N/L, respectively. The authors felt, though, that further optimization of the system was required to reduce the costs associated with the recirculation loop before the unit could be effectively applied in conventional wastewater treatment plants.

Bioreactors could find application in industries increasingly required to meet more stringent environmental standards for discharge of wastewaters, particularly where effluents contain a range of contaminants best treated by aerobic digestion. Tannery effluent contains high-strength organic matter and, often, the heavy metal chromium. A bioreactor incorporating an immersed hollow-fiber membrane filtration unit was tested using tannery wastewater (Yamamoto and Win, 1991). The polyethylene membrane was manufactured by the Mitsubishi Rayon Co. The unsupported membrane had an effective surface area of 0.27 m^2 and a pore size of $0.1 \text{ }\mu\text{m}$. The reactor had a working volume of 2.25 L.

Optimization of the operating conditions indicated a sludge retention time of 20 days, and a sludge concentration of 30,000 to 40,000 mg/L should be applied. Under these conditions, a removal efficiency of COD and total chromium of 93.7 to 96.3% and 95.4 and 97.7% were achievable. The effluent total chromium concentration was 0.41 to 1.2 mg/L throughout the 60-day experiment. The chromium content of the sludge was as low as 1 to 5% throughout the experiment, although some accumulation of chrome content in the reactor was noted. The authors noted that the sludge concentrations of 10,000 mg/L would not ordinarily have been achievable by sedimentation in 30 min. The unit could achieve very high organic loadings, making any full-scale treatment system comparatively compact.

Initial results with bioreactors, coupled with recirculation through membranes, seems promising. Good sludge retention characteristics, coupled with significant removal of ammoniacal nitrogen and small footprint, would appear to be advantages that will see the potential for bioreactor-membrane units grow.

A full-scale ADUF[R] (anaerobic digestion ultrafiltration) process was installed at a maize processing plant, African Products, in South Africa (Ross et al., 1992). Earth instability had resulted in the company shutting down three of its five 225-m^3 clarigesters. The remaining two clarigesters were unable to process the total factory flow, mainly because of the poor settling properties and because of the nongranular digester sludge. A treatment plant was designed that incorporated a clarigester volume of 2610 m^3 and an unsupported, tubular, polyethersulphone membrane area of 668 m^2. The design flow was 500 m^3/d with a COD of 15 $kgCOD/m^3$ (peaking as high as 60 $kgCOD/m^3$). The digester-suspended-solids concentration was 21 kg/m^3. After 15 months of operation, the process was shown to be producing a colloid-free effluent at a mean COD removal efficiency of 97 percent. The permeate flux varied in the range of 37 to 8.1 $L/m^2 \cdot h$ at 35°C. Periodic cleaning with ethylenediaminetetraacetic acid (EDTA) was commenced only after 13 months of operation. A mean plant space load of 3 $kgCOD/m^3 \cdot d$ guaranteed reliability to withstand high COD shock loadings caused by variation in the feed flow. The process was envisaged as having important economic advantages to the treatment of a wide spectrum of organic industrial effluents.

3.1.2 Sludge Dewatering

The treatment of raw water usually involves a coagulation and flocculation stage followed by clarification and filtration. Up to 10 percent of the feedwater to the plant may be lost in the resultant sludge. The solids content of the sludge may be between 3 and 30 g/L, depending on the clarification chemicals and equipment. Quite apart from the disposal difficulties associated with the sludge, it contains a significant portion of potentially usable water.

Traditionally, a plate-and-frame filter press or centrifuge would be used for the purpose of dewatering this sludge. However, the tubular filter press (TFP) arrangement described earlier in this chapter offers a number of benefits. First, the cake is formed inside the self-supporting tubes; hence, filter plates are not required. Second, the TFP produces a very much thinner cake (typically 3 mm) than other filter press processes so that filtration rates are correspondingly higher (Treffry-Goatley et al., 1987). Third, the process may be automated with high sludge velocities being used in conjunction with external roller pinchers to dislodge the internally accumulated sludge (Rencken and Buckley, 1992).

A prototype unit was constructed at a water treatment facility to investigate the full-scale operability and economics of the process. It was designed to treat the waste sludge arising from the treatment of 30,000 m^3/d of water. Coagulation with bentonite resulted in a clarifier sludge of concentration 23 g/L and flow of 50 m^3/d. The processes produced a sludge at an average concentration of 28% m/m. The process and its economics were considered attractive (Rencken et al., 1990).

3.2 MEMBRANE DISINFECTION

Many manufacturers of membranes and RO equipment claim that no virus should be found in product water from membrane-treated waters. This argument is based on the pore size of the membranes relative to the physical size of the viruses. In practice, viruses (specifically coliphage T2 and poliovirus) have been found to bypass or penetrate commercial CA membranes. Nevertheless, removal efficiencies are in the 99.20 to 100.00 percent range (Eisenberg and Middlebrooks, 1986). The releases may be caused by imperfections in the membrane surface and unit seals.

Membrane processes are attractive for use in disinfection processes as a nonadditive treatment. Chlorine and ozone may react with organic impurities in water to produce undesirable chlorinated organic products. The number of by-products covered by the World Health Organization's (WHO) drinking water quality guidelines rose sharply from 2 in the 1984 edition to 15 in the 1992 version. Modeling studies by the Water Division of the U.S. Environmental Protection Agency (EPA) indicated that treatment of surface water by coagulation, filtration, and chlorination may produce trihalomethane levels in the 25- to 100-μg/L range (Anon, 1993a). This concentration may be correlated with a few (28 to 40 people in a population of 140 million) premature deaths caused by cancer. While the number of fatalities is negligible by comparison with the risks associated with waterborne pathogens (several thousand people for the same population sample), U.S. legislation may force a reduction in total trihalomethanes to 25 μg/L at an estimated cost of $1 to 2 billion a year in the United States or double present water treatment costs in the United Kingdom. Predictions as to the method to be used to effect the by-product reduction appear to center on granular activated carbon addition prior to disinfection to remove trihalomethane precursors.

Technically, the membrane process best able to remove pathogens is reverse osmosis but researchers have had success with a range of processes. Ultrafiltration has been shown to remove all pathogens, including viruses, from contaminated feeds. The process makes disinfection unnecessary, apart from the addition of a small amount of residual chlorine prior to distribution. Lyonnaise des Eaux Dumez has experience in operating ultrafiltration plants treating 40 m^3 of water per hour. They estimate that, for larger plants, treatment costs would be more than twice those of ozone and peroxide followed by granular activated carbon (Anon, 1993a). Despite these drawbacks, experimentation on and refinement of membrane processes will proceed as a possible nonadditive means of water treatment.

Ellender and Sweet (1972) experimented with flat-sheet Amicon PM-30 UF membranes (constructed of polyelectrolyte coacetate material) with molecular cutoff weights of 30,000 daltons. Viral inputs of 30,000/mL were treated such that effluents with viral concentrations of 0.002 and 0.05% were produced. These experiments led to further research that demonstrated that membrane processes can be combined to produce drinking water.

Membranes are not necessarily superior to conventional treatment for disinfection. Cationic polyelectrolyte addition in pretreatment, prefiltration, and hypochlorination removed 99.99 percent of an f2 virus added to natural water. No virus removal resulted when polymer addition was omitted from conventional treatment. Addition of viruses after filtration resulted in only 97.3 percent removal, following membrane treatment (Eisenberg and Middlebrooks, 1986).

Endotoxins have been removed from feed streams with varying success by membrane processes. Karamian (1975) tested CA-dioxane membranes with molecular mass cutoffs exceeding 1,000,000 daltons. Maximum rejections of 91.8 percent were recorded but this decreased (along with permeate flux) to a low of 80.4 percent after eight days. Klein et al. (1983) tested reverse-osmosis membranes and found them to be superior to ultrafilters. RO spiral-wound membranes containing an interfacially polymerized polyamide and preceded by a polysulfone ultrafilter produced an endotoxin-free solution.

Microfiltration has been shown to be a successful treatment for disinfecting sewage water. Using a Biocarbone biofilter and a Memcor microfilter (pore size 0.2 μm) in a pilot plant arrangement, Langlais et al. (1992) eliminated all fecal germs, tenia, and ascaris eggs from a poor-quality secondary sewage effluent (COD between 100 and 400 mgO_2/L, BOD_5 between 30 and 15 mgO_2/L). Total removal of free ameba cysts was unconfirmed. A turbidity removal of 99 percent was achieved but the color remained between 50 and 150 mg Pt. Col/L. Tests were reported a year later (Langlais et al., 1993) that indicated the same degree of disinfection of a good-quality secondary effluent (COD between 14 and 40 mgO_2/L). A major difference experienced was in the operability of the unit. Whereas the poor-quality effluent allowed a minimum filtration cycle of 72 h for a minimum permeate flow of 80 $L/m^2 \cdot h$, filtration cycles in excess of 300 h were possible with the good-quality effluent. A further comparison of the unit run in cross-flow and dead-end mode showed that the cross-flow mode produced 9 percent more permeate for the same recirculation flow rate, although this tendency remained unconfirmed.

Tests have also been run on the same membranes in Sydney, Australia (Peters, 1992). A full-scale microfiltration system was installed at the Sydney Water Board's Blackheath site in Blue Mountains and had been running for a year. It provided tertiary sewage treatment for a community of 3500 people. It was designed to treat peak dry weather flows of up to 2950 m^3/d of effluent from an existing trickle bed filter plant. Again, the unit was found to remove all indicator bacteria and viruses and showed significant reductions in BOD, turbidity, oils and greases, and some reduction in phosphorus. Additional claimed benefits of the system were the low energy requirements at 0.15 kWh/m^3 of effluent treated and the modular construction that allowed for prefabrication, ease of expansion, and small footprint.

An EXXFLOW microfiltration unit has been operational at a potable water treatment plant in Durban, South Africa, since 1991. The unit is run continuously, treating the intake water at the head of the works. The unit is being tested to assess its capability in terms of producing potable water of comparable quality to traditional processes. The unit has consistently eliminated all measured bacteria and viruses and reduced turbidity to less than 0.3 NTU (Pillay, 1994). The unit reduces suspended solids by 99 percent. No figures were available for the conventional process under the same conditions.

A comparison has been made between conventional sand filtration and microfiltration as a polishing stage in the treatment of domestic effluent (Oesterholt and Bult, 1993). The effect of the treatment on the microbacteriological content of the feed was examined. The pore size cutoff of the membrane was the same as that in the Memtec examples (viz., 0.2 μm). The researchers found that, while sand filtration removed all of the 26,000 *Escherichia coli* microorganisms per 100 ml, there was still a positive count of 2 after the microfiltration. The following determinands were also investigated in the study: suspended solids, COD, Kjeldahl nitrogen, phosphorus, copper, and zinc. With the addition of iron flocculant, sand filtration removed 70 percent of the suspended solids and total phosphorus, while microfiltration removed 90 percent. The removal of COD, nitrogen, and zinc was less encouraging at between 10 and 40 percent, with iron addition generally having a beneficial effect. The authors concluded that microfiltration was preferable from a technical viewpoint because of its higher removal efficiency across the range of determinands. However, they felt that the microfiltration was not an economically feasible option since the inclusive treatment costs were almost 14 times those of sand filtration.

Successful disinfection of *E. coli*–containing solutions was effected in an electrodialysis system (Tanaka et al., 1986). Optimal germicidal efficiency was observed in the region where current density was greater than the limiting current density (i.e., in the vicinity of the ion-exchange membranes). The germicidal efficiency is considered to be a result of the synergistic effect of the H^+ and OH^- ions produced by water dissociation. At current densities of 1.63 A/dm^3, pHs around 4.6, and using ion-exchange membranes, the viable cell count in the unit effluent was reduced to zero in 20 min.

Membrane systems are also gaining favor as portable water purifiers. The 1993 winner of the Unique and Useful Consumer Plastics Product Award (sponsored by BASF) was the Explorer antimicrobial water purifier (Anon, 1993b). Materials used in the Explorer included polyurethane, woven nylon, nylon monofilament cloth, and polypropylene sintered porous plastic. The device permitted the manual production of water (up to a liter per minute) from almost any water source.

Membrane processes are finding increasing application in the field of water disinfection. This trend is likely to continue as disinfection by-products from additive treatment processes are more stringently regulated. Investigations into the applicability of the various membrane processes show, counterintuitively, that cross-flow microfiltration appears to be the most successful means of disinfection. The reported failure of ultrafiltration and reverse osmosis to completely disinfect contaminated waters is probably due to manufacturing defects, and these difficulties will need to be resolved with the manufacturers of the units. Finally, it would appear that the economics of membrane treatment will need to be addressed before the systems find widespread application in the field of water disinfection.

3.3 INORGANIC MATERIAL

3.3.1 Seawater and Brackish Waters

Membrane processes have been installed to treat seawater to produce potable water in regions where rainfall runoff is scarce. In producing potable water from sea- or brackwaters, typically the most difficult criterion to meet is to satisfy the World Health Organization Drinking Water Quality Guideline (WHO, 1984) for chloride concentration of less than 250 mg/L, rather than the guideline for total dissolved solids (TDS) of less than 1000 mg/L. To do this from seawater having a chloride ion concentration 19.8 g/L and using an RO system operating at 30 percent recovery requires a membrane chloride ion retention of about 99 percent. To ensure acceptable permeate quality for the duration of the membrane life, it is often necessary to design for a 99.5 percent salt retention.

Multistage flash distillation has been the predominant process for the treatment of seawater. Initially, membrane processes were precluded from consideration because the high operating pressures required to overcome the osmotic pressure of the water made the processes uneconomical (Gutman, 1987). Large brackwater RO plants were first built in the late 1960s. By the early 1970s, sufficient operational experience had been accumulated for large plants to be designed and installed.

The scale of membrane applications is now very large. Plants with a capacity in excess of 19,000 m³/d are commonplace. The Saudi Arabian city of Jeddah is served by six RO plants, all having a capacity in excess of 18,000 m³/d and a total capacity of 192,000 m³/d (Gutman, 1987). An RO plant commissioned in 1989 in Jeddah has a capacity of 56,800 m³/d (Abanmy and Al-Rashed, 1993). These plants treat borehole water to provide drinking water for the city. An even larger plant, with a capacity of nearly 450,000 m³/d is planned in the United States as part of the Colorado River desalting project (Gutman, 1987).

Recent advances in the state of the art have meant that RO desalination has now become economically attractive even at seawater concentrations, particularly because of improvements in the membrane materials and pretreatment. On Sanibel and Captiva, barrier islands on Florida's southwestern Gulf Coast, an RO plant was installed to provide an expandable capacity of 13,600 m³/d potable water (Smith and Derowitsch, 1994). After the original cellulose acetate membranes' performance deteriorated significantly toward the end of their design lifetime, the choice of Filmtec FT30 membranes as replacement significantly enhanced the performance and economics of the process. The membranes could operate at low pressure: for a 0.2% salt solution at 1.5 MPa, salt retention of 96 percent and fluxes of 6.7 m³/m²·d were routine. Furthermore, the membranes could operate over a pH range of 2 to 11, at temperatures up to 45°C, and were impervious to bacteriological attack. The economic implications of these process changes are as follows: Energy costs dropped to $0.05/m^3, nearly 25 percent less than predicted in 1980 dollars when the plant was installed. Acid costs (for hydrogen sulphide removal) have been reduced 60 percent because the more lenient pH operating range has allowed addition of acid to the product water directly. Finally, although replacement costs of the new membranes were expected to exceed those of the cellulose acetate (CA) membranes, this expense has also turned out to be lower because the membranes have been running effectively for seven years without sign of significant performance deterioration.

Various membrane processes have been tested as pretreatment to membrane desalination of sea- and brackwaters. Tests were performed on tubular and capillary polyethersulphone ultrafiltration modules as pretreatment for RO desalination (Strohwald and Jacobs, 1992). While UF had been shown to be economically competitive with processes such as combined media and carbon filtration, high foulant concentrations (particularly nutrients, suspended solids, and organics) posed particular problems that the UF units (developed by the authors) seemed to address. While no effective means of comparison could be found for the systems under investigation, the UF processes coped with the difficult treatment conditions and, on the basis of material cost, the capillary format proved competitive with the tubular technology, even though its permeate flux rates were somewhat lower.

In instances where the contaminants pose significant fouling problems for UF pretreatment, cross-flow filtration may be used to treat the seawater. Holdich and Zhang (1991) tested various membranes for pretreatment of seawater to be used in injection water in North Sea oil production operations. The presence of suspended solids in the injection water is considered problematic as these solids may clog the stone medium which the injection water must penetrate to displace the oil. An acrylonitrile plate-and-frame cross-flow filter was found to be one of the most suitable for this application. Difficulties associated with gel layer formation could be avoided by using an open pore structure which is too wide for the gel layer to form a strong mechanical bridge over. This open pore structure may allow fine particles to pass through the filter, possibly negating any possibility for RO filtering but allowing backwashing to sustain fluxes of 24 m^3/m^2·d. Lower fluxes would typically be expected where RO was to be used and backflushing would be less effective.

While being one of the oldest applications for industrial-scale reverse osmosis, desalination of brines and seawaters looks set to continue to represent the largest-scale application of the technology for some time yet. Significant experience has been gathered in the field. The use of more sophisticated membrane materials can significantly improve the performance of reverse-osmosis plants based on the traditional cellulose acetate format. Ultrafiltration and microfiltration may be used as effective pretreatments for the relatively more fouling-sensitive reverse osmosis.

3.3.2 Hardness

Cations and anions that cause water hardness have traditionally been removed by the precipitation of the ions concerned or the use of ion exchange for selective treatment.

Nanofiltration membranes are used to soften waters (and remove divalent ions as well). Because of their ability to reject high percentages of many dissolved components, nanofiltration membranes offer a single treatment alternative that will treat lower TDS waters for removal of specific contaminants. Nanofiltration (or ultra-low pressure membranes) typically provide 80 to 95 percent rejection of hardness and over 70 percent rejection of monovalent ions at only 500 to 700 kPa (Watson and Hornburg, 1989). Nanofiltration units have the advantage of removing tri-halomethanes without the necessity for activated carbon treatment or air stripping. Production costs for nanofiltration processes are reported to be lower than those associated with conventional lime softening and THM removal across the range of plant sizes.

Almost 450,000 m^3/d of nanofiltration facilities were planned for the Florida area (Conlon et al., 1989). Nanofiltration proved particularly effective in treating the high hardness and natural organics (such as humic and fulvic acids) encountered in the Florida waters. A three-stage process was typically used to effect 90 percent recovery. Typical operating pressures were between 620 and 825 kPa, but a booster pump was required between the second and third stage. Capital costs for the facilities were reported as less than $2.38 per m^3/d capacity ($0.90 per USgpd) for plants with over 100,000 m^3/d capacity (30 USmgd). Operation and maintenance costs are on the order of $0.10 per m^3 ($0.40 per 1000 US gal). It was found that blending waters treated with nanofiltration with waters treated via other techniques enhanced the quality of the final water. Using nanofiltration to treat waters with TDS values below 1000 ppm extended the economic treatability range for drinking water supplies.

A plant operating in Singapore produces 480 m^3/d of reclaimed water for industrial use (Chin and Ong, 1991). The process chain for water polishing comprises chemical coagulation and flocculation, multimedia sand filtration, and chlorination. A further reverse-osmosis stage uses spiral-wound cellulose acetate to remove most of the cations, anions, and heavy metals present in the water. Using the RO, more than 95 percent removal of chlorides, sulphates, calcium, sodium, and total hardness was achieved.

3.3.3 Nitrate Removal

Nanofiltration has also been proposed as a mechanism for nitrate removal from drinking water (Rautenbach and Gröschl, 1990). A number of municipal water-works face the problem of increasing nitrate concentration in their sources. At present, most concentration reduction is achieved by blending the high-nitrate waters. In cases where this is impossible, a reliable and cost-effective separation process must be installed. By treating the feed with sodium sulphate and using FilmTec NF 40 membranes, significant nitrate removal was effected. The nanofiltration stage produces a permeate containing sodium nitrate (and sodium chloride if chloride ions are present in the feed). By passing the permeate through an anion-exchanger bed, the nitrate ions are replaced with chloride ions, for example. The permeate and the retentate may then be recombined to produce a nitrate-free water. The recovery rate accorded to this system is high since the only losses would be those required for ion-exchange regeneration and flushing.

Furthermore, since the nanofiltration permeate is practically free of scaling components such as calcium sulphate and calcium carbonate, the permeate may be passed through a reverse-osmosis stage before blending, with an associated high recovery rate (on the order of 94 percent).

3.3.4 Heavy Metals and Other Inorganic Material

Heavy metals and other organics are not typically removed independently in general water treatment processes. In instances such as metal plating or wire drawing, membrane processes such as microfiltration and ultrafiltration have been successfully used to remove metal contaminants (including heavy metals) from specific process effluents. Often, the use of membrane processes on unmixed effluents allows the reuse of the concentrated retentate and cleaned permeate (Bradbury Ltd., 1993). However, the scope of application of such processes is considered too broad for a chapter such as this. Where an application might exist, it is advisable to consult the literature or contact a membrane manufacturer for specific information. Some examples of metal removal are discussed in other sections of this chapter where they illustrate the applicability of membrane processes in more general fields.

3.4 ORGANICS

The molecular size, shape, and chemical characteristics of an organic compound have been shown to influence the rate of permeation through a membrane by an organic compound. The stearic and polar effects specific to individual compounds have been shown to control the retention of compounds. Generalizations about functional groups may be made for similar membrane surface components. Furthermore, the retention of the compound may be related to its ability to form hydrogen bonds.

The majority of available information regarding solute-membrane interaction relates to cellulose acetate membranes even though fundamental research into contaminant-membrane interaction by polymer scientists has led to the refinement and development of a range of membrane materials. Nevertheless, little information has been made available on how contaminant-membrane interaction characteristics affect the retention of mixtures of natural or synthetic organic chemicals during plant scale-up operations (AWWA, 1989).

3.4.1 Pesticides and Synthetic Organic Compounds (SOCs)

Evidence of the use of membrane processes being used on an industrial scale for the removal of pesticides and SOCs is scarce. A number of studies have been undertaken to assess the feasibility of using membrane processes for the removal or concentration of specific pesticides, and some of these are discussed as follows. However, the reason that few of these processes are taken beyond feasibility or pilot-scale investigation probably involves the economics of such undertakings. With the exception of environmental remediation applications such as those governed by the U.S. Superfund program, membrane processes are likely to be considered as a last resort. Insufficient evidence regarding the persistence and long-term toxicity of most pesticides makes it difficult to assess the specific economic variables in each instance.

Membrane processes reviewed for the treatment of synthetic organic compounds (primarily pesticides) show significant disparity in the applicability of different membranes to the contaminants (AWWA, 1989). In certain instances, negative retention (the concentration of the contaminant was greater in the permeate than in the feed) of the organics occurred. This discussion will be restricted to examples of the application of membrane processes to the treatment of pesticides where positive results have been demonstrated.

Three RO systems have been studied at Water Factory 21 in Orange County, California (Reinhard et al., 1986). The aim of the study was to investigate organic removal using cellulose acetate and polyamide membranes. A full-scale (20,000 m^3/d) RO water treatment plant and two pilot plants (using spiral-wound polyamide membranes) were evaluated to determine total organic carbon (TOC) and SOC removal from lime-clarified secondary municipal wastewater.

The full-scale RO plant comprises spiral-wound cellulose acetate modules arranged in two units with three sections, each having 42 vessels 6 m long. They were connected in a 24-12-6 array to give a total membrane area of 4460 m^2. The plant was operated at 3170 kPa with an input feed flow of 18,150 m^3/d and gave an overall recovery of 84 percent. Pretreatment consisted of chlorination, sodium hexametaphosphate (SHMP) addition (to minimize scaling), and pH adjustment to 5.8 with sulphuric acid. The permeate was treated to remove carbon dioxide in an air-stripping tower. Analysis samples were taken after air stripping: a factor which may have skewed apparent performance of the RO plant.

The pilot plants comprised spiral-wound polyamide membrane modules arranged in a 2-1 array. Pretreatment consisted of acid and SHMP addition. The operating pressure was 1725 kPa. The average feed flow, recovery, and permeate flux of the first plant were 28 m^3/d, 52 percent, and 15 L/m^{2}h, respectively. The average feed flow, recovery, and permeate flux of the second plant were 45 m^3/d, 67 percent, and 22 L/m^{2}h, respectively.

Average TOC reduction was 89 percent to 0.95 mg/L at the full-scale RO plant, while both pilot plants removed 99 percent of the TOC to 0.1 mg/L. Statistically, the readings were significant to 75 percent and 83 percent in plants 1 and 2, respectively. COD was removed to below the detection limit (1 mg/L) by both pilot plants, but the full-scale plant achieved only 2.3 mg/L. The pilot plants also retained some 60 percent of the volatile organics, while the cellulose acetate appeared to have no effect on these compounds. Overall, the PA membranes retained more trace organics than the CA membranes.

There was a significant discrepancy between the performance of the two pilot units in terms of their effects on specific compounds (e.g., retention of chloroform at 71 and 22 percent, of naphthalene at 80 and 30 percent, and of 2,4,6 trichlorophenol at 100 and 59 percent for plants 1 and 2, respectively). There was no apparent reason for this difference and the explanation advanced by the authors centered around installation leaks, membrane defects, variation in the membrane manufacturing process, and sampling errors.

In the final analysis, the work demonstrated the difficulties associated with pilot plant work but showed the potential of the polyamide membranes for the treatment of pesticides and other SOCs. More than 86 percent of pentachlorophenol, 98 percent of trichloroethane, more than 67 percent of bromoform, and 79 percent of dibromochloromethane and bromodichloromethane were retained by the polyamide (PA) membranes. Not every analyte was best treated by the PA, though, a notable example being chlorobenzene where the PA membranes exhibited no or negative retention.

A contradictory set of results pertains to the treatability of lindane by CA RO. Laboratory-scale tests on a variety of pesticides were conducted at feed pressures of 1470 lb/in^2 (Hindin et al., 1969). Dichlorodiphenyltrichloroethane (DDT) feed concentrations of 910 to 7000 µg/L were reduced to less than 14 µg/L (more than 99.5 percent retention) at an average flux of 18.5 g/sfd. Tetrachlorodiphenylethane (TDE) feed concentrations of 23 to 532 µg/L were reduced to below 0.006 µg/L at fluxes of 15 g/sfd. Benzene hexachloride (BHC) feeds of 638 µg/L were reduced 52 percent to 306 µg/L at 24 L/m^{2}h. Finally, lindane feed concentrations between 50 and 500 µg/L were reportedly reduced to between 8 and 133 µg/L (representing an average retention of 78.7 percent) at average fluxes of 29 L/m^{2}h.

This is in contradiction to laboratory tests also conducted on CA membranes where lindane was noted to pass through and absorb on the membrane, only to be released during flushing (Malaiyandandi et al., 1980). A major portion of the lindane (40 percent) could not be accounted for. This was assumed to have been absorbed onto the membrane. Any membrane that absorbs and desorbs SOCs could not be used solely, if at all, for the treatment of potable water.

Alachor has been most successfully removed from contaminated waters by the use of thin film composite (TFC) membranes (Miltner et al., 1987). The TFC was compared to hollow-fiber amide membranes and spiral-wound CA membranes in single-pass configurations. The TFC delivered a 98.5 percent retention of the SOC. A pilot plant tested on 80-μg/L alachor-spiked groundwater successfully treated the pesticide to below detectable limits. A mass balance on the system indicated that no alachor had been absorbed by the membrane.

Alachor has also been successfully treated by nanofiltration (Taylor et al., 1989). In a one-month investigation, a membrane with a molecular weight cutoff of 300 (MWC 300), the following SOCs were completely retained: chlordane (MW 409), heptachlor (MW 373), methoxychlor (MW 346), and alachor (270). Mass balances conducted on the SOCs accounted for 92 to 100 percent, although complete recovery was evident for alachor only. The higher-weight SOCs may have been absorbed onto the membrane.

These few examples of the treatment of SOCs or pesticides by membrane processes show the need for adequate pilot-scale work in the treatment of such contaminated waters. The tendency of SOCs to be absorbed onto and subsequently desorb from membranes may make their use in the field of potable water treatment unlikely except for specific contaminant-membrane combinations. It would seem that the most likely application of membrane processes in the treatment of SOCs will be found in the field of environmental remediation where the waters can be passed through the membrane on a batch basis until the concentration meets the required specifications. Concerns about contaminant desorption and partial retention will not be strictly relevant when applied to batch treatment.

3.4.2 Oils and Oily Wastes

Treatment of contaminated effluents and groundwaters associated with the hydrocarbon and edible oil industries frequently involves the removal of free and emulsified oils. However, such wastes could arise in any industry where lubrication is required.

Ultrafiltration can be used to concentrate oily wastes (including oil emulsions). The permeate may be reused and the concentrate can be incinerated without the need for supplementary fuel (Eykamp, 1975). Feed concentrations of 1 to 5 percent oil have been successfully concentrated to 30 percent. Permeate fluxes of 60 to 120 L/m^{2}h are obtainable with permeate qualities of 30 to 400 mg/L of oil. By comparison, a 97 percent reduction of emulsified oil is achievable only after treatments of API separation, chemical coagulation, activated sludge, filtration, and carbon adsorption (Groves, 1976).

3.4.3 Solvents

The potential for the separation of organic solvents from water mixtures represents a challenging process with a wide range of potential applications. Such mixtures may be difficult to separate because of azeotropic points, the relatively high mobility of the solvents, and their associated health risks. Treatment of contaminated waters (particularly for environmental remediation purposes) to concentrate the organic content of the waters for disposal by energy-efficient incineration would seem to be one of the prime applications of such technology.

Separation of organic solvent–water mixtures has become possible with the advent of ceramic membrane filters. Tests conducted on organic polymer membranes tended to produce very low flux rates. Using a thin (less than 10 μm) silica-alumina porous membrane (average pore diameter of 1 μm) Sakohara et al. (1989) successfully separated binary water mixtures of methanol, ethanol, n-propanol, tetrahydrafuram (THF), and acetone. Both vapor permeation and pervaporation methods were tested with the membrane at temperatures of 50°C. Separation characteristics were markedly improved when the pinholes present in the membrane were plugged with organic material of relatively large molecular size, such as polyvinylacetate. With the improved membrane, high solvent retentions were achievable as well as good water permeation fluxes.

3.4.4 Textile Wastewater Treatment

The potential for membrane processes in the treatment at source of industrial effluents is discussed through the example of their use in the treatment of textile dyehouse effluents. The benefits of a holistic approach to effluent treatment combined with the realization of potential for resource recovery are indicated in the following examples.

Dyehouse effluents will vary according to the class of dye being used and the textile being dyed. As dyehouses are not restricted to single feedstocks or uniform product, there will typically be variation in the characteristics of the wastewater that arises from a single dyehouse. Any membrane system that is installed for the treatment of dyehouse efluents will need to accommodate this variation. To minimize the extent of any effluent treatment system, the efficiency of the process discharging the effluent must first be considered to identify opportunities for waste minimization at source. Concentrated wastes of distinct characteristics will generally prove easier to treat than dilute wastes containing a variety of different contaminants.

Membrane processes have the potential to treat dyehouse effluents to remove the dyestuff and allow the reuse of auxiliary chemicals used for dyeing or to concentrate the dyestuffs and auxiliaries and produce purified water. The membrane process selected will depend on the class of dyestuff used and the degree of effluent segregation within the dyehouse wastewater system.

Microfiltration will be effective in removing colloidal dyes from exhausted dyebaths and subsequent rinses. A 40-m³/d woven firehose microfiltration plant has been constructed for this purpose (Buckley, 1992). The feed to the plant was from a pressure jet used for dyeing 100 percent polyester fabric. Pretreatment of the effluent feed consisted of a wedge-wire screen to remove lint, alum dosing (50 mg/L as aluminum), and pH correction to 5.5 to 6.0. The average flux achieved was 50 L/m²h at an inlet pressure of 400 kPa and a tube velocity of 1.5 to 2.0 m/s. The permeate produced by the process was of low turbidity but was slightly colored. The residual color was due to the solubility of some of the disperse dyes used.

Reverse osmosis is suitable for removing ions and larger species from dyebath effluents. The presence of colloidal particles will cause fouling of some RO units. A 40-m³/d RO plant has been constructed to treat varied effluents, from cotton/synthetic and cotton to mixed dyehouse effluents. Because of the presence of disperse dyes, pretreatment consisted of flocculation with alum and adjustment of the pH to values between 5.0 and 6.0, followed by cross-flow microfiltration. The plant consisted of two stages, each comprising spiral-wound modules. The first stage was fitted with brackwater membranes and the second with seawater membranes. Operating pressure in stage 1 varied between 2.7 and 3.1 MPa. The brackwater membranes gave percentage rejections of over 96 percent for conductivity, total solids, and sodium ions. The percentage rejection for color was 90 percent and, for total carbon, 87 percent. The concentrate for stage 1 was used as the feed for stage 2. Operating pressure in stage 2 varied between 5.3 and 5.8 MPa, reflecting the higher osmotic pressure of the feed solution. The percentage rejection for conductivity, total solids, and sodium ions was over 98 percent and for color 94 percent. The quality of the composite permeate returned to the dyehouse compared favorably with tap water. In dyeing trials, 200,000 m of fabric were dyed using reverse-osmosis permeate with no discernible effect on dyeing quality. RO of untreated effluent from the dyeing of cotton fabric was not successful. Membrane fouling occurred due to the presence in the effluent of scouring chemicals and cotton waxes.

Unless the dyehouse dyes only one type of fabric, the selection of membranes suitable for the treatment of dyehouse effluents is limited by the variable nature of the effluents produced. The use of dynamic membranes may circumvent this problem. The loss of filtration efficiency may be countered by the improved service life of the membranes coupled with their rugged characteristics. An ultrafiltration plant consisting of 40 stainless steel modules with a dynamic membrane area of 256 m^2 was installed at a textile mill that dyed polyester/viscose fabric (Buckley, 1992). The dyehouse effluent contained both soluble and colloidal dyestuffs, acetate, alkali, salt, and organic auxiliary chemicals. The effluent varied in pH from 5 to 9, with temperatures up to 75°C and total dissolved solids up to 4500 mg/L. The color of the effluent often exceeded 10,000 ADMI color units. The pore size of the porous elements was modified by precoating the tubes with a precipitated zirconium hydroxide solution, prior to the formation of a hydrous zirconium (IV) oxide membrane, which was in turn converted to a polyacrylic acid. The inlet feed pressure was 6 MPa and the outlet pressure was 3.8 to 4.0 MPa. Cross-flow velocity was approximately 1.5 m/s and the operating temperature was 45°C. The average permeate flux for a water recovery of 85 percent was 33 L/m^2h. Color removal from the effluent was 95 percent or better. Ionic rejection (conductivity) was approximately 80 percent. The permeate has been reused since 1985. The concentrate is discharged through a deep marine outfall.

Soma et al. (1989) used mineral membrane microfiltration to treat the high-temperature, variable-pH feed. Alumina Ceraver membranes of 0.2-μm mean pore diameter were used in laboratory and industrial pilot plant trials. Pressures of 1 to 5 bar and cross-flow rates of 3 to 5 m/s were used in the trials. The trials showed that insoluble dyes were retained by virtue of their dimensions. Ionic dyes were retained when they were made insoluble by complexation with other dyes of opposite ionic charge (as a result of mixed baths, for example) or if they were fixed on micelles resulting from the agglomeration with oppositely charged surfactants and if the surfactants were subsequently absorbed onto the membrane surface. Ionic dye retention was aided by the formation of dynamic membranes on the membrane surface. The base of the dynamic membrane is normally present in the wastewater, especially in finishing effluents.

Nanofiltration (or charged ultrafiltration) may be used to decolorize effluent while retaining electrolyte. During the reactive dyeing of cellulosic fibers, electrolytes such as sodium chloride or sodium sulphate are added to the dye bath to exhaust the dye onto the fabric. Commercially available nanofiltration membranes have been used to treat cotton dyehouse effluents. The effluent contained reactive dyes and their hydrolyzed residuals, electrolyte (sodium chloride or sodium sulphate) ethylene diamine tetra-acetic acid (EDTA). The concentration of electrolyte was usually 20 to 100 g/L sodium chloride or 15 to 85 g/L sodium sulphate. The spent dye bath had a pH value of 11.5 to 12.5. Two types of test apparatus were installed at the dyehouse (Buckley, 1992). A flat-sheet membrane test cell was used to obtain point rejection data. Membrane flux and salt rejection data were obtained using a spiral-wrap element of area 0.56 m^2. As a first step, the electrolyte concentration in the dye bath was rationalized by choosing three standard solution strengths. Color rejection was found to be greater than 99 percent for blue, green, yellow, and red dyes at a dyeing concentration of 1 g/L dye and 50 g/L sodium chloride. Salt rejection increased with operating pressure and salt concentration. Color rejection increased with increasing operating pressure, so a balance had to be found between acceptable color rejection and salt rejection. Based on the results obtained, a full-scale nanofiltration plant has been designed to treat reactive dye liquors from a medium-sized (200 tons/month) cotton dyehouse. Based on a conservative design flux of 10 L/m^{2}h, the plant should recoup the initial capital outlay in under three years. The mass of electrolytes discharged to the environment will be reduced by more than 90 percent.

These examples of membrane processes applied in the textile industry indicate that, prior to investing in any effluent treatment/recycle process, the operation should be optimized to minimize or rationalize all waste generation. If the correct membrane process can be selected for a specific function, the opportunities for waste reuse and material recycle can be greatly enhanced. Adequate scope exists among membrane processes for manipulating the process to suit the requirements of the application if sufficient attention is applied to the management of the waste-generating process.

3.5 MIXED WASTEWATERS

A membrane process can be used to concentrate or extract residual organics and dissolved solids from complex industrial wastes. There is a danger, though, that it may be seen as a catchall solution to wastewater problems. In reality, the applicability of the process should be carefully tested in each instance. Theoretical studies can be used to assess the chemical composition of the waters. Pilot-plant studies on side streams or synthesized effluents can assist in physically realizing the effects of and the potential for membrane processes as well as potential operating difficulties that must be accounted for. Finally, care should be taken in pilot work to account for the range of concentrations that will require treatment through the process as species may precipitate once saturation limits have been exceeded in the feed.

Where membrane processes have been successfully installed, there can be considerable advantages over conventional treatment systems. The process can concentrate the contaminants for disposal, reducing the cost of treatment. Where selective treatment may be applied, a useful species may be removed from the stream for reuse. More costly treatments, such as evaporation or solidification can further reduce the retentate volume to minimize disposal costs again.

In this regard, reverse osmosis is now an important instrument in environmental engineering. Modern high-retention membranes have the ability to retain organic and inorganic contaminants dissolved in water. Open-channel membranes with the ability to treat a wide range of pH feeds are making the technology robust enough to be applied in the most demanding of conditions.

Using the ROCHEM Disc-Tube-Module DT, a 36-m^3/h reverse-osmosis unit was installed to treat leachate at a domestic refuse dump in Schönberg, Germany (Peters, 1992). The DT module combines an open-channel design with narrow gap technology and a new membrane cushion concept. The leachate treatment plant was installed in January of 1990 and had been operating without problems until the time of writing. A two-stage process was used to give an overall salts and inorganic contaminant rejection of about 99 percent. Operating pressures range between 3.5 and 6.0 MPa at ambient temperatures. The pressure of the feed is varied according to the feed temperature. A study performed on the first stage of the plant during the first year of operation showed the energy requirements to be 5.4 kWh/m^3 of permeate produced and a final cost for permeate (including capital and construction) of 12.11 DM/m^3 (1991).

A novel membrane has been designed for pollution control from a sea-based dredging solids disposal pond in Osaka North Port, Japan (Fukanaga et al., 1991). A landfill site, accepting refuse, incineration ash, sludge, and dredged solids from the adjacent river and port, has been established as a form of land reclamation. The landfill is sealed from the adjacent waters by seawalls. The water from the landfill flows back into the adjacent harbor through two consecutive outlets. A means had to be devised to improve the wastewater quality and, in particular, curb the exponential growth of phytoplankton in the water. A membrane of 30-m width and 2-m depth was suspended at the surface across each of the outlets of the ponds. The membrane was made of woven polyester with the porosity ratio decreasing down the membrane from 4 to 5 percent to 40 to 50 percent at the bottom. The membranes were varied according to the season. The membranes successfully retained up to 73 percent of the COD to below 20 mg/L during the two years that the site was being reclaimed. Furthermore, suspended solids and turbidity were reduced by about 75 and 68 percent, respectively. The membrane application represented the most economical wastewater treatment method under the circumstances and successfully reduced the concentration of conventional determinands tested.

Various innovative membrane solutions have been proposed to deal with difficult liquid wastes. A prime difficulty is fouling. Electrically enhanced filtration processes have been developed to control fouling of membranes and thereby maintain high permeation rates. Two treatment processes, able to treat solid-liquid suspensions over a wide range of concentrations, were developed at Harwell Laboratories (UK) to pilot scale (Macdonald, 1990). The processes are direct membrane cleaning (DMC) and electroosmotic dewatering (EOD). They have been demonstrated for the treatment of, for instance, water treatment flocs, ferric hydroxide suspensions, and dispersed absorbers for toxic species removal.

DMC enhances the filtration of finely divided solids at low concentrations from aqueous streams. A high permeation rate can be maintained at electrically conductive micro- or ultrafiltration membranes by flowing the fluid in the cross-flow mode and periodically applying short current pulses to the membranes. The electrical pulses generate gas bubbles which scour the membrane surface to remove the superficial fouling layer. The high permeation rates and the lower cross-flow velocities minimize plant wear, as well as reducing the plant size and pumping energy requirements.

EOD is a sludge-thickening process which can be used to treat feeds with an intermediate solids content (typically a few percent). As opposed to other membrane processes, the driving force is an electric field applied across a nonconducting membrane. The solid and liquid components are separated by their charges. As a result, it achieves an extremely high solids retention factor with rapid permeation rates (0.3 to 1.5 m^3/m^{2}h) with minimal membrane fouling. The high solids concentration is kept fluid by cross-flow recirculation (1 m/s). The energy consumption is low in comparison to techniques such as evaporation (approximately 1.5 to 7.0 percent), and coupled with DMC units, EOD can yield very clean permeate and minimize the volume of solids residue for disposal.

3.5.1 Wastewater Reclamation

A number of applications for wastewater recovery have been nullified. These generally pertain to areas where the rainfall and runoff is too low to support a steady supply of fresh water for distribution to the local population. Alternatively, the population in an area may be too large for the water resources available, making it essential to distribute potable water to people only and reclaim the waste for uses with less-stringent requirements. Various options can be considered depending on the needs of the community and the relative availability of potable water. In some instances, reclaimed water is used for nonpotable purposes such as irrigation or industrial use. In instances where water must be recirculated to provide potable water, the treatment requirements are significantly more stringent. It is in the latter instance that membrane processes may be used to provide high-quality water.

Japan embarked on a bold six-year R&D project, termed *Aqua Renaissance '90*, in 1986 (Kimura, 1991). The objective of the project was to develop low-cost and low-energy treatment processes by combining bioreactors with membrane units to produce reusable water from industrial and domestic wastewater. A range of membrane materials, both ceramic and polymeric, and configurations were tested. A prime consideration was to keep pretreatment to a minimum, balancing costs and membrane cleaning requirements. Polyvinylalcohol and polysulphone capillary modules are being used in large-scale plants (5.4-m^3 bioreactors and 100 m^2 of membrane) and polyethylene hollow-fiber modules are used with the small-scale or rural plants (1-m^3 bioreactors and about 50 m^2 of membrane). Permeate fluxes of 0.38 and $0.58 \text{ m}^3/\text{m}^2 \cdot \text{d}$ with COD retentions of 94.0 and 98.5 percent have been reported for the large- and small-scale units, respectively. Early results indicate that the treatment of concentrated wastewater with membrane separation units uses less than a third of conventional power requirements and that some of the membrane units are achieving their treatment targets.

Other tests on water reclamation systems seem to indicate that careful control of substances that are disposed as domestic effluents would be needed to meet stringent processing requirements prior to processing. A $400\text{-m}^3/\text{d}$ plant was installed at Port Elizabeth, South Africa, to treat secondary domestic municipal effluent (Odendaal, 1991). Flow reversal and spongeball cleaning were used to control excessive fouling. While the product water was found to be of high quality chemically, a number of problems were encountered. The reduction in ammonia nitrogen from 10 to 1 to 2.5 mg/L was insufficient, phenol removal was inadequate (30 to 10 μg/L), the bacteria removal was such that post-disinfection would still have been required, and bioassays indicated toxicity in the product water. This was subsequently found to be due to the presence of formaldehyde, which was used as a membrane preservative.

More success was achieved as part of a study to investigate the potential of reverse osmosis as a means of producing high-quality water for reuse for recreational purposes (Suzuki and Minami, 1991). Part of the study involved selecting membrane materials that successfully retained the key contaminants and were sufficiently resistant to fouling. A comparative chart was constituted that showed spiral-wound polyvinyl alcohol composite membranes and hollow-fiber cellulose triacetate membranes to be the most suitable. The treatment train mixed secondary effluent with sodium hypochlorite and sulphuric acid. Ferric chloride was used as a flocculant in the first tank. Anthracite and sand dual-media filters then pretreated the water prior to cartridge filtration with 3-μm filters. The stream was equally divided to treat the final water with one of the two membrane types. The design capacity of the plant was $40 \text{ m}^3/\text{d}$. The permeate produced by the plant was colorless, odorless, and transparent with an appearance equivalent to potable water. The retention rate of fecal coliform material was 100 percent and the COD of the permeate was 1 to 2 mg/L. The chromacity was 1 degree or less and the phosphorus content was 0.01 mg/L. Under the same pressure conditions, the polyvinyl alcohol membranes produced three times more permeate than comparable cellulose acetate membranes. Moreover, these membranes were sufficiently resistant to fouling, and cleaning produced a recovery to at least 100 percent permeate fluxes.

3.6 CONCLUSIONS

The range and scope of membrane applications in the field of wastewater treatment and water provision has grown rapidly during the past two decades. Water providers can make use of membrane processes for stand-alone units or in conjunction with traditional unit processes. In the field of desalination, the application of processes such as reverse osmosis is likely to increase, especially for smaller-scale plants. With more stringent disinfection requirements likely to be introduced, membrane processes also offer distinct advantages over chemical addition, although the economics of the processes do not presently look attractive.

As a wastewater treatment option, membrane processes can offer opportunities for recovery of materials at source as well as innovative treatment options for industrial and domestic wastewater. Source treatment with membrane processes has been used in the paint, motor manufacture, metal-plating, and wire-drawing industries for some time as a means of recovering valuable raw materials while treating waste streams to meet discharge limits. Bioreactors and sludge dewatering membrane units are offering new solutions for wastewater treatment works where space and efficiency requirements are at a premium.

In a field as new as that of membrane processes, refinements of systems and new applications are commonplace. Consultants, membrane scientists, and plant engineers would be well advised to consult the latest literature regarding the process envisaged when designing new plants or specifying replacement membranes. Membrane manufacturers can also supply information on product developments and process modifications. Three databases have been established to gather and disseminate information about membranes, modules, and modeling. These are MEMFO (managed by the Wastewater Technology Centre, Ontario, Canada), the Module Database (developed by Ralph Guenter at Harburg University, Germany), and EMILY (the Electronic Membrane-Information Library managed by the Pollution Research Group at the University of Natal, South Africa).

Finally, the ability of membranes, and reverse osmosis in particular, to remove a broad range of contaminants to very low levels should not be confused with the very real difficulties of applying what is essentially a sensitive process. Account should be taken of the operating conditions required of the membranes and the concentration variations possible down the plant. Realistic data gathered from pilot plant work that considers the range of stream concentrations can be indispensable in this regard.

Membrane processes represent an exciting development in the field of separation technology. They provide solutions to previously difficult problems. Properly considered and adequately designed, membrane processes can offer the user significant advantages over traditional processes. As the cost differential between the traditional and the membrane processes shrinks, the application of the technology is likely to become commonplace.

REFERENCES

Abanmy, A. B. A., and M. B. O. Al-Rashed, 1993, "Saline Water Conversion Corporation (S.W.C.C.) in the Kingdom of Saudi Arabia," *Desal. Wat. Reuse,* **3**(2):11–17.

American Water Works Association, 1989, *Assessment of Potable Water Membrane Applications and Research Needs,* AWWA Research Foundation and AWWA.

Anon, 1993a, "Water Industry Looks Ahead on Disinfection By-products," *Ends Report* **217**(February):11.

Anon, 1993b, "Unique and Useful Consumer Plastics Product Award," *Plastics Eng.* **51** (April).

Bradbury Ltd., 1993, *Pollution Control for Coating Processes, Printing and the Manufacture of Dyestuffs, Printing Ink and Coating Materials,* DOE Contract No 7/9/648 (in print).

Buckley, C. A., 1992, "Membrane Technology for the Treatment of Dyehouse Effluents," *Wat. Sci. Tech.* **25**(10):203–209.

Cartwright, P. S., 1992, "Industrial Wastewater Treatment with Membranes—A United States Perspective," *Wat. Sci. Tech.* **25**(10):373–390.

Chaize, S., and A. Huyard, 1991, "Membrane Bioreactor on Domestic Wastewater Treatment Sludge Production and Modelling Approach," *Wat. Sci. Tech.* **23**(Kyoto):1591–1600.

Chiemchaisri, C., Y. K. Wong, T. Urase, and K. Yamamoto, 1992, "Organic Stabilization and Nitrogen Removal in Membrane Separation Bioreactor for Domestic Wastewater Treatment," *Wat. Sci. Tech.* **25**(10):231–240.

Chin, K. K., and S. L. Ong, 1991, "A Study of Reclamation of Sewage for Industrial Waters," *Wat. Sci. Tech.* **23**(Kyoto):2181–2187.

Conlon, W. J., C. D. Hornburg, B. M. Watson, and C. A. Kiefer, 1990, "Membrane Softening: The Concept and Its Applicability to Municipal Water Supply," *Desalination* **78**:157–176.

Cross, J., 1992, "Membrane Processes: Versatile Technology for Cutting Costs and Protecting the Environment," *Filtration & Separation* **29**(5):386–390.

Eisenberg, T. N., and E. J. Middlebrooks, 1986, *Reverse Osmosis Treatment of Drinking Water,* Ann Arbor Science, Stoneham, Mass.

Ellender, R. D., and B. H. Sweet, 1972, "Newer Membrane Concentration Processes and Their Application to the Detection of Viral Pollution of Water," *Wat. Res.* **6**:741–746.

Eykamp, W., 1975, "Ultrafiltration of Aqueous Dispersions," *79th National A.I.Ch. E. Meeting,* Houston, Tex.

Fukunaga, I., K. Takamizawa, Z. Inoue, T. Hasebe, M. Konoae, and K. Hatano, 1991, "Effluent Pollution Control of Osaka North Port Dredged Solid Disposal Site (North Section-Site II & III)," *Wat. Sci. Tech.* **23**(Kyoto):1619–1628.

Groves, G. R., 1976, "Ultrafiltration of Industrial Effluents," *Symposium on Selected Studies on Demineralisation,* CSIR, Pretoria, South Africa.

Gutman, R. G., 1987, *Membrane Filtration: The Technology of Pressure-Driven Crossflow Processes,* Adam Hilger, Bristol, UK,

Hindin, E., P. J. Bennett, and S. S. Narayanan, 1969, "Organic Compounds Removed by Reverse Osmosis," *Wat. Sew. Works* **116**(2):466–470.

Holdich, R. G., and G. M. Zhang, 1991, "Seawater Cross-Flow Filtration," *Filtech Conference,* Filtration Society; Karlsruhe, October 15–17.

Karamian, N. A., and P. F. Waters, 1975, "A Membrane for Removing Endotoxins from Aqueous Solutions," *Desalination* **35**:21–38, **17**(3):329–338.

Kimura, S., 1991, "Japan's Aqua Renaissance '90 Project," *Wat. Sci. Tech.* **23**(Kyoto):1573–1582.

Klein, E., P. Feldhoff, and T. Turnham, 1983, "Molecular Weight Spectra of Ultrafilter Rejection," *J. Mem. Sci.* **15**:245–257.

Kolega, M., G. S. Grohmann, R. F. Chiew, and A. W. Day, 1991, "Disinfection and Clarification of Treated Sewage by Advanced Microfiltration," *Wat. Sci. Tech.* **23**(Kyoto):1609–1618.

Langlais, B., Ph. Denis, S. Triballeau, M. Faivre, and M. M. Bourbigot, 1992, "Test on Microfiltration as a Tertiary Treatment Downstream of Fixed Bacteria Filtration," *Wat. Sci. Tech.* **25**(10):219–230.

Langlais, B., Ph. Denis, S. Triballeau, M. Faivre, and M. M. Bourbigot, 1993, "Microfiltration Used as a Means of Disinfection Downstream: A Bacterial Treatment Stage on a Fixed-Bed Bacteria," *Wat. Sci. Tech.* **27**(7–8):19–27.

Macdonald, E. K., 1990, "Liquid Effluents: New Solutions to Old Problems," *Proc. of Symp. on Tech. Advances in Treating Difficult and Toxic Wastes;* N. J. D. Graham and C. J. Murphy (eds.), Imperial College, London, UK, Feb. 20.

Mackay, D., and T. Salusbury, 1988, "Choosing between Centrifugation and Cross-Flow Microfiltration," *Chemical Engineer,* April, pp. 45–50.

Magara, Y., K. Nishimura, M. Itoh, M. Tanaka, O. O. Hart, (ed.) and C. A. Buckley, (ed.), 1992, "Biological Denitrification System with Membrane Separation for Collective Human Excreta Treatment Plant," *Wat. Sci. Tech.* **25**(10):241–251.

Malaiyandandi, M., P. Blais, and V. S. Sastri, 1980, "Separation of Lindane from Its Aqueous Solutions by Reverse Osmosis," *Sep. Sci. Tech.* **15**(7):1483–1488.

Miltner, R. J., C. A. Fronk, and T. F. Speth, 1987, "Removal of Alachor from Drinking Water," in *Proc. of the Nat. Conf. on Env. Eng.,* Orlando, Fla., July, pp. 204–211.

Odendaal, P. E., 1991, "Wastewater Reclamation Technologies and Monitoring Techniques," *Wat. Sci. Tech.* **24**(9):173–184.

Oesterholt, F. I. H. M., and B. A. Bult, 1993, "Improving Municipal Waste Water Quality by Effluent Polishing: A Pilot Scale Experiment at Winterswijk, The Netherlands," *Wat. Sci. Tech.* **27**(5–6):277–286.

Peters, T., 1992, "Adding to the Armoury of Environmental Protection," *Wat. Qual. Int.* (2):22–23.

Pillay, V. L., 1994, private communication (Dr. Pillay is a post-doctoral researcher in the Dept. of Chemical Engineering, University of Natal, Durban, South Africa).

Pollution Research Group, Department of Chemical Engineering, University of Natal, Durban, 1983, A Guide for the Planning, Design and Implementation of Wastewater Treatment Plants in the Textile Industry, Water Research Commission, Pretoria.

Rautenbach, R., and A. Gröschl, 1990, "Separation Potential of Nanofiltration Membranes," *Desalination* **77**:73–84.

Reinhard, M., N. L. Goodman, P. L. McCarty, and D. G. Argo, 1986, "Removing Trace Organics by Reverse Osmosis Using Cellulose Acetate and Polyamide Membranes," *J. AWWA* **78**(4):163–174.

Rencken, G. E., and C. A. Buckley, 1992, "Dewatering Sludges Using a Novel Membrane Technology," *Wat. Sci. Tech.* **25**(10):41–54.

Rencken, G. E., K. Treffry-Goatley, and C. A. Buckley, 1990, "Operation and Theory of the Tubular Filter Press," *Proc. Vth World Filtration Cong.,* Nice, France, vol. 2, pp. 197–204.

Rencken, G. E., 1992, "Performance Studies of the Tubular Filter Press," Ph.D. thesis; University of Natal, Durban, South Africa.

Ross, W. R., J. P. Barnard, N. K. H. Strohwald, C. J. Grobler, and J. Sanetra, 1992, "Practical Application of the ADUF Process to the Full-Scale Treatment of a Maize-Processing Effluent," *Wat. Sci. Tech.* **25**(10):27–39.

Sakohara, S., S. Kitao, M. Ishizaki, and M. Asaeda, 1989, "Separation of Solvent/Water Mixtures by Porous Membrane," *First Int. Conf. on Inorganic Membranes,* Montpellier, France, July 3–6.

Smith, Jr., C. V., D. D. Gergorio, and R. M. Talcott, 1969, "The Use of Ultrafiltration Membranes for Activated Sludge Separation," *presented paper at the 24th Annual Purdue Industrial Waste Conference,* Ind.

Smith, T., and R. Derowitsch, 1994, "Membranes Decrease Energy Demand in RO System," *J. Am. Water Works Assoc.* (January):132–134.

Soma, C., M. Rumeau, and C. Sergent, 1989, "Use of Mineral Membranes in the Treatment of Textile Effluents," *First Int. Conf. on Inorganic Membranes,* Montpellier, France, July 3–6.

Strohwald, N. K. H., and E. P. Jacobs, 1992, "An Investigation into UF Systems in the Pretreatment of Seawater for RO Desalination," in *Membrane Technology in Wastewater Management, Wat. Sci. and Tech.* **25**(10):69–78.

Suzuki, Y., and T. Minami, 1991, "Technological Development of a Wastewater Reclamation Process for Recreational Reuse: An Approach to Advanced Wastewater Treatment Featuring Reverse Osmosis Membrane," *Wat. Sci. Tech.* **23**(Kyoto):1629–1638.

Swart, A. F., 1994, "Considerations in the Selection of Operating Regimes for Microfiltration," M.Sc.Eng. thesis, Pollution Research Group, University of Natal, Durban, South Africa.

Tanaka, T., Sato, T., and Suzuki, T., 1986, "Disinfection of *Escherichia coli* by Using Water Dissociation Effect on Ion Exchange Membrane," in E. Drioli and M. Nakagaki (eds.), *Membranes and Membrane Processes,* Plenum Press, New York, pp. 421–428.

Taylor, J. S., S. J. Duranceau, L. A. Mulford, W. M. Barrett, and D. K. Smith, 1989, *SOC Rejection by Nanofiltration,* U.S. EPA Water Engineering Research Laboratory, Cincinnati, Ohio.

Treffry-Goatley, K., M. I. Buchan, G. E. Rencken, W. J. Voortmann, and C. A. Buckley, 1987, "The Dewatering of Sludges Using a Tubular Filter Press," *Desalination* **67**:467–479.

Vial, D., R. Phan Tan Luu, and A. Huyard, 1992, "Optimal Design Applied to Ultrafiltration on Tertiary Effluent with Plate and Frame Modules," *Wat. Sci. Tech.* **25**(10):253–261.

Watson, B. M., and C. D. Hornburg, 1989, "Low-Energy Membrane Nanofiltration for Removal of Colour, Organics and Hardness from Drinking Water Supplies," *Desalination* **72**:11–22.

World Health Organization (WHO), 1984, *Guidelines for Drinking-Water Quality, Vol. 1, Recommendations,* Geneva, Switzerland.

Yamamoto, K., and K. M. Win, 1991, "Tannery Wastewater Treatment Using a Sequencing Batch Membrane Reactor," *Wat. Sci. Tech.* **23**(Kyoto):1639–1648.

CHAPTER 4
MASS TRANSPORT AND PERMEATE FLUX AND FOULING IN PRESSURE-DRIVEN PROCESSES

Mark R. Wiesner
Department of Environmental Science and Engineering
Rice University
Houston, Texas

P. Aptel
Laboratoire de Génie Chimique, CNRS
Université Paul Sabatier
Toulouse, France

The capital and operating costs of membrane systems typically scale directly as a function of the membrane permeate flux J.[1] Where it is possible to move more water across a unit area of membrane per unit time, less membrane area will be required to provide for the design flow. This results in a lower cost for membrane modules, peripheral piping and pumps, monitoring equipment, skids, foundations, and buildings. The cost of replacing membranes as reflected in the membrane life is often the single largest component of operating cost. By reducing the amount of membrane area to be replaced, a higher permeate flux also corresponds to a lower operating cost. Thus, permeate flux and the factors that influence it are central considerations in determining membrane performance and cost.

In this chapter, factors influencing mass transport and permeate flux are considered. Particular emphasis is placed on the role of materials in the feedwater, how they distribute themselves near the membrane, their interactions with each other and with the membrane, and their impact on permeate flux. We will build on material presented in this chapter when discussing principles of contaminant rejection in Chap. 5.

4.1 PERMEATION BEHAVIOR
OF CLEAN MEMBRANES

The flux J of very clean water across a membrane without materials deposited on its surface or within its pores is often described by Darcy's law:

$$J = \frac{\Delta p}{\mu R_m} \tag{4.1}$$

where Δp is the pressure drop across the membrane (the transmembrane pressure drop, or TMP), μ is the absolute viscosity (of the water), and R_m is the hydraulic resistance of the clean membrane with dimensions of reciprocal length. This expression is similar in form to the Kedem-Katchalsky equation derived from irreversible thermodynamics (see Chap. 5). If the flux of permeate J is much greater than the flux of solute J_s (in a dilute solution, for example), then

$$J \propto L_V(\Delta p - \sigma \Delta \Pi) \tag{4.2}$$

The variables in Eq. (4.2) are as defined in Chap. 5. For the time being, we will limit our interest in Eq. (4.2) to the observation that the dependence of permeate flux on the transmembrane pressure derived from irreversible thermodynamics is similar to a Darcy-type expression. By analogy, Eq. (4.1) may be modified to account for the reduction in the net transmembrane pressure (TMP$_{net}$) due to the effects of osmotic pressure:

$$J = \frac{(\Delta p - \sigma_k \Delta \Pi)}{\mu R_m} \tag{4.3}$$

where σ_k is treated as an empirical constant. The change in osmotic pressure across the membrane, $\Delta \Pi$, derives from the rejection of materials by the membrane and is, to a first approximation, inversely proportional to the molecular weight of the rejected species. The osmotic pressure of macromolecular and colloidal species is usually quite small. Since these are the only species rejected to any significant degree by UF and MF membranes, the correction for osmotic pressure can, as a first approximation, be neglected for these membranes.

If the bulk of permeate passes by advection through a network of capillary pores in the membrane, as in UF and MF, permeate flux through the clean membrane can be modeled as Poiseuille flow through a large number of these pores in parallel. Advective flow of permeate and the presence of pores through RO membranes has been questioned by some investigators.[2] For UF and MF membranes, permeate flux through a membrane characterized by an effective pore radius of r_{pore} is described as:

$$J = \frac{fr_{pore}^2 \Delta p}{8\mu\theta\delta_m} \tag{4.4}$$

where f is the fraction of open pore area on the membrane surface, θ is the pore tortuosity factor, and δ_m is the effective thickness of the membrane. It is significant that Eq. (4.4) predicts an increase in permeate flux with decreasing membrane thickness δ_m. This has been the principal consideration in developing asymmetric membranes with thin membrane skins. The pore size of the underlying support structure relative to its thickness should be large enough so that the support does not produce any appreciable resistance to permeate flow.

Like Eqs. (4.1) and (4.2), Eq. (4.4) predicts that permeate flux should increase proportionally with an increase in Δp. The presence of dissolved and colloidal materials in water can produce deviations from this linear behavior of permeate flux in Δp due to the accumulation of materials on the membrane. A discussion of such mass-transfer-limited behavior is reserved for the following section of this chapter. Equation (4.4) also predicts that with all other factors being equal, a membrane with a larger effective pore radius (or molecular weight cutoff, MWCO) should have a higher permeate flux. Thus, MF membranes are expected to have a higher permeate flux than UF membranes, and UF membranes should have a higher permeate flux than NF membranes, and so on. However, reductions in permeate flux with decreasing pore size are partially mitigated by the fact that f tends to increase and membrane thickness δ_m tends to decrease with decreasing pore radius. Approximate values for membrane resistance in RO, NF, UF, and MF membranes are listed in Table 4.1.

TABLE 4.1 Approximate Values of Membrane Resistance and Transmembrane Pressures (TMP) for Pressure-Driven Membrane Processes

Process	Approximate R_m (m^{-1})	Approximate TMP (kPa)
Reverse osmosis	10^{10}	800–8000
Nanofiltration	10^{8}	350–1000
Ultrafiltration	10^{7}	50–700
Microfiltration	10^{6}	30–300

In the presence of dissolved and colloidal materials, the effective pore size of the membrane may change over time due to adsorption of materials on pore walls leading to a constriction of the pore. Particles may also block the entrance to pores. A more detailed discussion of these issues is reserved for the following section on permeate flux decline.

4.1.1 Temperature Effects

Permeate flux, as calculated by Eq. (4.4), is predicted to increase with decreasing viscosity of the permeating water. For this reason, more favorable permeation rates can be achieved at higher temperatures. It is often useful to correct permeate flux to a reference temperature (e.g., 20°C) for purposes of comparison. One method of correcting for temperature is to multiply the permeation rate observed at a given temperature by the ratio of the viscosity of the permeate at that temperature to its viscosity at the reference temperature. However, changes in temperature result in a wide range of effects that go beyond the viscosity of the permeate alone. The effect of temperature on permeate flux is often described using an expression of the form of the Arrhenius equation:

$$J_T = J_{20} \exp\left(\frac{s}{T}\right) \tag{4.5a}$$

where J_T is the permeate flux at an arbitrary temperature T, J_{20} is the permeate flux measured at a reference temperature of 20°C, and s is an empirical constant that must be evaluated for each membrane.[3] Thus, the temperature correction factor

(TCF) by which permeate flux at some reference temperature (in this case, 20°C) is multiplied is the ratio of J_T to J_{20}. There is no special reason that 20°C must be used as the reference temperature. However, the use of a different reference temperature requires the evaluation of a corresponding value for the empirical constant s. Expressions of similar form have been developed for correcting permeate flux of RO membranes over a limited range of temperatures. In all cases, these corrections for permeate flux apply within the temperature ranges of tolerance for the membranes. Using a reference temperature of 25°C, Pohland[4] reports that the following expression for the temperature correction factor is correct within approximately 3 percent:

$$\frac{J_T}{J_{25}} = 1.03^{(T-25)} \tag{4.5b}$$

Marsh and Eriksson[5] give the following formulas for the TCF to be applied to a FilmTec FT30 membrane:

$$\frac{J_T}{J_{25}} = \exp\left[2640\left(\frac{1}{298} - \frac{1}{273+T}\right)\right] \qquad \text{for } T \geq 25°C \tag{4.5c}$$

$$\frac{J_T}{J_{25}} = \exp\left[3480\left(\frac{1}{298} - \frac{1}{273+T}\right)\right] \qquad \text{for } T \leq 25°C \tag{4.5d}$$

4.1.2 Effects of Membrane Charge

When the walls of membrane pores carry a charge, a range of complex phenomena may intervene in altering the effective pore size of the membrane. Solvation of the membrane surface and the resultant layer of structured water on the pore wall may also reduce the effective r_{pore}. Electrokinetic phenomena may exert an effect on membrane permeability that can be interpreted as a change in the net pressure applied (Δp) or in the viscosity of the permeating water μ. Typically, such deviations from ideality are attributed to changes in the viscosity of the permeating fluid near the capillary walls. These so-called electroviscous effects have their origin in the interaction between the permeating fluid and the ions in the diffuse layer near the pore walls.[6]

Primary electroviscous effects on flow through porous media, including membranes, have been recognized for some time. The flow of permeate through the pore displaces ions that have accumulated in the diffuse layer near the surface of the charged membrane pore. As a result, an electrical potential is established that produces an additional resistance to flow as backmigrating ions exert a drag force on the water moving through the membrane capillaries. The resultant permeate flux is lower than that predicted by Eq. (4.4). This reduced permeability can be interpreted as an increase in the apparent viscosity of the permeating water. The apparent viscosity μ_a of the water flowing through charged capillaries can be related to the zeta-potential ζ of the pore surface and the double layer thickness $1/\kappa$ by:

$$\frac{\mu_a}{\mu} = \left[\frac{1 - 8b\zeta^2(1-G)F}{\kappa r_{pore}}\right]^{-1} \tag{4.6}$$

where b is a dimensionless parameter that depends solely on electrolyte properties, given by:

$$b = \frac{(\varepsilon_o\varepsilon_w)^2 k^2 T^2 \kappa^2}{16\pi^2\mu k e_o^2} \tag{4.7}$$

T is the absolute temperature; k is the universal Boltzman's constant; ε_o and ε_w are the permittivity of a vacuum and the dielectric constant of water, respectively; e_o is the proton charge; k is the conductivity of the electrolyte; κ is the Debye-Huckel parameter; ζ is a dimensionless zeta-potential, given by:

$$\zeta = \frac{e_o\zeta}{kT} \tag{4.8}$$

G is the ratio of the mean electrostatic potential across to the ζ-potential, F is a factor used to correct Smoluchowski's equation for effects of high zeta-potential and fine capillaries, and r_{pore} is the radius of the pores. The ζ-potential is typically assumed to be a close approximation to surface potential Ψ_o. Ionic strength affects apparent viscosity through its effects on both the Debye constant κ and surface potential.

Electroviscous effects alone do not appear to adequately describe membrane permeability as a function of pH and ionic strength. Much greater changes in membrane permeability and very different overall trends as a function of pH compared with electroviscous theory have been reported for experiments with ceramic membranes.[7] The presence of humic substances in the water appears to reduce the sensitivity of permeate flux to pH while also reducing the permeate flux at all values of pH due to adsorptive fouling (discussed later).

4.2 REDUCTIONS IN PERMEATE FLUX

As materials accumulate near, on, and within the membrane, they may reduce the permeability of the membrane by blocking or constricting pores and by forming a layer of additional resistance to flow across the membrane (Fig. 4.1). Reductions in permeate flux over time may be substantial and represent a loss in the capacity of a membrane facility (Fig. 4.2). For the example illustrated in Fig. 4.2, a membrane facility with an installed membrane area calculated based on the initial permeability of these membranes would produce only one-third the calculated design flow after the first 80 minutes of operation. Thus, reductions in permeate flux and procedures for maintaining permeate flux must be considered in both the design and operation of membrane facilities. The characteristics and location of the deposited materials can play an important role in determining the extent and reversibility of permeate flux decline. Washing the membrane, either hydraulically or chemically, may remove some of the accumulated materials and partially restore permeate flux. A reduction in permeate flux that cannot be reversed is referred to as *membrane fouling*. Frequently, both reversible and irreversible decreases in permeate flux decline are referred to as membrane fouling, and materials in the water that produce reductions in permeate flux are collectively referred to as *foulants*.

The distinction between reversible and irreversible reductions in permeate flux is entirely dependent on the context in which membranes are operated and cleaned. That is to say that the process of permeate flux decline is extremely path dependent. The degree of irreversible fouling tends to reflect the "memory" of the membrane for extreme conditions it has been exposed to during operation such as the highest TMP or the worst feedwater quality. A different order of addition of chemical cleaning agents (e.g., acid wash followed by base versus base followed by acid) usually produces different degrees of permeate flux recovery. Also, hydraulic-based mea-

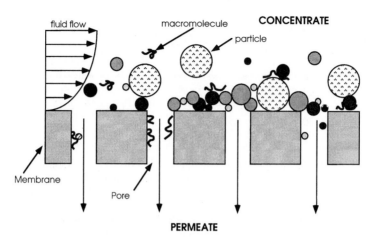

FIGURE 4.1 The accumulation of materials on, in, and near a membrane in the presence of a cross flow. Accumulated materials produce an additional layer of resistance to permeate flux. Some materials are small enough to enter membrane pores where they may be deposited, further reducing the permeation rate.

sures to reverse permeate flux decline, such as backflushing, tend to become less effective over time, to the extent that all of the permeate flux decline would be considered irreversible by this operation. Under these conditions, chemical cleaning may be required to restore permeate flux. Thus, it is probably more practical to consider an *irreversible* loss in permeate flux to be the difference between the permeate flux of the newly installed membrane and the permeate flux observed after applying the most rigorous cleaning procedure envisioned for a given membrane system. A loss in permeate flux that is truly irreversible, usually requiring replacement of the membrane, is sometimes termed membrane *poisoning*.

4.2.1 Silt Density Index

The tendency for a raw water to foul a membrane is sometimes assessed by means of an empirical test of filterability, the *silt density index* (SDI). The ASTM standard for this test (D4185) is conducted using a 0.45-μm membrane in a dead-end filtration cell. The SDI is calculated from the time t_i required to filter a fixed volume (typically 500 mL) of raw water through a clean membrane and the time t_f required to filter the same volume after the membrane has been used for a defined length of time. Standard conditions for determining the SDI include a 47-mm-diameter filter, and applied pressure of 2 bar (30 lb/in²), and a total test time T_t of 900 s. Thus, the SDI is calculated as:

$$\text{SDI} = \frac{100\left(1 - \dfrac{t_i}{t_f}\right)}{t_t} \tag{4.9}$$

where all the times are expressed in seconds. Upper SDI limits of approximately 3 to 5 have been found to apply for the successful operation of outside in hollow fibers

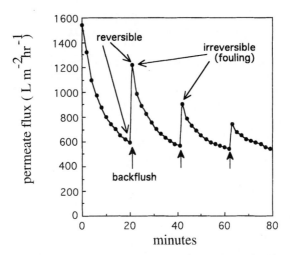

FIGURE 4.2   Reductions in permeate flux over time. In this case, periodic cleaning of the membrane by hydraulic backflushing restored a portion of the membrane permeability, but with decreasing efficacy when irreversible fouling occurs. Chemical cleaning may restore a portion of the permeate flux lost to irreversible fouling.

and spiral-wound RO membranes. While the SDI is often found to be a useful qualitative indicator of the need for pretreatment and feedwater treatability, the SDI is an inadequate indicator of membrane performance for design or control purposes. Limitations of the test include its inability to capture many of the interactions between foulants in the feedwater and the membrane that occur in practice. Morover, the rejection characteristics and fluid mechanics of the test cell differ substantially from actual operating membrane units. Thus, both pilot studies and a more fundamental understanding of the principles governing mass transport and permeate flux decline are required to assess the treatability of a given feedwater.

4.2.2   Causes of Permeate Flux Decline and Methods for Fouling Control

The accumulation of materials in the membrane, e.g., by adsorption to pore walls, results in an increase in the resistance of the membrane over time. Permeate flux decline can be described mathematically by generalizing Eq. (4.3) to the case where resistance to permeate flux is produced by both the membrane and materials accumulated near, on, and in the membrane. A layer or *cake* of materials deposited on the membrane surface and more loosely associated materials in the concentration boundary layer, or concentration-polarization layer, present additional resistances to permeation, designated as R_c and R_{cp}, respectively. These resistances vary as a function of the composition and thickness of each layer, which in turn are determined by the raw water quality and the characteristics of mass transfer in the membrane module. The resulting expression for permeate flux is

$$J = \frac{\Delta p - \sigma_k \Delta \Pi}{\mu \{R_m(t) + R_c[\delta_c(t), \ldots] + R_{cp}(D, J)\}} \tag{4.10}$$

where δ_c is the thickness of the cake (or gel) layer and D is the diffusivity of the material in the concentration-polarization layer. The resistance terms are all functions of time and are related in a complex fashion to the hydrodynamics of the membrane system and the raw water quality. By comparison of Eq. (4.10) with Eq. (4.3), the initial resistance of a clean membrane can be expressed as

$$R_m|_o = \frac{8\theta \delta_m}{f r_{\text{pore}}^2} \tag{4.11}$$

Over time, the membrane resistance may change due to fouling.

4.2.2.1 Cake Formation.

Cake formation, pore blockage, and adsorptive fouling appear to be the predominant causes of decreased permeate flux over time in UF and MF membranes. In most instances encountered in water and wastewater treatment, it appears that the concentration-polarization layer, if it is formed, contributes negligible resistance to permeate flux; i.e., $R_{cp} \ll R_c$ and, therefore, R_{cp} may be neglected. Nonetheless, concentration polarization plays a key role in the formation of cakes and gel layers.

The resistance of the cake R_c can be expressed as the product of the specific resistance of the cake $\hat{R}_c$ and the cake thickness δ_c. By the Kozeny equation, the specific resistance of an incompressible cake composed of uniform particles can be calculated as:

$$\hat{R}_c = \frac{180(1 - \varepsilon_c)^2}{d_p^2 \varepsilon_c^3} \tag{4.12}$$

where ε_c is the porosity of the cake and d_p is the diameter of particles deposited. This expression predicts that resistance to permeation by a deposited cake should increase as the particles composing the cake decrease in size. The resistance produced by RO and NF membranes is likely to be large in comparison with cakes of deposited colloidal materials. However, gel layers of macromolecular materials may produce significant resistance. Cake resistance may be small in comparison with the resistance of a UF or MF membrane if the particles deposited in the cake are large compared with the effective pore size of the membrane. In this case, permeate flux may be relatively independent of the concentration of particles in feed streams containing large particles (Fig. 4.3).

However, in most UF and MF applications to water and wastewater treatment, cake resistance exerts a significant influence on permeate flux. The cake itself may act as a filter and remove smaller particles or it may compact over time. Cake formation, in concert with pore blockage and adsorptive fouling, may result in resistances to permeate flux that eventually exceed the membrane resistance; the cake or gel layer deposited on a membrane essentially acts as second membrane through which the permeating water must pass. This is often observed in practice when raw waters are applied to the membranes with little pretreatment. One important implication of this observation is that the permeate fluxes observed when operating MF and UF facilities for some period without backflushing may be quite similar (Fig. 4.4).

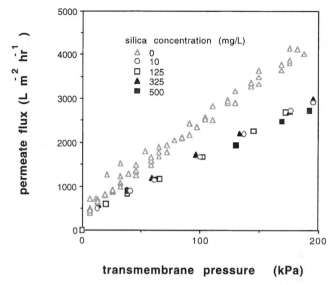

transmembrane pressure   (kPa)

FIGURE 4.3   Permeate flux of a 0.8-μm nominal pore diameter ceramic membrane as a function of mass particle concentration in the feed stream. Silica particles in the feed stream have a number-average diameter of 1.5 μm. Higher concentrations do not reduce permeate flux beyond the initial fouling produced by very low concentrations. *(Data adapted from Nazzal.[40])*

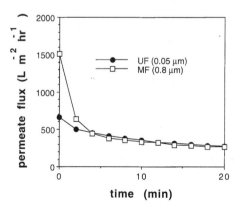

time   (min)

FIGURE 4.4   Permeate flux decline after placing UF and MF membranes into operation immediately after a backflush. The higher permeate flux of the more porous MF membrane decreases to that of the UF membrane as materials deposit on the membrane and resistances due to fouling dominate.

   When cake resistance is high relative to the resistance of the membrane, decreases in permeate flux may be correlated with the mass of material deposited on the membrane.[8] Over relatively short periods of time, permeate flux decreases until transport of particles to the cake deposited on the membrane is balanced by particle transport from the cake, and permeate flux attains an approximately constant value. This is illustrated in an experiment in which a UF membrane was used to filter a bentonite suspension (Fig. 4.5).[8] The decreases in permeate flux are mirrored by increases in the mass of bentonite deposited on the membrane. As the cross-flow velocity is increased, the plateau value of permeate flux also increases. However, during the early stages of permeate flux decline, the rate of flux decline and the mass deposited are nearly independent of the cross-flow velocity, suggesting that the drag force on particles associated with permeation is initially far higher than the sum of any backtransport processes. As the cake grows in thickness, backtransport becomes more significant.

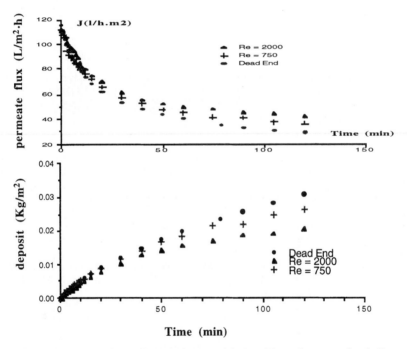

Time (min)

FIGURE 4.5   Effect of cross-flow velocity on particle deposition and permeate flux decline during ultrafiltration of 0.3-g/L suspension of bentonite at a transmembrane pressure of 50 kPa. *(Adapted from Gourgues et al.[8])*

   Factors affecting the permeability of clean membranes are also likely to be important for deposited layers. For example, primary and secondary electroviscous effects can affect cake permeability in an analogous fashion to that in the membrane. In addition, tertiary electroviscous effects, those resulting in a change in the geometry of the system, may affect the permeability of the colloidal deposits.[9] Data reported by Bacchin and co-workers[10] illustrate that, up to a point, the cake formed during the filtration of a suspension of bentonite particles reduces permeate flux to

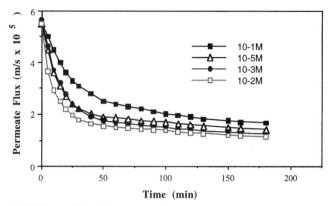

FIGURE 4.6   The effect of electrolyte strength on permeate flux during the ultrafiltration of suspension of bentonite particles at a transmembrane pressure of 100 kPa and a cross-flow velocity of 0.09 m/s. *(After Bacchin et al.[10])*

a greater extent as the ionic strength of the solution increases (Fig. 4.6). These decreases in permeate flux are attributed to a reduction in electrostatic repulsion between particles in the cake that leads to a more dense cake. However, when ionic strength increases beyond a critical point, coagulation of the particles occurs, resulting in larger particles and a more porous cake. Similar results have been reported by other investigators.[9,11]

Numerous strategies have been proposed to restore permeate flux caused by cake or gel layer formation. Hydrodynamic or gas cleaning of the membrane through backflushing (reversing the direction of flow across the membrane), fast-flushing (a short burst of high-velocity fluid to scour the membrane surface), or pulsing is implemented fairly easily and can be automatically initiated at regular and frequent intervals during operation. The introduction of particles or other mechanical scouring agents as well as chemical cleaning using detergents, enzymes, and solutions of base or acid, may be required when hydrodynamic measures become ineffective. However, these procedures tend to be more time consuming and costly. They are usually performed on an as-needed basis. Much consideration has been given to the design and operating conditions that may inhibit cake formation. These include pretreatment to alter particle size (see Chap. 16), to remove or complex foulants (see, for example, Chap. 15), or to optimize the fluid mechanics of membrane elements to promote scouring of the membrane surface.

4.2.2.2 Precipitative Fouling.   Precipitation or scale formation on the membrane may occur when salts in the raw water are concentrated beyond their solubilities. As a result, precipitative fouling is an important consideration in the operation of RO and NF membranes. Concentration of scale-forming species may occur due to two phenomena: (1) bulk concentration of the salts as water permeating through the membrane is removed from the salt solution and (2) concentration polarization.

Consider a sparing soluble salt consisting of anion A and cation B in equilibrium with its precipitated solid phase. Common foulants of concern are calcium, magnesium, iron, and other metals. The most common precipitates of these metals are as hydroxides, carbonates, and sulfates. In the absence of interactions with other salts, the product of the activities of A and B, a_A and a_B, have been found to be constant.

This solubility product K_{sp} can be expressed as a function of the concentrations of A and B:

$$K_{sp} = \gamma_A^x[A^{y-}]^x\gamma_B^y[B^{x+}]^y \tag{4.13}$$

where $\gamma_{A,B}$ are the free ion activity coefficients of A and B, [A] and [B], and x and y are, respectively, the molal concentrations in solution and the stoichiometric coefficients for the precipitation reaction of A and B. For example, if calcium and sulfate are present at sufficient concentrations (or, more correctly, activities) to initiate precipitation of $CaSO_4$, then at equilibrium at a temperature of 20°C, the product of the activities of calcium and sulfate in solution will be equal to the K_{sp} for $CaSO_4$, approximately 1.9×10^{-4}. The mean activity coefficients can be estimated as function of ionic strength I as

$$\log \gamma_{A,B} = -0.509 z_A z_B \sqrt{I} \tag{4.14}$$

For dilute solutions, as are typical of most natural waters, the activity coefficients are approximately equal to 1. However, as the feedwater is concentrated, the activity coefficients of solutes may decrease sufficiently to make concentration a poor approximation of activity. The presence of other electrolytes may further decrease ion activities through effects on ionic strength and ion pairing. The bulk concentration of the rejected salts (for example, the cation B) in the rejectate flow exiting a membrane module C_r increases over the feed concentration C_f as the recovery r and global rejection R increase:

$$C_r = C_f \frac{1 - r(1 - R)}{1 - r} \tag{4.15}$$

(Further discussion of rejection and recovery follow in Chap. 5.) When the ratio of the product of ion activities in the rejectate (right-hand side of Eq. (4.13) after substituting $[A]_r$ and $[B]_r$ calculated from the feed concentration) to the solubility product K_{sp} is greater than 1, there is a potential for precipitation of the salt. In fact, Eq. (4.15) underestimates the concentrations that may provoke scale precipitation since it does not account for concentration polarization. A more detailed discussion of concentration polarization, including the mathematics of solute transport, is presented later in this chapter. It suffices for the time being to note that as water permeates through salt-rejecting membranes, the rejected ions accumulate near the membrane (i.e., in a boundary layer) at concentrations higher than those in the bulk fluid far from the membrane. The ratio of concentration in the boundary layer to that in the bulk rejectate is termed the *concentration-polarization factor* PF. Typically, PF is estimated as an exponential function of the recovery r of the form:

$$PF = \exp(Kr) \tag{4.16}$$

where K is a semiempirical constant typically taking on values of 0.6 to 0.9 for commercially available RO modules. The constant K for ionic species is largely a function of the permeate flux and ion diffusivity.

Strategies for avoiding precipitative scaling focus on reducing the concentration of either the anion or the cation portion of the ion product. For example, acid may be added to reduce the concentrations of anionic species such as hydroxide and carbonate that may precipitate metal ions. Feedwater might also be pretreated by lime softening, precipitation, or ion exchange to remove scale-forming metals. Finally, antiscaling agents such as hexametaphosphate may also be added to impede precipitation.

EXAMPLE 4.1 *Nanofiltration membranes are being used to soften a groundwater. These particular spiral-wound membranes are known to reject 90 percent of the divalent ions. The empirical constant for the polarization factor can be taken to be 0.8. Neglecting the effects of other electrolytes, what do you estimate to be the maximum recovery that this system can be operated at without pretreatment if the groundwater contains 3×10^{-3} M Ca and 2.5×10^{-4} M sulfate? Assume that the solubility product for gypsum ($CaSO_4 \cdot 2H_2O$) controls calcium sulfate solubility and that it is equal to 2.5×10^{-5}.*

1. We know that, as water is filtered and ions are retained on the membrane, the concentration of ions will increase above those in the groundwater. The highest concentrations will be those near the membrane due to concentration polarization. So we will use concentrations near the membrane c_w as the worst case to estimate the limits of solubility. Near the membrane, when gypsum is just ready to precipitate, the ion product will be approximately equal to the solubility product for gypsum. So we want to ensure that the ion product does not exceed the solubility product [Eq. (4.13)]:

$$[Ca^{2+}]_w[SO_4^{2-}]_w \leq 2.5 \times 10^{-5}$$

(assuming the activity coefficients to be approximately equal to 1).

2. The concentrations at the wall can be calculated from the concentrations in the reject (the bulk) using the polarization factor since $c_w = c_r PF$. Thus, by Eq. (4.16),

$$(PF)[Ca^{2+}]_r(PF)[SO_4^{2-}]_r = [\exp (0.8r)]^2[Ca^{2+}]_r[SO_4^{2-}]_r \leq 2.5 \times 10^{-5}$$

3. The concentrations in the reject flow will themselves be functions of the recovery and the rejection. Substituting for the reject concentrations in terms of the feedwater concentrations [Eq. (4.15)], we obtain:

$$[\exp (0.8r)]^2\left[\frac{1 - r(1 - R)}{1 - r}\right]^2[Ca^{2+}]_f[SO_4^{2-}]_f \leq 2.5 \times 10^{-5}$$

Rearranging this expression,

$$[\exp (0.8r)]^2\left[\frac{1 - r(1 - R)}{1 - r}\right]^2 \leq \frac{2.5 \times 10^{-5}}{[Ca^{2+}]_f[SO_4^{2-}]_f} = \frac{2.5 \times 10^{-5}}{(3 \times 10^{-3})(2.5 \times 10^{-4})} = 33.33$$

$$[\exp (0.8r)]\left[\frac{1 - r(1 - R)}{1 - r}\right] \leq 5.77$$

By iteration (using the given value of R = 0.9), we obtain:

$$r_{max} = 0.71$$

4.2.2.3 Adsorptive Fouling. Adsorption or deposition of materials over time in membrane pores leads to a decrease in f or r_{pore} [Eq. (4.4)] and results in an increase in R_m which is often difficult to reverse. The adsorption of organic materials to membrane surfaces is often a controlling factor in determining membrane performance. Humic acids and other naturally occurring organic materials can have a much greater effect on permeate flux than clays or other inorganic colloids, even at lower mass concentrations (Fig. 4.7). The importance of adsorption as a fouling mechanism is further underscored by its effect on the permeate flux of two mem-

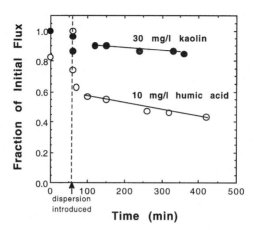

FIGURE 4.7 Reductions in permeate flux over time produced by dispersions of humic acid and kaolin particles in the feed stream. Humic acid reduces permeate flux to a much greater degree than do kaolin particles. (*Data adapted from Lahoussine-Turcaud.[18]*)

branes with very different initial permeabilities (Fig. 4.8). Tannic acid was observed to reduce the permeate flux of a membrane with twice the initial flux of a second membrane to a level below that of the second membrane treating an identical solution of tannic acid.

The characteristics of organic materials that determine their relative propensity to foul membranes include their affinity for the membrane, molecular weight, functionality, and conformation. Negatively charged functional groups on organic polyelectrolytes (such as humic and fulvic acids) may be electrostatically repulsed by negatively charged membrane surfaces. Typically, polysulfone, cellulose acetate, ceramic, and thin film composite membranes used for water and wastewater treatment all carry some degree of negative surface charge. Usually, greater charge density on the membrane surface is associated with greater membrane hydrophilicity. Hydrophobic interactions may increase the accumulation of naturally occurring organic matter on membranes, leading to more adsorptive fouling. Lâiné and co-workers found that lake water produced significantly less fouling of hydrophilic membranes than of hydrophobic membranes.[12] Adsorption of humic materials on membranes has been observed to reduce the apparent electrokinetic potential near membrane surfaces while increasing the hydrophilicity of the surface.[13] Conditions that render dissolved organic materials more hydrophobic should also augment adsorptive fouling of membranes.[14] The stability of natural organic matter is enhanced by negatively charged functional groups on these molecules. Calcium ions and protons may associate with these functional groups rendering the molecule more or less stable. Higher calcium concentration and lower pH values tend to favor the adsorption of humic materials on activated carbon and other solids. There is some experimental evidence that lower pH values may increase the hydrophobicity of humic materials as measured by the octanol-water partition coefficient.[13] While solution chemistry may affect the hydrophobicity of natural organic matter, the formation of salt bridges and direct precipitation of NOM may also reduce its stability in water. These trends have been observed in the fouling of membranes by proteins.[15] These

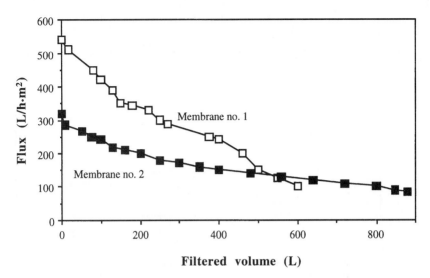

FIGURE 4.8 Adsorptive fouling of two membranes with differing initial permeabilities by a solution of tannic acid.

trends have also been observed experimentally using hydrophobic ultrafiltration membranes.[13] These laboratory observations support field experience where hard surface water has been observed to exhibit a greater tendency to foul RO membranes than soft surface water despite the fact that both waters produced the same SDI.[5]

Some progress has been made in identifying the specific fractions of natural organic matter, based on chemical composition, that may be responsible for membrane fouling. Mallevialle and co-workers[16] found evidence implicating the polysaccharide fraction of NOM as determined by pyrolysis-GC/MS, as being the principal fouling agent of cellulosic ultrafiltration membranes used in one pilot study. Polysaccharides and polyhydroxyaromatics have been identified as two dominant fractions of NOM in surface waters and karstic groundwaters using the pyrolysis-GC/MS fractionation technique. Grozes et al.[17] investigated permeate flux decline due to the adsorption of tannic acid (MW 700) and a dextran (MW 10,400) on different UF membranes. These two compounds were selected to represent, respectively, the polyhydroxyaromatic and polysaccharide fractions of NOM. In these experiments, adsorption phenomena were isolated from other phenomena such as cake formation and concentration polarization. The dextran solution was found to have little effect on permeate flux of a 100,000-MWCO membrane while a tannic acid solution, as reported by previous researchers,[18] produced significant fouling of the membrane (Fig. 4.9). Adsorption isotherms indicated that dextran adsorption on the membrane was 10 times less than that for tannic acid.

In addition to their direct effect on membrane fouling, the organic matter in natural waters has been shown to play a determinant role in the cohesion of colloids deposited on the membrane. Analysis of the organic foulants in natural waters and their relative concentrations in the deposited cake suggest that polyphenolic compounds, proteins, and polysaccharides bind together colloids that deposit on the membrane and may cement the cake to the membrane surface.[19] Thus, adsorptive fouling and stabilization of the cake (or formation of a gel layer) by organic matter

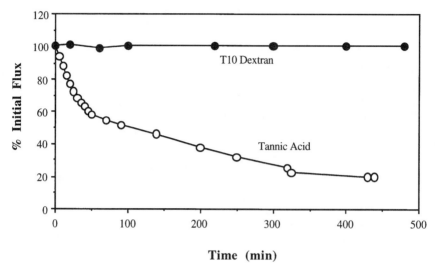

FIGURE 4.9   Permeate flux decline for a 100,000 molecular weight cutoff membrane produced by solutions of dextran and tannic acid.

in water impairs the efficiency of purely hydraulic cleaning methods such as back-flushing, fast pulsing, or cross-flow reversal. As a consequence, reagents used for chemical cleaning of the membrane must efficiently deteriorate or redissolve organic compounds. For this reason, solutions used to chemically wash membranes are typically composed of caustics and enzymes.

4.3   MASS TRANSPORT PROCESSES

In all pressure-driven membrane processes, the flow of permeate to and through the membrane also entails the transport of materials in the feedwater to the vicinity of the membrane. Intuitively, smaller materials such as ionic species and small organic compounds do not significantly affect bulk fluid flow in the region above the membrane. In the absence of a concentration gradient, these smaller materials are expected to move more or less with the fluid (neglecting self-diffusion). Larger particulate species may disturb the flow of fluid significantly and their trajectories may not follow fluid streamlines.

Transport to and subsequent rejection of materials by the membrane lead to an accumulation of these materials near the membrane surface and a concentration gradient for backdiffusion (Fig. 4.10). The nature of mass transport to the boundary layer region of the membrane and the balance of this transport by backdiffusion processes and advection along the membrane surface (in the presence of a cross flow) determine the degree to which mass accumulates near the membrane. This, in turn, may determine resistance to permeate flux in the event that a cake or gel layer is deposited. These transport processes may also determine the critical conditions for precipitation of dissolved materials on the membrane leading to membrane fouling.

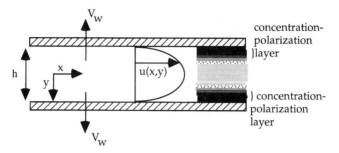

FIGURE 4.10   Fluid flow and concentration polarization for a system of two parallel membranes.

Considerable study has been made of mass transport in cross-flow membrane filtration in both laminar and turbulent flow. Due to the commercial importance of hollow-fiber, tubular, spiral-wound, and plate-and-frame membranes, most of these studies have considered two-dimensional flow in either cartesian or cylindrical coordinates. In all but the tubular membranes, flow is laminar.

4.4   TRANSPORT OF DISSOLVED SPECIES

The rejection of ionic species transported toward RO and NF membranes by advection associated with water permeating through the membrane results in an accumulation of these species near the membrane relative to the bulk fluid. This concentration gradient leads to brownian backdiffusion from the membrane (Fig. 4.11). The thickness of the concentration boundary layer is a function of the relative magnitudes of advective and diffusive transport. This boundary layer is called the *concentration-polarization layer*. The concentration profile of rejected species and, therefore, an estimate of the thickness of the concentration-polarization layer can be calculated from

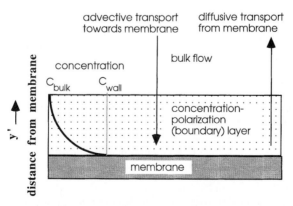

FIGURE 4.11   Changes in solute concentration within the concentration-polarization layer.

the advective-diffusion equation for any axial distance down the membrane element. In the absence of external forces and neglecting diffusion in the axial (x) direction, the advective-diffusion equation in cartesian coordinates for laminar flow and steady state reduces to

$$u \frac{\partial c}{\partial x} + v \frac{\partial c}{\partial y} = D_B \frac{\partial^2 c}{\partial y^2} \qquad (4.17)$$

A solution for the components of fluid velocity, u and v, as a function of x and y is required to solve Eq. (4.17). When the concentration boundary layer is much smaller than the momentum boundary layer (a requirement typically met in water filtration), the equations for fluid flow can be decoupled from the mass transport equation. Typical boundary conditions for Eq. (4.17) in the geometry depicted in Fig. 4.7 are:

$$c = c_o \qquad\qquad \text{at } x = 0 \quad \text{(uniform concentration at the inlet)} \qquad (4.18)$$

$$D_B \frac{\partial c}{\partial y} = V_{\text{wall}} c_{\text{wall}} \qquad \text{at } y = \frac{h}{2} \quad \text{(mass balance at the membrane)} \qquad (4.19)$$

$$\frac{\partial c}{\partial y} = 0 \qquad\qquad \text{at } y = 0 \quad \text{(symmetry)} \qquad (4.20)$$

The effects of concentration polarization can be estimated in a more approximate fashion by assuming that the permeation rate and concentration-polarization layer are constant with axial distance. A mass balance on the concentration-polarization layer then yields

$$-D_B \frac{\partial c}{\partial y} = V_{\text{wall}} c \qquad (4.21)$$

Integrating this expression over the thickness of the concentration-polarization layer δ_{cp} results in the following expression for the concentration of solute at the membrane surface c_{wall}

$$c_{\text{wall}} = c_{\text{bulk}} \exp \left(\frac{\delta_{\text{cp}}}{D_B} V_{\text{wall}} \right) \qquad (4.22)$$

For a given feed flow, the recovery of a membrane module is directly proportional to the permeation rate V_{wall}. Thus, the form of the equation given earlier for the polarization factor PF [Eq. (4.16)] can be derived directly from film theory by substituting for permeation rate (V_{wall}) in terms of recovery R in Eq. (4.22).

Equation (4.22) can be rearranged to express the limiting permeate flux V_w as a function of the bulk concentration c_{bulk}, the limiting wall concentration c_{wall}, the diffusion coefficient for the solute, and the concentration-polarization layer thickness

$$J = V_{\text{wall}} = \frac{D_B}{\delta_{\text{cp}}} \ln \frac{c_{\text{wall}}}{c_{\text{bulk}}} \qquad (4.23)$$

This so-called film layer model describes permeate flux under mass-transfer-limited conditions and ratio of diffusivity to concentration-polarization layer thickness is a mass transfer coefficient k

$$k = \frac{D_B}{\delta_{cp}} \tag{4.24}$$

Equation (4.23) predicts that under mass-transfer-limited conditions, permeate flux is independent of transmembrane pressure. In practice, permeate flux is often limited by mass transfer at higher transmembrane pressures (Fig. 4.12). At lower transmembrane pressures, permeate flux may be proportional to the transmembrane pressure drop as predicted by Eqs. (4.1) to (4.4). Pressure-independent permeation may also be interpreted within the context of Eq. (4.10); under conditions of significant concentration polarization and gel formation, increases in transmembrane pressure are countered by increases in the thickness and/or resistance of polarization and deposited gel layers. If permeate flux is mass transport limited and diffusion is brownian, permeate flux can be calculated using correlations for the mass transfer coefficient of the form:

$$Sh = \frac{kd_h}{D_B} = A(\text{Re})^{\alpha}(\text{Sc})^{\beta} \tag{4.25}$$

where $Re = u_{ave}d_h/\nu$, $Sc = \nu/D_B$ and ν is the kinematic viscosity, u_{ave} is the average cross-flow velocity, and d_h is the hydraulic diameter of the membrane element (e.g., diameter of the hollow fiber). If, in laminar flow, the velocity is fully developed and the concentration boundary layer is not, the Graetz-Leveque correlation (among others) can be used to estimate the mass transfer coefficient:

$$Sh = 1.86(\text{Re})^{0.33}(\text{Sc})^{0.33}\left(\frac{d_h}{L}\right)^{0.33} \tag{4.26}$$

The Linton and Sherwood correlation can be used to calculate the mass transfer coefficient in turbulent flow as, for example, may occur in tubular membranes

$$Sh = 0.023(\text{Re})^{0.83}(\text{Sc})^{0.33} \tag{4.27}$$

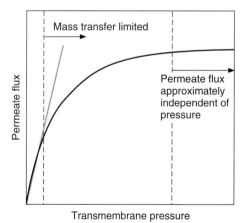

FIGURE 4.12 Permeate flux as function of transmembrane pressure; pressure-dependent and mass-transfer-limited permeate flux.

In contrast with Eq. (4.24), these correlations predict a larger value for the mass transfer coefficient and, therefore, permeate flux under mass-transfer-limited conditions, at higher velocities (Re). This is due to the effect of the cross flow in decreasing the boundary layer thickness as velocity increases.

An alternative approach to predicting permeate flux in the presence of concentration-polarization is to directly correlate permeate flux with parameters describing solute concentration and flow using a functional form derived from film theory. For example, Flemmer and co-workers[20] derived the following expression for permeate flux as a function of the bulk concentration c_{bulk} and flow exiting the membrane module Q_r (a surrogate for cross-flow velocity):

$$J = \alpha c_B^\beta Q_r^\gamma \ln \left(\frac{c_{wall}}{c_{bulk}} \right) \tag{4.28}$$

where the parameters α, β, and γ are empirical constants determined from operating the membrane over a range of values for c_B and Q_r. The concentration of material at the wall is also treated as a fitting coefficient. Equation (4.28) was derived under the assumption that the entire membrane module was gel-polarized, that is, that c_{wall} had reached some maximum value. Using a spiral-wound ultrafiltration module operated at constant inlet pressure, they determined values of $\alpha = 0.855$, $\beta = -0.108$, $\gamma = 0.478$, and $c_{wall} = 33.3$ expressed as percent mass fraction. Thus, increasing the cross-flow velocity (as captured in Q_r) is predicted to increase permeate flux, while increases in particle concentration c_{bulk} are predicted to decrease permeate flux. This correlation method has been found to yield accurate predictions of permeate flux for specific membrane modules, even for macromolecular solutes which may exhibit low brownian diffusivities. In subsequent work,[21] an expression was derived for the case in which only a fraction of the membrane module is gel-polarized.

While film theory and the semiempirical approaches based on film theory are adequate to describe permeate flux when true solutes accumulate near the membrane, film theory must be modified to describe the transport of larger materials such as particles and macromolecules that exhibit low brownian diffusivities. The transport of these materials and their impacts on permeate flux are discussed in the following section.

4.5 TRANSPORT OF PARTICULATE SPECIES

Colloidal materials transported to the membrane may form a cake that reduces permeate flux or obstructs the bulk flow over the membrane. Early efforts to describe colloidal transport to the membrane and the steady-state concentration profile near the membrane using mass transport expressions developed for solutes underpredicted permeate flux. Expressions for solute transport predict higher concentrations of colloids near the membrane, thus greater cake deposition and lower permeate fluxes than are actually observed in experiments and in practice. Numerous transport mechanisms for colloidal species were proposed that might lead to reduced concentrations of colloids near the membrane and thereby resolve the so-called flux paradox.[22, 23] In fact, it appears that several of these proposed transport mechanisms are operative in determining the performance of membranes used to treat suspensions of colloidal materials.[18, 24–26] In particular, inertial lift induced by wall effects has been shown to reduce the transport of larger colloids to the membrane surface, particularly at high cross-flow velocities. In addition, shear-induced diffusion and other orthokinetic effects tend to increase the backtransport of colloids from the membrane. Like brownian diffusion, inertial lift and shear-induced diffusion are functions of particle size; brownian diffusion decreases with increasing particle size, while inertial lift and shear-induced diffusion increase with particle size. Also, once deposited as a cake, larger particles tend to produce less resistance to permeation for the same mass deposited [see Eq. (4.12)]. Thus, the performance of membrane facilities and the success of pretreatment are likely to depend on the characteristics of the particle size distribution in the feedwater. In the following sections, factors controlling particle transport to the membrane and their backtransport from the membrane are briefly discussed.

4.5.1 Particle Transport to the Membrane

Particles, like solutes introduced in the feed to a membrane module, may be transported to the membrane surface due to the drag force associated with water permeating through the membrane. Backtransport of solutes and particles may occur due to turbulent or brownian diffusion. Unlike solutes, several additional forces may affect particle transport and must be considered in assessing the possibility that a particle entering a membrane element is transported to the membrane surface. Colloids may experience: (1) sedimentation, (2) inertial lift, (3) effects of virtual mass, (4) van der Waals attractions between particle or between particles and the membrane, (5) diffuse layer repulsion between particles or between particles and the membrane, and (6) shear-induced diffusion (or other orthokinetic transport mechanisms). Diffusive transport is of greater relative importance at low fluid velocities (small Peclet numbers). In the bulk flow over the membrane (Fig. 4.13), concentration gradients are likely to be small and cross-flow velocities are relatively high. Therefore, diffusive transport can typically be neglected in the bulk flow. Similarly, diffuse layer repulsion and Van der Waals attractions are significant only at small distances between the interacting surfaces—a condition that is likely to be met only in the boundary layer of the membrane. Virtual mass, the acceleration force associated with entrained water surrounding a particle, is typically small in membrane systems.[27] A force and torque balance on the particles in the bulk flow describes, to a great extent, the trajectories of these particles and their potential for transport from the bulk flow to the boundary layer of the membrane.

For flow conditions typical of most RO, NF, UF, and MF membranes, gravity and the inertial lift forces are the two dominant transport mechanisms in addition to per-

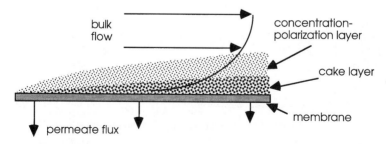

FIGURE 4.13   Bulk flow over a membrane surface and the associated boundary layer and a deposited cake.

meate drag (advection towards the membrane) that potentially determine particle trajectories in the bulk flow. Membrane modules that treat feedwaters with significant particle concentrations (e.g., UF or MF) are often installed vertically to avoid particle deposition on the membrane surface due to gravity. In this case, particle trajectories in the bulk flow are, in theory, determined primarily by the effects of permeate drag and inertial lift. Belfort and co-workers[23,28] adapted early works on particle transport in nonporous tubes[29-31] to describe inertial lift effects on particles in membranes and obtained some experimental validation for this theory.[32] The importance of inertial lift has been subsequently validated for conditions more closely approximating those of operating UF and MF membranes.[27,33]

Inertial lift is small for small particles. The behavior of smaller particles in the bulk flow is similar to that of solutes in that these smaller particles tend to follow fluid streamlines. For example, Fig. 4.14 depicts the trajectories calculated for particles 2 µm in radius in a channel 0.45 m long bounded by a membrane on the lower side and a solid wall on the upper side. These trajectories are virtually identical to the streamlines calculated for flow in this system. Particles entering near the membrane surface are carried to the membrane surface with water as it permeates through the membrane. Particles near the upper, solid wall remain in the bulk flow and exit with the concentrate. By tracing back from the exit to the entrance and interpolating, critical

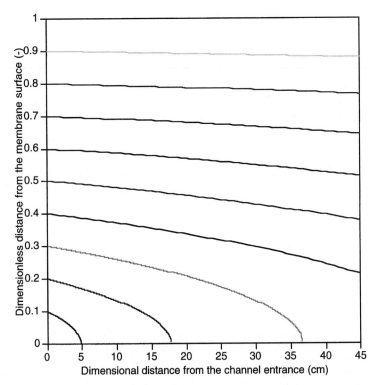

FIGURE 4.14   Position trajectories of small particles ($r_p = 2$ µm) in an ultrafiltration module. A channel geometry of 45 cm in length and 762 µm in height is assumed. Calculations also assume a mean cross-flow velocity of 33.33 cm/s and a constant permeate flux of 0.015 cm/s. *(Adapted from Chellam.[27])*

trajectory of a particle that just exits the element can be estimated. The critical trajectory in this case enters at a nondimensional distance of approximately 0.35. Thus, approximately 35 percent of the entering particles are predicted to be transported to the vicinity of the membrane and, because particle trajectories follow streamlines in this case, approximately 35 percent of the water entering the element would be expected to permeate through the membrane in a single pass.

The inertial lift on large particles is significant and, therefore, the transport of larger particles to the vicinity of the membrane may be reduced. In Fig. 4.15, trajectories for particles 10 μm in radius are shown for the same membrane geometry and flow conditions assumed in generating Fig. 4.14. In this instance, particle trajectories do not follow streamlines and all of the particles entering the membrane element are predicted to exit with the concentrate.

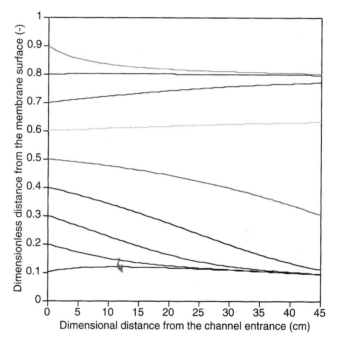

FIGURE 4.15   Position trajectories of large particles (r_p = 10 μm) in an ultrafiltration module. A channel geometry of 45 cm in length and 762 μm in height is assumed. Calculations also assume a mean cross-flow velocity of 33.33 cm/s and a constant permeate flux of 0.015 cm/s. *(Adapted from Chellam.*[27]*)*

The critical particle size below which smaller particles are predicted to be transported to the membrane increases with increasing permeation rate. That is, particles of larger and larger sizes are predicted to deposit on membranes as the permeation rate increases and permeation drag therefore becomes more important compared with inertial lift. This implies that membranes operating at higher permeation rates, such as MF and UF membranes, should draw more particles of larger size to the membrane, compared with NF or RO membranes. The critical particle size for transport to the membrane can be estimated by comparing the approximate magnitude of inertial lift, expressed as a velocity v_L to the magnitude of the permeate flux (Fig. 4.16).

The maximum inertial lift velocity can be estimated as[34]

$$v_L = 0.1182 \, \frac{u_{ave}^2 r_p^2}{\nu h^2} \tag{4.29}$$

where u_{ave} is the mean cross-flow velocity, r_p is the particle radius, ν is the kinematic viscosity of the fluid (water), and h is a characteristic length for the channel such as hollow-fiber radius or spacer thickness.

4.5.2 Diffusive Backtransport Within the Concentration-Polarization Layer

The previous discussion of solute transport considered the steady-state balance between the advective flux of materials to the membrane and diffusive backtransport. For solutes and small colloids, and macromolecules, diffusion is primarily brownian. The brownian diffusion coefficient D_B for small colloids and macromolecules can be calculated from the Stokes-Einstein equation

$$D_B = \frac{kT}{6\pi\mu r_p} \tag{4.30}$$

where k is the Botzmann constant and T is absolute temperature. Equation (4.30) predicts that the diffusion coefficient should decrease with increasing particle size. Numerous models have been proposed for shear-induced transport of particles

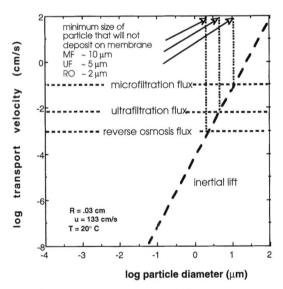

FIGURE 4.16 A comparison of the approximate magnitude of inertial lift as a function of particle diameter with permeation rates for several membrane processes.

along or away from the membrane surface. These models often share the property that particle transport is predicted to increase with the square of particle diameter and linearly with shear rate. One such solution to the flux paradox is that brownian diffusion is supplemented by a process of shear-induced or orthokinetic diffusion.[35] Correlations for the shear-induced diffusion coefficient of particles predict that the diffusion coefficient should increase with particle diameter squared and with the shear rate and particle volume fraction at the membrane.[35–37] Some experimental validation of the model has been presented in the literature.[38] If transport by brownian and orthokinetic diffusion are assumed to be additive, a minimum in particle diffusivity is predicted (Fig. 4.17).

Under the assumption of additivity, the film model for mass-transfer-limited permeate flux [Eq. (4.22)] can be modified to account for both brownian and shear-induced diffusivity:

$$J = \frac{D_B + \dot{\gamma} r_p^2 \tilde{D}_{sh}}{\delta_{cp}} \ln \frac{c_{wall}}{c_{bulk}} \tag{4.31}$$

where the second term in the numerator is the shear-induced diffusivity D_{sh}, $\dot{\gamma}$ is the shear rate, and $\tilde{D}_{sh}$ is a dimensionless function of particle volume fraction given by empirical relationships developed from laboratory data. Assuming a volume fraction of 0.15 near the cake and the functional form proposed by Leighton and Acrivos,[36] the second term in the numerator of Eq. (4.31), shear-induced diffusivity, can be approximated as

$$D_{sh} = 0.02 \frac{du}{dx} a_p^2$$

Analogous to the case of solute transport, a more rigorous estimate of particle transport and permeate flux in the presence of colloidal materials requires a numerical solution of the advective-diffusion equation

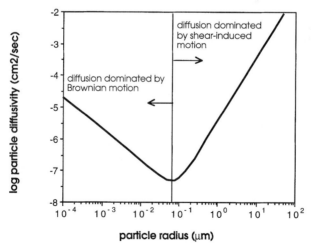

FIGURE 4.17 Brownian and orthokinetic diffusivity as a function of particle size for conditions typical of hollow-fiber UF membranes. (*Adapted from Wiesner.[25]*)

$$\frac{\partial u\phi}{\partial x} + \frac{\partial v\phi}{\partial y} = \frac{\partial}{\partial y}\left(D\,\frac{\partial \phi}{\partial y}\right) \tag{4.32}$$

where ϕ is the volume fraction of colloids with radius r_p in the water. The sum of brownian and shear-induced diffusivity D can be rewritten as (Sethi, in preparation #326)

$$D = \frac{kT}{6\pi\mu_o r_p} + \frac{\tau_{\text{wall}}}{\mu_o \eta(\phi)}r_p^2 \tilde{D}_{\text{sh}}(\phi) \tag{4.33}$$

where τ_{wall} is the shear rate at the membrane wall, μ_o is the absolute viscosity of the water at low particle volume fractions, and $\eta(\phi)$ is the relative viscosity which varies with volume fraction. Simultaneous consideration of brownian and shear-induced diffusivity leads to a minimum in particle backtransport for particles in the size range of several tenths of a micrometer and a corresponding minimum in permeate flux. It may follow that particles in this size range preferentially accumulate near the membrane. However, there is considerable evidence that larger particles may enhance the transport of smaller particles by sweeping or scouring them from the membrane surface. Theory for particle transport and membrane performance in the more realistic case of polydisperse suspensions of particles in the feedwater is poorly developed. For the purpose of obtaining very approximate calculations of membrane performance, models for single-size particles may be adapted to the polydisperse case by choosing an appropriate "average" particle radius. An average radius $\bar{r}_p$ that accounts for the relative propensity of particles to accumulate near the membrane has been suggested (Sethi, in preparation #326), in which

$$\bar{r}_{p,\text{ave}} = \frac{\displaystyle\sum_{i\in(\text{colloids}|p_i\,<1)} \frac{r_{p_i}\phi_{\text{bulk}_i}}{D_i}}{\displaystyle\sum_{i\in(\text{colloids}|p_i\,<1)} \frac{\phi_{\text{bulk}_i}}{D_i}} \tag{4.34}$$

where D_i and $\phi_{\text{bulk},i}$ are the total diffusivity and bulk volume fraction of particles, respectively, with particle of radius $r_{p,i}$, and p_i is the fraction of particles in size class i present in the feed that passes through the membrane.

4.6 MACROSCOPIC VIEW OF CHANGES IN PERMEATE FLUX

The preceding discussion considers the role of mass transport in determining permeate flux from first principles. Alternatively, permeate flux decline can be considered from a macroscopic perspective.[39] Permeate flux decline by adsorptive fouling, cake formation, pore blockage, etc., can be shown to imply a specific shape of the permeate flux curve over time. For example, when a membrane is first placed into service, the cake and concentration-polarization resistances in Eq. (4.9) are zero, and permeate flux is limited by the resistance of the membrane. The accumulation of materials on (and in) the membrane over time produces an increase in R_c and R_m over time. In the case where R_c is dominated by the formation of a cake of constant resistance per unit depth, R_{cp} is negligible, and membrane resistance is constant over time (e.g., no adsorptive fouling)

$$R_c = \hat{R}_c\delta_c \tag{4.35}$$

where R_c is the specific resistance of the cake (resistance per unit depth), and δ_c is the depth of the cake. Under ideal conditions, $\hat{R}_c$ can be calculated using Eq. (4.12). However, in many cases, $\hat{R}_c$ will increase over time as the cake compacts and as smaller colloidal materials are "filtered" by the cake. Cake growth at any time should be proportional to the rate at which materials are carried to the membrane surface. However, as the cake depth increases, shear-induced backtransport of deposited materials should occur at a greater rate. These assumptions may be expressed in the following differential equation describing the change in cake thickness with time:

$$\frac{\partial\delta_c}{\partial t} = k_1 J - k_2\delta_c \tag{4.36}$$

The rate constant k_1 describes the transport of cake-forming materials to the membrane and should increase with increasing feed concentration. Theoretical consideration of particle trajectories in membrane elements indicates that k_1 should decrease with increasing particle diameter and increasing cross-flow velocity due to the effects of inertial lift.[28] The constant k_2 indicates the relative importance of backtransport mechanisms in removing materials from the membrane and is also predicted to increase with increasing cross-flow velocity. If it is assumed that the cake is composed of a fluidized layer of particles, a minimum in k_2 with respect to particle size is predicted for colloids near 0.1 μm in diameter based on the combined effects of brownian diffusion and shear-induced diffusivity.[25]

Under the assumption of a constant membrane resistance, constant pressure drop, R_c, dominated resistance, and a specific resistance of the cake that is invariant with time, the time derivative of Eq. (4.10) yields,

$$\frac{\partial J}{\partial t} = \frac{-J\hat{R}_c(k_1 J - k_2\delta_{\text{cake}})}{R_m + \hat{R}_c\delta_{\text{cake}}} \tag{4.37}$$

During the early stages of cake formation, deposition of materials to the membrane may be much greater than loss by fluid shear ($k_1 J > k_2\delta_c$) and initial cake resistance may be less than the resistance of the membrane ($R_m \geq a\delta_c$). Under these assumptions, the time rate of change of permeate flux during the early stages of filtration can be approximated as a pseudo-second-order equation in J:

$$\frac{\partial J}{\partial t} \cong -k_3 J^2 \tag{4.38}$$

where $k_3^* = \hat{R}_c k_1 / R_m$. It is important to note that a higher value for the specific resistance of the deposit $\hat{R}_c$ results in a higher value of k_3^*. Cakes composed of larger particles should produce a lower specific resistance.[26] Thus, an increase in particle size may reduce k_3^* both through effects on particle transport and cake morphology. The slope of a linear plot of $(1/J)$ versus time yields the cake formation rate constant k_3^*. As cake formation proceeds, $\hat{R}_c\delta_c \geq R_m$. For the case of active cake formation where $k_1 J > k_2\delta_c$ and cake resistance controls permeate flux, the following rate expression is obtained

$$\frac{\partial J}{\partial t} \cong -k_3^* \left(\frac{\hat{R}_c}{\Delta p} \right) J^3 \tag{4.39}$$

and the cake formation rate constant is derived from a least-squares fit to a plot of $(1/J^2)$ versus time. The transmembrane pressure drop Δp often increases as J decreases due to a decreasing backpressure on the permeate side of the membrane. Also, membrane systems may be operated in a constant permeate flux, rather than a constant transmembrane pressure mode. When this is the case, transmembrane pressure increases over time. When Δp is not constant over time, equations similar to (4.38) and (4.39) for second- and third-order flux decay can be applied to the specific permeate flux $J^* = J/\Delta p$.

Expressions also can be derived, assuming other mechanisms for permeate flux decline such as the development of a concentration-polarization layer, the blockage of membrane pores, and the adsorption of materials with the membrane matrix. Some of these expressions are summarized in Table 4.2. The generic rate constant K in each of these expressions is an indication of the rate at which permeate flux declines. In all cases, K is directly proportional to a deposition, adsorption, or cake formation rate constant [e.g., k_1 in Eq. (4.36)].

Analysis of permeate flux data using the kinetic expressions in Table 4.2 allows for a comparison of the effect of operating conditions, raw water quality, and membrane characteristics on the rate, extent, and nature of permeate flux decline. The values of rate constants evaluated from field data assuming a single mechanism of flux decline can be used to compare the treatability of various raw waters as well as the impact of various pretreatments on permeate flux. Identification of rate expressions that best describe permeate flux may suggest mechanisms of membrane colmatage and/or fouling. They may also be useful in prescribing the frequency of procedures for permeate flux enhancement such as hydrodynamic or chemical cleaning. While different mechanisms sometimes lead to similar rate expressions, some of this ambiguity can be reduced by designing pilot studies to differentiate between reversible and irreversible reductions in permeate flux. However, in practice several rate expressions may be found to yield an acceptable fit to the data and it is likely that several mechanisms will play a role in reducing permeate flux.

TABLE 4.2 Kinetic Expressions for Permeate Flux Decline

Cause of permeate flux decline	Permeate flux equation	Linearized form of equation
Cake formation membrane-limited flux	$J = \dfrac{J_o}{1 + J_o K t}$	$\dfrac{1}{J} = \dfrac{1}{J_o} + K t$
cake-limited flux	$J^2 = \dfrac{J_o^2}{1 + J_o^2 K t}$	$\dfrac{1}{J^2} = \dfrac{1}{J_o^2} + K t$
Concentration polarization	$J = J_{SS}^* + B \exp(-Kt)$	$\ln(J - J_{SS}) = \ln B - Kt$
Adsorptive pore fouling	$J = J_{SS} + B \exp(-Kt)$	$\ln(J - J_{SS}) = \ln B - Kt$
Pore blockage (initial)	$J = J_o \exp(-Kt)$	$\ln\left(\dfrac{J}{J_o}\right) = -Kt$

* J_{SS} = steady-state permeate flux.
Adapted from Wiesner.[39]

REFERENCES

1. K. D. Pickering and M. R. Wiesner, 1993, "Cost Model for Low-Pressure Membrane Filtration," *Journal of Environmental Engineering, American Society of Civil Engineers,* **119** (5):000–000.

2. U. Merten, 1966, *Desalination by Reverse Osmosis,* MIT Press, Cambridge, Mass.

3. J. D. Henry, 1972, "Crossflow Filtration," in N. N. Li (ed.), *Recent Developments in Separation Science,* CRC Press Inc., Linden, N.J.

4. H. W. Pohland, 1988, "Theory of Membrane Processes," *Proceedings of the AWWA Annual Conference,* Orlando, Fla.

5. A. R. Marsh, and P. K. Eriksson, 1988, "Projecting R.O. Desalination System Performance with Filmtec Spiral-Wound Elements," *Proceedings of the Seminar on Membrane Processes,* Orlando, Fla.

6. R. J. Hunter, 1981, *Zeta Potential in Colloid Science,* Academic Press, New York.

7. F. Nazzal, and M. R. Wiesner, 1994, "pH and Ionic Strength Effects on the Performance of Ceramic Membranes in Water Filtration," *Journal of Membrane Science,* **93:**91–103.

8. C. Gourgues, P. Aimar, and V. Sanchez, 1992, "Ultrafiltration of Bentonite Suspensions with Hollow Fiber Membranes," **84:**61–77.

9. R. M. McDonogh, C. J. D. Fell, and A. G. Fane, 1984, "Surface Charge and Permeability in the Ultrafiltration of Non-Flocculating Colloids," *Journal of Membrane Science,* **21:**285–294.

10. P. Bacchin, P. Aimar, and V. Sanchez, "A Model for Colloidal Fouling of Membranes" (in press).

11. R. M. McDonogh, A. G. Fane, and C. J. D. Fell, 1989, "Charge Effects in the Crossflow Filtration of Colloids and Particulates," *Journal of Membrane Science,* **43:**69–85.

12. J.-M. Laîné, et al., 1989, "Effects of Ultrafiltration Membrane Composition," *Journal American Water Works Association,* **81**(11):61–67.

13. M. M. Clark, and C. Jucker, 1993, "Interactions Between Hydrophobic Ultrafiltration Membranes and Humic Substances," *Proceedings of the Membrane Technology Conference,* Baltimore, Md.

14. V. Lahoussine-Turcaud, et al., 1990, "Coagulation Pretreatment for Ultrafiltration of a Surface Water," *Journal American Water Works Association,* **82**(12):76–81.

15. U. Merin and M. Cheryan, 1980, "Ultrastructure of the Surface of a Polysulphone Ultrafiltration Membrane," **25:**21–39.

16. J. Mallevialle, C. Anselme, and O. Marsigny, 1989, "Effects of Humic Substances on Membrane Processes," in *Advances in Chemistry* (ed.), American Chemical Society, Denver, Colo.

17. G. Grozes, C. Anselme, and J. Mallevialle, 1993, "Effect of Adsorption of Organic Matter on Fouling of Ultrafiltration Membranes," *Journal of Membrane Science,* **84:**61–77.

18. V. Lahoussine-Turcaud, M. R. Wiesner, and J. Y. Bottero, 1990, "Fouling in Tangential-Flow Ultrafiltration: The Effect of Colloid Size and Coagulation Pretreatment," *Journal of Membrane Science,* **52:**173–190.

19. J. L. Bersillon, 1988, "Fouling Analysis and Control," in L. Cecille and J. C. Toussaint (eds.), *Future Industrial Prospects of Membrane Processes,* Elsevier Applied Science, Oxford.

20. R. L. C. Flemmer, C. A. Buckley, and G. R. Groves, 1982, "An Analysis of the Performance of a Spiral-Wound Ultrafiltration Membrane, With a Turbulence-Promoting Net," *Desalination,* **41:**25–32.

21. R. L. C. Flemmer, 1983, "Performance of a Spiral Wound Ultra-Filtration Membrane with Turbulence Promoting Net Under Conditions of Strong Pressure Gradient and Fouling," **48:**321–329.

22. W. F. Blatt, et al., 1970, "Solute Polarization and Cake Formation in Membrane Ultrafiltration: Consequences and Control Techniques," in J. E. Flinn (ed.), *Membrane Science and Technology, Industrial, Biological and Waste Treatment Processes,* Plenum Press, New York.

23. G. Green, and G. Belfort, 1980, "Fouling of Ultrafiltration Membranes: Lateral Migration and Particle Trajectory Model," *Desalination,* **35:**129–147.

24. M. R. Wiesner, M. M. Clark, and J. Mallevialle, 1989, "Membrane Filtration of Coagulated Suspensions," *Journal of Environmental Engineering, American Society of Civil Engineers,* **115**(1):20–40.

25. M. R. Wiesner, and S. Chellam, 1992, "Mass Transport Considerations for Pressure-Driven Membrane Processes," *Journal American Water Works Association,* **84**(1):88–95.

26. A. G. Fane, 1984, "Ultrafiltration of Suspensions," *Journal of Membrane Science,* **20:**249–259.

27. S. Chellam, and M. R. Wiesner, 1992, "Particle Transport in Clean Membrane Filters in Laminar Flow," *Environmental Science and Technology,* **26**(8):1611–1621.

28. F. W. Altena, and G. Belfort, 1984, "Lateral Migration of Spherical Particles in Porous Flow Channels: Application to Membrane Filtration," *Chemical Engineering Science,* **49**(2): 343–355.

29. R. G. Cox, and H. Brenner, 1968, "The Lateral Migration of Solid Particles in Poiseuille Flow—I Theory," *Chemical Engineering Science,* **23:**147–173.

30. R. G. Cox, and S. G. Mason, 1971, "Suspended Particles in Fluid Flow through Tubes," *Annual Review of Fluid Mechanics,* **3:**291–316.

31. P. G. Saffman, 1967, "The Lift on a Small Sphere in a Slow Shear Flow," *Journal of Fluid Mechanics,* **22**(2):386–400.

32. J. R. Otis, et al., 1986, "Measurements of Single Spherical Particle Trajectories with Lateral Migration in a Slit with One Porous Wall under Laminar Flow Conditions," *Experiments in Fluids,* **4:**1–10.

33. G. Belfort, R. H. Davis, and A. L. Zydney, 1994, "The Behavior of Suspensions and Macromolecular Solutions in Crossflow Microfiltration," *Journal of Membrane Science,* **96:**1–58.

34. G. Belfort, 1989, "Fluid Mechanic of Membrane Modules," *Journal of Membrane Science,* **40:**123–147.

35. R. H. Davis, and D. T. Leighton, 1987, "Shear-Induced Transport of a Particle Layer Along a Porous Wall," *Chemical Engineering Science,* **42**(2):275–281.

36. D. Leighton, and A. Acrivos, 1987, "The Shear-Induced Migration of Particles in Concentrated Suspensions," *Journal of Fluid Mechanics,* **181:**415–439.

37. C. A. Romero, and R. H. Davis, 1988, "Global Model of Crossflow Microfiltration Based on Hydrodynamic Particle Diffusion," *Journal of Membrane Science,* **39:**157–185.

38. C. A. Romero, and R. H. Davis, 1991, "Experimental Verification of the Shear-Induced Hydrodynamic Diffusion Model of Crossflow Microfiltration," *Journal of Membrane Science,* **62:**249–273.

39. M. R. Wiesner, S. Veerapaneni, and D. Brejchová, 1992, "Improvements in Membrane Microfiltration Using Coagulation Pretreatment," in H. Hahn and R. Klute (eds.), *Chemical Water and Wastewater Treatment II,* Springer-Verlag, Berlin.

40. F. F. Nazzal, K. D. Pickering, and M. R. Wiesner, 1991, "Particle Deposition on Tubular Ceramic Membranes," *Proceedings of the Membrane Technologies in the Water Industry,* Orlando, Fla.

CHAPTER 5

PRINCIPLES OF REJECTION IN PRESSURE-DRIVEN MEMBRANE PROCESSES

Mark R. Wiesner
Department of Environmental Science and Engineering
Rice University
Houston, Texas

Chris A. Buckley
Pollution Research Group
Department of Chemical Engineering
University of Natal
Durban, South Africa

Pressure-driven membrane processes such as reverse osmosis (RO), nanofiltration (NF), ultrafiltration (UF), and microfiltration (MF) resemble one another in many regards (geometry, flow configuration, etc.). However, the underlying principles governing rejection of contaminants by the membrane differ substantially. For example, rejection of contaminants by RO membranes is largely a function of the relative chemical affinity of the contaminant for the membrane vis-a-vis water. In contrast, rejection by MF membranes is due almost entirely to physical sieving of materials at the membrane surface. NF and UF membranes may reject contaminants to some extent based on combinations of physical sieving and contaminant-membrane chemistries. In the following sections, we provide an overview of processes governing membrane rejection in pressure-driven processes and the mathematical approaches to describing rejection in each case.

5.1 DEFINITIONS OF REJECTION

Rejection of a given material by a membrane is usually defined as one minus the ratio of the permeate concentration to the concentration "somewhere" else on the concentrate side of the membrane. In the most common usage, rejection across a membrane refers to the ratio of concentrations in the permeate and feed streams. This *global* rejection R, based on feed and permeate concentrations, is calculated as:

$$R = 1 - \left(\frac{c_p}{c_f}\right) \tag{5.1}$$

where c_p and c_f are the permeate and feed concentrations (Fig. 5.1).

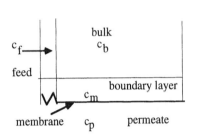

bulk
c_b

c_f

feed

boundary layer
c_m

membrane c_p permeate

FIGURE 5.1   Concentrations at various locations in a membrane system.

While this is the most commonly used definition of rejection, we might also consider an alternative, global measure of rejection, based on the fraction R_{mass}, of the mass of a specific material entering a membrane system that does not pass through the membrane. The following expression for mass rejection can be derived:

$$R_{\text{mass}} = 1 - \left(\frac{c_p}{c_f}\right) r \tag{5.2}$$

where r is the recovery of the membrane system (volume of permeate produced per volume of water introduced). These expressions for global rejection may yield different values as a function of time. Concentrations in the feed are often variable. Concentrations in the permeate may change as feed concentrations, mass transport, membrane condition, or other factors change over time. With the exception of physical sieving where contaminants are completely retained by the membrane, the concentration of a contaminant in the permeate is likely to increase as system recovery increases; i.e., $c_p = f(r)$.

The bulk concentration of a material increases as flow proceeds along the membrane due to the permeation of water through the membrane. Also, rejected materials tend to accumulate at higher concentrations near the membrane compared with the bulk flow due to concentration polarization. Thus, the effective rejection of materials by a membrane may be very different from that calculated based on the average feed and permeate concentrations. Occasionally, we are interested in the *local* rejection R_{local} of a given membrane for a specific contaminant. Local rejection is a function of the concentrations directly on either side of the membrane at a specific location along the membrane and is defined as

$$R_{\text{local}} = 1 - \left(\frac{c_p}{c_{\text{wall}}}\right) \tag{5.3}$$

where c_{wall} is the concentration at the membrane surface. The local rejection varies as a function of the mass transfer of contaminant to the membrane, which results in an elevated concentration relative to the bulk concentration ($c_{\text{wall}} \geq c_{\text{bulk}} \geq c_f$). The origin of the elevated concentration near the membrane (concentration polarization) as a consequence of mass transport to and from the membrane has been discussed in Chap. 4. The concentration near the membrane surface can be predicted as a function of the permeate flux and mass transfer coefficient by calculating a polarization factor PF such that $c_{\text{wall}} = (\text{PF})c_{\text{bulk}}$. The local rejection varies along the membrane since the local mass transfer coefficient and permeate flux may also vary with location. When rejection is expressed as a function of the bulk concentration rather than the concentration at the membrane surface, it is referred to as the *apparent rejection* $R_{apparent}$ and is given as:

$$R_{apparent} = 1 - \left(\frac{c_p}{c_{bulk}}\right) = 1 - (1 - R_{local})(PF) \qquad (5.4)$$

If a mass balance is performed over the membrane module, the following expression is derived relating the global to the apparent rejection:

$$R = 1 - \left(\frac{c_p}{c_f}\right) = 1 - \frac{1 - (1 - r)^{1 - R_{apparent}}}{r} \qquad (5.5)$$

5.2 MATHEMATICAL DESCRIPTION OF PARTICLE REJECTION

5.2.1 Mechanical Sieving at the Membrane Surface

Particulate materials are removed by membranes predominantly through physical sieving. This is true even for RO membranes; chemical factors play virtually no role in the rejection of colloidal materials by RO membranes. Membranes conceived specifically for removing or separating colloidal materials, such as microfiltration and ultrafiltration membranes, reject materials largely by mechanical sieving. However, electrostatic interactions, dispersion forces, and hydrophobic bonding may significantly affect the rejection of materials with dimension(s) similar to the size of pores in UF or MF membranes.

A quantitative analysis of mechanical sieving (or steric rejection) was first presented by J. D. Ferry in 1936.[1] Assuming pores of cylindrical geometry and spherical particles, he derived an expression for the fraction of particles p passing through a pore of radius R (Fig. 5.2). This expression was subsequently modified[2] to account

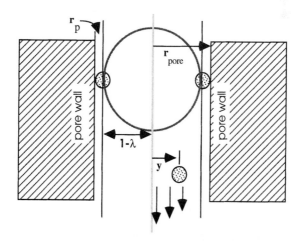

FIGURE 5.2 Definition of terms used to describe particle removal in a membrane pore.

for the lag velocity of particles with respect to the fluid permeating through the membrane pore. The rejection of particles by a membrane $(1 - p)$ can be estimated using this expression, as a function of the nondimensionalized particle radius, $\lambda = r_p/r_{pore}$, as

$$p = \begin{cases} (1 - \lambda)^2[2 - (1 - \lambda)^2]G & \lambda \le 1 \\ 1 & \lambda > 1 \end{cases} \qquad (5.6)$$

where G is the lag coefficient empirically estimated by Zeman and Wales[2] as

$$G = \exp(-0.7146\lambda^2) \qquad (5.7a)$$

Earlier work (e.g., Lakshminarayanaiah, 1965) considered the effect of a nonuniform velocity distribution in the membrane pores yielding the following expression for G:

$$G = 1 - 2.104\lambda + 2.09\lambda^3 - 0.95\lambda^5 \qquad (5.7b)$$

As particle radius approaches the pore radius, Eq. (5.7b) leads to substantially lower estimates of particle rejection than does Eq. (5.7a).

Rigorously derived, $(1 - p)$ corresponds to the local rejection of the membrane, R_{local}. Measurements of apparent rejection can be used to calculate a value p^*, which theoretically corresponds to the product of the polarization factor PF and the particle passage p. Removal of materials in deposited cake or gel layers may further alter the apparent rejection of the membrane.

Further modifications to this model account for electrostatic repulsion and dispersion forces near the pore wall,[3] a more rigorous treatment of drag on particles transported through membrane pores by advection and diffusion,[4] and the diffusion and convection of flexible macromolecules through cylindrical pores of molecular dimensions.[5] The latter modification results in rejection coefficients that are larger than those for solid spherical particles (Fig. 5.3).

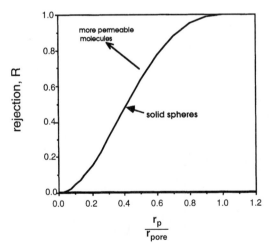

FIGURE 5.3 Rejection of particles and macromolecules as a function of the equivalent solid sphere radius of the molecule.

A simple approach for extending the physical sieving model for particle removal to describe the rejection of macromolecular compounds such as humic materials, involves substituting the molecule's hydrodynamic radius for particle radius. Empirical constants in an expression for the molecule's hydrodynamic radius are then manipulated to fit a given data set. For larger-molecular-weight materials, hydrodynamic radius can often be correlated with the compound's molecular weight $\overline{M}$. These correlations are typically of the form:

$$a_p = Z_1(\overline{M})^{Z_2} \tag{5.8}$$

where Z_1 and Z_2 are empirical constants. The constant Z_2 contains information on molecule geometry. For a perfectly spherical molecule, Z_2 takes on its theoretical minimum value of 1/3. The theoretical maximum value of 1 for Z_2 corresponds to a linear molecule. It is possible that Z_2 may take on values outside its theoretical range when Z_1 and Z_2 are treated as fitting constants within the context of the sieving model.

By the sieving mechanism, rejection of organic compounds is predicted to increase with molecular weight [assuming that molecular size also increases as described by Eq. (5.8)]. Such behavior has been observed with respect to the removal of the larger-molecular-weight fractions of natural organic matter (NOM) by ultrafiltration and nanofiltration membranes (e.g., see Refs. 6 and 7). NOM frequently reacts with chlorine added for disinfection or oxidation to form trihalomethanes (THMs) and other disinfection by-products. Removal of THM precursor materials closely follows removal of the dissolved organic carbon. The ability of NF membranes to remove significant fractions of DOC and the associated THM precursors make them attractive alternatives to other treatment processes for DOC removal such as granular activated carbon adsorption (see Fig. 5.4).

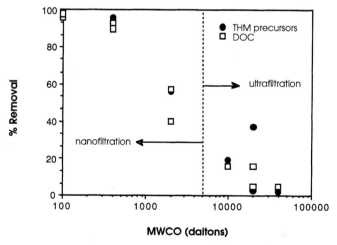

FIGURE 5.4   Removal of DOC and THM precursors as a function of membrane molecular weight cutoff. *(Data from Refs. 6 and 7.)*

5.2.2 Rejection of Deformable Drops

A rigid particle will not enter a membrane pore smaller than itself. However, deformable drops, such as are present in macro- or microemulsions, may squeeze through a membrane pore if the transmembrane pressure is great enough to overcome surface tension forces. An estimate of the maximum transmembrane pressure (TMP) below which drops of a given size will be rejected by the membrane can be derived from the Young-Laplace equation for capillary pressure across curved interfaces.[8] Idealizing the membrane pore as a straight cylindrical capillary (Fig. 5.5),

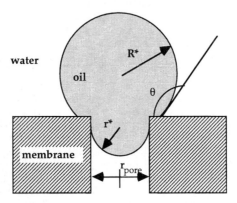

FIGURE 5.5 Passage of an oil drop through a membrane pore.

assuming that the fluids are incompressible and that hysteresis is negligible, the critical pressure p_c required to force an oil droplet through a pore is calculated as

$$p_c = \left(\frac{2\gamma_{o/w}}{r^*}\right) - \left(\frac{2\gamma_{o/w}}{R^*}\right) \tag{5.9}$$

where $\gamma_{o/w}$ is the interfacial tension between the oil and water, R^* is the radius of curvature of the portion of the drop outside the capillary, r^* is the radius of curvature of the portion of the drop inside the capillary ($r^* = -r/\cos\theta$), and θ is the contact angle measured through the oil phase. Equation (5.9) can be used as a starting point to estimate the pressure at which a membrane may become oil wet in the treatment of oil emulsions. When the transmembrane pressure exceeds a critical value, oil drops are expected to move through membrane pores.

5.2.3 Cake Removal

While the membrane itself determines the initial removal of materials from the permeate and serves as an absolute barrier to specific materials, the rejection characteristics of the membrane system may change over time as materials are deposited on the membrane surface. Indeed, this concept is exploited in the application of "dynamic" membranes, formed from materials deposited on a porous support, that are periodically removed and replenished. Mathematical treatment of this process has borrowed concepts from packed-bed filtration. Suspended colloidal materials transported toward the cake with the permeating water may be removed at the cake surface by simple sieving. Particles entering the cake may be removed by previously deposited particles that form a type of micropacked bed. Removal within the cake may be due to simple interception of mobile colloids if the streamline they follow brings them in contact with a previously deposited (immobile) colloid. Gravity and brownian motion may cause particles to deviate from their streamlines. As a result, gravity and brownian motion may also bring particles in contact with the immobile particles composing the cake. The previously deposited particles are termed *collectors* in the packed-bed filtration literature. Unlike mathematical descriptions of packed-bed filtration, the collectors may not be large compared with the particles being removed. While some progress has been made in describing cake growth and particle removal by cakes for the case of monodisperse suspensions, the general case of a suspension and cake composed of particles of many sizes, as is commonly encountered in practice, is considerably more complex and is poorly understood.

5.3 SOLUTE SEPARATION

Rejection of solutes by nanofiltration and reverse-osmosis membranes has been the subject of considerable research. Unlike the mechanism of physical sieving that dominates the rejection behavior of UF and MF membranes, rejection of solutes by RO and NF membranes depends in a complex fashion on the chemistry of the solute-membrane interactions.

5.3.1 Observed Trends in Inorganic Solute Removal

The rejection of ionic solutes by RO membranes has been observed, to an approximation, to follow the lyotropic series (increasing rejection with increasing hydrated radius). This is also consistent with the observation that ions of higher valency tend to be rejected more than are ions of lower valency. The lyotropic series predicts that the rejection of cations by RO membranes should obey the following order:

$$Mg > Ca > Sr > Ba > Ra > Li > Na > K$$

and similarly, anion rejection should occur in the following order:

$$SO_4^{2-} > Cl^- > Br^- > NO_3^- > I^-$$

However, this is, at best, a rough generalization. Exceptions to the lyotropic series are not uncommon in experiments with single solutes.[9] In addition, this order may not be observed due to ion pairing, complexation, and other solute-solute interactions (see subsequent sections). Nonetheless, the lyotropic rule serves as a useful device for envisioning the rejection of ionic species within the context of a number of models, including the extension of a sieving-type model for rejection to include rejection of ions with a specified hydrated radius, the rejection due to differential rates of diffusive transport, and the rejection of inorganic solutes due to negative adsorption at membrane surfaces.

5.3.2 Rejection of Organic Solutes

The rejection of organic compounds by RO and NF membranes appears to be somewhat more complex than the case of inorganic ions. It is likely that several mechanisms come into play in reducing the concentration of organic solutes in RO and NF permeates. Although considerable attention is given to molecular weight ($\overline{M}$) as a predictor of the rejection of organic compounds by NF and UF membranes, $\overline{M}$ appears to be of considerably less consequence for low-molecular-weight compounds removed by RO membranes, which include many of the synthetic organic chemicals (SOCs) of concern (Fig. 5.6).

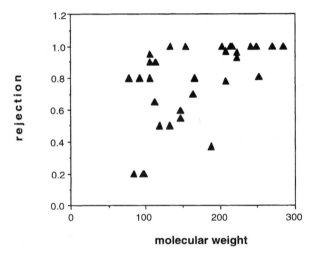

FIGURE 5.6   Rejection of various synthetic organic chemicals as a function of molecular weight. *(Data from Fronk and co-workers, 1990[23,24].)*

The composition of the membrane appears to play a critical role in determining the rejection of SOCs with molecular weights less than approximately 200 (Fig. 5.7), possibly supporting the role of sorbed water at the membrane interface and within the membrane matrix in determining rejection. However, comparisons between membranes of different composition are complicated by the fact that the molecular weight cutoff of the membranes may also differ.

If molecular weight does play some role in determining the rejection of these smaller compounds, this role may be in part related to a sieving phenomenon. But, in addition, the molecular weight of the compound will affect its diffusivity. Larger-molecular-weight materials may diffuse through the membrane at a slower rate, leading to lower concentrations in the membrane permeate. Other characteristics of these compounds that appear to affect their removal by membranes include molec-

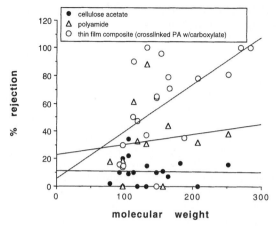

FIGURE 5.7 SOC rejection as a function of molecular weight for three types of membranes.[23]

ular conformation, polarity, dielectric constant, and hydrophobicity. In particular, it has been noted that the removal of many organic acids is greater when functional groups are charged at higher pH values. However, these trends are somewhat confusing in that very similar molecules differing only by the placement of the acidic group on the molecule are removed to a remarkably different degree for the same amount of dissociation.[10] The factors governing rejection are therefore apparently quite complex. However, the following trends in removal of organic solutes have been observed:

1. Merten and co-workers[11] analyzed data reported previously by Sourirajan and co-workers[12] and concluded that rejection of organic compounds increases with increased molecular weight in a homologous series and with increased branching for molecules of equal molecular weight.

2. Compounds without ionizable groups are rejected by RO (and probably NF) membranes to a lesser extent than are compounds with ionizable groups.

3. Compounds with ionizable groups are rejected to a greater extent when these groups are ionized.

4. Phenolic compounds are, as a class, poorly rejected. Low-molecular-weight chlorinated hydrocarbons (e.g., some insecticides and herbicides) also tend to be poorly rejected.[13]

5. Compounds that are capable of significant hydrogen bonding tend to be removed to a lesser extent (e.g., alcohols, aldehydes, acids, urea).[14]

6. Interactions with natural organic materials may significantly affect the rejection of specific SOCs.

7. Organic acids and amines are rejected better when present as a salt.[14]

8. VOCs, including THMs, most phenolic compounds, chlorinated hydrocarbons, and many pesticides, are poorly removed by RO membranes.[14]

9. Low-MW polar organics tend to be removed better by nonpolar membranes.[14]

5.4 MATHEMATICAL MODELS FOR SOLUTE TRANSPORT AND REJECTION

Models for separation of solutes may be grouped into three categories: (1) preferential adsorption/capillary flow, (2) irreversible thermodynamics, and (3) solution diffusion. In the following sections, these three categories of models are presented. The development of these models leads to expressions for water permeation rate as well as solute flux. An analysis of solute transport through membranes or, conversely, their rejection is closely related to the analysis of mass transport and permeate flux. Mass transport and permeate flux have been addressed in Chap. 4. Some of this material will be referenced in presenting models for separation. In addition to these three classes of models, models for membrane rejection have been developed based on the Donnan exclusion principle. This topic is briefly discussed as a special case for charged membranes. A more thorough treatment of the subject is given in Chap. 7, where models for selectivity in electrodialysis membranes are discussed.

5.4.1 Preferential Sorption–Capillary Flow

The term *preferential sorption* refers to a layer of water sorbed at the membrane surface. This layer may exist due to the repulsion of ions by membrane materials of low dielectric constant, the excessive energy required to strip ions of their hydration spheres as they move from the bulk water to the water-membrane interface, or the hydrophilic nature of the membrane surface that actively coordinates and orders water near the membrane surface. Preferential sorption of water at membrane surfaces corresponds to a negative sorption of the solute at membrane surfaces leading to a deficit of the solute in the membrane phase. Transport of the solute across the membrane is therefore impeded. Viewed from a different perspective, a layer of "pure" water sorbed to the membrane surface is "drained" off under an applied pressure. In the preferential sorption–capillary flow model and its variations, the solute is assumed to move through the membrane by diffusion, advection, or both.[9] Thus, the membrane is viewed as a microporous medium. Separation is assumed to be a function of both the surface chemistry of the membrane and the fashion in which water is transported through the membrane. Ions with large hydrated radii are impeded from entering the sorbed water layer due to the energy required to strip water from their hydration shells near the membrane. The preferential sorption model assumes that ions passing through the membrane must also diffuse through the adsorbed water film. Ions with smaller hydrated radii may be able to diffuse through the adsorbed water film or may be convected through pores (Fig. 5.8). The diffusivity of species in regions of structured water is not well understood. The sorbed layer of water imparts rejection characteristics that are independent of the size of pores in the membrane, up to the point that the thickness of the sorbed water layer is equal to or greater than the pore size.

Mathematically, permeate flux in the preferential sorption–capillary flow model is expressed as

$$J = B(\Delta p - \Delta\Pi) \tag{5.10}$$

where B is the pure water permeability of the membrane, Δp is the transmembrane pressure, and $\Delta\Pi$ is the change in osmotic pressure across the membrane. Because the membrane is assumed to transport water through small pores, the similarity of

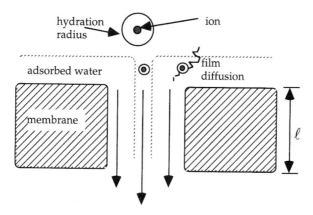

FIGURE 5.8 Qualitative description of ion transport and rejection by the preferential sorption model.

Eq. (5.10) to Poiseuille's law [see, for example, Eq. (4.4)] is not coincidental. However, osmotic pressure reduces the effective pressure drop through the membrane pores.

Solute flux J_s is expressed as

$$J_i = \frac{c_m K_{d,i} D_{i,m}}{\delta_m} (x_{i,m} - x_{i,p})$$ (5.11)

where c_m is the molar density; $K_{d,i}$ is the solute distribution coefficient for species i; $D_{i,m}$ is the diffusivity of the solute in the membrane; x_i is the mole fraction of the solute next to the membrane (m) on the concentrate side of the membrane and in the permeate (p); and δ_m is the membrane thickness.

Equations (5.10) and (5.11) predict that membrane performance, measured as permeate flux or as solute removal, should decrease with increasing feed concentration. As the feed concentration increases, the osmotic pressure also increases. Operating the membrane system at a higher recovery r also increases the concentration of solute applied to the membrane. To achieve the same permeate flux, a higher operating pressure is required to overcome the osmotic pressure. Elevated concentrations of solute near the membrane due to concentration polarization exacerbate this effect since the resultant osmotic pressures will be higher than those calculated from either the feed or bulk concentrations. A higher concentration of solute in the feed (and therefore at the membrane) increases the flux of solute across the membrane. Note that, to a first approximation, the flux of solute across the membrane is independent of the operating pressure while the permeate flux increases with increasing pressure. If Eq. (5.11) is divided by Eq. (5.10), we arrive at an expression that predicts a decrease in the concentration of solute in the permeate for an increase in the transmembrane pressure Δp. This corresponds to an increase in solute rejection. In fact, higher operating pressures do tend to increase the global rejection. The following empirical equation has been found to produce a satisfactory fit to data on rejection as a function of the transmembrane (operating) pressure:[9]

$$R = \frac{z_1 \Delta p}{(z_2 \Delta p + 1)}$$ (5.12)

where z_1 and z_2 are empirically determined constants. The form of this equation follows directly from the definition of rejection and the ratio of Eq. (5.11) to (5.10). It has been suggested[9] that the similarity of Eq. (5.12) to the Langmuir isotherm is consistent with the preferential sorption–capillary flow mechanism.

5.4.2 Irreversible Thermodynamics

Models based on irreversible thermodynamics yield general guidelines for membrane separation and insight into the underlying mechanisms of separation. However, specifics of the membrane structure and the transport processes are not accounted for.

Modeling membrane separation as a spontaneous irreversible process implies that the entropy of the system increases and that free energy is dissipated. Defining ϕ as the free-energy dissipation function, free-energy dissipation can be expressed as the sum of the fluxes of quantities multiplied by their driving forces:

$$\phi = \sum_{i=1}^{n} J_i F_i \qquad (5.13)$$

For a sufficiently slow process (i.e., small deviations from equilibrium), the fluxes of all quantities J_i can be expressed as linear combinations of the driving forces F_k:[15]

$$J_i = \sum_{k=1}^{n} L_{ik} F_k \qquad (5.14)$$

The phenomenological coefficients L_{ik} relate the flux of component i to force k. In the case $i = k$, or L_{ii}, these coefficients can sometimes be easily interpreted. For example, consider the case of pure water (the single component, indexed as 1) flowing across a membrane under pressure at a constant temperature. The only phenomenological coefficient for this system, L_{11}, relates the flux of water to the force of pressure. In this instance, Eq. (5.14) reduces to an expression that is equivalent to Eq. (4.1) and L_{11} can be interpreted as the membrane permeability. In the cases where there is more than one component and $i \neq k$, the L_{ik} describe coupled transport of these components. It has been shown both theoretically and experimentally that the effect of force k on the flux of component i is equivalent to considering the flux of component k on force i. Thus, the matrix of proportionality coefficients (L_{ik}) is symmetric (the Onsager reciprocal relation), reducing the number of coefficients from n^2 to $[n(n-1)/2] + n$. Again, this assumption is not valid far from equilibrium.

For isothermal conditions, the free-energy dissipation function can be written as,

$$\phi = \sum_{i=1}^{n} J_i \nabla(-\mu_i) \qquad (5.15)$$

where the gradient of μ_i, the chemical potential of component i, is the driving force for the flux of component i. Equation (5.15) can be integrated across the membrane of thickness δ_m (assuming one-dimensional flow) to yield an expression for the total free-energy dissipation across the membrane ϕ_{mem}:

$$\int_{0}^{\delta_m} \phi(z) \, dz = \int_{0}^{\delta_m} \sum_{i=1}^{n} J_i \nabla(-\mu_i) \, dz = \phi_{mem} \qquad (5.16)$$

The details of this integration require several simplifying assumptions or approximations regarding the local properties of transport within the membrane.[16]

If a two-component system (e.g., water w and one solute s) is considered, the energy dissipation function integrated across the membrane is written as

$$\phi_{mem} = J_w \Delta\mu_w + J_s \Delta\mu_s \tag{5.17}$$

Substituting the definition for $\Delta\mu$,

$$\Delta\mu_i = RT\Delta \ln a_i + V_i \Delta p \tag{5.18}$$

where R is the gas constant, T is absolute temperature, a_i is the activity of the solute or water, V_i is its molar volume, and Δp is the pressure differential across the membrane, and using the Lewis equation for osmotic pressure,

$$V_w \Delta\Pi_w = -RT\Delta \ln a_w \tag{5.19}$$

with the van't Hoff equation for dilute concentrations where the activity of the solute a_s is approximated by its concentration c_s,

$$\Delta\Pi_w = RT\Delta \ln c_s \tag{5.20}$$

the Kedem-Katchalsky model is derived:[17]

$$J_V = L_V \Delta p + L_{VD}\Delta\Pi_w \tag{5.21}$$

$$J_D = L_{DV}\Delta p + L_D \Delta\Pi_w \tag{5.22}$$

where J_V is the volume flux of everything across the membrane (water and solute); $J_D = J_s/c_{save} - J_V$; $c_{s,ave}$ is the log-mean solute concentration across the membrane; and J_s is the solute flux. Note that L_V corresponds to L_{11} in the one-component example previously given. The notation L_p is often used in place of L_V and is referred to as the *hydraulic permeability of the membrane*. The forces and conjugate fluxes in Eq. (5.21) and (5.22) are readily identified as follows:

Force	Conjugate flux
$\Delta\Pi$	$J_w V_w + J_s V_s$
Δp	$J_s/c_s - J_w V_w$

If we define the reflection coefficient $\sigma = -(L_{VD}/L_V)$, Eq. (5.20) can be further simplified as

$$J_V = L_V(\Delta p - \sigma\Delta\Pi_w) = J_w V_w + J_s V_s \tag{5.23}$$

For all practical purposes, J_V is proportional to the permeate flux by a factor of the molar volume of water, assuming that the solute volume flux is small. Examination of Eq. (5.23) reveals that the reflection coefficient can be estimated by determining the pressure drop Δp across the membrane as that at which the volumetric flux (or approximately the permeate flux) is zero. The solute concentrations on either side of the membrane can then be measured to estimate the differential osmotic pressure. The reflection coefficient is then calculated as $\sigma = \Delta p/\Delta\Pi$ and is equal to 1 in the absence of coupled fluxes.

An expression analogous to Eq. (5.21) can be derived for solute flux:

$$J_s = c_{s,ave}(1 - \sigma)J_V + \omega\Delta\Pi_w \tag{5.24}$$

where $\omega = (L_V L_D - L_{VD}^2)/L_V$. Equation (5.24) implies that there should be less rejection of the solute as osmotic pressure increases. Equation (5.23) indicates that permeation of water through the membrane will not occur until a mechanical pressure differential Δp is applied that exceeds the difference in osmotic pressure between the concentrate and the permeate. The applied pressure essentially reverses the direction of the flux of water that would occur under isobaric conditions (Fig. 5.9), leading to the term *reverse osmosis.*

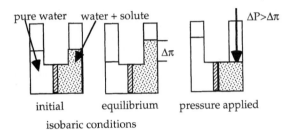

pure water   water + solute $\Delta P > \Delta \pi$

$\Delta \pi$

initial   equilibrium   pressure applied

isobaric conditions

FIGURE 5.9 Osmosis and reverse osmosis of water across semipermeable membrane.

Based on this requirement of overcoming the osmotic pressure of the feedwater, the theoretical energy required to desalinate 1 m³ of seawater at 27°C is on the order of 0.7 kWh. In practice, the applied pressure is 2 to 3 times the osmotic pressure. Assuming only a 25 percent recovery and taking account of mechanical losses, etc., the actual energy required to desalinate 1 m³ of seawater is approximately 10 times the theoretical value. (This is still substantially less than the approximately 10 kWh/m³ required for competing distillation processes.)

Fortunately, not all solutes in water exert high osmotic pressures. Higher-molecular-weight solutes produce less osmotic pressure than do small-molecular-weight compounds. This can be illustrated by considering the case of dilute concentrations for which the activity coefficients approach a value of 1 and Eq. (5.20) can be approximated as

$$\Pi \cong \frac{c}{\overline{M}} RT \qquad (5.25)$$

where c is the mass concentration of the solute, and $\overline{M}$ is the molecular weight of the solute. At the extreme, colloidal particles may be considered as very large "solutes" with very high molecular weights. Osmotic pressure can be neglected in calculating the permeate flux of MF and UF membranes since the larger colloidal and macromolecular materials rejected by these membranes exert very little osmotic pressure and the smaller-molecular-weight solutes pass through these membranes (Chap. 4).

5.4.3 Solution-Diffusion Model

The solution-diffusion model proposed by Lonsdale, Merten, and Riley[18,19] describes solute and solvent transport across membranes in terms of the relative affinities of these components for the membrane and their diffusive transport within the membrane "phase." The driving forces for transport are the differences in chemical potential across the membrane due to differences in concentration and pressure. The flux of water is given as:

$$J_w = -\frac{D_w c_{w,\text{mem}}}{RT}\frac{d\mu_w}{dz} \approx -\frac{D_w c_{w,\text{mem}}}{RT}\frac{\Delta\mu_w}{\delta_m} \tag{5.26}$$

Substituting the Lewis equation for osmotic pressure, the following equation is derived for water flux:

$$J_w = -K_w \frac{(\Delta p - \Delta\Pi_w)}{\delta_m} \tag{5.27}$$

where the specific hydraulic permeability $K_w = D_w c_{w,\text{mem}}\overline{V}_w/RT$. Comparison of this result with Eq. (5.23) reveals that these two approaches yield similar expressions for permeate flux. In fact, the thermodynamic model previously presented reduces to the solution-diffusion model if it is assumed that water and solute transport are not coupled ($\sigma = 1$). The flux of solute J_s is given as

$$J_s = -D_s K_d \frac{(c_{\text{bulk}} - c_p)}{\delta_m} \tag{5.28}$$

where K_d is the distribution coefficient which describes the relative affinity of the solvent for the membrane. Solute flux is predicted to be independent of the transmembrane pressure drop Δp. Increases in Δp increase J_w, and thus selectivity is predicted to increase with Δp.

Mathematically, these results are virtually identical to those arising from the preferential sorption–capillary flow model. However, the physical interpretation of the underlying phenomena and the associated model coefficients differ.

This model has been modified[20] to include advective transport through membrane pores as well as diffusion. This modification leads to the following expression for the local rejection of the membrane:

$$R_{\text{local}} = \left[1 + \left(\frac{D_s K_d}{K_w}\right)\left(\frac{1}{\Delta p - \Delta\Pi}\right) + \left(\frac{K_{\text{pf}}}{K_w}\right)\left(\frac{\Delta p}{\Delta p - \Delta\Pi}\right) \right]^{-1} \tag{5.29}$$

where K_{pf} is an additional fitting constant describing pore flow through the membrane.

5.5   DONNAN EQUILIBRIA AND ELECTRONEUTRALITY EFFECTS ON THE REJECTION OF IONS

Charged functional groups on membranes attract ions of opposite charge (counterions). This is accompanied by a deficit of like-charged ions (co-ions) in the membrane and results in a so-called Donnan potential. Although co-ions are not able to enter the membrane, it is possible to pass water through these membranes under pressure. The accumulation of co-ions in the membrane concentrate is accompanied by an accumulation of counter-ions as well, due to the need to preserve electroneutrality in the solution. By this mechanism, membrane rejection is predicted to increase with increasing membrane charge and with ion valence. This principle has been incorporated into calculations based on the extended Nernst-Planck equations to describe the selectivity of such charged membranes.[21] In particular, this approach has been applied to describe the rejection characteristics of NF as well as RO membranes.

Regardless of the actual mechanism of rejection, the requirement for electroneutrality in the concentrates and permeates of membranes that reject ions may lead to significant modification of a membrane's rejection characteristics in the presence of mixed solutes. For example, as calcium is rejected by an NF membrane, higher rejections of monovalent anions such as nitrate or chloride may occur. The passage of these anions through the membrane and into the permeate while calcium is rejected results in a reduced pH of the permeate as the dissociation equilibrium of water shifts to provide more protons to balance the permeating anions. The presence of two competing anions in the concentrate may lead to a lower rejection of one by NF membranes as the second increases in concentration. For example, sulfate, which is rejected to a large degree by NF membranes, reduces the rejection of both nitrate and chloride as the sulfate concentration is increased.[22]

REFERENCES

1. J. D. Ferry, 1936, "Statistical Evaluation of Sieve Constants in Ultrafiltration," *Journal of General Physiology*, **20**:95–104.

2. L. Zeman and M. Wales, 1981, "Polymer Solute Rejection by Ultrafiltration Membranes," in A. F. Turbak (ed.), *Synthetic Membranes, Vol. II. Hyper and Ultrafiltration Uses*, American Chemical Society, Washington D.C.

3. T. Matsuura and S. Sourirajan, 1983, *Journal of Colloid and Interface Science*.

4. W. M. Deen, 1987, "Hindered Transport of Large Molecules in Liquid-Filled Pores," *AIChE J.*, **33**:1409–1425.

5. M. G. Davidson and W. M. Deen, 1988, "Hydrodynamic Theory for the Hindered Transport of Flexible Macromolecules in Porous Membranes," *J. Mem. Sci.*, **35**:167–192.

6. J. S. Taylor, et al., 1988, "Comparison of Membrane Processes at Ground and Surface Water Sites," *Proceedings of the Annual Conference of the American Water Works Association*, Orlando, Fla.

7. J. S. Taylor, D. M. Thompson, and J. K. Carswell, 1987, "Applying Membrane Processes to Groundwater Sources for Trihalomethane Precursor Control," *J. AWWA*, **79**(8):72–82.

8. S. M. Santos, 1993, "Treatment of Produced Water from Oil and Gas Wells Using Crossflow Ultrafiltration Membranes," Master's thesis, Rice University, Houston, Tex.

9. S. Sourirajan, 1970, *Reverse Osmosis*. Academic Press, New York.

10. M. K. A. Agbekodo, 1994, "Elimination par Nanofiltration des Composés Organiques d'une Eau de Surface Prétraitée," Thèse de doctorat, Université de Poitiers, Poitiers, France.

11. U. Merten, I. Nusbaum, and R. Miele, 1968, *Organic Removal by Reverse Osmosis*, Gulf General Atomic Inc. San Diego, Calif.

12. S. Sourirajan, 1965, "Characteristics of Porous Cellulose Acetate Membranes for the Separation of Some Organic Substances in Aqueous Solutions," *I & EC Product Research and Development*, **4**(3):201–206.

13. H. K. Lonsdale, et al., 1969, *Study of the Rejection of Various Solutes by Reverse Osmosis Membranes*, Research and Development Report No. 447, USGPO, Washington, D.C.

14. T. N. Eisenberg and E. J. Middlebrooks, 1986, *Reverse Osmosis Treatment of Drinking Water*, Butterworth Publishers, Stoneham, Mass.

15. L. Onsager, 1931, *Phys. Rev.,* **37**:405.

16. K. S. Spiegler and O. Kedem, 1966, *Desalination,* **1**:311.

17. O. Kedem and A. Katchalsky, 1958, *Biochem. Biophys. Acta,* **27**:229.

18. H. K. Lonsdale, U. Merten, and R. L. Riley, 1965, *J. Appl. Polym. Sci.,* **9**:1341.

19. U. Merten, *Desalination by Reverse Osmosis.* MIT Press, Cambridge, MA (1966).

20. T. K. Sherwood, P. T. L. Brian, and R. E. Fisher, 1967, "Desalination by Reverse Osmosis," *Ind. Eng. Chem. Fundam.,* **6**(1):2–12.

21. L. Dresner, 1972, "Some Remarks on the Integration of the Extended Nernst-Planck Equations in Hyperfiltration of Multicomponent Solutions," *Desalination,* **10**:27.

22. R. Rautenbach and A. Groschl, 1990, "Separation Potential of Nanofiltration Membranes," *Desalination,* **77**:73–84.

23. C. A. Fronk, B. W. Lykins, Jr., and J. K. Carswell, 1990, "Membranes for Removing Organics from Drinking Water," *Proceedings of the Annual Meeting of the American Filtration Society,* Alexandria, Va.

24. C. Fronk, 1990, "In House Membrane Research," *Proceedings of the Membranes Workshop,* Cincinnati, Ohio.

CHAPTER 6
MEMBRANE BIOFOULING

Harry F. Ridgway, Ph.D.
Biotechnology Research Department
Orange County Water District
Fountain Valley, California

Professor Hans-Curt Flemming
Department of Microbiology
University of Duisburg
Duisburg, Germany

6.1 INTRODUCTION

Reverse osmosis (RO) and related membrane separations processes, such as micro-filtration (MF), ultrafiltration (UF), and nanofiltration (NF), play essential roles in modern water treatment practice (Brandt et al., 1993; Buros, 1989; Crossley, 1983; Ko and Guy, 1988; Pusch and Walch, 1982). In MF and UF processes, feedwater is forced through a microporous synthetic membrane that rejects macromolecular solutes and particulates primarily by size exclusion (Jacangelo et al., 1989; Kesting, 1985; Pusch and Walch, 1982). In RO and NF processes, water diffuses under an applied pressure across an ultrathin semipermeable membrane barrier which preferentially rejects solutes and suspended solids, including bacteria, virus, and other microorganisms (Allegrezza, 1989; Brandt et al., 1993; Pusch and Walch, 1982; Riley et al., 1964, 1973). Solute transport through RO/NF membranes is based on the specific molecular properties of the compound (e.g., atomic dimensions, ionization state, solubility, etc.) (Dickson, 1988; Pusch and Walch, 1982) and occurs in response to the trans-membrane solute concentration gradient (Brandt et al., 1993; Dickson, 1988). Water transport is driven by the osmotic gradient; thus, permeate flow may be increased in proportion to the applied feed pressure (Dickson, 1988). Membranes are often operated in a cross-flow regime to minimize accumulation of solute and colloids in the viscous sublayer (boundary layer) proximal to the membrane surface (concentration polarization) (Dickson, 1988; Lepore and Ahlert, 1988). Up to 90 percent of flow in RO systems is tangential to the membrane surface, whereas the remainder (permeate) passes through the membrane. In spite of the cross-flow component, a proportion of the feedwater colloids and microorganisms entering the module are transported to the membrane surface where they adsorb, forming a thin fouling layer (Lepore and Ahlert, 1988). Once attached, microorganisms may grow and multiply at the expense of feedwater nutrients, forming a biological film, i.e., a biofilm, which may compromise membrane performance and drive up operation and maintenance (O&M) costs (Figs. 6.1 to 6.3) (Argo and Ridgway, 1982; Flemming, 1991, 1993; Flemming and Schaule, 1988a, 1988b, 1988c, 1992a, 1992b, 1993; Ridgway, 1987, 1988; Ridgway et al., 1981, 1983, 1984a, 1984b, 1984c, 1985, 1986a, 1986b, 1992; Ridgway and Safarik, 1991; Schaule, 1992; Schaule et al., 1990, 1993a, 1993b).

6.2 PHENOMENON OF MEMBRANE BIOFOULING

Biofouling is operationally defined as biofilm formation resulting in an unacceptable degree of system performance loss (e.g., excessive loss of thermal conductivity

FIGURE 6.1 SEM photomicrograph of a typical biofilm formed on a cellulose acetate membrane fed with pretreated municipal wastewater (at Water Factory 21, California). Note ro-shaped bacteria and EPS. Bar = 5 μm. Inset = low magnification view of same biofilm.

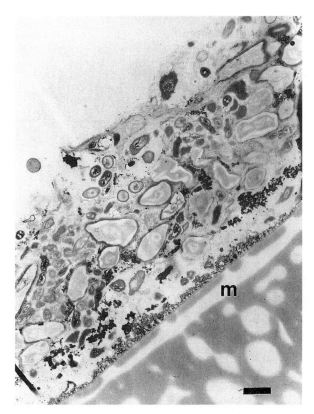

FIGURE 6.2 TEM photomicrograph of cross section through a biofilm formed on a polyetherurea UF membrane (m) fed with domestic tap water showing cells and unidentified electron-dense material embedded in the EPS matrix. Bar = 1 μm. *(Photo courtesy of Gabriela Schaule, University of Stuttgart, Germany.)*

in cooling systems or increased fluid frictional drag in flow systems, etc.) (Characklis, 1990b, 1991). Thus, biofilm formation invariably precedes biofouling, and a biofilm may exist in the absence of detectable biofouling. For a given performance parameter (e.g., water transport) (Flemming, 1993; Flemming et al., 1992, 1993), a threshhold value may be defined below which system performance is considered intolerable (the so-called pain threshold) (Flemming, 1991, 1993; Flemming et al., 1992). A biofilm detection threshold may also be defined as the minimum biofilm that can be detected by analytical methods, such as microscopic inspection.

Biofilm formation involves the accumulation of microorganisms (e.g., bacteria, fungi, microalgae) at a phase transition interface, which may be either solid-liquid, gas-liquid, or liquid-liquid. Transport of microorganisms to an interface may occur passively by diffusion (i.e., Brownian motion in the case of microorganisms), gravitational settling, or bulk fluid convection. Motile bacteria may actively seek surfaces via chemotactic processes (DeWeger et al., 1987). Biofilms may or may not uniformly cover the substratum and minimally consist of one, but usually multiple lay-

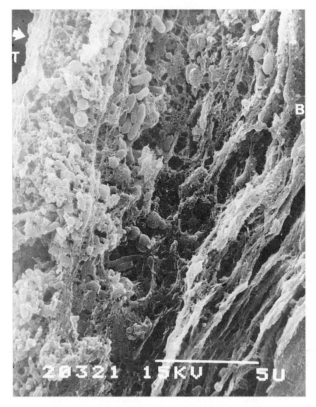

FIGURE 6.3   SEM photomicrograph of a biofilm formed on a cellulose acetate RO membrane fed with pretreated municipal wastewater (at Water Factory 21, California). Biofilm is displayed in edge view to illustrate the layering of cells/EPS. Layering effect may result from changes in compaction as influenced by the system operating pressure. T = top (feed-water) surface of biofilm; B = bottom of biofilm nearest to membrane; flux direction indicated by arrow. Bar = 1 μm.

ers of living and dead microorganisms and their associated extracellular products (e.g., heteropolysaccharides, glycoproteins, lipids, etc.).

Bacteria accumulate on membrane surfaces by two processes: (1) attachment (i.e., adhesion, adsorption) and (2) growth (i.e., cell multiplication). Adhesion is a prerequisite for biofilm formation and involves stereospecific or nonspecific bonding interactions between the synthetic polymer substratum and adhesive structures (e.g., flagella, fimbria) (De Graaf and Mooi, 1986; Wicken, 1985) or exposed macromolecules on the cell surface (e.g., peripheral membrane proteins, heteropolysaccharides, lipopolysaccharides, mycolic acids, etc.) (Belas and Colwell, 1982; Characklis, 1990a, 1990b; Costerton et al., 1985; Cowan et al., 1992; Marshall, 1985; Marshall and Blainey, 1991; Wicken, 1985). Primary adhesion refers to initial attachment of cells to a virgin substratum, while secondary adhesion involves attachment to a preestablished biofilm. Once attached, a cell may grow and multiply using nutrients transported to the surface from the bulk fluid.

The ratio of planktonic to attached microbes in membrane systems and other aquatic environments typically exceeds one million (Costerton et al., 1987; Geesey et al., 1978). The physiology of biofilm bacteria may differ from that of suspended (i.e., planktonic) bacteria due to the activation and expression of different sets of genes in the attached cells (Dagostino et al., 1991; Davies et al., 1993; Vandevivere and Kirchman, 1993).

Bacteria have evolved elaborate adhesion mechanisms, and virtually all natural and synthetic materials are susceptible to bacterial attachment and colonization. Biofilms are observed in natural aquatic ecosystems (Costerton et al., 1987; Geesey et al., 1978), on prosthetics and medical implant devices (Costerton, et al., 1987, 1994, 1985; Costerton and Irvin, 1981), and in industrial process streams (Flemming, 1991; Frith, 1988; Schaule et al., 1993b). Even materials exhibiting low surface free energies, such as Teflon (Dexter, 1978; Dexter et al., 1975) and hydrophobic plastics (Fletcher, 1976; Fletcher and Loeb, 1979; Fletcher and Marshall, 1982; Fletcher and McEldowney, 1983; Ludwicka et al., 1985; Paul and Loeb, 1983; Rosenberg and Doyle, 1990; Schoenen, 1990), are susceptible to colonization over a wide range of physicochemical conditions. Synthetic membrane polymers used in water treatment applications are no exception to this general rule (Flemming, 1991; Flemming and Schaule, 1988a, 1988b, 1988c; Ridgway et al., 1984b, 1985; Ridgway and Safarik, 1991; Schaule, 1992; Schaule et al., 1990). Treatment of large volumes of water in the shortest time requires maximizing physical contact between the feedwater and the membrane polymer. While this strategy optimizes water transport kinetics and solute concentration, it also increases the likelihood of contact between waterborne microorganisms and the membrane surface. If the contact time is of sufficient duration (seconds), bacteria may attach and grow into a biofilm (Characklis, 1990a, 1990b; Costerton et al., 1985; Flemming and Schaule, 1988b, 1988c; Lawrence et al., 1987; Marshall, 1985, 1991; Marshall and Blainey, 1991). Biofouling of membrane surfaces is more problematic than abiotic colloidal fouling or mineral scaling, since attached cells multiply in a geometric fashion using trace nutrients continuously supplied in the feedwater. A single bacterium entering a membrane module may result in extensive biofouling if the growth rate of the sessile population is high.

Biofilm bacteria exhibit dynamic adhesion in which bonding to the membrane increases with time due to the biosynthesis of adhesive extracellular biopolymers (Fig. 6.2). The biopolymers are frequently produced in amounts that completely envelop attached cells in a viscous hydrated gel (a hydrogel) (Allison and Sutherland, 1984; Characklis, 1990a; Christensen and Characklis, 1990; Costerton et al., 1987; Flemming, 1993; Flemming et al., 1992a, 1992b; Geesey et al., 1992). This gel-like matrix is referred to as the *glycocalyx* (Costerton and Irvin, 1981; Geesey, 1982), *slime layer, capsule,* or *extracellular polymeric substances* (EPS) (Geesey, 1982; Neu and Marshall, 1990). Bacterial EPS typically consists of acidic heteropolysaccharides with a high net-negative zwitterionic charge density (Christensen and Characklis, 1990; Costerton et al., 1987, 1994; Costerton and Irvin, 1981; Geesey, 1982; Marshall, 1991; Marshall and Blainey, 1991; Neu and Marshall, 1990; Wicken, 1985). The EPS enhances the survival and robustness of the biofilm microorganisms by serving as a chemically reactive diffusional transport barrier retarding convective flow and slowing the penetration of antimicrobial agents (e.g., chlorine and other microbicides) into the biofilm (Bryers, 1993; Cargill et al., 1992; Characklis, 1990a, 1990b, 1991; Costerton et al., 1987; Costerton and Irvin, 1981; Geesey, 1982; Geesey et al., 1992; LeChevallier, 1987, 1991; LeChevallier et al., 1984, 1988, 1990; Pyle and McFeters, 1990; Suci et al., 1994). In addition, the EPS matrix reinforces cellular bonding to membranes and other substrata (Bryers, 1993; Costerton and Irvin, 1981; Costerton et al., 1985; Marshall, 1985, 1991; Marshall and Blainey, 1991; Marshall et al., 1991;

Neu and Marshall, 1990) and mechanically cross-links and stabilizes the biofilm, thereby reducing its susceptibility to sloughing by hydrodynamic shear.

Biofilm bacteria obtain carbon and energy for growth from dissolved feedwater organics (assimilable organic carbon, or AOC) (Huck, 1990; van der Kooij, 1992), although adsorbed nutrients may also be mobilized and scavenged directly from surfaces (Kefford et al., 1982). Trace organics and other nutrients are concentrated within the EPS matrix due to its inherent electrostatic charge and large surface-to-volume ratio. The sorbed organics provide biofilm cells with higher nutrient levels than are present in the bulk fluid. Thus, the biofilm lifestyle enables microorganisms to survive and multiply even in extremely low nutrient environments (Bryers, 1993; Geesey et al., 1978), such as ultrapure water systems containing as little as 5 µg/L of AOC (Geesey et al., 1992; Mittleman, 1991; Parise et al., 1988; Patterson et al., 1991; Pearson et al., 1984; Pittner, 1993b). Thus, adhesion is recognized as an effective survival strategy evolved by microorganisms to cope with starvation conditions (Amy and Morita, 1983; Dawes, 1984; Kjelleberg and Hermansson, 1984; Novitsky and Morita, 1976, 1977). The AOC may be regarded as the potential biofilm; thus, AOC measurements provide an indication of the likelihood and probable extent of biofilm formation in membrane systems.

6.3 DETECTION AND MONITORING OF MEMBRANE BIOFOULING

Numerous methods are available for detecting and identifying membrane foulants (Amjad et al., 1988; Isner and Williams, 1993). These include optical microscopy, scanning and transmission electron microscopy (SEM, TEM, respectively), atomic force microscopy (AFM), x-ray fluorescence emission spectroscopy, attenuated total reflection Fourier transform infrared spectrometry (ATR-FTIR), energy-dispersive x-ray microanalysis (EDXM), and Auger spectroscopy.

Membrane biofilms may be detected directly by (1) microscopic inspection of exposed membrane surfaces (Paul and Loeb, 1983; Wolfaardt et al., 1991; Ridgway, 1987, 1988; Schaule, 1992; Schaule et al., 1990) or (2) biochemical/microbiological characterization of material scraped from membranes (e.g., heterotrophic plate count determinations) (Flemming, 1991; Ridgway et al., 1981, 1983, 1984a). Indirect methods of biofilm detection include (1) measurement of some effect of the biofilm on membrane function or system performance (e.g., loss of water flux, increased module differential pressure or solute transport) (Flemming, 1993; Flemming et al., 1992b, 1993; Ridgway 1987, 1988; see next section), (2) enumeration of greater numbers of bacteria in the brine than can be accounted for from the feedwater, implying biofilm growth and cell detachment (Ridgway et al., 1981, 1983), or (3) use of test coupons or flow cells placed on the same feed stream as the actual membrane modules. Of these techniques, direct microscopic examination of membrane surfaces is the most reliable and conclusive method for assessing biofilm development. Information obtained from optical microscopy can be extended and quantified by the use of organic dyes which preferentially react with bacterial EPS (Allison and Sutherland, 1984), fluorescent probes such as 2,4-diamidino-2-phenylindole (DAPI) or Hoechst 33278 that bind specifically to cellular nucleic acids (Paul and Loeb, 1983; Wolfaardt et al., 1991), fluorescent redox-sensitive probes such as 5-cyano-2,3-ditolyl tetrazolium chloride (CTC) that respond to a functional electron-transport system (Kuhlman et al., 1989; Rodriguez et al., 1992; Schaule et al., 1993a; Stellmach, 1984; Yu and McFeters, 1994), or fluorescent probes such as rhodamine 123 that respond to a cell membrane potential (Yu and McFeters, 1994).

Major improvements in imaging capability and quality of optical microscopy have occurred in recent years with the advent of scanning-laser and digitally based confocal microscopy, which permit the three-dimensional (3-D) reconstruction of virtual biofilms from a series of digitized 2-D optical sections acquired along the specimen z axis (Fig. 6.4) (Costerton et al., 1994; Jovin and Arndt-Jovin, 1989; Lichtman, 1994). Wolfaardt et al. (1994) recently applied confocal microscopy to analyze the 3-D multicellular organization of contaminant-degrading biofilm communities, and membrane biofilms have now been imaged using digital confocal microscopy (Fig. 6.5). A disadvantage of microscopic methods, however, is that they are usually destructive in nature, requiring sacrifice of a membrane module from the system. This problem may be avoided by employing in-line microscope flow cells or test coupons, but there must be assurance that coupons accurately simulate real membrane surfaces in terms of polymer chemistry, structure, hydrodynamic behavior, and fouling propensity.

Both TEM and SEM have been employed extensively to characterize the surfaces of biofouled RO membranes (Amjad et al., 1988; Isner and Williams, 1993; Ridgway et al., 1983). One concern with electron microscopic methods is the sus-

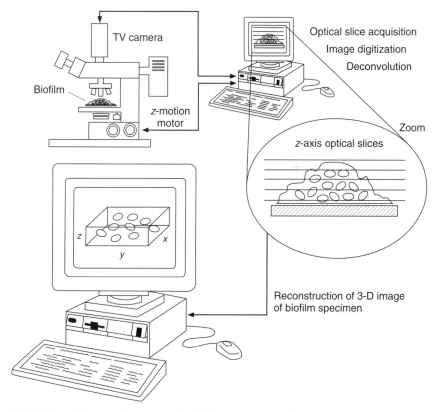

FIGURE 6.4 Schematic illustration of a DCM system. Optical slices are captured digitally along the specimen z axis. Hazy and defocused information in each section is subsequently removed digitally via deconvolution algorithms. The sections are then digitally recombined to produce a 3-D virtual image of the specimen (see Fig. 6.5).

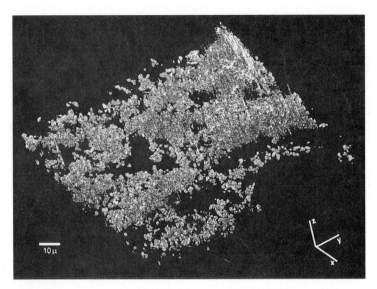

FIGURE 6.5 DCM image of a mycobacterial biofilm grown on a simulated membrane (cellulose acetate coated coupon). The bacteria were stained with DAPI, a fluorescent dye that intercalates into cellular DNA (see text). The entire image may be rotated and viewed in any orientation, allowing precise localization of any cell within the biofilm.

ceptibility of the membrane polymers to electron beam damage. Even at low accelerating voltages (10 to 15 kV), holes and other artifacts may be introduced into the membrane by the electron beam. Using a combination of ruthenium- and osmium-tetroxide fixation, Kutz et al. (1985a) were able to prevent electron beam damage to cellulose acetate RO membranes. Specimen dehydration and high vacuums required for electron microscopy may also introduce structural artifacts.

AFM avoids many of these concerns since membrane specimens can be viewed in a fully hydrated condition (Fig. 6.6) (Chi et al., 1993). The principle of operation of the AFM is based on measuring the defection and torsion of a cantilever mechanism that is rastered over the specimen surface within the zone of interatomic interactions (van der Waals forces). Thus, AFM has application in resolving molecular-scale interactions between bacteria and the membrane polymer surface (Fig. 6.7) (Bremmer et al., 1992).

White (1986), Nivens et al. (1993), and Bremmer and Geesey (1991) recently applied ATR-FTIR for the noninvasive monitoring of biofilm development on germanium internal reflection elements (IREs) (Fig. 6.8). Using ATR-FTIR, it is possible to nondestructively detect and quantify specific infrared (IR) absorption bands that are characteristic of proteins, heteropolysaccharides, and other biological macromolecules associated with developing biofilms (see Fig. 6.9). The kinetics of biofilm development on the surface of an IRE mounted in a flow cell may be monitored in pseudo-real time by programming the instrument to collect IR spectra at user-prescribed intervals. The IREs can be coated with thin films of membrane polymers, such as cellulose acetate or polysulfone, to simulate an actual membrane surface. Suci et al. (1994) recently employed ATR-FTIR spectrometry to investigate the dynamics of chemical biocide (ciprofloxacin) penetration into *Pseudomonas aeruginosa* biofilms formed on copper-coated IRE crystals.

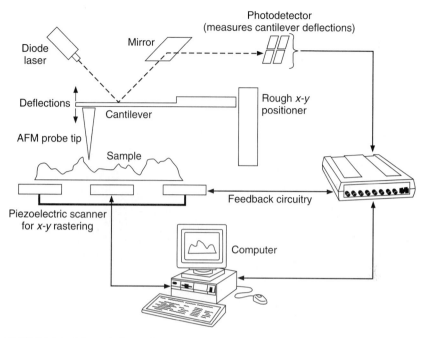

FIGURE 6.6   Schematic illustration of AFM system. A silicon nitride probe tip mounted on a cantilever is rastered in the x-y plane over the specimen surface. z-axis cantilever deflections and sideways torsion (measured by laser) signal tip-specimen interactions and surface topographic features. Digital data are recombined to generate a 3-D virtual image of the specimen topography (see Fig. 6.7).

6.4 SYMPTOMS AND CONSEQUENCES OF MEMBRANE BIOFOULING

Biofilms develop without regard for the benefit or liability to the system. Thus, biofilm formation on separation membranes often degrades system performance and drives up O&M costs (Flemming et al., 1992b; Pittner, 1993a; Taylor et al., 1989). These effects are most obvious in RO systems, which are dependent on the integrity and function of semipermeable membranes whose surfaces are exceedingly thin (<0.2 μm) and fragile (Brandt et al., 1993; Pusch and Walch, 1982; Riley et al., 1964, 1973) (Fig. 6.10; Table 6.1). The net effect of membrane biofouling is an elevated system energy consumption resulting primarily from a measurable decline in the *apparent* permeability of the RO membrane to water (i.e., membrane flux decline; Fig. 6.11) (Flemming, 1993; Flemming et al., 1992b, 1993; Lepore and Ahlert, 1988; Ridgway, 1987, 1988; Ridgway et al., 1984a). Flux loss due to biofouling is especially detrimental in large separation facilities (>4 × 10⁶ L/day capacity) where high electrical (pumping) costs are needed to maintain operating pressures and constant product output. Flux decline may be rapid (e.g., several days) or more gradual (weeks to months), depending on the physicochemical and microbiological properties of the feedwater, membrane polymer, and biofilm. Flux decline kinetics are often biphasic, i.e., an initial rapid loss followed by a slower decline (Flemming, 1993; Flemming et al., 1992b, 1993; Ridgway, 1987, 1988; Ridgway et al., 1981) (Fig. 6.11). Both decline

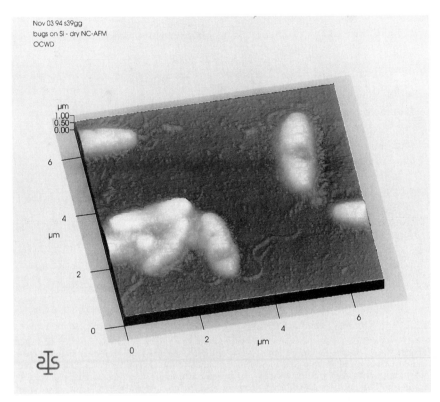

Nov 03 94 s39gg
bugs on Si - dry NC-AFM
OCWD

FIGURE 6.7   AFM image of bacteria attached to silicon surface. Note EPS matrix surrounding cells.

phases are typically curvilinear or polynomial functions with respect to elapsed time and can be only poorly approximated by linear regression models. The initial rapid flux decline is typically correlated with the early attachment and proliferation of microorganisms on the membrane surface, suggesting a possible causal relationship (Flemming, 1993; Flemming et al., 1992b, 1993; Ridgway et al., 1981). The biosynthesis and extrusion from attached cells of EPS presumably also contribute significantly to the initial rapid flux loss. The slow decline (plateau) phase results from establishment of an equilibrium condition in which biofilm growth and EPS production are balanced by biofilm loss (cell detachment or sloughing) caused by hydrodynamic shear at the solution-biofilm interface (Flemming, 1993; Flemming et al., 1992b, 1993; Lepore and Ahlert, 1988).

The molecular basis of flux decline due to biofouling is poorly understood. No quantitative models are available concerning the relationship between biofilm physicochemical properties (thickness, surface roughness, microbial composition, EPS structure and biochemistry, degree of hydration, etc.) and water flux decline. However, it is likely that flux decline is related to water transport impedance offered by the biofilm itself (i.e., attached cells plus EPS matrix), rather than to some modification of the inherent transport characteristics of the separation polymer. Given the intimate association of the EPS matrix with the synthetic polymer membrane (see Fig. 6.16), it is conceivable that individual EPS biopolymers could intercalate

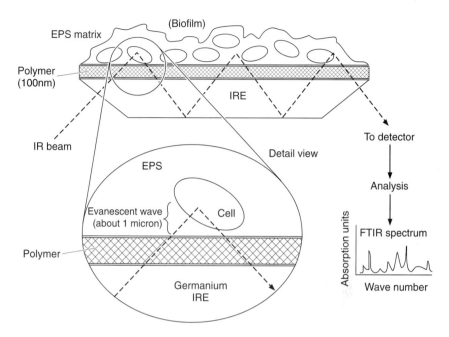

FIGURE 6.8   Schematic illustration of ATR-FTIR spectrometry technique for monitoring biofilm formation. The molecular composition of a biofilm growing on the surface of a polymer-coated germanium IRE is quantified by determining the attenuation of an IR signal (evanescent wave) which penetrates a short distance (about 1 μm) into the liquid phase at each reflection node. The polymer coating must be thin (about 100 nm) to allow biofilm detection (see Fig. 6.9).

within or otherwise chemically bond to the membrane and alter its water and solute permeation properties. Such polymer interactions notwithstanding, the biofilm functions as an independent diffusional transport barrier which retards convective fluid motion proximal to the membrane surface, a concept which is consistent with the hydrated gel nature of microbial biofilms (Bryers, 1993; Christensen and Characklis, 1990; Costerton et al., 1987; Flemming et al., 1992b; Geesey, 1982; Geesey et al., 1992; Neu and Marshall, 1990).

Leslie et al. (1993) and Hodgson et al. (1993) attributed low permeate fluxes and solute rejections of biofouled MF membranes to accumulated EPS rather than to the colloidal nature of bacterial cells comprised by the biofilm. Using *Pseudomonas* strain SW8, a gram-negative, hydrophobic, marine bacterium, they were able to enhance water flux and macromolecular solute (protein) passage through bacterial cakes formed on MF membranes by chemical modification of the EPS matrix. Modifications of the EPS were induced by treatment with ethylenediaminetetracetic acid (EDTA), a divalent-cation chelating agent, or pronase, a proteolytic enzyme (Hodgson et al., 1993; Leslie et al., 1993). These treatments did not change the packing density, compressibility, or 3-D arrangement of cells comprised by the biofilm (cake), but only the porosity and diffusional properties of the EPS matrix in which the cells were embedded.

Experiments reported by Flemming and co-workers (1992a, 1992b) are consistent with this EPS transport barrier model and suggest that the molecular structure

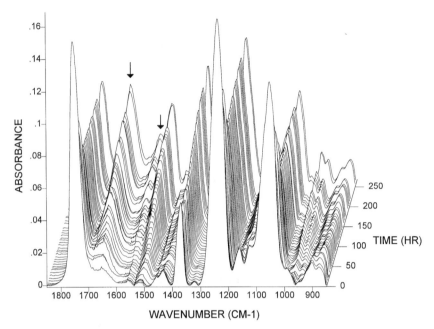

FIGURE 6.9   Time-resolved ATR-FTIR spectrum indicating growth of a *Mycobacterium* biofilm on a germanium IRE coated with an 80-nm-thick cellulose acetate film to simulate a membrane surface. IR bands associated with the cellulose acetate are located near 1750, 1371, 1238, and 1050 cm^{-1}. Bands at 1654 cm^{-1} (amide I) and 1547 cm^{-1} (amide II) indicate protein associated with biofilm cells. Note biofilm continued to accumulate for the entire experiment (about 262 h).

of the EPS can be chemically modified to increase its water and solute permeability. Experiments were described in which the water permeability of an artificial biofilm consisting of a 2-cm-thick agarose plug was increased following exposure to a mixture of proprietary anionic and nonionic surfactants. Under their experimental conditions, the permeability of the agarose layer increased more than fivefold without a corresponding decrease in plug thickness, implying that the agarose had assumed a more permeable molecular conformation. The authors speculated that such an approach could be used to enhance the flux of biofouled membranes without having to remove the biofilm.

Biofilm development increases fluid frictional resistance as water is transported tangential to the membrane surface (Bryers, 1993; Characklis, 1990a, 1990b, 1991), resulting in a higher module differential pressure (Δp_{mod}) (Flemming, 1993; Flemming et al., 1992b, 1993; Lepore and Ahlert, 1988; Ridgway, 1987, 1988). If the allowable Δp_{mod} is exceeded, adjacent membrane leaves within the module may shift relative to one another, causing telescoping, or the element may collapse along its longitudinal axis (Fig. 6.12). Even minor telescoping may cause abrasion of semipermeable membrane surfaces resulting in solute leakage.

The EPS matrix suppresses turbulent mixing at the membrane surface, resulting in enhanced concentration polarization, whereby solutes accumulate in the viscous sublayer (Brandt et al., 1993; Dickson, 1988; Lepore and Ahlert, 1988). When this occurs, solute transport through the membrane increases in response to greater ionic activity in the boundary layer, resulting in deterioration of permeate quality (Fig.

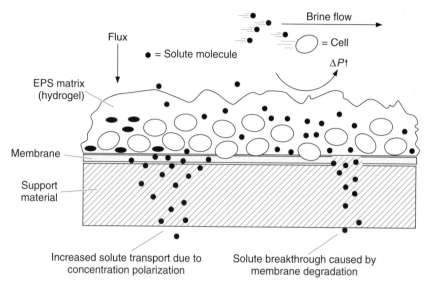

FIGURE 6.10 Schematic summarizing major effects of membrane biofouling.

6.11). Increased concentration polarization also drives up system energy costs, since greater operating pressures are required to overcome the higher osmotic gradient within the boundary layer.

Biofilms form on other module components, including the permeate surfaces of cross-flow membranes (Ridgway, 1987, 1988; Ridgway et al., 1983), the woven polyester support fabrics such as Texlon (Fig. 6.13), the permeate collection material, and the feed channel spacer materials (Schaule et al., 1993b). Microbial colonization of permeate surfaces of RO membranes is usually sparse compared to feedwater surfaces due to oligotrophic conditions on the former. Bacteria on the permeate surface may detach or slough, resulting in microbial contamination of the permeate which is of concern in semiconductor manufacturing (Frith, 1988; Mittleman, 1991), pharmaceuticals production (Mittleman, 1991; Parise et al., 1988), hemodialysis systems (Pearson et al., 1984; Vincent et al., 1989), and other ultrapure water applications (Mittleman, 1991; Pittner, 1993b). Adhesion and biofilm formation on the filtrate surfaces of MF and UF membranes could be expected to be more severe than that on the permeate surfaces of RO and NF membranes, since nutrients readily pass through MF and UF membranes.

Filamentous fungi belonging to the genera *Penicillium* and *Aspergillus* biodegrade and invasively colonize the polyester polyurethane glues used to seal adjacent membrane leaves together in RO, NF, UF, and some MF modules (Fig. 6.14). Growth of such fungi may occlude the feed channels (increasing the Δp_{mod}) and destroy the physical integrity and adhesive function of the glue lines, resulting in leakage of solutes and colloids into the permeate. The specific physiological growth requirements of the degradative fungi (required trace nutrients, pH and temperature optima, etc.) are unknown; however, similar biodeterioration of polyester polyurethane materials has been documented in other systems and broad spectrum biocidal agents, such as Vinyzene BP (active ingredient: 10,10'-oxybisphenoxarsine),

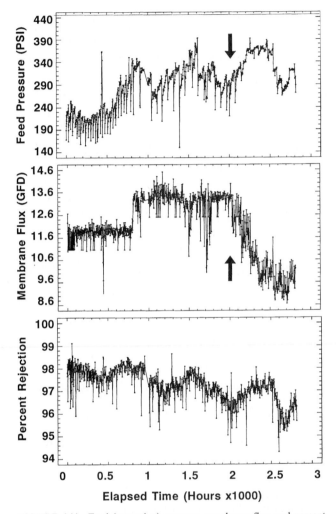

FIGURE 6.11 Feed (operating) pressure, membrane flux, and percent mineral rejection data for a TFC membrane fed with pretreated municipal wastewater at Water Factory 21, California. Flux data are uncorrected for pressure to illustrate the impact of operating pressure on productivity when a biofilm is present. Membranes were cleaned at about one-month (720-h) intervals, but cleaning was discontinued at about 2000 h of elapsed time (*arrows*). The system flux declined rapidly thereafter in spite of an increase in operating pressure. Solute rejection also gradually declined, probably in response to biofilm accumulation on the membrane surface.

effective against a range of fungi and bacteria, have been successfully incorporated into the plastics formulations (Kay et al., 1990; Seal, 1990).

Bacteria, fungi, and other microorganisms composing biofilms may directly (via enzymes) or indirectly (via localized pH or redox potential changes) degrade the membrane polymer (Flemming et al., 1992b; Ho et al., 1983; Kutz et al., 1985, 1986;

TABLE 6.1   Principal Adverse Effects of RO Membrane Biofouling

Observed biofouling effect	Description/comment
Membrane flux decline	Gradual loss of water flux due to biofilm accumulation; results in corresponding increase in system energy costs; flux may be partially restored by periodic chemical cleaning.
Reduced solute rejection	Results from enhanced opportunity for concentration polarization within the membrane biofilm, which has a hydrogel consistency; may also result from membrane biodegradation.
Enhanced mineral scaling	Biofilm formation enhances opportunity for mineral scaling due to increased concentration polarization or by providing nucleation sites for precipitate growth.
Increased module differential pressure	Results from increased fluid frictional drag (energy losses) associated with surface biofilm formation; also caused by physical blockage of feed channel spaces due to biological growth; effect may be severe; effect may or may not be reversible by cleaning or biocide treatment.
Permeate contamination	Caused by cellular detachment and/or biomass sloughing from colonized permeate surfaces of membrane, including polyester support fibers, glue lines, permeate collection materials, etc.
Membrane biodegradation	Caused by direct enzymatic hydrolysis of membrane polymers or by pH extremes associated with microcolonies on the membrane surface; may occur rapidly under physiologically favorable circumstances (e.g., warm temperatures, suitable nutrients); effect is essentially irreversible resulting in permanent loss of solute rejection properties.
Module component biodeterioration	Usually results from glue line biodegradation.
Reduced membrane life	Results from a combination of all of the above factors; inappropriate or excessive biocide applications and excessive cleaning frequencies contribute to shortened membrane life.

Ridgway, 1988; Sinclair, 1982), or other RO module components (e.g., glue lines, support fabrics, channel spacers). Biodegradation and/or biodeterioration of a membrane polymer decrease membrane life and accelerate loss of membrane structural integrity, which can lead to catastrophic system failure.

Evidence for direct biodegradation of cellulose acetate RO membranes has been reviewed by Ridgway (1988) and Sinclair (1982). Ho et al. (1983) were unable to demonstrate biodegradation of cellulose acetate membrane material by numerous cellulolytic fungi and bacteria, although some of the fungi degraded cellulose acetate following deacylation. The inability of the organisms to degrade native cellulose acetate was attributed to the high degree of acetate substitution of the cellulose polymer. Accordingly, Reese (1957, 1976) found that a number of pure strains of cellulolytic fungi including *Aspergillus, Aureobasidium, Cladosporium, Fusarium, Penicillium, Phialophora,* and *Tricoderma* completely degraded a water-soluble cellulose acetate with a degree of acetate substitution (DS value) of 0.76 using the esterase cellobiose octaacetase. None of 33 noncellulolytic bacteria tested degraded cellulose acetate. The fungi were unable to degrade cellulose triacetate with a DS

FIGURE 6.12 Longitudinal collapse of RO elements operated in excess of manufacturer's specified module differential pressure. Differential pressure increase resulted primarily from the combined effects of colloidal and microbial fouling.

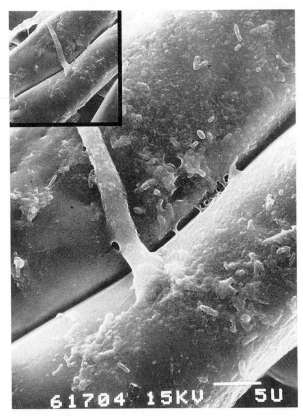

FIGURE 6.13 SEM photomicrograph of biofilm development on nonwoven polyester support fibers (inset) on permeate surface of RO membrane. Note partial occlusion of rod-shaped bacteria by EPS. Bar = 5 μm.

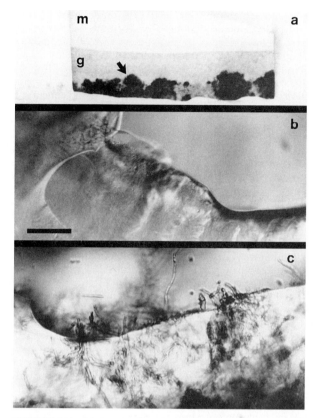

FIGURE 6.14 Biodeterioration of polyurethane glue line of an RO membrane element by filamentous fungi: (*Panel a*) Macroscopic view of glue line (approximately actual size); dark irregular patches indicate areas of penetration of fungal filaments (hyphae) into glue line (arrow); m = cellulose acetate membrane; g = polyurethane glue line. (*Panel b*) Microscopic view of uninfected (control) region of glue line. (*Panel c*) microscopic view of infected region of glue line showing invasion by fungal hyphae and loss of glue integrity. Bars = 100 μm.

value of 2.86, but degradation of blend cellulose acetates with intermediate DS values ($>0.76 < 2.86$) was not tested. None of the fungi or bacteria tested caused a decline in the solute rejection properties of the cellulose acetate membranes. Reese postulated that about one acetyl substitution per anhydroglucose unit (DS value = 1.0) was sufficient to confer protection from enzyme mediated biodegradation. However, because the DS value is an average, a value greater than 1.0 may be needed to assure at least one substitution per monosaccharide unit (Ho et al., 1983).

More recently, aerobic biodegradation of cellulose acetate of unspecified DS value by bacteria in activated sludge effluent was reported by Buchanan et al. (1993). Dorschel et al. (1993) similarly described isolation of cellulose acetate degrading bacteria with esterase activity from wastewater-activated sludge. Biodegradation was confirmed by $^{14}CO_2$ release from ^{14}C-labeled cellulose acetate, demonstrating complete mineralization of the polymeric substrate by the bacterial isolates. The authors speculated that some earlier pure culture studies may have failed by not accounting for the requisite paired action of microbial esterases and cellulases; i.e., initial esterase-mediated deacylation appeared to be a necessary prerequisite for biodegradation of the cellulose backbone structure by cellulases. Accordingly, Reese (1957, 1976) noted that among fungi that degraded cellulose acetate, the esterase cellobiose octaacetase was always found concurrently with cellulase activity. Ford (1989) similarly observed that the acyl-substituted cellulosic cell walls of Pangola grass could be rendered susceptible to biodegradation following deacylation with sodium borohydride.

Gross et al. (1993) reported decomposition of cellulose acetate in an anaerobic bioreactor environment as well as biodegradation by a single bacterium isolated from a simulated aerobic composter. Augusta et al. (1992) observed the release of carbon dioxide from *Comamonas* sp. cultures incubated with cellulose acetate of unspecified DS value, inferring mineralization of the substrate. Soviet researchers (Mamed-Yarov et al., 1987) have also described decomposition of cellulose acetete (unspecified DS value) by a number of fungi, including *Tricoderma lignorum, T. viride, T. koningii, Aspergillus niger, Penicillium* sp., *Chaetonium* sp., *Helminthosporium* sp., and *Cladosporium* sp. Furthermore, they found that incorporation of carbamyl groups into cellulose esters afforded protection from biodegradation. In contrast, solvent casted membranes of propionyl substituted esters of cellulose acetate were found by Gilmore et al. (1992) to be partially biodegraded after four months of incubation in activated sludge.

Perhaps the strongest evidence of cellulose acetate biodegradation was provided by Sinclair (1982). He reported biodegradation of cellulose acetate membranes (DS value = 2.5, 40% acetyl, 95.6 percent solute rejection) by 86 out of 90 primary cellulose acetate enrichment cultures that had been inoculated with raw or pretreated brackish surface feedwater supplying the 3×10^8 L/day U.S. Bureau of Reclamation desalting facility located in Yuma, Arizona. Biodegradation was noted by the appearance of holes or a weakening of the membrane surface specifically associated with the areas of microbial colonization. Seven pure bacterial isolates were obtained from the enrichment cultures which exhibited growth on cellulose acetate. Sinclair (1982) has speculated that microorganisms with appropriate esterase activity (e.g., cellobiose octaacetase) could attach (via a specific cell surface adhesin) to areas where microscopic structural imperfections exist on the membrane surface. Once attached, the organisms could proceed to remove cumbersome acetyl pendent groups with an appropriate esterase and further expand the diameter of the microflaws with cellulase to an extent sufficient to cause membrane failure (i.e., solute breakthrough).

Kutz et al. (1985b, 1986) described a *Seliberia* sp. that was associated with microscopic channels or tunnels in cellulose acetate RO membrane surfaces that had apparently been partially hydrolyzed or biodegraded (Fig. 6.15). *Seliberia* was also recovered from a raw groundwater source which supplied the RO system. Cellulose acetate membrane coupons incubated for four weeks at 25°C in a mineral salts basal medium with the *Seliberia* isolate showed significant hydrolysis of the membrane polymer, although solute rejection was not affected. No evidence was presented that linked the observed membrane degradation with a known enzymatic activity (e.g., cellobiose octaacetase). Thus, it is possible that abiotic physicochemical processes— e.g., localized extremes of pH—could have contributed to an acceleration of the observed membrane hydrolysis.

There are no reports documenting biodegradation of noncellulosic aromatic or aliphatic cross-linked polyamide, polyetherurea, polyvinylalcohol, or other thin film

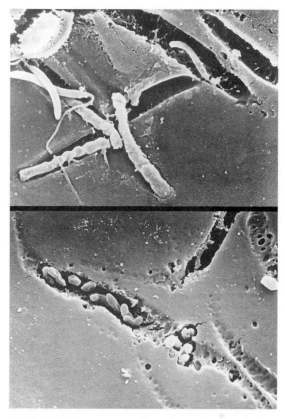

FIGURE 6.15   Apparent biodeterioration of cellulose acetate membrane surface by helical-shaped *Seliberia*-like microbe (*upper panel*) and unidentified rod-shaped bacteria (*lower panel*). (*SEM photos courtesy of Susan Bradford, Orange County Water District, Fountain Valley, California.*)

composite (TFC) membranes. The absence of such reports should not be interpreted to mean that slow biodegradation or biodeterioration of these synthetic polymeric materials does not occur. The TFC polymers contain aromatic ring and amide bond structures which closely resemble naturally occurring biomolecules and these moieties could represent potentially rich sources of nitrogen and reduced carbon for microbial growth.

Engineered water systems such as cooling towers, domestic water distribution systems, and membrane facilities often create highly specialized physicochemical microenvironments which may strongly select for the development of specific types of microorganisms. Health risk problems may occur when the growth of primary and/or opportunistic microbial pathogens is favored in these systems (Mistry and Woerkom, 1990). For example, the development of *Legionella pneumophila,* the etiological agent of legionellosis (Legionnaires' disease) is associated with certain water distribution systems, cooling towers, and evaporative condensers (Muraca et al., 1988). Du Moulin and Stottmeir (1986) and Du Moulin et al. (1985) reported the enrichment and isolation of *Mycobacterium* species from domestic household and hospital hot water supply systems. Many members of this genus are unusually resistant to antimicrobial agents such as chlorine and are considered to be opportunistic human and/or animal pathogens. Carson et al. (1988) similarly found a prevalence of nontuberculous mycobacteria in water supplies associated with renal dialysis centers that incorporated membrane systems. Finally, biofilms enriched in nontuberculous mycobacteria have been found to develop in certain RO membrane systems treating municipal wastewater for reclamation purposes (Ridgway, 1987, 1988; Ridgway et al., 1984a). Direct bodily contact with, ingestion of, or inhalation of aerosolized material from such biofilms could markedly increase disease risk for immunocompromised or otherwise susceptible individuals. Plant operations personnel who occasionally come into contact with membrane biofilms should exercise due caution. To minimize liability, biofilm analyses should be performed by trained technicians in a suitably equipped and controlled laboratory according to accepted microbiological practices and procedures.

6.5 OCCURRENCE OF MEMBRANE BIOFOULING

Biofouling is a widespread problem limiting the performance and application of RO and other membrane separation processes (Durham, 1989; Eisenberg and Middle-brooks, 1984; Flemming, 1991, 1993; Flemming and Schaule, 1988a, 1988b, 1988c; Flemming et al., 1992a, 1992b, 1993; Lepore and Ahlert, 1988; Paul, 1991; Ridgway 1987, 1988; Winters et al., 1983; Zeiher et al., 1991). The primary source of microbial contamination is typically the system feedwater, and surface waters in particular contain high numbers of microorganisms which lead to microbial problems (Applegate et al., 1989). The most adherent and rapidly growing microorganisms generally contribute most to nascent biofilm development. Biofouling is especially evident, and perhaps most critical, in large seawater desalination and wastewater reuse facilities where feedwaters contain high numbers of microorganisms and available nutrients (Bailey et al., 1974; Bettinger, 1981; Finken et al., 1979; Ridgway et al., 1983; Schaule et al., 1993b; Wechsler, 1976; Winfield, 1979a, 1979b).

   Reverse osmosis and other membrane systems are characterized by large areas of exposed membrane surface which offer a suitable habitat for adherent bacteria. In contrast to other surfaces, membranes are subject to flow normal to the membrane surface. While this component is usually only 10 to 15 percent of the total module flow in RO membranes, it is nevertheless sufficient to convectively transport colloids and microorganisms to the membrane surface, thereby increasing the probability of contact and successful adhesion (see Sec. 6.8).

   The piping and pretreatment systems prior to the RO system offer many surfaces suitable for biofilm development, such as ion exchangers (Flemming, 1991; Mittleman, 1991), granulated activated carbon filters (Camper et al., 1986; Geldreich et al., 1985; LeChevallier et al., 1984; Mittleman, 1991; Suffet and Pipes, 1986), degasifiers (Mittleman, 1991), cartridge filters (Malik et al., 1989; Parise et al., 1988; Patterson et al., 1991), holding tanks (Mittleman, 1991), and other materials (Schoenen, 1990). These biofilms tend to release microorganisms which colonize other surfaces of the system—among these, the membranes. The use of reinforced fiberglass piping often promotes surface biofouling by photosynthetic microalgae due to light penetration (Bergman et al., 1988). Overfeeding of organic polymeric flocculants (e.g., polyacrylates) used for suspended solids removal can result in accumulation of microbial aggregates on membrane surfaces, and the flocculants themselves may serve as potential nutrient sources. Ahmed and Alansari (1989) reported that addition of the calcium antiscalant sodium hexametaphosphate (SHMP) to the feedwater could serve both as a source of microorganisms, as well as a nutrient supply for biofilm growth, primarily because of orthophosphate formation resulting from SHMP hydrolysis. Accordingly, Mindler and Bateman (1980) reported enhanced microbial growth in SHMP system pipelines, tanks, and cartridge filters following dechlorination in a seawater RO pilot plant in St. Croix, U.S. Virgin Islands; however, there was no evidence that the observed biofilm growth adversely affected RO membrane performance. Sodium thiosulfate, used to neutralize chlorine, has also reportedly been found to serve as a nutrient source for membrane biofouling bacteria (Winters et al., 1983).

Numerous reports exist of biofouling problems in membrane systems. Among 70 U.S. reverse osmosis membrane facilities surveyed by Paul (1991), 58 reported having "above average" problems with membrane fouling, with biofouling representing the most common operational problem experienced. Eisenberg and Middlebrooks (1984) similarly reported that membrane fouling and flux decline were the most common problems reported among 28 RO water treatment plants surveyed. Membrane biofouling was believed to represent a major proportion of the reported problems. Membrane biofouling problems have been reported for nearly every type of feedwater and industrial process (Eisenberg and Middlebrooks, 1984), including brackish surface water (Bettinger, 1981; Kaakinen and Moody, 1984) and groundwater treatment (Ahmed et al., 1989; Bergman et al., 1988; Kutz et al., 1985b, 1986; Lepore and Ahlert, 1988; Malik et al., 1989), seawater desalination (Applegate et al., 1989; Crossley, 1983; Finken et al., 1979; Goodwyn, 1976; Hassan et al., 1989; Mindler and Bateman, 1980; Saad, 1992; Winters et al., 1983), wastewater reclamation (Argo and Ridgway, 1982; Bailey et al., 1974; Ridgway, 1987, 1988; Schaule et al., 1993b; Wechsler, 1976; Winfield, 1979a, 1979b), fully treated potable water (Payment, 1989; Payment et al., 1992), and ultrapure water production (Flemming, 1991; Mittleman, 1991; Patterson et al., 1991; Pittner, 1993b). Membrane biofouling is especially prevalent in the food and beverage industries where process streams often harbor high concentrations of soluble proteins and other organics which serve as nutrients for biofilm microorganisms. Biofouling occurs regardless of the nature of the membrane polymer type (e.g., cellulosic, polyamide, polysulfone, polypropylene, etc.) or the module configuration (spiral wound, hollow fiber, flat sheet), although clear differences in the magnitude and susceptibility to biofouling can be attributed to these parameters (see below). Factors that favor membrane biofouling include reduced feedwater cross-flow velocities (i.e., higher product recoveries) (Wechsler, 1976), elevated operating pressures and temperatures, high feedwater concentrations of organics (Applegate et al., 1989; Flemming et al., 1992b; Winfield, 1979a; Winters et al., 1983), use of small feed channel dimensions, and membrane polymers with enhanced bacterial affinities (Flemming, 1993; Flemming and Schaule, 1988a, 1988b, 1988c; Ridgway et al., 1984b, 1984c, 1985; Ridgway and Safarik, 1991; Schaule, 1992; Schaule et al., 1990).

Ridgway and co-workers (Ridgway, 1987, 1988; Ridgway et al., 1981, 1983) performed chemical and microbiological characterization of biofilms formed on blend cellulose acetate RO membranes used to demineralize municipal wastewater at Water Factory 21 in southern California. Clear evidence of biofouling occurred on both the feedwater and permeate (Texlon support) surfaces of these membranes in spite of extensive pretreatment measures, including high-pH lime clarification, polymer flocculation, multimedia filtration, continuous chlorine injection, and cartridge filtration (Argo and Ridgway, 1982; Nusbaum and Argo, 1984; Ridgway et al., 1983).

The 3 to 5 mg/L of free chlorine injected at this facility rapidly combines with an excess of ammonia normally present in the wastewater (typically 20 to 30 mg/L) to form monochloramine, which has been shown to be superior to free chlorine as a biofilm disinfectant (LeChevallier, 1987, 1991; LeChevallier et al., 1988, 1984, 1990). The monochloramine residual at the membrane surface was sufficient to prevent uncontrolled proliferation of bacteria, although certain types of resistant bacteria were still recovered from test membranes (see Sec. 6.7).

Membrane biofouling and biodeterioration of ancillary module components has been observed at the U.S. Bureau of Reclamation's large RO desalting facility in Yuma, Arizona. This facility has experienced biodeterioration of the cellulose acetate RO membranes and module components (e.g., membrane glue lines) which tended to occur when membrane modules were placed in long-term cold storage (weeks to months) during periods of plant inactivity. The biodeterioration resulted from the invasion and growth of filamentous fungi belonging to the genera *Aspergillus* and *Penicillium* (see Sec. 6.4). These fungi readily attacked the polyurethane-based glues used to seal adjacent membrane leaves together in the modules. Growth of fungi within the glue lines was, in some cases, sufficient to compromise the structural integrity of the module and caused direct leakage of solute into the product collection material (Fig. 6.14).

Schaule et al. (1993b) reported similar membrane biofouling at a municipal refuse site in Germany that incorporates TFC aromatic cross-linked polyamide membranes for removing ammonia and other substances from a leachate waste stream. Interestingly, most of the accumulated biofilm in the membrane modules appeared not to be associated with the membrane surface itself, but rather with the plastic feed channel spacer material (Vexar). Numerous experimental chemical cleaning formulations including tetramethylurea, periodate, and selected commercial biodispersants were only partially effective in removing the biofouling layers and restoring membrane performance. Some restoration of water flux was noted in the absence of any significant decline in the module differential pressure, suggesting that much of the biofilm was not actually removed during cleaning, but that its water permeation characteristics had improved (see Sec. 6.4). Membrane autopsies performed before and after chemical cleanings were consistent with this interpretation.

Amjad et al. (1991) reported severe biological fouling of fully aromatic cross-linked polyamide RO membranes used at a semiconductor manufacturing facility located in the midwestern United States. The polyamide TFC membranes were used in the final stages of an ultra-high-purity water system. Interestingly, obligately anaerobic sulfate reducing bacteria and aerobic (or facultative) heterotrophic bacteria were identified as the primary fouling microorganisms. The addition of sodium metabisulfite in the feed stream to inactivate residual chlorine prior to RO was thought to contribute to the anaerobic bacterial activity by lowering the system redox potential. Replacement of the metabisulfite feed with a granular activated carbon bed seemed to reduce the effects of membrane biofouling and enhance system production.

Despite very low microbial populations in a brackish anaerobic groundwater supply used to feed RO membranes at a desalting facility in Englewood, Florida, unacceptable flux declines were observed shortly after plant start-up (Bergman et al., 1988). While microorganisms were probably partly responsible for the membrane flux declines, medium- to low-molecular-weight organics (<10,000 daltons) were also suspected of contributing to flux losses at this facility. It was not determined whether these trace organics could serve as an assimilable organic carbon source for bacteria on the membrane surfaces.

Zeiher et al. (1991) reported three case histories of RO membrane facilities that experienced significant biological fouling problems. In a paper mill facility located in the southern United States, aromatic crossed-linked polyamide TFC membranes were fed with pretreated deep well water. While iron deposits found on the membrane surfaces partially explained observed flux losses, significant concentrations of viable bacteria were detected in the permeate and brine streams. The amount of bacteria rejected by the system was less than expected based on a 70 percent recovery system, indicating deposition of bacteria on the membrane surfaces. A similar biological problem was encountered at a second midwestern U.S. electronic component manufacturing facility also using polyamide TFC membranes. The feed source was groundwater disinfected with iodine and subsequently treated with sodium bisulfite to remove excess disinfectant. Extensive microbial fouling was also observed at a midwestern U.S. oil refinery incorporating TFC membranes fed with chlorinated groundwater treated with metabisulfite. Reduced system flow rates and increasing module differential pressures were correlated with the accumulation of primarily aerobic heterotrophic bacteria on the membrane surfaces.

Many seawater desalination facilities have been affected by membrane biofouling, including the large desalination plants at Ras Abu Jarjur, Bahrain (Ahmed et al., 1989) and at Saint Croix, U.S. Virgin Islands (Mindler and Bateman, 1980). The use of sodium hexametaphosphate (SHMP), a calcium carbonate antiscalant, at the Ras Abu Jarjur facility was associated with excessive microbial aftergrowth and biofouling of TFC hollow-fiber membranes, possibly by providing a source of microbially available orthophosphate (Ahmed et al., 1989). Addition of a minimum concentration of 0.25% (wt/vol) sodium metabisulfite to the SHMP tanks suppressed microbial growth without affecting SHMP performance or orthophosphate formation.

Biofouling is also a significant though manageable problem with polypropylene hollow-fiber MF membranes used for potable and wastewater treatment applications. Microfiltration membranes effectively remove bacteria and colloids, making them useful as an RO pretreatment to replace lime clarification in wastewater reclamation. Left unchecked, bacteria in the wastewater feed (typically activated sludge effluent) irreversibly foul MF membrane surfaces and clog pores, resulting in unacceptably high membrane cleaning frequencies in spite of backwash provisions. Continuous dosage of the wastewater feed with 3 to 5 mg/L monochloramine suppresses biofouling (without oxidative damage to the membrane), resulting in extended run times between chemical cleaning cycles.

Jacangelo et al. (1989) reported suspected biofouling of hollow-fiber UF membranes fed with native river water. The exact nature of the fouling material was not investigated. However, the fouling was noted only following the failure of a chemical feed pump used to supply a continuous dosage of disinfectant (3.5 mg/L free chlorine) to the membranes. Thus, biofouling was suspected.

Chlorine disinfection of system feedwaters does not necessarily ensure that membrane biofouling will not occur. Ridgway et al. (1984a) reported that biofouling of cellulose acetate RO membranes at a wastewater reclamation plant continued at close to control (unchlorinated) rates in the presence of high feedwater dosages of monochloramine (10 to 15 mg/L). Hassan et al. (1989) reported that biofouling of RO membranes was problematic at the Al-Birk, Saudi Arabia, seawater desalination plant where the feedwater was first chlorinated (4 mg/L) and then dechlorinated immediately before entering the membrane system. Bacteria surviving the disinfection process may have regrown on the membrane surfaces following removal of the chlorine residual; however, it is also possible that the chlorine may have oxidized complex trace organics present in the seawater to smaller compounds that were more easily assimilated by the fouling bacteria. Pilot-scale experiments performed by Applegate et al. (1989) using membrane biofouling bacteria isolated from Middle Eastern RO plants incorporating hollow-fiber TFC membrane modules indicated that humic acids and other trace organics in seawater were oxidized to smaller molecules by free chlorine which, in turn, served as nutrients for the membrane biofilms. Monochloramine was found to provide good biofilm disinfection with greatly reduced trace organic oxidation, resulting in less bacterial aftergrowth. Brief exposure of the TFC membranes to monochloramine did not adversely affect membrane performance. A similar release of low-molecular-weight assimilable organics might be expected to result from ozonation of feedwaters (Gilbert, 1988).

Interestingly, not all seawater desalination plants have experienced significant biofouling problems. Light et al. (1988) and Hassan et al. (1989) reported that RO plants in Saudi Arabia (Umm Lujj and Jeddah) produced high-quality potable water at better than design capacity from nonchlorinated Red Sea surface water for more than two years. During this time, no TFC membranes were replaced and chemical cleaning was evidently not required. However, algal growth was controlled in these membrane desalination systems by the addition of low concentrations (0.5 to 1.0 mg/L) of copper sulfate. It is possible that the copper sulfate also suppressed bacterial fouling in these membrane systems. Some corrosion of 316 stainless steel treatment and desalination materials was noted at the Jeddah facility (Hassan et al., 1989); however, whether this corrosion was microbially influenced is unknown.

6.6 MEMBRANE BIOFILM STRUCTURE AND CHEMISTRY

The microscale structure and composition of membrane biofilms determine their macroscale effects on system performance (Flemming, 1993; Flemming et al., 1992b, 1993). Flemming et al. (1992b), Ridgway (1987, 1988), and Ridgway and Safarik (1991) have reviewed the ultrastructural and physicochemical properties of membrane biofilms. In contrast to the strictly tangential flow experienced by biofilms on nonporous surfaces, a significant proportion (10 to 15 percent) of the water and solute mass is continuously transported under pressure vertically through membrane biofilms. Thus, membrane biofilms are exposed to greater nutrient fluxes than other types of biofilms, a factor which accelerates their growth kinetics accordingly. Biofilms on membranes are also subject to greater hydraulic compaction than other biofilms as a result of the applied transmembrane pressure. These differences may result in special ultrastructural features of membrane biofilms, such as higher density due to physical compaction. Membrane biofilm thickness may vary from a single monolayer of cells (about 1 μm thick) to several tens of micrometers in older biofilms (Ridgway et al., 1983, 1984a). Viewed in cross section by TEM or SEM, individual cells composing membrane biofilms are typically distributed throughout the EPS matrix (Ridgway, 1987, 1988; Ridgway et al., 1981, 1983; Schaule, 1992) (Figs. 6.2 and 6.3). Many of the biofilm cells are clearly lysed and inactive, while others appear structurally intact and possibly metabolically viable. The membrane biofilms comprise mainly EPS, dead (lysed) and viable cells, and other electron-dense debris, possibly entrained or precipitated inorganic material. Biofilm development is typically greatest on the feedwater surfaces of RO membranes due to greater nutrient fluxes.

Recent experimental evidence indicates that some biofilms are not structurally homogeneous masses, but rather possess a high degree of microscale complexity, including extensive networks of fluid-filled channels (voids) (Bryers, 1993; Costerton et al., 1994). Drury et al. (1993a, 1993b) observed latex bead penetration into *Pseudomonas aeruginosa* biofilms grown in laboratory annular reactors, inferring the presence of open pores and channels that extend to the substratum. Insertion of specially fabricated microelectrodes into experimental biofilms has indicated that the voids contain oxygen and solute concentrations more similar to those of the bulk fluid than the interior of the EPS matrix (Costerton et al., 1994). In addition, laminar flow has been demonstrated within biofilm channels using fluorescent latex beads as motion indicators (Stoodly et al., 1994). The concept of biofilm heterogeneity and void networks is relatively new and may help explain why biofilm permeability can be enhanced following certain chemical treatments (e.g., detergent treatment) without a corresponding loss of biofilm mass (Flemming et al., 1992a, 1992b; Schaule et al., 1993b). The occurrence and distribution of biofilm voids in membrane systems are unknown; however, voids might be expected to be less prevalent in membrane biofilms due to hydraulic compression effects, especially in RO and NF systems where operating pressures are generally high. Should they occur, however, the voids might provide a strategic avenue for the introduction of antimicrobial agents for biofilm control.

Biofouling of membrane surfaces rarely occurs in the absence of mineral deposition (mineral scaling). Because of the coincidence of mineral deposition and biofilm formation in membrane systems, it is often difficult to attribute performance decline exclusively to one process or the other. In some cases, however, there is a clear dominance of either mineral deposition or biofouling (e.g., in municipal wastewater reclamation applications where microbial fouling is predominant). Zeiher et al. (1991) and Schaule et al. (1993b) demonstrated significant iron scaling in association with RO membrane biofilms. Indeed, the very process of biofilm formation could be expected to promote mineral scaling in RO and NF systems due to enhanced solute concentration polarization (Flemming, 1993; Flemming et al., 1993; Ridgway, 1987, 1988). In addition, certain attached bacteria may themselves serve as specific nucleation sites for mineral precipitation or crystallization (Ridgway, 1988). Examples of such organisms would include iron- and manganese-oxidizing bacteria (Greene and Madgwick, 1991; Howsam and Tyrrel, 1990; Nealson et al., 1988). Many bacteria can produce intracellular deposits of polyphosphate, iron, or elemental sulfur (Deinema et al., 1980; Miguez et al., 1986). The high amphoteric charge density of the EPS associated with biofilms on membranes and other surfaces results in the concentration of heavy metals and other charged species within the biopolymer matrix (Bryers, 1994; Costerton and Irvin, 1981; Costerton et al., 1994; Geesey, 1982; Geesey et al., 1992). Techniques currently used to quantify mineral deposition on biofouled membrane surfaces (e.g., EDX, standard extraction/analytical techniques) cannot differentiate between mineral deposition by purely abiotic processes from biologically mediated mineral scaling.

Chemical analyses of biofilm material scraped from cellulose acetate RO membranes used to demineralize municipal wastewater at Water Factory 21, California, revealed significant levels of iron, calcium, sulfate, phosphate, chloride, and chromium (Nusbaum and Argo, 1984; Ridgway et al., 1981, 1983). Elemental species identified in similar RO membrane biofilms from the same facility scanned by EDXM included sodium, magnesium, aluminum, silicon, phosphorus, sulfur, chlorine, potassium, calcium, iodine, titanium, iron, and copper (Ridgway et al., 1981). The concentrations of these inorganic species (expressed as a percentage of sample dry weight) increased as a function of the stage from which the membrane module was recovered. Thus, membrane biofilms recovered from third-stage modules exhibited greater mineral deposits than first-stage biofilms. Higher mineral compositions were also found in later modules within individual pressure vessels (Nusbaum and Argo, 1984; Ridgway et al., 1981). These trends are correlated with increasing brine concentrations along the flow path of individual pressure vessels and between system stages.

6.7 MICROBIAL COMPOSITION
OF MEMBRANE BIOFILMS

Under actual plant operating conditions, biofouling may occur rapidly, with complete biofilm coverage of membrane surfaces occurring in less than two weeks (Ridgway, 1987, 1988). Numerous different bacteria participate in early biofilm development on RO membranes (Flemming and Schaule, 1988a, 1988b; Ridgway et al., 1983; Safarik et al., 1989; Schaule, 1992; Schaule et al., 1990). However, the microbial species that predominate at any given stage of biofilm development varies from one facility to another and ultimately depends on the interaction of a multitude of incompletely understood physicochemical, environmental, and biological factors related to the feedwater, the membrane substratum, and the microorganisms.

Ho and co-workers (1983) isolated numerous fungal and bacterial genera from biofilms formed on cellulose acetate RO membranes fed with domestic Roswell, New Mexico, city water consisting of blended surface and brackish groundwater. Pretreatment consisted of sand filtration followed by disinfection with either chlorine or sodium metabisulfite. Fungal genera recovered from the biofouled membranes included *Acremonium, Candida, Cladosporium, Cleistothecial ascomycetes, Fusarium, Geotrichum, Mucorales, Mycelia sterilia, Penicillium, Phialophora, Rhodotorula,* and *Trichoderma. Penicillium* and *Fusarium* were the most frequently isolated strains and most of the fungi were cellulolytic but did not degrade cellulose acetate under laboratory test conditions. Bacterial genera recovered from the same membranes included *Acinetobacter, Arthrobacter, Bacillus, Flavobacterium, Kurthia, Lactobacillus, Micrococcus, Micromonospora,* and *Pseudomonas. Bacillus, Lactobacillus,* and *Pseudomonas* were recovered most frequently. Payment (1989) reported isolation of *Chromobacterium, Moraxella, Acinetobacter, Alcaligenes,* and *Pseudomonas* species from the product collection reservoirs of 300 domestic RO membranes fed with fully treated and chlorinated surface water. Chlorine was removed by a GAC prefilter to protect the polyamide membranes from oxidative damage. This author speculated that the isolates may have been associated with the permeate surfaces of the RO units, although this was not proven. Sinclair (1982) described the isolation and identification of *Pseudomonas aeruginosa* and *Pseudomonas fluorescens* from cellulose acetate membrane enrichment cultures inoculated with raw and pretreated surface water from the U.S. Bureau of Reclamation desalting facility in Yuma, Arizona. Kutz et al. (1985b) recovered a species of the oligotrophic bacterium *Seliberia* from cellulose acetate membranes that had been reportedly biodegraded (see Fig. 6.15). The same *Seliberia* isolate was also recovered from raw groundwater used to feed the membrane system.

Few studies have been performed concerning temporal or successional changes in the microbial community structure of membrane biofilms. Ridgway et al. (1984a), Ridgway and Safarik (1991), and Safarik et al. (1989) reported that *Mycobacterium* spp. initially colonized cellulose acetate or aromatic cross-linked polyamide TFC membranes used to demineralize filtered and chlorinated activated-sludge effluent at Water Factory 21 in southern California. Acid-fast, rapidly growing mycobacteria were the predominant species isolated from biofouled membranes at this facility for several weeks to several months after placing new membranes into operation. Membranes at this facility received extensively pretreated secondary (activated-sludge) municipal effluent with continuous dosing of 3 to 5 mg/L chlorine. Although chlorine is typically added to the feedwater at this facility in the free form, i.e., as hypochlorous acid, it rapidly combines with an excess of ammonia in the wastewater (20 to 30 mg/L) to form monochloramine (Nusbaum and Argo, 1984; Ridgway et al., 1981). The microbial community structure of the membrane biofilm changed significantly over time, and mycobacteria were no longer detected after seven months of operation. Bacteria detected in older membrane biofilms included *Acinetobacter, Flavobacterium/Moraxella, Pseudomonas/Alcaligenes, Bacillus, Micrococcus, Serratia,* and other unidentified types, with *Acinetobacter* and *Flavobacterium/Moraxella* representing the predominant genera (Ridgway et al., 1981). Bacteria recovered from the permeate surfaces of the same membranes included *Acinetobacter, Pseudomonas/Alcaligenes, Bacillus/Lactobacillus, Micrococcus,* and *Serratia,* with *Acinetobacter, Bacillus/Lactobacillus,* with *Pseudomonas/Alcaligenes* representing the predominant types. Two factors may favor early membrane colonization by mycobacteria in this system. First, mycobacteria attach to RO membranes more rapidly and tenaciously than certain other bacteria, possibly because of their extraordinarily hydrophobic cell surfaces (see Sec. 6.8) (Bendinger et al., 1993; Katoh and Matsuo, 1983; Ridgway, 1987, 1988; Ridgway et al., 1984b, 1984c, 1985, 1986a, 1986b; Ridgway and Safarik, 1991). Second, since mycobacteria are typically more resistant to chemical biocides than other microbes (Haas and Engelbrecht, 1980), they may have survived better than their less-resistant competitors in the chloraminated feedwater at this facility.

Mycobacteria have not been detected in early biofilms from all RO facilities. Flemming (1993), Flemming and Schaule (1988a, 1988b, 1988c), Schaule (1992), and Schaule et al. (1993a), for example, found that *Pseudomonas vesicularis, Acinetobacter calcoaceticus, Staphylococcus warneri,* and *Staphylococcus simulans* were the predominant species in nascent biofilms from cellulose acetate and polyetherurea TFC membranes fed with fully treated and chlorinated domestic tap water in Stuttgart, Germany. These bacteria were considered to be "fast-adhering" strains, capable of attaching irreversibly to virgin membrane surfaces in large numbers within minutes to hours of first exposure to the water. These and other similar data (Flemming, 1993; Ho et al., 1983; Sinclair, 1982) suggest that the same membrane type can be colonized by different bacteria depending on geographic location, feedwater composition, and other factors.

These microbial identification studies were performed by conventional plating and colony isolation methods. While plating techniques are simple and inexpensive, they typically recover only a small proportion (0.01 to 10 percent) of the viable microbial community, since the range of nutrients and incubation conditions are inadequate or inhibitory to many environmental microbes. Future investigations of microbial diversity in membrane biofilms should employ microbial identification methods that do not depend on prior organism subculture, such as gene-probe analysis, ribosomal RNA sequence analysis, restriction fragment length polymorphism analysis, or signal lipid biomarker analysis (Guckert et al., 1986; McKinley et al., 1988; Welch, 1991; White, 1985).

6.8 BACTERIAL ATTACHMENT AND BIOFILM FORMATION

The principal events characterizing membrane biofouling are summarized in Fig. 6.16 and Table 6.2. The process of bacterial attachment in membrane systems is important for at least two reasons. First, attachment is the earliest step in biofilm formation (Costerton et al., 1985; Doyle, 1991; Marshall, 1985, 1991; Marshall and Blainey, 1991; Schaule, 1992; Schaule et al., 1990) and the initial attachment pattern contributes to the course of subsequent biofilm development and maturation (Costerton et al., 1994; Geesey et al., 1992). As mentioned previously, the earliest stages of microcolony development (e.g., the pattern of bacterial attachment to the substratum) influences the remaining course of biofilm development and determines the final microstructural features of the mature biofilm (e.g., the location and distribution of flow channels and void spaces). Second, the attachment stage represents a logical and potentially sensitive point at which to intervene in the biofouling process by the application of adhesion inhibitors. Bacteria readily attach to RO membranes even in the complete absence of transmembrane pressure differentials (Flemming and Schaule, 1988a, 1988b, 1988c; Ridgway, 1987, 1988; Ridgway et al., 1984b, 1984c, 1985; Ridgway and Safarik, 1991; Schaule, 1992; Schaule et al., 1990), suggesting an inherent affinity between the bacterium (or cell surface macromolecules) and the synthetic membrane polymer. The strength of the bacterium-membrane association appears to influence how rapidly a membrane undergoes biofouling under actual operating conditions. A comprehensive understanding of bacterial attachment mechanisms should allow prediction of membrane biofouling potentials from inexpensive and rapid laboratory adhesion tests, similar to that proposed by Winters et al. (1983). At this time, only a cursory knowledge exists of the subcellular and molecular mechanisms that mediate bacterial attachment to synthetic polymer membranes used in the membrane industry. Weak attachment usually commences within seconds following first contact of the bacterium with the membrane surface (Flemming and Schaule, 1988a, 1988b, 1988c; Ridgway et al., 1985; Ridgway and Safarik, 1991; Schaule, 1992). There is little or no detectable lag phase. Bacterial attachment to RO or other membranes is probably not substrate specific (i.e., stereospecific macromolecular binding sites are likely not involved), although the bacterium-membrane association is extremely intimate as evidenced from electron microscope images (Fig. 6.17), and certain fouling bacteria exhibit a strong preference for certain membrane materials (Flemming and Schaule, 1988b, 1988c; Ridgway et al., 1985; Ridgway and Safarik, 1991; Schaule, 1992; Schaule et al., 1990) (see below).

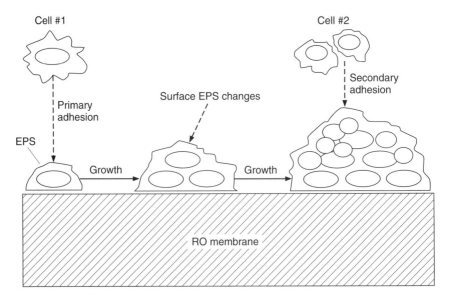

FIGURE 6.16   Schematic illustration of major events in membrane biofouling process.

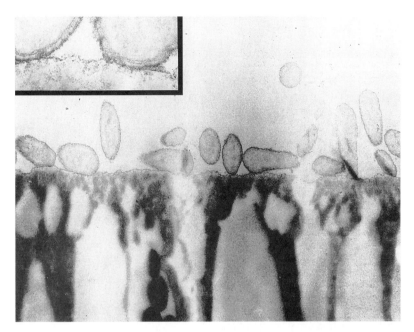

FIGURE 6.17   TEM photomicrograph of *Pseudomonas diminuta* cells attached to an aromatic cross-linked polyamide RO membrane. Note intimate association of bacterial surface macromolecules with the synthetic polymer surface (*inset*). Cells are about 1 μm in diameter. *(Courtesy of Gabriella Schaule, University of Stuttgart, Stuttgart, Germany.)*

Once a cell has attached, it may begin to grow and multiply at the expense of feedwater nutrients (Flemming, 1993). Initial attachment is governed by the physicochemical nature of the cell surface (Costerton et al., 1985; Marshall, 1991; Ridgway et al., 1985), the membrane polymer chemistry and structure (Flemming, 1993; Ridgway and Safarik, 1991; Schaule, 1992; Schaule et al., 1990), and the aqueous (i.e., feedwater) environment (Bryers, 1993; Characklis, 1990a; Leslie et al., 1993; Marshall, 1985; Marshall and Blainey, 1991; Marshall et al., 1971). The influence of the membrane on initial attachment may be further modified by formation of a macromolecular conditioning film formed by the adsorption of feedwater organics.

The bacterial attachment process itself is divided into several distinct phases (Costerton et al., 1985; Marshall, 1985, 1991; Marshall and Blainey, 1991), including (1) the close approach of the bacterium to the RO membrane surface via settling under the influence of gravity, diffusion (brownian motion), or convective fluid transport, or by motility and chemotaxis; (2) a reversible attachment phase in which the cell is transiently poised close to the membrane surface within the viscous sublayer (Lawrence et al., 1987) but still in a weak bonding state that can be easily disrupted by mild fluid shear; and (3) an irreversible attachment phase in which the cell cannot be removed by the moderate fluid shear forces associated with actual membrane systems (Fig. 6.18). Irreversible attachment is mediated by the biosynthesis and extrusion of EPS (Bryers, 1993; Characklis, 1983; Costerton et al., 1985; Geesey, 1982; Marshall, 1985; Vandevivere and Kirchman, 1993).

Figure 6.19 illustrates the type of laboratory adhesion assay that has been used to study bacterial attachment to membrane surfaces (see Ridgway, 1988). The assay is

TABLE 6.2 Principal Events in Membrane Biofouling Processes

Event	Time to onset*	Description/explanation
Primary organic film	Seconds/minutes	Typically referred to as the *conditioning* film; defined as the rapid adsorption of dissolved organic macromolecules and inorganic substances at the membrane/liquid interface.
Primary cell adhesion	Seconds/minutes	Refers to pioneer bacterial attachment; dependent on nature of cell surface, membrane type, feedwater chemistry, and system hydrodynamics; provides major contribution to early biofilm accumulation.
Cellular detachment	Seconds/minutes	Influences biofilm accumulation rate; detachment is sometimes enhanced by microbicidal agents, dispersants, etc.
Cell growth/multiplication	Minutes/hours	Occurs at expense of soluble and sorbed feedwater nutrients; may provide greatest contribution to biofilm formation where biocides are not present.
Biopolymer (EPS) synthesis	Minutes/hours	Provides for greater biofilm structural integrity; acts as a reactive transport barrier to chemical biocides; promotes nutrient concentration/storage.
Particle/colloid entrainment	Seconds/minutes	Secondary effect where suspended particles and colloidal material are passively entrained in the biopolymer matrix or within biofilm void spaces.
Secondary cell adhesion	Days/weeks	Commences after primary biofilm formation by pioneer cells; probably strongly influenced by surface properties and physiology of primary biofilm and leads to greater species diversity.
Biofilm sloughing	Days/weeks	Refers to cell and biomass detachment; occurs in response to changes in hydrodynamic shear or turbulence forces, or introduction of biocides, dispersants, etc.
Biofilm scenescence	Weeks/months	Refers to accelerated cell die-off in old biofilms; cell death is in equilibrium with biofilm growth in continuous flow systems; may result in release of soluble nutrients via cell lysis.

* Refers to time after a new membrane is placed into operation.

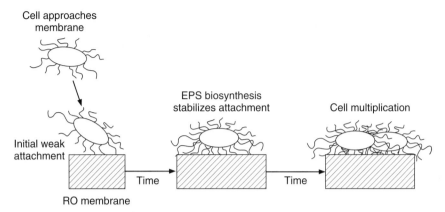

FIGURE 6.18 Schematic illustration of the bacterial adhesion process. Note that the irreversible adhesion phase is associated with EPS biosynthesis.

conducted in the absence of pressure differentials in order to isolate the effect of cellular affinity for the polymer membrane as a separate and distinct variable. The attachment of *Mycobacterium* and *Acinetobacter* species to cellulose acetate and polysulfone membranes was primarily mediated by hydrophobic rather than polar (i.e., ionic) interactions (Ridgway, 1988; Ridgway et al., 1985, 1986a; Ridgway and Safarik, 1991). Accordingly, surfactants such as the nonionic polyoxyethylene detergent Triton X-100, sodium dodecyl sulfate, and certain quaternary ammonium biocides interfered with these hydrophobic interactions and retarded mycobacterial adhesion kinetics to cellulose acetate membranes (Ridgway, 1988; Ridgway et al., 1985, 1986a). The inclusion of selected surfactants in some commercial membrane cleaning formulations designed to control biofouling may be rationalized on this

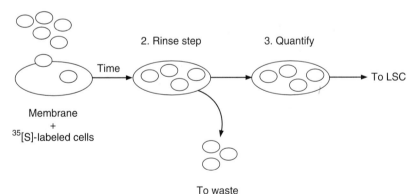

FIGURE 6.19 Schematic illustration of laboratory bacterial adhesion assay. Bacteria are allowed to come into contact with only one surface of the membrane coupon during the assay. No pressure differentials are applied. Potential inhibitors of cell adhesion may be added during the attachment step (1); potential cleaning agents may be added during the rinse step (2). Number of attached cells is quantified by determining bound radioactivity via liquid scintillation counting (LSC). (See Ridgway, 1988; Ridgway and Safarik, 1991.)

basis (Amjad, 1989; Amjad et al., 1993). However, the adhesion-blocking ability of detergents is highly specific and variable depending on subtle differences in the molecular structure of the particular compound under investigation (Ridgway, 1988; Ridgway et al., 1986a). Certain quaternary amine surfactants (e.g., dodecyltrimethyl ammonium chloride), for example, strongly enhanced mycobacterial attachment to cellulose acetate RO membranes, while very closely related compounds (e.g., hexa-decyltrimethyl ammonium chloride) inhibited attachment (Ridgway, 1988).

Adsorption of microorganisms at interfaces via a hydrophobic mechanism has been well documented in numerous systems (Busscher et al., 1984, 1986; Dahlback et al., 1981; Dexter et al., 1975; Doyle, 1991; Fattom and Shilo, 1984; Fletcher and Loeb, 1979; Fletcher and Marshall, 1982; Fletcher and McEldowney, 1983; Paul and Jeffrey, 1984, 1985; Rosenberg and Doyle, 1990; Rosenberg et al., 1980). Yet cell surface hydrophobicity does not always correlate with increased bacterial sorption to solid substrata (Paul and Jeffrey, 1984; Rosenberg and Doyle, 1990). Fletcher and McEldowney (1983) reported that both electrostatic and hydrophobic interactions mediated attachment of several different bacterial genera to native (hydrophobic) and modified (hydrophilic) polystyrene surfaces. However, these workers noted that the greatest degree of detachment from the plastic surfaces was produced by detergent treatment, suggesting some degree of hydrophobic interaction for all strains. In a study of the effects of surfactants on the attachment of estuarine and marine bacteria to different surfaces, Paul and Jeffrey (1984, 1985) found that attachment to glass, a hydrophilic material, was less affected by surfactants than adhesion to hydrophobic polystyrene, inferring an ionic mechanism of attachment to the former substrate. Low concentrations of Triton X-100 (0.002% wt/vol) having little or no effect on cell viability and growth strongly suppressed adhesion to hydrophobic polystyrene surfaces for 10 of the 11 bacterial isolates tested. This nonionic detergent also caused displacement from polystyrene of previously attached *Vibrio proteolyticus* cells.

Schaule (1992) and Schaule et al. (1990) reported that, whereas Triton X-100 inhibited attachment of *Pseudomonas vesicularis, Acinetobacter calcoaceticus, Staphylococcus warneri,* and a mixed native population to several different RO and UF membrane polymers (e.g., polysulfone, polyethersulfone, and polyetherurea), adhesion to polyamide membranes was not as strongly inhibited by this detergent. This observation suggested that adhesion forces other than hydrophobic interactions might be involved in bacterial attachment to the polyamide membrane surfaces. These same researchers demonstrated that there was a clear effect of nutrient starvation on the ability of the different fouling bacteria to adhere to these membrane polymers. In general, starved cells were less adherent than nonstarved cells, suggesting that the physiological status of bacteria can influence their sorptive properties (Flemming, 1993; Flemming and Schaule, 1988a, 1988b, 1988c).

Flemming (1993) and Flemming and Schaule (1988a, 1988b) reported that cells of *Pseudomonas diminuta* that had been metabolically inactivated with peracetic acid attached with essentially the same kinetics as viable (untreated) cells to aromatic cross-linked polyamide RO membranes. Ridgway et al. (1984a) reported similar kinetics of biofilm formation on cellulose acetate membranes fed with pretreated secondary effluent receiving either 3 to 5 mg/L monochloramine or 10 to 15 mg/L monochloramine. For several months, no viable bacteria were detected by plate count methods in membrane biofilm scrapings from the membranes that received the highly chloraminated feedwater, suggesting that dead cells may colonize membranes nearly as rapidly as live cells. Thus, initial adhesion to membranes may be a largely passive process, dependent primarily on abiotic physicochemical interactions between cell surface ligands and the substratum.

These observations may be subject to other interpretations and therefore do not rigorously exclude a role for cell physiology in influencing initial attachment events. Indeed, Schaule (1992) and Schaule et al. (1990) reported slight inhibition of adhesion to polyethersulfone membranes after treating *Pseudomonas diminuta* cells with various antibiotic substances (gentomycin, chloramphenicol, etc.) or following inactivation by ultraviolet irradiation. Furthermore, nutrient-starved cells harvested from stationary-phase batch cultures were found to be significantly less adhesive to polysulfone, polyamide, or polyetherurea membranes than nonstarved log-phase cells, implying that the physiological status of the cell can be a significant determinant of initial attachment kinetics (Flemming, 1993; Flemming and Schaule, 1988a, 1988b).

It is perhaps noteworthy that other researchers have similarly found that physiologically impaired or dead bacteria (e.g., antibiotic-treated cells) are less adherent than active cells (Gismondo et al., 1983; Salit, 1992; Scheld et al., 1981). Paul (1984) demonstrated that cells of the marine bacterium *Vibrio proteolytica* were less adhesive to hydrophobic polystyrene surfaces following exposure of the bacteria to various metabolic inhibitors or antimicrobial agents such as azide, dinitrophenol, chloramphenicol, puromycin, azauridine, rifampin, *p*-chloromercuribenzoate, cephalothin, formaldehyde, or mercuric chloride. *Vibrio proteolytica* cells inactivated by heating at 85°C for 10 min were also less adhesive, although such treatment could have denatured or otherwise altered the conformation of cell surface macromolecules involved in the adhesion process. Fluorodeoxyuridine and nalidixic acid, which interfere with DNA replication, had little or no effect on *Vibrio proteolytica* attachment to polystyrene. Paul (1984) concluded that energy metabolism and protein synthesis (including transcription) were probably involved in some aspect of the adhesion process, whereas DNA synthesis was not.

Leslie and co-workers (1993) reported that bacterial cakes formed on MF membranes by the hydrophobic marine *Pseudomonas* sp. strain SW8 redispersed (by applied fluid shear) less effectively than those of a hydrophilic *Escherichia coli* strain. This observation is consistent with the findings of Ridgway et al. (1984c, 1985) regarding the enhanced adhesion of hydrophobic mycobacteria to RO membranes compared to hydrophilic *E. coli* and *Acinetobacter* strains. Yet Leslie et al. (1993) also demonstrated that cake stability was not exclusively determined by colloidal properties such as cell surface hydrophobicity or charge. In experiments with two populations of strain SW8, cake stability depended on the molecular composition of the cell surface, i.e., the nature of the EPS matrix. The redispersion of cakes of SW8, cultivated under carbon-limited conditions, decreased as the electrolyte concentration of the suspending medium increased; conversely, redispersion of cakes of SW8 cultivated under nitrogen-limited conditions was independent of the electrolyte concentration of the suspending medium. The observed differences in redispersion were not correlated with cell surface hydrophobicity as determined by hydrocarbon partitioning or with cell surface charge as measured by electrophoretic mobility. These observations imply that factors other than the bulk colloidal properties of the cell (i.e., hydrophobicity and charge) also contribute to adhesion and cake stability. A role for the molecular composition of the EPS matrix in cake stability was suggested in experiments using carbon-limited SW8 where differences between redispersion of carbon- and nitrogen-limited SW8 at low electrolyte concentrations was negated by substituting calcium for the sodium cation. It was hypothesized that the divalent cation caused bridging between polymers on adjacent cell surfaces in the cake, which prevented redispersion of the carbon-limited bacteria at low ionic strength. Interestingly, adhesion of carbon- and nitrogen-limited SW8 cells to MF membranes was reduced at low ionic strength, presumably due to electrostatic repulsion between the bacterium and the substratum.

Thus, different surface molecules may be responsible for adhesion and cake stabilization. The authors speculated that the specific molecular composition and/or conformation of the EPS matrix of strain SW8 determined the strength of cell-cell interactions in bacterial cakes. The physiological status of the cell (e.g., carbon or nitrogen limitation) was presumed to influence the EPS structure/conformation in strain SW8. The observations discredit the use of simple colloidal models based on inert particles to model and predict bacterial adhesion and fouling in MF and RO systems.

6.9 INHIBITION OF BACTERIAL ATTACHMENT TO MEMBRANES

Inhibition of bacterial attachment by physical or chemical means represents a potential strategy for controlling biofilm formation in membrane systems. As previously indicated, certain surfactants and other chemical agents interfere with bacterial attachment to synthetic polymer membranes (Ridgway, 1988; Ridgway et al., 1985, 1986a; Schaule, 1992; Schaule et al., 1990). Selected chemical substances which interfere with bacterial attachment to membrane polymers are listed in Table 6.3. Interestingly, many of the most effective inhibitors of bacterial adhesion to membranes are surfactants, such as Triton X-100, sodium dodecyl sulfate, and quaternary amines, which disrupt or otherwise weaken hydrophobic bonds; or chaotropic agents, such as dodecylguanidineacetate, which disrupt intramolecular hydrogen bonds involved in the stabilization of secondary and tertiary structure of proteins and other macromolecules involved in adhesion. These effects are consistent with the involvement of hydrophobic interactions in bacterial attachment to synthetic polymer membranes and other similar materials. The mechanism of inhibition of adhesion by proteolytic enzymes is less clear (Danielsson et al., 1977) but might involve proteolytic cleavage of hydrophobic surface polypeptides or exposure of hydrophobic moieties following removal of hydrophilic protein masking groups.

It may be possible to adapt or reformulate adhesion inhibitors (adhesion blocking agents) to serve as feedwater additives for the purpose of controlling membrane biofouling (see Sec. 6.12). Surfactants hold special promise as adhesion blocking agents in UF systems where many of the membrane polymers currently in use are hydrophobic (e.g., polysulfone, polyethersulfone, sulfonated polysulfone) and, as a consequence, tend to favor microbial attachment. A factor constraining the general use of surfactants for controlling microbial colonization and biofouling of RO membranes, however, is the occasional unfavorable interaction of the surfactant with the membrane polymer. Even neutral surfactants like Triton X-100, for example, react irreversibly with some polyamide TFC membranes, rapidly destroying their flux and/or mineral-rejection properties. Certain quarternary amine surfactants (e.g., benzalkonium chloride) exhibit similar behavior in which they bind irreversibly to membrane surfaces and dramatically diminish flux. Analysis of surfactant effects on bacterial adhesion in RO and other membrane systems should be undertaken, since the number of detergents so far examined represents a fraction of those available for exploration.

The Synperonic detergents described by Marshall and Blainey (1991) are somewhat unusual surfactants that interfere with bacterial attachment and retard surface biofouling of hydrophobic polymeric plastics. Synperonic detergents are water-soluble copolymers of poly[ethylene oxide] and poly[propylene glycol] that inhibit bacterial attachment to hydrophobic polystyrene surfaces under conditions of defined hydrodynamic shear. The Synperonic detergents rapidly sorb to hydrophobic surfaces placed in aqueous environments via a central hydrophobic section of the molecule which is symmetrically bracketed by two hydrophilic polymer moieties (*tail groups*). The hydrophilic tail groups, whose chain lengths can be readily varied, tend not to associate with the substratum, but rather extend freely into the aqueous phase for some distance. The inhibition of bacterial attachment caused by these agents is related to the overall length of the dual hydrophilic pendent groups which bind water molecules in an icelike fashion and prevent biopolymer (EPS) bridging to the substratum by the bacteria. Using a strain of *Pseudomonas*, Marshall and Blainey (1991) demonstrated that Synperonic F-108 impeded attachment to polystyrene for one to two days, but had little or no inhibitory effect on biofilm development over a period of several weeks. The lack of a long-term inhibitory effect on biofilm growth may reflect gradual desorption of the surfactant copolymer, masking by biotic and abiotic foulants, or surfactant biodegradation.

6.10 BACTERIAL ATTACHMENT TO DIFFERENT MEMBRANES

A given bacterium will exhibit different adsorption kinetics to different membrane polymers (Fig. 6.20) (Flemming, 1993; Flemming and Schaule, 1988a, 1988b, 1988c; Ridgway and Safarik, 1991; Schaule, 1992; Schaule et al., 1990). Thus, the physicochemical nature of the membrane polymer is an important and influential factor in initial bacterial attachment and early biofilm formation (Laîné et al., 1989). Manufacturers of RO membranes traditionally have not taken advantage of this fact and optimize membranes solely on the basis of water flux and solute rejection. The biofouling potential of the membranes is rarely, if ever, considered during membrane design and manufacturing.

Membrane polymer properties such as surface microtopography, polymer homogeneity, surface charge, charge distribution, surface free energy, membrane polymer density, and membrane porosity all influence bacterial affinity. Although more definitive data are needed, preliminary findings indicate that differences observed in bacterial attachment to RO membranes under laboratory conditions correlate qualitatively with the observed flux declines of membranes operated in pilot systems under actual field conditions (Flemming et al., 1993; Ridgway and Safarik, 1991). Thus, laboratory adhesion assays, which are considerably more rapid and inexpensive to perform than extended field trials, may be capable of predicting full-scale membrane performance. Indeed, rapid bacterial adhesion tests might be employed in the future to screen and rank new membrane polymers for their *biofouling potentials* (Flemming and Schaule, 1988a; Ridgway, 1988).

The relative influence of different membrane properties (membrane hydrophobicity, surface charge, etc.) on bacterial adhesion and biofouling kinetics remains unclear; thus, it is not possible to predict the potential for membrane biofouling based on such variables. The physicochemical composition of the feedwater, as well as the specific surface properties of the bacterial cell, are also critical variables in determining membrane biofouling potential.

Schaule (1992), Schaule et al. (1990), and Flemming and Schaule (1988a, 1988b) described the adhesion of three different bacteria to four different UF and RO membrane polymers. *Pseudomonas vesicularis, Acinetobacter calcoaceticus,* and *Staphylococcus warneri* readily attached to polysulfone, polyethersulfone, aromatic cross-linked polyamide, and polyetherurea membranes, with the latter membrane

TABLE 6.3 Selected Compounds Inhibiting Bacterial Adhesion to RO/UF Membranes and Related Synthetic Polymers

Substance	Concentration range	Bacteria/membrane	Reference
Dodecylguanidineacetate	0.05%	Adhesion of *Ps. vesicularis*, *Ac. calcoaceticus* to polysulfone partially inhibited; no inhibition using polyamide.	Flemming and Schaule, 1988a, 1988b, 1988c
Triton-X series (poly[oxyethylenephenoxy-N-oxide]*n*)	0.01–0.1%	• Inhibition of *Ac. calcoaceticus* to polysulfone by Triton-X-100; no inhibition of *Ps. vesicularis* or *Staphylococcus warneri* to polysulfone; adhesion inhibition of all strains tested with polyethersulfone or polyetherurea; no inhibition observed with polyamide. • Inhibition of *Mycobacterium* adhesion to cellulose acetate. Less inhibition observed as detergent chain length increased.	Flemming and Schaule, 1988a, 1988b Ridgway et al., 1984c, 1985, 1986a
Sodium bisulfite	0.1–1.0%	Little inhibition of adhesion of above bacterial strains to polysulfone, polyethersulfone, polyamide; some inhibition with polyetherurea.	Flemming and Schaule, 1988a, 1988b
Extremes of ionic strength	>1.0 molar	Partial inhibition of *Mycobacterium* observed at NaCl, LiCl, or LiBr concentrations of 5 molar.	Ridgway et al., 1984c, 1985
Synperonic F-108 (poly[ethylene oxide]*n* + poly[propylene glycol]*n* copolymer)	Monomolecular film	Inhibition of adhesion of *Pseudomonas* strain EK20 to hydrophobic polystyrene.	Marshall and Blainey, 1991
N,N-dimethylamide	Monomolecular film	Partial adhesion inhibition of several marine bacteria to hydrophobic polystyrene. Possible biofilm removal activity.	Marshall and Blainey, 1991
Polyethyleneimine	2.0%	Partial inhibition of *Mycobacterium* adhesion to cellulose acetate membrane.	Ridgway et al., 1984c, 1985

Agent	Concentration	Effect	Reference
Sodium dodecyl sulfate	0.1%	• Partial inhibition of *Mycobacterium* adhesion to cellulose acetate. • Strong inhibition of adhesion of *Ps. diminuta* to polyethersulfone.	Ridgway et al., 1984c, 1985 Schaule, 1992
Brij 35 (poly[oxyethene ether]*n*)	0.1%	Strong inhibition of adhesion of *Ps. diminuta* to polyethersulfone.	Schaule, 1992
Tween 20 (poly[ethoxysorbitanlaurate]*n*)	0.1%	Strong inhibition of adhesion of *Ps. diminuta* to polyethersulfone membrane.	Schaule, 1992
Tween 80	0.1%	Strong inhibition of adhesion of *Vibrio proteolytica* to hydrophobic polystyrene.	Paul and Jeffrey, 1985
Pluronic 64 (poly[propylene glycol ethoxylate]*n*)	0.1%	Strong inhibition of adhesion of *Ps. diminuta* to polyethersulfone.	Schaule, 1992
Teric PE series (poly[propylene glycol ethoxylate]*n*)	0.1%	Partial inhibition of adhesion of *Ps. diminuta* to polyethersulfone.	Schaule, 1992
Proteolytic enzymes (pronase, trypsin, chymotrypsin)	10 μg/mL	Partial inhibition of attachment of *V. proteolytica* to hydrophobic polystyrene.	Paul and Jeffrey, 1985
Cetyltrimethylammonium bromide	0.1%	Partial inhibition of attachment of *V. proteolytica* to hydrophobic polystyrene.	Paul and Jeffrey, 1985
N-lauroylsarcosinate	0.1%	Partial inhibition of *Mycobacterium* adhesion to cellulose acetate.	Schaule, 1992
Hexadecyltrimethylammonium bromide	0.1%	Partial inhibition of *Mycobacterium* adhesion to cellulose acetate membrane. Adhesion promoted by dodecylbenzyldimethylammonium chloride or dodecyltrimethylammonium bromide.	Ridgway et al., 1984c; Ridgway, 1988

6.37

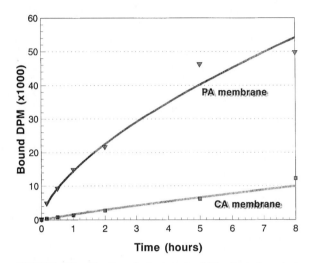

FIGURE 6.20 Kinetics of attachment of *Mycobacterium* strain BT2-4, a hydrophobic isolate, to two different polymer membranes, cellulose acetate (CA) blend and aromatic cross-linked polyamide (PA). The adhesion assay was performed using a phosphate buffer at pH 7.0 according to the method illustrated in Fig. 6.19. *y* axis indicates the number of attached cells per unit area of membrane surface expressed as disintegrations per minute (DPM).

type displaying the lowest bacterial affinity. The attachment of starved cells was less than that of unstarved bacteria, suggesting that the physiological status of the bacteria could influence their adhesion. Furthermore, starved bacteria colonized the membrane surfaces in a more patchy fashion compared to unstarved cells.

Ridgway and Safarik (1991) observed that the attachment kinetics of *Mycobacterium* strain BT2-4 was correlated with the hydrophobicity of the membrane polymer. Strain BT2-4 is involved in early biofilm development on cellulose acetate RO membranes used to demineralize pretreated secondary municipal effluent at Water Factory 21, California (Ridgway et al., 1984b). Mycobacteria adhered most rapidly to dimethylsilicone membranes, which exhibited an extremely hydrophobic surface. The least amount of adhesion was observed for comparatively hydrophilic cellulose acetate membranes or cellulose acetate membranes that contained a preestablished biofilm. It was also reported that mycobacterial adhesion was independent of the net negative charge associated with a series of polysulfone membranes containing different degrees of sulfonation (0 to 15 percent sulfate by weight). In contrast, amine substituted cellulose acetate membranes carrying a net positive charge and fully aromatic cross-linked polyamide (TFC) membranes stimulated mycobacterial attachment. Increased adherence to amine substituted membranes is consistent with the net negative surface charge of most bacteria and colloids (Cowan et al., 1992; Marshall, 1985, 1991; Marshall and Blainey, 1991; Marshall et al., 1971; Wicken, 1985).

6.11 PHYSIOLOGY OF MEMBRANE BIOFILMS

The metabolic state and activity of bacteria composing membrane biofilms determine the nature and extent of biofilm development and the effects of biofouling on membrane performance. Membrane biofilms are capable of assimilating exogenously supplied organic compounds (e.g., ^{3}H-glucose), indicating the bacteria composing the biofilms are metabolically active in situ (Ridgway et al., 1981). Membrane biofilms also contain high levels of adenosine 5'-triphosphate (ATP), indicating the presence of metabolically active cells (Ridgway et al., 1981, 1983). The amount of ATP in certain biofilms has been shown to be distributed asymmetrically through the depth of the biofilm, reflecting physiological differences in cells located near the top and bottom of the biofilm (Kinniment and Wimpenny, 1992). More recently, Ridgway et al. (1992), Rodriguez et al. (1992), Schaule et al. (1993a), and Stewart et al. (1994) applied the redox-sensitive dye 5-cyano-2,3-ditolyl tetrazolium chloride (CTC) to explore the electron-transport activity of biofilm microorganisms (Fig. 6.21). Oxidized CTC is nearly colorless and nonfluorescent; however, viable bacteria possessing a functional electron-transport (respiratory) system reduce CTC to the insoluble CTC-formazan (CTF), which fluoresces at 620 nm when excited with far blue light of 420 nm (see Seidler [1991] for a review of tetrazolium salt chemistry and reactions in biological systems). The CTC molecule competes with molecular oxygen for reduction by dehydrogenases (e.g., succinate dehydrogenase) or cytochromes (Kuhlman et al., 1989; Stellmach, 1984). The CTF is deposited intracellularly (similar to other nonfluorescent tetrazolium dyes such as INT) (Zimmermann et al., 1978), permitting localization of viable respiring bacteria in biofilms when examined by epifluorescence microscopy. The 3-D positions of CTC-reducing bacteria in these biofilms may be determined by using confocal microscopy. The accumulated CTF provides a stable chemical record of the respiratory history of the cell. Using CTC, active bacteria on optically opaque or translucent surfaces can be readily imaged.

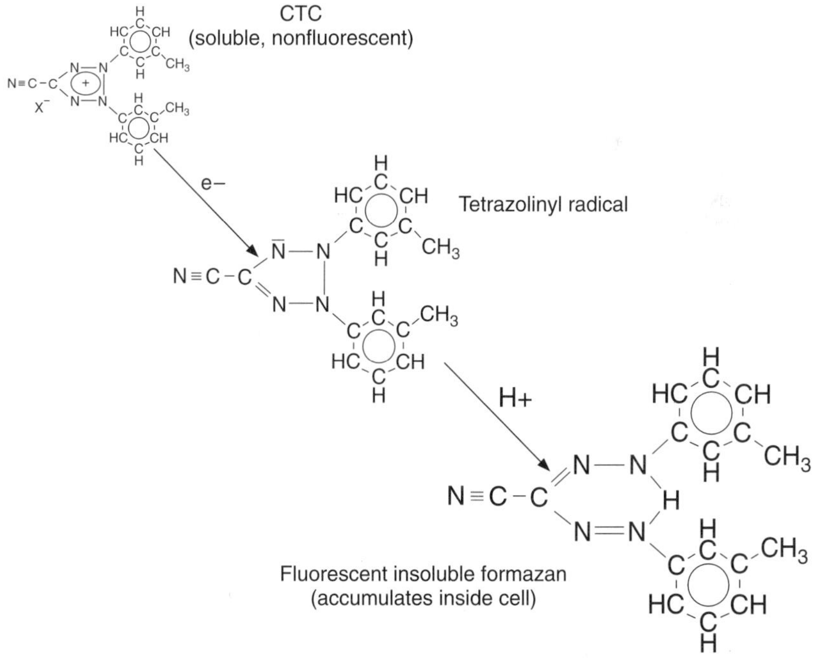

FIGURE 6.21 Schematic illustration of CTC reduction reactions. The oxidized form of CTC is soluble and nearly colorless. The reduced compound (formazan) is insoluble and deposits intracellularly as red fluorescent granules (refer to Seidler, 1991, for a review of tetrazolium chemistry).

Ridgway et al. (1992) demonstrated that brief nutrient starvation resulted in suppression of cellular respiration (and, hence, CTC-reducing activity) in nascent biofilms grown on cellulose acetate–coated coupons in the laboratory. Monospecies membrane biofilms composed of *Mycobacterium* strain BT2-4 were less susceptible to nutrient starvation than comparable *Pseudomonas putida* biofilms. As shown in Figs. 6.22 and 6.23, there was a dramatic decline in average mycobacterial cell respiration (i.e., the amount of CTF deposited per cell) as a function of the starvation time. Poorly respiring cells cannot process reduced carbon for energy production, resulting in slowing or cessation of growth. Thus, nutrient deprivation may represent a potentially useful strategy for suppression of biofilm growth in membrane systems.

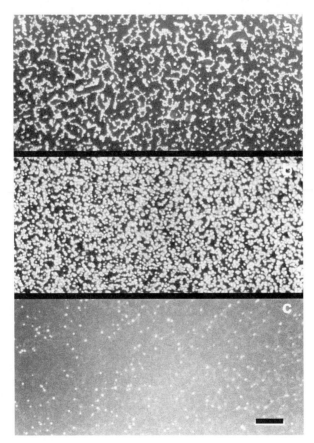

FIGURE 6.22   Influence of nutrient starvation on CTC reduction by *Mycobacterium* strain BT2-4 in biofilm. (*Panel a*) Biofilm on cellulose acetate–coated coupon surface stained with DAPI to indicate locations of all cells (respiring and nonrespiring). (*Panel b*) unstarved control preparation of same biofilm as in panel *a* showing distribution of respiring CTC-reducing bacteria. (*Panel c*) decrease in CTC-reducing cells in same biofilm following a 60-min starvation period prior to CTC addition. Refer to Rodriguez et al. (1992) and Schaule et al. (1993a) for CTC methods. Bar = 10 μm.

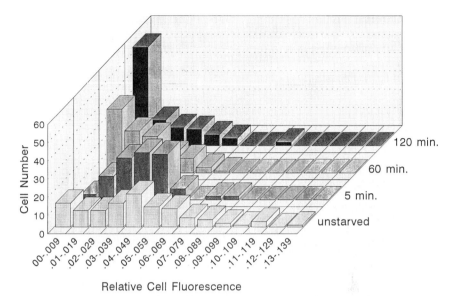

Relative Cell Fluorescence

FIGURE 6.23 Frequency histograms showing the respiratory activities of *Mycobacterium* strain BT2-4 cells in a simulated membrane biofilm before and after nutrient starvation for different times. For each experimental condition (i.e., 0 to 120 min of starvation prior to CTC addition), the relative amount of CTC formazan in 100 randomly chosen individual cells was determined by digital image analysis. Note that the longer the starvation period, the lower the proportion of cells that exhibit CTC reducing activity.

CTC has also been successfully applied for the rapid evaluation of the effects of chemical biocides on the respiratory activity of bacteria in biofilms reared on simulated membrane surfaces (polymer-coated coupons). In such experiments, CTC is applied following biocide treatment to determine the number and location of surviving bacteria within the biofilm. The series of DCM images presented in Fig. 6.24 indicates that the biocide isothiazolinone (Kathon; Rohm and Haas, Philadelphia) partially inhibited electron-transport activity in model biofilms composed of *Pseudomonas putida* strain 54g (similar data for *Mycobacterium* BT2-4 not shown). Physiological indicators such as CTC may prove useful in future investigations directed at identifying and evaluating new antimicrobial agents for the control of membrane biofouling.

Two other methods that have been employed to investigate in situ microbial activity in biofilms include (1) ATR-FTIR (Bremmer and Geesey, 1991; Suci et al., 1994) to noninvasively track surface biofilm development and EPS biosynthesis (see Sec. 6.3 and Figs. 6.8 and 6.9) and (2) the introduction of plasmid vectors harboring reporter genes (e.g., beta-galactosidase) placed under the control of promoter sequences involved in the regulation of EPS biosynthesis into specific biofilm microbes (Davies et al., 1993). Suci et al. (1994) described the use of ATR-FTIR spectrometry for monitoring in situ biofilm growth rates and the transport kinetics of ciprofloxacin, a fluoroquinoline antibiotic, into *Pseudomonas aeruginosa* biofilms. Shifts in the IR vibrational modes of biofilm DNA/RNA indicated possible interactions of ciprofloxacin with these macromolecular targets. Both the ATR-FTIR and reporter gene approaches are nondestructive and provide in situ response measure-

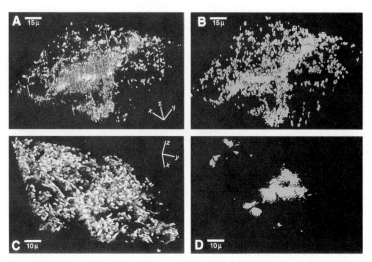

FIGURE 6.24   Series of DCM images of a membrane biofilm (*Pseudomonas putida* strain 54g) before and after treatment with the biocide isothiazolinone (0.1 percent wt/vol) (Kathon; Rohm and Haas). (*Panel A*) Control (untreated) biofilm stained with DAPI to indicate 3-D locations of all cells in biofilm. (*Panel B*) Same image area as in panel *A* showing 3-D locations of CTC-reducing cells. Note that most cells are actively respiring. (*Panel C*) DAPI stained biofilm image after 20-min exposure to biocide. (*Panel D*) Same image area as in panel *C* showing 3-D distribution of CTC-reducing cells after biocide treatment. Note that most cells have been inactivated by the biocide.

ments of EPS biosynthesis under different environmental conditions. Results of these studies indicate that perturbation of the chemical microenvironment (e.g., by nutrient depletion or antibiotic addition) results in significant changes in the physiological responses of biofilm microorganisms.

6.12   BIOFOULING CONTROL STRATEGIES

Control of membrane biofouling is necessary not only during continuous plant operation, but also during extended periods of plant inactivity due to system repair or modifications. Biofouling must also be controlled when newly manufactured membrane modules are packaged and stored for long periods prior to shipping or installation.

Potential points of intervention in the membrane biofouling process are shown schematically in Fig. 6.25 and summarized in Table 6.4. Biofouling control strategies fall into four general categories: (1) selection and optimization of effective feedwater pretreatment methods, (2) optimization of effective membrane cleaning methods or other biofilm modifications, (3) optimization of system operating pressure and recovery, and (4) selection of the most appropriate membrane. Feedwater pretreatment by high-pH lime clarification, chemical coagulation, multimedia filtration, cartridge microfiltration, and biocide applications impedes biofouling and typically extends membrane life, but these measures also escalate O&M costs (Bates et al., 1988; Bettinger, 1981; Brandt et al., 1993; Nusbaum and Argo, 1984).

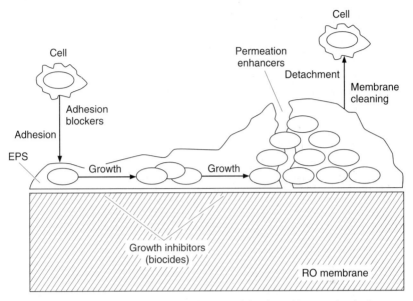

FIGURE 6.25 Schematic illustration showing potential points of intervention in the membrane biofouling process (see text for further explanation).

TABLE 6.4 Potential Points of Intervention in the Membrane Biofouling Process

Intervention point or process	Approach or strategy
Bacterial adhesion	Addition to feedwater of biodispersants, surfactants, chaotropic agents, chelating agents, biocides, and other chemical compounds or formulations that interfere with bacterial attachment and colonization processes.
Growth and cell division	Introduction of nonmetabolizable nutrient analogs, nutrient limitation (starvation), application of antimicrobial agents, biocides, electrical fields, or other conditions or substances that retard biofilm growth.
Feedwater bacteria	Use of physicochemical processes such as continuous microfiltration, affinity columns, bioflocculants, etc., that physically remove or precipitate bacteria from the feedwater.
Biofilm structure	Treatment of biofilm with physicochemical cleaning agents, chelating agents, chaotropic agents, ultrasound, dispersants, electric fields, air backwashes (in the case of MF), or other agents/conditions which tend to disrupt biofilm integrity.
Biofilm permeation	Addition of physicochemical agents/conditions or application of processes which enhance biofilm permeability to water and solute molecules; avoids need to remove biofilm.
Membrane polymer	Design and synthesis of new membrane polymers with low affinity surfaces, surfaces that can be regenerated in situ, membranes with delayed-release antimetabolites or biocides incorporated, chlorine-resistant membranes, etc.

The most widely employed biofouling control measures include the use of disinfectants (Alashri et al., 1992; Brandt et al., 1993; Characklis, 1990b; Flemming, 1993; Flemming et al., 1992b; LeChevallier, 1987, 1991; LeChevallier et al., 1984, 1988, 1990; McCoy, 1987; Rowley, 1992; Saad, 1992), membrane cleaning agents (Amjad, 1989; Amjad et al., 1993; Flemming, 1993; Whittaker et al., 1984), and special liquid- or air-backwash procedures in the case of MF systems. Recent experimental evidence indicates certain commercial membrane cleaning agents increase the water permeation of artificial "biofilms" (e.g., hydrated agarose gels) as well as bacterial filter and sludge cakes (Flemming et al., 1992a; Leslie et al., 1993). In these experiments, the agarose gel or filter cake was not removed by the cleaning agent, but its water permeation characteristics were improved. Thus, in practice, it may be possible to increase biofilm permeation in situ without having to actually remove the biofilm. An increase in biofilm permeability would allow better penetration of chemical biocides to target cells buried within the EPS matrix.

Dhar et al. (1981) and Gordon et al. (1981) observed an antibacterial effect of applied electric currents at surfaces. More recently, Blenkinsopp et al. (1992) demonstrated that laboratory-grown biofilms of *Pseudomonas aeruginosa* are most susceptible to certain antimicrobial agents (e.g., antibiotics such as Tobramycin) when they are exposed to reversing direct-current (DC) electric fields. While the exact mechanism of the enhanced susceptibility has not been fully elucidated, it appears that the DC electric field may greatly increase biofilm permeability to solutes, including antimicrobial agents. It is probably safe to speculate that a similar approach might someday be adapted for biofouling control in membrane systems.

Lozier and Sierka (1985) investigated the use of ozone and ultrasound to reduce membrane biofouling. Zips et al. (1990) applied ultrasonic energy to physically remove attached cells of *Pseudomonas diminuta* from modified polysulfone UF membranes. Bacterial detachment was found to depend on the ultrasonic intensity, exposure time, and the distance and position of the transducer relative to the bio-fouled membrane surface. Biofilm cells detached according to a first-order exponential decay model. More than 95 percent biofilm removal was achieved within 60 s at an applied ultrasonic energy of 2 W and a frequency of 38 kHz (biofilm positioned 3 cm distance from transducer). The authors speculated that bacterial attachment and membrane biofouling might be impeded by the application of continuous or intermittent acoustic energy before biofilm formation.

Nutrient starvation represents a potentially effective strategy for inhibiting biofilm development on membranes. However, biofilm formation presumably will occur to a limited extent even under the most stringent starvation conditions, since starved cells could be expected to adhere to membrane surfaces (especially if operated in a dead-end filtration mode), though perhaps not as rapidly or tenaciously as unstarved cells (Flemming and Schaule, 1988a, 1988b; Schaule, 1992; Schaule et al., 1990). Feedwater pretreatment by granular activated carbon (GAC) effectively removes trace organic nutrients; however, GAC particles (with their sorbed layer of organics) provide a favorable habitat for biofilm formation (Burlingame et al., 1985; Camper et al., 1986). Once colonized, bacteria are shed in high numbers from GAC systems, and these cells may contaminate downstream processes (Geldreich et al., 1985).

Little research has been reported on the design of anti-biofouling membranes, possibly because of a general lack of understanding concerning the molecular mechanisms of bacterial attachment to synthetic separation membranes. Yet, it is known that bacteria sometimes display large differences in their affinity for different membrane polymers (see Fig. 6.20). In certain instances, the actual field performance of RO membranes appears to be correlated with their relative bacterial affinities as determined in the laboratory (Ridgway and Safarik, 1991). Such observations suggest that it may be possible to limit biofouling by designing new membrane polymers or surface coatings that retard bacterial attachment and subsequent colonization.

The application of chemical cleaning agents is perhaps the most commonly practiced means to combat biofouling in membrane systems. Variations in cleaning effectiveness can be achieved by altering (1) the formulation of the cleaning solution, (2) the frequency of cleaning, and (3) the cleaning protocol. Cleaning solutions specifically formulated for biologically fouled membranes usually contain one or more surfactants (detergents) (Table 6.5). The purpose of the detergents is to denature macromolecules (cell wall polypeptides, lipopolysaccharides, EPS, etc.) involved in maintaining cell envelope integrity and/or stabilizing biofilm EPS structure. In addition, surfactants assist in disrupting or otherwise weakening hydrophobic interactions involved in the adhesion of bacteria to membrane surfaces. There are more than 100 different detergents that are commercially available. Which type is selected for incorporation into a particular cleaning solution is critical, since some detergent/membrane combinations are incompatible and will result in irreversible loss of flux or solute rejection properties. For example, the nonionic polyoxyethylene n-oxide detergents (e.g., Triton X-100) are incompatible with TFC aromatic cross-linked polyamide membranes. Use of this surfactant with polyamide TFC membranes causes a rapid and irreversible loss of membrane flux, presumably due to strong hydrophobic bonding of the surfactant to the membrane. Cationic surfactants such as quaternary ammonium compounds (*quats*) exhibit similar bonding behavior to TFC membranes, which is unfortunate since many of the quats exhibit excellent microbicidal activity (Block, 1991b) and inhibit bacterial attachment (Paul and Jeffrey, 1985; Ridgway et al., 1986a; Ridgway, 1988). The positively charged quaternary amine group of benzalkonium chloride (dodecylbenzyldimethylammonium chloride) and related quats undergo strong ionic interactions with negatively charged carboxyl and carbonyl groups in the polyamide backbone. Secondary hydrophobic interactions with the membrane polymer involving the hydrocarbon moiety of the detergent may serve to further reinforce the ionic bonding. Reaction of quaternary amines with the membrane surface causes a dramatic and irreversible reduction in water flux and solute rejection. The bound quat cannot be removed by high concentrations of lithium chloride or other salts, even when these are applied in the presence of anionic surfactants such as sodium dodecyl sulfate, which does not affect membrane function.

In general, anionic surfactants are compatible with the most commonly used separation membranes, such as cellulose acetate and polyamide. Examples of inexpensive commercially available anionic detergents include sodium dodecyl sulfate (SDS) and sodium dodecyl benzenesulfonate (SDBS). These anionic detergents typically exhibit less biocidal activity than quaternary ammonium compounds, but they are still very effective agents for the disruption of cytoplasmic membrane structure and function in many, but not all, bacteria (Block, 1991a, 1991b; Dychdala and Lopes, 1991; Merianos, 1991; Moriyon and Berman, 1982), especially when they are used at elevated temperatures (>30°C). Mycobacteria and bacterial spores are typically resistant to these detergents. Detergents such as SDS and SDBS are generally most effective within the concentration range of about 0.01 to a few percent by weight. Destabilization of cell envelopes by detergents may be enhanced by the addition of EDTA or related divalent cation chelating agents (Block, 1991a).

Amphoteric (zwitterionic) detergents such as phosphatidylethanolamine, lecithin, betaines (e.g., N-dodecyl-N-N-dimethylglycine), and the CHAPS series have not yet been applied as cleaning agents for membranes. However, because they possess both positive (amine) and negative (carboxyl) ligands, their degree of effective-

TABLE 6.5 Some Generic Components in Cleaning Solutions Designed to Treat Biofouled Separations Membranes

Component	Examples	Membrane compatibility*	Concentration[†]	Function/comment
Detergent	SDS SDBS Triton series Quaternary amines Many others	CA, PA, PS CA, PA, PS CA, PS, PE CA, PS, PE	0.01–2.0%	Detergents function to (1) disrupt hydrophobic bonds involved in bacterial attachment and EPS stabilization, and (2) denature proteins and other macromolecules involved in stabilizing cell envelopes of bacteria. Detergents weaken biofilm structural integrity and exhibit biocidal activity in many instances (e.g., the quats). Compatibility of untested detergents with membranes must be determined empirically.
Chaotropic agents	Urea Guanidium HCl	CA, others unknown CA, others unknown	6–8 molar 1–2 molar	Chaotropic agents disrupt hydrogen bonds involved in stabilizing secondary/tertiary structure of biopolymers. Most effective when used in combination with denaturing agents such as SDS and/or chelating agents such as EDTA.
Chelating agent	Citrate EDTA	All All	0.1–1.0% 0.1–1.0%	Chelate divalent cations such as calcium and magnesium, which are involved in stabilizing cell envelopes of bacteria. EDTA also exhibits broad spectrum biocidal activity. Chelating agents are especially effective when applied in combination with detergents and other antimicrobial agents and enzymes.
Enzyme(s)	Proteases Esterases Lipases Polysaccharidases Lysozyme	All	10–100 mg/L 10–100 mg/L 10–100 mg/L 10–100 mg/L	A combination of enzymes may assist in breaking down the biofilm EPS matrix which typically consists of complex heteropolysaccharides, lipopolysaccharides, and proteins.
Biocide	Sodium bisulfite Quaternary amines Formaldehyde Glutaraldehyde Isothiozolinone (Kathon) Sodium benzoate Monochloramine EDTA Many others	All CA, PS, PE All All All CA, PA, PS All	10–100 mg/L 0.1–1.0% 0.1–5.0% 0.1–5.0% 0.1–1.0% 0.5–5.0 mg/L 0.1–1.0%	Broad spectrum microbial biocides are useful in inhibiting or slowing regrowth of biofilm microorganisms following chemical cleaning of membranes. Compatibility of untested biocides with membranes must be determined empirically.

* CA = cellulose acetate; PA = polyamide; PS = polysulfone; PE = polyetherurea.
[†] Percentages are wt/vol.

ness and compatibility with different membranes might be adjusted by changing the solution pH accordingly.

Whittaker et al. (1984) demonstrated that a solution of urea and SDS resulted in greater biofilm removal than either compound alone. The compounds were tested against native mixed-species biofilms formed on the surfaces of cellulose acetate RO membranes operated on a pretreated municipal wastewater feed at Water Factory 21 in southern California. High concentrations of urea and related compounds (e.g., guanidinium hydrochloride) disrupt hydrogen bonds involved in stabilizing the secondary and tertiary structure of proteins and other exposed biopolymers. These so-called chaotropic agents enhance the denaturing activity of detergents such as SDS, possibly by making detergent binding sites more accessible.

Other components which may be added to membrane cleaning solutions include enzymes (e.g., proteases, esterases, polysaccharidases, lysozyme) to aid in breaking down the EPS matrix and cell walls of bacteria (Sutherland, 1990; Whittaker et al., 1984), divalent cation chelating agents such as EDTA for weakening cell envelope integrity, and chemical biocides for inhibiting cell proliferation during or following cleaning.

A typical membrane cleaning protocol consists of (1) an initial 10- to 20-min membrane rinse with product water, (2) recirculation of the cleaning solution at maximum system differential pressure (usually about 60 lb/in^2) for one to several hours at elevated temperature (30 to 40°C), (3) an extended soaking period lasting from one hour to overnight, (4) final recirculation for 20 to 30 min, and (5) a final rinse with permeate (10 to 20 min). The pH of the cleaning solution should be adjusted to a value that is compatible with the type of membrane being used (pH 6 to 8 for cellulose acetate; pH 3 to 12 for polyamide and polysulfone). The cleaning solution should be continuously filtered for solids removal during all recirculation events.

Stenstrom et al. (1982) investigated the use of combined chemical (citric acid, enzyme detergent) and physical (spongeball) cleaning on tubular cellulose acetate RO membranes used to demineralize pretreated municipal wastewater (trickling filter effluent). Interestingly, enzyme detergent-spongeball cleaning without citric acid was relatively ineffective. It may be speculated that citric acid helped to solubilize or chelate iron salt precipitates from the membrane surface which could not be removed effectively solely by physical processes.

Another factor which influences the effectiveness of membrane cleaning is the application frequency. If the cleaning frequency is too low, the biofilm will have an opportunity to completely regrow and stabilize before the next cleaning cycle, thereby extending the time of suboptimal system operation. Low cleaning frequencies (e.g., once every few months) are also less effective because older biofilms are typically more difficult to remove via chemical cleaning than nascent biofilms (Argo and Ridgway, 1982), presumably due to a thicker biofilm and more extensive EPS biosynthesis. Thus, cleaning at very frequent intervals (e.g., every few weeks), that is, well before extensive biofilm/EPS accumulation, is generally preferable and will tend to maximize system performance and membrane/module life. Criteria used to establish the optimum cleaning frequency for any given system include (1) the observed flux decline kinetics, (2) the decline in mineral rejection, (3) the increase in the module differential pressure, and (4) the rate of biofilm development as determined by regular membrane autopsy or coupon analysis.

A wide range of potentially effective disinfectants and microbicidal agents (biocides) are available for inhibiting the growth of microorganisms on membrane surfaces, although most have not been tested for their compatibility with membrane systems (see Block, 1991a). As a general rule, chemical biocides are superior to ultraviolet (UV) irradiation as feedwater disinfectants, since the latter does not provide a continuous biocide residual at the aqueous/membrane interface where the biofilm develops. Thus, if a single feedwater bacterium escaped UV disinfection, it could result in extensive biofouling following attachment to the membrane surface. Key issues concerning biocide use for membrane systems are (1) the microbicidal effectiveness of the biocide, particularly against biofilms; (2) the short- and long-term compatibility of the biocide with the synthetic membrane polymer and other module components (e.g., polyurethane glues, plastics, rubber O-ring seals); (3) the chemical stability and breakdown products of the biocide; (4) the susceptibility of the biocide to biotransformation and/or biodegradation; (5) the toxicity of the biocide for humans and the environment; (6) the ability of the biocide to penetrate the membrane polymer to effect disinfection on both sides of the membrane; (7) ease of biocide handling and disposal issues; and (8) cost of the biocide.

Biocides may be broadly categorized as either oxidizing or nonoxidizing (Table 6.6). Examples of oxidizing-type biocides that are useful for membrane systems include free chlorine, monochloramine, chlorine dioxide, iodine, peracetic acid, ozone, N-halo compounds such as 1-bromo-3-chloro-5,5-dimethylhydantoin, and the recently developed N-halamines such as 3-chloro-2,2-dialkyl-4,4-dimethyl-1,3-oxazolidine and trichloromelamine (Worley and Williams, 1988). Nonoxidizing biocides include aldehydes (e.g., formaldehyde, glutaraldehyde), sodium bisulfite, EDTA, benzoic acid, organobromine compounds (e.g., 2,2-bromo-3-nitropropionamide), organosulfur compounds (e.g., 2-methyl-4-isothiazolin-3-one), organonitrogen compounds (e.g., quaternary amines, dodecylguanidine hydrochloride) (Block, 1991b), organometallics (e.g., tributyltin oxide), and metal salts such as copper sulfate (see Block [1991a] for a comprehensive review of biocide categories and industrial applications).

The use of potent broad-spectrum biocides, such as the free and combined forms of chlorine, to control biofouling on membranes and other surfaces (LeChevallier, 1987, 1991; LeChevallier et al., 1988, 1984, 1990; McCoy, 1987) is becoming less acceptable because of the potential for disinfection by-product formation, oxidative damage to the membranes (Glater and McCray, 1983; Glater et al., 1981, 1983), stricter environmental regulations, and associated transport and handling hazards, especially in urban settings. For these reasons, there has been much interest in recent years in evaluating the effectiveness of food- and cosmetic-type preservatives (EDTA, sorbate, benzoate, etc.) for membrane applications.

There are two main applications of biocides in membrane systems. First, a biocide may be added continuously or intermittently (e.g., via a metering pump) to the system feedwater in an attempt to suppress or otherwise control the unrestricted growth of biofilm microorganisms on the membrane surfaces. Inexpensive biocides that are fully compatible with the membrane over extended periods of contact time are most appropriate in this type of application. Examples of effective feedwater biocides are monochloramine for cellulose acetate membranes and sodium bisulfite for polyamide TFC membranes. A second general application of biocides is in the preservation of the polymer membranes and related module components (e.g., glues, plastic spacers, other materials of construction) during extended periods of membrane storage or plant shutdown. Here, too, the biocide selected must be fully compatible with the membrane polymer and other system components, since contact times are typically long (weeks to years). However, biocide cost becomes a secondary consideration, since there is typically only a single application of the compound (e.g., a solution of the biocide may be pumped into pressure vessels prior to plant shutdown). Other considerations for membrane storage are the safety and ease of handling of large volumes of the biocide solu-

TABLE 6.6 Disinfecting Agents Commonly Used to Control Biofouling in Membrane Systems

Biocide category	Examples	Useful concentration range	Membrane compatibility	Comments
Oxidizing	Chlorine	0.1–1.0 mg/L	CA, PS	The oxidizing biocides listed are used primarily as feedwater additives. Because monochloramine exhibits reduced oxidizing activity compared to free chlorine, it does not harm performance of PA membranes. It is an excellent biocide, especially with respect to biofilms (see text).
	Monochloramine	0.5–5.0 mg/L	All	
	Peracetic acid	0.1–1.0 mg/L	CA, PS	
	Hydrogen peroxide	0.1–1.0 mg/L	All	
	Iodine			
Nonoxidizing	Formaldehyde	0.5–5.0%	All	Sodium bisulfite is the only nonoxidizing biocide currently used as a feedwater additive. All others listed are used primarily to preserve membranes from biodegradation/biodeterioration during plant inactivity.
	Glutaraldehyde	0.5–5.0%	All	
	Bisulfite	1.0–100 mg/L	All	
	2-methyl-4-isothiazolin-3-one	0.01–1.0%	All	
	Quaternary amines	0.01–1.0%	CA, PS	
	Benzoate	0.1–1.0%	All	
	EDTA	0.01–1.0%	All	
Irradiation	Ultraviolet	1–2 MR*	All (conditional)[†]	Disinfection by irradiation is very effective but leaves no biocide residual following exposure; thus, surviving bacteria may regrow on membrane surfaces. Gamma irradiation is particularly suited for disinfection of new membrane modules (sealed in plastic wrap) prior to long-term storage.
	Gamma		All (conditional)	

* MR = megarads.
[†] Compatibility depends on radiation dosage, temperature, pH, redox potential, and other factors.

tion, as well as the ability to chemically or physically inactivate the biocide prior to disposal.

Increased operating pressure (higher flux) results in greater biofouling kinetics in RO systems. This relationship is partly due to increased mass transport (including microorganisms) toward the membrane surface at higher flux values. However, an additional factor is that at higher flux values nutrients become more concentrated near the membrane surface, which, in turn, accelerates biofilm growth. Thus, biofouling may be retarded to some extent by consistently operating the membranes at the lowest possible flux or system recovery.

There are sometimes large performance differences when different membrane types are used to treat the same feedwater. For example, polyamide membranes fed with pretreated municipal wastewater experience more rapid and extensive biofouling than cellulose acetate membranes (Ridgway and Safarik, 1991). Furthermore, the biofilm that develops is bound more strongly to the polyamide surface, decreasing the effectiveness of chemical cleaning agents. A similar relationship is observed when the bacterial adhesion characteristics of these membranes are compared under laboratory conditions, i.e., mycobacteria attach more rapidly to the polyamide membrane than to the cellulose acetate membrane. Thus, it is essential that the appropriate membrane is selected to minimize biofouling. Because it is not yet possible to reliably predict membrane biofouling potentials solely from laboratory tests, it is usually necessary to compare different membrane types, module configurations, and pretreatment alternatives in pilot-scale tests using the actual site feedwater before making a final selection.

6.13 CONCLUDING REMARKS AND FUTURE TRENDS

Biofouling is a pervasive membrane systems problem. The economic impact on plant performance can be substantial (Paul, 1991; Taylor et al., 1989; Durham, 1989; Pittner, 1993a) and the useful life of the membranes is often dramatically reduced. The biofouling process involves adhesion and growth of microorganisms on the membrane surface; however, very little is understood about the fundamental nature of the adhesion and growth processes. Additional research in these areas could lead to the identification and development of novel biofouling control strategies. For example, many bacteria compete with one another for surface adhesion sites (Belas and Colwell, 1982; Bibel et al., 1983); thus, it may be possible to direct biofilm development by the addition of selected bacteria to the feedwater that can be subsequently more easily removed by chemical cleaning than the native biofilm. The influence of the membrane as substratum also plays an important role in the rate and extent of biofouling. The observation that bacterial attachment and colonization occurs more rapidly on some membranes than others suggests that it may be possible to create a new generation of membrane polymers that combine superior flux and solute rejection properties with low biofouling potential. Efforts to create anti-biofouling membranes by applying monolayer assemblies of fluorinated pyridimium bromide have met with some success in laboratory experiments (Speaker, 1985; Speaker and Bynum, 1983). Certain quaternary ammonium salts can be immobilized on surfaces and still retain antimicrobial activity (Nakagawa et al., 1984a, 1984b). Kawaguchi and Tamura (1984) described the synthesis of novel halogen-resistant polyamide membranes which could be used in the presence of continuous chlorine dosage to suppress biofilm formation. Thus, it may be possible to design and fabricate new membrane polymers that are not only optimized with respect to flux and rejection properties, but also retard adhesion and biofilm growth by virtue of their surface chemistries.

ACKNOWLEDGMENTS

The authors wish to acknowledge Dr. Greg Leslie for contributing the DCM images used in this chapter, Dr. Kenneth Ishida for contributing the ATR-FTIR spectra of biofilm growth on membrane polymer-coated IREs, and Ms. Grisel Rodriguez, Ms. Andrea Kern, Mrs. Hanna Rentscher, and Ms. Jana Safarik for their contributions concerning the CTC reactions of nutrient-deprived biofilms. Don Phipps kindly proofread the manuscript. These individuals are members of the Biotechnology Research Department, Orange County Water District, Fountain Valley, California, and of the Institut für Siedlungswasserbau, Wassergüte, und Abfallwirtschaft, University of Stuttgart, Germany. We also thank Mr. Katsumi Ishiguro, Kurita Water Industries, Tokyo, Japan, for providing expert technical assistance with some of the bacterial attachment studies and Park Scientific Instruments, Inc., Sunnyvale, California, for use of the AFM image of bacteria on a silicon surface.

REFERENCES

Ahmed, S. R., M. S. Alansari, and T. Kannari, 1989, "Biological Fouling and Control at Ras Abu Jarjur RO Plant: A New Approach," *Desalination* **74**:69–84.

Alashri, A., M. Valverde, C. Roques, and G. Michel, 1992, "Sporocidal Properties of Peracetic Acid and Hydrogen Peroxide, Alone and in Combination, in Comparison with Chlorine and Formaldehyde for Ultrafiltration Membrane Disinfection," *Can. J. Microbiol.* **39**:52–60.

Allegrezza, A. E., 1988, "Commercial Reverse Osmosis Membranes and Modules," in B. S. Parekh (ed.), *Reverse Osmosis Technology: Applications for High-Purity Water Production,* Marcel Dekker, Inc., New York, pp. 53–120.

Allison, D. G., and I. W. Sutherland, 1984, "A Staining Technique for Attached Bacteria and Its Correlation to Extracellular Carbohydrate Production," *J. Microbiol. Methods* **2**:93–99.

Amjad, Z., 1989, "Advances in Membrane Cleaners for Reverse Osmosis Systems," *Ultrapure Wat.* **6**:38–42.

Amjad, Z., J. Hooley, and J. Pugh, 1991, "Reverse Osmosis Failures: Causes, Cases, and Prevention," *Ultrapure Wat.* **8**:14–21.

Amjad, Z., J. D. Isner, and R. C. Williams, 1988, "Reverse Osmosis: The Role of Analytical Techniques in Solving Reverse Osmosis Fouling Problems," *Ultrapure Wat.* **5**:20–26.

Amjad, Z., K. R. Workman, and D. R. Castete, 1993, "Considerations in Membrane Cleaning," in Z. Amjad (ed.), *Reverse Osmosis: Membrane Technology, Water Chemistry, and Industrial Applications,* Van Nostrand Reinhold, New York, pp. 210–236.

Amy, P. S., and R. Y. Morita, 1983, "Starvation-Survival Patterns of Sixteen Freshly Isolated Open-Ocean Bacteria," *Appl. Environ. Microbiol.* **45**:1109–1115.

Applegate, L. E., C. W. Erkenbrecher, and H. Winters, 1989, "New Chloramine Process to Control Aftergrowth and Biofouling in Perasep B-10 RO Surface Seawater Plants," *Desalination* **74**:51–67.

Argo, D. G., and H. F. Ridgway, 1982, "Biological Fouling of Reverse Osmosis Membranes at Water Factory 21," *Proc. 10th Annual Conference and Trade Fair of the Water Supply Improvement Association,* vol. III, Special Interest Presentations, July 25–29, 1982, Honolulu, Hawaii.

Augusta, J., R. J. Mueller, and H. Widdecke, 1992, "Biodegradation and Biocorrosion of Synthetic Materials: A Comparison Using *Pseudomonas aeruginosa* as Test Microorganism," *Chem. Eng. Technol.* **64**:67–68.

Bailey, D. A., K. Jones, and C. Mitchell, 1974, "The Reclamation of Water from Sewage Effluents by Reverse Osmosis," *J. Wat. Pollu. Contrl. Fed.* **73**:353–364.

Bates, W. T., B. L. Coulter, and D. J. Thomas, 1988, "Pretreatment Guidelines for Reverse Osmosis," *Ultrapure Wat.* **5**:34–41.

Belas, M. R., and R. R. Colwell, 1982, "Adsorption Kinetics of Laterally and Polarly Flagellated *Vibrio," J. Bacteriol.* **151**:1568–1580.

Bendinger, B., H. H. M. Rijnaarts, K. Altendorf, and A. J. B. Zehnder, 1993, "Physicochemical Cell Surface and Adhesive Properties of Coryneform Bacteria Related to the Presence and Chain Length of Mycolic Acids," *Appl. Environ. Microbiol.* **59**:3973–3977.

Bergman, R. A., H. W. Harlow, and P. E. Laverty, 1988, "Characterizing and Controlling Microbial Activity at the Englewood RO Plant," *Proc. National Water Supply Improvement Association,* July 31–August 4, San Diego, Calif.

Bettinger, G. E., 1981, "Controlling Biological Activity in a Surface Water Reverse Osmosis Plant," Du Pont Corp. report, Haskell Laboratory for Toxicology and Industrial Medicine, Wilmington, Delaware, 1981, pp. 1–9.

Bibel, D. J., R. Aly, C. Bayles, W. G. Strauss, H. R. Shinefield, and H. I. Maibach, 1983, "Competitive Adherence as a Mechanism of Bacterial Interference," *Can. J. Microbiol.* **29**:700–703.

Blenkinsopp, S. A., A. E. Khoury, and J. W. Costerton, 1992, "Electrical Enhancement of Biocide Efficacy against *Pseudomonas aeruginosa* Biofilms," *Appl. Environ. Microbiol.* **58**:3770–3773.

Block, S. S., 1991a, *Disinfection, Sterilization, and Preservation,* Lea and Febiger, Philadelphia.

Block, S. S., 1991b, "Surface-Active Agents: Amphoteric Compounds," in S. S. Block (ed.), *Disinfection, Sterilization, and Preservation,* Lea and Febiger, Philadelphia, pp. 263–273.

Brandt, D. C., G. F. Leitner, and W. E. Leitner, 1993, "Reverse Osmosis Membranes State of the Art," in Z. Amjad (ed.), *Reverse Osmosis: Membrane Technology, Water Chemistry, and Industrial Applications,* Van Nostrand Reinhold, New York, pp. 1–36.

Bremmer, P. J., and G. G. Geesey, 1991, "An Evaluation of Biofilm Development Utilizing Nondestructive Attenuated Total Reflectance Fourier Transform Infrared Spectroscopy," *Biofouling* **3**:89–100.

Bremmer, P. J., G. G. Geesey, and B. Drake, 1992, "Atomic Force Microscopy Examination of the Topography of a Hydrated Bacterial Biofilm on a Copper Surface," *Curr. Microbiol.* **24**:223–230.

Bryers, J. D., 1993, "Bacterial Biofilms," *Curr. Opinion Biotechnol.* **4**:197–204.

Buchanan, C. M., R. M. Gardner, R. J. Komarek, S. C. Gedon, and A. W. White, 1993, "Biodegradation of Cellulose Acetates," in C. Ching, D. L. Kaplan, and E. L. Thomas (eds.), *Biodegradation of Polymer Packaging,* Technomic, Lancaster, Pa., pp. 133–140.

Burlingame, G. A., I. H. Suffet, and W. O. Pipes, 1985, "Predominant Bacterial Genera in Granular Activated Carbon Water Treatment Systems," *Can. J. Microbiol.* **32**:226–230.

Buros, O. K., 1989, "Desalting Practices in the United States," *J. Amer. Water Works Assoc.* **81**:38–42.

Busscher, H. J., M. H. W. J. C. Uyen, A. W. J. van Pelt, A. H. Weerkamp, and J. Arends, 1986, "Kinetics of Adhesion of the Oral Bacterium *Streptococcus sanguis CH3* to Polymers with Different Surface Free Energies," *Appl. Environ. Microbiol.* **51**:910–914.

Busscher, H. J., A. H. Weerkamp, H. C. vam der Mei, A. W. J. van Pelt, H. P. de Jong, and J. Arends, 1984, "Measurement of the Surface Free Energy of Bacterial Cell Surfaces and Its Relevance for Adhesion," *Appl. Environ. Microbiol.* **48**:980–983.

Camper, A. K., M. W. LeChevallier, S. C. Broadway, and G. A. McFeters, 1986, "Bacteria Associated with Granular Activated Carbon Particles in Drinking Water," *Appl. Environ. Microbiol.* **52**:434–438.

Cargill, K. L., B. H. Pyle, R. L. Sauer, and G. A. McFeters, 1992, "Effects of Culture Conditions and Biofilm Formation on the Iodine Susceptibility of *Legionella pneumophila,"* *Can. J. Microbiol.* **38**:423–429.

Carson, L. A., L. A. Bland, L. B. Cusick, M. S. Favero, G. A. Bolan, A. L. Reingold, and R. C. Good, 1988, "Prevalence of Nontuberculous Mycobacteria in Water Supplies of Hemodialysis Centers," *Appl. Environ. Microbiol.* **54**:3122–3125.

Characklis, W. G., 1990a, "Biofilm Processes," in W. G. Characklis and K. C. Marshall (eds.), *Biofilms,* John Wiley & Sons, New York, pp. 195–232.

Characklis, W. G., 1990b, "Microbial Biofouling Control," in W. G. Characklis and K. C. Marshall (eds.), *Biofilms,* John Wiley & Sons, New York, pp. 585–634.

Characklis, W. G., 1991, "Biofouling: Effects and Control," in H.-C. Flemming and G. G. Geesey (eds.), *Biofouling and Biocorrosion in Industrial Water Systems,* Springer-Verlag, Berlin, pp. 7–27.

Characklis, W. G., and K. E. Cooksey, 1983, "Biofilms and Microbial Fouling," *Adv. Appl. Microbiol.* **29**:93–103.

Chi, L. F., M. Anders, H. Fuchs, R. R. Johnston, and H. Ringsdorf, 1993, "Domain Structures in Langmuir-Blodgett Films Investigated by Atomic Force Microscopy," *Sci.* **259**:213–216.

Christensen, B. E., and W. G. Characklis, 1990, "Physical and Chemical Properties of Biofilms," in W. G. Characklis and K. C. Marshall (eds.), *Biofilms,* John Wiley & Sons, New York, pp. 93–130.

Costerton, J. W., K. J. Cheng, G. G. Geesey, T. I. Ladd, J. C. Nickel, M. Dasgupta, and T. J. Marrie, 1987, "Bacterial Biofilms in Nature and Disease," *Ann. Rev. Microbiol.* **41**:435–464.

Costerton, J. W., and R. T. Irvin, 1981, "The Bacterial Glycocalyx in Nature and Disease," *Ann. Rev. Microbiol.* **35**:299–324.

Costerton, J. W., Z. Lewandowski, D. DeBeer, D. Caldwell, D. Korber, and G. James, 1994, "Biofilms, the Customized Microniche," *J. Bacteriol.* **176**:2137–2142.

Costerton, J. W., T. J. Marrie, and K. J. Cheng, 1985, "Phenomena of Bacterial Adhesion," in D. C. Savage and M. Fletcher (eds.), *Bacterial Adhesion: Mechanisms and Physiological Significance,* Plenum Press, New York, pp. 3–43.

Cowan, M. M., H. C. van der Mei, P. G. Rouxhet, and H. J. Busscher, 1992, "Physicochemical and Structural Investigation of the Surfaces of Some Anerobic Subgingival Bacteria," *Appl. Environ. Microbiol.* **58**:1326–1334.

Crossley, I. A., 1983, "Desalination by Reverse Osmosis," in A. Porteous (ed.), *Desalination Technology, Developments and Practice,* Applied Science Publishers Ltd., London, pp. 205–248.

Dagostino, L., A. E. Goodman, and K. C. Marshall, 1991, "Physiological Responses Induced in Bacteria Adhering to Surfaces," *Biofouling* **4**:113–119.

Dahlback, B., M. Hermansson, S. Kjelleberg, and B. Norkrans, 1981, "The Hydrophobicity of Bacteria—An Important Factor in Their Initial Adhesion at the Air-Water Interface," *Arch. Microbiol.* **128**:267–270.

Danielsson, A., B. Norkrans, and A. Bjornsson, 1977, "On Bacterial Adhesion—the Effect of Certain Enzymes on Adhered Cells of a Marine *Pseudomonas* sp.," *Bot. Marina* **XX**:13–17.

Davies, D. G., A. M. Chakrabarty, and G. G. Geesey, 1993, "Exopolysaccharide Production in Biofilms: Substratum Activation of Alginate Gene Expression by *Pseudomonas aeruginosa,*" *Appl. Environ. Microbiol.* **59**:1181–1186.

Dawes, E. A., 1984, "Stress of Unbalanced Growth and Starvation in Microorganisms," in M. H. E. Andrew and A. D. Russel (eds.), *The Revival of Injured Microbes,* Academic Press, New York, pp. 19–43.

De Graaf, F. K., and F. R. Mooi, 1986, "The Fimbrial Adhesins of *Escherichia coli,*" *Adv. Microbial Physiol.* **28**:65–143.

Deinema, M. H., L. H. A. Habets, J. Scholten, E. Turkstra, and H. A. A. M. Webers, 1980, "The Accumulation of Polyphosphate in *Acinetobacter* spp.," *FEMS Microbiol. Letters* **9**:275–279.

De Weger, L. A., C. I. M. van der Vlugt, A. H. M. Wijfjes, P. A. H. M. Bakker, B. Schippers, and B. Lugtenberg, 1987, "Flagella of a Plant-Growth-Stimulating *Pseudomonas fluroescens* Strain Are Required for Colonization of Potato Roots," *J. Bacteriol.* **169**:2769–2773.

Dexter, S. C., 1978, "Influence of Substratum Critical Surface Tension on Bacterial Adhesion—in situ Studies," *J. Colloid Interface Sci.* **70**:346–354.

Dexter, S. C., J. D. Sullivan, J. Williams III, and S. W. Watson, 1975, "Influence of Substrate Wettability on the Attachment of Marine Bacteria to Various Surfaces," *Appl. Microbiol.* **30**:298–308.

Dhar, H. P., D. H. Lewis, and J. O. Bockris, 1981, "The Electrochemical Diminution of Surface Bacterial Concentration," *Can. J. Microbiol.* **27**:998–1010.

Dickson, J. M., 1988, "Fundamental Aspects of Reverse Osmosis," in B. S. Parekh (ed.), *Reverse Osmosis Technology: Application for High-Purity Water Production,* Marcel Dekker, New York, pp. 1–51.

Dorschel, D., C. Buchanan, D. V. Strickler, R. J. Komarek, A. J. Matosky, and R. M. Gardner, 1993, "Isolation of Cellulose Acetate-Degrading Microorganisms," *Abstracts of the 205th Meeting of the American Chemical Society,* Part 1, Tech 22.

Doyle, R. J., 1991, "Strategies in Experimental Microbial Adhesion Research," in N. Mozes, P. S. Handley, H. J. Busscher, and P. G. Rouxhet (eds.), *Microbial Cell Surface Analysis,* VCH Publishers, New York, pp. 292–316.

Drury, W. J., W. G. Characklis, and P. S. Stewart, 1993a, "Interactions of 1 Micron Latex Particles with *Pseudomonas aeruginosa* Biofilms," *Water Res.* **27**:1119–1126.

Drury, W. J., P. S. Stewart, and W. G. Characklis, 1993b, "Transport of 1-Micron Latex Particles in *Pseudomonas aeruginosa* Biofilms," *Biotechnol. Bioeng.* **42**:111–117.

Du Moulin, G. C., I. H. Sherman, D. C. Hoaglin, and K. D. Stottmeier, 1985, "*Mycobacterium avium* complex, an emerging pathogen in Massachusetts," *J. Clin. Microbiol.* **22**:9–12.

Du Moulin, G. C., and K. D. Stottmeier, 1986, "Waterborne Mycobacteria: An Increasing Threat to Health Demographic Patterns, Larger Immunodeficient Populations, and the Prevalence of Mycobacteria in Water Systems Contribute to a Rising Problem," *ASM News* **52**:525–529.

Durham, L., 1989, "Biological Growth Control Problem in RO Systems," *Ultrapure Wat.* **6**:30–37.

Dychdala, G. R., and J. A. Lopes, 1991, "Surface-Active Agents: Acid-Anionic Compounds," in S. S. Block (ed.), *Disinfection, Sterilization, and Preservation,* Lea and Febiger, Philadelphia, pp. 256–262.

Eisenberg, T. N., and E. J. Middlebrooks, 1984, "A Survey of Problems with Reverse Osmosis Water Treatment," *J. Amer. Water Works Assoc.:* 44–49.

Fattom, A., and M. Shilo, 1984, "Hydrophobicity as an Adhesion Mechanism of Benthic Cyanobacteria," *Appl. Environ. Microbiol.* **47**:135–143.

Finken, H., J. Kaschemekat, and J. Mohn, 1979, "Reverse Osmosis: Field Trials of a GKSS Plate Module Plant on Baltic Sea Water," *Desalination* **29**:285–294.

Flemming, H.-C., 1991, "Biofouling in Water Treatment," in H.-C. Flemming and G. G. Geesey (eds.), *Biofouling and Biocorrosion in Industrial Water Systems,* Springer-Verlag, Berlin, pp. 47–80.

Flemming, H.-C., 1993, "Mechanistic Aspects of Reverse Osmosis Membrane Biofouling and Prevention," in Z. Amjad (ed.), *Reverse Osmosis: Membrane Technology, Water Chemistry, and Industrial Applications,* Van Nostrand Reinhold, New York, pp. 163–209.

Flemming, H. C., and G. Schaule, 1988a, "Biofouling on Membranes—a Microbiological Approach," *Desalination* **70**:95–119.

Flemming, H. C., and G. Schaule, 1988b, "Investigations on Biofouling of Reverse Osmosis and Ultrafiltration Membranes: Part I: Initial Phase of Biofouling," *Vom Wasser* **71**:207–223.

Flemming, H., and G. Schaule, 1988c, *Proc. Imtec '88 International Membrane Technology Conference, Biofouling on Membranes—a Microbiological Approach,* 1988, pp. D34, D37.

Flemming, H.-C., G. Schaule, and R. McDonogh, 1992a, "Some Second Thoughts about Biofouling," *Proc. International Membrane Science and Technology Conference,* November 10–12, 1992, University of New South Wales, Sydney, Australia, pp. 244–247.

Flemming, H.-C., G. Schaule, and R. McDonogh, 1993, "How Do Performance Parameters Respond to Initial Biofilm Formation on Separation Membranes?" *Vom Wasser* **80**: 177–186.

Flemming, H.-C., G. Schaule, R. McDonogh, and H. F. Ridgway, 1992b, "Effects and Extent of Biofilm Accumulation in Membrane Systems," in G. G. Geesey, Z. Lewandowski, and H.-C.

Flemming (eds.), *Biofouling and Biocorrosion in Industrial Water Systems,* CRC Press, Boca Raton, pp. 63–89.

Fletcher, M., 1976, "The Effects of Proteins on Bacterial Attachment to Polystyrene," *J. Gen. Microbiol.* **94**:400–404.

Fletcher, M., and G. I. Loeb, 1979, "Influence of Substratum Characteristics on the Attachment of a Marine Pseudomonad to Solid Surfaces," *Appl. Environ. Microbiol.* **37**:67–72.

Fletcher, M., and K. C. Marshall, 1982, "Bubble Contact Angle Method for Evaluation Substratum Interfacial Characteristics and Its Revelance to Bacterial Attachment," *Appl. Environ. Microbiol.* **44**:184–192.

Fletcher, M., and S. McEldowney, 1983, "Microbial Attachment to Nonbiological Surfaces," in M. J. Klug and C. A. Reddy (eds.), *Current Perspectives in Microbial Ecology, Proceedings of the Third International Symposium on Microbial Ecology,* American Society for Microbiology, Washington, D.C., pp. 124–129.

Ford, C. W., 1989, "Deacylation and Cleavage of Cell Walls of Pangola Grass with Borohydride," *Phytochem.* **28**:393–398.

Frith, C. F., 1988, "Electronic-Grade Water Production Using Reverse Osmosis Technology," in B. S. Parekh (ed.), *Reverse Osmosis Technology: Applications for High-Purity Water Production,* Marcel Dekker, New York, pp. 279–310.

Geesey, G. G., 1982, "Microbial Exopolymers: Ecological and Economic Considerations," *ASM News* **48**:9–14.

Geesey, G. G., R. Mutch, and J. W. Costeron, 1978, "Sessile Bacteria: An Important Component of the Microbial Population in Small Mountain Streams," *Limmol. Oceanogr.* **23**:1214–1223.

Geesey, G. G., M. W. Stupy, and P. J. Bremer, 1992, "The Dynamics of Biofilms," *International Biodeterioration & Biodegradation* **30**:135–154.

Geldreich, E. E., R. H. Taylor, J. C. Blannon, and D. J. Reasoner, 1985, "Bacterial Colonization of Point-of-Use Water Treatment Devices," *J. Amer. Water Works Assoc.* **77**:72–80.

Gilbert, E., 1988, "Biodegradability of Ozonization Products as a Function of COD and DOC Elimination by the Example of Humic Acids," Water Res., **22**:123–126.

Gilmore, D. F., N. Lotti, R. W. Lenz, M. Scandola, and R. C. Fuller, 1992, "Biodegradability of Blends of Poly(beta-hydroxybutyrate-co-hydroxyvalerate) (PHBV) with Ester Substituted Cellulose: Poly-beta-hydroxybutyrate Degradation and Poly-beta-hydroxyvalerate Degradation When Blended with Butyryl or Propionyl Ester of Cellulose Acetate," *Proc. American Society for Microbiology, 92d Annual Meeting,* New Orleans, La.

Gismondo, M. R., M. A. Romeo, G. Tempera, D. Iannello, and M. Carbone, 1983, "Comparative Effects of Subinhibitory Concentrations of Miokamycin and Other Antibiotics on Oral Streptococci," *Drugs Under Exptl. Clin. Res.* **9**:801–808.

Glater, J., and S. McCray, 1983, "Changes in Water and Salt Transport During Hydrolysis of Cellulose Acetate Reverse Osmosis Membranes," *Desalination* **46**:389–397.

Glater, J., J. W. McCutchan, S. B. McCray, and M. R. Zachariah, 1981, "The Effect of Halogens on the Performance and Durability of Reverse-Osmosis Membranes," in A. F. Turbak (ed.), *Synthetic Membranes: Desalination,* American Chemical Society, Washington, D.C., pp. 185–190.

Glater, J., M. R. Zachariah, S. B. McCray, and J. W. McCutchan, 1983, "Reverse Osmosis Membrane Sensitivity to Ozone and Halogen Disinfectants," *Desalination* **48**:1–16.

Goodwyn, P. P., 1976, "One Pass Seawater Desalting RO Plant Evaluation," Office of Water Research and Technology Report No. OWRT 14-34-0001-6507; NTIS No. PB-262-207.

Gordon, A. S., S. M. Gerchakov, and L. R. Udey, 1981, "The Effect of Polarization on the Attachment of Marine Bacteria to Copper and Platinum Surfaces," *Can. J. Microbiol.* **27**:698–703.

Greene, A. C., and J. C. Madgwick, 1991, "Microbial Formation of Manganese Oxides," *Appl. Environ. Microbiol.* **57**:1114–1120.

Gross, R. A., J. D. Gu, D. T. Eberiel, M. Nelson, and S. P. McCarthy, 1993, "Cellulose Acetate Biodegradability in Simulated Aerobic Composting and Anaerobic Bioreactor Environments as Well as by a Bacterial Isolate Derived from Compost," in C. Ching, D. L. Kaplan, and E. L. Thomas (eds.), *Biodegradation of Polymer Packaging,* Technomic, Lancaster, Pa., pp. 257–279.

Guckert, J. B., M. A. Hood, and D. C. White, 1986, "Phospholipid Ester-Linked Fatty Acid Profile Changes During Nutrient Deprivation of *Vibrio cholerae:* Increases in the *trans -cis* Ratio and Proportions of Cyclopropyl Fatty Acids," *Appl. Environ. Microbiol.* **52**:794–801.

Haas, C. N., and R. S. Engelbrecht, 1980, "Chlorine Dynamics During Inactivation of Coliforms, Acid-Fast Bacteria and Yeasts," *Water Res.* **14**:1749–1757.

Hassan, A. M., S. Al-Jarrah, T. Al-Lohibi, A. Al-Mamdan, and L. M. Bakheet, 1989, "Performance Evaluation of SWCC SWRO Plants," *Desalination* **74**:37–50.

Ho, L. C. W., D. D. Martin, and W. C. Lindemann, 1983, "Inability of Microorganisms to Degrade Cellulose Acetate Reverse-Osmosis Membranes," *Appl. Environ. Microbiol.* **45**:418–427.

Hodgson, P. H., G. L. Leslie, R. P. Schneider, A. G. Fane, C. J. D. Fell, and K. C. Marshall, 1993, "Cake Resistance and Solute Rejection in Bacterial Microfiltration: The Role of the Extracellular Matrix," *J. Membr. Sci.* **79**:35–53.

Howsam, P., and S. F. Tyrrel, 1990, "Iron Biofouling in Groundwater Abstraction Systems: Why and How?," in P. Howsam (ed.), *Microbiology in Civil Engineering,* Chapman and Hall, London, pp. 192–197.

Huck, P. M., 1990, "Measurement of Biodegradable Organic Matter and Bacterial Growth Potential in Drinking Water," *J. Amer. Water Works Assoc.* **82**:78–86.

Isner, J. D., and R. C. Williams, 1993, "Analytical Techniques for Identifying Reverse Osmosis Foulants," in Z. Amjad (ed.), *Reverse Osmosis: Membrane Technology, Water Chemistry, and Industrial Applications,* Van Nostrand Reinhold, New York, pp. 237–274.

Jacangelo, J. G., E. M. Aieta, K. E. Carns, E. W. Cummings, and J. Mallevialle, 1989, "Assessing Hollow-Fiber Ultrafiltration for Particulate Removal," *J. Amer. Water Works Assoc.* **81**:68–75.

Jovin, T. M., and D. J. Arndt-Jovin, 1989, "Luminescence Digital Imaging Microscopy," *Annu. Rev. Biophys. Biophys. Chem.* **18**:271–308.

Kaakinen, J. W., and C. D. Moody, 1984, *Proc. Characteristics of Reverse Osmosis Membrane Fouling at the Yuma Desalting Test Facility,* presented at Symposium on Reverse Osmosis and Ultrafiltration American Chemical Society, Philadelphia, Pennsylvania, 1984, pp. 1, 21.

Katoh, M., and Y. Matsuo, 1983, "Adherence of *Mycobacterium lepraemurium* to Tissue Culture Cells," *Hiroshima J. Med. Sci.* **32**:285–290.

Kawaguchi, T., and H. Tamura, 1984, "Chlorine-Resistant Membrane for Reverse Osmosis. 1. Correlation between Chemical Structures and Chlorine Resistance of Polyamides," *J. Appl. Polymer Sci.* **29**:3359–3367.

Kay, M. J., L. H. G. Morton, and E. L. Prince, 1990, "The Biodeterioration of Polyester Polyurethane in Soil/Marine Contact," in P. Howsam (ed.), *Microbiology in Civil Engineering,* Chapman and Hall, London, pp. 109–120.

Kefford, B., S. Kjelleberg, and K. C. Marshall, 1982, "Bacterial Scavenging: Utilization of Fatty Acids Localized at a Solid-Liquid Interface," *Arch. Microbiol.* **133**:257–260.

Kesting, R. E., 1985, *Synthetic Polymeric Membranes: A structural Perspective,* 2nd ed., John Wiley & Sons.

Kinniment, S. L., and J. W. T. Wimpenny, 1992, "Measurements of the Distribution of Adenylate Concentrations and Adenylate Energy Charge across *Pseudomonas aeruginosa* Biofilms," *Appl. Environ. Microbiol.* **58**:1629–1635.

Kjelleberg, S., and M. Hermansson, 1984, "Starvation-Induced Effects on Bacterial Surface Characteristics," *Appl. Environ. Microbiol.* **48**:497–503.

Ko, A., and D. B. Guy, 1988, "Brackish and Seawater Desalting," in B. S. Parekh (ed.), *Reverse Osmosis Technology: Applications for High-Purity Water Production,* Marcel Dekker, Inc., New York, pp. 185–278.

Kuhlman, V. U., E. Severin, J. Stellmach, C. Wiezorek, and K. Echsler, 1989, "Fluoreszierende Formazane in der Durchflubzytometrie Untersuchungen iher Sauerstoffempfindlichkeit," *Acta Histochemica* XXXVII (supplement): 221–230.

Kutz, S. M., D. L. Bentley, and N. A. Sinclair, 1985a, "Improved Fixation of Cellulose-Acetate Reverse-Osmosis Membrane for Scanning Electron Microscopy," *Appl. Environ. Microbiol.* **49**:446–450.

Kutz, S. M., D. L. Bentley, and N. A. Sinclair, 1986, "Morphology of a Seliberia-like Organism Isolated from Reverse Osmosis Membranes," *Proc. IVth Inter. Sym. Microbial Ecol:* 584–587.

Kutz, S. M., N. A. Sinclair, and D. L. Bentley, 1985b, "Characterization of *Seliberia* and Its Effect on Reverse Osmosis Membrames," *Abstracts of the Annual Meeting of the American Society for Microbiology*, 85th Annual Meeting, Las Vegas, Nevada, 1985, pp. 223, 223.

Laine, J.-M., J. P. Hagstrom, M. M. Clark, and J. Mallevialle, 1989, "Effects of Ultrafiltration Membrane Composition," *J. Amer. Water Works Assoc.* **81**:61–67.

Lawrence, J. R., P. J. Delaquis, D. R. Korber, and D. E. Caldwell, 1987, "Behavior of *Pseudomonas fluorescens* within the Hydrodynamic Boundary Layers of Surface Microenvironments," *Microbial Ecol.* **14**:1–14.

LeChevallier, M. W., 1987, "Disinfection of Bacterial Biofilms," *Proc. of the Sixth Conference on Water Disinfection: Environmental Impact and Health Effects*, May 3–8, pp. 1–20.

LeChevallier, M. W., 1991, "Biocides and the Current Status of Biofouling Control in Water Systems," in H.-C. Flemming and G. G. Geesey (eds.), *Biofouling and Biocorrosion in Industrial Water Systems*, Springer-Verlag, Berlin, pp. 113–132.

LeChevallier, M. W., C. D. Cawthon, and R. G. Lee, 1988, "Inactivation of Biofilm Bacteria," *Appl. Environ. Microbiol.* **54**:2492–2499.

LeChevallier, M. W., T. S. Hassenhauer, A. K. Camper, and G. A. McFeters, 1984, "Disinfection of Bacteria Attached to Granular Activated Carbon," *Appl. Environ. Microbiol.* **48**: 918–923.

LeChevallier, M. W., C. D. Lowry, and R. G. Lee, 1990, "Disinfecting Biofilms in a Model Distribution System," *J. Amer. Water Works Assoc.* **82**:87–99.

Lepore, J. V., and R. C. Ahlert, 1988, "Fouling in Membrane Processes," in B. S. Parekh (ed.), *Reverse Osmosis Technology: Application for High-Purity Water Production*, Marcel Dekker, New York, pp. 141–184.

Leslie, G. L., R. P. Schneider, A. G. Fane, K. C. Marshall, and C. J. D. Fell, 1993, "Fouling of a Microfiltration Membrane by Two Gram-Negative Bacteria," *Colloids and Surfaces A: Physicochemical and Engineering Aspects* **73**:165–178.

Lichtman, J. W., 1994, "Confocal Microscopy," *Sci. Amer.* **271**:40–45.

Light, W. G., J. L. Perlman, A. B. Reidinger, and D. F. Needham, 1988, "Desalination of Non-Chlorinated Surface Seawater Using TFC Membrane Elements," *Desalination* **70**:N1–3.

Lozier, J. C., and R. A. Sierka, 1985, "Using Ozone and Ultrasound to Reduce RO Membrane Fouling," *J. Amer. Water Works Assoc.* **77**:60–65.

Ludwicka, A., L. M. Switalski, A. Lundin, G. Pulverer, and T. Wadstrom, 1985, "Bioluminescent Assay for Measurement of Bacterial Attachment to Polyethylene," *J. Microbiol. Methods* **4**:169–177.

Malik, A. L. A., N. G. Younan, B. J. R. Rao, and K. M. Mousa, 1989, "Skid Mounted Mobile Brackish Water Reverse Osmosis Plants at Different Sites in Kuwait," *Desalination* **75**:341–361.

Mamed-Yarov, M. A., R. S. Alimardanov, Z. H. M. Mamedova, and R. L. Akopyan, 1987, "Biological Stability of Carbamyl Group-Containing Cellulose Esters," *Izvestiya Akademii Nauk Azerbaidzhanskoi Ssr Seriya Biologicheskikh Nauk* **0**:113–117.

Marshall, K. C., 1985, "Mechanisms of Bacterial Adhesion at Solid-Water Interfaces," in D. C. Savage and M. Fletcher (eds.), *Bacterial Adhesion: Mechanisms and Physiological Significance*, Plenum Press, New York, pp. 133–162.

Marshall, K. C., 1991, "The Importance of Studying Microbial Cell Surfaces," in N. Mozes, P. S. Handley, H. J. Busscher, and P. G. Rouxhet (eds.), *Microbial Cell Surface Analysis*. VCH Publishers, New York, pp. 4–19.

Marshall, K. C., and B. L. Blainey, 1991, "Role of Bacterial Adhesion in Biofilm Formation and Biocorrosion," in H.-C. Flemming and G. G. Geesey (eds.), *Biofouling and Biocorrosion in Industrial Water Systems*, Springer-Verlag, Berlin, pp. 29–46.

Marshall, K. C., R. Stout, and R. Mitchell, 1971, "Mechanism of the Initial Events in the Sorption of Marine Bacteria to Surfaces," *J. Gen. Microbiol.* **68**:337–348.

McCoy, W. F., 1987, "Strategies for the Treatment of Biological Fouling," in M. W. Mittelman and G. G. Geesey (eds.), *Biological Fouling of Industrial Water Systems: A Problem Solving Approach,* Water Micro Associates, San Diego, pp. 247–268.

McKinley, V. L., J. W. Costerton, and D. C. White, 1988, "Microbial Biomass, Activity, and Community Structure of Water and Particulates Retrieved by Backflow from a Waterflood Injection Well," *Appl. Environ. Microbiol.* **54**:1383–1393.

Merianos, J. J., 1991, "Quaternary Ammonium Antimicrobial Compounds," in S. S. Block (ed.), *Disinfection, Sterilization, and Preservation,* Lea and Febiger, Philadelphia, pp. 225–255.

Miguez, C. B., T. J. Beveridge, and J. M. Ingram, 1986, "Lipopolysaccharide Changes and Cytoplasmic Polyphosphate Granule Accumulation in *Pseudomonas aeruginosa* During Growth on Hexadecane," *Can. J. Microbiol.* **32**:248–253.

Mindler, A. B., and S. T. Bateman, 1980, "Pilot Plant Study on Marine Microorganisms and Organic Matter in Seawater Desalination by Reverse Osmosis," prepared for Office of Water Research and Technology, Washington, D.C., pp. 2, 126.

Mistry, G. J., and R. van Woerkom, 1990, "Engineered Water Systems and Waterborne Disease," in P. Howsam (ed.), *Microbiology in Civil Engineering,* Chapman and Hall, London, pp. 75–79.

Mittleman, M. W., 1991, "Bacterial Growth and Biofouling Control in Purified Water Systems," in H.-C. Flemming and G. G. Geesey (eds.), *Biofouling and Biocorrosion in Industrial Water Systems,* Springer-Verlag, Berlin, pp. 133–154.

Moriyon, I., and D. T. Berman, 1982, "Effects of Nonionic, Ionic, and Dipolar Ionic Detergents and EDTA on the *Brucella* Cell Envelope," *J. Bacteriol.* **152**:822–828.

Muraca, P. W., V. L. Yu, and J. E. Stout, 1988, "Environmental Aspects of 'Legionnaires' Disease," *J. Amer. Water Works Assoc.* **80**:78–86.

Nakagawa, Y., H. Hayashi, T. Tawaratani, H. Kourai, T. Horie, and I. Shibasaki, 1984a, "Disinfection of Water with Quaternary Ammonium Salts Insolubilized on a Porous Glass Surface," *Appl. Environ. Microbiol.* **47**:513–518.

Nealson, K. H., B. M. Tebo, and R. A. Rosson, 1988, "Occurrence and Mechanisms of Microbial Oxidation of Manganese," *Adv. Appl. Microbiol.* **33**:279–318.

Neu, T. R., and K. C. Marshall, 1990, "Bacterial Polymers: Physicochemical Aspects of Their Interactions at Interfaces," *J. Biomaterials Applications* **5**:107–133.

Nivens, D. E., J. Schmit, J. Sniatecki, T. Anderson, J. Q. Chambers, and D. C. White, 1993, "Multichannel STR-FTIR Spectrometer for On-line Examination of Microbial Biofilms," *Appl. Spec.* **47**:668–671.

Novitsky, J. A., and R. Y. Morita, 1976, "Morphological Characterization of Small Cells Resulting from Nutrient Starvation of a Psychrophillic Marine Vibrio," *Appl. Environ. Microbiol.* **32**:617–622.

Novitsky, J. A., and R. Y. Morita, 1977, "Survival of a Psychrophillic Marine Vibrio under Long-Term Nutrient Starvation," *Appl. Environ. Microbiol.* **33**:635–641.

Nusbaum, I., and D. G. Argo, 1984, "Design, Operation, and Maintenance of A 5-mdg Wastewater Reclamation Reverse Osmosis Plant," in G. Belfort (ed.), *Synthetic Membrane Processes, Fundamental and Water Applications,* Academic Press, Harcourt Brace Jovanovich, New York, pp. 377–436.

Parise, P. L., B. S. Parekh, and R. T. Smith, 1988, "Reverse Osmosis for Producing Pharmaceutical-Grade Waters," in B. S. Parekh (ed.), *Reverse Osmosis Technology: Applications for High-Purity Water Production,* Marcel Dekker, New York, pp. 347–398.

Patterson, M. K., G. R. Husted, A. Rutkowski, and D. C. Mayette, 1991, "Isolation, Identification, and Microscopic Properties of Biofilms in High-Purity Water Distribution Systems," *Ultrapure Wat.* **8**:18–24.

Paul, D. H., 1991, "Reverse Osmosis: Scaling, Fouling, & Chemical Attack," *Desal. & Water Reuse* **1**:8–11.

Paul, J. H., 1984, "Effects of Antimetabolites on the Adhesion of an Estuarine *Vibrio* sp. to Polystyrene," *Appl. Environ. Microbiol.* **48**:924–929.

Paul, J. H., and W. H. Jeffrey, 1984, "The Effect of Surfactants on the Attachment of Estuarine and Marine Bacteria to Surfaces," *Can. J. Microbiol.* **31**:224–228.

Paul, J. H., and W. H. Jeffrey, 1985, "Evidence for Separate Adhesion Mechanisms for Hydrophilic and Hydrophobic Surfaces in *Vibrio proteolytica*," *Appl. Environ. Microbiol.* **50**:431–437.

Paul, J. H., and G. I. Loeb, 1983, "Improved Microfouling Assay Employing a DNA-Specific Fluorochrome and Polystyrene as Substratum," *Appl. Environ. Microbiol.* **46**:338–343.

Payment, P., 1989, "Bacterial Colonization of Domestic Reverse Osmosis Water Filtration Units," *Can. J. Microbiol.* **35**:1065–1067.

Payment, P., E. Franco, L. Richardson, and J. Siemiatycki, 1992, "Gastrointestinal Health Effects Associated with the Consumption of Drinking Water Produced by Point-of-Use Domestic Reverse Osmosis Filtration Units," *Appl. Environ. Microbiol.* **57**:945–948.

Pearson, F. C., J. Bohon, W. Lee, G. Bruszer, M. Sagona, G. Jakubowski, R. Dawe, D. Morrison, and C. Dinarello, 1984, "Characterization of *Limulus* Amoebocyte Lysate-Reactive Material from Hollow-Fiber Diaylzers," *Appl. Environ. Microbiol.* **48**:1189–1196.

Pittner, G. A., 1993a, "The Economics of Desalination Processes," in Z. Amjad (ed.), *Reverse Osmosis: Membrane Technology, Water Chemistry, and Industrial Applications,* Van Nostrand Reinhold, New York, pp. 76–103.

Pittner, G. A., 1993b, "High Purity Water Production Using Reverse Osmosis Technology," in Z. Amjad (ed.), *Reverse Osmosis: Membrane Technology, Water Chemistry, and Industrial Applications,* Van Nostrand Reinhold, New York, pp. 334–363.

Pusch, W., and A. Walch, 1982, "Synthetic Membranes—Preparation, Structure, and Application," *Angew. Chem.* **21**:660–685.

Pyle, B. H., and G. A. McFeters, 1990, "Iodine Susceptibility of Pseudomonads Grown Attached to Stainless Steel Surfaces," *Biofouling* **2**:113–120.

Reese, E. T., 1957, "Biological Degradation of Cellulose Derivatives," *J. Indust. Eng. Chem.* **49**:89–93.

Reese, E. T., 1976, "History of the Cellulase Program at the U. S. Army Natic Development Center," *Biotechnol. Bioeng.* **6**:9–20.

Ridgway, H. F., 1987, "Microbial Fouling of Reverse Osmosis Membranes: Genesis and Control," in M. W. Mittelman and G. G. Geesey (eds.), *Biological Fouling of Industrial Water Systems: A Problem Solving Approach,* Water Micro Associates, San Diego, Calif., pp. 138–193.

Ridgway, H. F., 1988, "Microbial Adhesion and Biofouling of Reverse Osmosis Membranes," in B. S. Parekh (ed.), *Reverse Osmosis Technology: Applications for High-Purity Water Production,* Marcel Dekker, New York, pp. 429–481.

Ridgway, H. F., D. G. Argo, and B. H. Olson, 1981, "Factors Influencing Biofouling of Reverse Osmosis Membranes at Water Factory 21: Chemical, Microbiological, and Ultrastructural Characterization of the Fouling Layer," Vol. III-B, final report prepared for U.S. Department of the Interior, Office of Water Research and Technology, contract No. 14-34-0001-8520, Washington, D.C.

Ridgway, H. F., C. A. Justice, C. Whittaker, D. G. Argo, and B. H. Olson, 1984a, "Biofilm Fouling of RO Membranes—Its Nature and Effect on Treatment of Water for Reuse," *J. Amer. Water Works Assoc.* **79**:94–102.

Ridgway, H. F., A. Kelly, C. Justice, and B. H. Olson, 1983, "Microbial Fouling of Reverse-Osmosis Membranes Used in Advanced Wastewater Treatment Technology: Chemical, Bacteriological, and Ultrastructural Analyses," *Appl. Environ. Microbiol.* **45**:1066–1084.

Ridgway, H. F., M. G. Rigby, and D. G. Argo, 1984b, "Adhesion of a *Mycobacterium* sp. to Cellulose Diacetate Membranes Used in Reverse Osmosis," *Appl. Environ. Microbiol.* **47**: 61–67.

Ridgway, H. F., M. G. Rigby, and D. G. Argo, 1984c, "Biological Fouling of Reverse Osmosis Membranes: The Mechanism of Bacterial Adhesion," *Proc. Water Reuse Symposium III, The Future of Water Reuse,* vol. 3 (published by AWWA Research Foundation, Denver, Col.), August 26–31, 1984, San Diego, Calif., pp. 1314–1350.

Ridgway, H. F., M. G. Rigby, and D. G. Argo, 1985, "Bacterial Adhesion and Fouling of Reverse Osmosis Membranes," *J. Amer. Water Works Assoc.* **77**:97–106.

Ridgway, H. F., D. M. Rogers, and D. G. Argo, 1986a, "Effect of Surfactants on the Adhesion of Mycobacteria to Reverse Osmosis Membranes," *Proc. 5th Annual Semiconductor Pure Water Conference,* January 16–17, 1986, San Francisco, Calif., pp. 133–164.

Ridgway, H. F., D. M. Rogers, and D. G. Argo, 1986b, "Biofilm on Reverse Osmosis Membranes," *Wat. Waste Treat.* **63**:45–48.

Ridgway, H. F., and J. Safarik, 1991, "Biofouling of Reverse Osmosis Membranes," in H.-C. Flemming and G. G. Geesey (eds.), *Biofouling and Biocorrosion in Industrial Water Systems,* Springer-Verlag, Berlin, pp. 81–111.

Ridgway, H. F., J. Safarik, and G. Rodriguez, 1992, "Life in an RO Biofilm: Breathing and Other Problems of Bacteria in Very Attached Relationships," *Proc. International Membrane Science & Technology Conference,* November 10–12, 1992, University of New South Wales, Sydney, Australia, pp. 248–250.

Riley, R. L., J. O. Gardiner, and U. Merten, 1964, "Cellulose Acetate Membranes: Electron Microscopy of Structure," *Sci.* **143**:801–803.

Riley, R. L., G. R. Hightower, and C. R. Lyons, 1973, "Thin-Film Composite Membrane for Single-Stage Seawater Desalination by Reverse Osmosis," in *Applied Polymer Symposium,* no. 22, John Wiley & Sons, pp. 255–267.

Rodriguez, G. G., D. Phipps, K. Ishiguro, and H. F. Ridgway, 1992, "Use of a Fluorescent Probe for Direct Visualization of Actively Respiring Bacteria," *Appl. Environ. Microbiol.* **58**:1801–1808.

Rosenberg, M., and R. J. Doyle, 1990, "Microbial Cell Surface Hydrophobicity: History, Measurement, and Significance," in R. J. Doyle and M. Rosenberg (eds.), *Microbial Cell Surface Hydrophobicity,* American Society for Microbiology, Washington, D.C., pp. 1–38.

Rosenberg, M., D. Gutnick, and E. Rosenberg, 1980, "Adherence of Bacteria to Hydrocarbons: A Simple Method for Measuring Cell-Surface Hydrophobicity," *FEMS Microbiol. Letters* **9**:29–33.

Rowley, L. H., 1992, "A Screening Study of 12 Biocides for Potential Use with Cellulose Acetate Reverse Osmosis Membranes," paper presented at *Biennial Conference of the National Water Supply Improvement Association,* August 23–27, 1992, Newport Beach, Calif.

Saad, M. A., 1992, "Biofouling Prevention in RO Polymeric Membrane Systems," *Desalination* **88**:85–105.

Safarik, J., J. Williams, and H. F. Ridgway, 1989, "Analysis of Biofilm from Reverse Osmosis Membranes by Computer-Programmed Polyacrylamide Gel Electrophoresis," *Proc. of the American Society for Microbiology,* May 14–18, New Orleans, La.

Salit, I. E., 1982, "Effect of Subinhibitory Concentrations of Antimicrobials on Meningococcal Adherence," *Can. J. Microbiol.* **29**:369–376.

Schaule, G., 1992, *Bakterielle Adhäsion und Aggregation auf inerten Oberflachen: ihre Rolle bei der Bildung von Biofilmen auf Umkehrosmose-Membranen,* Kommissionsverlag R. Oldenbourg.

Schaule, G., H.-C. Flemming, and K. Poralla, 1990, "Primary Biofouling in Membrane Processes—Influence of Various Factors," *Dechema Biotechnology Conference,* pp. 410–413.

Schaule, G., H.-C. Flemming, and H. F. Ridgway, 1993a, "Use of 5-cyano-2,3-ditolyl Tetrazolium Chloride for Quantifying Planktonic and Sessile Respiring Bacteria in Drinking Water," *Appl. Environ. Microbiol.* **59**:3850–3857.

Schaule, G., A. Kern, and H.-C. Flemming, 1993b, "RO Treatment of Dump Tricking Water: Membrane Biofouling," *Desal. & Water Reuse* **3**:17–23.

Scheld, W. M., O. Zak, K. Vosbeck, and M. A. Sande, 1981, "Bacterial Adhesion in the Pathogenesis of Infective Endocarditis: Effect of Subinhibitory Antibiotic Concentrations on Streptococcal Adhesion *in-vitro* and the Development of Endocarditis in Rabbits," *J. Clin. Invest.* **68**:1381–1384.

Schoenen, D., 1990, "Influence of Materials on the Microbiological Colonization of Drinking Water," in P. Howsam (ed.), *Microbiology in Civil Engineering,* Chapman and Hall, London, pp. 121–145.

Seal, K. J., 1990, "Biodeterioration of Materials Used in Civil Engineering," in P. Howsam (ed.), *Microbiology in Civil Engineering*, E. & F.N. Spon, London, pp. 39–52.

Seidler, E., 1991, "The Tetrazolium-Formazan System: Design and Histochemistry," *Prog. Histochem. Cytochem.* **24**(1).

Sinclair, N. A., 1982, *Microbial Degradation of Reverse Osmosis Desalting Membranes, Operation and Maintenance of the Yuma Desalting Test Facility*, vol. IV, U.S. Department of the Interior, Bureau of Reclamation, Yuma, Arizona.

Speaker, L. M., 1985, "Aft (Anti-Fouling Technology) for Membranes and Nonpermeable Surfaces," *Proc. at the Second International Conference on Fouling and Cleaning in Food Processing*, July 14, 1985, Madison, Wis. pp. 1–12.

Speaker, L. M., and K. R. Bynum, 1983, "Oriented Monolayer Assemblies to Modify Fouling Propensities of Membranes," in K. L. Mittal (ed.), *Physicochemical Aspects of Polymer Surfaces*, Plenum Press, New York, pp. 817–841.

Stellmach, J., 1984, "Fluorescent Redox Dyes: 1. Production of Fluorescent Formazan by Unstimulated and Phorbol Ester- or Digitonin-Stimulated Ehrlich Ascites Tumor Cells," *Histochem.* **80**:137–143.

Stenstrom, M. K., J. R. Davis, J. G. Lopez, and J. W. McCutchan, 1982, "Municipal Wastewater Reclamation by Reverse Osmosis—a 3-Year Case Study," *J. Wat. Pollu. Contrl. Fed.* **54**:43–51.

Stewart, P. S., T. Griebe, R. Srinivasan, C.-I. Chen, F. P. Yu, D. DeBeer, and G. A. McFeters, 1994, "Comparison of Respiratory Activity and Cuturability During Monochloramine Disinfection of Binary Population Biofilms," *Appl. Environ. Microbiol.* **60**:1690–1692.

Stoodley, P., D. DeBeer, and Z. Lewandowski, 1994, "Liquid Flow in Biofilm Systems," *Appl. Environ. Microbiol.* **60**:2711–2716.

Suci, P. A., M. W. Mittleman, F. P. Yu, and G. G. Geesey, 1994, "Investigation of Ciprofloxacin Penetration into *Pseudomonas aeruginosa* Biofilms," *Antimicrobial Agents and Chemotherapy* **38**:2125–2133.

Suffet, I. H., and W. O. Pipes, 1986, "Predominant Bacterial Genera in Granular Activated Carbon Water Treatment Systems," *Can. J. Microbiol.* **32**:226–230.

Sutherland, I. W., 1990, "Enzymes Degrading Exopolysaccharides," in I. W. Sutherland (ed.), *Biotechnology of Microbial Exopolysaccharides*, Cambridge University Press, Cambridge, pp. 39–53.

Taylor, J., L. A. Mulford, S. J. Duranceau, and W. M. Barrett, 1989, "Cost and Performance of a Membrane Pilot Plant," *J. Amer. Water Works Assoc.* **81**:52–60.

van der Kooij, D., 1992, "Assimilable Organic Carbon as an Indicator of Bacterial Regrowth," *J. Amer. Water Works Assoc.* **84**:57–65.

Vandevivere, P., and D. L. Kirchman, 1993, "Attachment Stimulates Exopolysaccharide Synthesis," *Appl. Environ. Microbiol.* **59**:3280–3286.

Vincent, F. C., A. R. Tibi, and J. C. Darbord, 1989, "A Bacterial Biofilm in a Hemodialysis System: Assessment of Disinfection and Crossing of Endotoxin," *Trans. Amer. Soc. Artif. Intern. Organs* **XXXV**:310–313.

Wechsler, R., 1976, "Reverse Osmosis on Secondary Sewage Effluent: The Effect of Recovery," *Water Res.* **11**:379–385.

Welch, D. F., 1991, "Applications of Cellular Fatty Acid Analysis," *Clin. Microbiol. Rev.* **4**:422–438.

White, D. C., 1985, "Quantitative Physico-Chemical Characterization of Bacterial Habitats," in J. Poindexter and E. Leadbetter (eds.), *Bacteria in Nature*, Plenum Publishing, New York, pp. 177–203.

White, D. C., 1986, "Environmental Effects Testing with Quantitative Microbial Analysis: Chemical Signatures Correlated with in situ Biofilm Analysis by FTIR," *Tox. Assess.* **1**:315–338.

Whittaker, C., H. F. Ridgway, and B. H. Olson, 1984, "Evaluation of Cleaning Strategies for Removal of Biofilms from Reverse-Osmosis Membranes," *Appl. Environ. Microbiol.* **48**:395–403.

Wicken, A. J., 1985, "Bacterial Cell Walls and Surfaces," in D. C. Savage and M. Fletcher (eds.), *Bacterial Adhesion: Mechanisms and Physiological Significance,* Plenum Press, New York, pp. 45–70.

Winfield, B. A., 1979a, "A Study of the Factors Affecting the Rate of Fouling of Reverse Osmosis Membranes Treating Secondary Sewage Effluents," *Water Res.* **13**:565–569.

Winfield, B. A., 1979b, "The Treatment of Sewage Effluents by Reverse Osmosis—pH Based Studies of the Fouling Layer and Its Removal," *Water Res.* **13**:561–564.

Winters, H., I. Isquith, W. A. Arthur, and A. Mindler, 1983, "Control of Biological Fouling in Seawater Reverse Osmosis Desalination," *Desalination* **47**:233–238.

Wolfaardt, G. M., R. E. M. Archibald, and T. E. Cloete, 1991, "The Use of DAPI in the Quantification of Sessile Bacteria on Submerged Surfaces," *Biofouling* **4**:265–274.

Wolfaardt, G. M., J. R. Lawrence, R. D. Robarts, S. J. Caldwell, and D. E. Caldwell, 1994, "Multicellular Organization in a Degradative Biofilm Community," *Appl. Environ. Microbiol.* **60**:434–446.

Worley, S. D., and D. E. Williams, 1988, "Halamine Water Disinfectants," *CRC Crit. Rev. Environ. Contrl.* **18**:133–175.

Yu, F. P., and G. A. McFeters, 1994, "Physiological Responses of Bacteria in Biofilms to Disinfection," *Appl. Environ. Microbiol.* **60**:2462–2466.

Zeiher, E. H. K., C. C. Pierce, and D. Woods, 1991, "Biofouling of Reverse Osmosis Systems: Three Case Studies," *Ultrapure Wat.* **8**:50–63.

Zimmermann, R., R. Iturriaga, and J. Becker-Birck, 1978, "Simultaneous Determination of the Total Number of Aquatic Bacteria and the Number Thereof Involved in Respiration," *Appl. Environ. Microbiol.* **36**:926–935.

Zips, A., G. Schaule, and H.-C. Flemming, 1990, "Ultrasound as a Means of Detaching Biofilms," *Biofouling* **2**:322–333.

CHAPTER 7
MODELS FOR SELECTIVITY IN ELECTRODIALYSIS

Japie J. Schoeman
Environmentek, CSIR
Pretoria, South Africa

7.1 SELECTIVITY IN ELECTRODIALYSIS— AN INTRODUCTION

7.1.1 Introduction

Electromembrane processes are finding their place more and more in industry along with microfiltration (MF), ultrafiltration (UF), and reverse osmosis (RO). Some of the similarities between electromembrane processes and these other three processes are:

• Operate by permeation of materials through selective membranes.

• Use many membranes arrayed in parallel.

• Boundary-layer control is important.

• Membrane fouling is a problem.

However, electrodialysis (ED) differs from these other processes in that it is electrically driven rather than pressure driven. Thus, in ED only ions and associated water are transferred.

Electromembrane processes make use of ion-exchange membranes and the processes are usually continuous. The most valuable property of ion-exchange membranes is their permselectivity, which is the ability to allow passage of certain ionic species while preventing the passage of other species (Davis and Brockman, 1972). Because this property of the membranes is more important than their ion-exchange capacity, they are often called *permselective* or *ion-selective* membranes. Permselective membranes have the ability to discriminate between positively and negatively charged species. Permselectivity in ED membranes forms the basis of the ED and electrodialysis reversal (EDR) processes.

Ion selectivity in ED/EDR also has to do with the discrimination between ions of the same charge by ion-exchange membranes (Nishiwaki, 1972; Kedem et al., 1992).

Some membranes have the ability to preferentially transport monovalent ions to divalent ions. Another form of selectivity in ED/EDR is the ability of certain anion-exchange membranes which have been developed to have a low permeability for hydrogen ions (Boudet-Dumy, Lindheimer, and Gavach, 1991). These membranes can be effectively applied for acid concentration, and these membranes are known as *acid-blocking membranes*.

Permselectivity, ion selectivity, models describing selectivity in ED/EDR, and models that can be used to predict membrane performance are described in the literature (Malherbe and Mandersloot, 1960; Strathmann, 1992; Davis and Brockman, 1972; Korngold, 1984; Hoffer and Kedem, 1972; Garza, 1973; Garza and Kedem, 1976; Kedem, Bar-On, and Warshawsky, 1986; Schoeman, 1992; Rubinstein, 1990; and others). Selectivity in ED/EDR, the physical chemistry of ion-selective membranes, and models for selectivity in ED/EDR will be considered in this chapter.

7.1.2 Ion-Exchange Membranes

Ion-exchange membranes are used in most electrodialytic processes. They are ion exchangers in the form of films. There are two basic types: cation-exchange and anion-exchange membranes. Ion-exchange membranes are selective in that they are permeable to cations, but not to anions, or vice versa. Polymer chains are shown in Fig. 7.1 that have negatively charged groups chemically attached to them. The polymer chains are intertwined and also cross-linked at various points. Positive ions are shown freely dispersed in the voids between the chains. However, the fixed negative charges on the chains repel negative ions that try to enter the membrane and exclude them. Thus, because of the negative fixed charges, negative ions cannot permeate the membrane, but positive ones can. If positive fixed charges are attached to the polymer chains instead of negative fixed charges, selectivity for negative ions is achieved. This exclusion as a result of electrostatic repulsion is termed *Donnan exclusion* (Helfferich, 1962).

Selectivity on its own is not enough to make an ion-exchange membrane that is practical for low-cost processing. In addition, the resistance of the membrane to ion transfer must be low. To decrease the resistance, the degree of cross-linking is decreased so that the average interchain distances and the lengths of polymer segments that are free to move are increased. However, if this enlargement of void spaces between polymer chains is carried too far, it can result in volumes in the center of the voids that are not affected by the fixed charges on the chains (the repulsion effect of fixed charges decreases rapidly with distance). Volumes that are unaffected by the fixed charges result in ineffective repulsion of the undesired ions and lowered selectivity. For this reason, a compromise between selectivity and low resistance must usually be made. Membranes now available combine excellent selectivity with low resistance, high physical strength, and long lifetimes.

There are two general types of commercially available ion-exchange membranes. Heterogeneous and homogeneous membranes. Heterogeneous membranes, in which ion-exchange particles are incorporated in film-forming resins, are made by:

1. Calendering mixtures of ion-exchange and film-forming materials
2. Casting films from dispersions of ion-exchange materials in solutions of film-forming materials and allowing the solvent to evaporate
3. Casting films of dispersions of ion-exchange material in partially polymerized film-forming polymers and completing the polymerization

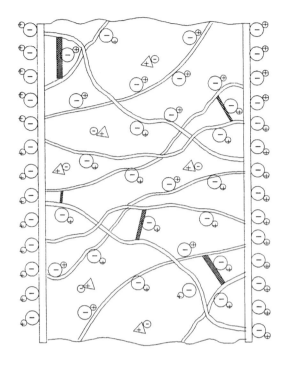

⊖   Fixed negatively charged exchange site: SO_3^-

⊕   Mobile positively charged exchangeable cation: Na^+

⊂▭   Polystyrene chain

▰▰▰   Divinylbenzene crosslink

FIGURE 7.1   Diagram of cation-exchange membrane.

Homogeneous ion-exchange membranes have been made by several methods:

1. Polymerization of mixtures of reactants that can undergo condensation polymerization. At least one of the reactants must contain a moiety that is ionic or can be made to be so charged.
2. Polymerization of mixtures of reactants (one of which is anionic or cationic) that can undergo additional polymerization.
3. Graft polymerization of moieties that are anionic or cationic (or can be made to be) into preformed films.
4. Casting films from a solution of a linear film-forming polymer and a linear polyelectrolyte, and allowing the solvent to evaporate.

Of the theoretical models that have been used to describe ion-exchange membranes, the most popular have been homogeneous models (Davis and Brockman,

1972). Mackay and Meares (1955) present the following six assumptions as the basis for a homogeneous model of cation-exchange membranes:

1. The membrane consists of an isotropic three-dimensional network of polymer chains to which anionic groups are chemically bound. This is swollen by water or an electrolyte solution which constitutes an internal aqueous phase.

2. The anionic groups are completely ionized and are uniformly distributed through the polymer.

3. The internal aqueous phase contains free cations (termed counter-ions because their charges are opposite to the charges affixed to the membrane matrix) to balance the electrical charge of the fixed ions. Sorbed electrolyte, made up of an equal number of free cations and anions, may also be present. The internal aqueous phase is qualitatively similar to an ordinary aqueous electrolyte solution.

4. The fixed anions are surrounded by, or projected into, the internal aqueous phase. The polymer network to which these ions are fixed has the thermal vibrations and rotations characteristic of a swollen rubberlike gel, and there is a definite volume concentration of fixed ions in the system in any particular state.

5. The average distance between junction points of the rubberlike polymer network is large compared to the dimensions of the free ions in the internal solution. Therefore, the network exerts no mechanical sievelike effect on the movement of the free ions.

6. The volume concentration of fixed ions does not vary appreciably with the electrolyte concentration in the external solution with which the membrane is in contact.

Since ion-exchange membranes contain an appreciable concentration of mobile ions, they display good electrolytic conductivity, usually about the same as that of 0.01 to 0.05 mol/L salt solutions (Davis and Brockman, 1972).

7.1.3 Transport Process Occurring During Electrodialysis

7.1.3.1 Transport Processes. A number of transport processes occur simultaneously in multicompartment electrodialysis, and these are illustrated in Fig. 7.2 (Wilson et al., 1960).

The term *ion-selective* or *permselective* is used to describe membranes which, if interposed between two electrolyte solutions, show selective permeability toward ions of a particular sign (Malherbe and Mandersloot, 1960). Such membranes are ion exchangers in sheet form, a cation exchanger being selectively permeable to cations, and an anion exchanger showing the same behavior toward anions; the two types are sometimes designated as "negative" and "positive," respectively, according to the sign of the *fixed ions* in each case.

Counter-ion transport constitutes the major electrical movement in the process; the counter-ions transport with them by electroosmosis a certain quantity of water. Co-ion transport is comparatively small and is dependent on the quality of the ion-selective membrane and on the brine concentration. Water is also transported electroosmotically with the co-ions. Diffusion of electrolyte occurs from the brine to the diluate compartment because in the electrodialysis process the brine stream is usually more concentrated than the diluate stream. Water transport is also associated with electrolyte diffusion. Water transport due to osmosis takes place from the low-concentration diluate compartment into the higher-concentration brine compartment.

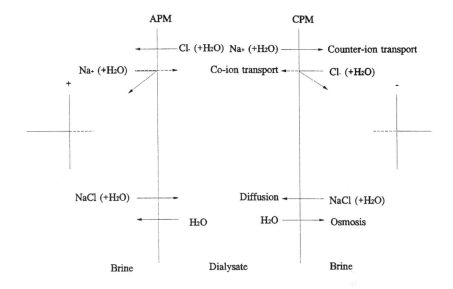

APM: Anion permeable membrane
CPM: Cation permeable membrane

FIGURE 7.2   Illustration of transport processes which can occur simultaneously during the electrodialysis process. *(From Wilson et al., 1960. Reproduced by permission of the authors.)*

The efficiency of demineralization of the liquid in the diluate compartment may be considerably reduced by the countereffects of co-ion transport, diffusion, water transport associated with counter-ion movement, and osmosis. The effect of these unwanted transfer processes can, however, be reduced by the correct selection of membranes and by the selection of the optimum operational procedure for a particular application. Osmosis and electroosmosis are effects which limit the usefulness of electrodialysis as a method of concentrating electrolyte solutions.

Besides having such qualities as chemical stability, mechanical strength, and dimensional stability under conditions of use, an ion-selective membrane should (Wilson et al., 1960):

1. Have a high ion selectivity, that is, a negligible permeability toward ions of the same sign as the fixed ions of the membrane. The selectivity, which may be expressed in terms of the transport number of the "counter-ions" in the membranes, decreases as the salt concentration of the solution in contact with the membranes increases.

2. Have a high electrical conductance when in equilibrium with the most dilute solution to be encountered in the process under consideration. This factor influences the ohmic resistance of an electrodialysis unit.

3. Permit a negligible rate of free electrolyte diffusion under the conditions of concentration difference expected in the process. Salt diffusion through the membrane counteracts electrodialytic salt transport, which is the object of the process, and, hence, causes a loss in efficiency. Free diffusion is controlled by the salt con-

centration difference across the membrane; as this difference increases, so the rate of diffusion increases. This is affected by the membrane selectivity; the higher the membrane selectivity, the smaller the relative increase in the diffusion for a given increase in concentration difference.

4. Have a low osmotic permeability. Ion-selective membranes exhibit anomalous osmotic behavior in electrolyte systems, the most common effect being the abnormally high osmotic flow which takes place in electrolyte systems as compared with nonelectrolyte systems.

7.1.3.2 Coulomb Efficiency.

At the salt concentrations encountered in the practical application of the electrodialysis process, membranes do not exhibit ideal ion selectivity, and the resulting deviation from ideal performance is expressed by the current or coulomb efficiency, which is the ratio of equivalents of electrolyte displaced to faradays of electricity passed in the process. Coulomb efficiency expresses the performance of the process with respect to current and differs from the thermodynamic or minimum work efficiency, which expresses performance with respect to energy consumption.

In practice, the coulomb efficiency of an electrodialytic process is often determined by an inexact method. The flow rate of an effluent stream is determined, and samples are taken of the influent and effluent stream. The quantity of electrolyte displaced is obtained by multiplying the effluent flow rate by the change in concentration found. The coulomb efficiency is then calculated from these values as in Eqs. (7.1) and (7.2). This procedure, however, is subject to an error due to water transfer, as a result of which the effluent and influent flow rates of a single stream are not identical; in addition, there is a minor volume change resulting from the salt displacement. When coulomb efficiencies are determined in this way, it often occurs that the values obtained for the brine (concentrate) and diluate streams individually are unequal.

Thus, in a continuous process, the coulomb efficiency (η_c) is (Wilson et al., 1960):

$$\frac{\text{Number of equivalents transported}}{\text{number of faradays passed} \times \text{number of membrane pairs employed}}$$

and the apparent values are:

$$\eta_{cd} = \frac{Q_d(c_{od} - c_{td})F}{n_{pr}\,I} \tag{7.1}$$

$$\eta_{cb} = \frac{Q_b(c_{tb} - c_{ob})F}{n_{pr}\,I} \tag{7.2}$$

where　　η_{cd}, η_{cb} = apparent coulomb efficiencies based on diluate and brine streams, respectively

Q_d, Q_b = diluate and brine efflux rates, respectively, mL/s

c_{od}, c_{ob} = diluate and brine influent concentrations at time zero, respectively, equiv/mL

c_{td}, c_{tb} = diluate and brine effluent concentrations at time t, respectively, equiv/mL

F = Faraday's constant, coulomb/equiv

n_{pr} = number of membrane pairs

I = current passed, A

Determinations of η_{cd} and η_{cb} enable both the true coulomb efficiency and the mols water transported per equivalent of salt transfer to be calculated.
    The following data were obtained in a continuous ED/EDR process:

Feedwater concentration:	0.0480 ge/L
Product water concentration:	0.0078 ge/L
Diluate flow rate:	9765 L/h
Electrical current (I):	25.5 A
Number of cell pairs:	450
Faraday's constant:	26.8 A-h

Therefore

$$\eta_{cd} = \frac{9765\,(0.0480 - 0.0078)\,26.8}{450 \times 25.5}$$

$$= 91.7\%$$

In a batch process, the coulomb efficiency is:

$$\eta_{cd} = \frac{(c_0 V_0 - c_t V_t)F}{n_{pr}\displaystyle\int_0^t I(t)\,dt} \tag{7.3}$$

$$\eta_{cb} = \frac{(c_t V_t - c_0 V_0)F}{n_{pr}\displaystyle\int_0^t I(t)\,dt} \tag{7.4}$$

where c = concentration of salt solution, equiv/L
         V = volume, L
         $_0$ = concentration and volume at time zero
         $_t$ = concentration and volume at time t

    The following data were obtained during a batch ED experiment with sodium chloride solution:

Feedwater concentration:	2934 mg/L
Product water concentration:	374 mg/L
Initial feed volume:	30 L
Final product volume:	29.6 L
Coulombs (A-s) used:	1852
Number of cell pairs:	75
Faraday's constant:	96,500 C/mol

Therefore

$$\eta_{cd} = \frac{[30\,(0.0502) - 29.6\,(0.00640)]\,96{,}500}{75 \times 1852}$$

$$= 91.5\%$$

Factors which limit the true coulomb efficiency, η_c, are as follows.

Imperfect Selectivity. The penetration of co-ions into a membrane not only reduces the transport number of the counter-ions but leads to the membrane's permitting electrolyte diffusion. Considering the transport numbers only, the coulomb efficiency of an electrodialysis process is (Wilson et al., 1960):

$$\eta_c = 1 - (t_-^c + t_+^a) \tag{7.5}$$

in which t_-^c is the anion transport number in the cation-permeable membrane, and t_+^a is the cation transport number in the anion-permeable membrane.

The effect of backdiffusion on coulomb efficiency can in principle be described by adding a term to this equation, i.e.:

$$\eta_c = 1 - (t_-^c + t_+^a) - \frac{FJ}{i} \tag{7.6}$$

in which i is the current density and J the rate of electrolyte diffusion occurring under the influence of the concentration difference across the membranes which exists in practice. This concentration difference is generally greater than the concentration difference of the bulk solutions because of the existence of polarized layers at the membrane surfaces. Furthermore, the rate of diffusion through ion-selective membranes varies in a complex way, both with the concentration difference and with the absolute level of concentration; this probably results from swelling effects and the variation of ion selectivity with concentration. It is generally possible to estimate the value of the diffusion term only approximately from the measurements made in normal laboratory characterization of membranes (Wilson et al., 1960).

Unwanted Ion Transfer. The effect of unwanted ion transfer on coulomb efficiency may be neglected since practical installations must be operated under substantially depolarized conditions (Wilson et al., 1960). Unwanted ion transfer can influence the coulomb efficiency obtained in a three-compartment arrangement because of the relatively high hydrogen and hydroxyl ion permeabilities of most anion-permeable and cation-permeable membranes, respectively.

Electrical Leakage and Short-Circuiting. The partial short-circuiting of a membrane stack by conduction through manifolds, or by the interconnection of electrolyte feed systems to the electrodes, reduces the effective coulomb efficiency (Wilson et al., 1960). Electrical leakage is, therefore, one of the important considerations in multicell design; in general, such leakage should not exceed 2 percent. The extent of electrical leakage is a complex function of the internal dimensions of a multicell, of the concentration of liquids handled, and of the electrical conductivity of the membranes.

7.2 PHYSICAL CHEMISTRY OF ION-SELECTIVE MEMBRANES

The physical chemistry of ion-exchange membranes is described by Malherbe, Le Roux, and Mandersloot (1960); Helfferich (1962); and others.

7.2.1 Phenomenon of Ion Selectivity in Light of the Fixed-Charge Theory and the Donnan Concept

The chemical structure of a typical commercial cation-exchange resin, sulphonated cross-linked polystyrene, is indicated in Fig. 7.3. It consists of a network of carbon chains and benzene nuclei, with sulphonic acid groups attached to the nuclei. If this structure is immersed in an aqueous medium, it absorbs water and swells due to the affinity between the water molecules and the sulphonic acid groups. The sulphonic acid groups dissociate to yield negatively charged *fixed* ions, chemically bound to the hydrocarbon matrix, and mobile cations, in this case hydrogen ions. The latter are free to move under the influence of an applied electrical field or to exchange with other cations diffusing into the exchanger from the external solution.

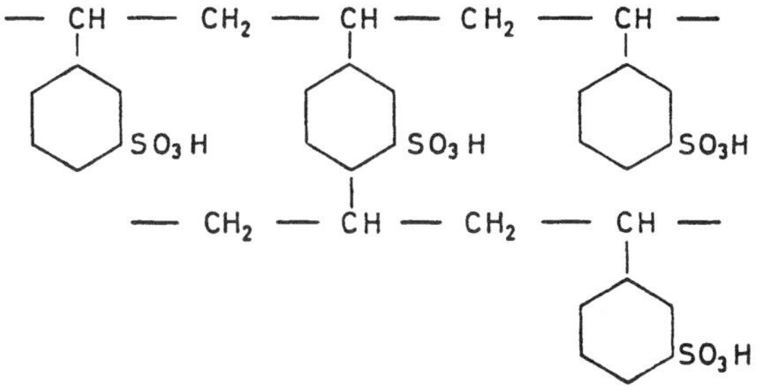

FIGURE 7.3 Structure of sulphonated cross-linked polystyrene. *(From Malherbe and Mandersloot, 1960. Reproduced by permission of the authors.)*

Suppose the cation-exchanger in Fig. 7.3 is in sheet form and in equilibrium with a solution of hydrochloric acid, as indicated diagrammatically in Fig. 7.4 (Malherbe and Mandersloot, 1960). There are present in the swollen exchanger water molecules, fixed-negative sulphonate ions, and an equivalent number of hydrogen ions. In addition to these, equal numbers of hydrogen and chloride ions enter the exchanger by diffusion from the external solution.

If electrical current is passed through the system by applying an external field in a direction perpendicular to the membrane-liquid interface, both hydrogen and chloride ions take part in the transport of electricity through the membrane. It is found, however, that the transport number of the hydrogen ions in the membrane exceeds the value in free solution. In qualitative terms, it may be stated that the electrostatic repelling forces between the fixed-negative ions in the membrane and the chloride ions are the ultimate cause of selective permeability. These forces prevent chloride ions from entering the membrane phase in appreciable number; consequently, they cannot take part in the transport of current in the membrane to the same extent as in free solution.

A clearer picture may be obtained by applying to the system the ideas embodied in the *fixed-charge theory* of membrane phenomena, due originally to Teorell (1935, 1953) and Meyer and Sievers (TMS) (1936a, 1936b). According to this theory, the membrane is viewed as a quasihomogeneous phase in which fixed ions, counter-ions, and co-ions (the last two terms denote mobile ions possessing charges opposite and similar in sign, respectively, to that of the fixed ions) are uniformly dispersed in the internal aqueous phase pervading the membrane. The idea of membrane pores is given less prominence than in many of the older treatments since in modern synthetic ion exchangers, with fixed ion concentrations ranging from about 1 to 8 mol/L and higher, the average distance between the more or less randomly distributed functional groups is small and of the same order as the sizes of the interstices between the units constituting the hydrocarbon matrix. The idea of a membrane

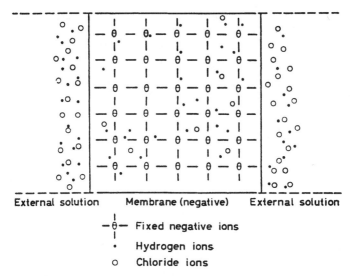

External solution   Membrane (negative)   External solution

$-\overset{|}{\underset{|}{\theta}}-$ Fixed negative ions

•   Hydrogen ions

○   Chloride ions

FIGURE 7.4   Diagrammatic representation of a negative membrane in equilibrium with a solution of hydrochloric acid. *(From Malherbe and Mandersloot, 1960. Reproduced by permission of the authors.)*

involving pores with charged walls and electrical double layers is, therefore, no longer utilized. The Donnan concept is applied to the membrane-liquid interface by regarding it as a *semipermeable membrane* separating two phases: the external one consists of water molecules and mobile ions of both signs that can diffuse across the interface, whereas the internal or membrane phase contains the same components plus the fixed ions, which cannot diffuse.

Although the Donnan equation in its simplest form (which neglects activity coefficients and relates internal to external ionic concentrations) is only a rough approximation for real systems, it is sufficient for a qualitative description of the variation of internal with external concentration at membrane-liquid interfaces.

For the system of Fig. 7.4, the electroneutrality condition is

$$[H^+] = \overline{X} + [Cl^-] \tag{7.7}$$

where $[H^+]$, $[Cl^-]$, and $\overline{X}$ are the hydrogen, chloride, and fixed ion concentrations, respectively, in the solution pervading the membrane. Treating both internal and external solutions as ideal, equality of chemical potentials according to the Gibbs-Donnan principle requires that at equilibrium

$$[H^+][Cl^-] = c^2 \tag{7.8}$$

where c is the concentration of the external solution. Combination of Eqs. (7.7) and (7.8) yields

$$[H^+] = \left[\left(\frac{\overline{X}}{2}\right)^2 + c^2\right]^{1/2} + \frac{\overline{X}}{2} \quad \text{and} \quad [Cl^-] = \left[\left(\frac{\overline{X}}{2}\right)^2 + c^2\right]^{1/2} - \frac{\overline{X}}{2} \tag{7.9}$$

It can readily be seen that for membranes of high fixed ion concentration in equilibrium with relatively dilute external solutions

$$[H^+] \cong \overline{X} \gg c \gg [Cl^-] \tag{7.10}$$

i.e., the concentration of co-ions in the membrane is low compared with both that of the counter-ions and that of the ions in the external electrolyte. This *ion exclusion effect* is largely responsible for the ion-selective action of ion-exchange membranes, since the preponderance of counter-ions over co-ions naturally causes them to carry a greater fraction of the total current in the membrane than in free solution, where the concentrations of the two types of ions are equal.

Reference is often made to an "ideally selective membrane." This is a hypothetical membrane with a fixed ion concentration so high in comparison with the concentration of the external solution that co-ions are completely excluded. If current is passed through such a membrane, it is carried by the counter-ions only; hence, their transport number equals unity.

Consideration of Eq. (7.9) shows that at sufficiently low external concentrations any charged membrane behaves as an ideally selective one.

The variation of internal with external ionic concentration, as predicted by Eq. (7.9), is illustrated in Fig. 7.5. As the concentration of the external solution is increased, more electrolyte diffuses into the membrane, causing an increase in $[Cl^-]$, $[H^+]$, and the ratio $[Cl^-]/[H^+]$, which means that relatively more co-ions become available for the transport of electricity through the membrane, and that the membrane becomes less selective. This decrease of selectivity with increasing external concentration is observed in practice (see Fig. 7.6) and, in the light of the preceding considerations, it can be seen that it is due largely to the gradual weakening of the ion exclusion effect as the concentration of the external solution increases.

7.2.2   A Quantitative Measure of Ion Selectivity

The transport number of the counter-ions is sometimes used as a measure of the selectivity of a membrane. A more rational measure would be one which indicated the

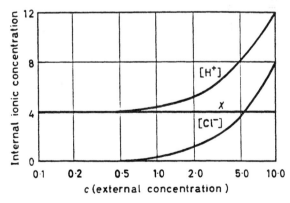

FIGURE 7.5   Dependence of internal concentration upon external concentration as predicted by the Donnan equation for a cation-selective membrane $\overline{X} = 4$. *(From Malherbe and Mandersloot, 1960. Reproduced by permission of the authors.)*

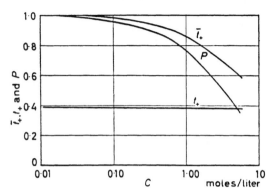

FIGURE 7.6 Dependence of counter-ion transport numbers and permselectivity upon external concentration. *(From Malherbe and Mandersloot, 1960. Reproduced by permission of the authors.)*

increase in transport number over the value in free solution, due to the presence of the membrane. To meet this requirement, the permselectivity S has been defined as:

$$S = \frac{\text{(transport number of counter-ions in membrane)} - \text{(transport number in free solution)}}{1 - \text{(transport number in free solution)}}$$

$$= \frac{\bar{t} - t}{1 - t} \tag{7.11}$$

where the barred symbol refers to the membrane phase.

The permselectivity therefore expresses the increase in transport number over the value in free solution as a fraction of the maximum possible increase, i.e., the increase that would be observed in the case of an ideally selective membrane, for which $\bar{t}$, and hence S, is equal to 1.

The dependence of counter-ion transport number and permselectivity on external concentration, observed for a cation-selective membrane in equilibrium with NaCl solution, is illustrated in Fig. 7.6.

7.2.3   Permselectivity and Ion-Exchange Membrane Equilibria

Permselectivity and ion-exchange membrane equilibria are described by Strathmann (1992), Korngold (1984), and Davis and Brockman (1972).

The permselectivity of ion-exchange membranes results from the exclusion of co-ions from the membrane phase (Fig. 7.1). For a cation-exchange membrane in the dilute solution of a strong electrolyte, this can be illustrated using the model of Gregor (1951) according to Strathmann (1992). Here, the concentration of the cations is higher in the exchanger than in the solution, because the cations are attracted by the negatively charged fixed ions. On the other hand, the concentration of mobile anions in the solution is higher than in the exchanger. An equilibrium in the concentration of electrically neutral-charged particles could take place via diffusion. Because electroneutrality is required here as well, the process is limited if charged particles are involved. The passing of cations into the solution and of anions into the cation-exchange membrane leads to a counteracting space charge. Therefore, the so-called Donnan exclusion equilibrium is established between the attempt of diffusion on one side and the establishment of an electrical potential difference on the other. The activity of the counter-ions in the exchanger is higher than in the solution and the activity of the co-ions is smaller, respectively. The thermodynamical treatment by Donnan and Guggenheim (1932) is based on an equilibrium in the membrane phase (indicated by absence of a superscript) and in the outer phase (indicated by super-script $'$) of the electrochemical potential μ_i of all ions that are able to permeate through the membrane:

$$\mu_i = \mu_i' \tag{7.12}$$

In both of the adjacent phases, the concentrations as well as the osmotic pressure and the electrical potential can be different. For the distribution of a special ion, an established electrical potential difference $\psi - \psi'$, the Donnan potential $\Delta\psi_{\text{don}}$, can be described as a function of the different activities a_i and the swelling pressure π_i (Strathmann, 1992):

$$\Delta\psi_{\text{don}} = \psi - \psi' = \frac{1}{z_i F} \left[RT \ln \left(\frac{a_i}{a_i'} \right) - \overline{V}_i \pi_i \right] \tag{7.13}$$

where z_i = valency of ion species i
        F = Faraday's constant
        R = gas constant
        T = absolute temperature
        $\overline{V}_i$ = partial molar volume of component i
        π_i = swelling pressure

The numerical value of $\Delta\psi_{\text{don}}$ is negative for the cation-exchange membrane and positive for the anion-exchange membrane. Unfortunately, the Donnan potential cannot be evaluated by direct measurement; however, it presents the starting point for the calculation of the distribution of the co-ions between the solution and the membrane and, hence, for the determination of the permselectivity. The swelling pressure, which is directly proportional to the concentration of the fixed ions and inversely proportional to the concentration of the electrolyte, can be substituted, and for 1:1 electrolytes, an approximate relation for the concentration of the co-ions in the membrane c_{co} is obtained (Strathmann, 1992):

$$c_{\text{co}} = \frac{c'^2}{\overline{X}} \left(\frac{\gamma'}{\gamma} \right)^2 \tag{7.14}$$

where c' = electrolyte concentration in solution
        $\overline{X}$ = concentration of fixed ions in membrane
        γ' and γ = average activity coefficients of salt in solution and membrane, respectively

In a typical situation in which a membrane with a fixed ion concentration of 6×10^{-3} eq/cm^3 is immersed in a solution of 0.01 N NaCl, the co-ion concentration in the membrane is only 1.7×10^{-8} eq/cm^3.

Equation (7.14) is based on the theory of Teorell (1951) and Meyer and Sievers (1936). However, the more complex structure of modern membranes cannot be adequately described exclusively by this theory (Strathmann, 1992). The remaining differences in the observed and expected membrane behavior are primarily a result of the nonuniformity in the distributions of molecular components in the membrane. This results from structural irregularities on a molecular level and from the influence of the electric field. Additionally, the practical application of the thermodynamic relations is rather limited by the difficulties in the experimental measurement of independent interaction, diffusion, resistance, and frictional coefficients.

The Donnan exclusion equilibrium as well as the selectivity increase depend on (Strathmann, 1992):

1. The concentration of the fixed ions
2. The increasing valence of the co-ions
3. The decreasing valence of the counter-ions
4. The decreasing concentration of the electrolyte solution
5. The decreasing affinity of the exchanger with respect to the counter-ions

Further important parameters for the characterization of ion-exchange membranes are the density of the polymer network, the hydrophobic and hydrophilic properties of the matrix polymer, the distribution of the charge density, and the morphology of the membrane itself (Strathmann, 1992). All these parameters do not only determine the mechanical properties, but they also have a considerable influence on the sorption of the electrolytes and the nonelectrolytes and therefore on the swelling.

The most desirable properties for ion-exchange membranes are (Strathmann, 1992):

• *High permselectivity:* An ion-exchange membrane should be highly permeable to counter-ions, but should be impermeable to co-ions.

• *Low electrical resistance:* The permeability of an ion-exchange membrane for the counter-ions under the driving force of an electrical potential gradient should be as high as possible.

• *Good mechanical and form stability:* The membrane should be mechanically strong and should have a low degree of swelling or shrinking in transition from dilute to concentrated ionic solutions.

• *High chemical stability:* The membrane should be stable over a pH range from 0 to 14 and in the presence of oxidizing agents.

Optimization of the properties of ion-exchange membranes is often difficult because the parameters determining the different properties often have opposing effects (Strathmann, 1992). For instance, a high degree of cross-linking improves the mechanical strength of the membrane but also increases its electrical resistance. A high concentration of fixed ionic charges in the membrane matrix leads to a low electrical resistance but, in general, causes a high degree of swelling combined with poor mechanical stability. The properties of ion-exchange membranes are determined by two parameters: the basic polymer matrix and the type and concentration of the fixed ionic moiety. The basic polymer matrix determines to a large extent the mechanical, chemical, and thermal stability of the membrane. Very often, the matrixes of ion-exchange membranes consist of hydrophobic polymers such as polystyrene, polyethylene, and polysulphone. Although these basic polymers are insoluble in water and show a low degree of swelling, they may become water soluble by the introduction of the ionic moieties. Therefore, the polymer matrixes of ion-exchange membranes are very often cross-linked. The degree of cross-linking then determines to a large extent the degree of swelling and the chemical and thermal stability, but it also has a large effect on the electrical resistance and the permselectivity of the membrane.

The type and concentration of the fixed ionic charges determine the permselectivity and electrical resistance of the membrane, but they also have a significant effect on the mechanical properties of the membrane (Strathmann, 1992). The degree of swelling, especially, is affected by the concentration of the fixed charges. The following moieties are used as fixed charges in cation-exchange membranes:

$$SO_3^- \quad -COO^- \quad -PO_3^{2-} \quad -HPO_2^- \quad -AsO_3^{2-} \quad SeO_3^-$$

In anion-exchange membranes, fixed charges may be:

$$-NH_3^+ \quad -RNH_2^+ \quad -R_3N^+ \quad =R_2N^+ \quad -R_3P^+ \quad -R_2S^+$$

These different ionic groups have significant effects on the selectivity and electrical resistance of the ion-exchange membrane. The sulphonic acid group, e.g., $-SO_3^-$, is completely dissociated over nearly the entire pH range, while the carboxylic acid group, $-COO^-$, is virtually undissociated when the pH range is <3. The quaternary ammonium group $-R_3N^+$ again is completely dissociated over the entire pH range, while the primary ammonium group $-NH_3^+$ is only weakly dissociated. Accordingly, ion-exchange membranes are referred to as being weakly or strongly acidic or basic in character. Most commercially available cation-exchange membranes have $-SO_3^-$ or $-COO^-$ groups, and most anion-exchange membranes contain $-R_3N^+$ groups.

7.2.4   Diffusion of Electrolytes in and Through Ion-Selective Membranes

Diffusion of a single electrolyte under a concentration gradient is described by Malherbe and Mandersloot (1960).

Diffusion is of importance in electrodialysis, as backdiffusion across the membranes tends to decrease the efficiency of the process. It is therefore instructive to consider the case of an ion-selective membrane separating two solutions of the same electrolyte (say, NaCl) at different concentrations, and to investigate the relationships between the electrolyte flux J, the concentrations c_1 and c_2 of the solutions, and membrane properties such as the fixed ion concentration $\overline{X}$.

Consider the three cases illustrated in Fig. 7.7: (1) a nonselective membrane with no ionizable groups, $\overline{X} = 0$; (2) a negative membrane of intermediate selectivity, say $\overline{X} = 4$ (arbitrary units); and (3) an "ideally selective" negative membrane, i.e., one with a fixed ion concentration which is so high that co-ions are completely excluded. It is assumed that the external solutions are agitated sufficiently vigorously to break up diffusion layers.

In the case of the nonselective membrane, the relation between flux and concentration difference takes the form of Fick's law, i.e.

$$J = K_{ff}D\frac{(c_1 - c_2)}{\delta_m} \tag{7.15}$$

where D is the diffusion coefficient of the electrolyte in free solution ($cm^2\,s^{-1}$), δ_m is the membrane thickness (cm) and K_{ff} is a *form-factor* characteristic of the membrane, which takes into account the tortuosity of the paths along which the diffusing ions move, and the fact that only a certain fraction of the membrane area is available

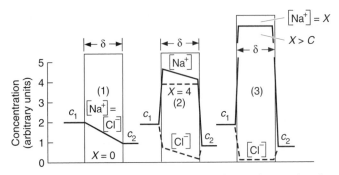

FIGURE 7.7   Concentration profiles across negative membranes of varying fixed ion concentration separating sodium chloride solutions of concentration 2 and 1, respectively. *(From Malherbe and Mandersloot, 1960. Reproduced by permission of the authors.)*

for the diffusion of ions, J is the amount of electrolyte diffusing through 1 cm^2 of membrane in one second, and c_1, c_2 are concentrations of the external solutions. It should be remembered that, with most electrolytes, D is not a constant but shows minor variations with concentration (Robinson and Stokes, 1955).

In case 2, Eq. (7.15) does not hold. If the membranes of cases 1 and 2 have the same internal geometry—i.e., their form factors (K_{ff}) are identical—the flux observed will be smaller in case 2 than in case 1; also, the relation between flux and concentration is more complicated than the proportionality observed with nonselective membranes. The reason for this is that in the case of ion-selective membranes in contact with electrolyte solutions, there are concentration discontinuities at the membrane-liquid interfaces as a result of the Donnan equilibrium. Since these concentration discontinuities correspond to equilibrium conditions, they have no direct effect on the electrolyte flux, but they determine the concentration gradient in the interior of the membrane, which in turn determines the flux. For the nonselective membrane, the absolute value of the concentration gradient is $(c_1 - c_2)/\delta_m$, whereas in the case of the ion-selective membrane of fixed ion concentration X, its absolute average value is, according to the simple Donnan Eq. (7.9)

$$\frac{[Cl^-]_1 - [Cl^-]_2}{\delta_m} = \frac{\{(\tfrac{1}{2}\overline{X})^2 + c_1^2\}^{1/2} - \{(\tfrac{1}{2}\overline{X})^2 + c_2^2\}^{1/2}}{\delta_m} < \frac{(c_1 - c_2)}{\delta_m} \qquad (7.16)$$

where $[Cl^-]_1$ and $[Cl^-]_2$ denote the concentrations of co-ions (and hence of the Donnan-sorbed NaCl) in the membrane phase at the two interfaces. The inequality in Eq. (7.16) is consistent with the experimental observation that the electrolyte flux across an ion-selective membrane is less than that across a nonselective membrane of equal thickness and similar internal geometry.

As the membrane becomes more selective (i.e., as $c/\overline{X} \to 0$, corresponding to case 3 in Fig. 7.7), the concentration gradient tends to zero. For an ideally selective membrane, the flux is therefore zero as a result of the complete exclusion of co-ions, and such a membrane is impermeable to diffusing electrolyte under the conditions postulated.

On account of their complexity, the exact theoretical description of membrane diffusion phenomena is difficult. Nevertheless, Teorell (1935, 1953) has indicated a very successful approach to the problem. He used the concepts of the fixed-charge theory and a kinetic treatment, starting from the general principle

$$\text{Flux} = \text{mobility} \times \text{concentration} \times \text{total driving force} \qquad (7.17)$$

and following the procedure of Nernst (1888, 1889), he considered the total driving force as being composed of an osmotic and an electrical term:

$$\text{Total driving force per gram ion} = -\left[RT\frac{d(\ln \bar{c})}{dx} + zF\frac{d\phi}{dx}\right] \qquad (7.18)$$

where $\bar{c}$ = concentration of ion in membrane phase
  x = distance in membrane, extending from $x = 0$ at interface (1) to $x = \delta_m$ at other interface (2)
  z = valency (positive for cations and negative for anions)
  ϕ = electrical potential within membrane (usually taken $\phi_1 = 0$ at $x = 0$)

A brief elucidation of Eq. (7.18) may be of value here. According to modern views, the driving force giving rise to diffusion of component i in solution is (in the absence of electrical effects) the negative gradient of its chemical potential, $d\mu_i/dx$ (Robinson and Stokes, 1955). By definition

$$\mu_i = \mu_i^0 + RT \ln a_i = \mu_i^0 + RT \ln c_i\gamma_i$$

where a_i, c_i, and γ_i denote the activity, concentration, and activity coefficient, respectively, of component i, and μ_i^0 is its chemical potential in the standard state. Since this is constant,

$$\frac{d\mu_i}{dx} = RT\frac{d(\ln c_i\gamma_i)}{dx}$$

If electrical forces (e.g., an applied electrical field, or a diffusion potential, or both) are also operative, the total force includes an electrical term $zF\,d\phi/dx$. As pointed out by Teorell (1953), some workers prefer to combine the osmotic and electrical terms and to regard the driving force as the negative gradient of the *electrochemical potential* $d\bar{\mu}/dx$,

where $\bar{\mu}_i = \bar{\mu}_i^0 + RT \ln \bar{a}_i$

$\bar{a}_i$ being referred to as the *electrochemical activity* of species i. Equation (7.18) then simplifies to:

$$\text{Total driving force per gram ion} = -RT\,d\ln\frac{\bar{a}_i}{dx}$$

$$= -RT\frac{d\ln(c_i\gamma_i\xi_i)}{dx}$$

where $$\xi_i = \exp\left(\frac{z_iF\phi}{RT}\right)$$

In Teorell's treatment, solutions are assumed ideal and activity coefficients equated to unity; also, the effect of liquid flow due to osmosis is neglected. The differential equation for the flux of a monovalent cation is therefore

$$J_+ = -\overline{U}_+ \overline{c}_+ \left(RT \frac{d(\ln \overline{c}_+)}{dx} + F \frac{d\phi}{dx} \right) \tag{7.19}$$

where $\overline{U}_+$ is the absolute mobility (diffusion) of the cation in the membrane. A similar relation holds for monovalent anions, with J_-, $\overline{U}_-$, $\overline{c}_-$, and $-F(d\phi/dx)$ replacing the corresponding terms for the cation. Equation (7.19) is, from the point of view of steady-state thermodynamics, incomplete and approximate since mutual interactions between the various species of migrating particles are neglected (Malherbe and Mandersloot, 1960). For an ion-selective membrane of fixed ion concentration $\omega \overline{X}$ ($\omega = +1$ for positive and -1 for negative membranes), separating two solutions of a uni-univalent electrolyte at different concentrations, integration of Eq. (7.19) yields for the electrolyte flux

$$J = -\frac{RT}{\delta_m} \frac{2\overline{U}_+ + \overline{U}_-}{\overline{U}_+ + \overline{U}_-} \left[(\overline{c}_2 - \overline{c}_1) - \frac{F}{2RT} (\phi_2 - \phi_1)\omega \overline{X} \right] \tag{7.20}$$

where $\overline{c}_1$ and $\overline{c}_2$ are the concentrations of the Donnan-sorbed electrolyte at the membrane faces, and $\phi_2 - \phi_1$ is the diffusion potential existing in the interior of the membrane. The diffusion potential can be calculated by the Henderson formula (Malherbe and Mandersloot, 1960) and is caused by the difference in the mobilities of counter-ions and co-ions. Since, in order to preserve electroneutrality, the two types of ions must move with equal velocities, the effect of the diffusion potential is to retard the diffusion of the faster ion and to accelerate that of the slower ion. The factor

$$\frac{2RT \, \overline{U}_+ \overline{U}_-}{\overline{U}_+ + \overline{U}_-} = \overline{D} \tag{7.21}$$

is the diffusion coefficient of the electrolyte in the membrane; since the mobilities in the membrane phase are not the same as those in free solution, the diffusion coefficients for the two media are also different. They are related by a form factor similar to the one defined in Eq. (7.15), i.e.

$$\overline{D} = K_{ff} D \tag{7.22}$$

For membranes of high fixed ion concentration ($\overline{X} > c$), the diffusion potential is small, and the electrical term in Eq. (7.20) can be neglected in comparison with the osmotic term. Introducing this simplification and expressing internal concentrations $\overline{c}$ in terms of external concentrations c by means of the simple Donnan equation, Eq. (7.20) becomes

$$J \cong -\frac{\overline{D}}{\delta_m} (\overline{c}_2 - \overline{c}_1)$$

$$= -\frac{\overline{D}}{\delta_m} \left\{ \left[\left(\frac{\omega \overline{X}}{2} \right)^2 + c_2^2 \right]^{1/2} - \left[\left(\frac{\omega \overline{X}}{2} \right)^2 + c_1^2 \right]^{1/2} \right\} \qquad \text{cf. Eq. (7.16)}$$

$$\cong -\frac{\overline{D}}{\delta_m} \frac{(c_2^2 - c_1^2)}{\overline{X}} \tag{7.23}$$

The last step is obtained by making use of the expansion

$$\left[1 + \left(\frac{2c}{\omega \overline{X}}\right)^2\right]^{1/2} = 1 + \frac{2c^2}{(\omega \overline{X})^2} + \cdots$$

For the case in which diffusion takes place from a solution of concentration c through the membrane into pure water (i.e., $c_1 = c$, $c_2 = 0$), Eqs. (7.15) and (7.23) reduce to:

$$\text{Nonselective membrane:} \qquad J = \overline{D}\,\frac{c}{\delta_m}$$

$$\text{Ion-selective membrane:} \qquad J \cong \frac{\overline{D}}{\overline{X}}\,\frac{c^2}{\delta_m}$$

$$(7.24)$$

indicating that for high fixed ion concentrations, and within the limits of validity of the Teorell treatment, the electrolyte flux is approximately proportional to the square of the external concentration.

Equation (7.23) represents the dependence of diffusion flux on external concentrations as $c/\overline{X} \to 0$ (i.e., as case 3, Fig. 7.7, is approached) and contrasts with Eq. (7.16) which holds for a nonselective membrane (case 1, $\overline{X} = 0$). Membranes of intermediate properties may be expected to show a behavior intermediate between the linear and quadratic dependences on external concentration indicated by Eqs. (7.16) and (7.23). Some isolated observations have served to illustrate the fact that membranes at ordinary concentrations (~0.2 N) can show a complex dependence of diffusion flux on external concentration (Malherbe and Mandersloot, 1960). Diffusion was measured in systems in which membranes separated well-stirred NaCl solutions and distilled water. The fluxes obtained with NaCl concentrations of 1.5 and 0.5% were compared. The ratio of these fluxes was found with most parchment-based membranes to lie between 4.0 and 5.0. With some highly cross-linked membranes, values of the flux ratio of less than 3.0 were found, a value of 1.81 being the lowest observed ratio. Such a serious deviation from the behavior predicted by Eq. (7.23) suggests that factors such as anomalous osmosis leading to incongruent diffusion and variation of swelling with external concentration are important in determining the diffusion flux for membranes of case 2.

Teorell's basic assumptions are those usually made in applications of the fixed-charge theory, i.e., absence of sieve effects, instantaneous establishment of permanent Donnan equilibria at the interfaces, and a constant fixed ion concentration throughout the membrane. The most serious limitations of his theory are perhaps the use of concentrations instead of activities and the neglect of osmotic flow. Schlögl (1955), in his theoretical treatment of anomalous osmosis, has indicated that in certain cases a strong positive osmosis (i.e., from the dilute to the concentrated solution) can have the effect of carrying electrolyte along in the same direction by convection. This results in so-called *incongruent diffusion*, or movement of electrolyte from a lower to a higher concentration. Hirsch-Ayalon (1951) has observed this phenomenon for the diffusion of Na_2SO_4 across an oxidized cellophane membrane.

Mackie and Meares (1955) investigated the diffusion of several electrolytes (HCl, NaCl, NaBr, $MgCl_2$, $MgSO_4$) through a disc of Zeo-Karb 315 (a sulphonated resin of the phenolformaldehyde type) and compared experimental fluxes with those calculated from theory. Their basic equations were the same as those of Teorell, except that activity coefficients and an osmotic flow term were included, e.g., for the cation flux

$$J_+ = -\overline{U}_+ \, \overline{c}_+ \left[RT \frac{d(\ln c_+ \gamma_+)}{dx} + z_+ F \frac{d\phi}{dx} \right] + \overline{c}_+ v \qquad (7.25)$$

where γ_+ = molar ionic activity coefficient
         z_+ = valency
         v = rate in cm/s of osmotic flow

The other terms have the same significance as in Eq. 7.19. In order to integrate the equation, the authors established empirical relationships between experimentally determined activity coefficients and ionic concentrations, and also measured the osmotic flow, calculating its influence on the concentration profile by making the reasonable assumption that it would affect only the distribution of the Donnan-sorbed electrolyte. Ionic mobilities in the resin phase were calculated from those in free solution, using a form factor based on considerations of the structure of the swollen resin and its water content.

The integrated flux equation contained three terms representing the contributions of concentration gradient, the variation of activity coefficients, and the osmotic flow, respectively. It was found that each of these terms made a significant contribution to the total theoretical flux. Agreement between theoretical and experimental curves of flux versus ingoing concentration was good.

7.2.5   Membrane Potentials

Membrane potential is the potential difference which exists across a membrane separating two electrolyte solutions of different composition. If the solutions contain the same electrolyte at different concentrations, it is called a *concentration potential;* in cases where the two solutions contain more than one species of counter-ion, the term *multi-ionic potential* (MIP) is used, of which a *bi-ionic potential* (BIP) represents a special case (Malherbe and Mandersloot, 1960). In using these terms, it is usually understood that the system referred to is at constant temperature and pressure.

Like liquid junction potentials, membrane potentials cannot be calculated from the electromotive forces (emf's) of cells without making nonthermodynamic assumptions concerning the values of single ion activities. In experiments designed to measure membrane potentials directly, use is made of salt bridges, and the liquid junction potentials are either assumed to be zero, or their approximate values can be calculated (Malherbe and Mandersloot, 1960). In spite of these uncertainties, membrane potentials have been studied extensively because of their fundamental interest, their bearing on certain biological processes, and their importance in the characterization of membranes.

7.2.5.1   *Concentration Potentials*
The Nernst Equation.   Consider the following cell in which $a_1 > a_2$:

$$\overline{\text{Ag}} \mid \text{AgCl} \mid \text{NaCl aq.}(a_1) \mid \frac{\text{ion-selective}}{\text{membrane}} \mid \text{NaCl aq.}(a_2) \mid \text{AgCl} \mid \text{Ag}^+ \qquad (7.26)$$

This is a concentration cell "with transference," in which a membrane replaces the liquid junction, and its emf may be calculated along quasi-thermodynamical lines by considering the changes brought about by the passage of one faraday of electricity under the driving force of its own potential. This results in the transfer of $\bar{t}_+$ equivalents of NaCl from solution 1 to solution 2, and the movement of electrons through the external connecting wire from the electrode on the left to that on the right. The quantity $\bar{t}_+$ is an average transport number of the sodium ion in the membrane. The reactions taking place at the electrodes are $Ag + Cl^- = AgCl + e^-$ at the left, and the reverse one at the right. If the process is carried out reversibly, consideration of the free energy change involved leads to the following expression for the emf of the cell:

$$
E_{cell} = \bar{t}_+ \left(\frac{RT}{F}\right) \ln \left(\frac{a_1}{a_2}\right)
$$

$$
= 2\bar{t}_+ \frac{RT}{F} \ln \left[\frac{(a_\pm)_1}{(a_\pm)_2}\right] \tag{7.27}
$$

where $a_\pm$ denotes the mean ionic activity in each solution. $[a = (a_\pm)^2 = a_+ a_-]$. Equation (7.27) is exact, provided that $\bar{t}_+$ is taken as an average transport number, with the water as frame of reference.

The emf of the cell can be regarded as the algebraic sum of three components: the potentials between the electrodes and the solutions with which they are in contact (the algebraic sum of these will be denoted by E_{el}), and the membrane potential E_m. The Nernst expression for the former is

$$
E_{el} = \left(\frac{RT}{F}\right) \ln \left[\frac{(a_-)_1}{(a_-)_2}\right] \tag{7.28}
$$

where a_- is the activity of the chloride ion, so that the membrane potential

$$
E_m = E_{cell} - E_{el}
$$

$$
= \frac{RT}{F} \left[\bar{t}_+ \ln \frac{(a_+)_1(a_-)_1}{(a_+)_2(a_-)_2} - \ln \frac{(a_-)_1}{(a_-)_2}\right]
$$

$$
= \frac{RT}{F} \left[\bar{t}_+ \ln \frac{(a_+)_1}{(a_+)_2} - \bar{t}_- \ln \frac{(a_-)_1}{(a_-)_2}\right] \tag{7.29}
$$

since $\bar{t}_+ + \bar{t}_- = 1$.

Equation (7.29) contains the single ion activities a_+ and a_-, which cannot be evaluated, except in the Debye-Hückel range, i.e., for concentrations less than 10^{-3} molar, and it is therefore of very limited practical use in this form. Usually the non-thermodynamic and approximate assumption is made that (Malherbe and Mandersloot, 1960)

$$
\frac{(a_+)_1}{(a_+)_2} = \frac{(a_-)_1}{(a_-)_2} = \frac{(a_\pm)_1}{(a_\pm)_2} \tag{7.30}
$$

so that

$$
E_m = \frac{RT}{F} (\bar{t}_+ - \bar{t}_-) \ln \frac{(a_\pm)_1}{(a_\pm)_2} \tag{7.31}
$$

which gives the membrane potential in terms of transport numbers and mean ionic activities, both of which can be evaluated experimentally. Equation (7.31) is a special case of the Henderson equation for liquid junction potentials (Malherbe and Mandersloot, 1960).

If Equation (7.31) is regarded as exact, it is clear that an ideally selective negative membrane $(\bar{t}_+ = 1, \bar{t}_- = 0)$ placed in cell Eq. (7.26) would give rise to a total emf,

$$E_{\max} = \frac{2RT}{F} \ln \frac{(a_{\pm})_1}{(a_{\pm})_2} \tag{7.32}$$

the membrane potential being half this value because the membrane and resultant electrode potentials are equal and reinforce each other. In the case of an ideally selective positive membrane $(\bar{t}_+ = 0, \bar{t}_- = 1)$, the two last-named potentials are again equal in magnitude, but opposite in sense; the total emf is therefore zero. In actual membranes, the transport number of the co-ions is not zero, so that for cation-selective membranes E lies somewhere between $t_+E_{\max}$ and $E_{\max}$, whereas for anion-selective membranes E has a positive value less than $t_+E_{\max}$ (t_+ being the transport number of the cation in free solution). The transport number of the sodium ion in the membrane may be calculated from the experimentally measured value of E according to the formula

$$\bar{t}_+ = \frac{E}{E_{\max}} \tag{7.33}$$

Correction of the Nernst Equation for Water Transport.   Staverman (1952) and Scatchard (1953) have indicated the effect of water transport on the membrane potential. Scatchard pointed out that the transport numbers $\bar{t}_+$ and $\bar{t}_-$ in Eq. (7.31) are those relative to the liquid, which moves with respect to the membrane. If it is desired to express the membrane potential in terms of "true" transport numbers, i.e., using the membrane as frame of reference, a correction for water transfer must be introduced. Scatchard's equation for the membrane potential may be written as

$$\frac{E_m F}{RT} = -\int_{(1)}^{(2)} \Sigma_i \bar{t}_{\text{tri}} \, d \ln a_i \tag{7.34}$$

where t_{tri} is the "true" transference number of species i, i.e., the number of moles per faraday of species i transferred across the membrane in the direction of positive current. According to this definition, the transference number of a cation is positive, and that of an anion negative. The transference number of water, $\bar{t}_{\text{trw}}$, may be either positive (in the case of cation-exchange membranes) or negative (for anion-exchangers). Applying Scatchard's equation to a membrane separating two solutions of a uni-univalent electrolyte, and integrating

$$\frac{E_m F}{RT} = \bar{t}_{\text{tr+}} \ln \frac{(a_+)_1}{(a_+)_2} + \bar{t}_{\text{tr-}} \ln \frac{(a_-)_1}{(a_-)_2} + \bar{t}_{\text{trw}} \ln \frac{(a_w)_1}{(a_w)_2} \tag{7.35}$$

where a_w is the activity of the water, and the $\bar{t}$'s now denote mean "true" transference numbers. Equation (7.35) is similar to Eq. (7.31) (note that $t_{\text{tr-}}$ is negative), except for the water correction term $\bar{t}_{\text{trw}} \ln (a_w)_1/(a_w)_2$. Consideration of the signs of the transference numbers and logarithmic terms shows that the water correction term has the effect of decreasing the absolute value of the membrane potential.

Scatchard showed that, by making use of the Gibbs-Duhem equation, the water correction term can be obtained in a more useful form:

$$-\int_{(1)}^{(2)} \bar{t}_{\text{trw}} \, d \ln a_w = 0.018 \int_{(1)}^{(2)} c\bar{t}_{\text{trw}} \, d \ln a_{\pm}^2$$

$$= -0.036c' \, \bar{t}_{\text{trw}} \ln \frac{(a_{\pm})_1}{(a_{\pm})_2} \tag{7.36}$$

In the second expression, c denotes a concentration intermediate between c_1 and c_2. This intermediate concentration is that of a hypothetical solution with which a certain thin slice of the membrane can be considered to be in equilibrium; in the final expression, $\bar{t}_{trw}$ and c' represent mean values. Introducing this correction into Eq. (7.27), the following expression is obtained for the emf of the cell Eq. (7.26)

$$E = \frac{2RT}{F}(\bar{t}_{tr+} - 0.018\, c'\, \bar{t}_{trw}) \ln \frac{(a_\pm)_1}{(a_\pm)_2} \qquad (7.37)$$

from which it can be seen that the relationship between the electrometric transport number $\bar{t}_+$ (calculated according to Eq. (7.33)) and the "true" transference number $\bar{t}_{tr+}$ is

$$\bar{t}_{tr+} = \bar{t}_+ + 0.018\, c'\, \bar{t}_{trw} \qquad (7.38)$$

Equation (7.38) has been tested, and experimental values of $\bar{t}_{tr+}$ determined by a Hittorf method showed reasonable agreement with those calculated from $\bar{t}_+$ and $\bar{t}_{trw}$ (Malherbe and Mandersloot, 1960).

Using cation-selective membranes, Stewart and Graydon (1957) made careful measurements of water transfer and membrane potentials, and calculated emf corrections for water and co-ion transfer, respectively. They generally found that the water correction term was greater than the co-ion correction term.

The Teorell-Meyer-Sievers Equation. The Nernst equation expresses the membrane concentration potential in terms of average transport numbers in the membrane and mean ionic activities in the external solutions. Using the concepts of the fixed-charge theory and, introducing certain simplifying assumptions, Teorell (1953) and, independently, Meyer and Sievers (1936, 1936) derived an equation for the membrane potential in terms of the external concentrations, ionic mobilities in the membrane, and the fixed ion concentration. The most important simplifications introduced were:

1. Solutions were considered to be "ideal," i.e., concentrations were used instead of activities.
2. The mobility ratio $\overline{U}_+/\overline{U}_-$ was assumed to be the same throughout the membrane.
3. The fixed ion concentration $\overline{X}$ was assumed constant throughout the membrane.
4. The effect of water transport was neglected.
5. The distribution of ions at the membrane-liquid interface was assumed to be a Donnan equilibrium.

Teorell used a kinetic method, while Meyer and Sievers used a quasi-thermodynamical one. The equations obtained were identical and may be expressed in the following form for a uni-univalent electrolyte at external concentrations c_1 and c_2:

$$E_m = \frac{RT}{F}\left[\omega \ln \frac{c_2(x_1 + 1)}{c_1(x_2 + 1)} + \overline{U} \ln \frac{x_1 - \omega\overline{U}}{x_2 - \omega\overline{U}} \right] \qquad (7.39)$$

where $\omega = +1$ or -1 for an anion or cation selective membrane, respectively;

$$x = \left[\left(\frac{2c}{\overline{X}} \right)^2 + 1 \right]^{1/2} \quad \text{and} \quad \overline{U} = \frac{\overline{U}_+ - \overline{U}_-}{\overline{U}_+ + \overline{U}_-}$$

The fact that the right-hand side of Eq. (7.39) consists of two terms has a physical significance: the first is known as the *Donnan term,* because it is the algebraic sum of two potential *jumps* occurring at the membrane faces, and the second is a diffusion potential existing in the interior of the membrane, sometimes referred to as the *Henderson term,* since it is identical with the expression derived by Henderson (1907) for diffusion potentials, of which Eq. (7.31) is a special case. The Donnan term is an equilibrium potential having its origin in the concentration discontinuities at the membrane-liquid interfaces and has no direct effect on ionic fluxes through the membrane. In this respect, it differs from the *Henderson potential,* which is caused by a diffusion process and is closely connected with transport phenomena.

For a highly selective membrane ($\overline{X} \gg 2c$), the Donnan potential is the main term, and in the limit (i.e., for an ideally selective membrane), the Henderson term vanishes. Equation (7.39) then reduces to

$$E_m = \omega \, \frac{RT}{F} \, \ln \frac{c_2}{c_1} \tag{7.40}$$

If $\overline{X} \ll 2c$, the membrane has low selectivity, and the diffusion potential is the important term. For $\overline{X} = 0$ (i.e., a nonselective membrane), the Donnan term vanishes; i.e., the normal liquid junction potential is observed.

It may be of value to consider the Donnan and diffusion potentials a little more closely by applying Eq. (7.39) to a hypothetical case of a cation-selective membrane ($\omega \overline{X} = -10$ in arbitrary units) separating two NaOH solutions with concentrations 5 and 2, respectively. The concentration and potential profiles are indicated in Fig. 7.8.

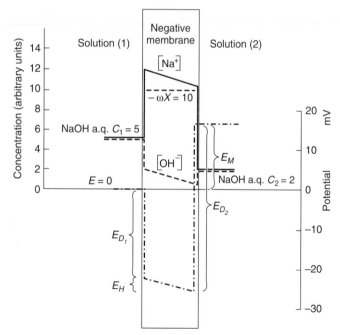

FIGURE 7.8 Concentration and potential profiles across a negative membrane separating solutions of NaOH. *(From Malherbe and Mandersloot, 1960. Reproduced by permission of the authors.)*

Concentrations in the membrane have been calculated according to the simple Donnan equation; in the membrane interior, the profiles are represented as linear, although this is generally not the case (Teorell, 1953).

In calculating the diffusion potential, the assumption was made that the mobility ratio is the same as in free solution, i.e., $\overline{U}_+/\overline{U}_- = 0.254$ and $\overline{U} = -0.596$. The values of the two Donnan potentials are calculated according to the formulas

$$E_{D1} = \frac{RT}{F} \ln \frac{(c_+)_1}{[\text{Na}^+]_1}$$

and

$$E_{D2} = \frac{RT}{F} \ln \frac{[\text{Na}^+]_2}{(c_+)_2}$$

where c_+ represents the concentration of Na^+ in each external solution, and $[\text{Na}^+]_1$, $[\text{Na}^+]_2$ the concentrations of Na^+ in the membrane at each interface (for an exact calculation the concentrations must be substituted by single ion activities). The membrane potential is given by:

$$E_m = E_{D1} + E_H + E_{D2}$$
$$= -22.7 - 2.8 + 42.3$$
$$= 16.8 \text{ mV} \tag{7.41}$$

It will be noticed that E_m is opposite in sign to E_H; this is often (but not always) so when the co-ions in a membrane are more mobile than the counter-ions, as in the present case. On the other hand, within the limits of the validity of Eq. (7.39), E_H and E_m must always be of the same sign if the counter-ions move faster than the co-ions. These qualitative rules can easily be verified by inspection of the signs of the various terms in Eq. (7.39), which shows that:

1. If $\overline{U}_{\text{counter-ion}} > \overline{U}_{\text{co-ion}}$, the signs of E_H and $(E_{D1} + E_{D2})$, and hence also E_m, are the same.

2. If $\overline{U}_{\text{co-ion}} > \overline{U}_{\text{counter-ion}}$, E_H and $(E_{D1} + E_{D2})$ differ in sign. Thus, E_H and E_m may have either similar or opposite sign, depending on whether $|E_H| > |E_{D1} + E_{D2}|$, or vice versa.

3. If $\overline{U}_{\text{counter-ion}} = \overline{U}_{\text{co-ion}}$, the diffusion potential vanishes, and the membrane potential is simply the sum of the two Donnan potentials.

In spite of the limitations arising from the simplifying assumptions made in the derivation of the Teorell-Meyer-Sievers equation, experiment has shown that, if experimental conditions do not depart too far from the "ideal" ones postulated, the equation is essentially correct. It has in fact proved to be of great value in the interpretation of membrane potential data, and the basic concepts embodied in the theory have aided in the understanding of other membrane phenomena, such as electrical conductance, electroosmosis, streaming potentials, and anomalous osmosis (Malherbe and Mandersloot, 1960).

Equation (7.39) predicts a falling off of the membrane potential with increasing external concentration, from a theoretical maximum value given by Eq. (7.40), to the normal liquid junction potential, Eq. (7.31). This decline is observed in practice (Malherbe and Mandersloot, 1960).

7.2.6 Transport Number and Conductance and Their Relation to Flows and Forces in Electrodialysis

Transport numbers and conductance and their relation to flows and forces in ED/EDR are described by Schoeman (1992).

Consider a system consisting of two aqueous solutions containing only one permeable electrolyte separated by a membrane. Different concentrations, pressures, and electrical potentials are allowed on both sides of the membrane. Envisage further operation of two forces with two conjugated flows which may pass from one side of the membrane to the other. The simplest choice of flows and forces would be the flow of cation J_1, driven by the difference in electrochemical potential $\Delta\tilde{\mu}_1$, and the flow of anion J_2, driven by the corresponding force $\Delta\tilde{\mu}_2$. The following simple phenomenological equations can then be set up (Katchalsky and Curran, 1967).

$$J_1 = L_1 \Delta\tilde{\mu}_1 \tag{7.42}$$

$$J_2 = L_2 \Delta\tilde{\mu}_2 \tag{7.43}$$

where L_1 and L_2 are the phenomenological coefficients which characterize the system. The chemical potential of the electrolyte $\Delta\mu_s$ is equal to the electrochemical potentials of the cation and the anion.

$$\Delta\mu_s = \Delta\tilde{\mu}_1 + \Delta\tilde{\mu}_2 \tag{7.44}$$

The electrical current I through a membrane is related to the ionic flows by the relationship (Katchalsky and Curran, 1967).

$$I = (z_1 J_1 + z_2 J_2)F \tag{7.45}$$

where z_1 = valence of cation
z_2 = valence of anion
F = Faraday's constant

When $I = 0$, then $J_1 = J_2$.

The electromotive force E acting on the system can be determined by introducing a pair of electrodes reversible to one of the ions, say ion 2, and measuring the potential difference. The value of E is related thermodynamically to the difference in electrochemical potential of ion 2 (Katchalsky and Curran, 1967).

$$E = \frac{\Delta\tilde{\mu}_2}{z_2 F} \tag{7.46}$$

for NaCl, $z_2 = -1$

$$\text{and} \quad E = \frac{\Delta\tilde{\mu}_2}{-F} \tag{7.47}$$

$$\text{or} \quad EF = -\Delta\tilde{\mu}_2 \tag{7.48}$$

Membrane conductance is usually carried out under isothermal, isobaric conditions with constant salt concentration across the membrane.

The electric current I through the membrane is:

$$\text{when } \Delta\mu_s = 0, \quad \text{then } \Delta\tilde{\mu}_1 = -\Delta\tilde{\mu}_2 \tag{7.49}$$

$$I = F(J_1 - J_2) \tag{7.50}$$

Substituting Eqs. (7.42) and (7.43) into Eq. (7.50) gives

$$I = F(L_1 \Delta \tilde{\mu}_1 - L_2 \Delta \tilde{\mu}_2) \tag{7.51}$$

But $\quad\quad\quad\quad \Delta \tilde{\mu}_1 = -\Delta \tilde{\mu}_2$

$$\therefore I = F(-L_1 \Delta \tilde{\mu}_2 - L_2 \Delta \tilde{\mu}_2) \tag{7.52}$$

$$= -F \Delta \tilde{\mu}_2 (L_1 + L_2) \tag{7.53}$$

But $\quad\quad\quad\quad EF = -\Delta \tilde{\mu}_2$

$$\therefore I = F^2 E (L_1 + L_2)$$

$$\therefore \left[\frac{I}{E} \right]_{\Delta \mu_s = 0, Jv = 0} = F^2 (L_1 + L_2) = \text{conductance} \tag{7.54}$$

when $I = 0$, then

$$J_1 - J_2 = 0 \tag{7.55}$$

Substituting Eqs. (7.42) and (7.43) into Eq. (7.55) gives

$$L_1 \Delta \tilde{\mu}_1 - L_2 \Delta \tilde{\mu}_2 = 0 \tag{7.56}$$

But $\quad\quad\quad\quad \Delta \tilde{\mu}_1 = \Delta \mu_s - \Delta \tilde{\mu}_2$

$$L_1 (\Delta \mu_s - \Delta \tilde{\mu}_2) - L_2 \Delta \tilde{\mu}_2 = 0 \tag{7.57}$$

and

$$\Delta \tilde{\mu}_2 = \frac{L_1}{L_1 + L_2} \Delta \mu_s \tag{7.58}$$

or

$$-EF = \frac{L_1}{L_1 + L_2} \Delta \mu_s \tag{7.59}$$

$$\therefore \left[\frac{EF}{\Delta \mu_s} \right]_{I = 0, Jv = 0} = -\frac{L_1}{L_1 + L_2} \tag{7.60}$$

Consider

$$\left[\frac{J_1}{I} \right]_{\Delta \mu_s = 0} = \frac{L_1 \Delta \tilde{\mu}_1}{F^2 E (L_1 + L_2)} \tag{7.61}$$

But $\quad\quad\quad\quad \Delta \tilde{\mu}_1 = -\Delta \tilde{\mu}_2 \quad\quad \text{and} \quad\quad EF = -\Delta \mu_2$

$$\therefore \left(\frac{J_1}{I} \right)_{\Delta \mu_s = 0} = \frac{L_1(-\Delta \tilde{\mu}_2)}{-\Delta \mu_2 (L_1 + L_2) \cdot F} \tag{7.62}$$

$$= \frac{1}{F} \cdot \frac{L_1}{L_1 + L_2} \tag{7.63}$$

$$\therefore \left[\frac{JF}{I}\right]_{\Delta\mu_s = 0; Jv = 0} = \frac{L_1}{L_1 + L_2} \tag{7.64}$$

$$= \Delta t \text{ (transport number)}$$

$$= -\left(\frac{EF}{\Delta\mu_s}\right)_{I = 0; Jv = 0} \tag{7.65}$$

Note: The membrane potential $\Delta\psi$ is related to the electromotive force measured between reversible electrodes by the expression (Katchalsky and Curran, 1967).

$$\Delta\psi = E - \frac{\Delta\bar{\mu}_2}{z_2 F} \tag{7.66}$$

7.2.7   Measurement of Transport Number

The efficiency with which a membrane transports selectively any particular ionic species may be inferred by measuring the transport number of the species in the membrane. Two methods are normally used to determine membrane transport number. They are:

- The EMF method (Lakshminarayanaiah, 1969)
- The Hittorf's method (Lakshminarayanaiah, 1969)

In these methods different concentrations of electrolyte exist on either side of the membrane, even though in the Hittorf's method one might start initially with the same concentration. Therefore, the transport number values derived by these methods cannot be directly related to a definite concentration of the external solution.

Membrane potentials measured using concentrations c' and c'' on either side of the membrane may be used in the following equation to derive an average transport number:

$$\frac{E}{E_{max}} = 2\bar{t}_+ - 1 \qquad \bar{t}_+ = \left(\frac{E}{E_{max}}\right) + 0.5 \tag{7.67}$$

If Ag-AgCl electrodes immersed in two chloride solutions are used, $\bar{t}_+$ is derived from (Lakshminarayanaiah, 1969):

$$E = 2\bar{t}_{+(app)} \frac{RT}{F} \ln \frac{a'}{a''} \tag{7.68}$$

The derived transport number value has been called the *apparent* transport number because, in this type of measurement, water transport has not been taken into account. This apparent value will be close to the true value when very dilute solutions are used.

In the Hittorf's method a known quantity of electricity is passed through the membrane cell containing two chambers filled with the same electrolyte separated by a membrane. Cations migrate to the cathode and anions migrate to the anode. The concentration change brought about in the two chambers, which is not more than about 10 percent, is estimated by the usual analytical methods. The transport number is calculated from

$$t_i = \frac{FJ_i}{I} \tag{7.69}$$

The determination of meaningful transport numbers for any membrane-electrolyte system calls for careful control of a number of factors. The important factors for the control of the concentration of the donating or receiving side are (Lakshminarayanaiah, 1969):

- External concentration
- Current density
- Difference in concentration on either side of the membrane

The effect of current density on the values of $\bar{t}_i$ has been demonstrated by Kressman and Tye (1956) using multicompartment cells and by Lakshminarayanaiah and Subrahmanyan (1964) using simple cells. When external concentrations are small (<0.1 mol/L) an increase of current density decreases $\bar{t}_i$ values. This is attributed to polarization effects at the membrane-solution interface facing the anode.

The amount of polarization decreases as the concentration is increased. When the external concentration is 0.1 mol/L, $\bar{t}_i$ exhibits a maximum at a certain current density below which the $\bar{t}_i$ values decrease as the current density is decreased and above which also $\bar{t}_i$ values decreased as the current density is increased. The decrease as the current density is lowered is attributed to backdiffusion of the electrolyte.

When external concentrations >0.1 mol/L are used, polarization effects are negligible but backdiffusion becomes dominant. As the quality of backflux due to diffusion is determined by the concentration differences allowed to build up during electrodialysis, it should be made as small as possible to derive meaningful values for $\bar{t}_i$.

7.3 SELECTIVITY MODELS

Different selectivity models will be considered in this section.

7.3.1 Current Efficiency and Brine Concentration in Electroosmotic Pumping (EOP)

Electroosmotic pumping in a sealed-cell ED stack is described by Kedem, Bar-On, and Warshawsky (1986); Garza (1973); and Schoeman (1992).

7.3.1.1 The Stationary State.   In the unit cell flow regime, ED/EDR becomes a three-port system, like RO. The feed solution is introduced; it passes between the membranes and leaves the system. The permeate composition is completely determined by the membrane performance under the conditions of the process.

The membranes and flows in a cell pair are shown schematically in Fig. 7.9. The amount of salt J, leaving at e per unit time and unit area of membrane pair, is given by:

$$J = J_1^C - J_1^A = J_2^C - J_2^A \tag{7.70}$$

where C denotes the cation-exchange membrane, A the anion-exchange membrane, 1 the cation, and 2 the anion. For simplicity, only uni-univalent electrolytes will be discussed here.

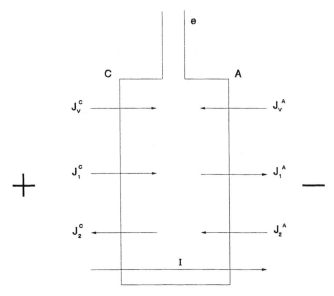

FIGURE 7.9 Flows in the brine cell. *(Kedem, Bar-On, and Warshawsky, 1986.)*

The total salt flow entering into the brine cell through the membrane may be written as a sum of two separate contributions from both membranes by the following definition:

$$J^C = \frac{J_1^C + J_2^C}{2} \quad \text{and} \quad J^A = \frac{J_1^A + J_2^A}{2} \tag{7.71}$$

where

$$J = J^C - J^A \tag{7.72}$$

The current efficiency η_c is also a sum of two contributions.

$$\eta_c = \eta_c^C + \eta_c^A = \frac{J^C - J^A}{I/F} \tag{7.73}$$

The total volume of solution leaving a unit cell per unit area A is:

$$J = J_V^C - J_V^A \tag{7.74}$$

hence, the concentration of the brine c_b is given by

$$c_b = \frac{J}{J} = \frac{J^C - J^A}{J_V^C - J_V^A} \tag{7.75}$$

7.3.1.2 Electroosmotic Pumping in Single Cell. Electroosmotic pumping (EOP) was studied in a single cell pair (Garza and Kedem, 1976), with special precautions to ensure effective stirring. Polarization is largely suppressed in this setup, and it was possible to study high current densities. The flows obtained were characterized by two remarkably simple features: current efficiency was nearly constant in

The fo
0.1 mol/L
Selemion

J:

J_{osm}:

J_{elosm}:

c_b:

26:

The m

an

Fo
ex

It has l
predict m
(Schoema
rent effici
density us
AMV an
branes); (
possible t
with an a
port numl

Correl
efficiency
apparent
than curre
Backdiffu
rent effici
(Δt^A) gave
tion. Ratie
Israeli Al
quently, i
hydrochle
from the
Correl
efficiency

Th
ne

hig
inf
W
me
flo
is i
tia
ap

fol
de
cri
tra
Th
pr
kn

η_w
the

---

a wide range of current densities, and the brine concentration at high current density attained a constant value, independent of current density and only slightly dependent on feed concentration. The current efficiencies were close to the value expected from transport numbers. For the less permselective membrane pairs, the observed η_c was even higher than the value predicted from the membrane potential measured between feed and brine solutions, neglecting backdiffusion.

7.3.1.3 Flux Equations, Membrane Potentials, Current Efficiency, and Brine Concentration.

The total ED/EDR process comprises three independent flows and forces: electric current and potential, volume flow and pressure/osmotic pressure, salt flow and concentration difference. For small flows and gradients linear equations can be written for each of the flows, including the influence of all gradients (Kedem and Katchalsky, 1963; Staverman, 1952).

In practical ED/EDR, especially in EOP, flows and forces are large and one cannot expect linear equations to hold, even if the usually defined membrane transport coefficients are constant. In fact, transport coefficients may vary considerably in the concentration range between feed and brine. For an adequate discussion of flows under these conditions, the analysis given previously for RO would be followed (Spiegler and Kedem, 1966). In the schematic representation in Fig. 7.10 the membrane is broken down into differential elements, separated by uniform solution segments which are in equilibrium with the two contiguous membrane faces. All fluxes going from left to right are counted positive. The gradient of a scalar Y, dy/dx, is taken as the value of the scalar on the right (double prime) minus the value on the left (single prime), divided by the distance. On the other hand, the operator Δ is defined with the opposite sign, in order to bring the notation in line with that of previous publications (Johnson, Dresner, and Kraus, 1966; Kedem and Katchalsky, 1961). $\Delta c = c' - c''$. Thus, $y = \int''_{} dy$.

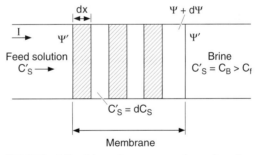

FIGURE 7.10 Schematic representation of cation-exchange membrane. (*Kedem, Bar-On, and Warshawsky, 1986.*)

Salt flow across a differential layer of a cation-exchange membrane can be written as a function of electric current, volume flow, and concentration gradient (Kedem and Katchalsky, 1963; Staverman, 1952; Kedem, Bar-On, and Warshawsky, 1986).

$$J^C = \frac{J_1^C + J_2^C}{2} = c_s(1 - \sigma^C)J_V^C - \overline{P}^C \frac{dc_s}{dx} + \frac{\Delta t}{2}\frac{I}{F} \qquad (7.76)$$

where

$$\Delta t = t_1 - t_2 = 2t_1 - 1 \qquad (7.77)$$

Th
cussec
causin
Th
ric uni

where

Bri

where

Th
0.1 mc
Selem

2J:

Δī:

Memt

Th

Maxin
man, 1

where

tory. Ratios of $\overline{\Delta t}/\eta_c$ of 1.0 to 1.1 (0.05 mol/L, Ionac membranes); 0.9 to 1.0 (0.1 mol/L, Ionac membranes); 0.9 (1.0 mol/L, Selemion AMV and CMV); 1.1 (0.5 mol/L, Selemion AMP and CMV) were obtained. Therefore, it should be possible to predict current efficiency for concentration of caustic soda solutions with an accuracy of approximately 20 percent and better from the apparent transport number of the membrane pair.

Brine concentration in EOP can be predicted from the apparent transport numbers ($\overline{\Delta t}$'s) and water flows through the membranes [Eq. (7.85)]. The ratio $c_{b,calc}/c_{b,exp}$, however, decreased with increasing feed concentration (Schoeman, 1992). Maximum brine concentration c_b^{max} can be predicted from two simple models [Eqs. (7.86) and (7.87)]. A very good correlation was obtained by the two methods. Maximum brine concentration increased with increasing feed concentration and appeared to level off at high feed concentration (0.5 to 1.0 mol/L) (Schoeman, 1992).

7.3.2 Discrepancy Between Transport Numbers Derived from Potential Measurements and Current Efficiency Actually Obtained

The correct relationships to be used when measuring membrane potential for the prediction of desalting in ED/EDR, are as follows (Schoeman, 1992):

$$\left[\frac{J}{I}\right]_{\Delta\mu_s; J_V = 0} = \frac{\Delta t}{F} = -\left[\frac{\Delta\psi}{\Delta\mu_s}\right]_{I = 0; J_V = 0} \tag{7.88}$$

The correct Onsager relationship for potential measured is at zero current and at zero volume flow, and for the transport number, at zero concentration gradient and zero volume flow. In practical ED/EDR, measurements are conducted at zero pressure and in presence of concentration gradients and volume flows. These factors will influence the results considerably in all systems in which volume flow is important and where the concentration factor is high, as is encountered in EOP. In the measurement of membrane potential, the volume flow is against the concentration potential and, in general, will decrease the potential. In ED/EDR water flow helps to increase current efficiency, but the concentration gradient acts against current efficiency.

7.3.3 Permselectivity of a Membrane in Working Systems

Permselectivity of a membrane in a working system is described by Narebska and Koter (1988).

By the permselectivity of a membrane some authors understand the transport number of counter-ions itself, $S = \bar{t}_1$, while others mean the ratio:

$$S = \frac{(\bar{t}_1 - t_1)}{(1 - t_1)} \tag{7.89}$$

In this formula, $\bar{t}_1$ is the transport number of the same ion 1 in an aqueous solution of concentration close to that within the membrane. Since for computing the permselectivity defined in such a way that the corresponding data for aqueous electrolytes should be known, it is a somewhat more complicated and a relative value. Narebska and Koter use the term *permselectivity* in the simplest meaning, i.e., as the transport number of counter-ions in the membrane.

$$\psi_{\text{theoretical}} = \frac{RT}{F} \times 2.303 \, \frac{\log \gamma'' c''}{\log \gamma' c'}$$

$$= 0.059158 \log \left[\frac{0.668}{0.657} \times \frac{2}{1} \right]$$

$$= 18.2 \text{ mV}$$

$$\Delta t^C = \frac{\psi_{\text{measured}}}{\psi_{\text{theoretical}}}$$

$$= \frac{16.9}{18.2}$$

$$\Delta t^C = \bar{t}_1^{\ C} - \bar{t}_2^{\ C}$$

$$= 0.93$$

$$\bar{t}_1^{\ C} + \bar{t}_2^{\ C} = 1$$

and $$\bar{t}_1^{\ C} = 0.97$$

The average transport number (Δt^C) is closer to the value for c_f than for c_b. For concentration-dependent transport numbers the actual current efficiency is expected to be less than that predicted from membrane potentials, i.e.

$$\eta_c < \frac{\Delta\psi_{\text{measured}}^C + |\,\Delta\psi_{\text{measured}}^A\,|}{|\,2\Delta\psi_{\text{theoretical}}\,|} \tag{7.83}$$

The correlation given by Eq. (7.83) is valid only if the influence of volume flow is negligible.

It was found that η_c for a single membrane pair was equal to this value or even higher (Garza and Kedem, 1976; Schoeman, 1992). This is due to the substantial influence of the electroosmotic and osmotic flow into the brine cells, increasing η_c. Water flow, both osmotic and electroosmotic, enters the brine cell through both membranes. It increases the flows of counter-ions into the cell and decreases the flows of co-ions leaving the brine. The total effect of volume flow into the brine cell is increased salt flow J. There is also a slight influence of osmotic flow on the potential measurements, in the opposite direction, decreasing $\Delta\psi_{\text{measured}}$ and, thus, the apparent transport number.

The importance and mechanism of this effect was shown in model calculations following the TMS model (Garza, 1973). It was shown that in EOP at high current density a plateau appears in the concentration profile near the feed surface or at a critical concentration in the membrane, and not near the brine surface. The concentration gradient near the concentrated solution is counteracted by the volume flow. The overall efficiency of the ED/EDR process η_d is given by the amount of desalted product solution and not by the salt transferred through the membranes. As is well known,

$$\eta_d = \eta_c \, \eta_w$$

η_w allows for the volume of water passing through the membranes and, thus, lost to the brine.

This detrimental influence of water flow is often emphasized. The effect discussed here deals with a different phenomenon: the coupling of water and salt flows, causing an increasing in η_c itself.

The amount of salt transferred per faraday of current passed through a symmetric unit cell is given by (Garza, 1973; Schoeman, 1992):

$$\eta_c = \frac{2Jc_b}{i/F} \equiv \overline{\Delta t} \tag{7.84}$$

where η_c = current efficiency
J = volume flow through membrane
c_b = brine concentration
i = applied current density
F = Faraday's constant
$\overline{\Delta t}$ = effective transport number of membrane pair

Brine concentration (c_b) in EOP is given by (Garza, 1973; Schoeman, 1992):

$$c_b = \frac{I\overline{\Delta t}}{2FJ} \tag{7.85}$$

where

$$\overline{\Delta t} = \frac{\Delta t^C + \Delta t^A}{2}$$

The following data were obtained during a bench-scale ED experiment when a 0.1 mol/L NaCl solution was electrodialyzed at a current density of 15 mA/cm^2 using Selemion AMV and CMV membranes:

2J:	0.152 cm/h
$\overline{\Delta t}$:	0.82
Membrane area:	7.55 cm^2

The brine concentration can be calculated as follows:

$$c_b = \frac{i\overline{\Delta t}}{2FJ}$$

$$= \frac{15 \times 0.82}{96,500 \times 4.22 \times 10^{-5}}$$

$$= 3.02 \text{ mol/L}$$

Maximum brine concentration in EOP is given by (Garza and Kedem, 1976; Schoeman, 1992):

$$c_b^{max} = \frac{1}{2F\beta} \tag{7.86}$$

where β is the electroosmotic coefficient and

$$c_b = \frac{c_b^{max}}{1 + J_{osm}/J_{elosm}} \tag{7.87}$$

The following data were obtained during a bench-scale ED experiment when a 0.1 mol/L NaCl solution was electrodialyzed at a current density of 15 mA/cm² using Selemion AMV and CMV membranes:

J:	0.152 cm/h
J_{osm}:	0.06770 cm/h
J_{elosm}:	0.08430 cm/h
c_b:	2.83 mol/L
$2\bar{b}$:	0.198 L/F

The maximum brine concentration can be calculated as follows:

$$c_b^{max} = c_b\left(1 + \frac{J_{osm}}{J_{elosm}}\right)$$

$$= 2.83\left(1 + \frac{0.06770}{0.08430}\right)$$

$$= 5.10 \text{ mol/L}$$

$$c_b^{max} = \frac{1}{2\beta F}$$

$$= \frac{1}{0.198}$$

$$= 5.05 \text{ mol/L}$$

It has been found that a simple potential measurement can be used effectively to predict membrane performance for salt, acid, and base concentration with ED/EDR (Schoeman, 1992). The ratio between the apparent transport number $(\overline{\Delta t})$ and current efficiency (η_c), however, depends on the feed concentration and the current density used. Ratios of $\overline{\Delta t}/\eta_c$ varied between 1.0 and 1.07 (0.1 mol/L feed, Selemion AMV and CMV, salt concentration); 0.95 to 1.09 (0.5 mol/L feed, Ionac membranes); 0.95 and 1.02 (0.5 mol/L, Ionics membranes). Consequently, it should be possible to predict current efficiency for concentration of sodium chloride solutions with an accuracy of approximately 10 percent and better from the apparent transport number of the membrane pair.

Correlations obtained between the apparent transport number $(\overline{\Delta t})$ and current efficiency for membranes used for acid concentration were unsatisfactory. The apparent transport number of the membrane pair $(\overline{\Delta t})$ was from 1.5 to 4 times higher than current efficiency in the feed acid concentration range from 0.05 to 1.0 mol/L. Backdiffusion of hydrochloric acid through the membranes caused the lower current efficiency. However, the apparent transport number of the anion membrane (Δt^A) gave a much better indication of membrane performance for acid concentration. Ratios of $\Delta t^A/\eta_c$ of 1.1 to 1.2 (1.0 mol/L, Selemion AAV); 0.97 to 0.84 (1.0 mol/L Israeli ABM-2); 0.92 to 0.97 (0.1 mol/L, Israeli ABM-1) were obtained. Consequently, it should be possible to predict current efficiency for concentration of hydrochloric acid solutions with an accuracy of approximately 20 percent and better from the apparent transport number of the anion membrane.

Correlations obtained between the apparent transport number $(\overline{\Delta t})$ and current efficiency of membranes investigated for caustic soda concentration were satisfactory. Ratios of $\overline{\Delta t}/\eta_c$ of 1.0 to 1.1 (0.05 mol/L, Ionac membranes); 0.9 to 1.0 (0.1 mol/L, Ionac membranes); 0.9 (1.0 mol/L, Selemion AMV and CMV); 1.1 (0.5 mol/L, Selemion AMP and CMV) were obtained. Therefore, it should be possible to predict current efficiency for concentration of caustic soda solutions with an accuracy of approximately 20 percent and better from the apparent transport number of the membrane pair.

Brine concentration in EOP can be predicted from the apparent transport numbers ($\overline{\Delta t}$'s) and water flows through the membranes [Eq. (7.85)]. The ratio $c_{b,\text{calc}}/c_{b,\text{exp}}$, however, decreased with increasing feed concentration (Schoeman, 1992). Maximum brine concentration c_b^{max} can be predicted from two simple models [Eqs. (7.86) and (7.87)]. A very good correlation was obtained by the two methods. Maximum brine concentration increased with increasing feed concentration and appeared to level off at high feed concentration (0.5 to 1.0 mol/L) (Schoeman, 1992).

7.3.2 Discrepancy Between Transport Numbers Derived from Potential Measurements and Current Efficiency Actually Obtained

The correct relationships to be used when measuring membrane potential for the prediction of desalting in ED/EDR, are as follows (Schoeman, 1992):

$$\left[\frac{J}{I}\right]_{\Delta\mu_s;\,J_V=0} = \frac{\Delta t}{F} = -\left[\frac{\Delta\psi}{\Delta\mu_s}\right]_{I=0;\,J_V=0} \tag{7.88}$$

The correct Onsager relationship for potential measured is at zero current and at zero volume flow, and for the transport number, at zero concentration gradient and zero volume flow. In practical ED/EDR, measurements are conducted at zero pressure and in presence of concentration gradients and volume flows. These factors will influence the results considerably in all systems in which volume flow is important and where the concentration factor is high, as is encountered in EOP. In the measurement of membrane potential, the volume flow is against the concentration potential and, in general, will decrease the potential. In ED/EDR water flow helps to increase current efficiency, but the concentration gradient acts against current efficiency.

7.3.3 Permselectivity of a Membrane in Working Systems

Permselectivity of a membrane in a working system is described by Narebska and Koter (1988).

By the permselectivity of a membrane some authors understand the transport number of counter-ions itself, $S = \bar{t}_1$, while others mean the ratio:

$$S = \frac{(\bar{t}_1 - t_1)}{(1 - t_1)} \tag{7.89}$$

In this formula, $\bar{t}_1$ is the transport number of the same ion 1 in an aqueous solution of concentration close to that within the membrane. Since for computing the permselectivity defined in such a way that the corresponding data for aqueous electrolytes should be known, it is a somewhat more complicated and a relative value. Narebska and Koter use the term *permselectivity* in the simplest meaning, i.e., as the transport number of counter-ions in the membrane.

By definition, the transport number $\bar{t}_1$ represents the ratio of the current transported by counter-ions to the total current passing through the membrane:

$$S = \bar{t}_1 = \left[\frac{z_1 F J_1}{i}\right]_{\Delta c = 0} \tag{7.90}$$

where z_1 = valence of ion 1
F = Faraday's constant
J_i = flux of species i, mol/m²/s
i = electric current density, A/m²
Δc = concentration difference, mol/dm³

When defining the permselectivity of the membrane in working systems, it is enough to account for the phenomena produced by the concentration and electric potential difference, leaving the main formula unchanged, i.e.,

$$S_d = \left[\frac{z_1 F J_1}{i}\right]_{\Delta c, i, J_v \neq 0} \tag{7.91}$$

where J_v = volume flux, m/s.

Since the so-defined permselectivity accounts for all the fluxes in the membrane at Δc, $i \neq 0$, it is called dynamic as opposed to the static one. For 1:1 electrolyte, Eq. (7.91) simplifies to:

$$S_d = \left[\frac{J_1}{J_1 - J_2}\right]_{\Delta c, i, J_v \neq 0} \tag{7.92}$$

To solve Eq. (7.91) the phenomenological transport equation $J_i = L_{i1}\Delta\bar{\mu}_1 + L_{i2}\Delta\bar{\mu}_2 + L_{iw}\Delta\mu_w$, $i = 1, 2, w$ should be applied first (μ_i = chemical potential of species i [J/mol]; and L_{ik} = conductance coefficients (mol²/J/m²/s). Rearranging this equation (at zero pressure difference), the equation for permselectivity can be derived composed of two terms:

$$S_d = \left[\frac{J_1}{J_1 - J_2}\right]_{\Delta c = 0} + \left[\frac{J_s}{J_1 - J_2}\right]_{\Delta c \neq 0} \tag{7.93}$$

The first term relates the permeation of counter-ions to diffusion, and the second one, to the electric transport phenomena. The final rearrangement of Eq. (7.93) gives the following for the dynamic permselectivity:

$$S_d = \bar{t}_1 + \left[\bar{t}_1\bar{t}_2\frac{k}{F^2} + L_{i2} - 0.018\overline{m}\left(\bar{t}_w\bar{t}_2\frac{k}{F^2} + L_{2w}\right)\right]\frac{F\Delta\mu_s}{i} \tag{7.94}$$

where $\overline{m}$ = mean molality
k = conductivity of membrane, Ω^{-1}/m²

For convenience, Eq. (7.94) can be abbreviated to the form:

$$S_d = \bar{t}_1 + \alpha(\overline{m})\frac{F\Delta\mu_s}{i} \tag{7.95}$$

where $\alpha(m)$ is the total bracket.

As intended, S_d is related to the difference in the molarities of the electrolytes bathing the membrane (through $\Delta\mu_s$) and the current density i, which is even more useful than the equation with the electric potential difference.

The comparison of Eqs. (7.94) and (7.95) to the one defining selectivity as $S = \bar{t}_1$ makes the difference between the static and dynamic permselectivities distinct.

At $\Delta\mu_s \rightarrow 0$ and $i \rightarrow 0$, $S_d \rightarrow \bar{t}_1$, i.e., at zero condition the dynamic permselectivity approaches the static one. At nonzero parameters $S_d < S$.

The dynamic permselectivity is a complex function of the transport numbers of ions and water, and also of the membrane conductivity k_m and the two conductance coefficients, L_{12} and L_{2w}.

When applying irreversible thermodynamics, some authors assume the interactions between the co- and counter-ions to be close to zero and neglect the L_{12} coefficient. It has been proved here that by neglecting L_{12} and L_{2w} the results obtained differ by up to 10 percent from S_d computed while applying the full equation. The results for the Nafion 120 membrane separating sodium chloride solutions or sodium hydroxide of the molarity ratio $m'/m'' = 10$ and at $i = 500 \text{ A/m}^2$ are presented in Table 7.1.

TABLE 7.1   Permselectivity of Nafion 120 membrane

Static (S), dynamic (S_d), and the dynamic one while computed with the simplified equation (S_d')

$\bar{m}$	$S = \bar{t}_1$	S_d*	$S_d'{}^\dagger$
	NaCl (298 K)		
0.1	0.999	0.995	0.998
0.5	0.98	0.95	0.96
1.0	0.96	0.88	0.91
	NaOH (298 K)		
0.1	0.983	0.968	0.965
0.5	0.87	0.75	0.73
1.0	0.80	0.56	0.60

* At $m'/m'' = 10, I = 500 \text{ A/m}^2$

$\dagger$ $S_d' = S_d - (L_{12} - 0.018\,\bar{m}\,L_{2w})\dfrac{F\Delta\mu_s}{i}$ at m'/m'' and I as above.

Source: From Narebska and Koter, 1988. Reproduced by permission of the authors.

The dynamic permselectivity computed either using Eq. (7.94) (column 3) or basing on the same equation, but neglecting the L_{12} and L_{2w} coefficients (column 4), shows the effect of simplification. It should be admitted, however, that at still higher concentration, neglecting the coefficients will result in increasing deviations between the simplified and real values.

The computed results of the dynamic permselectivity obtained when applying Eq. (7.93) and the experimental data for the Nafion 120 membrane can be seen in Figs. 7.11 and 7.12. Here the permselectivity is shown as a function of the molarity of solute in one compartment and the ratio of this molarity to the molarity on the opposite side of the membrane (x axis). The graphs present the results obtained for the regions characterized by low, mean, and high molarities, i.e., 0.1, 0.5, and 1.0.

The curves in the z-x plane reflect the separation ability of the membrane as desalination proceeds. In every diagram, the starting point on the z axis corresponds

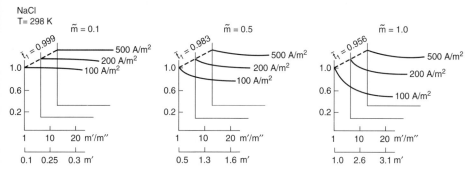

FIGURE 7.11   Dynamic selectivity of Nafion 120 membrane in contact with aqueous sodium chloride solutions at logarithmic mean concentrations; $\tilde{m} = 0.1; 0.5$ and 1; $T = 298$ K. *(From Narebska and Koter, 1988. Reproduced by permission of the authors.)*

to $m'/m'' = 1$ referred to the static permselectivity $\bar{t}_1$. Because the static permselectivity does not depend on the current density, it has the same value for the family curves, irrespective of i. Every other curve (in the z-y plane) shows how much increasing the current improves the membrane permselectivity.

The conclusions summarizing the results presented in Figs. 7.11 and 7.12 are as follows:

1. The permselectivity of ion-exchange membranes in working systems with $\Delta c, i \neq 0$, called *dynamic*, is below that found in the zero condition *(static)*. With Δc approaching zero, the dynamic permselectivity S_d reaches a limiting value, which is the static permselectivity.

2. The increasing molarity ratio of the separated solutions (with progressing desalination) strongly depresses the membrane permselectivity. For sodium hydroxide solutions at high molarity, it is almost impossible to get uphill transport of sodium ions.

3. A high current density limits the effects of diffusion and osmotic flows favoring the effective transport of counter-ions.

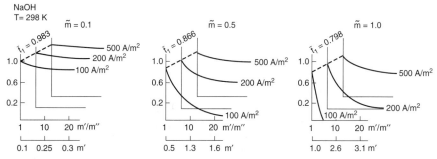

FIGURE 7.12   Dynamic selectivity of Nafion 120 membrane in contact with aqueous sodium hydroxide solutions; $\tilde{m} = 0.1; 0.5$ and 1 M; $T = 298$ K. *(From Narebska and Koter, 1988. Reproduced by permission of the authors.)*

7.3.4 Hyperfiltration in Charged Membranes—Prediction of Salt Rejection from Equilibrium Measurements

Hoffer and Kedem (1972) established a quantitative correlation between co-ion uptake and salt rejection for polylysine-collodion membranes. For this purpose, the simple TMS model had to be modified to take into account the electrostatic interaction between the charged matrix and the counter-ions. They assumed, as in the TMS model, that the interstitial solution is an ideal one, but the counter-ion concentration is given by the "free" counter-ions.

The Donnan equilibrium between solution and membrane, at the interface is then given by:

$$\frac{\bar{c}_1}{\phi_w} \cdot \frac{\bar{c}_2}{\phi_w} = \frac{\overline{X} + \bar{c}_s}{\phi_w} \cdot \frac{\bar{c}_s}{\phi_w} = c_s^{\ 2} \tag{7.96}$$

where $\bar{c}_2$ is the co-ion concentration per cubic centimeter of membrane phase, identified with the salt concentration $\bar{c}_s$, $\bar{c}_1$ is the concentration of free counter-ions per cubic centimeter, $\overline{X}$ is the effective charge density, and c_s is the salt concentration in the external solution. Here, as in the treatment of polyelectrolyte solutions, it is assumed that the activity coefficient of the co-ions is unity.

Equation (7.96) may be written in terms of the co-ion distribution coefficient between external and interstitial solution K_d, defined by:

$$K_d = \frac{\bar{c}_s}{c_s \phi_w}$$

where ϕ_w = volume fraction of water.

$$\left(\frac{\overline{X}}{(c_s \phi_w) + K_d} \right) K_d = 1 \tag{7.97}$$

Hence, the experimental determination of the co-ion distribution coefficient also gives the effective charge density or free counter-ion concentration.

The effectiveness of hyperfiltration is described by the salt rejection R, defined as:

$$R = 1 - \frac{c_s''}{c_s'} \tag{7.98}$$

where c_s' and c_s'' are the salt concentrations in feed and product solutions, respectively. At stationary hyperfiltration the concentration of the product is given by the ratio between salt flow and volume flow

$$c_s'' = \frac{J_s}{J_V} \tag{7.99}$$

The calculation of salt rejection for membranes represented by the TMS model showed that, at high flow rates, the concentration profile of the salt in the membrane flattens in the vicinity of the feed boundary. Then, the whole process is governed by the salt exclusion at feed boundary, and the rejection may be expressed as a function of the membrane charge density and the relative mobilities of the ions. A consideration of the Donnan equilibrium at the feed interface and the maintenance of electroneutrality during hyperfiltration led to a simple expression for the salt rejection at high flow rates R_∞ where $t_1 = u_1/(u_1 + u_2)$, u_1 and u_2 are the mobilities of the counter-ion and co-ion, respectively, and thus t_1 is the transport number in free solution.

$$R_\infty = 1 - \frac{c_s'\phi_w}{\overline{c_s'} + t_1\overline{X}} \qquad (7.100)$$

Regarding $\overline{X}$ as the effective charge density, Eqs. (7.97) and (7.100) give the relation between salt rejection and the measured equilibrium distribution of the co-ion, at the external concentration c_s'

$$R_\infty = 1 - K_d' \frac{1}{t_2 K_d'^2 + t_1} \qquad (7.101)$$

where $t_2 = 1 - t_1$.

The validity of Eq. (7.101) is limited to the simple association model in which a single effective charge density is used to describe both equilibrium and transport properties.

Three polylysine membranes were equilibrated with a number of sodium chloride solutions and co-ion concentration was determined in each membrane, at each external concentration. The measured values of the co-ion concentration, as well as the distribution coefficient and R computed from Eq. (7.101), are given in Table 7.2 (typical). Salt rejection was determined in hyperfiltration experiments with feeds of the same concentration as the equilibrating solutions. Since there is no flow rate dependence of salt rejection in these experiments, the measured salt rejection may be identified with R_∞. The results showed in the last column of Table 7.2 show fair agreement with the rejections computed from equilibria data. Thus, from relatively simple models one may predict the hyperfiltration behavior of charged membranes. Alternatively, the permselectivity of charged membranes may be predicted from simple models.

TABLE 7.2   Electrolyte Distribution Coefficients and Rejection in a Polylysine-Collodion Membrane*

Concentration of equilibrating solution	Co-ion concentration equiv/L $\times 10^3$	κ'	R_∞[†]	R(exptl)
0.0048	0.85	0.18	71.3	71.5
0.0098	2.36	0.24	61.3	59
0.023	10.2	0.43	35.6	39.3
0.048	29.8	0.62	18.4	21.3
0.095	70.7	0.74	10.3	10.3

* Water content of the membrane 44.4%; rejection measured at 0.02 cm/min at 200 lb/in².
[†] See Eq. (7.101).
Source: Hoffer and Kedem, 1972.

7.3.5   Selectivity Between Ions of the Same Charge

7.3.5.1   *Factors Affecting Permselectivity.*   The specific permselectivity between ions of the same charge (ions A and B), S_B^A, is defined as (Nishiwaki, 1972):

$$S_B^A = \frac{t_A c_B}{t_B c_A} \qquad (7.102)$$

where t_A = transport number of A ion
         t_B = transport number of B ion
         c_A = concentration of A ion in dilute solution
         c_B = concentration of B ion in dilute solution

In currently available univalent-selective membranes, the specific permselectivity of magnesium ions relative to sodium ions $S_{Na^+}{}^{Mg2+}$ is in the range of 0.2 to 0.3, that for calcium ions relative to sodium ions $S_{Na^+}{}^{Ca2+}$ is 0.3 to 0.4, and that for sulphate ions relative to chloride ions $S_{Cl^-}{}^{SO_4{}^{2-}}$ is 0.02 to 0.08. Yawataya et al. (1962) have shown that the transport numbers of divalent cations through univalent-selective cation-exchange membranes increase with increases in current density, as shown in Fig. 7.13. These changes in transport numbers cause changes in the specific permselectivities in accordance with the definition of $S_B{}^A$ given previously. It was also found that the concentration of ions in the dilute solution and the linear velocity of the dilute solution affect $S_B{}^A$ slightly. Therefore, the current density, the concentrations of ions, and the velocity of the dilute solution must all be considered in selecting the conditions for operation of the electrodialytic concentration process with univalent-selective membranes.

7.3.5.2   Application of Monovalent-Ion-Permselective Membranes to the Treatment of an Industrial Wastewater by Electrodialysis.

Saracco, Zanetti, and Onofrio (1993) studied the selective transport of ions (Cl$^-$ and C$_2$O$_4{}^{2-}$) through monovalent-ion-permselective membranes. Modeling studies were based on a kinetic approach by use of an Arrhenius-type expression relating the separation fac-

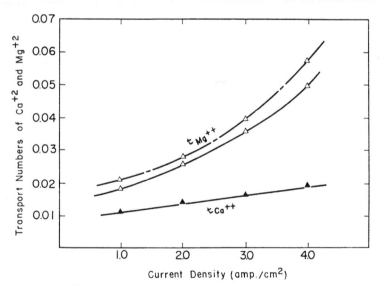

Current Density (amp./cm^2)

Solid lines: solution 0.45N NaCl + 0.02N CaCl$_2$ + 0.10N MgCl$_2$

Broken lines: solution 0.45N NaCl + 0.10N MgCl$_2$

FIGURE 7.13   Effect of current density on transport numbers for Mg^{2+} and Ca^{2+} in univalent-selective membranes. *(From Nishiwaki, 1972. Reproduced by permission of the publishers.)*

tor to current density, temperature, and concentrations in the diluate. The selective transport of ions through the monovalent-ion-permselective membranes is of the nonequilibrium type. A satisfactory correlation was obtained between experimental and modeling results for determination of the separation factor.

Modeling was conducted as follows: the separation factor (α, $Cl^-/C_2O_4^{2-}$) was evaluated by use of the following expression:

$$\alpha = \frac{J_s c_{ox}}{J_{ox} c_s} \tag{7.103}$$

where J_s, J_{ox} = molar fluxes of sodium chloride and sodium oxalate through membrane, $mol \cdot h^{-1}$.
c_s, c_{ox} = concentration of sodium chloride and sodium oxalate in bulk of dilute solution, $mol \cdot L^{-1}$.

The Donnan equilibrium expression is typically adopted as an approach for modeling in ion-exchange membranes. Diffusive transport is described by the Nernst-Planck equation and by the Einstein expression, which links diffusion coefficients to electric mobilities (Flett, 1983). However, this approach fails when mass fluxes through the membrane became relatively high (Timashev, 1991).

Sata (1973) has pointed out the prevalence of nonequilibrium phenomena in the permeation of Na^+ and Ca^{2+} ions through cationic membranes modified by the presence of a layer of surface-active agent. Similarly, it is likely that, due to the strong dependency of the separation factor on the imposed current density, the highly cross-linked thin layer present on both sides of the employed anionic membranes governs the $C_2O_4^{2-}/Cl^-$ separation according to mainly kinetic mechanisms. Therefore, the problems of formulating an adequate kinetic expression, quantitatively accounting for the dependency of the separation factor on the main operating parameters, arises. The Donnan equilibrium formulation, for diluted solutions of a monovalent electrolyte is as follows (Lacey, 1972):

$$\bar{c}_i = K_{eq}\, c_i^2 \quad \text{with} \quad K_{eq} = \frac{1}{X}\left(\frac{\gamma_\pm}{\bar{\gamma}_\pm}\right)^2 \tag{7.104}$$

where $\bar{c}_i, c_i$ = concentration of ion i inside membrane and in bulk of solution, $equiv \cdot m^{-3}$,
K_{eq} = equilibrium constant
X = concentration of fixed charges in membrane, $equiv \cdot L^{-1}$
$\bar{\gamma}_\pm, \gamma_\pm$ = activity coefficients inside and outside membrane

From this it can be assumed, in nonequilibrium conditions (whereas current density remarkably influences the separation of ions), that

$$\bar{c}_i = K_i\, c_i^n \tag{7.105}$$

where ion-concentration variations in the boundary layers have been neglected (K_i = kinetic constant related to ion i). In this expression, K_i is no longer an equilibrium constant but accounts for the kinetic phenomena which govern the transport of ions through the membrane.

The following Arrhenius-type kinetic expression was suitable for data fitting:

$$K_i = K_{\infty,i}\, \exp\left(-\frac{E_i - |z_i|\, FR_m l_e}{RT}\right) \tag{7.106}$$

where $K_i, K_{\infty,i}$ = kinetic constant related to ion i
E_i = apparent activation energy for entrance of ion i in membrane, $kJ \cdot mol^{-1}$
z_i = valence of ion i
F = Faraday's constant = 96,480 C $\cdot$ equiv^{-1}
R_m = electrical resistance of highly cross-linked layer, Ω
l_e = effective current, A
R = ideal gas constant = 8.314 J $\cdot$ mol^{-1} $\cdot$ K^{-1}
T = temperature, K

This expression denotes that, other things being equal, the electric energy possessed by the ions (proportional to the potential difference $R_m l_e$ imposed over the thin selective layer) helps to overcome the energetic barrier (apparent activation energy) that the layer itself imposes to the permeating ions.

Supposing the monovalent-ion-selectivity to be exclusively determined by passing through the highly cross-linked layer at the diluate side of the membrane, the following equation can be assumed:

$$\frac{\overline{c}_s}{\overline{c}_{ox}} = \frac{J_s}{J_{ox}} \tag{7.107}$$

Therefore, considering the expression (7.103), it can be written:

$$\alpha = \frac{K_s}{K_{ox}}\left(\frac{c_s}{c_{ox}}\right)^{n-1}$$

$$= \left[\frac{K_{\infty,s}}{K_{\infty,ox}}\exp\left(-\frac{E_s - E_{ox}}{RT}\right)X\exp\left(\frac{|z_s| - |z_{ox}|\,FR_m l_e}{RT}\right)\right]\left(\frac{c_s}{c_{ox}}\right)^{n-1} \tag{7.108}$$

The selectivity toward chloride ions is assured by the fact that E_s and E_{ox} have different values; i.e., the membrane imposes two different potential barriers to the passing of the two ions. On the other hand, as the current density increases, the electric energy available for the two ions becomes increasingly larger than these apparent activation energies, thus determining a reduction of the separation factor.

This behavior was quantitatively followed by experimental data. By expressing Eq. (7.108) in logarithmic terms, the following linear formulation was obtained:

$$\ln \alpha = A' + B' \cdot \frac{1}{T} + C' \cdot \frac{l_e}{T} + D' \cdot \ln\left(\frac{c_s}{c_{ox}}\right) \tag{7.109}$$

By least-square fitting of experimental data (Sarocco, Zanetti, and Onofrio, 1993), the following values were obtained.

$$A' = \ln\frac{K_{\infty,s}}{K_{\infty,ox}} = -1.255;\ B' = \frac{E_{ox} - E_s}{R} = 1523\ K;$$

$$C' = \frac{(|z_s| - |z_{ox}|)\,FR_m}{R} = -74.7\ K\cdot A^{-1};\ D' = n - 1 = 0.270$$

Several conclusions can be drawn from these results. Though the temperature range in the different tests was unavoidably narrow, the difference $(E_{ox} - E_s)$ equal to 12.7 KJ/mol deducible from the B' value is reasonable. Tanaka and Seno (1981) obtained comparable results for Na$^+$/Ca^{2+} separation through modified cationic membranes on the grounds of a similar modeling approach. On the basis of the c'

coefficient value, it can be easily calculated that R_m is approximately equal to 6.4 mΩ, which is nearly half the absolute resistance of the ACS anionic membranes, which have, indeed, two such layers (Sarocco, Zanetti, and Onofrio, 1993).

Finally, the n value, equal to 1.27 as attainable from D', may account for the partial presence of equilibrium phenomena in combination with kinetic ones, according to which unitary value would be intuitively expected. On the other hand, the equilibrium condition expressed by Eq. (7.104) entails a value of 2 for the exponent n. An intermediate condition between a quasi-equilibrium state and a completely kinetic controlled one may be active.

The model showed good agreement with experimental data in the estimation of the separation factor. Separation factors are summarized in Table 7.3.

TABLE 7.3 Separation Factors

Run	T, °C	l_e, A	α exp	α model	Error, %
1	25	1.9	48.1	50.2	+4.43
3	25	2.86	40.0	40.8	+2.02
5	25	4.48	27.9	29.0	+4.11
7	30	2.85	43.5	41.9	−3.73
9	30	3.43	30.5	28.9	−4.64
11	35	2.34	47.2	47.8	+1.23
13	35	3.29	34.4	33.7	−1.99
15	35	4.80	21.0	21.0	+0.16

Source: Sarocco, Zanetti, and Onofrio, 1993.

7.3.5.3 *Electrodialysis in Advanced Wastewater Treatment.*   Selectivity in electrodialysis application is described by Smith and Eisenmann (1967).

The following flux equations constitute a mathematical basis for the quantitative treatment of selectivity in an electrodialysis system.

$$J_i = \frac{-\bar{u}_i \bar{c}_i}{|z_i|F} \left[RT \text{ grad } (\ln \bar{c}_i \bar{\gamma}_i) + z_i F \text{ grad } \phi \right] + \omega \, \bar{u}_o \bar{c}_i \text{ grad } \phi \qquad (7.110)$$

The notation $\bar{u}_i, \bar{c}_i, \bar{\gamma}_i$ refers to the mobility, concentration, and activity coefficient of species i in the membrane phase; R is the universal gas constant; T is the absolute temperature; z_i is the electrochemical valence of species i; ϕ is the electrical potential (V); ω is the electrochemical valence of fixed ionic groups; $\bar{u}_o$ is the "mobility" of pore liquid within the membrane; F is Faraday's constant. The flux Eq. (7.110) applies to all mobile species in the system. The set of flux equations, one equation for each species, is subject to the restriction of electroneutrality:

$$\Sigma_i = z_i \bar{c}_i + \omega \bar{X} = 0 \qquad (7.111)$$

where $\bar{X}$ is the concentration of fixed ionic groups. The set of flux equations is related to the current density (assuming 100 current efficiency) by:

$$\frac{i}{A} = F \Sigma_i z_i J_i \qquad (7.112)$$

The flux equations must be solved simultaneously to arrive at the flux J_i of any one species. This is, in fact, a greatly simplified treatment. The most important simplifications are:

1. An idealized permselective membrane has been assumed, i.e., one with a fixed ion concentration that is so high that co-ions (in this case cations) are completely excluded. As waste-to-product concentration ratios increase, this becomes an increasingly poor assumption; flux equations, therefore, should be included for all co-ion species as well as for co-ion concentrations in the electroneutrality relationship.

2. Steady-state conditions are assumed; i.e.:

$$\operatorname{div} J_i = 0 \tag{7.113}$$

3. The flux equations, in the form given here, imply the validity of the diffusion and electrical mobility relationship. This essentially is equivalent to neglecting "coupling" of individual fluxes by other than electrical forces and convection.

4. Any differences in interaction of the various species with structural parts of the membrane have been neglected.

5. It is also assumed that species cross the membrane as such (no complexing with other species within the membrane) and that their motion is not geometrically coupled (as in single-file diffusion).

6. Complete membrane diffusion control has been assumed. Equations must be solved simultaneously.

In practice, ionic transference by electrodialysis is neither completely membrane-diffusion nor film-diffusion controlled, but lies somewhere between. With the simplifications noted, Eqs. (7.110), (7.111), and (7.112) constitute a mathematical basis for the quantitative treatment of selectivity in an electrodialysis system.

The rate of removal (flux) of any particular species is dependent on the rates of removal of all other species and, in this manner, the rate becomes a function not only of the particular species concentration and mobility, but also of all other species' concentrations and mobilities. In an attempt to obtain usable information regarding the rates of removal of the various ionic species present in a secondary sewage effluent, without becoming embroiled in unjustifiable mathematics, three separate approaches have been used (Smith and Eisenmann, 1967):

1. Some effects of certain external solution parameters have been examined by studying relative transfer rates across an anion-permeable membrane.

2. Possible effects of certain membrane parameters have been determined by estimating relative transfer rates from membrane equilibrium data.

3. The overall effects of various operating conditions, including membrane fouling and scaling, have been estimated from information collected through exhaustion tests.

Separation Factor.   The concept of membrane selectivity involves the ability of an ion-exchange membrane to transfer ionic species preferentially. This selectivity may be described, for most practical purposes, by a separation factor α:

$$\alpha_k{}^i = \frac{dc_i}{c_i} \cdot \frac{c_k}{dc_k} \tag{7.114}$$

where $\alpha_k{}^i$ = the relative separation of species i over species k.
This expression obviously may also be written:

$$\alpha_k{}^i = \frac{J_i/c_i}{J_k/c_k} \tag{7.115}$$

By the preceding definitions, a completely nonselective membrane would have a separation factor equal to unity for all combinations of species; i.e., the overall flux of all species would be identically proportional to the external solution concentration of that species:

$$\frac{J_i}{c_i} = \frac{J_j}{c_j} = \frac{J_k}{c_k} = \frac{J_i}{c_i} \ldots = \frac{J_n}{c_n} \tag{7.116}$$

In view of flux Eq. (7.110), the theoretical significance of the separation factor defined by

$$\alpha_T^i \doteq \frac{\% \text{ species } i \text{ removed}}{\% \text{ total meq removed}}$$

and Eqs. (7.114) and (7.115) is questioned. This separation factor is quite useful, however, in estimating the overall effects of various solution and operating parameters on relative rates of removal. It is, moreover, easily adapted to engineering calculations to estimate removal rates and to determine limiting ionic species for design purposes.

Effect of Solution Parameters on a Separation Factor. Field experience in electrodialysis and general fundamentals of ion-exchange have shown that separation factors are dependent on such parameters as total solution concentration, solution pH and temperature, fractional concentration of the species of concern, and the nature of fractional concentration of competing species. The flux Eq. (7.110) shows such a dependency, the individual membrane concentrations $\bar{c}_i$ and valences z_i being related to external solution characteristics. The effects of pH and fractional concentration were examined in this study for a phosphate-chloride system. Experiments were conducted in an electrodialysis stack.

The results of the study indicate that the relative rate of phosphate removal can be increased by raising the solution pH, and that phosphate will move faster than chloride, through the anion membrane tested, only at very low (<0.1) fractional phosphate concentrations and at high pHs (>8.0).

Effect of Membrane Parameters on Separation Factor. Examination of flux Eq. (7.110) shows that the rate of transfer of any species i to be proportional to the product of its membrane concentration $\bar{c}_i$, its mobility in the membrane phase $\bar{u}_i$, and the total driving force acting on the species (assuming complete membrane control and ignoring convection)

$$J_i \doteq -\frac{\bar{u}_i \bar{c}_i}{|z|F} [\text{driving force}] \tag{7.117}$$

If, in an electrodialysis application, the electrical components of the driving force are much more significant than the diffusional component, i.e.:

$$\text{grad} (\ln \bar{c}_i \gamma_i) \ll z_i F \, \text{grad} \, \phi \quad \text{and} \quad J_i \doteq \pm \bar{u}_1 \bar{c}_i \, \text{grad} \, \phi$$

(γ_i = molar activity coefficient of species i)

then the ratio of fluxes of any two species approaches the ratio of the products of their mobilities and concentrations:

$$\frac{J_i}{J_k} \doteq \frac{\bar{u}_i \bar{c}_i}{\bar{u}_k \bar{c}_k} \frac{\text{grad} \, \phi}{\text{grad} \, \phi} = \frac{\bar{u}_i \bar{c}_i}{\bar{u}_k \bar{c}_k} \tag{7.118}$$

This relationship might lead to a relatively simple method of examining membranes for their utility in separating charged ionic species.

The specific conductance of an ion-exchange membrane may be expressed (Smith and Eisenmann, 1967):

$$\bar{\kappa} = \frac{d}{R_m A} \doteq 0.001 F \Sigma \bar{u} \, \bar{c}' \tag{7.119}$$

so that the specific conductance of the membrane completely in the form of species i would be

$$\bar{\kappa}_i = \frac{d_{mi}}{R_{mi}A} \doteq 0.001 F \bar{u}_i \bar{c}_i' \tag{7.120}$$

The mobility of species i would then be

$$\bar{u}_i \doteq \frac{1000 d_{mi}}{F \bar{c}_i R_{mi} A} \tag{7.121}$$

where d_{mi} = thickness of swollen membrane equilibrated with species i, cm
$\bar{c}_i'$ = milliequivalents of species i per mL of swollen membrane
R_{mi} = electrical resistance of membrane equilibrated with species i, Ω
A = area of measuring probes, cm^2

The ratio of the mobilities then of any two species will be:

$$\frac{\bar{u}_i}{\bar{u}_k} = \frac{d_{mi}}{d_{mk}} \frac{\bar{c}_k'}{\bar{c}_i'} \frac{R_{mk}}{R_{mi}} \tag{7.122}$$

If the membrane is equilibrated with only one species i, the concentration $\bar{c}_i'$ may be expressed in terms of membrane capacity:

$$\bar{c}_i' = \frac{(\text{capacity, meq})}{(\text{area, cm}^2) \times (d_{mi}, \text{cm})} \tag{7.123}$$

Combining Eqs. (7.122) and (7.123):

$$\frac{\bar{u}_1}{\bar{u}_k} = \left(\frac{R_{mk}}{R_{mi}}\right)\left(\frac{d_{mi}}{d_{mk}}\right)^2 \tag{7.124}$$

Thus, the ratio of the ionic mobilities of any two species in a particular membrane may be determined by measurements of the electrical resistance and thickness of the membrane equilibrated first with one species and then with another.

If now a membrane sample is equilibrated in a mixed solution of two ionic species and the relative equilibrium concentrations are determined, an ion-exchange separation factor may be described.

$$(\alpha_k^{\ i})_{IE} = \frac{\bar{c}_i}{\bar{c}_k} \cdot \frac{c_k}{c_i} \tag{7.125}$$

$\bar{c}_i, \bar{c}_k$ = equilibrium membrane concentrations of species i and k
c_i, c_k = equilibrium solution concentrations of species i and k
$(\alpha_k^{\ i})_{IE}$ = ion-exchange separation factor

This *ion-exchange separation factor* Eq. (7.125) is different from the separation factor described by Eq. (7.115) in that the former deals with static equilibrium membrane concentrations while the latter is concerned with changes in solution concentration due to ionic transport across the membrane. Combining Eqs. (7.115), (7.118), and (7.124) will yield an expression for the separation factor described by Eq. (7.126).

$$(\alpha_k^{\ i}) = \frac{J_i/c_i}{J_k/c_k} \doteq \frac{\overline{u}_i\overline{c}_i/c_i}{\overline{u}_k\overline{c}_k/c_K} \doteq \frac{R_{mk}}{R_{mi}} \left(\frac{d_{mi}}{d_{mk}}\right)^2 (\alpha_k^{\ i})_{IE} \tag{7.126}$$

Thus, it appears that separation factors might be estimated from static equilibrium measurements. The derivation of Eq. (7.126) necessitated the following assumptions:

1. All similarly charged species are subjected to essentially the same driving force. This implies that the log-activity gradients of all similarly charged species are either similar or negligible compared to the potential gradient. If none of the separation factors differ greatly from unity, all log-activity gradients will, in fact, be similar. That the electrical potential gradient is much more significant than activity gradients in an electrodialysis system appears to be a reasonable assumption, but has not been verified experimentally.

2. Individual ionic mobilities remain constant in the presence of other species. Although this is not strictly true, the differences are negligible compared to other possibilities of error.

3. Convective transport is ignored while the derivation does not specifically consider convection, the mobilities determined by resistance measurements certainly include the effects of convective transport.

4. Static or ion-exchange equilibrium concentrations do not differ from the dynamic membrane equilibrium concentrations found in an electrodialysis system. This assumption, while adequate as a first approximation, does not consider the effects of different solutions on either side of the membrane.

Until the assumptions noted are verified experimentally, separation factors calculated from static equilibrium measurements cannot be regarded quantitatively. Nevertheless, these measurements offer a qualitative description of the effects of modification of various membrane parameters on selectivity.

Test Procedure.   Membrane samples were equilibrated for 24 h in the more concentrated solutions (1.0 mol/L Cl^-, NO_3^-, SO_4^{2-}, etc.). This was followed by three washings in the more dilute solutions to remove any excess salt concentration from the surface or interior of the membranes. Samples were then equilibrated in the dilute solutions (0.01 mol/L) an additional 24 h before resistance and thickness measurements were undertaken.

For the static equilibrium measurements, mixed solutions were selected that approximated the concentration ratios found in a secondary sewage effluent used in exhaustion runs and product-to-waste ratio studies. Total solution concentrations were approximately 0.01 mol/L.

Membrane samples were first placed in the Cl^- form by use of the equilibrium procedure described for the mobility ratio determinations. They were then washed in distilled water and equilibrated for 7 days in the mixed solutions. These solutions were analyzed both before and after the 7-day equilibration period, and the membrane-to-solution concentration ratios calculated.

Selectivity data calculated with Eq. (7.125) is shown in Table 7.4.

TABLE 7.4 Estimated Separation Factors

	Membrane type			
	AR110 DYG-067	AR110 CZG-067	AR111 BZL-183	AR111 GWL-183
$\alpha_{Cl^-}^{NO_3^-}$	0.5	0.8	5	8
$\alpha_{Cl^-}^{SO_4^{2-}}$	3	3	8	7
$\alpha_{Cl^-}^{HPO_4^{2-}}$	0.6	0.7	0.2	0.4
$\alpha_{Cl^-}^{PO_4^{3-}}$	2	1	26	(large)

Source: Smith and Eisenmann, 1967.

The separation factors (Table 7.4) are subject to the assumptions discussed in the derivation of Eq. (7.126). While inherent membrane selectivity can be adequately described by this equation, the selectivity obtained in operation may be quite different (Smith and Eisenmann, 1965). The separation factors obtained from the membrane parameter study and from the solution parameter study are quite different (phosphate-chloride separation factors for membrane AR111 BZL-183). This is because the condition of negligible activity gradients assumed in the derivation of Eq. (7.126) obviously cannot be applied to the system that was used for the solution parameter study. In fact, approximately two-thirds of the combined phosphate and chloride transfer was due to diffusion because of backdiffusion of nitrate ions. The results of the two studies indicate that actual operation in a concentrating-diluting (electrodialysis) system will tend to depress selectivities inherent in the membranes; i.e., the system will not exhibit the degree of selectivity indicated by a study of membrane parameters alone.

7.3.6 Some New Developments in Ion-Exchange Membranes

Kedem et al. (1992) studied nitrate selectivity in ion-exchange membranes based on preferential uptake. The specific binding site will not lead to an ion-specific membrane. However, if preferred uptake is based on the character of the matrix in complete analogy to extraction, the mobility of the preferred ion may be preserved. Using this principle, they were able to obtain preferred transport of nitrate. For the separation of nitrates from other univalent salts, preferential extraction of nitric acid may serve as a model. In the extraction by amine solutions the sequence

$$ClO_4^- > NO_3^- > Cl^- > Br^- > SO_4^{2-}$$

was found. A versatile polymer which allows variation of the nature and concentration of amino groups in an anion-exchange membrane was developed using a halomethylation procedure.

The aminated polysulphone (APS) membranes were found to have good permselectivity. Since aminated polysulphone membranes are prepared from mixed solutions of bromomethylated polysulphone (BMPS) and polysulphone (PS) and aminated, a number of parameters for a graded change in membrane properties have been possible: the ratio of substitution of BMPS, the ratio of BMPS/PS, and the nature of the amino groups. All these change the hydrophilic/hydrophobic balance. In addition, different interactions with chloride and nitrate ions may be expected. From the extraction results previously mentioned, it may be inferred that the ratio between tertiary and quaternary amino groups is of special importance. Therefore, membranes with different ratios of tertiary and quaternary amino groups were prepared.

With all quaternary polyamines no selectivity between chloride and nitrate was observed. Increasing the fraction of tertiary groups increased the transport number of nitrate in the presence of excess chloride. This is shown in Table 7.5.

TABLE 7.5   Transport Numbers t_i for Varying Proportion of Tertiary Groups

Proportion of amino groups, %				
Tertiary	Quaternary	t Na$^+$	t NO$_3^-$	t Cl$^-$
10	90	0.08	0.22	0.70
30	70	0.08	0.61	0.31
47	53	0.17	0.65	0.18

Source: Kedem et al., 1994

In the chosen membrane composition, the quaternary groups were polymer – $N(C_2H_5)_3$ and the tertiary groups polymer – $N(C_4H_7)_2$. The increase of hydrophobicity by the butyl groups also contributes to nitrate selectivity.

The equilibrium distribution of chloride and nitrate in basic pH was determined in the same membranes. These results showed that the preferential transport of NO_3^- closely follows preferential uptake. This means that the mobility of NO_3^- is not hampered by specific binding. Indeed, the resistance of the membranes in chloride form or in nitrate form are similar.

7.3.7   Transport Properties of Anion-Exchange Membranes (Acid Blocking) in Contact with Hydrochloric Acid Solutions

Boudet-Dumy, Lindheimer, and Gavach (1991) have used a gel model to describe transport properties of anion-exchange membranes in contact with hydrochloric acid solutions. The membrane is assimulated to a polyelectrolyte solution. However, it was found that the gel model did not fit the experimental data–membrane considered as a gel phase and complete dissociation. It was found that HCl molecules sorbed to a significant amount in the membranes with respect to water. These data strongly suggest an incomplete ionic dissociation inside the membrane phase. A fraction of the sorbed HCl probably remains in the neutral molecule form.

In order to analyze the results of transport number measurements in Selemion AAV and Morgane ARA acid-blocking membranes, a modified gel model for the membrane phase was used by Boudet-Dumy, Lindheimer, and Gavach (1991). The membrane is assimilated to a polyelectrolyte solution; considering that the moving electrical species are only H_3O^+ and Cl^- (membranes in contact with hydrochloric acid solution). For an incomplete ionic dissociation inside the membrane phase, the following expression applies:

$$t_{Cl^-} = \frac{(\overline{X} + \gamma\overline{m}\,\overline{u}_{Cl^-})}{(\overline{X} + \gamma\overline{m})\overline{u}_{Cl^-} + \gamma\overline{m}\,\overline{u}_{H_3O^+}} \qquad (7.127)$$

where γ is the dissociation coefficient of sorbed HCl; $\overline{m}$ is, at the sorption equilibrium, the molality of the electrolyte which is present in the membrane; $\overline{X}$ is the molality of fixed charges; and $\overline{u}$ is the apparent mobility of ions in the membrane;

$$\beta = \frac{\overline{m}}{\overline{X}} = \frac{n_{HCl}}{n_{sites}}$$

Equation (7.127) can be rearranged into the following form:

$$\frac{1}{t_{Cl^-}} = 1 + \frac{\overline{u}_{H_3O^+}}{\overline{u}_{Cl^-}} \cdot \frac{\gamma\beta}{1 + \gamma\beta} \tag{7.128}$$

Equation (7.128) shows that the values of the transport number of the moving ions inside the membrane phase are dependent not only on the ratio of their mobilities and the amount of sorbed acid but also on the dissociation coefficient of sorbed acid.

If the ratio of the mobilities of the free ions $\overline{u}_{H_3O^+}/\overline{u}_{Cl^-}$ maintains a constant value, $1/t_{Cl^-}$ will have the same variations as $\gamma\delta/(1 + \gamma\delta)$. In Fig. 7.14, the calculated variations of this expression are plotted as a function of β for four different values of γ. From this curve, it follows that $1/t_{Cl^-}$ is lowered when the dissociation coefficient γ is low. Therefore, selectivity for chloride is higher when the dissociation coefficient is lower.

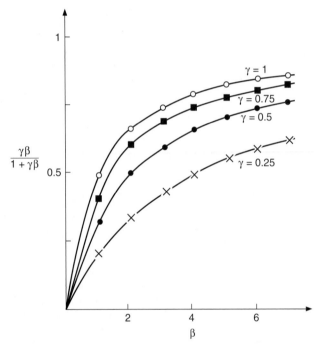

FIGURE 7.14 Variation of $\gamma\beta/(1 + \gamma\beta)$ as a function of β for several values of γ. *(Boudet-Dumy, Lindheimer, and Gavach, 1991.)*

It can be concluded that the lower dissociation of sorbed HCl is a factor which decreases the proton leakage of the anion-exchange membrane.

As for the variation of the ratio of the mobilities of H_3O^+ and Cl^-, no direct information could be deduced from the experimental data. A maximum was observed in the curves of the membrane conductivity when conductivity was plotted against β (Boudet-Dumy, Lindheimer, and Gavach, 1991). Considering again an incomplete dissociation of the sorbed HCl in the membrane phase, it can be concluded that the mobilities of the free ions have decreased because the expression of the membrane conductance is:

$$\Lambda = F[(\overline{X} + \gamma\overline{m})\overline{u}_{Cl^-} + \gamma\overline{m}E\overline{u}_{H_3O^+}] \tag{7.129}$$

The decrease of the values of free ion mobilities has two distinct origins: the increase of the local viscosity of the hydrophilic pathways in the membrane phase and the increase of the inter-ionic interactions, these two effects being due to high value reached by the mobility of HCl in the membrane phase.

For a current density of 30 mA-cm^{-2}, the unidirectional flux J of Cl$^-$ through the ARA membrane from the anode to the cathode steadily increased as the amount of sorbed electrolyte increased (Boudet-Dumy, Lindheimer, and Gavach, 1991). This result shows that chloride ions are associated with the movement of positively charged species. This fact may be due to the formation of an aggregate form such as (H$_4$OCl)$^+$ resulting from the solvation of a proton by a water molecule and an HCl molecule, ion association inside the membrane overcoming the state of a neutral HCl molecule. This result confirms the role of ion association in this membrane.

7.3.8 Concentration-Polarization Effects in Electrodialysis on Counter-Ion Selectivity of Ion-Exchange Membranes with Differing Counter-Ion Distribution Coefficients

This work (Rubenstein, 1990) is concerned with theory of the effect of concentration polarization under direct current on the counter-ion selectivity of an ion-exchange electrodialysis membrane. The equilibrium selectivity is induced by the differing uptake of the counter-ions by the membrane, owing to the difference in their distribution coefficients between the membrane and the solution phases.

Numerical calculations are supplemented by an asymptotic analysis for a certain range of bulk concentration and distribution coefficients, which provides an analytical expression for the dependence of selectivity on the electric current in the entire current range.

Rubinstein (1990) shows that most of the selectivity reduction in electrodialysis occurs for currents much lower than the limiting value. This reduction occurs "sooner" for lower concentrations of the preferred ion.

If near-equilibrium electrodiffusion in the membrane is the rate-determining step in ion transfer, the equilibrium ionic composition of the membrane determines its selectivity. In particular, this implies high selectivity for the preferentially extracted counter-ion. For a higher current, preferential transfer of the preferred counter-ion results in its enhanced depletion in the unstirred layer, so that the membrane is equilibrated with a solution containing an appreciably lower relative amount of the preferred ion than the bulk solution. These changes in the ionic contents of the membrane result in a rapid reduction of counter-ion selectivity, while the voltage-current curve still remains nearly linear at still higher voltage. When the Nernst layer gets appreciably depleted of both ions, the limiting stage of counter-ion transfer shifts from the membrane to the unstirred layer. As a result, the selectivity approaches its low limiting value, while the voltage-current curve exhibits typical nonlinearity with ultimate saturation.

ACKNOWLEDGMENT

The authors would like to thank René Cassidy for typing the manuscript and André Steyn for drawing the graphs.

REFERENCES

Boudet-Dumy, M., A. Lindheimer, and C. Gavach, 1991, "Transport Properties of Anion-Exchange Membranes in Contact with Hydrochloric Acid Solutions: Membranes for Acid Recovery by Electrodialysis," *Journal of Membrane Science,* **57**:57–68.

Davis, T. A., and G. F. Brockman, 1972, "Physiochemical Aspects of Electromembrane Processes," in R. E. Lacey and S. Loeb (eds.), *Industrial Processing with Membranes,* Wiley-Interscience, New York.

Donnan, F. G., and E. A. Guggenheim, 1932, "Exact Thermodynamics of Membrane Equilibrium," *Z. Physik. Chemie,* **A162**:346–360.

Flett, D. S., 1983, *Ion-Exchange Membranes,* Ellis Horwood, Chichester.

Garza, G., 1973, "Electrodialysis by Electro-Osmotic Pumping and Ion Separation with Charged Membranes," Ph.D.diss., Weizmann Institute of Science, Rehovot, Israel.

Garza, G., and O. Kedem, 1976, "Electro-Osmotic Pumping in Unit-Cells," *5th International Symposium on Fresh Water from the Sea,* **13**:79–87.

Gregor, H. P., 1951, "Gibbs-Donnan Equilibria in Ion-Exchange Resin Systems," *J. Am. Chem. Soc.,* **73**:642–650.

Helfferich, F. G., 1962, *Ion-Exchange,* McGraw-Hill, New York.

Henderson, F., 1907, *Z. Phys. Chem.,* **59**:118.

Hirsch-Ayalon, P., 1951, "Over Abnormal Osmose: Proeven over Waterbeweging Daar Een Geladen Membraan Tijdens Diffusie van Opgeloste Stof," *Chem. Weekbl.,* **47**:1025.

Hoffer, E., and O. Kedem, 1972, "Hyperfiltration in Charged Membranes: Prediction of Salt Rejection from Equilibrium Measurements," *Journal of Physical Chemistry,* **76**(24):3638–3641.

Johnson, J. S., Jr., L. Dresner, and K. A. Kraus, 1966, chapter 8, *Principles of Desalination,* Academic Press, New York.

Katchalsky, A., and P. F. Curran, 1967, *Non-Equilibrium Thermodynamics in Biophysics,* Harvard University Press, Cambridge, Mass.

Kedem, O., and A. Katchalsky, 1961, "Physical Interpretation of Phenomenological Coefficients of Membrane Permeability," *J. Gen. Physiol.,* **45**:143.

Kedem, O., and A. Katchalsky, 1963, "Permeability of Composite Membranes, Part I: Electric Current, Volume Flow and Flow of Solute Through Membranes," *Trans. Faraday Soc.* **59**:1918–1930.

Kedem, O., Z. Bar-On, and A. Warshawsky, 1986, "Electro-Osmotic Pumping in a Sealed-Cell ED Stack," *AIChE Symposium Series, Industrial Membrane Processes,* **82**(248):19–27.

Kedem, O., A. Warshawsky, I. Rubinstein, and A. Eyal, 1992, *Some New Developments in Ion-Exchange Membranes,* Weizmann Institute of Science, Rehovot, Israel.

Korngold, E., 1984, "Electrodialysis—Membranes and Mass Transport," in *Synthetic Membrane Processes,* G. Belfort (ed.), Academic Press, New York.

Kressman, T. R. E., and F. L. Tye, 1956, "The Effect of Current Density on the Transport of Ions Through Ion-Selective Membranes," *Discussions Faraday Soc.,* **21**:185–220.

Lacey, R. E., 1972, "Basis of Electromembrane Processes," in R. E. Lacey and S. Loeb (eds.), *Industrial Processing with Membranes,* Wiley-Interscience, New York.

Lakshminarayanaiah, N., 1969, *Transport Phenomena In Membranes,* Academic Press, New York.

Lakshminarayanaiah, N., and V. Subrahmanyan, 1964, "Measurement of Membrane Potentials and Test of Theories," *J. Polymer Sci.,* **Pt. A2**(10):4491–4502.

Mackay, D., and P. Meares, 1955, "The Diffusion of Electrolytes in a Cation-Exchange Resin Membrane," *Proc. Roy. Soc. (London),* **A232**:498–509.

Mackie, J. S., and P. Meares, 1955, "The Sorption of Electrolytes by a Cation-Exchange Resin Membrane," *Proc. Roy. Soc.,* **232A**(1191):485–518.

Malherbe, P. Le Roux, and G. B. Mandersloot, 1960, "The Physical Chemistry of Ion-Selective Membranes," in J. R. Wilson (ed.), *Demineralization by Electrodialysis,* Butterworths Scientific Publication, London.

Meyer, K. H., and J. F. Sievers, 1936a, "La Perméabilité des Membranes, I. Thérie de la Perméabilité Ionique," *Helv. Chim. acta,* **19**:649.

Meyer, K. H., and J. F. Sievers, 1936b, "La Perméabilité des Membranes, II. Essais Avec des Membranes Sélectives Artificielles," *Helv. Chim. acta,* **19**:665.

Narebska, A., and S. Koter, 1988, "From Theory to Practice: Transport Phenomena in Working Membrane Systems," in Alicja M. Mika and Thomasj Z. Winnicki (eds.), *Advances in Membrane Phenomena and Processes,* lecture textbook of the ESMST Summer School, June 6–12th, 1988, Gdansk-Sobieszewo. Wroclaw Technical University Press, Wroclaw, 1989.

Nernst, W., 1888, *Z. Phys. Chem.,* **2**:613.

Nernst, W., 1889, *Z. Phys. Chem.,* **4**:129.

Nishiwaki, T., 1972, "Concentration of Electrolytes Prior to Evaporation with an Electromembrane Process," in R. E. Lacey and S. Loeb (eds.), *Industrial Processing with Membranes,* Wiley-Interscience, New York.

Robinson, R. A., and R. H. Stokes, 1955, *Electrolyte Solutions,* Butterworths, London.

Rubinstein, I., 1990, "Theory of Concentration Polarization Effects in Electrodialysis on Counter-Ion Selectivity of Ion-Exchange Membranes with Differing Counter-Ion Distribution Coefficients," *J. Chem. Soc. Faraday Trans.,* **86**(10):1857–1861.

Sarocco, G., M. C. Zanetti, and M. Onofrio, 1993, "Novel Application of Mono-Ion-Permselective Membranes to the Recovery Treatment of an Industrial Wastewater by Electrodialysis," *Industrial and Engineering Chemistry Research,* **32**(4):657–662.

Sata, T., 1973, "Properties of a Cation-Exchange Membrane Adsorbed or Ion-Exchanged with Hexadecylpyridinium Chloride," *Electrochim. Acta,* **18**(20):199–203.

Scatchard, G. J., 1953, "Ion-Exchanger Electrodes," *J. Amer. Chem. Soc.,* **75**:2883.

Schlögl, R., 1955, "Theory of Anomalous Osmosis," *Z. Phys. Chem.,* **3**:73–102.

Schoeman, J. J., 1992, "Electrodialysis of Salts, Acids and Bases by Electro-Osmotic Pumping," Ph.D.diss., University of Pretoria, South Africa.

Smith, J. D., and L. Eisenmann, 1967, "Electrodialysis in Advanced Waste Treatment," Report submitted in fulfillment of Contract No. SAph 76690 between the Public Health Service, Inc. submitted February 1, 1965, U.S. Department of the Interior, Federal Waste Pollution Control Administration, Cincinnati, Ohio, February 1967.

Spiegler, K. S., and O. Kedem, 1966, "Thermodynamics of Hyperfiltration (Reverse Osmosis): Criteria for Efficient Membranes," *Desalination,* **1**:311–326.

Staverman, A. J., 1952, "Non-Equilibrium Thermodynamics of Membrane Processes," *Trans. Faraday Soc.,* **48**:176–185.

Stewart, R. J., and W. F. Graydon, 1957, "Ion-Exchange Membranes: III Water Transfer," *J. Phys. Chem.,* **61**:164–168.

Strathmann, H., 1992, in W. S. Ho. Winston and K. Kamalesk, Sirkar (eds.), *Membrane Handbook,* Van Nostrand-Reinhold, New York.

Tanaka, G., and M. Seno, 1981, "Treatment of Ion-Exchange Membranes to Decrease Divalent Ion Permeability," *J. Membr. Sci.,* **8**:115–127.

Teorell, T., 1935, *Proc. Soc. Exp. Biol.,* **33**:282.

Teorell, T., 1951, "Quantitative Treatment of Membrane Permeability," *Z. Electrochem.,* **55**:460–469.

Teorell, T., 1953, "Transport Processes and Electrical Phenomena in Ionic Membranes," *Prog. Biophys.,* **3**:305.

Timashev, S. F., 1991, *Physical Chemistry of Membrane Processes,* Ellis Horwood, New York.

Wilson, J. R., B. A. Cooke, W. G. B. Mandersloot, and S. G. Wiechers, 1960, "The Electrodialysis Process," in J. R. Wilson (ed.), *Demineralization by Electrodialysis.* Butterworths Scientific Publication, London.

Yawataya, T., H. Hani, Y. Oda, and A. Nishihara, 1962, *Dechema Monograph,* **47**:501.

CHAPTER 8
MULTIPHASE MEMBRANE PROCESSES

P. Aptel
Laboratoire de Génie Chimique, CNRS
Université Paul Sabatier
Toulouse, France

M. J. Semmens
Department of Civil Engineering
University of Minnesota
Minneapolis, Minnesota

8.1 INTRODUCTION

In the other chapters in this book, it is assumed that water is present on both sides of the membranes that are used for water treatment. In this chapter, processes will be addressed in which the membrane is used to expose the water to a different phase to facilitate the removal of particular contaminants or the transfer of gases. Examples of the other phases include a gas, a vacuum, a solvent, or a chemically reactive solution. Water may therefore only be in contact with one side of the membrane.

There are several important differences from the other processes that are specific to multiphase separations. First, the driving force for separation is provided by maintaining a concentration gradient across the membrane. This is usually accomplished by exploiting the chemical characteristics of the contaminants that need to be removed (e.g., volatility, polarity, charge, dissociation constant, etc.). Second, the common and desired characteristic of these membranes is that they are largely impermeable to water. For example, in the pervaporation process a polymeric membrane is chosen to be selective for and highly permeable to the organic compounds that we wish to remove, but as impermeable as possible for water.

8.2 MEMBRANE TYPES

There are three different types of membrane that may be used for phase contact processes:

1. Dense membranes
2. Hydrophobic porous membranes
3. Composite membranes

The dense membranes are made of a solid nonporous polymer. A common example is silicone rubber which is permeable to low-molecular-weight organic compounds and gases. Wilderer, Braeutigam, and Sekoulof (1985); Côté et al. (1988 and 1989); and Hirasa, Ichijo, and Yamauchi (1991) have used these membranes to transfer oxygen into water for wastewater treatment applications, and numerous investigators have demonstrated the use of silicone or other elastomeric membranes for VOC removal by pervaporation processes (Nguyen and Nobe, 1987; Psaume et al., 1988; Nijhuis, 1990; Böddeker, Bengston, and Bode, 1990; and Neel, 1992).

Mesoporous and microporous membranes are widely used in water treatment applications for ultrafiltration and nanofiltration. However, the membranes used in these applications are made from functionalized polymers that render the membranes hydrophilic so that when the membrane is contacted with water the pores wet and water can flow through the membrane convectively. For phase contact applications it is important that the membrane is not hydrophilic and that the pores do not wet. The membrane is used to create an interface for mass transfer and, as such, the membrane must be hydrophobic and the pores sufficiently small to avoid wetting under the conditions of operation. Porous polyolefin membranes (e.g., polypropylene and polyethylene) with pore sizes below 0.1 μm are effective membranes for many phase contact processes.

The presence of high concentrations of certain organic solvents and the presence of oils or surfactants may reduce the surface tension of the water and/or modify the surface of the membrane such that wetting occurs. This is a major disadvantage of using porous membranes, since once wetting occurs the desirable transport properties of the membrane are lost. These problems are not experienced with dense membranes.

Composite membranes combine the advantages and selectivity of dense polymers with the faster transport kinetics of porous membranes. Composite membranes can be created by coating a very thin dense polymer layer on the surface of a porous membrane support. This dense layer physically covers and seals the pores of the membrane so that wetting is impossible even under the most adverse conditions.

Example applications of multiphase separations are listed in Table 8.1, which also includes processes in which the membrane itself presents a different phase to the water and is exploited to separate contaminants.

In this chapter, our discussion is limited mainly to processes of potential significance to the treatment of potable water. Some discussion is also included of processes that offer promise for the remediation of groundwaters that may be used as a source of water supply and the treatment of wastes generated during water treatment.

8.3 MASS TRANSPORT CONSIDERATIONS

The driving force for the mass transport across the membrane is the chemical potential gradient across the membrane. The chemical potential of a specie is given by the expression:

$$\mu = \mu^{\circ}(T) + RT \ln (a) \tag{8.1}$$

where μ° is the standard chemical potential of the pure compound and a is its activity.

TABLE 8.1 Multiphase Separations—Applications

Phases	Membrane types	Process
Water-gas	Dense, composite, and hydrophobic porous	Gas transfer
Water-vacuum	Dense and composite	Pervaporation
Water-vacuum or air	Hydrophobic porous	Membrane distillation and membrane stripping
Water-water	Dense and hydrophobic porous	Biologically or chemically mediated separations
Water-oil	Dense, composite, and hydrophobic porous	Organics separation
Water-solvent	Dense, composite, and hydrophobic porous	Solvent extraction

The chemical potential across the membrane is defined as the chemical potential on the permeate side minus the chemical potential on the feed side. If we let the subscripts f stand for feed side and p for permeate side, then we obtain:

$$\mu_p - \mu_f = [\mu°(T) + RT \ln (a_p)] - [\mu°(T) + RT \ln (a_f)] = RT \ln \left(\frac{a_p}{a_f} \right) \qquad (8.2)$$

The chemical potential gradient must be negative for the process to be thermodynamically feasible and thus the term $\ln (a_p/a_f)$ must be negative. This requires that $a_p < a_f$.

In most water treatment applications we are concerned with the removal of specific contaminants. In such cases the value of a_f is fixed, and the separation can be accomplished only by providing conditions that lower the value of a_p. The maximum driving force that may be provided under such circumstances corresponds to the condition under which the value of a_p approaches zero.

In applications where a membrane is being used to add a chemical to the water, the value of a_f represents the activity of the chemical on the feed side of the membrane and a_p represents the activity of the chemical in the water phase. In this case, the value of a_p is determined by the concentration needed in the water, and the driving force is controlled by increasing the value of a_f. Clearly, in these applications, higher concentration gradients may be maintained since large values of a_f may be achieved by using high concentrations of the chemical on one side of the membrane.

8.3.1 Kinetics of Mass Transport

Mass transport across a membrane may involve as many as five steps. For example, for a dense membrane, the steps include:

1. Diffusion from the water phase to the surface of the membrane
2. Selective partitioning into the membrane phase
3. Selective transport (diffusion) through the membrane
4. Desorption from the permeate side of the membrane
5. Diffusion away from the membrane and into the bulk fluid of the extracting phase

As with all mass transfer operations the slowest step in this sequence will limit the overall rate of mass transfer. The slowest step is determined by the membrane characteristics, the fluid flow regimes that are maintained on each side of the membrane, the properties of the chemical being separated, and the properties of the phases that are involved. The partitioning and desorption steps are generally not considered to be rate limiting; indeed, it is usually assumed that an equilibrium condition prevails at the interface between the membrane and the fluid phases. Therefore, one or more of steps 1, 3, and 5 may control the rate of mass transfer.

The overall mass transfer coefficient may be calculated based on the resistance-in-series model. In this model, a stagnant film is presumed to exist in the fluid phases on either side of a dense membrane, and chemical transfer occurs by molecular diffusion through these films and the membrane itself. The thickness of these stagnant films, or fluid-phase boundary layers, is determined by the hydrodynamic conditions in each phase, and they will become thinner as the Reynolds number increases.

8.4 RESISTANCE IN SERIES MODEL

The membrane with its fluid boundary layers is depicted in Fig. 8.1. If the membrane is a dense polymer, three different phases are involved in this analysis. The picture presented in Fig. 8.1 is representative of the kind of profiles that might prevail in a membrane air stripping or pervaporation process.

The mass transfer coefficient for each phase (k_l, k_m, and k_g) may be estimated by dividing the diffusion coefficient of the contaminant in that phase by the distance through which the molecules must diffuse. In practice, however, the individual mass transfer coefficients cannot be calculated directly since the boundary layer thick-

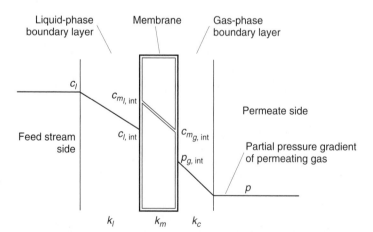

FIGURE 8.1   An illustration of the resistances to mass transfer that are considered in the resistance-in-series model for pervaporation. In this figure, the flux across the membrane is necessarily equal to the flux across each of the fluid boundary layers.

nesses are dependent on the local mixing conditions in the water and air, and the diffusion path of a contaminant across a membrane is not necessarily equal to the thickness of the membrane. As a result, the individual mass transfer coefficients have to be evaluated empirically and much experimental work has been conducted to obtain the dimensionless correlations needed to calculate the individual mass transfer coefficients under different operating conditions. The calculation of the individual mass transfer coefficients is discussed at length as follows.

There is no way to accurately measure the interfacial concentrations under different operating conditions; however, we can reasonably assume that the concentrations in each phase are in equilibrium at an interface. Thus, we need to know the partition coefficient that describes the equilibrium partitioning of the chemical between the membrane and the water K_d and Henry's law constant H_u for the distribution of the contaminant between the water and air phases. Where

$$K_d = \frac{c_m}{c_l} \quad \text{and} \quad H_u = \frac{p}{c_l}$$

Notice that K_d is dimensionless since it is simply derived from the ratio of the concentrations in the two phases. The value of the Henry's constant will depend on the way in which it is defined; for the preceding definition the units would be $Pa \cdot m^3 \cdot mol^{-1}$. Sometimes it is convenient to use a dimensionless Henry's constant H_u, which is defined as the ratio of the gas-phase and water-phase concentrations at equilibrium:

$$H_u = \frac{c_g}{c_l}$$

In practical applications for water treatment, it is most common to express mass transfer in terms of the liquid-phase (water) concentrations as shown in Eq. (8.4).

$$J = k_l(c_l - c_{l,\text{int}}) = k_m K_d(c_{m_{l,\text{int}}} - c_{m_{g,\text{int}}}) = k_g H_u(p_{g,\text{int}} - p) \tag{8.3}$$

where
$\qquad J = \text{flux}$
$\qquad c_l = \text{bulk liquid concentration}$
$\qquad c_{l,\text{int}} = \text{interface liquid concentration}$
$\qquad c_{m,l,\text{int}} = \text{membrane – water interface concentration}$
$\qquad c_{m,g,\text{int}} = \text{membrane – air interface concentration}$
$\qquad p_{\text{int}} = \text{interface gas partial pressure}$
$\qquad p = \text{bulk gas partial pressure}$
$\qquad k_l = \text{liquid film mass transfer coefficient}$
$\qquad k_m = \text{membrane mass transfer coefficient}$
$\qquad k_g = \text{gas film mass transfer coefficient}$

This equation can be rearranged and simplified so that the flux is expressed in terms of the bulk concentrations in the air and water phases and an overall mass transfer coefficient. If the bulk concentrations are expressed in terms of liquid-phase concentrations, then the overall liquid-phase mass transfer coefficient K_{OL} is used:

$$J = K_{\text{OL}}(c_l - c_l^*) \tag{8.4}$$

where the value of c^* is assumed to be in equilibrium with the partial pressure in the gas phase on the other side of the membrane (i.e., $c^* = p/H_u$). The overall liquid-phase mass transfer coefficient can then be expressed in terms of the mass transfer coefficients for each phase and the relationship has the following form:

$$\frac{1}{K_{OL}} = \frac{1}{k_l} + \frac{1}{K_d k_m} + \frac{1}{H_u k_g} \tag{8.5}$$

The relationship of the overall mass transfer coefficient K_{OL} to the film transfer coefficients k_l and k_g will depend on the type of membrane used. The preceding relationship was derived for a dense membrane which is treated as a separate phase; however, if a microporous membrane is used, then one of the fluid phases will fill the pores of the membrane, and a different relationship must be used. The relationships for different configurations of the membrane are presented in Table 8.2. In these relationships, the value of K_d is determined by the phase into which the transferring solute is partitioning; in Table 8.2, for example, K_d represents the ratio of the concentration in the membrane to the liquid-phase concentration under equilibrium conditions.

TABLE 8.2 Relationships for the Overall Mass Transfer Coefficient for Different Types of Membrane and Conditions of Operation

(Based on the liquid-phase concentrations)

Membrane	Fluid-phase occupying membrane pores	Overall mass transfer relationship
Dense	None	$\frac{1}{K_{OL}} = \frac{1}{k_l} + \frac{1}{K_d k_m} + \frac{1}{H_u k_g}$
Microporous	Water	$\frac{1}{K_{OL}} = \frac{1}{k_l} + \frac{1}{k_m} + \frac{1}{H_u k_g}$
Microporous	Gas	$\frac{1}{K_{OL}} = \frac{1}{k_l} + \frac{1}{H_u k_m} + \frac{1}{H_u k_g}$
Microporous	Solvent	$\frac{1}{K_{OL}} = \frac{1}{k_l} + \frac{1}{K_d k_m} + \frac{1}{K_d k_s}$

The overall mass transfer coefficient can also be calculated based on the concentration gradients being expressed as gas-phase concentrations. For example, from Eq. (8.4) we may write:

$$J = K_{OG}(p^* - p)$$

where K_{OG}, the overall gas-phase mass transfer coefficient, is calculated as follows:

$$\frac{1}{K_{OG}} = \frac{H_u}{k_l} + \frac{1}{K_d' k_m} + \frac{1}{k_g}$$

where K_d' represents the ratio of the concentration in the membrane to the gas-phase concentration under equilibrium conditions.

EXAMPLE 8.1 *A membrane is to be selected to extract a gas or volatile organic compound (VOC) from water. Assume the conditions of operation fix the values of k_l and k_g at 10^{-6} and 2×10^{-2} m/s, respectively, and that the Henry's constant is 0.15 (dimensionless concentration ratio). Calculate the overall liquid-phase transfer coefficient for the transfer of the VOC across the following three membranes and identify which membrane would give the best transfer rate.*

1. A dense membrane, given $k_d = 103$ and $k_m = 5 \times 10^{-9}$ m/s

2. A porous membrane in which the pores are gas filled, given $k_m = 10^{-2}$ m/s

3. A porous membrane in which the pores are filled with water, given $k_m = 10^{-7}$ m/s

Answer:

1. Dense membrane:

$$\frac{1}{K_{OL}} = \frac{1}{k_l} + \frac{1}{K_d k_m} + \frac{1}{Hk_g}$$

$$\frac{1}{K_{OL}} = \frac{1}{10^{-6}} + \frac{1}{103 \times 5 \times 10^{-9}} + \frac{1}{0.15 \times 2 \times 10^{-2}}$$

$$\therefore K_{OL} = 3.4 \times 10^{-7} \text{ m/s}$$

2. Porous membrane with air-filled pores:
   In this case,

$$\frac{1}{K_{OL}} = \frac{1}{k_l} + \frac{1}{Hk_m} + \frac{1}{Hk_g}$$

$$\frac{1}{K_{OL}} = \frac{1}{10^{-6}} + \frac{1}{0.15 \times 10^{-2}} + \frac{1}{0.15 \times 2 \times 10^{-2}}$$

$$\therefore K_{OL} \cong 10^{-6} \text{ m/s}$$

3. Porous membrane with water-filled pores:

$$\frac{1}{K_{OL}} = \frac{1}{k_l} + \frac{1}{k_m} + \frac{1}{Hk_g}$$

$$\frac{1}{K_{OL}} = \frac{1}{10^{-6}} + \frac{1}{10^{-7}} + \frac{1}{0.15 \times 2 \times 10^{-2}}$$

$$\therefore K_{OL} \cong 9.09 \times 10^{-8} \text{ m/s}$$

From this analysis it is clear that the porous membrane with air-filled pores has the largest value of K_{OL} and will give the fastest rate of mass transfer for VOC removal.

It is the additive character of the reciprocals of the mass transfer coefficients as expressed in Eq. (8.5) that lends itself to analogy with electrical resistance theory. When resistances are aligned in series, their overall resistance is equivalent to the sum of the individual resistances. Likewise, in the case of mass transfer across a membrane, the overall resistance is equal to the sum of the liquid film resistance, the membrane resistance, and the gas film resistance, the three terms on the right-hand side of Eq. (8.5). We discuss these individual resistances at length as follows.

8.5 RESISTANCE IN THE FLUID BOUNDARY LAYERS

The fluid mass transfer coefficients may be estimated using dimensionless correlations. These correlations are available for a variety of configurations and for different ranges of operation (Reynolds numbers), and it is important to select a correlation that matches as closely as possible the conditions that you wish to model. The correlations typically express the Sherwood number (Sh) as a function of the Reynolds number (Re) and the Schmidt number (Sc). These correlations generally take the following form:

$$Sh = p \cdot Re^q \cdot Sc^r \cdot \left(\frac{d_e}{L}\right)^s \tag{8.6}$$

The values for p, q, r, and s are dependent on the operating conditions and the design of the membrane contactor, and different values can be found from literature. A summary of several correlations for tubular and spiral-wound membrane contactors is presented in Table 8.3. The fluid film mass transfer coefficients that govern the rate of mass transfer can therefore be determined as a function of the unit dimensions and the Sh, Re, Sc, and Gr or Pe numbers. These in turn are defined as:

$$Sh = \frac{kd_e}{D} \qquad Re = \frac{vd_e}{v} \qquad Sc = \frac{v}{D} \qquad \text{and} \qquad Gr = \frac{d_e^2 v}{DL},$$

TABLE 8.3 Dimensionless Correlations Used to Determine Mass Transfer Coefficients*

Tubular and hollow fine-fiber reactors		
Flow inside the tubes (fibers)		
Straight tubes		
Sh = 1.62 Gr$^{0.33}$	Re < 2000	Lévêque, 1928
Sh = 0.026 Re$^{0.8}$ Sc$^{0.33}$	Re > 4000	Yang & Cussler, 1986
Helically wound tubes		
Sh = 0.142(De′)$^{0.75}$ Sc$^{0.33}$	Re 180–2000	Moulin et al., 1994
Flow outside the tubes (fibers)		
Parallel flow—fluid flow parallel to the axis of the tubes		
Sh = β[d_e(1 − φ)/1] Re$^{0.6}$ Sc$^{0.33}$	φ is packing factor	Prasad and Sirkar, 1988
Sh = 0.022 Re$^{0.6}$ Sc$^{0.33}$	—	Knudsen and Katz, 1958
Sh = 1.25(Red_e/1)$^{0.93}$ Sc$^{0.33}$	Re: 5–3500, φ = 0.25	Yang and Cussler, 1986
Sh = 0.0104 Re$^{0.806}$ Sc$^{0.33}$	Re: 600–46000	Ahmed and Semmens, 1992
Sh = 0.61 Re$^{0.363}$ Sc$^{0.33}$	Re: 0.6–49	Côté et al., 1989
Sh = (0.53–0.58 φ)Re$^{0.53}$ Sc$^{0.33}$	Re: 20–350; 30< >76	Costello et al., 1993
Cross flow—fluid flow normal to the axis of the tubes		
Sh = 0.90 Re$^{0.4}$ Sc$^{0.33}$	External void fraction of 0.93	Yang and Cussler, 1986
Sh = 1.38 Re$^{0.34}$ Sc$^{0.33}$	External void fraction of 0.30	Yang and Cussler, 1986
Sh = 0.32 Re$^{0.61}$ Sc$^{0.31}$	Widely spaced tube banks	Kreith, 1973
Sh = 0.39 Re$^{0.59}$ Sc$^{0.32}$	Closely spaced tube banks	Kreith, 1973
Fluid flow at an angle other than 90 degrees to the axis of the tube		
Shβ = CβSh	Cβ = 0.00216(β)$^{1.38}$	Ahmed, 1991
Shβ = cos^2 (90 − β)Sh	β is degree of angle	Wickramasinghe, 1992
Spiral-wound contactors		
Sh = 0.065 Re$^{0.875}$ Sc$^{0.25}$	Re < 1000	Schock and Miquel, 1987

* For additional correlations and a more complete analysis, the reader is referred to V. Gekas and B. Hallström (1987), S. R. Wickramasinghe (1992), and Hickey and Gooding (1994).

where k = film mass transfer coefficient
v = fluid velocity
v = kinematic viscosity of fluid considered
d_e = characteristic length (e.g., inside diameter for flow inside tube)
D = diffusivity of chemical in fluid considered
L = channel or fiber length

The characteristic length d_e that is used in these correlations is defined in different ways for different fiber-flow configurations. For example, for flow outside the fibers, when a cross-flow design is used, the characteristic length is equal to the outside fiber diameter. When the flow is parallel to the axis of the fibers, however, an equivalent diameter is used. The equivalent diameter is equal to four times the hydraulic radius, where the hydraulic radius is defined as the cross-sectional area for fluid flow divided by the wetted perimeter. For parallel flow in a module of shell diameter d_s containing n fibers having diameter d_f, the value of d_e is calculated as follows:

$$d_e = \frac{d_s^2 - nd_f^2}{d_s + nd_f} \qquad (8.7)$$

Several semiempirical relations have been developed to determine the mass transfer coefficients across the fluid boundary layers next to the membrane. Table 8.3 gives some of the semiempirical correlations that have been developed to determine the fluid film boundary layer resistances in membrane contactors. These correlations have been determined by different investigators and apply to spiral-wound and hollow fine-fiber membrane modules. For the hollow fine-fiber modules, different correlations exist for flow inside and outside the fibers. In addition, for flow outside the fibers, different correlations exist for fluid flow parallel to the fiber axis and for fluid flow patterns that cross the fibers either normally or at some angle to the fiber axis.

The Reynolds number exponent in Eq. (8.6) is an indication of the mass transfer regime involved. For example, when q has a value of 0.33, it indicates a laminar mass transfer with fully developed velocity profile and a developing concentration profile. The exponent q takes on a value of 0.5 to 0.6 when both the velocity and concentration profiles are developing, as might be expected for entrance region conditions. The value of q approaches 1 for mass transfer in the turbulent flow regime (Iverson, 1994).

The mass transfer coefficient for flow inside straight tubes is well described (Psaume et al., 1988; Kreulen et al., 1993; and Cussler, 1994). However, when tubes are wound in a helical or spiral configuration, the curved tubes create centrifugal forces that induce secondary flows which make the fluid flow more complex and improve mass transfer. Dean (1928) first studied flow in curved pipes using a concentric toroidal coordinate system. In the case of a laminar flow through a cylindrical tube, the axis of which forms a helix of small pitch, the primary (axial) flow is accompanied by a secondary flow field which acts in a plane perpendicular to the tube axis and which is symmetrical about the plane of curvature of the tube. This results in the formation of vortices. To take into account the curvature of the tubes in a helical contactor, the flow regime in the tubes is characterized by the Dean number (De), which is a modified Reynolds number and is defined as follows:

$$\text{De} = \text{Re}\left(\frac{d_s}{d_i}\right)^{0.5} \qquad (8.8)$$

where d_i = internal diameter of tube
           d_s = coil diameter

Vortex formation is influenced both by the diameter of the helical coil and the pitch of the coil (i.e., the distance between adjacent coils in the spiral). To account for the pitch of the helix, a slightly different definition of the Dean number is used:

$$De' = Re \sqrt{\frac{d_s[1 + (p/(\pi \cdot d_s))^2]}{d_i}} \tag{8.9}$$

where the pitch of the helical coil is equal to $2\pi p$.

The value of a film mass transfer coefficient can be determined by selecting a correlation that best describes the flow conditions in a contactor and expanding the dimensionless correlations to express the mass transfer coefficient in terms of the design and operating conditions.

The characteristic length d_e that is used in these correlations is defined in different ways for different fiber-flow configurations. For example, when a cross-flow design is used, for flow outside the fibers, the characteristic length is equal to the fiber diameter. When the flow is parallel to the axis of the fibers, however, an equivalent diameter is used. The equivalent diameter is equal to four times the hydraulic radius where the hydraulic radius is defined as the cross-sectional area for fluid flow divided by the wetted perimeter. For flow in a module of shell diameter d_s containing n fibers having diameter d_f, the value of d_e is calculated as follows:

$$d_e = \left(\frac{d_s^2 - nd_f^2}{d_s + nd_f} \right) \tag{8.10}$$

EXAMPLE 8.2   *A shell and tube hollow-fiber module is 0.3 m long and has an internal shell diameter of 60 mm. If the module contains 3000 hollow fibers that have an external diameter of 400 μm, and a water is pumped through the shell at a flowrate of 4 L/s, calculate the value of the liquid film coefficient k_l on the outside of the fibers. Assume the kinematic viscosity of the water is 10^{-6} m²/s, the diffusivity of the contaminant in water is 10^{-8} m²/s, and that flow is parallel to the hollow fibers.*

1. Calculate the Reynolds number for the shell side of the module:

   a. Calculate the equivalent diameter:

$$d_e = \left(\frac{d_s^2 - nd_f^2}{d_s + nd_f} \right)$$

$$d_e = \left(\frac{(0.06)^2 - 3000(400 \times 10^{-6})^2}{0.06 + 3000(400 \times 10^{-6})} \right)$$

$$d_e = 0.0025 \text{ m}$$

   b. Calculate the velocity of the water outside the fibers:

$$v = \frac{4Q}{\pi(d_s^2 - nd_f^2)}$$

$$v = \frac{4(4.0 \times 10^{-3})}{\pi\{(0.06)^2 - 3000(400 \times 10^{-6})^2\}}$$

$$v = 1.618 \text{ m/s}$$

c. *Calculate the Reynolds number:*

$$\text{Re} = \frac{d_e v}{v} = \frac{1.632*0.0025}{10^{-6}} = 4800$$

d. *Select a dimensionless correlation that was evaluated for a Reynolds number range that embraces the value we calculated above. In this case, the correlation of Knudsen and Katz is chosen from Table 8.3.*

$$\text{Sh} = 0.022 \text{ Re}^{0.6} \text{ Sc}^{0.33}$$

$$\frac{k_l d_e}{D} = 0.022(4800)^{0.6}\left(\frac{v}{D}\right)^{0.33}$$

$$k_l = 0.022(4800)^{0.6}\left(\frac{v}{D}\right)^{0.33}\frac{D}{d_e}$$

$$k_l = 0.022(4800)^{0.6}\left(\frac{10^{-6}}{10^{-8}}\right)^{0.33}\frac{10^{-8}}{0.0025}$$

$$k_l = 5.9*10^{-5} \text{ m/s}$$

8.6 MASS TRANSFER IN THE MEMBRANE

The membrane mass transfer coefficient k_m is calculated by dividing the diffusion coefficient of the compound through the membrane D_m by the thickness of the membrane δ as shown in Eq. (8.11). Where ε is the porosity of the membrane and θ is the tortuosity of the membrane pores.

$$k_m = \frac{\varepsilon D_m}{\delta\theta} \tag{8.11}$$

In the case of a dense membrane, the value of D_m is the value for the diffusion of the compound through the solid polymer matrix. But for microporous membranes, the value of D_m will vary depending on the phase that occupies the pores. If a gas phase occupies the pores, then the diffusion rate of a molecule through the pores is about five orders of magnitude faster than if the pores are filled with water or another liquid phase. Since microporous membranes are made with different porosities and since the pores through the membrane are tortuous, these factors will naturally influence the mass transfer rate. Usually the porosity and tortuosity are incorporated into an apparent diffusion coefficient D_m^* and Eq. (8.11) reduces to the following form:

$$k_m = \frac{D_m^*}{\delta} \tag{8.12}$$

For a membrane contactor operating with gas-filled pores, Iverson (1994) has shown that the tortuosity factor can be considered a predictable parameter. He has found good agreement with experimental results for sulfur dioxide diffusion, by dividing the pore structures into basic geometric shapes. For a pore structure similar to the interstices between loosely packed spheres (e.g., membrane formation by sintering and stretching), the tortuosity factor can be predicted from the following factor:

$$\theta = \frac{1}{\varepsilon} \tag{8.13}$$

For phase inversion membranes, the basic structure is similar to closely packed spheres and the tortuosity factor can be predicted from

$$\theta = \frac{(2 - \varepsilon)^2}{\varepsilon} \tag{8.14}$$

The first structure results in the highest possible mass transfer rates and is the best choice; however, for very high porosities (>0.85) the difference between the two expressions is small.

8.6.1   The Effect of Chemical Reactions on Mass Transfer

8.6.1.1   Chemical Speciation in the Aqueous Phase.   When the transport behavior of acid gases such as sulfur dioxide and carbon dioxide, and basic gases such as ammonia are considered, the chemistry of these gases can profoundly influence their rate of mass transfer. Consider, for example, the case of ammonia. Ammonia is a basic gas and when dissolved it reacts with water to form ammonium ions according to the following reaction:

$$NH_3 + H_2O \Leftrightarrow NH_4^+ + OH^- \qquad K_b = 10^{-4.7} \tag{8.15}$$

The speciation of the ammonia in water is therefore a function of pH with the fraction of the dissolved ammonia in the ammonium form increasing as the pH decreases. Since the ammonium ion is nonvolatile, only the free ammonia can cross the membrane and, as a result, the performance of the membrane in separating ammonia from water is a strong function of pH.

When molecules of NH_3 are removed from the water by the membrane, it disturbs the local chemical equilibrium between the species represented in Eq. (8.15). In this case, some of the ammonium ions will react with hydroxide to form free ammonia and to reestablish a local equilibrium. The resulting reduction in hydroxide concentration may reduce the local pH if the water is not well buffered. These effects must be taken into consideration if we wish to characterize the removal of ammonia under different operating conditions.

As a result of these shifts in equilibrium in the stagnant film near the membrane, concentration gradients are established with respect to ammonia, ammonium, and hydroxide, and all of these ions effectively diffuse toward the membrane as a result of the removal of ammonia. The conventional film theory must be modified to allow for these nonlinear chemical reactions. Assuming the concentrations of ammonia, ammonium, and hydroxide to be represented by c_1, c_2, and c_3, respectively, then a mass balance on each specie near the membrane yields the following three equations:

$$0 = D \frac{d^2c_1}{dz^2} + r \tag{8.16}$$

$$0 = D \frac{d^2c_2}{dz^2} - r \tag{8.17}$$

$$0 = D \frac{d^2c_3}{dz^2} - r \tag{8.18}$$

in which r represents the rate of formation of free ammonia, and the value of the diffusion coefficient D is assumed to be equal for all three species. Semmens, Foster, and Cussler (1990) solved these differential equations and showed that the overall mass transfer coefficient is a mathematically complex expression that incorporates the ionization constant for ammonia as well as the bulk concentrations of ammonium and hydroxide. In this chapter we deal only with a simplified result that represents the behavior of a system that has a low ammonium concentration and is well buffered.

In this situation the overall rate of mass transfer is related to the concentration of free ammonia c_1 in solution and, since the system is well buffered, this is a constant proportion of the total ammonium concentration c (i.e., $c = NH_3 + NH_4^+$). Using Eq. (8.15), we can express this fraction as a function of the pH and the acid dissociation constant of ammonia ($K_a = 10^{-9.3}$)

$$c_1 = \frac{c}{\{1 + ([H^+]/K_a)\}} \tag{8.19}$$

Thus, the overall transfer rate of ammonia across a membrane is given by the following equation:

$$J = \frac{K_{OL}(c - c^*)}{\{1 + ([H^+]/K_a)\}} \tag{8.20}$$

When this equation is compared with Eq. (8.4), it is clear that this correction for ionization effectively modifies the overall mass transfer coefficient; i.e.

$$K_{OL,measured} = \frac{K_{OL}}{\{1 + ([H^+]/K_a)\}} \tag{8.21}$$

This equation can be linearized by taking the reciprocals and plotting the data in the form of Fig. 8.2. In this figure, the reciprocal of the overall mass transfer coefficient measured at different pH values is plotted as a function of the fraction of the ammonium that is in the ammonia form. The slope and the intercept of the plot are equal to the reciprocal of the overall mass transfer coefficient for free ammonia.

$$\frac{1}{K_{OL,measured}} = \frac{1}{K_{OL}}\left(1 + \frac{[H^+]}{K_a}\right) \tag{8.22}$$

The approach illustrated here for ammonia is also applicable to other gases that participate in acid-base reactions (Howe and Lawler, 1989; Chang, 1988; Semmens, Foster, and Cussler, 1990). In all cases, the overall mass transfer coefficient is modi-

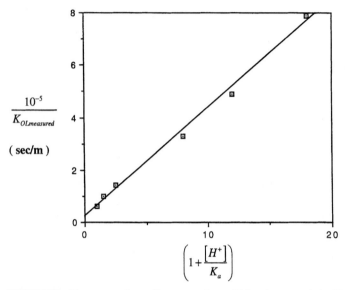

FIGURE 8.2   The measured overall mass transfer coefficient for ammonia is pH dependent. When the data are plotted in the form of Eq. (8.22), the data yield a linear plot that verifies this model.

fied by the fraction of the gas that is present as a volatile specie. The simple nature of the pH correction means that if a mass transfer coefficient for a gas is measured at any pH, the behavior under different pH conditions can be estimated. It is unimportant whether the mass transfer is controlled by the membrane or the boundary layers on either side of the membrane; the correction is the same.

EXAMPLE 8.3   *The overall mass transfer coefficient for hydrogen sulfide gas in a hollow-fiber membrane module is estimated to be 5×10^{-5} m/s at a pH of 5.8. Calculate the effective overall mass transfer coefficient of hydrogen sulfide at a pH of 8.0.*

Unlike the preceding example of ammonia, in this case the gas is acidic. Only the dissolved hydrogen sulfide gas can pass through the membrane; dissociated sulfide ions are nonvolatile. It is therefore necessary to calculate the fraction of the total sulfide that is present as hydrogen sulfide.

Hydrogen sulfide is a weak diprotic acid and can lose one proton to form HS^- or two protons to form $S^=$. The total sulfide concentration in water is the sum of these different species, i.e.

$$\text{Total sulfide concentration} = [H_2S] + [HS^-] + [S^=]$$

However, the first pKa of hydrogen sulfide is 7.1, and the second pKa is 1.0. This means that in the pH range of 5.8 to 8.0 we need only consider the first dissociation constant, and we may ignore the formation of $S^=$ ions since their concentration will be negligible. Considering only the HS^- ion, we can write the following equations to express the fraction of the sulfide that is present as hydrogen sulfide gas:

$$\frac{[H^+][HS^-]}{[H_2S]} = 10^{-7.1}$$

Total sulfide concentration $= [HS^-] + [H_2S]$

Fraction of sulfide present as gaseous hydrogen sulfide $= \dfrac{[H_2S]}{[HS^-] + [H_2S]}$

Dividing through by [H₂S] and simplifying, the fraction becomes

$$\frac{1}{10^{-7.1}/[H^+] + 1}$$

hence, for this acid gas the correction factor for calculating the overall mass transfer coefficient at any pH becomes

$$K_{OL,measured} = \frac{K_{OL}}{(1 + 10^{-7.1}/[H^+])}$$

For this example, the value of K_{OL} can be determined from the measured value at pH 5.8, and then the expected value of the overall mass transfer coefficient at pH 8.0 can be calculated:

$$K_{OL} = K_{OL\,at\,pH\,=\,5.8}\left(1 + \frac{10^{-7.1}}{[10^{-5.8}]}\right) = 5 \times 10^{-5} \times 1.05 = 5.25 \times 10^{-5} \text{ m/s}$$

$$K_{OL\,at\,pH\,8.0} = \frac{K_{OL}}{\{1 + (10^{-7.1}/[10^{-8.0}])\}} = \frac{5.25 \times 10^{-5}}{8.94} = 5.87 \times 10^{-6} \text{ m/s}$$

8.6.1.2 Chemical Reactions in Scavenging Phase. Sometimes it is useful to exploit the chemical character of a gas to enhance its separation from water. This is accomplished by providing a nonvolatile chemical scavenger on the other side of the membrane. The scavenger must react with the volatile gas to form species that are nonvolatile. For example, ammonia transfer can be encouraged by maintaining sulfuric acid on one side of the membrane so that ammonia crossing the membrane is rapidly converted to ammonium sulfate. While the acid is present, the effective ammonia concentration is essentially zero, and the mass transfer rate across the membrane is maximized.

For removal with chemical reaction, the microporous membrane can have liquid phases on both sides of the membrane and, therefore, two liquid-phase mass transfer coefficients contribute to the overall mass transfer coefficient. However, the mass transfer correlations that have been developed here cannot be used to predict the mass transfer coefficient in the scavenging phase since the chemical reaction between the transferring gas and the scavenging chemical reduces the effective boundary layer and increases the apparent mass transfer coefficient.

When a scavenger is selected to remove a gaseous component, the reaction is typically fast and irreversible. The reaction is second order, and the concentrations of both the scavenger and the reacting gas are important in determining the rate of reaction. However, if the concentration of the scavenging chemical c_s is present in excess, the removal of c_1 may be treated as a pseudo first-order reaction:

$$-\frac{dc_1}{dt} = \kappa_1 c_1 \qquad \text{where } \kappa_1 = \kappa c_s$$

then the mass transfer coefficient for the scavenging phase is simply:

$$k_L = \sqrt{\kappa_1 D}$$

The impressive feature of this relationship is that the mass transfer coefficient is independent of the velocity of the scavenging liquid phase and simply dependent on the square root of the concentration of the scavenger.

For a more complete discussion of the influence of chemical reactions on interfacial mass transfer, the reader is referred to Cussler (1984). This reference also presents a variety of instructive problems and solutions related to the calculation of mass transfer coefficients for first- and second-order reactions.

8.6.2 Transport Limitations

From the preceding discussion it is apparent that the overall mass transfer coefficient for a contaminant may be limited either by transport across a membrane or through one of the fluid boundary layers. It is axiomatic that if we wish to take advantage of the selectivity of a membrane, the mass transfer must be limited by the membrane. If mass transfer is limited by one or another of the fluid boundary layers, then the character and selectivity of the membrane is less important or unimportant.

By way of example, consider that the module described in Example 8.2 is to be used for the removal of trace organics from water by pervaporation. Assume the fibers are made of a dense membrane and that the membrane mass transfer coefficient k_m is equal to 2×10^{-5} m/s. Assume further that the inside of the fibers is evacuated so that mass transfer is controlled by either the membrane or the liquid film on the outside of the fibers (i.e., the resistance of the gas phase or vacuum on the other side of the membrane is negligible). The liquid film mass transfer coefficient can be determined as a function of water flow rate through the module, following the procedure illustrated in Example 8.2. If the liquid film resistance is plotted as a function of $Re^{-0.6}$, the data yield a linear relationship that passes through the origin when the value of Reynolds number approaches infinity. The membrane resistance is represented as a horizontal line on the same figure since the value is unaffected by the water flow rate. The overall mass transfer resistance is equal to the sum of the two resistances. Figure 8.3 illustrates such a graph.

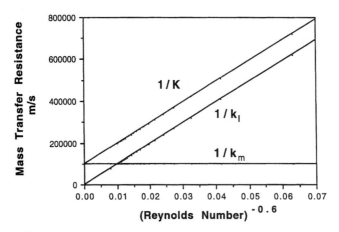

FIGURE 8.3 The values of the film, membrane, and overall mass transfer resistances as a function of Reynolds number to the negative 0.6 power.

An examination of Fig. 8.3 reveals that the membrane resistance is greater than the liquid film resistance at high liquid flow rates (i.e., low values of $Re^{-0.6}$) and that the liquid film resistance dominates the overall mass transfer coefficient above a value of $Re^{-0.6} = 0.01$ (i.e., below a Reynolds number of about 2150).

This effect is presented in Fig. 8.4 in a somewhat different manner. The overall mass transfer coefficient is shown as a function of Reynolds number for three different membrane transfer coefficients. The curves illustrate the effect, for example, of improving the selectivity of the membrane by a factor of two and three. We assume that the improvement in selectivity is specific to the transport of the volatile organic compounds and that the water mass transfer coefficient is constant. Under these conditions, as the selectivity is improved, the membrane resistance for the organic compounds is decreased and the overall mass transfer coefficient is increased. However, inspection of Fig. 8.4 shows that the benefit of this improved membrane selectivity is gained only when the module is operated at high Reynolds numbers. For example, at a Reynolds number of 3000, the overall mass transfer coefficient of the most selective membrane is only about 80 percent greater than that of the least selective membrane even though there is a 200 percent difference in membrane selectivity. The situation is even worse when the Reynolds number is only 100; the improvement is less than 10 percent. The point is that improvements in membrane selectivity or performance may not necessarily translate into significant improvement process performance unless the module design and conditions of operation are manipulated to take proper advantage of the improvements.

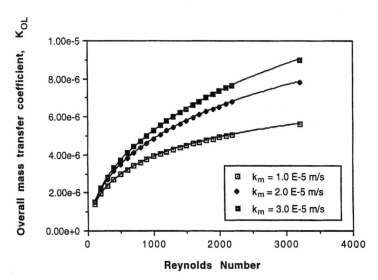

FIGURE 8.4 The effect of Reynolds number on the overall mass transfer coefficient for three different membrane resistances.

A second and unusual consequence of the importance of the liquid film resistance at low Reynolds numbers is that the selectivity can be increased by simply using thicker membranes of the same material. This will not cause a significant decrease in the flux of the organic compounds; however, it will reduce the water flux across the membrane and, as a result, the permeate will be richer in organics.

Finally, if the rate of mass transfer in a membrane module is governed by the fluid-phase boundary layers, then the flux of the volatile organic compounds is insensitive to the type of membrane used (i.e., the membrane may be porous or dense, isotropic or composite, etc.) This means that different membranes may be selected to meet different objectives. For example, if the objective is simply to remove organics from water, then a low-cost microporous membrane could be used. However, if the objective is to recover the organic, then a more expensive, dense, thick, and "waterproof" membrane will produce a more concentrated permeate.

REFERENCES

Ahmed, T., 1991, "Gas Transfer Through Membranes," Ph.D. thesis, Department of Civil and Mineral Engineering, University of Minnesota.

Böddeker, K. W., G. Bengston, and E. Bode, 1990, "Pervaporation of Low Volatility Aromatics from Water," *J. Membrane Sci.,* **53:**143–158.

Chang, Y. Y.,

Costello, M. J., A. G. Fane, P. A. Hogan, and R. W. Schofield, 1993, "The Effect of Shell-Side Hydrodynamics on the Performance of Hollow Fiber Modules," *J. Membrane Sci.,* **80:**1–11.

Côté, P. L., J.-L. Bersillon, A. Huyard, and J.-M. Faup, 1988, "Bubble-Free Aeration Using Membranes: Process Analysis," *Journal WPCF,* **60**(11):1986–1992.

Côté, P., J.-L. Bersillon, and A. Huyard, 1989, "Bubble-Free Aeration Using Membranes: Mass Transfer Analysis," *J. Membrane Sci.,* **47:**91–106.

Cussler, E. L., 1984, *Diffusion,* Cambridge University Press, London.

Cussler, E. L., 1994, "Hollow Fiber Contactors," in J. G. Crespo and K. W. Böddeker (eds.), *Processes in Separation and Purification, Nato ASI Series E—vol. 272,* Kluwer Academic Publishers, Dordrecht, pp. 375–394.

Dean, W. R., 1928, "The Streamline Motion of Fluid in a Curved Pipe," *Phil. Mag.,* **5**(4): 674–695.

Gekas, V., and B. Hallström, 1987, "Mass Transfer in the Membrane Concentration Polarization Layer Under Turbulent Cross Flow, I: Critical Literature Review and Adaptation of Existing Sherwood Correlations to Membrane Operations," *J. Membrane Sci.,* **30:**153–170.

Hickey, P. J. and C. H. Gooding, 1994, "Mass Transfer in Spiral Wound Pervaporation Modules," *J. Membrane Sci.,* **92:**59–74.

Hirasa, O., H. Ichijo, and A. Yamaguchi, 1991, "Oxygen Transfer from Silicone Hollow-Fiber Membrane to Water," *J. Ferment. Biol.,* **71:**206–207.

Howe and Lawler, 1989,

Iverson, S. R., 1994, "Membrane Contactors to Replace Conventional Scrubbers," Ph.D. thesis, Danish Academy of Technical Sciences, Denmark.

Knudsen, J. G., and D. L. Katz, 1958, *Fluid Dynamics and Heat Transfer,* McGraw-Hill, New York.

Kreith, F., 1973, *Principles of Heat Transfer,* Intext Educational Publishers, New York.

Kreulen, H., C. A. Smolders, G. F. Versteeg, and W. P. M. van Swaaij, 1993a, "Microporous Hollow Fiber Membrane Modules as Gas-Liquid Contactors, Part I: Physical Mass Transfer Processes," *J. Membrane Sci.,* **78:**197–216.

Kreulen, H., C. A. Smolders, G. F. Versteeg, and W. P. M. van Swaaij, 1993b, "Microporous Hollow Fiber Membrane Modules as Gas-Liquid Contactors, Part II: Mass Transfer with Chemical Reaction," *J. Membrane Sci.,* **78:**217–238.

Lévêque, M. A., 1928, "Les lois de la transmission de chaleur par convection," *Ann. des Mines* **13.**

Moulin, P., C. Serra, J. C. Rouch, and Ph. Aptel, 1994, "Mass Transfer Improvements by Secondary Flows in Gas-Liquid Contactors," in *ESMST XIth Annual Summer School,* University of Glasgow, Glasgow, U.K.

Neel, J., 1992, "Current Trends in Pervaporation," in P. L. T. T. Napoli (ed.), *CEE Brazil Workshop on Membrane Separation Processes,* Rio de Janeiro, May 3–8, pp. 182–198.

Nguyen, Q. T., and K. N. Nobe, 1987, "Extraction of Organic Contaminants in Aqueous Solutions by Pervaporation," *J. Membrane Sci.,* **30:**11–22.

Nijhuis, H. H., 1990, "Removal of Trace Organics from Water by Pervaporation," Ph.D. thesis, Twente University, The Netherlands.

Pankhania, M., T. Stephenson, and M. J. Semmens, 1994, "A Hollow Fiber Bioreactor for Wastewater Treatment using Bubbleless Membrane Aeration," *Water Research,* **28**(10): 2233–2236.

Prasad, R., and K. K. Sirkar, 1988, "Dispersion Free Solvent Extraction with Microporous Hollow Fiber Modules," *AIChE J.,* **34**(2):177–188.

Psaume, R., Ph. Aptel, Y. Aurelle, J. C. Mora, and J.-L. Bersillon, 1988, "Pervaporation: Importance of Concentration Polarization in the Extraction of Trace Organics from Water," *J. Membrane Sci.,* **36:**373–384.

Schock, G., and A. Miquel, 1987, "Mass Transfer and Pressure Loss in Spiral-Wound Modules," *Desalination,* **64:**339.

Schwarz, S. E., M. J. Semmens, and K. Froehlich, 1991, "Membrane Air Stripping: Operating Problems and Cost Analysis," *45th Purdue Industrial Waste Conference Proceedings,* Lewis Publishing, Chelsea, Mich.

Semmens, M. J., D. M. Foster, and E. L. Cussler, 1990, "Ammonium Removal from Water Using Microporous Hollow Fibers," *J. Membrane Sci.,* **51:**127–140.

Sieder, E. N., and G. E. Tate, 1936, "Heat Transfer and Pressure Drop in Liquids in Tubes," *Ind. Eng. Chem.,* **28**(12):1429–1435.

Voss, M., 1994, "Membrane Gas Transfer: Practical Applications," M.S. thesis, University of Minnesota, Minneapolis, Minn.

Wickramasinghe, S. R., 1992, "The Best Hollow Fiber Membrane," Ph.D. thesis, Department of Chemical Engineering, University of Minnesota, Minneapolis, Minn.

Wilderer, P. A., J. Brautigan, and I. Sekoulov, 1985, "Application of Gas Permeable Membranes for Auxiliary Oxygenation of Sequencing Batch Reactors," *Conser. Recycl.,* **8**(1–2):181–192.

Yang, M. C., and E. L. Cussler, 1986, "Designing Hollow-Fiber Contactors," *AIChE Journal,* **32**(11):1910–1916.

CHAPTER 9
REVERSE OSMOSIS AND NANOFILTRATION

James S. Taylor
University of Central Florida
Orlando, Florida

Ed P. Jacobs
University of Stellenbosch
Stellenbosch, South Africa

9.1 INTRODUCTION

Reverse osmosis and nanofiltration are the most commonly used membrane processes for potable water treatment in the United States. Like any water treatment process, RO and NF are selected on a cost and quality basis, with cost having the primary significance most of the time. Reverse osmosis is capable of rejecting contaminants or particles with diameters as small as 0.0001 μm, whereas nanofiltration can reject contaminants as small as 0.001 μm. Both reverse osmosis and nanofiltration can be described as diffusion-controlled processes in that mass transfer of ions through these membranes is diffusion controlled. Consequently, these processes can remove salts, hardness, pathogens, turbidity, disinfection by-product (DBP) precursors, synthetic organic compounds (SOCs), pesticides, and most all potable water contaminants known today. No process represents universal treatment for all contaminants (Taylor, 1995; Taylor et al., 1990). Most dissolved gases such as hydrogen sulfide and carbon dioxide and some pesticides will pass through RO or NF membranes. However, membrane technology can be used to treat a broader range of potable water contaminants than any other treatment technology.

This chapter deals with applications of reverse osmosis and nanofiltration for the production of drinking water. The chapter is written emphasizing application. The main sections in the chapter discuss the targeted contaminants that can be advantageously treated by RO or NF, source water selection, process design, operational guidelines, and cost.

9.1.1 Targeted Contaminants

One of the first membrane applications for the utilization of membrane technology was the conversion of seawater into drinking water by reverse osmosis. The targeted contaminant in this application would be dissolved salts or, primarily, sodium chloride. The initial cellulose acetate membranes were much less permeable than modern membranes and required a much higher driving force or pressure. The high pressure was in excess of 1000 lb/in^2 (68 atm) and only 10 to 25 percent of the raw water was recovered as product. Power cost alone was in excess of $4/kgal. However, there were areas where desalination by reverse osmosis was the most economical method of drinking water production. However, the development of thin film composites provided greater saline rejection and greater productivity. Development of coastal communities and the associated brackish water reserves led to the utilization of low-pressure reverse-osmosis applications for removing the same dissolved solids as in the desalination of seawater by reverse osmosis. The primary difference between a low-pressure and a high-pressure reverse-osmosis membrane is the water and salt permeability. The high-pressure reverse-osmosis membrane is much less permeable than the nanofiltration membrane because the salt concentration in seawater is typically an order of magnitude higher than the salt concentration in brackish water. As noted in the previous chapters on mass transfer, the amount of salt that passes through a diffusion-controlled membrane is directly related to the concentration in the feed stream. A high-pressure reverse-osmosis membrane must be much "tighter" than a low-pressure reverse-osmosis membrane to achieve the same concentration of dissolved solids or salt in the permeate. The use of reverse osmosis for rejection of dissolved solids or salt may be the membrane application of the moment, but it is the most commonly utilized membrane process for drinking water production in the world today. There are more than 100 drinking water plants in the United States that are using reverse osmosis to produce drinking water from brackish water. There are no desalination processes treating seawater in the United States; however, there are several major facilities using reverse osmosis for desalination of seawater in the Middle East.

DBPs are a major targeted contaminant that can be controlled by reverse osmosis and nanofiltration. DBPs are formed when the natural organic material (NOM) in drinking water reacts with chlorine or other chemical oxidizing agents used for disinfection. Not all of the NOM can be assumed to be DBP precursors, but reverse osmosis and nanofiltration effectively control DBPs by rejecting the NOM or DBP precursors (Taylor et al., 1986, 1987, 1989a, 1989b; Jones et al., 1992; Tan et al., 1991). Reverse-osmosis and nanofiltration membranes that have molecular weight cutoffs of 500 or less have been found to reject more than 90 percent of the DBP precursors from highly organic surface and groundwaters. The removal of NOM not only reduces DBP formation but decreases chlorine demand in the distribution system. Consequently, the minimum disinfectant residual concentrations required by regulation are more easily maintained with these systems. The increased removal of NOM would remove a bacteriological food source and very likely reduce biological activity and regrowth in the distribution system.

Turbidity, hardness, and color are targeted contaminants that can be removed by nanofiltration or reverse osmosis. Turbidity is regulated as a primary standard, color as a secondary standard, and hardness is not regulated but is often a targeted contaminant, as in a softening process. Nanofiltration was primarily developed as a membrane softening process to offer an alternative to chemical softening and is used as a softening process that also controls DBP precursors in Collier County and Fort Meyers, Florida (Boyle Engineering, 1995). Typically, nanofiltration will remove

60 to 80 percent of the total hardness, more than 90 percent of the color, and virtually all of the turbidity in the feed stream. However, nanofiltration is not a process that can be used successfully for turbidity removal because of fouling. If the particles producing the turbidity are not removed by pretreatment, they will plug or foul the membrane and make the process uneconomical.

Pesticides are targeted contaminants that can also be removed by nanofiltration or reverse osmosis. Reverse-osmosis and nanofiltration membranes have been shown to completely remove some pesticides under certain operating conditions (Duranceau et al., 1990, 1992; Hofman et al., 1993; Baier et al., 1987). Many European countries have significant pesticide contamination in surface waters such as the Rhine and are actively pursuing the use of reverse osmosis and nanofiltration for removal of pesticides to meet regulatory limits (Hopman et al., 1993). Pesticides with molecular weights greater than 190 were removed beyond detectable limits using nanofiltration; however, EBD completely passed the nanofilter. Reverse osmosis has been shown to be very effective for the removal of some pesticides, but pilot studies are required on specific source waters before treatment criteria could be established.

Reverse osmosis and nanofiltration membranes have been shown to reject all viruses, bacteria, cysts, and other pathogenic organisms, but are seldom used for removal of biological pathogens because ultrafiltration and microfiltration are less costly membrane processes which can achieve the same degree of pathogenic rejection (Karamian, 1975). The point is sometimes lost that reverse osmosis and nanofiltration will reject everything out of the ionic size range. Bacteria, however, have been found in permeate samples from reverse-osmosis and nanofiltration pilot plants and can proliferate in the discharge lines. This does not mean that pathogens are not rejected by these membranes but that sterile operating conditions cannot be maintained. Bacteria have also been shown to penetrate membranes during growth and pass through defects in membranes. Consequently, membranes are not always completely effective for biological control and are commonly succeeded by a disinfection process. However, membranes are a very effective method of removing pathogens.

Reverse-osmosis and nanofiltration systems can be designed to produce a very noncorrosive water. Corrosive waters are generally characterized by a high-conductivity water that is not stabilized with respect to calcium carbonate. Both reverse-osmosis and nanofiltration systems can be designed to produce a well-buffered, low-TDS (total dissolved solids) water that is easily stabilized, which will be described in Sec. 9.4.5. Although not obvious, control of lead and copper corrosion can be accomplished by a membrane process.

The major issues in the '90s in water treatment have involved both water supply and quality. Reverse osmosis has increased the raw water supply by making possible the use of brackish waters for potable water supply. Significant quality issues have involved DBPs, disinfection, regrowth, SOCs, and corrosion. Reverse osmosis and nanofiltration will remove and reduce the potential for contamination in potable water supplies. There is simply no other treatment process that can produce the same water quality from a highly saline or organic groundwater source as economically as reverse osmosis and nanofiltration (Taylor et al., 1990, 1995). These processes will play an increasingly important role in the future of worldwide water supply.

9.2 MEMBRANE CHARACTERISTICS

This section describes the configuration of the common forms of RO/NF membranes that are used in potable water production, characteristics of active surface materials, and selection considerations. A detailed description of spiral-wound and hollow fine-fiber (HFF) elements is presented in order to describe the flow characteristics through the membrane. Common materials used for active membrane surfaces are discussed relative to mass transfer characteristics. Membrane selection criteria are also discussed relative to desired production characteristics and finished water quality criteria.

9.2.1 Configuration

There are many different configurations of membranes that are used in industrial practices. However, the hollow fine-fiber and spiral-wound elements are the most common by far to the production of drinking water.

9.2.1.1 Hollow Fine-Fiber Elements.   Membranes for producing drinking water are either in spiral-wound or HFF configurations. The spiral-wound configuration is by far the most common used for the production of drinking water, although the HFF configuration is used extensively for desalination of seawater in the Middle East.

The manufacturers of spiral-wound membranes and many professionals involved with membranes have stated that spiral-wound membranes maintain a constant production, foul less, and can be cleaned more thoroughly than HFF membranes. Du Pont developed the HFF technology and held the original patents. The development of the spiral-wound membrane may have been partially driven by Du Pont's initial position in the field. Although there are no definitive studies known to the author that document the fouling characteristics of HFF and spiral-wound membranes, the current availability and performance of several different spiral-wound membranes offers a more fertile research opportunity than investigation of HFF membranes.

HFF modules consist of a pressure vessel, inside which is a cartridge containing the HFF membrane bundle. An HFF membrane bundle is shown in Fig. 9.1. The inside and outside diameters of the HFF are 1.33E-4 ft (41 μm) and 2.9E-4 ft (90 μm), respectively. A 4-in (10.2-cm) Du Pont B-10 membrane contains 650,000 HFF, which are approximately 4 ft (3.28 m) long, and have 1500 ft^2 (139 m^2) of surface area. The bundle is created by folding a group of HFF in a U-form and epoxy-casting the one end in a tube sheet. The feed stream enters the HFF element from a perforated feeder tube placed in the center of the membrane bundle. The feed stream therefore flows radially from the center feed tube to the brine collection channel at the outside of the element.

The highest feed stream velocity is where the feed stream enters into the module, and the lowest velocity is at the outer edge of the membrane bundle. The recovery from an HFF element ranges from 10 to 50 percent and is typically higher than that from a spiral-wound element. The radial feed-stream velocity along the outside surface of the HFF varies from approximately 0.01 to 0.001 ft/s (0.003 to 0.0003 m/s) which give a Reynolds number ranging from 100 to 500 through the membrane.

There are other sizes of HFF elements, but the feed-stream velocities, recoveries, and Reynolds numbers are similar to those of the B-10 element. The feed flow is in the laminar region and is most likely to produce chemical or colloidal fouling near the brine collector and fouling from the collection of filtered particles near the feed tube in an HFF filter. The configuration of the HFF element is physically such that it is prone to fouling by particulate matter, which is retained by entrapment within the fiber mass; the HFF membrane bundle is difficult to clean.

9.2.1.2 Spiral-Wound Elements.   Spiral-wound elements are manufactured from flat-sheet membranes. A spiral-wound element is shown in Fig. 9.2. Spiral-wound

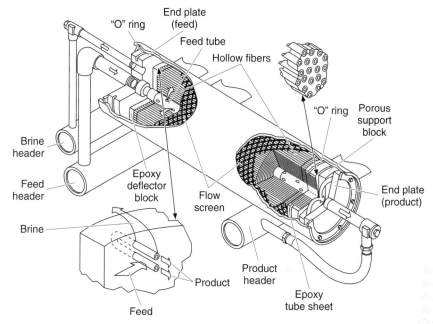

FIGURE 9.1 Diagram of a hollow fine-fiber reverse-osmosis membrane element.

elements typically consist of membrane envelopes attached to a center tube which collects the permeate stream. The design of a spiral-wound element differs according to manufacturer; however, the following description is applicable to FilmTec, Hydranautics, TriSep, Desal, and Fluid Systems spiral-wound designs.

A membrane envelope is one flat-sheet membrane which has been folded over a permeate stream spacer. The flat sheet consists of two integral layers and a permselective membrane layer supported on a porous support fabric. The active layer of the membrane is on the outside of the fold. The envelope is glued along three open sides and near the fold. The permeate spacer is completely enclosed in the fold. The fold

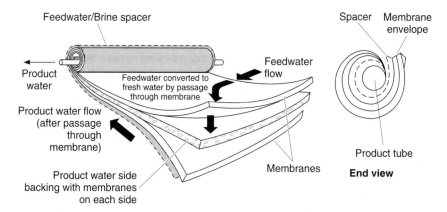

FIGURE 9.2 Diagram of a spiral-wound reverse-osmosis membrane.

end is glued a short distance away from the fold because the fold end is attached to the center collection tube. The fold end glue line stops the flow of the feed stream and allows the remaining pressure in the permeate stream to drive the permeate stream into the center collection tube. A feed stream spacer is attached to each envelope ahead of the fold end glue line. Several envelopes and feed stream spacers are attached to and wrapped in a spiral around the center collection tube. An epoxy shell is applied around the envelopes to complete the spiral-wound element.

The feed stream enters the open end of the spiral-wound element in the channel created by the feed-stream spacer. The feed stream can flow either in a path parallel to the center collection tube or through the active membrane film and membrane supports into a channel created by the permeate stream spacers. The permeate stream follows a spiral path into the center collection tube and is taken away as product water. As with HFF, the feed stream becomes more concentrated and is passed to a succeeding element. A 4-in (1.2-cm) FilmTec NF70 membrane of approximately 90 ft^2 (8.33 m^2) surface area contains four 3-ft (0.92-m) by 3.75-ft (1.14-m) envelopes. The total element is 3.33 ft (1.01 m) long, but the feed-stream path along the active membrane film is approximately 3 ft (0.91 m) because of the impervious glue lines.

The recovery in a spiral-wound element ranges from approximately 5 to 15 percent. The maximum feed and concentrate stream flow in a 4-in (10.2-cm) element are approximately 16 gpm (6.1E-2 m^3/min) and 3 gpm (1.1E-2 m^3/min), respectively. Neglecting the effect of the feed-stream spacer, the Reynolds number would typically be greater than 100 and less than 1000. The feed-stream spacer would create additional turbulence and increase the Reynolds number.

The physical configuration of the spiral-wound element is such that a turbulent feed flow is produced, with a further result that the membrane is more easily accessible to cleaning agents. The highest and lowest feed-stream velocities are at the entrance and exit ends of the element, respectively. The feed flow is in the laminar region and is most likely to produce chemical or colloidal fouling near the exit and fouling from the collection of filtered particles near the entrance of the spiral-wound element.

9.2.2 Active Surface Materials

As noted previously, the active layer on a membrane surface is very thin. Precise measurement of this layer is very difficult and typically is not available to users. The active membrane surfaces are made from organic polymers. The active surfaces of RO and NF membranes can be made from cellulose acetate and cellulose acetate derivatives, polyamides and polyamide derivatives, and other combinations of organic polymers. Cellulose acetate membranes are more hydrophilic than polyamide membranes and less likely to be fouled during operation. Ideally, the membrane surface would be hydrophilic and not attracted to hydrophobic particles which could be in the feedwater stream. Hydrophobic particles have a greater affinity for a nonhydrophilic surface and would be more likely to foul the membrane during operation. Diffusion-controlled membranes, like RO/NF membranes, need to be selective, permeable, mechanically stable, and resistant to chemical and temperature change. These surfaces are made to be asymmetric and homogeneous and as composites. The asymmetric homogeneous layer is typically 0.1 to 0.5 mm thick and produced in a phase-inversion process which controls the nature of the surface.

Composite membranes provide support and active surface layers from different materials, which can be advantageously selected separately. Typically, the support layer is made from a polysulfone which is layered on a fabric base. The active layer is generally produced from a phase interface polymerization which produces a cross-linked polyamide. The polysulfone layer is saturated with a polyethyleneimine (PEI) in a toluene disocyanate (TDI) in hexane under heat. As PEI is insoluble in hexane and TDI is insoluble in water, the polymerization occurs quickly at the interface.

9.2.3 Selection

Membranes are manufactured from a wide range of materials. The materials from which RO and NF membranes are produced are essentially hydrophilic, which means that water is able to associate with the membrane material; this is not a necessary requirement for materials from which UF and MF membranes are made, and these membranes are normally, therefore, hydrophobic in nature. This imparts a certain robustness to UF membranes which, unlike RO and NF membranes, are less susceptible to physical, mechanical, or chemical degradation.

Nonetheless, high on the list of important properties for RO and NF membranes are qualities such as mechanical and temperature stability and chemical and hydrolytic resistance. The first is very important, as RO membranes are operated at high pressures and compaction of the membrane due to the viscoelastic properties of polymer materials is one reason for irreversible loss of flux. Other important considerations are good selectivity and high productivity.

Seawater RO membranes must have very high salt-retaining properties to produce water of a potable standard in a single pass. Seawater has a total dissolved solids content of about 35,000 mg/L, and to produce water at the World Health Organization (WHO) standard of 500 mg/L at recovery ratios of more than 35 percent, the membranes must have salt-removal efficiencies more than 99 percent. Seawater membranes are exposed to very high operating pressures due to the high osmotic pressure of seawater feed, which is in the region of 2400 kPa (24.5 atm), which is approximately 10 psig per 1000 mg/L TDS. To ensure adequate productivity, seawater RO membranes are typically operated at pressures between 700 to 1000 lb/in^2 (48 to 68 atm). The operating pressures are much lower for brackwater applications.

Pressure-driven desalting membranes may be grouped into two classes: the *asymmetric membranes* (produced by wet-phase inversion techniques) and *thin film composite membranes* (e.g., produced by performing an interfacial cross-linking reaction on the surface of a non-desalting support membrane).

Asymmetric cellulose acetate (CA) RO membranes, which have been in use since the inception of membrane technology in the 1960s, are still commercial, despite certain shortcomings. CA membranes can tolerate continuous exposure to low concentration levels of residual chlorine (0.1 to 0.5 mg/L at 25°C), which is important in certain applications. They are, however, very susceptible to hydrolysis, a process which is accelerated outside an operating pH range of 4 to 6.5. The material has a low temperature threshold (<30°C), mainly due to hydrolysis and viscoelastic flow. The CA material is also susceptible to biological degradation. Cellulose triacetate (CTA) and CA/CTA blends are hydrolytically more stable than CA membranes, and the CA/CTA blend membranes outperform membranes from any of the two homopolymers.

The Du Pont Company tested a wide range of polymer materials in its quest for a material superior in performance to CA. Its final choice fell on a polyamide, and today their asymmetric HFF membranes are still being fabricated from an aromatic polyamide. These materials have excellent retention properties, but their flux performances tend to be lower than that of CA. Although the membranes can be operated over a wider pH range than CA can (pH 4 to 9 continuous) and are stable toward bacterial degradation, they are susceptible to oxidative degradation. This holds true for all nitrogen-containing membranes based on amide chemistry (i.e., membranes with –NH-CO functionality). Polyamides are sensitive to residual chlorine and, although the rate at which membranes succumb to attack by free chlorine is a function of pH, their performance will gradually (sometimes catastrophically) deteriorate upon exposure. Aramid membranes also have operating temperature limitations ($<40°C$); they undergo compaction as a result of creep, and permanent loss of flux performance is one reason for this limitation.

There are a number of commercial thin film composite membranes, the most widely known being the FT30 FilmTec membrane. These membranes are also based on amide chemistry and therefore suffer the same disadvantages regarding chlorine susceptibility. However, because of the method of fabrication, the membranes have near-perfect desalting barriers and very high salt-retention capabilities. Although the membrane skin layer which is deposited on the substrate membrane during fabrication is essentially dense, the film is so thin (<1 μm) that the flux performance of these membranes is very high. The membranes have the further advantage that their materials of construction can be chosen individually, and certain chemical features can therefore be designed into the membrane. Thin film membranes have good chemical and mechanical integrity and can withstand high operating pressures and temperatures as high as $50°C$.

Membranes are classified primarily by solute exclusion size, which is sometimes referred to as pore size. This is something of a misnomer in the case of thin film composite membranes because these NF and RO membranes do not have pores and pass solutes and solvent through a nonporous film by diffusion and not by convection.

As noted in the previous chapters, there are primary membrane classifications of interest in drinking water treatment based on the size of solute exclusion. An RO membrane retains species as small as 0.0001 μm, which is in the ion or molecular size range. An NF membrane retains species as small as 0.001 μm, which is also in the ionic and molecular size range. Solute mass transport in these processes is diffusion controlled. Ultrafiltration and microfiltration membranes have a minimum solute retention size of 0.01 and 0.1 μm, respectively. These membranes retain solutes, such as very large colloids, bacteria and suspended solids, by sieving and are not diffusion controlled. The transport of water (the solvent) through all of these membrane processes is pressure driven and follows Darcy's law.

As is the case with ultrafiltration membranes, RO and NF membranes are not always categorized by molecular weight cutoff (MWC). Although low-molecular-weight organic species such as alcohols, sugars, acids, and sometimes dyes are also used to categorize RO and NF membranes, these membranes are normally evaluated against solutions of sodium chloride and magnesium sulphate.

In aqueous solution, ions are surrounded by primary, secondary, and more layers of water bound to the ions by electrostatic forces. At low concentration levels, the retention of ionic species can, to some extent, be related to their hydration enthalpy. The hydration enthalpy is a measure of the hydration capacity; that is, the higher the hydration enthalpy, the larger the hydrated diameter and the higher the retention of that species. At high salt concentration levels, however, (e.g., >1 mol/L) the hydration diameter decreases as the secondary layers become less strongly attached, the ions become surrounded by the primary layers only, and the salt passage increases. Typically, an RO and NF membrane would make the following distinction in retention when challenged with the following feed solutions at low concentrations:

$$Na_2SO_4 > MgCl_2 > MgSO_4 > NaCl > NaBr$$

The distinction between RO and NF lies in the fact that a 50 percent NaCl retention NF membrane would, for example, retain more than 90 percent $MgSO_4$, whereas the distinction in the case of RO would not be as great. NF membranes are sometimes referred to as *loose* RO membranes.

Rautenbach (1989) provides the following guidelines for RO membrane selectivity:

- Multivalent ions are retained to higher levels than monovalent ions.
- Dissolved polar gases permeate well.
- pH has a marked effect on the retention of weak inorganic acids (boric acid) and organic acids.
- Within a homologous series, retention increases with increasing molecular mass.
- Isomer retention increases with increased branching, that is, tertiary > iso > secondary > primary.
- Components with molecular mass >100 dalton are retained irrespective of charge.

Almost all solutes of concern for potable water can be classified into three approximate size ranges: an ionic range from 0.0001 to 0.01 µm, a macromolecular range from 0.01 to 1 µm, and a fine particle range from 1 to 100 µm. These divisions are not exact but are useful to relate membrane applications to potable water treatment. The removal of turbidity or biological species, which is required for disinfection, can be accomplished by any pressure-driven membrane process, but it is accomplished most economically by ultrafiltration and microfiltration covering size ranges of approximately 0.003 to 10 µm and 0.04 to 20 µm, respectively. RO and NF will remove the same pathogens in addition to ions and macromolecules from feedwaters of approximately 0.0005 to 0.006 µm and 0.0007 to 0.01 µm, respectively.

Reverse-osmosis and nanofiltration membranes can also be categorized by molecular weight cutoff and mass transfer coefficients. The molecular weight cutoff is measured as daltons and is the degree of exclusion of a known solute and is determined for a given set of test conditions in the laboratory. Typical known solutes used for determination of molecular weight cutoff are sodium chloride, magnesium sulfate, dextrose, and some dyes. Solute and solvent permeability can be described by mass transfer coefficients and percent solute rejection. The most common method of describing membranes is by name, i.e., reverse osmosis or nanofiltration. Other chemical terms are molecular weight cutoff, percent rejection, and active surface. Although qualitative, immediate implication of desalination of sea- or brackish water or softening is realized. The mass transfer coefficients would be much more quantitative but are very seldom used.

Treatment objective will determine selection of membranes for RO/NF processes. NF will provide color removal, softening, or DBP control. RO will provide the same treatment as NF, in addition to greater TDS removal. Low-pressure RO is commonly used to produce drinking water from brackish raw waters with TDS concentrations to 12,000 mg/L. High-pressure RO can produce drinking water (TDS < 500 mg/L) from seawater (TDS > 34,000 mg/L). RO/NF membranes are used for a variety of reasons. A recently completed AWARF investigation of concentrate disposal found that the 137 responding utilities with membrane processes gave 21 different reasons for selecting membrane treatment, as summarized in Table 9.1 (Mickley et al., 1993). It is obvious from the distribution of responses that membrane processes are capable of removing many contaminants from drinking water sources. The only contaminants that would not be amenable to rejection by RO/NF membrane processes would be gaseous. Typically, dissolved gases will readily permeate a

TABLE 9.1 Distribution of Reasons for Selecting Membrane Process

Treat	No.	Treat	No.	Treat	No.	Treat	No.
Bacteria	1	F	11	NO_3	4	SO_4	19
HCO_3	1	Soften	17	Organics	2	TDS	100
Ca	5	Fe	11	Ra	16	THMs	5
Cl	55	Mn	3	Na	19	Turbidity	2
Desalt	6	Water R.	1				

membrane and no rejection will be realized. However, RO/NF membranes are capable of rejecting pesticides, heavy metals, DBP precursors, salts, bacteria, viruses, cysts (if they pass the prefilter), and numerous other contaminants. RO/NF processes can remove a broad spectrum of contaminants and can be used beneficially to treat potable water.

9.3 MEMBRANE SYSTEMS

The design of membrane systems requires an understanding of membrane terminology, element configuration, pretreatment, membrane treatment, and posttreatment. A detailed description of spiral-wound and hollow fine-fiber elements is presented in order to describe the flow characteristics through the membrane. Some models are discussed so that operational effects on finished water quality can be explained. Finally, a conventional system is described in divisions of pretreatment, membrane treatment, and posttreatment from a systems and unit operations basis.

9.3.1 Concentration Polarization

The most common method of describing a membrane is first by name, that is, reverse osmosis or nanofiltration, and then by its performance. Although qualitative, the description has an immediate implication of desalination of sea- or brackish water, or water softening. Unfortunately, there is no set standard against which the performances of membranes are measured (i.e., feedwater solution composition and concentration, pressure, temperature, Reynolds number) or reported by the different suppliers operating in the field.

Resistance to transport is a problem typical of membrane separation, and in this sense the efficiency of pressure-driven membrane processes depends not only on the properties of the membrane, but also very much on how the process is operated. System hydrodynamics has a direct influence on reducing transport resistances at the membrane interface, and module designs, which allow good fluid-flow distribution and mixing at the membrane interface, helps to alleviate the destructive effect which concentration polarization has on membrane performance. Most membrane suppliers will recommend a minimum linear feed flow rate and a maximum water recovery rate to minimize the effects of concentration polarization.

A method often used to compare the performance of (cellulose acetate) membranes graphically is through the log-log relationship of their pure-water permeability coefficient (k_w) versus their salt permeability coefficient (k_i), where the effects of concentration polarization (Denisov, 1994) are taken into consideration. The basic diffusion transport equations used for water and salt flux are given as follows (Merten et al., 1964; Lonsdale et al., 1965; Rosenfeld and Loeb, 1967):

water flux:
$$J = k_w[\Delta P - \Delta \Pi] \tag{9.1}$$

salt flux:
$$J_i = k_i \Delta C \tag{9.2}$$

In the operation of RO and NF membranes, solutes and other species in the feed stream are transported toward the membrane surface by bulk convective flow at a velocity equivalent to that of the permeating solvent. Because of the semipermeable nature of the membrane, the solute present in the feed stream will be retained at the membrane interface. The cross-flow velocity which prevails in the boundary region is essentially laminar and backtransport of retained solute into the bulk stream can therefore be only by diffusion. In order for the backdiffusion of the solute $-D(dC/dx)$ to balance the convective flow of solute toward the membrane surface (JC), the concentration of solute at the membrane (C_s) must be much higher than that in the bulk (C_b). (Figure 9.3 presents this case, assuming that no fouling layer is present on the membrane/feed interface.)

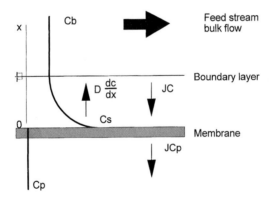

FIGURE 9.3 Boundary layer conditions at membrane/feed interface.

As membranes have imperfections, some solute will diffuse through the semipermeable barrier. At steady state, the convective transport of solute toward the membrane will therefore be equivalent to the diffusive backtransport of solute plus that which passes with the permeate (JC_p). From the boundary conditions given in Fig. 9.3, the following mass balance can be constructed.

$$JC - \left(-D\frac{dC}{dx}\right) = JC_p$$

$$x = 0 \Rightarrow C = C_s$$

$$x = \delta \Rightarrow C = C_b \tag{9.3}$$

Integration of Equation (a) results in

$$\ln \frac{C_s - C_p}{C_b - C_p} = \frac{J\delta}{D} \tag{9.4}$$

which may also be rewritten as

$$\frac{C_s - C_p}{C_b - C_p} = \exp\left[\frac{J\delta}{D}\right]$$

The mass transfer coefficient k_i relates to the ratio between the diffusion coefficient D and the thickness of the boundary layer δ:

$$k_i = \frac{D}{\delta} \tag{9.5}$$

The solute rejection of the membrane can be expressed as:

$$R = \left(\frac{C_s - C_p}{C_s}\right) = \left(1 - \frac{C_p}{C_s}\right) \tag{9.6}$$

which allows an expression for the concentration-polarization ratio to be developed from Eq. (9.4):

$$\frac{C_s}{C_b} = \frac{\exp(J/k_i)}{R + (1 - R)\exp(J/k_i)} \tag{9.7}$$

The product flux and retention performance characteristics of the membrane affect the concentration-polarization ratio; higher salt concentration at the membrane-solution interface increases the salt flux as well as the osmotic pressure, thereby reducing the net pressure driving force and permeate flow.

The concentration-polarization ratio for tubular membranes in the turbulent region has been given by Brian (1965) (Rosenfeld and Loeb, 1967; Goel and McCutchan, 1977) as:

$$\frac{C_s}{C_b} = \frac{\exp(K')}{R + (1 - R)\exp(K')} \qquad \text{where } K' = J/k_i$$

$$K' = \frac{JS_c^{0.67}}{v} \tag{9.8}$$

$$R = 1 - \frac{C_p}{C_s}$$

$$j_D = 0.023 \, \text{Re}^{-0.17} = \frac{k^o Sc^{0.67}}{v} \tag{9.9}$$

In terms of the solute rejection value R and the desalination ratio D_r, which describes the passing membrane on the bulk solute rejection of the membrane in dimensionless form:

$$R = \left(\frac{C_p}{C_b}\right)\left(\frac{C_b}{C_s}\right) \tag{9.10}$$

and

$$D_r = \frac{C_b}{C_p} \tag{9.11}$$

the concentration-polarization ratio can be expressed as follows:

$$\frac{C_s}{C_b} = \exp{(k)}\left[1 - \frac{1}{D_r}\right] + \frac{1}{D_r} \tag{9.12}$$

By incorporating concentration polarization into the transport equations for water and salt flux and simplifying, the following relationships result, from which the membrane water and salt flux coefficients can be calculated:

$$J = k_w\left[\Delta P - \Delta\Pi\left(\frac{C_s}{C_b}\right) + \frac{\Delta\Pi}{D_r}\right] \tag{9.13}$$

$$J_i = k_i C_b\left[\frac{C_s}{C_b} - \frac{1}{D_r}\right] \tag{9.14}$$

These equations allow the performance of membranes to be judged on the log-log relationship of their salt and water permeability coefficients without considering recovery.

where　　k_w = mass transfer coefficient for water, cm/s-atm
　　　　k_i = membrane permeability coefficient for salt, cm/s
　　　　C_b = salt concentration in brine bulk feed, g/cm^3
　　　　C_s = salt concentration at surface, g/cm^3
　　C_s/C_b = concentration-polarization ratio
　　　　C_p = salt concentration in product, g/cm^3
　　　　D = diffusion coefficient
　　　　D_r = desalination ratio
　　　　J = solvent flux, cm^3/cm^2·s
　　　　J_i = solute(i) flux, cm^3/cm^2·s
　　　　k = mass transfer coefficient, cm/s
　　　　Sc = Schmidt number for salt diffusion
　　　　Re = dimensionless Reynolds number
　　　　v = feed velocity, cm/s
　　　　R = intrinsic salt retention
　　　　j_D = Chilton-Colburn mass transfer factor
　　　　P = osmotic pressure, atm
　　　　v = kinematic viscosity
　　　　μ = dynamic viscosity
　　　　d_h = hydraulic radius

9.3.1.1　Sample Calculation. Sample calculation of the concentration-polarization ratio and membrane coefficients for a 12.7-mm (0.5-in) tubular membrane system operating under the following conditions:

Feed stream flow rate Q	(11.4 L/min) 190.5 cm^3/s
Product flux J	(1166 L/m^3·d) 1.35E-3 cm^3/cm^2·s
Membrane rejection R	(88.9%) 0.889
NaCl concentration in feed solution C_b	5200 mg/L
Operating pressure ΔP	(4 MPa) 39.5 atm
Operating temperature	25°C
Membrane inside diameter d	1.27 cm

Kinematic viscosity of water ν 8.93E-3 cm^2/s

NaCl diffusion coefficient D 1.48E-5 cm^2/s

Cross-section area of membrane A_c $(\pi/4)\,d = 1.27$ cm^2

Reynolds number   Re $= vd/\nu = (150 \times 1.27)/(8.93\text{E} - 3) =$ 21300

Chilton-Coburn factor $j_d = 0.023\,\text{Re}^{-0.17} = 4.22\text{E-3}$

Schmidt number   Sc $= \nu/D = 603$

Desalination ratio $D_r = 1/(1 - R) = 8.85$

From Eq. (9.12), the concentration-polarization ratio is calculated as:

$$\frac{C_s}{C_b} = \frac{1}{8.85} + \left[1 - \frac{1}{8.85}\right]\exp\left[\frac{1.35\text{E} - 3 \times 603^{0.67}}{150 \times 4.22\text{E} - 3}\right] = 1.15$$

This example demonstrates the concentration polorization at the membrane surface for the given conditions. Recovery is not considered in test and only one solute (NaCl) is considered. General estimates are often used to estimate concentration polorization and will be illustrated later in the text.

9.3.2 Process Operation

The following sections illustrate the effects of pressure, recovery, and hydrodynamics on the performance of a membrane process. The effects are related to the fundamental phenomena associated with membrane processes.

9.3.2.1 Pressure and Recovery Effects.

According to the model shown in Eq. (9.1), the product flow (i.e., water flux) through a membrane is proportional to the net pressure differential across the membrane. The salt flow is not affected by the applied pressure, but rather is proportional to the concentration difference. In an indirect way, however, the applied pressure does have an effect on the retention performance. At low operating pressures, the resulting water flux is lower while the salt flux remains at a constant level. The net effect is that the salt concentration in the permeate rises due to the lower dilution rate of the permeate. Figure 9.4 illustrates the effect of operating pressure on retention and flux performance.

The percentage recovery r, or conversion,

$$r = \left[\frac{Q_p}{Q_f}\right]100 = \left[\frac{Q_p}{Q_p + Q_c}\right]100 \tag{9.15}$$

also has an effect on membrane performance, as it affects both the salt and product flows. As conversion rises, the salt concentration in the feed stream increases, which causes an increase in the driving force for salt flow or salt passage. Furthermore, higher salt concentration levels in the feed stream increase the osmotic pressure, which reduce the net pressure driving force and, therefore, the product flow. Figure 9.5 illustrates the effect conversion levels have on membrane performance.

The concentration factor, on the other hand, can be related to the conversion by:

$$CF = \left[\frac{1}{1 - r}\right] \tag{9.16}$$

where r is expressed as a decimal fraction. This expression means simply that the concentration of a solute in the feed stream will double (if 100 percent retention is assumed), if the membrane plant is operated at a recovery of 50 percent (i.e., $r = 0.5$).

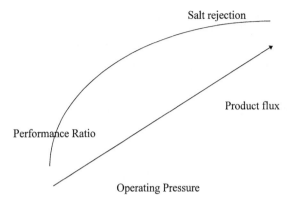

FIGURE 9.4 Membrane performance versus operating pressure.

Seawater RO plants are typically operated at recovery ratios of 30 to 50 percent, but in brackish water installations the recovery ratios can be as high as 85 percent or even greater. Effective scale control becomes very important at these high conversion rates.

At these high concentration factors, the negative effect of concentration polarization on membrane performance can be very serious in the sense that not only can the quality of the permeate produced decline, but also the product flux is decreased. Deposition of solute on the surface of membranes can change the separation characteristics, and the high concentration of solute at the membrane interface increases the risks of change in composition of the membrane material due to chemical attack (Matthiasson and Sivik, 1980).

9.3.2.2 Hydrodynamic Effects. Mass transfer coefficients (as previously shown) are related to the Sherwood number through semiempirical relationships, typically of the form:

$$\text{Sh} = a\text{Re}^b\text{Sc}^c = \frac{k i d_h}{D} = a\left(\frac{v d_h}{\nu}\right)^b\left(\frac{\nu}{D}\right) \tag{9.17}$$

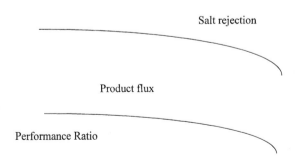

FIGURE 9.5 Membrane performance versus recovery.

where a, b, and c are constants. A number of these relationships exist in the literature for both laminar and turbulent flow regimes, as well as for flow in tubular devices and in channels between plates (Rautenbach, 1989; Porter, 1990).

From Eq. (9.17), it follows that higher mass transfer rates (k_i) will reduce the effect of concentration polarization. The concentration-polarization ratio can also be reduced by lowering the membrane flux and increasing the temperature to increase diffusion rates. However, from Eq. (9.17), higher rates of mass transfer are best achieved by changing the hydrodynamics of the membrane system. This is accomplished by operational techniques (optimization of pressure and linear velocity) or design practice (i.e., system design and module development). Lopez-Leiva (1980) stated that good design of an RO/UF module requires, among others, achieving concentration polarization as low as possible. To reduce the concentration-polarization magnitude requires the hydrodynamic conditions of high turbulent flow or high shear stress laminar flow.

The first condition is achieved in tubular membrane systems such as PCI, Abcor, Wafilin, and Membratek, while high shear stress is achieved with very small channels such as in hollow fibers and plate-and-frame devices such as DDS, Rhone-Poulenc, and GKSS.

The mass transfer rate can be improved by incorporating turbulence promoters within the feed flow channel. The best-known turbulence promoter in use is probably the brine-side spacer material used in spiral-wrap elements to maintain an open flow channel. The apparent Reynolds number for flow within the brine-side channel of these elements falls within the laminar flow region. However, due to the spacer design, excellent mixing and good mass transfer are achieved at low linear flow rates and low pressure differentials. Schock and Miguel (1987) developed correlations for typical RO conditions in a spacer-filled channel and showed that a flat-channel test cell could be used to simulate the conditions in a spiral-wound element. Using a flat-channel test-cell, Da Costa et al. (1991) examined the effect of spacer design on flux, mass transfer, and pressure loss for typical UF conditions and identified spacer-design factors which produce optimal performance. They obtained new correlations for mass transfer and friction factor and fresh insight into the role of spacers.

Mechanical and physical techniques have also been used in tubular membrane devices to improve mixing. A typical case is to install static mixers in the flow passage (Pitera and Middleman, 1973) or, in yet another approach, to employ secondary induced flow profiles such as Taylor vortices, which are established under conditions of curved flow, to improve mass transfer rates (Belfort, 1989). Lately, pulsatile flow and air sparging received much attention as techniques by which to improve membrane performance (Gupta et al., 1993).

9.4 DESIGN CRITERIA

This section introduces design and associated material, including subsections covering fouling indices, pretreatment, array configuration, design options, posttreatment, instrumentation and concentrate disposal. A conventional RO/NF treatment system includes pretreatment, membrane filtration, and posttreatment and is shown in Fig. 9.6. Common terminology for membrane processes is shown in Table 9.2.

Any raw water stream used as a feed stream to a membrane process must undergo either conventional or advanced pretreatment. Conventional pretreatment includes acid or antiscalant addition to prevent precipitation of salts during

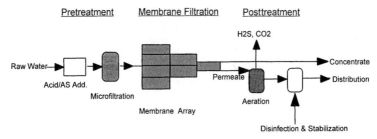

FIGURE 9.6   Conventional RO/NF membrane process.

membrane filtration. Advanced pretreatment occurs before conventional pretreatment and is necessary when the raw water contains excessive fouling materials. Membrane filtration is passage of the pretreated water through an active RO/NF membrane surface with a pore size of 0.001 to 0.0001 μm. Posttreatment includes many unit operations common to drinking water treatment such as aeration, disin-

TABLE 9.2   Membrane Terminology

Raw	Input stream to the membrane process.
Conventional RO/NF process	A treatment system consisting of acid or antiscalant addition for scaling control (pretreatment), RO/NF membrane filtration and aeration, chlorination, and corrosion control for posttreatment.
Feed	Input stream to the membrane array.
Concentrate, reject, retentate, residual stream	The membrane output stream that contains higher TDS than the feed stream.
Brine	Concentrate stream containing total dissolved solids greater than 36,000 mg/L.
Permeate or product	The membrane output stream that contains lower TDS than the feed stream.
Membrane element	A single membrane unit containing a bound group of spiral-wound or hollow fine-fiber membranes to provide a nominal surface area.
Pressure vessel	A single tube that contains several membrane elements in series.
Stage or bank	Parallel pressure vessels.
Array or train	Multiple interconnected stages in series.
High recovery array	Arrays where concentrate stream from succeeding arrays is fed to succeeding arrays to increase recovery.
System arrays	Several arrays that produce the required plant flow.
Rejection	Percent solute concentration reduction of permeate stream relative to feed stream.
Solute	Dissolved solids in raw, feed, permeate, and concentrate streams.
Solvent	Liquid containing dissolved solids, usually water.
Flux	Mass ($mL^{-2}t^{-1}$) or volume (Lt^{-1}) rate of transfer through membrane surface.
Scaling	Precipitation of solids in the element due to solute concentration on the feed stream of the membrane.
Fouling	Deposition of existing solid material in the element on the feed stream of the membrane.
Mass transfer coefficient (MTC)	Mass or volume unit transfer through membrane based on driving force.

fection, and corrosion control. In all potable water membrane processes there will be one feed stream entering the process and two exit streams leaving the process. One of the streams will be a concentrated stream and the remaining stream will be a product stream. As shown in Fig. 9.6, separating the process into advanced pretreatment, conventional pretreatment, membrane filtration, and posttreatment, and one entering and two exiting streams is helpful to gain an overview of RO/NF design.

9.4.1 Fouling Indices

Membrane fouling is an important consideration in the design and operation of membrane systems. Cleaning frequencies, pretreatment requirements, operating conditions, cost, and performance are affected by membrane fouling. The *silt density index* (SDI), *modified fouling index* (MFI), and *mini plugging factor index* (MPFI) are the most common fouling indices. In this article, the SDI, MFI, and the MPFI are defined using the basic resistance model and are quantitatively related to water quality and reverse osmosis and nanofilter membrane fouling.

Fouling indices are determined from simple membrane tests and are similar to mass transfer coefficients for membranes used to produce drinking water. The water must be passed through a 0.45-µm Millipore filter with a 47-mm internal diameter at 30 psig to determine any of the indices. The time required to complete data collection for these tests varies from 15 min to 2 h, depending on the fouling nature of the water. Although similar data are collected for each index, there are significant differences among these indices. Because of the effects of different filtration equipment, only a Millipore filter apparatus can be used to generate accurate index measurements.

9.4.1.1 SDI. The SDI is the most widely used fouling index and is shown in Eq. (9.1). The Millipore test apparatus is used to determine three time intervals for calculation of the SDI. The first two time intervals are the times to collect an initial 500 mL and final 500 mL. The third time interval is 5, 10, or 15 min and is the time between the collection of the initial and final sample. The 15-min interval is used unless the water is so highly fouling that the filter plugs before the 15-min interval is realized. The interval between the initial and final sample collection is decreased until a final 500-mL sample can be collected.

$$SDI = \frac{100[1 \ (t_i/t_f)]}{t} \tag{9.18}$$

where t_i = time to collect initial 500 mL of sample
t_f = time to collect final 500 mL of sample
t = total running time of test

The SDI is a static measurement of resistance which is determined by samples taken at the beginning and the end of the test. The SDI does not measure the rate of change of resistance during the test and is the least sensitive of the fouling indices. The SDI is not dynamic, measures only static resistance, and is not reflective of a continuously operated membrane process.

9.4.1.2 MFI. The MFI is determined using the same equipment and procedure used for the SDI, except that the volume is recorded every 30 s over a 15-min filtration period (Schippers and Verdouw, 1980). The development of the MFI is consistent with Darcy's law in that the thickness of the cake layer formed on the membrane surface is assumed to be directly proportional to the filtrate volume. The total resistance is the sum of the filter and cake resistance. The MFI is derived in Eqs. (9.19) to (9.21) and is defined graphically as shown in Fig. (9.7) as the slope of an inverse flow versus cumulative volume curve.

$$\frac{dV}{dt} = \frac{\Delta P}{\mu} \frac{A}{(R_f + R_k)} \tag{9.19}$$

$$t = \frac{\mu V R_f}{\Delta P A} + \frac{\mu V^2 I}{2\Delta P A^2} \tag{9.20}$$

$$\frac{1}{Q} = a + \text{MFI} \times V \tag{9.21}$$

where
R_f = resistance of filter, L^{-1}
R_k = resistance of cake, L^{-1}
I = measure of fouling potential, $L^2 t$
Q = average flow, v/t^{-1}
a, b = constants
μ = absolute viscosity, $ML^{-1}t^{-1}$

Based on the work of Schippers and Verdouw, Morris (1990) determined an instantaneous MFI by calculating the ratio of flow over volume in 30-s increments to increase fouling sensitivity. An MFI plot is shown in Fig. 9.7. Typically, the cake formation, cake buildup, and cake compaction or failure can be seen in three distinct regions on an MFI plot. The regions corresponding to blocking filtration and cake filtration represent productive operation, whereas compaction would be indicative of the end of a productive cycle.

9.4.1.3 MPFI.   The MPFI is defined as the ratio of flow versus time and is shown in Fig. 9.8 for the same tap water as was used for the MFI. The MPFI is stated math-

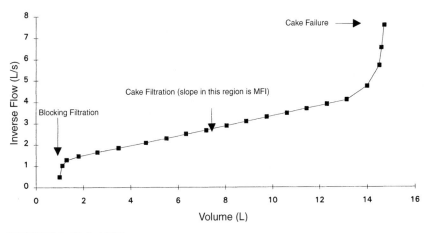

FIGURE 9.7 Typical MFI curve.

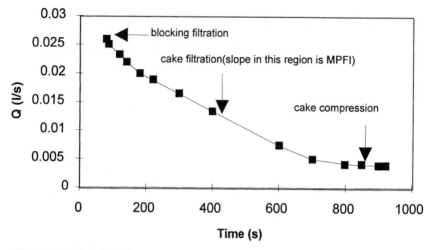

FIGURE 9.8 Typical MPFI curve.

ematically in Eq. (9.22). The MPFI curve ideally shows regions of blocking filtration, cake filtration, and cake compaction, as does the MFI curve. The MPFI is actually the water MTC showing change as a function of time and is equivalent to a direct determination of dJ/dt, since membrane pressure and area are constant. The MPFI would seemingly be the best indicator of membrane fouling, as dJ/dt is the exact measurement of productivity loss. However, since there is very little flow when fouling occurs, it is very difficult to collect flow and time data that accurately reflect fouling. The MFI is much more sensitive to fouling because both volume and flow are equally sensitive to fouling.

$$Q' = a + \text{MPFI}(t) \tag{9.22}$$

where Q' = flow at 30-s increments
   t = time of operation
   a = constant

9.4.1.4 Index Guidelines. Some approximations for required indices prior to conventional membrane treatment are given in Table 9.3 (Morris, 1990; Sung, 1994). The indices apply to the raw or advanced pretreated water before conventional pre-

TABLE 9.3 Fouling Index Approximations for RO/NF

Fouling index	Range	Application
MFI	$0–2$ s/L^2	Reverse osmosis
	$0–10$ s/L^2	Nanofiltration
MPFI	$0–3 \times 10^{-5}$ L/s^2	Reverse osmosis
	$0–1.5 \times 10^{-4}$ L/s^2	Nanofiltration
SDI	$0–2$	Reverse osmosis
	$0–3$	Nanofiltration

treatment unit operations. These numbers are only approximations and do not replace the need for pilot studies. Pretreatment requirements cannot be determined for most installations without a pilot study unless actual plant operating data can be obtained on a very similar water. Pilot studies have been omitted in the design of some brackish-water reverse-osmosis plants. However, the raw waters for these plants typically have TOCs and SDIs less than 1.

9.4.2 Pretreatment

Membrane operations require some measure of feedwater pretreatment upstream of the process. First, however, it is important to realize that pretreatment is process- and feedwater-specific and will therefore differ from application to application and site to site.

Pretreatment is the first step in controlling membrane fouling and can be quite involved. In its simplest form pretreatment involves microstraining with no chemical addition. However, when a surface water is treated, the pretreatment procedure may be much more involved and include pH adjustment, chlorination, addition of coagulants (e.g., alum, polyelectrolytes), sedimentation, clarification, dechlorination (e.g., addition of sodium bisulphite), adsorption onto activated carbon, addition of complexing agents (e.g., EDTA, SHMP), pH adjustment, and final polishing.

Factors that are important and must be considered when pretreatment is contemplated are:

• Material of membrane construction (asymmetric cellulosic or noncellulosic membranes, thin film ether, or amidic composite membranes)
• Module configuration (spiral wrap, hollow fine fiber, tubular)
• Feedwater quality
• Recovery ratio
• Final water quality

Figure 9.9 gives some indication of substances that are known to affect membrane performance and whose concentration and/or presence in the feedwater must be controlled.

Turbidity levels stipulated by membrane manufacturers are normally attained by conventional clarification techniques such as coagulation followed by sedimentation and sand filtration. In seawaters that are notoriously rich in suspended nutrients, such as along the west coast of Southern Africa, and for treating surface water with high TOCs as are common in the United States, ultrafiltration has been advocated as a viable pretreatment option capable of reducing the suspended solids concentration down to acceptable standards in a simple operation (Strohwald and Jacobs, 1992; Metcalf, 1992; Taylor, 1989a).

The pretreatment process shown in Fig. 9.6 consists of antiscalant and/or acid addition and microfiltration. These pretreatment processes are used to control scaling and to protect the membrane elements and are required for conventional reverse-osmosis or nanofiltration systems. The membranes can be fouled or scaled during operation. Fouling is caused by materials such as colloids that are present in the raw water and will reduce the productivity of the membrane. Scaling is caused by the precipitation of a salt within the membrane because of feed-stream concentration. If a raw water is excessively fouling, additional or advanced pretreatment is required.

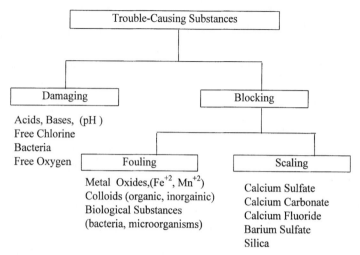

FIGURE 9.9   Substances potentially harmful to membranes. *(Rautenbach and Albrecht, 1989.)*

9.4.2.1 Advanced Pretreatment.   Advanced pretreatment would be unit operations that precede scaling control and static microfiltration. Examples of advanced pretreatment would be coagulation, oxidation followed by green sand filtration, groundwater recharge, continuous microfiltration, and GAC filtration. Any other unit operations that preceded conventional pretreatment would, by definition, be advanced pretreatment. In some pretreatment unit operations such as alum coagulation, the feedwater is saturated with a salt such as alum hydroxide. In such instances, the solubility of such salts must be accounted for in the feedwater stream to avoid precipitation in the membrane.

Clarification.   Clarification is standard practice in the production of potable water in which conventional clarification techniques such as sand filtration are employed. Colloidal species are negatively charged and remain in suspension because of charge repulsion. Their size range, 0.3 to 1.0 μm, make them difficult to remove by sand filtration only.

However, stringent standards set by membrane manufacturers more often than not necessitate that flocculation and coagulation steps be introduced prior to the polishing step. Flocculation and coagulation may be followed by sedimentation (if the turbidity levels are very high) before final filtration. The bed filters that are used may be stratified, single-medium sand filters or multilayered, with the layers consisting of coarse/fine sand, activated carbon, and anthracite.

The SDI of a feedwater to an RO plant should be less than 2 to minimize the rate of colloidal fouling, but surface waters treated by conventional clarification techniques often have higher SDIs. This is because the sand filters used in the final polishing step are not absolute, as separation is based on a mechanism of adsorption due to surface effects rather than sieving as in membrane filtration. The particle size of the sand used to pack these filters is normally larger than 500 μm. Furthermore, the filters are operated on a cyclic basis, in which they are backwashed with air before any breakthrough occurs. When even higher quality water is required, filters are precoated with still finer materials such as diatomaceous earth

(DE). It is possible to reduce the SDI from 4 to 5 down to less than 2 by means of DE filtration. Pressure medium filtration, in which filter media such as sand, carbon, or some other medium is used, offers another option for the reduction of the SDI from 6 to 3.

Flocculation and coagulation agents are added to water to neutralize the electrical charge on colloids and create *nuclei* onto which colloidal and suspended material in the water can adsorb, thus creating floc of larger dimensions and mass and which can be removed by sedimentation followed by sand filtration. Alum and ferric chloride are commonly used inorganic coagulants. Although coagulation is in itself a very complex physical chemistry process, the chemical reactions which institute flocculation are simple:

Alum added to waters with natural bicarbonate alkalinity:

$$Al_2(SO_4)_3 \; 18H_2O + 3 \; Ca(HCO_3)_2 \rightarrow 2 \; Al(OH)_3 \downarrow + 3 \; CaSO_4 + 18 \; H_2O + 6 \; CO_2$$

Alum added to waters with added lime alkalinity (lime [CaO]):

$$Al_2(SO_4)_3 \; 18H_2O + 3 \; Ca(OH)_2 \rightarrow 2 \; Al(OH)_3 \downarrow + 3 \; CaSO_4 + 18 \; H_2O$$

Ferric chloride added to waters with natural bicarbonate alkalinity and lime added for alkalinity:

$$2 \; FeCl_3 + 3 \; Ca(HCO_3)_2 \rightarrow 2 \; Fe(OH)_3 \downarrow + 3 \; CaCl_2 + 6 \; CO_2$$

$$2 \; FeCl_3 + 3 \; Ca(OH)_2 \rightarrow 2 \; Fe(OH)_3 \downarrow + 3 \; CaCl_2$$

Polymers of molecular mass typically 10^4 to 10^6 dalton are also used as coagulating agents. With respect to charge, polyelectrolytes may be anionic (negatively charged), cationic (positively charged), or nonionic (neutral). These coagulants have the advantage that the pH of the feedwater is not critical to ensure their proper functioning. However, as with the mixing of alum, the initial mixing step is most critical to achieve maximum effectivity. Polyelectrolytes are often injected into the feed flow line. Mixing is accomplished by means of static mixers and filtration by means of multimedia in-line pressure filters.

Other.   Several other types of pretreatment are possible for RO or NF membrane systems. Membrane fouling is not a well-understood phenomenon and many different pretreatments are possible for different types of fouling mechanisms. For example, if biological fouling was a problem, disinfection with a bactericide that is not harmful to the membranes would be a very realistic possibility. If foulants were biodegradable, biologically activated carbon would be feasible. Microfiltration or ultrafiltration would also be useful for removing some fouling contaminants. There are few limitations to the pretreatment processes that could be investigated in new applications of membrane processes.

The most serious fouling problems may exist when highly organic surface waters are treated by NF or RO for TOC or DBP control. Such applications may require cleaning frequencies of less than one month which are not typical to conventional RO or NF processes. However, membrane applications for organic control are not typical and will have to be developed with time. One possibility for controlling fouling is to highly automate the membrane cleaning process. Currently, conventional RO plants treat nonfouling groundwaters, and cleaning frequencies are typically six months or more. Consequently, the cleaning operation is not automated and is disruptive to normal operation. Development of nondamaging membrane cleaning agents and automation of the cleaning process should not be overlooked as potential answers to fouling control.

9.4.2.2 Scaling Control. Scaling control is essential in RO/NF membrane filtration. Three different methods of controlling precipitation or scaling within the membrane element are discussed with the identification of a limiting salt, acid addition for prevention of $CaCO_3$, and an example of an output from a computer program for prediction of antiscalant dose to control scaling.

Limiting Salt. The amount of antiscalant or acid addition is determined by the limiting salt. A diffusion-controlled membrane process will naturally concentrate salts on the feed side of the membrane. If excessive water is passed through the membrane, this concentration process will continue until a salt precipitates and scaling occurs. Scaling will reduce membrane productivity and, consequently, recovery is limited by the allowable recovery just before the limiting salt precipitates. The limiting salt can be determined from the solubility products of potential limiting salts and the actual feed-stream water quality. Ionic strength must also be considered in these calculations because the natural concentration of the feed stream during the membrane process increases the ionic strength, allowable solubility, and recovery. Calcium carbonate scaling is commonly controlled by sulfuric acid addition; however, sulfate salts are often the limiting salt. Commercially available antiscalants can be used to control scaling by complexing the metal ions in the feed stream and preventing precipitation. Equilibrium constants for these antiscalants are not available, which prohibits direct calculation. However, some manufacturer-provided computer programs are available for estimating the required antiscalant dose for a given recovery, water quality, and membrane. General equations for the solubility products and ionic strength approximations are given as follows.

$$A_n B_m \Leftrightarrow n A^{+p} + m B^{-q}$$

$$K_{sp} = (A^{+p})^n (B^{-q})^m$$

$$K_{sp} = \left(a\, \frac{A^{+p}}{x} \right)^n \left(b\, \frac{B^{-q}}{x} \right)^m$$

(9.23)

where x = fraction remaining
         $1 - x$ = fraction recovered
         K_{sp} = solubility product
         a = fraction of cation retained
         b = fraction of anion retained

$$u = 0.5\Sigma C_i Z_i^2 \cong (2.5 \times 10^{-5})(\text{TDS})$$

(9.24)

where u = ionic strength
         C = mol/L
         Z = ion charge
         TDS = mg/L

$$\log \gamma \cong -0.5 Z^2 \frac{\sqrt{u}}{1 + \sqrt{u}}$$

(9.25)

where γ = activity coefficient

A limiting salt calculation is shown in the following example. Rejection of ions will vary by operation conditions, molecular weight, and charge. The solubility product has been modified by a and b coefficients, as shown, for consideration of less than complete rejection. In the example, 0.95 rejection has been assumed for divalent ions. A more exact estimate of fraction rejected can be obtained from the manufacturer's literature or pilot studies. A raw water quality is given with the solubility products of limiting salts at 25°C. The limiting salt is identified by determining which salt allows the least recovery or minimizes $1 - x$. The solubilities of all possible salts are considered here. The first calculations are intended to demonstrate that both $CaCO_3$ and $BaSO_4$ are supersaturated. This would initially seem to indicate that recovery was not possible; however, when ionic strength is considered, the allowable recovery is 60 percent. $CaCO_3$ precipitation will be controlled by acid addition.

EXAMPLE 9.1 *Limiting salts.*

	Raw water quality used for examples				
	Concentration,		Concentration,	Salts to consider	
Parameter	mg/L	Parameter	mg/L	Salt	pK_{sp} at 25°C
pH	8.0	Cl^-	730	$CaCO_3$	8.3
Na^+	695	NO_3^-	0	$Ca_3(PO_4)$	6.8
K^+	8	F^-	1.1	CaF_2	10.3
Ca^{+2}	8	SO_4^{-2}	79	$CaSO_4$	4.7
Mg^{+2}	2	O-PO_4	0.7	$BaSO_4$	9.7
Fe^{+2}	0.5	HCO_3^-	631	$SrSO_4$	6.2
Mn^{+2}	0.02	SiO_2	24	SiO_2	2.7
Cu^{+2}	0	TDS	2200		
Ba^{+2}	0.04	H_2S as S	1		

For
$$CaCO_3 \rightarrow Ca^{2+} + CO_3^{2-} \quad K_{sp} = [Ca^{2+}][CO_3^{2-}] = 10^{-8.3}$$
$$Ca^{2+} = 8 \text{ mg/L} = 0.0002 \text{ M}$$
$$CO_3^{2-} = HCO_3^- \frac{\alpha_2}{\alpha_1} = HCO_3^- \frac{K_2}{H^+} = 0.00005 \text{ M} = 3 \text{ mg/L}$$

Then
$$\left(\frac{0.0002}{X}\right)\left(\frac{0.00005}{X}\right) = 10^{-8.3} \qquad X = 1.41 \text{ L}$$
$$r = 1 - X = -0.41 \text{ L}$$

For
$$BaSO_4 \rightarrow Ba^{2+} + SO_4^{2-} \qquad K_{sp} = [Ba^{2+}][SO_4^{2-}] = 10^{-9.7}$$

if
$$Ba^{2+} = 0.04 \text{ mg/L} = 3 \times 10^{-7} \text{ M}, SO_4^{2-} = 79 \text{ mg/L} = 0.0008 \text{ M}$$

Then
$$\left(\frac{3 \times 10^{-7}}{X}\right)\left(\frac{0.0008}{X}\right) = 10^{-9.7} \qquad X = 1.09 \text{ L}$$
$$r = 1 - X = -0.09 \text{ L}$$

These calculations show that both $CaCO_3$ and $BaSO_4$ are supersaturated, as indicated by the negative decimal fraction for r. A positive r would indicate the fraction of the feed stream that could be recovered.

Assuming 95 percent mass rejection

$$\text{TDS} = \left(\frac{0.95}{0.4}\right) = 5225$$

$$\mu = (2.5 \times 10^{-5})\text{TDS} = 0.13$$

$$\log \gamma = -0.5(2)^2 \cdot \frac{\sqrt{0.12}}{1 + \sqrt{0.12}} = -0.53$$

$$pK_c = pK_{sp} - (m)p\gamma - (n)p\gamma = 9.7 - (1)0.53 - (1)0.53 = 8.64$$

Recalculate X from:

$$K_c = \left(a\frac{[A^{+p}]}{X}\right)n\left(b\frac{[B^{-q}]}{X}\right)m = \left[(0.95)\frac{3 \times 10^{-7}}{X}\right]\left[(0.95)\frac{0.0008}{X}\right] = 10^{-8.64}$$

$$X = 0.31 \; L \qquad r = 1 - X = 0.69 \; L \text{ or } 69\%$$

Iterate calculations from beginning but use 0.31 for X in place of 0.4 to determine new TDS concentration in concentrate stream. After two iterations the recovery converges at 72 percent.

Acid Addition. Acid is usually added to allow $CaCO_3$ recovery to the limiting salt recovery, which is illustrated in the example. In this example, an assumption was made of 90 percent rejection for monovalent, 95 percent rejection for divalent, and 98 percent rejection for trivalent. It is important to realize the scaling calculation should be made using the concentrate stream. The equilibrium constants for the carbonate system, alkalinity, pH, calcium and recovery are required to make these calculations. In the example, pH 6.3 was required not to scale $CaCO_3$ which could be achieved by the addition of 206 mg/L H_2SO_4. The additional sulfate from acid addition should be considered for calculation of the limiting salt recovery.

EXAMPLE 9.2 *Acid addition to control $CaCO_3$ based on ionic strength.*
Given: $Ca^{2+} = 8$ mg/L, $HCO_3^- = 631$ mg/L, pH = 8.0, R = 72%, $\mu = 0.19$
Find: H_2SO_4 dose to prevent scaling of calcium carbonate.
Solution: Determine pH required on the feed side of membrane using concentrate stream.

For $\qquad CaCO_3 \rightarrow Ca^{2+} + CO_3^{2-} \quad K_{sp} = [Ca^{2+}][CO_3^{2-}] = 10^{-8.3}$

$$\log\gamma = -0.5(1)^2 \frac{\sqrt{0.19}}{1 + \sqrt{0.19}} = -0.15 \qquad p\gamma_{\pm 1} = 0.15 \qquad p\gamma_{\pm 2} = 0.60$$

$$pK_c = pK_{sp} - (m + n)p\gamma = 8.3 - (1)0.60 - (1)0.60 = 7.1$$

$$K_c = \left(a\frac{[A^{+p}]}{X}\right)^n\left(b\frac{[B^{-q}]}{X}\right)^m = \left[(0.95)\frac{Ca^{2+}}{X}\right]\left[(0.95)\frac{CO_3^{2-}}{X}\right] = 10^{-7.1}$$

$$Ca^{2+} = \left(\frac{0.95}{0.28}\right)\left(\frac{8}{40000}\right) = 0.00068 \; M$$

for $\qquad HCO_3^- \rightarrow H^+ + CO_3^{2-} \qquad K_2 = \frac{[H^+][CO_3^{2-}]}{[HCO_3^-]} = 10^{-10.3}$

$$K_2 = \frac{[H^+][\gamma_2 CO_3^{2-}]}{[\gamma_1 HCO_3^-]} \quad pK_{c2} = pK_2 + p\gamma_1 - p\gamma_2 = 10.3 + 0.15 - 0.60 = 9.85$$

$$CO_3^{2-} = \frac{[K_{c2}][HCO_3^-]}{[H^+]} = \frac{[10^{-9.85}](0.9/0.28)[(631/61000)]}{[H^+]} = \frac{10^{-11.3}}{[H^+]}$$

$$K_c = \left[\left(\frac{0.95}{0.28}\right)Ca^{2+}\right]\left[\left(\frac{0.95}{0.28}\right)CO_3^{2-}\right]$$

$$= \left(\frac{0.95}{0.28}\right)^2 (0.00068)\left(\frac{10^{-11.3}}{[H^+]}\right) = 10^{-7.1}$$

$$[H^+] = 10^{-6.3}$$

or   pH = 6.3 not to scale calcium carbonate in concentrate stream at 72 percent recovery.

The preceding calculation has determined the pH that will not scale $CaCO_3$ at 75 percent recovery. As is shown in the example, an assumption of 90 percent rejection of calcium and carbonate ions was assumed. This calculation is approximate but does illustrate the effect of ionic strength and the control of $CaCO_3$ solubility by adjusting pH. Control of $CaCO_3$ solubility is required in almost every RO/NF plant that will use groundwater.

Once the required pH has been determined for $CaCO_3$ scaling, the required acid dose can be calculated.

EXAMPLE 9.3   *Determine dose of H_2SO_4 to achieve pH.*

$$K_1 = \frac{[H^+][\gamma HCO_3^-]}{[H_2CO_3]} = 10^{-6.3} \qquad pK_{c1} = pK_1 - p\gamma$$

$$\log\gamma = -0.5(1)^2 \frac{\sqrt{0.19}}{1 + \sqrt{0.19}} = -0.15 \qquad p\gamma = 0.15$$

$$pK_{c1} = pK_1 - p\gamma = 6.3 - 0.15 = 6.15$$

$$K_{c1} = \frac{[H^+][HCO_3^-]}{[H_2CO_3]} = 10^{-6.15} = \frac{[10^{-6.3}][(0.9/0.28)(631/61000) - Y]}{[0 + X]}$$

Where Y is the H^+ needed to react with HCO_3^-.

$$10^{-9.85} + 10^{-5.64}Y = -10^{-5.79}Y + 10^{-7.78}$$

$$Y = 0.0042 \text{ M } H^+ = \frac{(0.0042 * 98000)}{2} = 206 \text{ mg/L } H_2SO_4$$

This is the H_2SO_4 that needs to be added in the last element of the last stage.

Manufacturers' Programs.   Antiscalants can also be used to determine allowable recovery. As noted, manufacturers such as Hydranautics, Desal, Fluid Systems, Dow-FilmTec, Du Pont, and TriSep provide computer programs that are available for estimating recovery. Antiscalants complex the metal cations, making them unavailable for precipitation. Unfortunately, the equilibrium constants are not available and the predicted recoveries cannot be checked. However, there is no doubt that antiscalants work, because they are used worldwide in membrane plants to increase recovery. Many membrane plants use both sulfuric acid and antiscalants. A manufacturer-provided antiscalant project is shown in Figure 9.10. Note that, even though the $BaSO_4$ is 608 percent saturated, a dose of 4.2 mg/L is indicated to control scaling. There is a design choice here, as all scaling could possibly be controlled by antiscalant addition. Antiscalants are typically an order of magnitude more costly than acid, but less is needed to control scaling. It is not uncommon for antiscalants to become ineffective if they react preferentially with a metal other than the limiting salt metal. High iron concentrations have been known to greatly decrease the effectiveness of certain antiscalants. These factors and others point out the need for pilot studies.

FLOCON 100 Dose Projection

Prepared by:		FEED WATER ANALYSIS	Date: 3/7/94 RECOVERY = 75%
Cations		Anions	
Ca	8	HCO_3	583.5
Mg	2	SO_4	202.2
Na	695	Cl	730
K	8	F	1.1
Ba	.04	NO_3	0
Sr	0	PO_4	.7
Fe	0.5	CO_3	.72
Silica	24	CO_2	45.12
Temperature	25	Pressure	150

The above feedwater analysis reflects the addition of sulphuric acid.
H_2SO_4 (100%) added, ppm = 45.09

Ionic strength, feed	.035	
Total cation as $CaCO_3$	1554	
Total anion as $CaCO_3$	1723	
TDS (feed)	2267	TDS (brine)
9070		
pH, raw water 8	pH, FEED 7.2	pH, brine 7.72

CAUTION! Anion/cation balance exceeds 5% error.
A more accurate water analysis is desirable.

FLOCON 100 dose for $CaCO_3$ control	Minimum	Recommended
	2.5 ppm	3.8 ppm
FLOCON 100 dose for $BaSO_4$ control	Minimum	Recommended
	3 ppm	4.2 ppm

Degree of saturation of potential scales, %

$CaCO_3$ (LSI)	$CaSO_4$	$SrSO_4$	$BaSO_4$
.97	2	0	767

To ensure operation in a scale-free position, FLOCON 100 should be dosed consistent with the greatest scale potential in the system as indicated by the highest dosage given above.

Ion product, $CaCO_3$ = 3.833975E-08   Ion product, $CaSO_4$ = 6.721338E-06
Ion product, $SrSO_4$ = 0   Ion product, $BaSO_4$ = 9.805316E-06
Ion product, CaF_2 = 4.28174E-11

FIGURE 9.10   Manufacturer-provided antiscalant project.

This program was run using 75 percent recovery but could have been run using a higher recovery, the point being that the use of an antiscalant will provide a higher recovery than acid alone. The antiscalant complexes the cations to prevent precipitation, whereas the acid destroys the anion to prevent precipitation. The mechanisms are different and can be optimized to maximize recovery and lower cost.

9.4.2.3 Static Microfiltration. Microfilters typically are sieving filters with pore diameters of 5 to 20 μm. The pressure drop through a microfilter does not usually exceed 5 lb/in² in nanofiltration applications and 10 lb/in² in reverse-osmosis applications. Microfiltration with a nanofiltration or a reverse-osmosis process protects only against foulants or materials in the solid phase before they enter the membrane process. There is no protection against scaling by microfiltration. Microfiltration alone will not be adequate to remove foulants from a feed stream with a high turbidity or suspended solids concentration. Generally, the size distribution of the foulants will include particles with diameters smaller than the microfilter pore size. Consequently, a maximum value for the feed-stream fouling indices is specified prior to microfiltration. A microfilter in a reverse-osmosis or nanofiltration process should be thought of as a necessary means of protecting the membrane elements.

9.4.3 Arrays

This section describes different models and design techniques used to size arrays. A linear solution diffusion model, a film theory solution diffusion model, and a coupling model are described. Single- and multistaged array equations for predicting permeate water quality are developed and a simple design example is presented.

9.4.3.1 Linear Solution Diffusion Model. There are many different theories and models describing mass transfer in diffusion-controlled membrane processes (see Fig. 9.11); however, a few basic principles or theories are used to develop most of these models. These are convection, diffusion, film theory, and electroneutrality. These principles or theories could be used to group models into linear diffusion models, exponential diffusion models, and coupling models. Most of the modeling efforts have been developed using very small test cells, have not incorporated product recovery into the model, and are of limited practical use.

The basic equations used to develop these models are shown in Eqs. (9.26) through (9.30) with reference to the membrane element shown in Fig. 9.11.

$$J = k_w(\Delta P - \Delta \Pi) = \frac{Q_p}{A} \tag{9.26}$$

$$J_i = k_i \Delta C = \frac{Q_p C_p}{A} \tag{9.27}$$

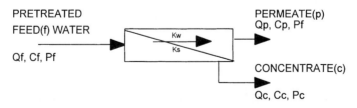

FIGURE 9.11 Basic diagram of mass transport in a membrane.

$$r = \frac{Q_p}{Q_f} \tag{9.28}$$

$$Q_f = Q_c + Q_p \tag{9.29}$$

$$Q_f C_f = Q_c C_c + Q_p C_p \tag{9.30}$$

where J = water flux, $L^3/L^2 t$
         J_i = solute flux, $M/L^2 t$
         k_w = solvent mass transfer coefficient, $L^2 t/M$
         ki = solute mass transfer coefficient, L/t
         ΔP = pressure gradient, L, $[(P_f + P_c)/2 - P_p]$
         $\Delta \Pi$ = osmotic pressure, L, $[(\Pi_f + \Pi_c)/2 - \Pi_p]$
         ΔC = concentration gradient, M/L^3, $[(C_f + C_c)/2 - C_p]$
         Q_f = feed stream flow, L^3/t
         Q_c = concentrate stream flow, L^3/t
         Q_p = permeate stream flow, L^3/t
         C_f = feed stream solute concentration, M/L^3
         C_c = concentrate stream solute concentration, M/L^3
         C_p = permeate stream solute concentration, M/L^3
         r = recovery
         A = membrane area, L^2
         Z = combined mass transfer term

If ΔC is defined as the difference of the average feed and brine stream concentrations and the permeate stream concentration, then Eq. (9.31) can be derived from Eqs. (9.26) and (9.30). This model can be described as a linear homogeneous solution diffusion model in that it predicts that solute flow is diffusion controlled and solvent flow is pressure (convection) controlled. Equation (9.31) can be simplified by including a Z term which incorporates the effects of the mass transfer coefficients, pressure and recovery, into a single term.

$$C_p = \frac{k_i C_f}{k_w (\Delta P - \Delta \Pi)[(2 - 2r)/(2 - r)] + k_i} = Z_i C_f \tag{9.31}$$

This is the simplest model, but it does allow the effect of five independent variables on permeate water quality to be considered, as shown in Fig. 9.12. If pressure is increased and all other variables are held constant, then permeate concentration will decrease. If recovery is increased and all other variables are held constant, then permeate concentration will increase. These effects may be hard to realize if an existing membrane array is considered, for it is impossible in such an environment to increase pressure without increasing recovery. However, it is very possible to increase pressure without varying recovery when arrays are designed. Decreasing feed-stream concentration consequently, pretreatment may be an option for decreasing the permeate stream concentration. Different membranes may have different mass transfer characteristics. Using a membrane with a lower molecular weight cutoff would probably decrease the permeate concentration, although the solvent and solute MTCs must be considered before such a result could be expected.

9.4.3.2   Film Theory Model.   The linear model can be modified by the incorporation of film theory which assumes that the solute concentration exponentially increases from the center of the feed-stream channel toward the surface of the mem-

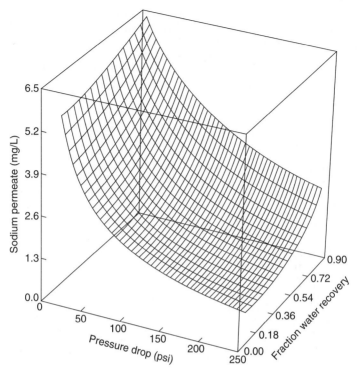

FIGURE 9.12 Three-dimensional plot of sodium permeate concentration as a function of pressure drop and water recovery.

brane and diffuses back into the bulk stream. Mathematically, this is shown in Eq. (9.3) and has been solved as shown in Eq. (9.16). The backdiffusion constant is introduced in Eq. (9.4).

$$J_i = -D_s \frac{dC}{dx} + C_i J \tag{9.3}$$

where D_s = diffusivity
C_i = concentration from the bulk to the membrane interface
z = path length or film thickness

$$\left[\frac{C_s - C_p}{C_b - C_p} \right] = e^{J/k_b} \tag{9.4}$$

where C_p = solute concentration in membrane permeate
C_b = solute concentration in membrane bulk
C_s = solute concentration at membrane surface
J = water flux through membrane
$k_b = D_s/x$ = diffusion coefficient from surface to bulk

Incorporation of Eq. (9.4) into Eqs. (9.26) through (9.30) results in Eq. (9.32), which is a development of the homogeneous solution diffusion model using concentration polarization. This model predicts that concentration at the membrane surface is higher than in the bulk of the feed stream. Such an effect is documented in the literature (Sung, 1993; Hofman, 1995) and the model shown in Eq. (9.32) accounts for this phenomenon. The backdiffusion coefficient k_b represents solute diffusion from the membrane surface to bulk in the feed stream, which is different than the mass transfer coefficient k_w, which represents solute diffusion through the membrane to the permeate stream.

$$C_p = \frac{C_f k_i e^{J/k_b}}{k_w(\Delta P - \Delta\pi)[(2 - 2r)/(2 - r)] + k_i e^{J/k_b}}$$ (9.32)

If solutions containing ions are assumed to be electrically neutral due to the presence of counter-ions, then solutes do not pass through a membrane in a charged state but a coupled state. Such phenomena can be described as electroneutrality and are observed in diffusion-controlled membrane processes when the permeate stream becomes more concentrated than the feed stream. Relatively high concentrations of sulfates have been observed to increase the rejection of calcium and decrease the rejection of chlorides. One interpretation of such observations is that the strong calcium sulfate couple is retained in the feed stream, and the weaker sodium chloride couple is forced to pass through the membrane to maintain equilibrium.

9.4.3.3 Coupling.   The coupling effect on mass transfer through membranes has been modeled using a statistical modification of free energy for single-solute systems, as shown in Eq. (9.33) (Rangarajan, 1976; Sung, 1993). The free-energy term is assumed to be different in the bulk solution and at the membrane surface because of a difference of ion concentration at those two locations. Consequently, the energy required to bring the ions to the surface is shown in Eq. (9.34) and is the difference of the free energy in bulk and surface solutions. The $\Delta\Delta G$ values for each ion are determined experimentally.

$$\frac{1}{\Delta G} = -\frac{1}{E} r - \frac{\Delta}{E}$$ (9.33)

where ΔG = free energy of coupled ion
   E = solution-dependent constant
   r = coupled ion radius
   Δ = statistical constant

$$\Delta\Delta G = \Delta G_I - \Delta G_B$$ (9.34)

where $\Delta\Delta G$ = difference of coupled ion free energy at interface I and bulk B

Membrane-specific solute mass transfer coefficients for single-solute systems have been determined experimentally and related to free energy by Eq. (9.35). Once the membrane specific constant $\ln (C^*)$ has been determined for a reference solute (e.g., NaCl), it is possible to determine k_s for any given solute in a single-solute system. Once k_s is known, the mass transport of any single solute in a diffusion-controlled membrane can be predicted.

$$\ln (k_i) = \ln (C^*) + \Sigma\left(-\frac{\Delta\Delta G}{RT}\right)$$ (9.35)

where C^* = membrane-specific constant
k_i = mass transfer coefficient
R = gas constant
T = temperature

Coupling can be used to model mass transport in a multisolute system. As with the single-solute system, electroneutrality is assumed in all phases (bulk, surface, permeate, film). A coupling model can incorporate either a homogeneous or an exponential solution diffusion model with the free-energy model. The model requires the mass transfer coefficients for water and a reference couple, the backtransfer coefficient, and the $\Delta\Delta G$ values for all ions to predict flux and concentration of the permeate stream. The mass transfer models shown in Eqs. (9.31) and (9.32) can both be used as coupling models as long as the k_i term is determined using $\Delta\Delta G$ or feed-stream solute composition. However, this term is membrane specific and must be for a given membrane. There is no model that accounts for the effects of the membrane surface on a fundamental basis.

The determination of $\Delta\Delta G$ is based on free energy and is therefore affected by feed-stream composition. The determination of free energy for any given reaction is a good method of determining the likelihood of the reaction proceeding, but this method relies on the consideration of all the free energies for the major chemical interactions. As the solutes and free energies are not known for many of the constituents in natural water, a basic free-energy model has not been developed for practical membrane design.

9.4.3.4 Array Modeling. The design will be illustrated using the linear model because of simplicity and a membrane manufacturer's computer program. The mass transfer coefficients for the linear model can be developed from field criteria, and the model can easily be used to predict permeate water quality for given changes in operation. Variations in sodium permeate concentration are shown in Fig. 9.12 for varying pressure and recovery using the linear model. There is a good deal of flexibility in reverse-osmosis or nanofiltration processes as to permeate water quality. Designers can vary flux and recovery to vary water quality. Anyone involved in the design or operation of membrane facilities should be aware of the coupling phenomenon and concentration polarization. The mass transfer coefficients for the film theory model are more difficult to develop but have been shown to more accurately describe pilot plant data. However, the best design criteria would come from site-specific pilot studies. Pilot studies may not be required for desalination of some brackish waters with very low organic concentration and no potential foulants, but there are many physical, chemical, and biological factors that are not realized initially and can cause serious operational problems. The only way to avoid these types of unrealized problems in design is to conduct a pilot study.

Linear Array. Equations (9.31) and (9.32) are very useful in that the effect of ΔP, C_f, r, $\Delta\Pi$, k_w, and k_i on C_p can be determined without experimentation if the six variables are known. ΔP, C_f, r, and $\Delta\Pi$ can be accurately assumed by the user. k_w and k_i should be measured experimentally for a given membrane and source but can be taken from the literature for a given membrane and a similar source. As shown in Eq. (9.36), a Z_i term is a modified MTC, which represents the effect of solvent mass transfer, solute mass transfer, recovery, and pressure on permeate stream concentration for any single membrane element and simplifies Eq. (9.31). If Eq. (9.32) were used, the modified MTC would be represented by Z_{cpi} which would include consideration of backdiffusion.

$$C_{p,i} = \frac{k_{i,i}C_{f,i}}{k_{w,i}\Delta P_i[(2-2r_i)/(2-r_i)] + k_{i,i}} = Z_i C_{f,i} \tag{9.36}$$

where Z_i = modified mass transfer coefficient
$\Delta P_i = \Delta P - \Delta\Pi$ = net average driving pressure including hydraulic losses

The subscript i in Eq. (9.36) and further equations is used to denote any stage in a multistage membrane array. Membrane systems are configured in arrays, which consists of stages. Consequently, Eq. (9.36) can be expanded to describe an entire membrane system. A typical two-stage membrane system is shown in Fig. 9.13. The first

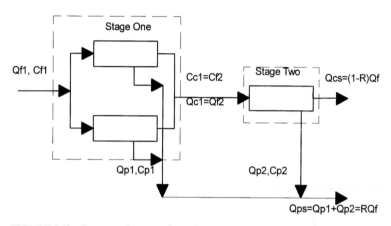

FIGURE 9.13 One-array (two-stage) membrane system.

stage shown in Fig. 9.13 consists of two pressure vessels, which typically contain six membrane elements. The second stage consists of one pressure vessel, which also typically contains six elements. The combination of these two stages is a 2-1 array, which indicates a two-stage array with two pressure vessels in stage one connected to one pressure vessel in stage two. The model can be used to predict the system permeate concentration or the effect of any of the six independent variables on permeate concentration if MTCs, pressures, and recoveries for each stage are known.

As noted in Fig. 9.13, the succeeding-stage feed-stream flow and concentration is always equal to the preceding concentrate stream flow and concentration in a high recovery system that is used to produce potable water. The interstage flow and solute concentration in a multiarrayed membrane system can be related by an interstage mass balance, as shown in Eqs. (9.37) through (9.40). The resulting X_i term is used in Eq. (9.40) to determine the feed-stream concentration for any stage from the initial feed-stream concentration and can be thought of as a concentration factor of stage i.

$$C_{C,i} = \frac{Q_{f,i}C_{f,i} - Q_{p,i}C_{p,i}}{Q_{c,i}} = \frac{C_{f,i} - r_i C_{p,i}}{1 - r_i} \tag{9.37}$$

$$C_{c,i} = C_{f,i+1} = C_{f,i}\left(\frac{1 - Z_i r_i}{1 - r_i}\right) \tag{9.38}$$

$$C_{c,i} = X_i C_{f,i} \tag{9.39}$$

$$X_i = \frac{1 - Z_i r_i}{1 - r_i} \; ; \qquad \text{Define } X_0 \equiv 1 \tag{9.40}$$

Consequently, an interstage equation for permeate concentration can be developed by interrelating all stages in any diffusion-controlled membrane system. Equation (9.40) interrelates all stages and is simply a flow-weighted concentration equation which represents the total solute mass from the permeate stream in each stage, as shown in the numerator of Eq. (9.40) and the total volume of solvent from the permeate stream in each stage as shown in the denominator of Eq. (9.40). Equation (9.40) becomes less complex when the stage Z_i and X_i terms are used as shown in Eq. (9.41). This equation can be used to describe membrane system performance and is as versatile as Eq. (9.31) or (9.32), which describes a single stage or element. The subscript j as i denotes the stage number in a multistaged membrane array, as shown in Eq. (9.42).

$C_{p,\text{system}}$

$$= \frac{C_f \displaystyle\sum_{i=1}^{n} \left[A_i k_{i,i} \Delta P_i \frac{k_{i,i}}{k_{W,i} \Delta P_i [(2 - 2r_i)/(2 - r_i)] + k_{i,i}} \prod_{j=0}^{i-1} \frac{1 - \dfrac{k_{i,j} r_j}{k_{W,j} \Delta P_j [(2 - 2r_j)/(2 - r_j)] + k_{i,j}}}{1 - r_j} \right]}{\displaystyle\sum_{i=1}^{n} A_i k_{W,i} \Delta P_i} \tag{9.41}$$

$$C_{p,\text{system}} = \frac{C_f \displaystyle\sum_{i=1}^{n} \left(A_i k_{i,i} \Delta P_i Z_i \prod_{j=0}^{i-1} X_j \right)}{\displaystyle\sum_{i=1}^{n} A_i k_{i,i} \Delta P_i} \tag{9.42}$$

In Eq. (9.42), the term $\prod_{j=0}^{i-1} X_j$ is the product of the concentration factors of all stages before stage i. The concentration factor is therefore mathematically defined as 1 before stage 1 ($X_O = 1$) (Duranceau, 1992).

The exponent J/k_b used in the film theory model, as shown in Eq. (9.4), can be approximated by a constant times the ratio of the flow through the membrane to the flow across the membrane, which is shown in Eq. (9.43). This is an approximation which compensates for the effects of multisolute solutions in a high-recovery or plant application.

$$\frac{J}{k_b} = k \left(\frac{r_e}{1 - r_e} \right) \tag{9.43}$$

where k_b = statistically derived coefficient
 r_e = average membrane recovery per element

The system permeate equation can also be modified to incorporate the film theory model as shown in Eqs. (9.44) and (9.45).

$$C_p = \frac{k_i e^{J/k_b} C_f}{k_W(\Delta P - \Delta \Pi) [(2 - 2r)/(2 - r)] + k_i e^{J/k_b}} = Z_{i,\text{cp}} C_f \tag{9.44}$$

where $Z_{i,\text{cp}}$ = modified film theory feed-stream mass transfer coefficient

$$C_{p,\text{system}} = \frac{C_f \sum_{i=1}^{n} \left(A_i k_{W,i} \Delta P_i Z_{i,\text{cp}} \prod_{j=0}^{i-1} X_{j,\text{cp}} \right)}{\sum_{i=1}^{n} A_i k_{W,i} \Delta P_i} \tag{9.45}$$

where $X_{j,\text{cp}}$ = modified film theory concentrate stream concentration factor

9.4.4 Array Design

The conventional membrane system shown in Fig. 9.6 consists of a membrane pre-treatment system, membrane arrays, and a posttreatment process. The membranes are typically described as elements and are defined in Table 9.2 with several basic terms that represent common membrane terminology. Unfortunately, these terms are not universally accepted and may have different meanings. For example, a membrane array can be defined as stage 1, or as the total of stages 1, 2, and 3. Care must be taken to ensure that parties discussing membrane systems are using the same membrane terminology.

9.4.4.1 Simple Sizing.   A simplistic application of a 4-2-1 array is shown in the following example to illustrate design of membrane configurations. Basic assumptions in this design are 8-in (20-cm) membrane elements with a 350-ft^2 (32.5-m^2) surface area operating at a flux of 15 gsfd (0.7 m^3/m^2·day) and an 85 percent recovery. One element will produce 5250 gpd. As seven elements are in each pressure vessel and seven pressure vessels are in one array, an array will produce approximately 250,000 gpd. Four such arrays would provide 1 mgd of permeate and 0.18 mgd of concentrate. Osmotic pressure has not been illustrated and will continuously decrease the recovery in each succeeding element.

EXAMPLE 9.4   *Assume:*

1. *Membrane element flux = 15 gal/ft^2 day (0.7 m/s)*
2. *Designing a 4-2-1 array*
3. *Each pressure vessel can contain 6 to 8 elements*
4. *Element surface area = 350 ft^2 (32.5 m^2)*
5. *Recovery = 85%*

Water 1 element can produce:

$Q_p = JA = $ (15 gal/ft^2·day)(350 ft^2/element)

$\qquad\qquad$ = 5250 gal/day·element (22.75 m^3/day·element)

Number of elements on 1 array:

$\qquad\qquad$ (7 PV/array)(7 elements/PV) = 49 elements/array

Gallons per day produced per array:

(5250 gal/day·element)(49 elements/array)

$\qquad\qquad$ = 257,250 gal/train·day (1115.75 m^3/train·day)

Trains needed to supply 1 mgd (3786 m³/day):

$$(10^6 \text{ gal/day})(\text{array day}/257{,}250 \text{ gal}) = 3.9 \text{ arrays}$$

Total number of elements needed:

$$(4 \text{ arrays})(49 \text{ elements/array}) = 196 \text{ elements}$$

Flow in and out of 1 array:

$$Q_f = \frac{Q_p}{r} = \frac{1 \text{ mgd}}{0.85} = 1.18 \text{ mgd } (4453 \text{ m}^3/\text{d})$$

$$\text{Flow in per array} = \frac{1.18 \text{ mgd}}{4} = 0.30 \text{ mgd} = 205 \text{ gpm } (0.78 \text{ m}^3/\text{min})$$

$$\text{Flow out per array} = \frac{1 \text{ mgd}}{4} = 0.25 \text{ mgd} = 174 \text{ gpm } (0.66 \text{ m}^2/\text{min})$$

9.4.4.2 Pressure and Flow Determination. The pressure and flow for a membrane array or system can be estimated for design by using the average element flux and system recovery. The first steps in determining the operating pressure and feedstream flow of an RO or NF plant is to determine the water produced in each element by using flux and membrane area, as shown in Eq. (9.46).

$$Q_{p-e} = JA_e \tag{9.46}$$

where Q_{p-e} is the permeate flow rate produced by a single element, J is the system design flux, and A_e is the area of a single element.

The permeate flow from each pressure vessel can be determined by multiplying the number of elements per pressure vessel by the permeate flow produced by a single element using Eq. (9.46).

$$Q_{p-v} = N_e Q_{p-e} \tag{9.47}$$

where Q_{p-v} is the permeate flow rate produced by a pressure vessel and N_e is the number of elements in one pressure vessel.

The permeate flow from each stage is calculated by multiplying the number of pressure vessels in each stage by the flow produced by a single pressure vessel, using Eq. (9.48).

$$Q_{p-s} = N_v Q_{p-v} \tag{9.48}$$

where Q_{p-s} is the permeate flow rate produced by a stage, and N_v is the number of pressure vessels in the stage.

The stage permeate flows can be added together to calculate the permeate flow produced by the system, using Eq. (9.49).

$$Q_{p-\text{sys}} = \sum_{i=1}^{n} Q_{p-s(i)} \tag{9.49}$$

where $Q_{p-\text{sys}}$ is the permeate flow rate produced by the system which is equal to the sum of the permeate flows from all stages in a system with n stages.

The required feed-stream flow to a pressure vessel, a stage, an array, or a system can now be estimated by dividing the associated permeate flow by the associated recovery, as shown in Eq. (9.50). As previously described, equal flow distribution among pressure vessels is assumed in a given stage.

$$Q_{F-(i)} = \frac{Q_{P-(i)}}{R_{(i)}} \tag{9.50}$$

where $Q_{F-(i)}$ is the feed-stream flow to either a pressure vessel (Q_{F-V}), a stage (Q_{F-s}), or the system (Q_{F-sys}). $Q_{p-(i)}$ is the permeate produced by the pressure vessel (Q_{p-v}), stage (Q_{p-v}), or system (Q_{p-sys}) and $R_{(i)}$ is the respective recovery for either the pressure vessel, stage, or system in decimal fraction.

The concentrate waste flow rate from the system is calculated as the difference between the feed flow rate to the system and the permeate flow produced by the system, as shown in Eq. (9.51).

$$Q_{W-sys} = Q_{F-sys} - Q_{p-sys} \tag{9.51}$$

where Q_{W-sys} is the concentrate waste flow rate from the system.

The manufacturer's recommended maximum and minimum element flows can be used to determine if recycle is required to overcome a minimum element flow violation or if the feed flow must be reduced. The need for recycle is made by determining the flow into the last element of a pressure vessel for each stage. The last element is used because the lowest feed flow is delivered to the last element of a pressure vessel because of the reduction due to the removal of permeate in the previous elements in the series, as shown in Eq. (9.52).

$$Q_{F-le} = Q_{F-V} - (N_e - 1)Q_{p-e} \tag{9.52}$$

Where Q_{F-le} is the feed-stream flow into the last element of a pressure vessel, Q_{F-V} is the feed flow into the respective pressure vessel, and Q_{p-e} is the permeate produced per element in the pressure vessel preceding the last element.

The required recycle flow is determined by the positive difference between the minimum manufacturer's recommended flow and the flow entering the last pressure vessel, Q_{F-le}, as shown in Eq. (9.53). If the difference is negative, the flow into the last element is greater than the minimum required flow and meets this criterion. This determination must be made for each stage in the array; however, the most critical element is typically the last element in the first stage.

$$Q_{Rmin-V(i)} = Q_{Fmin-e} - Q_{F-le} \tag{9.53}$$

where $Q_{R-V(i)}$, if positive, is the recycle flow required to meet the minimum feed flow Q_{Fmin-e} recommended by the manufacture.

The minimum system influent stream flow $Q_{Imin-sys}$ can be determined by adding the feed flow Q_{F-sys} based on element flux and recovery to the required recycle flow for each stage $Q_{R-S(i)}$, as shown in Eq. (9.54).

$$Q_{Imin-sys} = Q_{F-sys} + \sum_{i=1}^{n} Q_{R-S(i)} \tag{9.54}$$

The flow into the first element of each stage must also be compared to the maximum element flow specified by the manufacturer Q_{Imax-e}. Equation (6.10) determines the maximum feed-stream flow into a stage $Q_{Imax-s(i)}$ based on the manufacturer's maximum element flow, which is equally divided into each pressure vessel of stage i, $N_{V-s(i)}$.

Equation (9.55) can be used to determine the maximum feed flow into each preceding stage $Q_{I\max - s(n-1)}$. The subscript n refers to the last stage in the membrane system. Consequently, the maximum feed-stream flow is determined by working backward through the membrane array. The feed-stream flow into stage n is the concentrate stream flow from stage $n - 1$ and that equal distribution of feed-stream flow among stage pressure vessels is assumed.

$$Q_{I\max - s(i)} = Q_{I\max - e} N_{V - s(i)} \tag{9.55}$$

$$\frac{Q_{I\max - s(n-1)} \le Q_{I\max - s(n)}}{1 - R_{S(n-1)}} \tag{9.56}$$

Where $R_{S(n-1)}$ is the recovery of the preceding stage.

The permissible system influent flow range can then be determined from the minimum and maximum permissible flows and must fall between the flows determined by Eqs. (9.54) and (9.56).

Equation (9.57) calculates the influent flow rate to each stage as the sum of the feed and recycle flow rates to the stage, as calculated by Eq. (9.50), and the recycle flow rate.

$$Q_{I - s} = Q_{F - s} + Q_R \tag{9.57}$$

where $Q_{I - s}$ is the influent flow rate to a stage and Q_R is the recycle flow rate. If the recycle flow rate for a stage is zero, then the influent flow rate is equal to the feed flow rate for that stage. The recycle flow rate should be added to the feed flow rate of every stage downstream of the point where the recycle stream is blended with the feed stream. For example, if recycled concentrate is blended with feedwater prior to the first stage, the recycle flow rate should be added to the feed flow rate of each stage in the system to obtain the influent flow rate to each stage.

Equation (9.58) is used to calculate the recycle ratio r.

$$r = \frac{Q_R}{Q_F} \tag{9.58}$$

The required feed-stream pressure can be determined once the osmotic pressure gradient for the system is estimated and hydraulic losses through stage hardware and membrane elements are accounted for. The osmotic pressure gradient is estimated from the TDS of the system influent, waste and permeate streams, as shown in Eq. (9.59).

$$\Delta\pi_{sys} = \left[\left(\frac{TDS_I + TDS_W}{2} \right) - TDS_p \right] \times 0.01 \tag{9.59}$$

where $\Delta\pi_{sys}$ is an estimate of the average osmotic pressure gradient for the system in lb/in^2, TDS_I is the influent TDS concentration in mg/L, TDS_W is the concentrate waste TDS concentration in mg/L, TDS_p is the permeate TDS concentration in mg/L, and 0.01 is an approximation factor for converting TDS (mg/L) to pressure (lb/in^2).

Once the osmotic pressure has been estimated, the flux equation as shown in Eq. (9.60) can be used to estimate the feed-stream pressure P_F. This pressure is the pressure that the high-pressure pump must supply to drive the membrane system. The first stage losses must be considered in the second stage because P_F is the feed-stream pressure coming into the system. P_F is reduced in the first stage by the first-stage losses and those losses must be deducted from P_F when it appears in the second stage. Consequently, the loss term in each stage must be a summation of all the losses in the preceding stages in addition to the losses in that stage.

$$J = K_W \times \sum_{i=1}^{n} \Gamma_{(i)} \left\{ \left[\frac{2P_F - (2 \times i - 1) \times L}{2} \right] - P_p - \Delta\Pi_{(i)} \right\} \qquad (9.60)$$

$$\Gamma_i = \frac{Q_{p-s(i)}}{Q_{p-sys}} \qquad (9.61)$$

$$L = P_e + N_{el} \times P_{el} \qquad (9.62)$$

The array flow weighted factor Γ_i as shown in Eqs. (9.60) and (9.61) is the permeate stream flow produced by an individual stage divided by the total permeate stream flow produced by the system. If the numbers of elements in each pressure vessel and element flux in all stages are equal, then Γ_i is just the ratio of the pressure vessels in a given stage to the total pressure vessels in the system. In a 2-1 array with the same number of elements and element flux, Γ_i would be two-thirds and Γ_i would be one-third. K_W is the membrane mass transfer coefficient in gsfd/lb/in². As shown in Eq. (9.60), L in lb/in² is the sum of the mechanical head losses which consists of the per-element pressure loss P_{el} in lb/in² times the number of elements in the pressure vessel N_{el} as well, the pressure vessel and stage entrance and exit losses $P_{e(j)}$ in lb/in². $\Delta\Pi_{(i)}$ is the osmotic pressure for each stage, and P_p is the desired permeate pressure for downstream transmission.

The following example demonstrates the determination of pressure and flow requirements for a multistage NF membrane system. The flow information determined in the previous example for simple sizing is used in the example calculations.

EXAMPLE 9.5 *Assumptions:*

Parameter	Criteria
No. of stages	3
No. of pressure vessels per stage	4-2-1
No. of elements per pressure vessel	7
% recovery for system	85%
Design flux	15 gsfd
Surface area per element	350 sf
Mass transfer coefficient for water	0.30 gsfd/lb/in²
Maximum and minimum element flow	96 gpm, 24 gpm
Energy loss per element	3 lb/in²
Energy loss in stage entrance and exit	5 lb/in²
Feed-stream TDS	300 mg/L
% TDS rejection	70%

The water produced per element, pressure, array, and system can be determined from the average flux, number of elements, and array configuration.

Once the pilot plant production is determined, the need for recycle can be determined from the minimum and maximum flow restrictions for the element. This solution is based on the assumption that each pressure vessel will produce the same permeate flow and will make an equal contribution to the overall system recovery.

Determine feed stream flow based on recovery

$$\text{System recovery for first stage} = \frac{0.85}{7PV} \cdot \frac{4PV}{\text{stage 1}} = 0.486 = 48.6\%$$

$$\frac{174 \text{ gpm}}{4 + 2 + 1} = 24.9 \text{ gpm, water produced per PV}$$

$$\frac{24.9}{7} = 3.56 \text{ gpm/element}$$

$$Q_f = \frac{Q_{pv}}{R_{s1}} = \frac{24.9}{0.486} = 51.2 \text{ gpm/day PV}$$

Determine flow into last element in PV

$$Q_{fle} = Q_{fpv} - \sum_{i=1}^{n-1} Q_{pe_i}$$

$$Q_{fle} = 51.2 - 3.56 \cdot 6 = 29.9 \text{ gpm}$$

Determine if recycle is required for each pressure vessel

$$Q_{\text{min recycle}} = Q_{\text{mfg min}} - Q_{fle}$$

$$Q_{\text{min recycle}} = 24 \text{ gpm} - 29.9 \text{ gpm} = -5.9 \text{ gpm, no recycle is required}$$

$$Q_{\text{max recycle}} = Q_{\text{mfg max}} - Q_f$$

$$Q_{\text{max recycle}} = 96 \text{ gpm} - 51.2 \text{ gpm} = 44.8 \text{ gpm/PV} > 0, \text{ maximum flow is acceptable}$$

Determine stage 2 recycle

$$Q_{s2 \text{ feed}} = Q_{s1 \text{ conc}}$$

$$Q_{s2 \text{ feed}} = Q_{s1 \text{ feed}} - Q_{s1 \text{ permeate}}$$

$$Q_{s2 \text{ feed}} = (51.2 \text{ gpm} - 24.9 \text{ gpm})(4) = 105.2 \text{ gpm}$$

Adjusting for 2 PV in stage 2

$$Q_{fle, \, s2} = \frac{105.2}{2} - 3.56 \cdot 6 = 31.3 \text{ gpm}$$

$$Q_{\text{min recycle}} = 24 \text{ gpm} - 31.3 \text{ gpm} = -7.3 \text{ gpm} = 0 \text{ gpm/PV, no recycle is required}$$

$$Q_{\text{max recycle}} = 96 \text{ gpm} - \frac{105.2}{2} \text{ gpm} = 43.4 \text{ gpm/PV} > 0, \text{ maximum flow is acceptable}$$

Determine stage 3 recycle

$$Q_{s3 \text{ feed}} = Q_{s2 \text{ conc}}$$

$$Q_{s3 \text{ feed}} = Q_{s2 \text{ feed}} - Q_{s2 \text{ permeate}}$$

$$Q_{s2 \text{ feed}} = (52.6 \text{ gpm} - 24.9 \text{ gpm})(2) = 55.4 \text{ gpm}$$

$$Q_{fle, \, s3} = 55.4 - 3.56 \cdot 6 = 34.1 \text{ gpm}$$

$Q_{\text{min recycle}} = 24 \text{ gpm} - 34.1 \text{ gpm} = -10.1 \text{ gpm} = 0 \text{ gpm/PV, no recycle is required}$

$Q_{\text{max recycle}} = 96 \text{ gpm} - 55.4 \text{ gpm} = 40.6 \text{ gpm/PV} > 0, \text{ maximum flow is acceptable}$

Determine array recycle

$$Q_{\text{min} \times \text{recycle}} = 0 \text{ gpm}$$

$$Q_{\text{max recycle}} = 96 \cdot 4 - 205 = 179 \text{ gpm or } 44.8 \text{ gpm/PV}$$

Determine minimum array feed

$$Q_{\text{system}} = 205 \text{ gpm or } 51.2 \text{ gpm/PV}$$

Determine maximum array feed

$$Q_{\text{system}} = 96 \cdot 4 = 384 \text{ gpm or } 96 \text{ gpm/PV}$$

These calculations show that, for the given flux, recovery, and array, the input flow for each array can be between 205 and 384 gpm. Once the system flow has been determined, the pressure requirements can be determined for the membrane system using the basic flux equation on a flow weighted basis and accounting for the entrance and exit, element, and osmotic pressure energy losses. Osmotic pressure is estimated based on single element per pressure vessel. This assumption will be very close to the osmotic pressure of seven elements per pressure vessel.

Accounting for osmotic pressure:

$$\Delta\pi = \left(\frac{C_{f\text{TDS}} + C_{c\text{TDS}}}{2} - C_{p\text{TDS}}\right)(0.01 \text{ lb/in}^2/\text{mg·L})$$

$$C_{c\text{TDS}} = \frac{Q_f C_{f\text{TDS}} - Q_p C_{p\text{TDS}}}{Q_c}$$

$$C_{P1} = 300 \text{ mg/L} (1 - 0.7) = 90 \text{ mg/L}$$

$$C_{C1} = \left[\frac{(300)(1) - (0.486)(90)}{0.514}\right] = 499 \text{ mg/L}$$

$$\Delta\pi_1 = \left(\left(\frac{300 + 499}{2} - 90\right)\text{mg/L}\right)(0.01 \text{ lb/in}^2/\text{mg·L}) = 3.1 \text{ lb/in}^2$$

$$C_{P2} = 499 \text{ mg/L} (1 - 0.7) = 150 \text{ mg/L}$$

$$C_{c2} = \frac{(0.514)(499) - (0.243)(0.30)(499)}{0.514 - 0.243} = 812 \text{ mg/L}$$

$$\Delta\pi_2 = \left(\left(\frac{499 + 812}{2} - 150\right)\text{mg/L}\right)(0.01 \text{ lb/in}^2/\text{mg·L}) = 5.1 \text{ lb/in}^2$$

$$C_{P3} = 812 \text{ mg/L} (1 - 0.7) = 244 \text{ mg/L}$$

$$C_{c3} = \frac{(0.27)(812) - (0.12)(0.30)(812)}{0.27 - 0.12} = 1266 \text{ mg/L}$$

$$\Delta\pi_3 = \left(\left(\frac{812 + 1266}{2} - 244\right)\text{mg/L}\right)(0.01 \text{ lb/in}^2/\text{mg·L}) = 8.0 \text{ lb/in}^2$$

Solving for feed-stream pressure

$$J_w = \sum_{i=1}^{n} \Gamma_i K_{w_i} \left(\frac{2P - 2(i-1)L_i - L_i}{2} - P_{sys} - \Delta\Pi_i \right)$$

$$L_1 = L_2 = 7E\cdot3 \, \frac{\text{lb/in}^2}{E} + 5 \text{ lb/in}^2 + 5 \text{ lb/in}^2 = 31.0 \text{ lb/in}^2$$

$$15 \text{ gsfd} = 0.30 \text{ gsfd/lb·in}^2 \left(\frac{4}{7}\right)\left(\frac{2P - 31 \text{ lb/in}^2}{2} - 30 \text{ lb/in}^2 - 3.1 \text{ lb/in}^2 \right)$$

$$+ 0.30 \text{ gsfd/lb·in}^2 \left(\frac{2}{7}\right)\left(\frac{2P - 2\cdot31 \text{ lb/in}^2 - 31 \text{ lb/in}^2}{2} - 30 \text{ lb/in}^2 - 5.1 \text{ lb/in}^2 \right)$$

$$+ 0.30 \text{ gsfd/lb·in}^2 \left(\frac{1}{7}\right)\left(\frac{2P - 4\cdot31 \text{ lb/in}^2 - 31 \text{ lb/in}^2}{2} - 30 \text{ lb/in}^2 - 8 \text{ lb/in}^2 \right)$$

$$15 \text{ gsfd} = 0.17 \text{ gsfd/lb·in}^2 \left(\frac{2P - 31 \text{ lb/in}^2}{2} - 33.1 \text{ lb/in}^2 \right)$$

$$+ 0.085 \text{ gsfd/lb·in}^2 \left(\frac{2P - 93 \text{ lb/in}^2}{2} - 35.1 \text{ lb/in}^2 \right)$$

$$+ 0.043 \text{ gsfd/lb·in}^2 \left(\frac{2P - 155 \text{ lb/in}^2}{2} - 38 \text{ lb/in}^2 \right)$$

$$15 \text{ gsfd} = 0.17 \text{ gsfd/lb·in}^2 \, (P - 48.6 \text{ lb/in}^2) + 0.085 \text{ gsfd/lb·in}^2(P - 81.6 \text{ lb/in}^2)$$
$$+ 0.043 \text{ gsfd/lb·in}^2 \, (P - 115.5 \text{ lb/in}^2)$$

$$P = \frac{15\text{gsfd} + (8.26 + 6.94 + 4.97)\text{gsfd}}{0.30 \text{ gsfd/lb·in}^2} = 117.2 \text{ lb/in}^2$$

The systems design for the NF pilot plant has been completed. The high-pressure pump should be selected so that optimum operation occurs at approximately 117 lb/in² and from 205 to 384 gpm if there is one pump per array. The remaining design criteria have been shown earlier. However, this is a very basic illustration of sizing an RO/NF pilot plant. It may be desirable to use flow restriction on the permeate or concentrate lines or to use an interstage booster pump to increase flux. The flux will decrease in each element in a given pressure vessel due to increasing hydraulic and osmotic pressure losses which have not been accounted for on an element basis but on a pressure vessel basis in this example. The mass transfer coefficient would have to be adjusted for temperature if the operating temperature is different than the temperature referenced for the mass transfer coefficient.

9.4.4.3 Determination of Permeate Water Quality. The film theory model can be used to estimate permeate water quality. In the following example, MTCs were assumed to be determined from a pilot study or given by the membrane manufacturer. The concentration-polarization factor is also given. The example calculation shows the chloride concentrate in the permeate stream from each stage and the array [Hofman et al., 1993].

EXAMPLE 9.6   *Given:*   *A 2-1 array with 6 elements/pv, pv surface area = 30 m² (323 ft²), 5-bar driving force, $k_w = 1.46 \times 10^{-6}$ m/s − bar(0.2135 gsfd/lb/in²), $k_i = 1.87 \times 10^{-6}$ m/s(0.53 ft/day), k' = 0.75, r = 0.5 stage, $r_e = 0.083$/element*

Using the array film theory model:

$$C_{p,\text{system}} = \frac{C_f \sum\limits_{i=1}^{n} \left(A_i k_{w,i} \Delta P_i Z_{i,\text{cp}} \prod\limits_{j=0}^{j-1} X_{j-1,\text{cp}} \right)}{\sum\limits_{i=1}^{n} A_i k_{w,i} \Delta P_i}$$

$\sum\limits_{i=1}^{2} A_i k_{w,i} \Delta P_i$ = stage 1 (2 vessels/stage 1)(6 elements/p. vessel)(30 m²/element)

$\times$ (1.46 $\times$ 10⁻⁶ m/s·bar)(5 bar) + stage 2(1 vessel/stage 2)(6 elements/p. vessel)

$\times$ (30 m²/element)(1.46 $\times$ 10⁻⁶ m/s·bar)(5 bar) = 2.628 $\times$ 10⁻³ m³/s + 1.324 $\times$ 10⁻³ m³/s

$$= 3.942 \times 10^{-3} \text{ m}^3/\text{s} \ (89{,}999 \text{ gal/day})$$

$$Z_{i,\text{cp}} = \frac{k_{s,i} e^{k[r_e/(1-r_e)]}}{k_{w,i}\Delta P_i[(2-2r_i)/(2-r_i)] + e^{k[r_e/(1-r_e)]} k_{s,i}} = Z_1 = Z_2$$

$$= \left(\frac{(1.87 \times 10^{-6} \text{ m/s}) e^{0.75[0.083/(1-0.083)]}}{(1.46 \times 10^{-6} \text{ m/s·bar})(5 \text{ bar})\left[\dfrac{2-2(0.5)}{2-0.5}\right] + (1.87 \times 10^{-6} \text{ m/s}) e^{0.75[0.083/(1-0.083)]}} \right)$$

$$= 0.29$$

$$\prod_{j=0}^{1} X_j = (1)\left(\frac{1-Z_1 R_1}{1-R_1}\right) = (1)\left(\frac{1-(0.29)(0.50)}{1-0.5}\right) = 1.71$$

$C_{p,\text{system}}$

$$= \frac{100 \text{ mg/L } (2.628 \times 10^{-3} \text{ m}^3/\text{s})(0.29) + 100 \text{ mg/L } (1.314 \times 10^{-3} \text{ m}^3/\text{s})(0.29)(1.71)}{5.4 \times 10^{-3} \text{ m}^3/\text{s}}$$

= 100 mg/L (0.67)(0.29) + 171 mg/L (0.33)(0.29)

= 29.0 mg/L (0.67) + 49.6 mg/L (0.33) = 19.4 mg/L + 16.4 mg/L = 35.8 mg/L

The following comments are for clarification of model application. The five-bar feed-stream pressure gradient through each stage has been assumed to include the effects of osmotic pressure and the hydraulic losses. These effects can be estimated by assuming that 1 mg/L TDS is assumed to produce 0.7 millibars of osmotic pressure, a 0.20-bar pressure loss across each element, and a 0.34-bar stage entrance loss. The hydraulic effects on pressure gradient are accounted for during design and are typically controlled by high-pressure pumps and flow restrictors. The first- and second-stage Z terms are identical because stage recovery, pressure gradient, and membranes are identical. The Z_1 term is used to calculate the X_2 term as the stage-1 concentrate stream is the stage-2 feed stream. The denominator of Eq. (9.30) is the total permeate flow produced by the two-stage system and is shown as the first calculation in the example application. The final calculation is the calculation of the system permeate Cl⁻ concentration, which was estimated to be 35.8 mg/L. The stage-1 and -2 Cl⁻ concentrations are 29.0 and 49.6 mg/L, which are shown in the calculation of the flow-weighted system concentration. Since recovery, pressure, and the membranes are identical from stage 1 to stage 2, the second-stage permeate stream Cl⁻

concentration will always be higher than the first stage because the feed-stream Cl^- concentration is higher in stage 2 (171 mg/L) than in stage 1 (100 mg/L).

9.4.4.4   Manufacturers' Programs.   Dow-FilmTec, Hydranautics, Fluid System, and TriSep are manufacturers who provide computer programs for developing RO plant design criteria. These programs are not final design specifications and no manufacturer accepts responsibility for their use by anyone not authorized specifically by the manufacturer. These programs are only a tool to develop and test various system configurations and should not be regarded as complete. The programs do provide a means of estimating water production and quality from given parameters. The following is an example of a Dow-FilmTec computer program that was used to develop an initial design of an RO array. Feed flow rate, TDS concentration, feed temperature, recovery, array type (i.e., 4-2-1 array), number of elements per pressure vessels, and element type are required program inputs. The program outputs pressure, flux, flow, and water quality for all process streams by stage and element for each array. Scaling calculations are provided for calcium sulfate. See Fig. 9.14.

9.4.5   Posttreatment

Posttreatment consists of several different unit operations. The choice and the sequence of these unit operations are functions of the designer's choice and water quality. The primary posttreatment unit operations are sulfide removal, alkalinity recovery, aeration, disinfection, and stabilization. A systems view of posttreatment can be helpful to realize what a designer may wish to accomplish. The membrane process at this point has removed essentially all of the pathogens and the majority of the DBP precursors, salts, and other solutes in the feed stream. Solute removal has included carbonate alkalinity; however, all dissolved gases including carbon dioxide and hydrogen sulfide have passed through the membrane. The designer has to take care not to produce a finished water after posttreatment that has almost no alkalinity and contains significant sulfur turbidity as well as meeting disinfection requirements. The following example of posttreatment uses the permeate stream water quality illustrated in the previous membrane section. The water quality resulting from each posttreatment unit operation is illustrated in Table 9.4.

The changes in water quality will be discussed in the following sections. The sequence of unit operations assumed here is H_2S removal/disinfection, alkalinity recovery, and aeration/stabilization. There are different sequences of posttreatment but this sequence has the advantage of minimizing equipment, as H_2S removal/disinfection and aeration/stabilization are conducted simultaneously in separate unit operations.

9.4.5.1   Hydrogen Sulfide Removal and Disinfection.   Many of the groundwaters used for feed streams to reverse osmosis or nanofiltration plants contain hydrogen sulfide. Neither the conventional pretreatment process (microfiltration, acid, or antiscalant addition) nor the membrane process will remove hydrogen sulfide. Aeration and oxidation are the two primary means of removing hydrogen sulfide; however, the involved chemical reactions are not well defined. An often neglected problem in these hydrogen sulfide removal processes is the formation of elemental sulfur. Both entrained oxygen and chlorine can react with hydrogen sulfide to form elemental sulfur, as shown here.

FILMTEC RO SYSTEM ANALYSIS, SEPT 93 VERSION

PREPARED FOR: AWWARF
PREPARED BY: J.M.H.
DATE: 8-2-94

FEED:	204.71	1933 MG/L	25 DEG C
RECOVERY:	85%		

ARRAY:	1	2	3
NO. OF PV:	4	2	1
ELEMENT:	SW8040	SW8040	SW8040
NO. EL/PV:	7	7	7
EL. TOTAL:	28	14	7

FOULING FACTOR: 0.85

	FEED	REJECT	AVERAGE
PRESSURE (PSIG)	355.0	248.1	311.2
OSMOTIC PRESSURE (PSIG)	18.7	122.9	51.2

NDP(MEAN) = 260.1 PSIG
AVERAGE PERMEATE FLUX =   17.3 GFD   PERMEATE FLOW = 174.05 GPM

ARRAY	EL. NO.	RECOVERY	PERMEATE GPD	MG/L	FEED GPD	MG/L	P (PSIG)
1	1	0.089	6531	11	51.2	1933	350
	2	0.094	6337	12	46.6	2120	343
	3	0.101	6155	14	42.2	2339	336
	4	0.109	5978	17	38.0	2601	331
	5	0.119	5801	20	33.8	2918	327
	6	0.131	5614	24	29.8	3310	323
	7	0.145	5404	29	25.9	3805	320
2	1	0.081	5150	33	44.3	4445	313
	2	0.084	4938	38	40.7	4833	307
	3	0.088	4728	44	37.3	5274	302
	4	0.092	4514	51	34.0	5780	298
	5	0.097	4292	61	30.8	6362	294
	6	0.101	4058	73	27.9	7036	288
	7	0.106	3807	90	25.0	7819	288
3	1	0.055	3549	101	44.8	8731	281
	2	0.054	3322	115	42.3	9234	275
	3	0.054	3107	131	40.0	9759	269
	4	0.053	2897	147	37.9	10307	264
	5	0.052	2694	165	35.9	10877	260
	6	0.051	2496	185	34.0	11467	255
	7	0.050	2304	209	32.3	12073	252

ARRAY:	TOTAL	1	2	3
REJECT GPM:		88.5	44.8	30.7
REJECT MG/L:		4445	8731	12692
PERM GPD:	250627	167283	62975	20369
PERM MG/L:	37	18	54	145

FIGURE 9.14

PERMEATE, MG/L AS ION

NH$_4$	0.0	0.0	0.0	0.0
K	0.3	0.1	0.5	1.3
Na	8.5	3.9	12.3	34.0
Mg	0.6	0.3	0.8	2.1
Ca	4.1	2.1	6.0	15.3
HCO$_3$	0.8	0.4	1.2	3.3
NO$_3$	1.5	0.7	2.2	6.5
Cl	19.7	9.4	28.5	76.7
F	0.0	0.0	0.0	0.0
SO$_4$	1.7	0.9	2.5	6.2
SiO$_2$	0.0	0.0	0.0	0.0

FEED/REJECT, MG/L AS ION

NH$_4$	0.0	0.0	0.0	0.0
K	10.0	22.9	44.8	64.9
Na	343.0	788.0	1545.1	2242.2
Mg	40.0	92.1	181.2	263.9
Ca	300.0	690.9	1359.5	1979.6
HCO$_3$	28.7	65.8	128.9	186.8
NO$_3$	15.0	33.8	64.6	91.5
Cl	1056.0	2429.2	4772.5	6938.9
F	0.0	0.0	0.0	0.0
SO$_4$	139.9	322.4	634.6	924.6
SiO$_2$	0.0	0.0	0.0	0.0

TO BALANCE: 4.9 MG/L SODIUM AND 0.0 MG/L CHLORIDE ADDED TO FEED
FEED WATER IS WELL OR SOFTENED WATER (BW) SDI < 3

SCALING CALCULATIONS

	FEED	ACIDIFIED FEED	REJECT
HCO$_3$ (MG/L)	193.8	28.7	186.8
CO$_2$ (MG/L)	10.0	129.2	129.2
SO$_4$ (MG/L)	10.0	139.9	924.6
Ca (MG/L)	300.0	300.0	1979.6
TDS (MG/L)	1967.8	1932.6	12692.4
pH	7.50	5.56	6.37
LSI	0.63	−2.14	0.26
IONIC STRENGTH (MOLAL)	0.043	0.044	0.289
IP CaSO$_4$ (SQ. MOLAR)	0.78E-06	0.11E-04	0.48E-03

IP CaSO$_4$ AT SATURATION (SQ. MOLAR)   = 0.57E-03
SULFURIC ACID DOSING (MG/L, 100%)   = 132.6
TEMPERATURE (DEGREE C)   = 25.0
RECOVERY (PERCENT)   = 85.0
ESTIMATED PERMEATE pH IS 4.0
SULFURIC ACID CONSUMPTION (100% CONCENTRATION) = 147.8 KG/DAY

FIGURE 9.14   (*Continued*)

TABLE 9.4 Posttreatment Water Quality Changes by Unit Operation

	Permeate, mg/L	Disinfection H_2S destruction, mg/L	Alk recovery, mg/L	Aeration, mg/L
pH	4	3.43	7.6	7.6
H_2CO_3	129	130	6	0.6
HCO_3^-	0.8	0	122	122
H_2S as S	1	0	0	0
SO_4^{-2}	1.7	4.7	4.7	4.7
Cl^-	19.7	30.1	30.1	30.1
Ca^{+2}	4.1	4.1	50.9	50.9
TDS	34	51	220	220
DO	0	0	0	8.24 (25°C)
Cl_2	0	3	3	3

$$2H_2S + O_2 \Rightarrow 2S + 2H_2O$$

$$2H_2S + O_2 + 2H_2O \Rightarrow 2H_2SO_4$$

$$H_2S + Cl_2 \Rightarrow S + 2HCl$$

$$H_2S + 4Cl_2 + 4H_2O \Rightarrow H_2SO_4 + 8HCl$$

The pK for hydrogen sulfide is 7, and H_2S gas can essentially be completely *removed* at pHs below 6.5 without turbidity formation in an air-stripping process. Aeration or draft aeration, however, involves much less surface area for air stripping, and elemental sulfur can form. Elemental sulfur produced by biological or chemical oxidation is commonly seen on aeration towers at water plants treating a sulfide containing groundwater.

While the air-stripping process generally will avoid sulfur formation, all of the CO_2 is also lost during the volatilization process. Consequently, unless a carbonate salt is added or a significant amount of alkalinity passes through the membrane, there will be no carbonate buffering in the finished water. The final pH can be increased by the addition of sodium hydroxide but the finished water will have a very low buffering capacity and will be corrosive. Some nanofilters may pass significant amounts of bicarbonate but reverse-osmosis membranes typically reject more than 98 percent of the bicarbonate ions. If a nanofilter passes enough bicarbonate, then air stripping can be used to remove hydrogen sulfide. The pH can be reduced prior to air stripping with carbon dioxide to avoid alkalinity destruction. If the membrane does not pass adequate alkalinity, the chlorination at low pH is possible. Elemental sulfur has been shown not to form during the chlorination of hydrogen sulfide if the reaction pH is less than 3.7. Acid addition in the feed stream is common to avoid calcium carbonate scaling permeate stream, and pH is commonly between 5.0 and 6.0. The chlorination of H_2S will produce enough protons to lower the pH to 3.7 on most occasions. However, increased acid addition to the feed stream or direct addition to the permeate stream to achieve a permeate stream pH of 3.7 can be used. The advantage of direct feed-stream addition is that fewer sulfate ions (reduced corrosivity) are in the finished water, but the disadvantage is higher cost relative to permeate stream addition. The permeate-stream pH is lower than the feed-stream pH because the bicarbonate ions remaining after acid addition are rejected by the membrane and the dissolved carbon dioxide passes through the membrane. The amount of acid to add can be determined by determining the amount of acid to completely destroy the alkalinity plus the amount of acid to reduce the pH from the point of complete alkalinity destruction (pH 4.5 to 4.2) to pH minimum 3.7. The amount of acid to add to the permeate stream can be estimated using the equations as follows.

$$2HCO_3^- + H_2SO_4 \Rightarrow 2H_2CO_3 + SO_4^{-2}$$

$$pH_c = \frac{1}{2}(pK_1 + pC_{H_2CO_3})$$

$$(H^+)_{min} = 10^{-3.7} - 10^{-pH_c}$$

The pH of the permeate stream following H_2S destruction/disinfection is calculated in the following example. The initial water quality in this stream is the permeate water quality following membrane filtration. The pH after H_2S destruction/disinfection is determined to be 3.43, which will prohibit the formation of sulfur turbidity.

EXAMPLE 9.7 *H_2S destruction and disinfection.*

$pH = 4.0, H_2CO_3 = 129$ mg/L, $HCO_3^- = 0.8$ mg/L, H_2S

$$= 1 \text{ mg/L}, 3 \text{ mg/L } Cl_2 \text{ for disinfection}$$

$$H_2S + 4Cl_2 + 4H_2O \Rightarrow H_2SO_4 + 8HCl$$

Cl_2 required for H_2S destruction:

$$1 \text{ mg/L } H_2S \left| \frac{1 \text{ mmol}}{32 \text{ mg}} \right| \frac{4 \text{ mmol } Cl_2}{1 \text{ mmol } H_2S} \right| 71 \text{ mg} = 8.9 \text{ mg/L } Cl_2$$

$$H^+ \text{ generated} = 1 \text{ mg/L } H_2S \left| \frac{1 \text{ mmol}}{32 \text{ mg}} \right| \frac{10 \text{ mmol } HCl}{1 \text{ mmol } Cl_2} \right| = 0.31 \text{ mM} = 3.1 \times 10^{-4} \text{ M } H^+$$

$$Cl_2 + H_2O \Rightarrow HOCl + HCl$$

$$H^+ \text{ generated} = 3 \text{ mg/L } Cl_2 \left| \frac{1 \text{ mmol}}{71 \text{ mg}} \right| \frac{1 \text{ mmol } HCl}{1 \text{ mmol } Cl_2} \right| = 0.04 \text{ mM} = 4 \times 10^{-5} \text{ M } H^+$$

$$\text{Total } H^+ \text{ generated} = 3.1 \times 10^{-4} \text{ M} + 4 \times 10^{-5} \text{ M} = 3.5 \times 10^{-4} \text{ M}$$

pH after alk. destruction: $pH = 0.5(pK + p[H_2CO_3])$

$$H_2CO_3 = 129 \text{ mg/L} = 10^{-2.68} \text{ M}$$

$$pH = 0.5(6.3 - \log(10^{-2.68})) = 4.5$$

$$3.5 \times 10^{-4} \text{ M} - \frac{0.8}{61000} \text{ M} = 10^{-3.47} \text{ M } H^+$$

$$\text{Total } H^+ = 10^{-4.5} + 10^{-3.47} = 10^{-3.43}, pH = 3.43$$

Air stripping could be used to successfully remove all the sulfides, but only very little of bicarbonate alkalinity would remain. A better application would be to increase the pH to 6.3 prior to air stripping with calcium hydroxide or sodium hydroxide to recover 1 to 2 meq/L of alkalinity. Air stripping has the advantage that no chlorides or sulfates are added to the water. Both sulfates and chlorides will be increased by chlorination of H_2S; however, there is no required capture of the off gas as with air stripping of H_2S.

9.4.5.2 Alkalinity Recovery. If acid addition is used for scaling control, all the alkalinity in the raw water will be destroyed but not lost. The membrane is a closed system and the carbon dioxide will remain under pressure until exposed to an open system. Alkalinity recovery needs to be considered during scaling control and depends on how much carbon dioxide and bicarbonate is in the raw water. Normally, finished waters with 1 to 3 meq/L of bicarbonate alkalinity are considered highly desirable for corrosion control. Since carbon dioxide will pass unhindered through the membrane, the desired amount of alkalinity can be recovered in the permeate by acidifying the desired amount, passing it through the membrane, and adding the desired amount of base to convert the carbon dioxide back to its original bicarbonate form. The reactions are shown as follows.

$$HCO_3^- + H^+ \Rightarrow H_2CO_3$$

$$H_2CO_3 + OH^- \Rightarrow HCO_3^- + H_2O$$

EXAMPLE 9.8 *Alkalinity recovery.*
pH = 3.43, 130 mg/L H_2CO_3 for feed stream
Adjust pH to 4.5, destroy $10^{-3.43} - 10^{-4.5} = 10^{-3.47} H^+$

CaO required = $10^{-3.47}$ |28,000 mg/eq CaO| = 9.5 mg/L

Given 2 meq/L HCO_3^- desired:

CaO required = 2 meq/L |28 mg/meq CaO| = 56 mg/L

Total CaO required = 9.5 + 56 = 65.5 mg/L

Finished alkalinity = 2 meq/L |61 mg HCO_3^-/meq| = 122 mg/L

$$pH = pKa + \log \frac{[HCO_3^-]}{[H_2CO_3]}$$

$$= 6.3 + \log \left\{ \frac{[(122 \text{ mg/L})/(61 \text{ mg } HCO_3^-/\text{mmol})]}{[(6 \text{ mg/L})/(62 \text{ mg } H_2CO_3/\text{mmol})]} \right\} = 6.3 + 1.3 = 7.6$$

The following example calculation illustrates the alkalinity recovery using the water quality following H_2S destruction/disinfection. The pH is essentially determined by a two-step process. First, the amount of base added to get to a pH where alkalinity recovery can begin is determined, and then the amount of base necessary to recover the desired amount of carbonate alkalinity is determined. In the example shown in Table 9.4, both calcium and TDS are also observed to increase.

9.4.5.3 Aeration and Stabilization. These unit processes can be discussed simultaneously in membrane processes because they will be accomplished simultaneously. Chlorination could also be done here if desired. The pH calculation following aeration is determined by equilibrium with $CaCO_3$ and can be determined by the pH_s calculation shown as follows.

$$pH_s = pK_2 - pK_{sp} + pCa + pHCO_3^-$$

EXAMPLE 9.9 *Aeration.*

$$CaCO_3 = Ca^{+2} + CO_3^{-2} \qquad K_{sp} = 10^{-8.3}$$

$$HCO_3^- = H^+ + CO_3^{-2} \qquad K_2 = 10^{-10.3}$$

$$pH = pK_2 - pK_{sp} + p[Ca^{+2}] + p[Alk]$$

$$[Ca^{+2}] = \left| \frac{50.9 \text{ mg/L}}{40 \text{ mg/mmol Ca}} \right| = 1.27 \text{ mM} = 10^{-2.9} \text{ M}$$

$$[HCO_3^-] = \left| \frac{122 \text{ mg/L}}{61,000 \text{ mg/mol}} \right| = 10^{-2.7} \text{ M}$$

$$pH_s = pK_2 - pK_{sp} + p[Ca^{+2}] + p[Alk]$$

$$pH_s = 10.3 - 8.3 + 2.9 + 2.7 = 7.6$$

Note in Table 9.4, the H_2CO_3 following aeration is shown to decrease, and the dissolved oxygen is shown to increase.

If chlorine and a base are added to the process stream before aeration, disinfection, oxygen addition, and stabilization will occur. Almost no chlorine demand will remain following a reverse-osmosis or nanofiltration process. The chlorine will convert some of the recovered alkalinity to carbon dioxide which will be lost during aeration; however, the pH should return to the stabilization pH as carbon dioxide will tend to be at equilibrium with the atmospheric carbon dioxide. The pH will closely approach pH_s with respect to $CaCO_3$. The basic equations are shown as follows.

$$Cl_2 + H_2O \Rightarrow HOCl + HCl$$

$$HCl \Rightarrow H^+ + Cl^-$$

$$HOCl \Leftrightarrow H^+ + OCl^- \qquad pK = 7.4$$

$$pH = pK_{H_2CO_3} + \log \left| \frac{HCO_3^- - \alpha_{OCL^-} C_{T_{Cl_2}}}{H_2CO_3 + \alpha_{OCL^-} C_{T_{Cl_2}}} \right|$$

$$OH^- \text{ addition} = OH^- \text{ mg/meq} |meq/mmol| C_{T_{Cl_2}} |1 + \alpha_{OCL^-} C_{T_{Cl_2}}|$$

Chlorine addition to water will produce equal moles of hypochlorous acid and hydrochloric acid. The hypochlorous acid will partially ionize to hypochlorite ions and protons. The hydrochloric acid will completely ionize, producing protons and chloride ions. One mole of protons will be produced for every mole of hydrochloric acid and every mole of hypochlorite ion produced. Consequently, the complete proton production during chlorination would be canceled by the addition of OH^-, as shown previously. The pH during chlorination can be solved by an iteration process. Typical chlorine doses following a reverse-osmosis or nanofiltration process range from 5 to 10 mg/L. The following example is not taken from the water quality illustrated in the previous calculations. The assumptions have no relation to the previous examples and are used to illustrate chlorination following aeration.

EXAMPLE 9.10
Given: 2 meq $HCO_3^-/L = 10^{-2.7}$ mol HCO_3^-/L, pH = 8.3, $Cl_2 = 7.1$ mg $Cl_2/L = 10^{-4}$ mol HCO_3^-/L

Assume: pH = 7.0 after chlorination

$$a_{OCl} = \frac{K}{K + H^+} = \frac{10^{-7.4}}{10^{-7.4} + 10^{-7}} = 0.28$$

$$pH = 6.3 + \log \left| \frac{10^{-2.7} - (1 + 0.28)10^{-4}}{0 + (1 + 0.28)10^{-4}} \right| = 7.5 \neq 7.0, \text{ assume pH} = 7.4$$

$$a_{OCl} = \frac{K}{K + H^+} = \frac{10^{-7.4}}{10^{-7.4} + 10^{-7.4}} = 0.50$$

$$pH = 6.3 + \log \left| \frac{10^{-2.7} - (1 + 0.50)10^{-4}}{0 + (1 + 0.50)10^{-4}} \right| = 7.4 = 7.4 \text{ check}$$

$$Ca(OH)_2 = 1.5 \times 10^{-4} \text{ mol OH}^-/\text{L} \left| \frac{Ca(OH)_2}{2 \text{ OH}^-} \right| 74{,}000 \text{ mg/mol} = 5.5 \text{ mg/L}$$

The iteration will close very quickly. The preceding example shows how to correctly calculate pH following chlorination and how to return the pH to the pH before chlorination, which is the stabilization pH. The pH could be increased to a more encrusting pH.

9.4.6 Pilot Studies

Common questions are when should pilot studies be conducted and what information should be gathered during the pilot study. The cost of pilot studies can vary greatly depending on the size of the pilot plant, period of operation, data collection, and analysis. However, the costs of pilot studies are usually insignificant when compared to the cost of the actual plant, and pilot studies are a very cost effective way of avoiding problems after construction.

The need for conducting a pilot study can be shown by determining the number of RO/NF plants operating on similar waters in similar geographic areas. For example, brackish-water RO plants are common in Florida and are typically designed and built without a pilot study if the Floridian aquifer is the source. NF plants are more common in Florida than in other locations but NF pilot studies are still being conducted to determine design criteria. A pilot study should be conducted when surface waters are used as the source. A pilot study should be conducted if anything other than conventional pretreatment (acid or antiscalant addition and static microfiltration) is indicated from the raw water analysis or required in the preliminary systems design.

Minimum pilot study membrane requirements are a single-stage 4-in element. Smaller elements have the disadvantage of not being able to operate at 10 to 20 percent recoveries and are capable of being operated at higher recoveries with process modifications. Higher single-element recoveries have been obtained by using recycle flows to maintain a minimum flow velocity across the membrane. Multistage pilot plants operating at the same recoveries, fluxes, and pressures as would be experienced in actual operation are the best source for establishing design criteria. Typically, these pilot plants are two- or three-stage single-array systems that operate at 50 to 90 percent recovery with the capability to monitor pressure, flow, and water quality in and out of each pretreatment unit operation and stage. Posttreatment unit operations could also be monitored but typically are not.

Two different logs should be kept during pilot plant operations. The first is a record of operation, and the second is a form for recording data by pilot plant operators. The latter is not necessary if all data are collected and recorded using automated data-recording systems. All entries in either log should always be dated and clearly signed or initialed. The record of operation documents regular observations by operators and any changes in pilot plant conditions. Examples of such observations would be changes in chemical feed rates, operating pressures, chemical cleanings, flushings, operation stoppages, and any other happenings which normally or unexpectedly occur during pilot plant operation. Recording errors will invariably occur and data review is necessary.

Examples of suggested data collection are shown in Table 9.5. The most common parameters to monitor during operation are temperature, start and stop times, pressure, and flow in and out of each stage. Conductivity can be monitored in and out of each pressure vessel and can be used as a surrogate for other water quality parameters and an indicator for pressure vessel, seal, membrane, etc., integrity. Conductivity, pressure, and flow can be recorded continuously in more elaborate pilot plants or recorded in the pilot plant log at least twice by shift operators. Water quality parameters are typically measured in and out of each stage no more than weekly and at least monthly. The notation in Table 9.4 is consistent with model notation shown previously. P, Q, and C represent pressure, flow, and concentration, and the subscripts f, p, and c represent feed, permeate, and concentrate streams. Additional stages may need to be added to Table 9.5 if a multistage system is used.

These data recorded in Table 9.5 can be used for initial treatment and changes in performance during time of operation. Simple graphs of pressure, flux, and water quality versus time of operation will illustrate pilot plant performance. It is important that changes in operating conditions are noted on such graphs, so that any change during operation may be analyzed. Such changes might be a change of feedwater quality, pressure, recovery, or membrane. The data can also be utilized to calculate MTCs which can be plotted for the same operating conditions to compare membranes or pilot plant performance. Two important parameters are permeate water quality and productivity. Water quality goals are established by whoever conducts the study. Productivity declines will determine frequency of cleaning. Operating RO/NF plants have cleaning frequencies that range from quarterly to several years. Consequently, excessive fouling is occurring if the pressure drop through the membrane has to be increased by 10 percent in a three-month period to sustain the initial flux or production. The MTC could be monitored in the same manner if pressure cannot be increased.

9.4.7 Controls and Energy

This section briefly surveys the controls, instrumentation, power requirements, and pumps normally required for an RO/NF plant. The process variables that were previously described are the parameters used to control an RO/NF plant. These vari-

TABLE 9.5 Suggested Data Recording for RO/NF Pilot Plant Operation

Date	Shift	Oper	on	off	T	Stage 1								
						P_f	P_p	P_c	Q_f	Q_p	Q_c	C_f	C_p	C_c

ables are usually monitored by computer systems and the process is automatically controlled, but RO/NF processes can be monitored manually in very small plants. There are many different types of control systems that can be used in RO/NF plant design. These would include manual, on-off, proportional (P), integral (I), derivative (D), PI, and PID. Automatically controlled RO/NF plants use different types of control at different points within the process. For example on-off controls over a proportional band can be used to control levels in chemical feed tanks and reservoirs.

9.4.7.1 Power Supply.   Three-phase power is usually supplied for RO/NF plants. Rectifiers and transformers can be used to convert DC energy to three-phase or single-phase. However, estimating power requirements for an RO/NF plant can be done by determining the power requirement for the membranes and adding 10 percent for NF plants, 5 percent for low-pressure RO plants, and 1 to 2 percent for high-pressure RO plants. Equation (9.63) is a general equation used for calculating energy. Power requirement calculation is shown in the following example. This calculation is for a 10-mgd plant but could be easily made for any size plant, given the recovery, pump efficiency, and motor efficiency

$$E = \frac{(H)(0.00315)}{\eta_r \eta_p \eta_m} \tag{9.63}$$

where $E = $ kWh/kgal
$H = $ head in ft
$\eta_i = $ decimal efficiency of recovery, pump, and motor

EXAMPLE 9.11
Assume: 10-mgd (0.44-m^3/s) NF plant operating at 150-lb/in^2 (10.2-bar) feed pressure and 85% recovery; the efficiency of the pump and motor are both 85%.

100 lb/in^2|144 in^2/ft^2|ft^3/62.4 lb = 230.7 ft (70.3 m)

$$\text{Energy consumption} = \frac{230.7 \text{ ft} \times 1.5 \times 0.00315}{(0.85)^3} = \frac{1.09}{0.614}$$

$$= 1.774 \text{ kWh/kgal (468.5 kWh/km}^3\text{)}$$

NF energy consumption for 10 mgd = 1.774 kWh/kgal|1000 kgal/mgd|10 mgd

$$= 17,774 \text{ kWh}$$

Total energy consumption for 10-mgd NF plant = 17,774 kWh × 1.1 = 19,524 kWh

9.4.7.2 Controls.   RO/NF plants are operated with control based on the monitoring of flow, pressure, conductivity, temperature, pH, liquid level, specific chemical analyses, and lapsed time indicators. The comments here are only with regard to the pretreatment and membrane arrays

- Flow is monitored for control of water production in arrays and may be controlled automatically with on-off or PID controllers.

- Pressure is monitored for control of water production in arrays and may be controlled manually or automatically by regulating pump operation and valve settings through a computer system. Pressure is typically increased or decreased to increase or decrease production in each stage.

- Temperature is monitored for protection of pumps and membranes, with complete plant or array shutdown from on-off temperature control.
- pH can be monitored and controlled automatically by feedback control. pH monitoring is essential if acid is used to control scaling and will cause shutdown or alarms to activate if range is violated.
- Liquid levels are monitored and controlled by on-off proportional controllers in chemical feed systems for RO/NF plants.
- Specific chemical analyses such as conductivity or turbidity may be monitored automatically by feedback or manually to ensure membrane, pressure vessel, and array integrity.
- Lapsed-time meters are recorded continuously for plant analyses and maintenance. Actual time of operation is essential for accurate evaluation of plant performance.

9.4.7.3 Pumps. Pumps for RO/NF processes are typically centrifugal pumps that are constructed of noncorrosive materials, i.e., nonmetallics or alloys such as stainless steel. NF pumps are usually single-stage pumps but can be multistaged on occasion and would be expected to operate from 75 to 200 lb/in^2. The pressure range for RO pumps varies as to the TDS of the feedwater. Low-pressure RO pumps would be expected to operate over a pressure range of 175 to 400 lb/in^2. High-pressure RO pumps would be expected to operate to pressures of 1200 lb/in^2 or more. Because of the higher energy requirement, RO pumps are multistaged or very high frequency with increased impeller speeds. Seawater RO pump applications can utilize positive displacement pumps.

9.5 OPERATION

Operation of an RO/NF plant requires constant monitoring of water production and quality. There are several different operating parameters that can be successfully monitored to gauge overall plant performance. The most common parameters for monitoring are pressure drop, flux, conductivity, and water quality.

9.5.1 Monitoring

Each unit operation in an RO/NF plant should be monitored for function. Typical flow processes in an RO/NF plant include static microfiltration, acid or antiscalant addition, membrane filtration, aeration, disinfection, and corrosion control. Flow, pressure, and applicable water quality are general monitoring parameters that are applicable to each unit operation in an RO/NF process. Recording the performance of an RO/NF plant should begin immediately during startup and continue during operation. A startup report should include a complete plant description, pretreatment unit performance, membrane performance, and posttreatment performance:

- Complete plant descriptions can be described by a flow diagram, equipment and instrumentation literature provided by engineers and manufacturers, instrumentation literature and operating manuals, and material lists. The plant description should show water source, all pretreatment unit operations, RO/NF array configuration, and all posttreatment unit operations.

- Actual initial performance of pretreatment and RO/NF sections should be monitored. The pretreatment unit operations typically include static microfiltration and acid and/or antiscalant additions, which are described by SDI or some other fouling index, pH, flow, and pressure.
- All gauges and meters should be calibrated based on manufacturers' recommendations as to method. The expected process performance can be used to check gauges and meters. Flow gauges can also be checked by mass balances around stages or pressure vessels. Pressure gauges and conductivity meters can be checked by comparison to similar readings before and after the desired gauge or meter.

Operating reports must also be kept as part of monitoring. The operating data describe the performance of the RO and NF systems. They must be collected routinely during the life of the plant. These data, along with periodic water analyses, provide the necessary input for the evaluation of the plant's performance. Suggested monitoring and frequency of operating data are shown in Table 9.6. The information includes:

- Flows (product and brine of each stage)
- Pressures (feed, product, and brine of each stage and permeator brine sample port)
- Temperature (feed after RO/nanofilter pump or product)
- pH (feed, product, and brine)
- Conductivities/TDS of each stage and each permeator (feed, product, and brine)
- SDI of feed after static microfiltration
- Unusual incidents (upsets in pH, SDI, and pressure, shutdowns, etc.)
- Calibration of all gauges and meters

Complete water analysis of the feed, permeate, and concentrate streams for the plant and the raw water should be obtained at startup, quarterly, and as required by regulation thereafter: changes in conductivity can be correlated to TDS and possibly to other major ions. Suggested water quality monitoring is shown in Table 9.7.

TABLE 9.6 Typical Data Collection Frequency for RO/NF Plants

	8-h shift	Daily	Weekly	Monthly
Pressure:				
System (by stages)	X	X		
Pressure vessels (concentration sample port)			X	X
Flow	X	X		
Temperature	X	X		
TDS:				
System (by stages)	X	X		
Pressure vessels			X	X
pH	X	X		
SDI	X	X		
Water analysis (feed, concentration, product)				X
LSI (brine)			X	
Unusual incidence—on occurrence				

TABLE 9.7 Suggested Water Quality
Monitoring During Startup and by Quarter

Ca^{2+}	HCO_3^-
Mg^{2+}	SO_4^{2-}
Na^+	Cl^-
K^+	F^-
Ba^+	PO_4^{3-}
Fe(total and Fe^{2+})	SiO_2
Al^{3+}	TOC
pH	TDS
H_2S	THMs
Conductivity	HAAs

Partial water analyses of the feed, product, and brine for each stage should be obtained once a month and should include TDS, chloride, and conductivity. The chloride content and conductivity should be obtained twice a week. Conductivity can be monitored continuously throughout the process.

Records must be kept of the operating characteristics of the pretreatment equipment. Chemical consumption and performance must be recorded. Since pretreatment is site dependent, specific recommendations for all data record keeping are not given. Pretreatment operating data should include at least the following:

- SDI before and after acid addition once per day
- Inlet and outlet pressure of static microfilters twice a day
- Acid consumption once per day
- Discharge pressure of well pumps twice per day
- Calibration of all gauges and meters no less than once every three months

Maintenance records on the plant must also be kept. They provide further information about the performance of the permeators and the mechanical equipment in the plant. The logs include:

- Routine maintenance
- Mechanical failures
- Permeator replacements
- Cleanings (cleaning agents and conditions used)
- Chemical posttreatments (type and conditions used)
- Sterilization of freeze protection applicant
- Frequency of changing 5-μm filters
- Calibration of all gauges and meters

9.5.2 Quantity

Water production is typically monitored by pressure drop through and across pressure vessels. There are general approximations of pressure drop per element and pressure vessel entrance and exit losses which are usually identified in the manual of operation. Typically, entrance and exit losses to pressure vessels will not exceed

5 lb/in². Element losses average 2 to 3 lb/in². A 10- to 20-lb/in² loss in the feed stream through a pressure vessel is not uncommon; however, the change in the pressure losses across a pressure vessel can indicate excessive deposition on the feed side of the membrane. Pressure losses through a membrane are much higher than across the membrane. NF membranes typically require 50- to 75-lb/in² driving force during normal production. RO membranes in brackish-water applications may require a driving force from 125 to 250 lb/in² for adequate production. Seawater desalination driving forces are considerably higher and can exceed 1000 lb/in². RO or NF membranes require three to eight weeks to set, but once production is normalized, a normal pressure loss through and across the membranes for the required production will be established. Membranes will naturally build up resistance to water passing through the active membrane layer. The membranes need to be cleaned when this resistance causes an increase of 10 percent or more from the initial pressure loss. Membrane cleaning frequencies vary in typical plant applications from three months to more than two years, averaging approximately six months.

Comparison of plant flux with other plant performance requires that the data be normalized with respect to temperature. The viscosity of water decreases as temperature increases, which causes the MTC or flux to increase as temperature increases. Several equations are available for normalization of flux with respect to temperature. Equation (9.64) gives the temperature correction factor for flux adjustment and can be used directly for flow adjustment also. The normalization temperature is 25°C or 77°F and is the reporting standard worldwide (PEM, 1982).

$$\text{TCF} = \frac{J_{T^\circ C}}{J_{25^\circ C}} = 1.03^{(T-25)} \tag{9.64}$$

where TCF = temperature correction factor
J_T = flux at actual temperature
J_T = flux at 25°C
T = actual temperature

The following shows the adjustment of flow and flux for a membrane process operating at 50°F. A good rule of thumb is that the flux of water will increase about 3 percent for every 1°C or 1.8°F increase in temperature.

EXAMPLE 9.12
Given: $Q_p = 2000$ gpm, $T = 50°F$, $J = 15$ gsfd
Change temperature units:

$$°C = \tfrac{5}{9}(°F - 32)$$

$$= \tfrac{5}{9}(50 - 32) = 10°C$$

$$\text{TCF} = 1.03^{(10-25)} = 0.64$$

$$J_{25^\circ C} = \frac{J_{T^\circ C}}{0.64} = \frac{15}{0.64} = 23.4 \text{ gsfd}$$

$$Q_{25^\circ C} = \frac{Q_{T^\circ C}}{0.64} = \frac{2000}{0.64} = 3125 \text{ gpm}$$

9.5.3 Quality

Water quality in RO/NF membrane plants is typically monitored by conductivity measurements every 4, 8, or 24 h. Membranes would be expected to deteriorate naturally from use; however, the rate of water quality deterioration would normally be no more than 2 to 3 percent annually. Spikes or sudden increases in conductivity would indicate seal or O-ring failure as opposed to failure of the membrane surface.

Temperature will also affect the mass transfer of solutes. As shown in the Stokes-Einstein Eq. (9.65), diffusivity will increase and viscosity will decrease as temperature increases. The resulting effect is that as temperature increases the rejection of diffusion-controlled solutes will decrease even though flux will increase under the same conditions. Unfortunately, there are no convenient equations for adjusting solute MTCs with respect to temperature. However, it is important that water temperature is monitored during operation of RO/NF plants and that an awareness of the temperature effects on mass transfer is maintained when comparisons of plant data are made.

$$D_1 = \frac{D_2 v_2 T_1}{T_2 v_1} \tag{9.65}$$

where D_i = diffusivity at T_i
v_i = viscosity at T_i
T_i = temperature

9.5.4 Flux/MTC

A good method of monitoring productivity is monitoring the MTC for water over timer of production. The MTC is sometimes described as *specific flux* because it is the ratio of flux to pressure loss through the membrane. The MTC is a good measurement of efficiency and can be monitored, as is pressure, to determine acceptability of production. The pressure monitoring is acceptable as long as a constant production is maintained, which amounts to an equivalent measurement to the MTC. Flux is acceptable for production monitoring as long as the pressure is held constant, which is not acceptable for meeting a required production goal. As with flux, membrane MTCs must also be adjusted for temperatures for comparison.

9.5.5 Membrane Life

Membrane rejection of solutes and productivity (flux) will deteriorate over time due to compaction, chemical reactions, or biodegradation of the membrane surface. Membranes made of CA and derivatives of CA are biodegradable and must use pretreatment, such as chlorination, for microorganism control. TFC and PA membranes are resistant to biodegradation but can be damaged by oxidants such as chlorine. The normal life of membranes is reported to be three to five years. Cellulose acetate membranes can undergo hydrolysis reactions, can be degraded biologically, and are seldom warranted for more than three years. Thin film composite membranes do not readily undergo hydrolysis reactions, are very resistant to bacterial degradation, and typically have a five-year warranty. However, there are installations where cellulose acetate membranes have not been replaced over a 10-year period. The actual life of membranes is greater than three to five years. But if membranes are not cared for during operation or pretreatment systems are not properly designed, membrane replacement could be necessary in less than one year.

9.5.6 Membrane Cleaning

Membranes must be regularly cleaned during the operating life. The frequency and type of cleaning depends on the quality of the feed stream. A high-pH organic cleaner is used to remove organic solutes deposited on the membrane surface. EDTA in NaOH solutions is common for organic cleaning. Phosphoric acid is commonly used to remove inorganic deposits such as calcium carbonate. However, elemental sulfur and colloidal clays are almost impossible to remove from the membrane surface once deposited. There are several different solutions and chemicals used to clean membranes. Common phosphate-based detergents have been successfully used to remove organics from membrane surfaces. Manufacturers commonly recommend cleaning conditions such as temperature, pH range, frequency, and duration. Failure to abide within the manufacturer's recommended cleaning conditions could void membrane warranties.

9.6 CONCENTRATE DISPOSAL

Concentrate disposal from RO/NF plants is a significant problem for all RO/NF drinking water treatment plants because (1) concentrate disposal is highly regulated by state and/or federal governments, (2) there is a significant amount of concentrate stream to disposal, and (3) the concentrate stream contains a high concentration of salts and other chemicals added before membrane filtration.

The quality and quantity of concentrate streams from RO/NF plants can be easily estimated using simply recovery and rejection. A 10-mgd plant operating at 90 percent recovery will produce a concentrate stream of 1.11 mgd, which is determined by dividing the permeate flow, 10 mgd, by the recovery, 0.9. The quality of the concentrate-stream can be determined by dividing the decimal fraction solute rejection into the feed-stream recovery. The same plant producing 10 mgd at 90 percent recovery and rejecting 80 percent of 100 mg/L of the feed-stream chlorides would produce 20 mg/L of chlorides in the permeate stream and 820 mg/L of chlorides in the concentrate stream. On occasion, the rejection and recovery are confused. The calculations are illustrated as follows.

$$Q_c = \frac{Q_p(1-R)}{R} = \frac{10\,\text{mgd}(1.0-0.9)}{0.9} = \frac{1.0}{0.9} = 1.11\,\text{mgd}$$

$$C_p = (\text{Re } j)(C_f) = (1-0.8)(100\,\text{Cl}^-\,\text{mg/L}) = 20\,\text{Cl}^-\,\text{mg/L}$$

$$C_c = \frac{Q_f C_f - Q_p C_p}{Q_c} = \frac{(1\,\text{L})(100\,\text{Cl}^-\,\text{mg/L})\,(0.9\text{L})(20\,\text{Cl}^-\,\text{mg/L})}{0.1\,\text{L}} = 820\,\text{Cl}^-\,\text{mg/L}$$

These calculations can be made from the mass transfer equations presented in the section on array modeling. A similar set of equations could be developed for C_c as were developed for C_p.

TABLE 9.8 Concentration Disposal Technique Distribution by Plants and Flow

Technique/plant size, mgd	<.3	.3–1	1–3	>3	Total
Surface	34	12	9	11	66
Land application*	14	2	0	1	17
Sewer	18	8	3	3	32
Deep well injection	3	1	5	5	14
Evaporation pond	8	0	0	0	8
Total	77	23	17	20	137

* None planned in future.

The techniques used for concentrate disposal in the United States are shown by plant size and number in Table 9.8 (Mickley, 1993). There are five basic techniques used by the 137 plants Mickley surveyed in an AWWARF concentrate disposal study. These techniques show surface discharge as the most common technique of concentrate disposal, accounting for nearly half of techniques surveyed. Land application was not planned for any of the to-be-built plants surveyed. However, land application of NF concentrates is possible in some locations because of the low TDS concentration, typically 1000 to 2000 mg/L, relative to RO concentrates. Sewer discharge is usually an option only for very small plants and was the second most common technique for plants under 0.3 mgd. Deep well injection was more common in Florida than in any other state because of the regulatory environment in Florida and other states.

Concentrate discharge is regulated through programs in the Clean Water Act (CWA) and the Safe Drinking Water Act (SDWA). Surface waters fall under the National Pollutant Discharge Eliminate System (NPDES). Groundwaters fall under the Underground Injection Control program. Regulatory environments for concentrate disposal are controlled both by states and the USEPA, depending on primacy. Although these programs have existed since the 1970s, the regulatory environment is becoming more stringent. Bioassay testing is becoming more common in Florida and California and can be used to require additional treatment or stop planned disposal of concentrate streams.

9.7 COST

Cost of RO/NF plants has been estimated as a function of capacity and TDS or pressure by Morin (Morin, 1994). These estimated costs are shown in Table 9.9 for a low-pressure (LP) NF plant that was operating at 100 lb/in^2 and 85 percent recovery. There are no costs for concentrate disposal considered in this development.

All indirect costs were estimated as a fixed percentage of the direct cost, which can be determined from Table 9.8. In developing this cost analysis, a conventional RO/NF treatment system was assumed. Pretreatment included only acid and antiscalant addition for scaling control and microfiltration for element protection. The operating costs were estimated using \$0.075/kWh, and \$10/h for labor. Membranes were assumed to have a life of five years, be thin film composites, and have a purchase cost of \$1000 for an 8-in-diameter membrane. The amortization cost was determined using a 10 percent interest rate and 20 years.

TABLE 9.9 Process—LP (reference) TDS = 400 mg/L, Plant Capacity (mgd), Plant Recovery at 85%

	1.00	5.00	10.00	20.00	40.00
A. Direct construction costs (1000 $)					
1. Process equipment cost	622.00	2312.00	4227.00	7449.00	13,941.00
2. Building	178.00	891.00	1783.00	3566.00	7132.00
3. Feedwater supply, site dev., aux. equip.	182.00	316.00	527.00	912.00	1677.00
Subtotal direct cost	982.00	3519.00	6537.00	11,927.00	22,750.00
B. Indirect capital costs ($K/yr)					
4. Construction overhead	216.04	774.18	1438.14	2623.94	5005.00
5. Owners' costs	78.56	281.52	522.96	954.16	1820.00
6. Contingency	98.20	351.90	653.70	1192.70	2275.00
Subtotal indirect cost	392.80	1407.60	2614.80	4770.80	9100.00
Total construction cost	1443.54	5172.93	9609.39	17,532.69	33,442.50
Unit capital cost ($/gpd)	1.44	1.03	0.96	0.88	0.84
C. Annual operating cost ($K/yr)					
7. Electricity	53.28	265.20	530.50	1060.10	2117.70
8. Steam	0.00	0.00	0.00	0.00	0.00
9. Labor	281.00	312.00	343.40	374.60	437.00
10. Labor overhead	112.40	124.80	137.36	370.50	174.80
11. Membrane replacement	18.50	92.60	185.00	370.50	741.10
12. Chemicals	23.50	126.80	254.30	508.50	1013.80
13. Other Matl's and insurance	21.65	77.59	144.14	262.99	501.64
Annual O&M costs	510.33	998.99	1594.70	2947.19	4986.04
Amortization cost	147.02	526.86	978.72	1785.70	3406.12
Total annual cost	657.36	1525.86	2573.42	4732.89	8392.16
14. Yearly production (kgal)	310,250	1,551,250	3,102,500	6,205,000	12,410,000
15. O&M cost ($/kgal)	1.64	0.64	0.51	0.47	0.40
16. Total cost of water ($/kgal)	2.12	0.98	0.83	0.76	0.68

The costs show the decreased unit cost for large-scale plants as the total cost of water decreased from $2.12/kgal at 1 mgd to $0.68/kgal at 40 mgd. Significant total costs savings are realized at 5 mgd relative to 1 mgd as the total cost decreased from $2.12/kgal to $0.98/kgal.

The capital costs and O&M costs for RO and NF plants are shown in Figs. 9.15 and 9.16 for varying salt concentrations or feed-stream pressures. The higher TDS or salt concentration requires a tighter membrane and more energy to produce permeate or water. The 400-, 2000-, 4000-, and 34,000-mg/L TDS concentration correspond to feed-stream pressures of 100, 200, 300, and 1000 lb/in^2, respectively.

These figures show that the unit cost of water produced by RO/NF processes decreases as capacity increases and that the greatest incremental in percent of unit cost reduction is from 1 to 5 mgd. Additionally, the cost analysis indicates that there is a significant increase in the cost of treating seawater (34,000 mg/L) by RO relative to treating brackish or fresh water by low-pressure RO or NF.

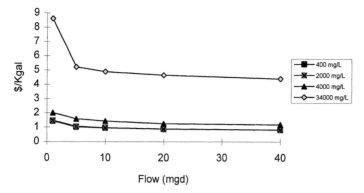

FIGURE 9.15   Capital cost of RO and NF plants by flow and TDS (mg/L).

9.8   CASE STUDIES

There are in excess of 140 membrane plants in the United States. RO/NF plants are selected for all types of water treatment issues, as shown in Table 9.1 (Mickley, 1993). The most common treatment issue was desalination or the removal of TDS; however, softening, color removal, and DBP precursor removal are also major reasons for selecting an RO/NF process. Two membrane plant case studies of one NF and one RO plant are presented in the following sections.

9.8.1   Fort Meyers

Fort Meyers is located on the southwest gulf coast of Florida. Previous to 1985, the City of Fort Meyers operated two softening plants that utilized a well field in a surficial aquifer for a raw water supply. The well field is recharged with water from the Caloosahatchee River, which is supplied by Lake Okeechobee. The Caloosahatchee River is highly colored and has moderate hardness. In the wet season the raw water

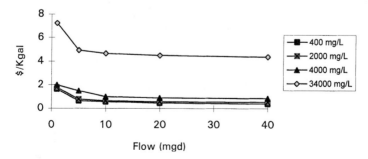

FIGURE 9.16   RO and NF plant operation and maintenance costs by flow and TDS (mg/L).

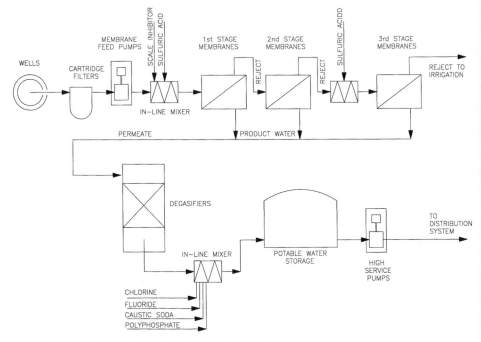

FIGURE 9.17   Fort Meyers Nanofiltration Plant flow diagram.

color, TOC, and hardness are approximately 200 CPU, 20 mg/L, and 150 mg/L $CaCO_3$. In the dry season the raw water color, TOC, and hardness are approximately 75 CPU, 10 mg/L, and 250 mg/L $CaCO_3$. In 1985, as a result of a regulatory consent order, the City of Fort Meyers selected an NF membrane process to reduce color, THMs, and hardness in the finished water.

The flow diagram and the plant information for the City of Fort Meyers NF water treatment plant is shown in Fig. 9.17 and Table 9.10 (Boyle Engineering, 1995). Raw water is delivered to the plant through a raw water transmission header. The well field is adjacent to the plant and includes 20 shallow wells equipped with submerged pumps. A telemetry system is used to control well field pumping. The raw water is passed through a 5-μm prefilter. Sulfuric acid and anti-scalant are added before a single set of high-pressure membrane pumps which pass the process stream to a three-stage membrane array. As shown in Fig. 9.18, each membrane array includes two first-stage membrane control blocks, one second-stage control block, and a third-stage control block. All pressure vessels contain seven 8.5-in-diameter membranes. Acid can also be added before the third stage.

Posttreatment consists of packed tower aeration for H_2S and CO_2 removal, followed by NaOH addition for stabilization, fluoride addition, chlorination, and the addition of a blended phosphate for corrosion control. The concentrate stream is discharged to a small lake that is used as an irrigation source for a municipal golf course. Historically, the lake had been recharged with water from the well field.

TABLE 9.10 Physical Parameters and Water Quality of Case Studies

	Ft. Meyers	Vero Beach
Startup	8/3/92	11/1/92
Design flow, mgd	20	6
Recovery, %	90	85
Membrane data		
Manufacturer	Hydranautics	Hydranautics
Model	NCM-1 and PVD-1	NCM-1
Surface	TFC-CLPA	TFC-CLPA
MWC daltons	200	<100
Process data		
Array/stages	3	1/2
Pressure vessels/stage	96	36/15
Elements/vessel	7	7
Flux, gsfd	14.7	15
Area/element, ft^2	350	350
Feed pressure, lb/in^2	145–155	175
Concentrate disposal	Irrigation	Surface canal
Water quality	Raw/finished	Raw/finished
pH	7.4/7.8	7.5/8.4
Calcium hardness, mg/L $CaCO_3$	164/64	327/2
Total hardness, mg/L $CaCO_3$	212/79	335/3
Total alkalinity, mg/L $CaCO_3$	170/77	220/20
Sulfate, mg/L	26/12	<5/<5
TDS, mg/L	377/195	5736/360
Color, CPU	93/<1	<5/<1
Turbidity, NTU	0.2/<0.1	7.2/<0.1
NPDOC, mg/L	39.7/1.1	<2.0/<0.5
THM, ug/L	537/15	<10/<10
Chloride, mg/L	73/65	2978/193
Iron, mg/L	0.64/0.20	<0.1/<0.01
Manganese, mg/L	BDL	BDL
Sodium, mg/L	27/21	1725/120

9.8.2 Vero Beach

The City of Vero Beach is located on the east coast of Florida. The city water supply was taken from a surficial and deep aquifer and treated in a 13-mgd lime-softening process. Although the wells were operated to optimize quality, salts and natural organic matter caused the raw water quality to vary significantly. In 1988 the city decided to meet future demand and improve drinking water quality by building an RO plant and blending the RO and lime-softened finish waters. The finished water goals were to produce a water with a hardness of 125 mg/L $CaCO_3$ and a TDS of 275 mg/L.

The Vero Beach RO flow diagram is shown in Fig. 9.19. The pretreatment system consists of antiscalant addition, sulfuric acid addition, and static 5-μm microfiltration. The membrane treatment units consist of high-pressure feed pumps, skids for housing several arrays, and a membrane cleaning system. Each skid contains a two-stage array with 36 first-stage pressure vessels and 15 second-stage pressure vessels.

FIGURE 9.18 Fort Meyers Nanofiltration Plant membrane arrays.

Each pressure vessel holds seven 8.5-in elements. The membrane cleaning system consists of a cleaning tank, cartridge filter, and recirculation pump.

The posttreatment system consists of degasifiers and acid addition for H_2S and CO_2 removal, and chloramination for disinfection. Chloramination is still used in the lime-softening process and was retained in the RO process for balance in the finished water. Corrosion control is accomplished by pH control through NaOH addition and blending of the finished waters. The off-gas from both aerators is treated for odor control by first passing the H_2S-laden gaseous stream through an NaOH solution for sulfide conversion and then treating the basic solution with NaOCl to precipitate the sulfide as sulfur. The waste stream following NaOCl treatment is small and is discharged to the lime-softening sludge settling ponds or the sewer (Boyle Engineering, 1995).

9.8.3 Son Tugores, Spain

The Son Tugores plant is located in Palma de Mallorca, Spain. The supply water is a brackish groundwater with a variable TDS concentration ranging from 2 to 10 g/L. The total plant capacity, 30,000 m^3/d (8 mgd), is obtained by varying the feedwater recov-

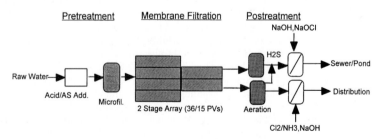

FIGURE 9.19 Vero Beach reverse-osmosis plant flow diagram.

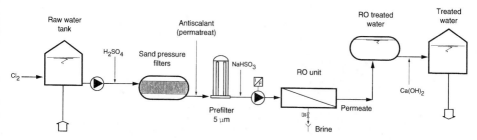

FIGURE 9.20 Simplified process schematic.

ery using variable-speed pumps depending on the TDS levels. The feedwater recovery ranges from 70 to 82 percent and the permeate TDS obtained is less than 500 mg/L.

Figure 9.20 shows a simplified process schematic. The pretreatment includes a prechlorination, a pH adjustment at 6.7 to 6.8 with sulfuric acid, a sand pressure filtration, a scaling control using PermaTreat 191 (Houseman Limited, Burnham Slough, UK) a 5-μm cartridge filtration, and an addition of sodium metabisulfite. The biofouling of the RO membranes is controlled through the prechlorination step; the sodium metabisulfite is used for dechlorination purposes. The RO unit is composed of six racks of 49 pressure vessels having each seven FilmTec BW-330 (Dow Europe Separation Systems, Rheinmünster, Germany), spiral-wound elements (i.e., 2058 membrane elements). Each rack is arranged in two-stage arrays, including 35 and 14 pressure vessels for the first and second array, respectively. Figure 9.21 pre-

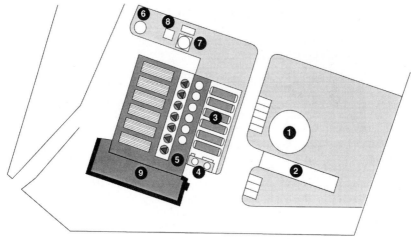

1. Raw water tank (prechlorination)
2. Pretreatment pumping station
3. Sand pressure filters
4. Chemical dosing station (H_2SO_4, antiscalant and $NaHSO_3$)
5. RO building
6. RO flushing water storage tank
7. Lime saturator
8. Lime storage tank
9. Administration building

FIGURE 9.21 Plant layout general overview.

FIGURE 9.22*a*   General RO building view.

FIGURE 9.22*b*   RO rack view.

sents a general plant overview, including the pretreatment, the RO building layout, the posttreatment, and the administration building. Figures 9.22*a* and *b* show views of the RO units inside the building, and the plant building including the administration office on the left, and the pressure sand filters in the right upper corner. The plant was started in April 1995, and the membranes had not been chemically cleaned at the time this report was prepared.

ACKNOWLEDGMENTS

The following people are acknowledged for their contributions to this chapter. The text materials were reviewed and edited by Shiao-Shing Chen, Charles Johnson,

Todd Shaw, and Luke A. Mulford of the University of Central Florida. Material for the chapter was supplied by Steven Duranceau of Boyle Engineering, Stuart A. McCellan of Dow Chemical U.S.A., Jean-Michel Laîné of the Centre of International Research for Water and Environment, Lyonnaise Des Eaux, The City of Fort Meyers, Florida, and the City of Vero Beach, Florida.

REFERENCES

Baier, J. H., B. W. Lykins, Jr., C. A. Fronk, and S. J. Kramer, 1987, "Using Reverse Osmosis to Remove Agricultural Chemicals from Groundwater," *Journal of American Water Works Association,* **79**(8):55–60.

Belfort, G., 1989, "Fluid Mechanics in Membrane Filtration: Recent Developments," *Journal of Membrane Science,* **40**:123–147.

Boyle Engineering Inc., 1995, Vero Beach and Fort Meyers Water Treatment Plant documentation, via Dr. S. J. Duranceau, February 1995, Orlando, Fla.

Brian, P. L. T., October 3–9, 1965, "Influence of Concentration Polarization on Reverse Osmosis Systems Design," *1st International Symposium on Water Desalination,* Washington, D.C.

Da Costa, A. R., A. G. Fane, C. J. D. Fell, and A. C. M. Franken, 1991, "Optimal Channel Spacer Design for Ultrafiltration," *Journal of Membrane Science,* **62**:275–291.

Denisov, G. A., 1994, "Theory of Concentration Polarization in Cross-Flow Ultrafiltration: Gel-Layer Model and Osmotic-Pressure Model," *Journal of Membrane Science,* **91**:173–187.

Duranceau, S. J., and J. S. Taylor, 1990, "Investigation and Modeling of Membrane Mass Transfer," *1990 Proceedings of the National Water Improvement Supply Association,* Orlando, Fla.

Duranceau, S. J., J. S. Taylor, January 1992, "A SOC Removal in a Membrane Softening Process," *Journal of American Water Works Association,* **84**(1)(11):68–78.

Goel, V., and J. W. McCutchan, January 1977, "Systems Design of a Tubular RO Plant," Desalination Report No. 64, Water Resources Center, USA.

Gupta, B. B., B. Zaboubi, and M. Y. Jaffrin, 1993, "Scaling up Pulsatile Filtration Flow Methods to a Pilot Apparatus Equipped with Mineral Membranes," *Journal of Membrane Science,* **80**:13–20.

Hofman, J. A. M. H., J. C. Kruithof, Th. H. M. Noij, and J. C. Schippers, March 1993, "Removal of Pesticides and Other Contaminants with Nanofiltration" (in Dutch), *H2O,* **7**.

Hofman, J. A. M. H., J. S. Taylor, J. C. Schippers, S. J. Duranceau, and J. C. Kruithof, January 1995, "Vereenvoudigd model voor de beschrijving van de weking van nano-en hyperfiltratie-installaties," *H₂O Journal* (Europe), **1**.

Hopman, R., C. G. E. M. Van Beek, H. M. J. Janssen, and L. M. Puiker, 1993, "Pesticides and the Drinking Water Supply in the Netherlands," Report No. 113, KIWA N.V., Nieuwegein, Netherlands.

Jones, P. A., and J. S. Taylor, 1992, "DBP Control by Nanofiltration, Cost and Performance," *Journal of American Water Works Association,* **84**:104–116.

Karamian, N. A., 1975, "A Membrane for Removing Endotoxins From Aqueous Solutions," *Desalination,* **17**:329–338.

Lonsdale, H. K., U. Merten, and R. L. Riley, 1965, "Transport Properties of CA Osmotic Membranes," *Journal Applied Polymer Science,* **9**:1341–1362.

Lopez-Leiva, M., 1980, "Ultrafiltration at Low Degrees of Concentration Polarization—Technical Possibilities," *Desalination,* **35**:115–128.

Matthiason, E., and B. Sivik, 1980, "Concentration Polarization and Fouling," *Desalination,* **35**:59–103.

Merten, U., H. K. Lonsdale, and R. L. Riley, August 1964, "Boundary-Layer Effects in Reverse Osmosis," *Industrial Engineering Chemistry Fundamentals,* **3**(3):210–213.

Metcalf, P. J. et al., 1992, "Water Science and Technology," *Proceedings from IAWPRC Conference,* Cape Town, South Africa.

Mickley, M., R. Hamilton, and J. Truesdall, 1993, *Membrane Concentrate Disposal,* AWWA Research Foundation, Denver.

Morin, O. J., September 1994, *Proceedings of the IDA Annual Conference,* West Palm Beach, Fla.

Morris, K. M., 1990, "Predicting Fouling in Membrane Separation Processes," Master's thesis, University of Central Florida, Orlando, Fla.

PEM Products Engineering Manual, 1982, E. I. Du Pont de Nemours & Co., Wilmington, Del.

Pitera, E. S., and S. Middleman, 1973, "Convection Promotion in Tubular Desalination Membranes," *Industrial Engineering Chemistry, Process Design and Development,* **12**:52–56.

Porter, M. C., 1990, "Ultrafiltration," in M. C. Porter (ed.), *Handbook of Industrial Membrane Processes,* Noyes Publications, New Jersey, p. 175.

Rangarajan, R., E. C. Matsuura, and S. Sourirajan, 1976, "Free Energy Parameter for Reverse Osmosis Separation of Some Inorganic Ions and Ion Pairs in Aqueous Solutions," *I & EC Process Design and Development,* **15**(4):529–534.

Rautenbach, R., and R. Albrecht, 1989, *Membrane Processes,* John Wiley, New York, p. 82.

Rosenfeld, J., and S. Loeb, January 1967, "Turbulent Region Performance of Reverse Osmosis Desalination Tubes," *I & EC Process Design and Development,* **6**(1):123–127.

Schippers, J. C., and J. Verdouw, 1980, "The Modified Fouling Index, a Method of Determining the Fouling Characteristics of Water," *Desalination,* **32**:137–148.

Schock, G., and A. Miguel, 1987, "Mass Transfer and Pressure Loss in Spiral-Wound Modules," *Desalination,* **64**:339–352.

Strohwald, N. K. H., and E. P. Jacobs, 1992, "An Investigation into UF Systems in the Pretreatment of Sea Water for RO Desalination," *Water Science and Technology,* **25**(10):69–78.

Sung, Larry, August 1994, "Modeling Mass Transfer in Nanofiltration," Ph.D. dissertation, University of Central Florida, Orlando, Fla.

Sung, L. K., K. E. Morris, and J. S. Taylor, Nov–Dec 1994, "Predicting Colloidal Fouling," *International DeSalination and Water Reuse Journal,* **4/3.**

Tan, Lo, and Gary L. Amy, 1991, "Comparing Ozonation and Membrane Separation for Color Removal and Disinfection By-product Control," *Journal of American Water Works Association,* **83**:74–79.

Taylor, J. S., D. Thompson, B. R. Snyder, J. Less, and L. Mulford, 1986, "Cost and Performance Evaluation of In-Plant Trihalomethane Control Techniques," EPA/600/52-85/138, U.S. EPA Water Engineering Research Laboratory, Cincinnati, Ohio.

Taylor, J. S., D. M. Thompson, and J. Keith Carswell, 1987, "Applying Membrane Processes to Groundwater Sources for Trihalomethane Precursor Control," *Journal of American Water Works Association,* **79**(8):72–82.

Taylor, J. S., L. A. Mulford, W. M. Barrett, S. J. Duranceau, and D. K. Smith, 1989a, "Cost and Performance of Membrane Processes for Organic Control on Small Systems," U.S. EPA Water Engineering Research Laboratory, Cincinnati, Ohio.

Taylor, J. S., L. A. Mulford, S. J. Duranceau, and W. M. Barrett, 1989b, "Cost and Performance of a Membrane Pilot Plant," *Journal of American Water Works Association,* **81**(11):52–60.

Taylor, J. S., Feb-March 1995, "Drinking Water Regulations and Membrane Applications," *International DeSalination and Water Reuse Journal,* **4/4.**

Taylor, J. S. et al., 1990, "Assessment of Potable Water Membrane Application and Research Needs," *AWWA Research Foundation Report,* Denver, Colo.

CHAPTER 10
ULTRAFILTRATION

C. Anselme
CIRSEE
Lyonnaise des Eaux
France

E. P. Jacobs
Institute for Polymer Science
University of Stellenbosch
Stellenbosch, South Africa

10.1   INTRODUCTION

By the 1930s, polymer ultrafiltration (UF) membranes with pores of various sizes had been developed; important work was done by William Elford, who cast membranes of cellulose acetate and of cellulose nitrate, and who also elucidated most of the principles of UF (Gregor and Gregor, 1978). During World War II the Germans used membranes to analyze for bacteria in drinking water and after the war this technology was taken to the United States. In the succeeding years, the U.S. government became worried about possible shortages of water before the end of the century, and the U.S. Department of the Interior set up the Office of Saline Water (OSW) and later the Office of Saline Water Research and Technology (OSWRT) which provided financial support for research into the desalination of water; a considerable sum was used for development of membrane separation processes for desalination.

The first move toward the realization of large-scale membrane separation was done by Reid and his colleagues at the University of Florida in the late 1950s (Reid and Breton, 1959), and the first practical RO membranes were developed in the early 1960s for desalination of seawater. These membranes resulted from the discovery by Loeb and Sourirajan at UCLA of a method for making an asymmetric cellulose acetate membrane (Loeb and Sourirajan, 1963). The resulting membranes had a dense skin layer integrally supported by a substructure progressively more porous with distance from the skin.

One method used to improve the flux performance of these initial asymmetric RO membranes was to reduce the thickness of the dense skin section of the RO membrane. This led to the development of a new-generation membrane, the thin film composite membrane, which consists of a very thin, dense, permselective layer overlying an ultraporous membrane support base. This development held major importance for ultrafiltration, as the support base that proved most appropriate for the task proved to be an ultrafilter, and major effort was expended to advance its development.

10.2 WATER TREATMENT OBJECTIVES

The principal risks associated with drinking water stem from biological sources. There are about two dozen infectious diseases whose incidence depends on water quality. These diseases may be caused by viruses, bacteria, protozoa, or worms. Other microorganisms present in water include fungi, algae, worms, rotifers, and crustaceans.

Viruses are minute infectious agents (pathogens) that are stable in the environment and commonly transferred by water. They are not cells but particles composed of a protein sheath surrounding a nucleic-acid core. They range in size from 10 to 25 nm. Hepatitis A, a particularly virulent disease, may be spread by water. Although free-chlorine dosages of 0.6 mg/L will inactivate viruses, it is difficult to monitor viral presence (Brock and Brock, 1978).

Protozoa are unicellular, colorless, generally motile organisms that lack a cell wall, and several are pathogenic to man. Some pathogenic protozoa which occur in drinking water are capable of forming spores or cysts that are highly resistant to common methods of disinfection and can be a source of waterborne infection. *Giardia lamblia* and *Cryptosporidium parvum* are of interest in drinking water.

Bacteria are single-cell organisms that live on soluble food in water. They are the basic units of plant life and range in size from 0.5 to 5 μm. Some form resistant spores that can remain dormant in unsuitable environmental conditions and become revitalized when conditions change. They can be grouped into four categories by form: spheroid, rod, curved-rod, and spiral.

Fortunately, these organisms can generally be removed by two operationally simple processes: filtration followed by disinfection. Table 10.1 gives a summary of the relative sizes of some of the smaller types of microorganisms.

10.2.1 Drinking Water Regulations

A general criterion for the quality of potable water is that there be no correlation between the amount of water consumed and its adverse effects on health. As a first requirement, this would require water to be free of pathogens and any harmful or objectionable biological species. Another is that water should not contain contaminants or chemicals that may be physiologically harmful or objectionable to the user. A further requirement is that there must be a minimum aesthetically acceptable standard relating to clarity, color, and objectionable tastes or odors.

Water is an excellent solvent and a medium which supports all forms of biological life. It is obvious that the quality of potable water should be ensured and controlled by means of *safe water* specifications, guidelines, and regulatory standards. However, the growth in population, the increasing costs of treatment and distribution, the quality of the raw water, and the sophistication of the end user, etc., place stresses on available resources, and questions arise regarding the possible risk of

TABLE 10.1 Relative Sizes of Various Forms of Microorganisms

Microorganism	Size, μm
Protozoans	
Giardia lamblia	$5-15 \times 10-20$
ovoid cyst	6×10
Entamoeba histolytica	15×25
cyst	10×15
Yeasts, fungi	$1-10$
Bacteria (*Salmonella, Shigella, Legionella,* etc.)	
Spherical (cocci)	$0.5-4$
Rod-shaped (bacilli)	$0.3-1.5 \times 1-10$
Escherichia coli (human feces)	0.5×2.0
Rod-shaped, curved (vibrios)	$0.4-2 \times 1.0-10$
Spiral-shaped (spirilla)	<50 in length
Filamentous	>100 in length
Viruses	$0.01-0.025$
Hepatitis A	
Proteins (10^4-10^6 kdalton)	$0.002-0.1$
Enzymes	$0.002-0.005$
Antibiotics, polypeptides	$0.0006-0.0012$

Source: Tchobanoglous and Schroeder (1985); Brock and Brock (1978); Tate and Arnold (1990); Gelman and Williams (1983).

consuming tap water. The booming market in bottled water and undersink treatment units are evidence of the reaction from the public.

Whatever standards or regulations are imposed will depend on political, practical, and regional considerations, and it is therefore to be expected that regulatory standards will differ from region to region. All decisions regarding regulatory standards of treatment must, in the end, reflect economic and technological feasibility, and risk assessment and management undoubtedly provide the key to drinking water treatment strategies (Cotruvo and Vogt, 1990; Mallevialle and Fiessinger, 1991).

Table 10.2 gives an excerpt of comparisons of U.S. primary regulations with EEC and WHO guidelines for water quality (Sayre, 1988) and with those of the South African Bureau of Standards (SABS 241, 1984) on some contaminants that could be affected by UF treatment of raw water.

A more in-depth discussion of drinking water regulations in relation to the development of low-pressure membrane filtration technologies in the water industry is presented in Chap. 11, on microfiltration.

10.2.2 Drinking Water Treatment

10.2.2.1 Conventional Treatment Operations (Efficiency and Reliability). The production of clear and sparkling water that is safe as far as disease is concerned requires chemical precipitation, adsorption, sedimentation, and filtration to remove biological forms, colorants, tastes, odors, iron, silicates, and manganese (see Table 10.3). The minimum treatment in the production of water of a potable standard is direct filtration through granular beds, but coagulation, sedimentation, and filtration are the principal unit operations more commonly encountered in the production of clarified potable water.

TABLE 10.2 Standards and Guidelines for Potable Water

Substance	USEPA maximum contamination level*	EEC or French maximum admissible concentration[†]	WHO guideline	SABS 241-1984 (recommended and maximum allowable levels)
Organics				
Trihalomethanes (µg/L)	100	1[‡]	30 (CHCl$_3$ only)	—
Phenolic compounds (as phenol) (µg/L)	—	0.5	—	5–10
Microbial and bacteriological				
Standard plate count (cfu/mL)				
22°C	—	<10 (plant) <100 (distribution system)	0	100–NS
37°C	—	<2 (plant) <10 (distribution system)	—	—
Coliforms (cfu/100 mL)	<1	<1, 95% of samples	0	0–5
Fecal coliform (cfu/100 mL)	<1 (most probable number)	<1, 100% of samples	—	0
Coliphages (pfu/50 mL)	—	<1, 100% of samples	—	—
Enteric viruses (pfu/10 L)	4-log removal	<1, 100% of samples	—	—
Giardia lamblia	3-log removal	—	—	—
Legionella	absent in 1 L	—	—	—
Clostridium perfringens (cfu/20 mL)	—	<1, 100% of samples	—	—
Salmonella (cfu/51)	—	<1, 100% of samples	—	—
Cryptosporidium parvum	3-log removal	—	—	—
Physical and organoleptical	[§]	—	—	—
Turbidity—NTU	1–5	0–4	<1	1–5

	U.S. secondary maximum contaminant level[†]	EEC guide level[†] 80/778/EEC		
Color	15 CU	15 mg/L Pt	15 CU	20 mg/L Pt
Odor at 25°C	3 TON	3 TON	—	Not objectionable
Suspended solids, mg/L	—	25	—	—
Biological oxygen demand, mg/L	—	<3	—	—

CU = color units, TON = threshold odor number, NTU = nephelometric turbidity units, NS = not specified
* Enforceable
[†] Nonenforceable
[‡] Guide level for chlorinated compounds except for carbone tetrachloride (3 µg/L), 1,2 dichloroethane and tetrachloroethylene (10 µg/L), trichloroethylene and chloroform (30 µg/L)
[§] Expected enforced surface water treatment rule: 6-log removal.
Source: Adapted from Pieterse, 1989.

TABLE 10.3 Techniques for the Removal of Suspended Solids

Technique	Unit operation
Physical straining	Cartridge filters
	Diatomaceous earth filtration
	Screening
	Ultrafiltration
Granular media filtration	Downflow sand filters
	Green sand filters
	Multimedia filters
	Upflow sand filters
Flotation	Dissolved air flotation
Gravity separation	Chemical sedimentation
	Clarification
	Primary sedimentation
	Secondary sedimentation

Many impurities in water are present as colloidal species that will not settle naturally or which cannot be removed by direct filtration. With the addition of coagulants (usually requiring 30- to 100-mg/L doses) of commercial products such as ferric chloride or alum, such particles can be caused to agglomerate to the point at which clarification can be achieved by sedimentation followed by filtration. However, the correct dose of coagulant, or optimum pH, must be determined experimentally since it cannot be calculated. Short-term upsets and seasonal changes in the quality of the intake water will affect the quality of the filtrate if dose rates and coagulants used are not corrected timeously (Metcalf and Eddy, 1991), or the mean of an automatic adjustment of the coagulant dose using a sensor such as a streaming current detector (SCD) (Hubele, 1992).

Figure 10.1 gives an example of the influence of an interruption in coagulant feed of a dual-media filter on total coliform removal efficiency. Starting from a 2-log removal efficiency, increasing to a 3- to 4-log removal with maturation of the filter bed, one can obtain no removal at all in the worst-case scenario of an interruption in chemical feed. When coagulant feed is resumed, initial 4-log removal efficiency is reached instantaneously.

Figure 10.2 illustrates the effects of filtration flow rate (velocity) and backwashing sequences on particle count in the size range of 0.5 to 5 μm, at the outlet of a first-stage GAC filter.

Floc formation of colloidal matter at low concentrations is improved by the addition of coagulating aids such as clay particles, which act as nuclei for polyelectrolyte and hydroxide (iron and aluminum) coagulating agents. Floc particles offer a large surface area onto which dissolved material in the water can be adsorbed. This surface-action effect assists in the reduction of colloidal turbidity and of dissolved colorants. About 90 percent of the turbidity and color is removed by the coagulation and sedimentation operations, but a certain amount of floc is carried over from the settling stages and must be removed by filtration.

A number of different conventional filters are used in the production of potable water. The different granular bed filtration systems that have been proposed and built may be classed according to (1) type of operation, (2) filtering medium used, (3) direction of flow during operation, (4) method of backwash, and (5) method to control flow rate. The mechanism by which suspended solids are removed is a com-

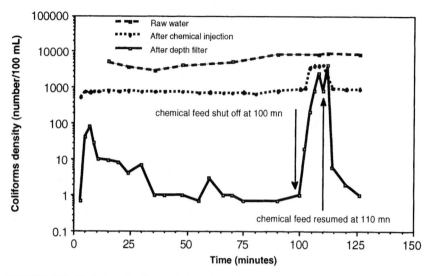

FIGURE 10.1   Evolution of coliforms during a filtration run using dual-stage filtration: effect of an interruption in chemical feed.

plex combination of various phenomena such as straining, interception, adhesion, impaction, sedimentation, flocculation, and physical and chemical adsorption, all with or without the addition of chemicals (flocculants and pH adjustment) upstream of the operation. The end of a filtration cycle becomes evident when the amount of suspended solids that break through from the feed into the filtrate reaches unac-

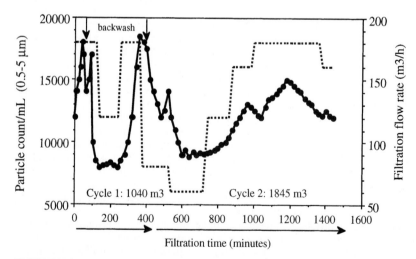

FIGURE 10.2   Treatment of surface water: effect of filtration flow rate on particle count at the outlet of first-stage GAC filter (Prechlorination + coagulation + settling + filtration).

ceptable limits or when the pressure drop across the filter bed reaches limiting values. Figure 10.3 shows how the duration of a filtration run is affected by these two factors (Metcalf and Eddy, 1991). The clarification process (coagulation, sedimentation, and filtration) is complex, and the filtration behavior of a suspension is not easily predicted mathematically.

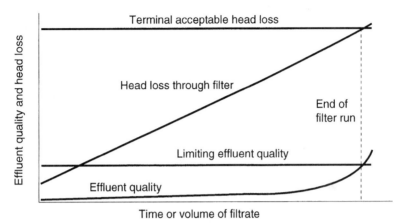

FIGURE 10.3   Definition sketch for the length of filtration run based on effluent quality and head loss. *(Metcalf & Eddy, 1991.)*

Filtration is not a straightforward matter of straining or sieving in granular bed–based systems (Amirtharajah, 1988). Once transported into the voids in a packed bed, suspended solids are attracted to the bed by the same physicochemical forces that act during coagulation. A granular bed is therefore able to remove particles that are considerably smaller than the voids in the bed. However, the small size of microorganisms such as enteric viruses means that their complete removal by coagulation and filtration cannot be guaranteed, even in the presence of measurable levels of free chlorine, and also, even if coliform bacteria (1/100 mL) and turbidity (1 NTU) standards have been met by the final product (Rose, 1985). Poor filter backwash procedures, filter channeling, or improperly operated filtration systems have been associated with the release of large numbers of entrapped *Giardia* and *Cryptosporidium* cysts (Longsdon, 1987; AWWA, 1980).

Postchlorination is required to disinfect filtered water and to ensure a residual free chlorine in the distribution network. However, some questions regarding the efficacy of disinfection and the formation of trihalomethanes arise (Van Steenderen et al., 1989). For example, there are indications that the free-chlorine concentration required to inactivate *Giardia lamblia* is much higher than the levels recommended for the inactivation of polio viruses and coliform bacteria (Wickramanayake et al., 1988; Wierenga, 1985). Chlorination leads to the formation of carcinogenic trihalomethanes, although there have been indications that the removal of humic matter TOC by adsorption, alum coagulation, and lime softening generally results in a reduction in the formation of trihalomethanes upon subsequent chlorination. This reduction was not proportional over the entire range of TOC removal (Jodellah and Weber, 1985).

The CT relationship is used to determine the requirements for inactivation of bacteria and protozoa. Table 10.4 shows the relationship between chlorine concentration and contact time to ensure a 99 percent inactivation of pathogenic agent and the influence of pH and temperature on the CT needed to ensure this limited efficiency. The viral CT values shown in the table are based on the inactivation of Coxsackie A2 viruses which are more resistant than polio virus or any pathogenic vegetative bacteria.

TABLE 10.4 Disinfection CT* for 99 Percent Inactivation of Pathogenic Agents

	Viral CT		Protozoan CT		
pH	0–5°C	10°C	5°C	15°C	25°C
6	—	—	80	25	15
7	12	8	100	35	15
7–7.5	20	15	—	—	—
7.5–8	30	20	—	—	—
8	35	22	150	50	15

* CT = product of chlorine concentration (mg/L) and contact time (min).
Source: From Lippy (1986).

10.2.2.2 Membrane Operations. Membrane separation, in conjunction with activated carbon adsorption, is a proven technology capable of achieving the low levels for total organic carbon (TOC) and reduction of submicron particle levels required of ultrapure water for electronic industry or municipal drinking water (Adham et al., 1993; Anselme et al., 1995). The UF process, part of the train of processes, therefore has many advantages over conventional clarification and filtration operations, and thus new ground may be broken in the field of water purification, dominated until now by chemical coagulation and depth-filtration technology.

The main advantages of low-pressure UF membrane processes when they are compared with conventional clarification (direct filtration, settling/rapid sand filtration, or coagulation/sedimentation/filtration) and disinfection (postchlorination) processes are (see Table 10.5):

• No need for chemicals (coagulants, flocculants, disinfectants, pH adjustment)
• Size-exclusion filtration as opposed to media depth filtration
• Good and constant quality of the treated water in terms of particle and microbial removal, regardless of the raw feedwater quality
• Process and plant compactness
• Simple automation

Ultrafiltration is a pressure-driven process by which colloids, particulates, and high-molecular-mass soluble species are retained by a mechanism of size exclusion, and, as such, provides means for concentrating, fractionating, or filtering dissolved or suspended species (Amy et al., 1987a). UF generally allows most ionic inorganic species to pass and retains discrete particulate matter and nonionic and ionic organic species, depending on the molecular weight cutoff (MWCO) of the membrane.

The MWCO is a specification used by membrane suppliers to describe the retention capabilities of a membrane, and it refers to the molecular mass of a macrosolute

TABLE 10.5 General Efficacy of Water Treatment Processes to Remove Contaminants

Contaminant category	Coagulation processes, sedimentation, filtration	Lime softening	Reverse osmosis	Ultrafiltration	Chemical oxidation, disinfection
Total coliforms	G-E	G-E	E	E	E
Giardia lamblia	G	G	E	E	G
Viruses	G-E	G-E	E	E	E
Legionella	G-E	G-E	E	E	E
Turbidity	E	G	E	E	P
Organics					
VOCs	P	P-F	F-G	P*	P-G
SOCs	P-G	P-F	F-E	P*	P-G
Pesticides	P-G	P-F	F-E	P*	P-G
THMs	P	P	F-G	P*	P
THM precursors	F-G	F-G	G-E	P-F*	P
Color	F-G	F-G	G-E	F	F-E
Iron	F-E	E	G-E	G	G-E
Manganese	F-E	E	G-E	G	F-E
Taste and odor	P-F	P-F	—	—	F-E

P—poor (0 to 20% removal); F—fair (20 to 60% removal); G—good (60 to 90% removal); E—excellent (90 to 100% removal); —insufficient data.
* G-E in case of combined use of PAC.
Source: Modified from Hamann, McEwan, and Myers (1990).

(typically, polyethylene glycol, dextran, or protein) for which the membrane has a retention capability greater than 90 percent. The MWCO can therefore be regarded as a measure of membrane pore dimensions. The retention–versus–molecular-mass curve does not show a sharp or absolute cutoff limit; instead, the shape of the curve depends not only on membrane pore-size distribution, but also on the conformation of the macrosolute being tested (globular, branched, or linear flexible) and on the operational conditions during evaluation (Cheryan, 1986; Gelman and Williams, 1983). There is no set international standard for the determination of MWCO, which means that membranes from different suppliers cannot be compared only on the manufacturers' MWCO specifications.

The diagram in Fig. 10.4 shows the cross sections of stratified granular filter beds after backwashing, and some resemblance between the asymmetry in those structures and that shown in Fig. 10.5 of the cross section of a UF membrane is evident. Some of the main differences which distinguish membrane separation from granular bed filtration are (1) the size of the openings or pores in the membrane surface, (2) that separation takes place at the surface of the membrane, and (3) that feed flow in membrane filtration is across the face of the membrane compared with normal (dead-end) flow with granular bed filtration.

An important feature of asymmetric UF membranes is the presence of a discernible thin skin at the filtration surface. This skin typically has a thickness of 0.1 to 1.0 μm and is supported by a more open substructure. The skin is highly permeable to water and retains suspended and dissolved solids by size exclusion. The minimum diameter of the pores is at the skin, so that, once a solute enters a pore, it will permeate with the filtrate and not be trapped in the membrane where it can cause fouling; this is quite different from the mechanism of separation by conventional depth filters.

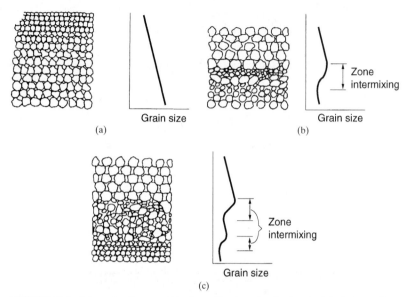

FIGURE 10.4 Schematic diagram of bed stratification after backwash: (*a*) single-medium bed; (*b*) dual-medium bed; (*c*) trimedium bed. *(Metcalf & Eddy, 1991.)*

Membrane filtration is a single process that is very effective in removing many water-soluble organic materials as well as microbiological contaminants. Since all UF membranes are capable of effectively straining protozoa and bacteria from water, the process offers a disinfected filtered product with little load on any post-treatment sterilization method such as UV radiation, ozone treatment, or even chlorination. The same applies to most viruses if as low as 100-kdalton cutoff UF membranes are used.

Table 10.6 shows some relationships between membrane pore size and various MWCOs. These, combined with the data shown in Fig. 10.6, suggest relative filtration size ranges.

Removal of VOCs. Disinfection by chlorine is simple and very efficient, but it leads to the formation of carcinogenic trihalomethanes (THMs) (Van Steenderen et al., 1989). Alternative methods that will reduce the incidence of VOCs in water either by removing the THMs or by removing their higher-molecular-mass precursors before chlorination are under consideration. For this second case, ultrafiltration is obviously able to remove 20 to 50 percent of THM precursors (Laîné et al., 1989, 1990; Jacangelo et al., 1989; Anselme et al., 1991).

THMs are low-molecular-mass organic compounds with sizes smaller than those of pores of UF membranes. However, it has been proved that UF membranes will remove such compounds from water in a stripping operation rather than in a filtration operation. In stripping volatile organics from water, the aqueous phase is brought into contact with a countercurrent gas stream across an ultraporous or microporous membrane interface (Semmens et al., 1989; Bessarabov et al., 1994). The same technique has also been used effectively in bubblefree oxygenation of water. In this case, the UF membrane was first coated with a thin silicone rubber layer to transport air into water by diffusion rather than by convection (Bessarabov et al., 1994).

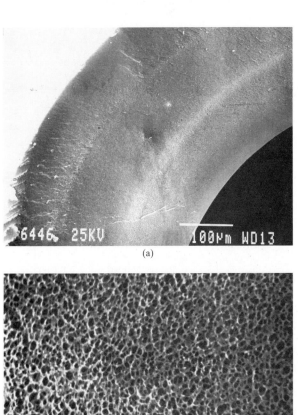

(a)

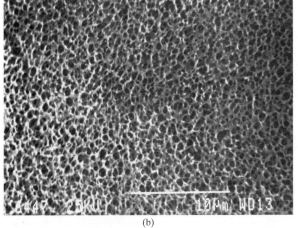

(b)

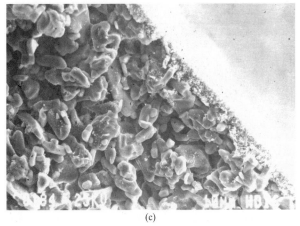

(c)

FIGURE 10.5 Modular structure of a UF polymeric: (*a*) and (*b*) Aquasource; and a ceramic membrane (*c*) Techsep.

10.11

TABLE 10.6 Relationship Between MWCO Ratings and Pore Size for Different Commercial Membranes

Membrane	Supplier	Material	MWCO, dalton	Average pore radius, nm
Nova 1	Filtron	PES*	1,000	2.7
Nova 5	Filtron	PES	5,000	4.78
Nova 10	Filtron	PES	10,000	12.7
Nova 10	Filtron	PES	30,000	8.48
Nova 50	Filtron	PES	50,000	1.28
PM10	Amicon	PSf†	10,000	3.8
PTGC M10	Millipore	PSf	10,000	3.05
PTTK M30	Millipore	PSf	30,000	6.81
gs90	DDS	Sulphonated PSf	1,500	1.1
gs81	DDS	Sulphonated PSf	6,000	2
gs61	DDS	Sulphonated PSf	20,000	3.14
BCDA	Aquasource	Cellulosic derivative	100,000	<10

* polyethersulphone (PES)
† polysulphone (PSf)
Source: Adapted from Tweddle et al. (1992).

Removal of Particulate Matter. UF will reduce the concentration of high-molecular-mass colorants, TOC, and turbidity of feedwater. The process is ideal for the removal of small particles from drinking water. However, in comparison with the number of studies on other applications of the process, relatively little attention has been shown in the literature for this purpose, even though it has been reported that

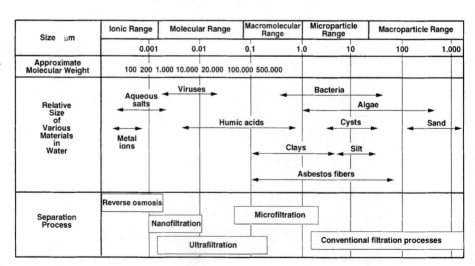

FIGURE 10.6 Selected separation processes used in water treatment and size ranges of various materials found in raw waters (from Jacangelo et al., 1989).

turbidities as low as 0.1 to 0.4 NTU are attainable from feedwaters with turbidity levels of up to 100 NTU and higher (Jacangelo et al., 1989; Bersillon et al., 1989).

The tables on membrane pore size and dimensions of microorganisms indicate that use of an ultrafiltration membrane will guarantee a disinfected product. Theoretically, this is attainable, and practice has shown that UF membranes are quite capable of eliminating or reducing bacterial and viral concentrations in the permeate as low-permeate turbidity may indicate. The process suffers, however, from two disadvantages: membrane and module imperfections and regrowth of bacteria. The membrane fabrication process does not produce a membrane with a skin layer free of imperfections. These imperfections contribute to the broad pore-size distribution characteristic of wet-phase inversion membranes. It is also difficult to achieve a 100 percent success rate in producing the glue lines of a spiral-wrap element. Sufficient coliforms may pass through such imperfections to contaminate the large product-side surfaces of the membrane devices on which regrowth may then take place. Fortunately, all UF membranes and modules are not facing the same imperfection problems and it has been proved, as explained in the following, that one can find industrial UF modules which are able to disinfect water with reliability. In fact, hollow-fiber configuration has demonstrated (Mandra et al., 1994) a higher efficiency and reliability in the disinfection of drinking water as compared to other configurations such as spiral-wound, plate-and-frame, and tubular ones. This is due to the epoxy potting construction of the hollow-fiber modules, which excludes any potential leak of seals which are used for other module constructions, as shown in Fig. 10.7.

In an ultrafiltration study conducted on two surface waters in Northern California, very low turbidity levels were obtained with regenerated cellulosic derivative hollow-fiber (Aquasource, France) capillary membranes with an MWCO of 100 kdaltons. Water from the Mokelumne River was treated after prefiltration to remove material greater than 100 μm. The turbidities of the influent ranged from 0.3 to 0.82 NTU. The turbidity of the recycled water rose to levels of up to 2.54 NTU, but that of the permeate remained at 0.03 to 0.04 NTU. The turbidity of the water influent treated in the other study, that from the Delta River, was much higher and ranged from 11.5 to 24.8 NTU. The turbidity of the recycled water reached levels as high as 55.3 NTU, although the same high-quality, low-turbidity permeate was produced (Jacangelo et al., 1989).

Conclusions from this study were that capillary membranes were useful in reducing turbidity, particle counts, levels of suspended solids, and the numbers of selected indicator bacteria in the two waters tested, as well as viruses seeded in the feedwater. The differences in quality of the two intake waters did not appear to affect the quality of the product water. Removal of microorganisms as high as 8 log for viruses and 5 log for *Giardia* and *Cryptosporidium* was demonstrated. No microorganisms were detected in the permeate, even for sample volume as large as 100 or 1000 L.

Similar results were reached from another study in which membranes were tested in dead-end filtration of untreated surface water. The effects of membrane materials and MWCO on TOC and turbidity removal were studied. It was concluded that a reduction of 40 percent in the TOC content of the water was achieved mainly as the result of the large concentration of low-MW organic components in the water, but a 98 percent reduction in turbidity was achieved (Laîné et al., 1989).

In another study (Bersillon et al., 1989), the first UF plant in operation in Amoncourt (France) in 1987 for municipal drinking water was carefully studied during its first six months of operation. Permeate water was found to exhibit turbidity lower than 0.1 NTU even for resource turbidity as high as 200 NTU.

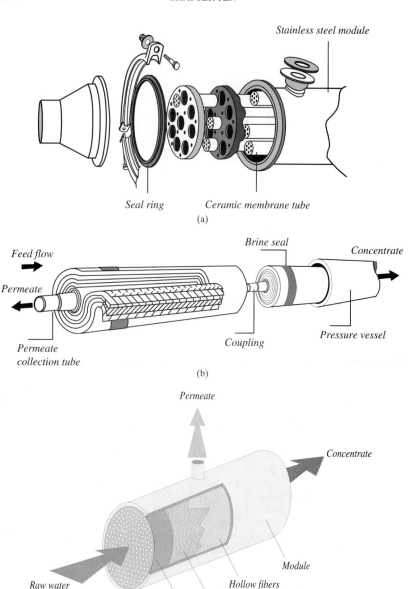

Stainless steel module

Seal ring   Ceramic membrane tube

(a)

Feed flow

Brine seal

Concentrate

Permeate

Permeate collection tube

Coupling

Pressure vessel

(b)

Permeate

Concentrate

Module

Raw water

Hollow fibers

Netting

Potting

(c)

FIGURE 10.7 Comparison of construction characteristics for three types of modules: (*a*) connection of a tubular ceramic module; (*b*) assembly of the spiral-wound RO membrane elements; (*c*) hollow-fiber module—epoxy potting assembly.

10.2.3 UF as Pretreatment to RO/NF Membrane Operations

Pretreatment of process water before RO is very important for membrane life and the economical operation of the RO plant. Conventional methods followed by microfiltration are usually used to produce a feedwater of very low turbidity. A number of cases are reported in which UF was used ahead of RO in the treatment of wastewater streams (Sinisgalli and McNutt, 1986; Cowan et al., 1992), and cases have been reported in which UF pretreatment was used before RO treatment of seawater. The replacement of RO membranes is prohibitively expensive, and the view is held that overdesign of the conventional stage of pretreatment is a better alternative to regular replacement of RO membranes.

Raw water pollutants include suspended solids (e.g., silt, clay, organic material, size range > 1 μm), microorganisms and marine organisms, soluble organic material (oil, fatty acids, polysaccharides, lipids), colloids (0.2 to 1.0 μm), and, sparingly, soluble inorganic matter—most of which can be ultrafiltered. The clarity of the feedwater to the RO plant depends on the type of RO membrane and module used. Hollow fine-fiber modules that produce permeate of the highest quality also require a high-quality feedwater, whereas the quality requirements for feed to spiral-wrap modules are less stringent.

The important features of UF pretreatment are (Heyden, 1985):

- Continuous and easily automated operation
- No breakthrough as occurs in granular media filtration
- Good downstream protection of RO membranes
- No addition of chemicals
- Simple chemical shock disinfection treatment
- Compact design of pretreatment equipment

UF modules of less-complicated design and which allow higher water-recovery rates would have an impact on the economic acceptability of UF pretreatment (Ericsson and Hallmans, 1991). Examples of such devices are the low-cost SWUF tubular units that have been tested with success on the nutrient-rich seawater along the west coast of Namibia. These 13-mm tubular units are operated under conditions of intermittent axial-flow reversal, with simultaneous release of cleaning foam balls. The 40-kdalton MWCO polyethersulphone membranes gave permeates showing SDI-values of 0.5 to 0.7 and turbidity levels of 0.09 to 0.2 NTU. This was achieved routinely with intake seawater that often had turbidity levels higher than 150 NTU. The Du Pont B-10 permeators that were operated on this water showed no decline in flux (Strohwald, 1992; Strohwald and Jacobs, 1992).

Other examples of such application of ultrafiltration are the pilot tests run in France with the Aquasource hollow-fiber module used as pretreatment of spiral-wound thin film composite FilmTec RO and NF modules when treating both a clarified Seine River water and a secondary treated municipal wastewater. In this last example, the membrane promising a market for water reuse is the objective.

The fouling index (Degremont, 1988) of Seine clarified water was consistently measured at between 6 and 10, as after-ultrafiltration values below 0.2 were obtained.

Nonmeasurable fouling indices (greater than 25) were determined for the secondary treated wastewater effluent; this value decreased to below 0.5 after ultrafiltration, allowing a three- to four-month operation duration before the FilmTec NF70 nanofiltration module required chemical cleaning. In that case, microfiltration at

0.2-μm pore size allows a consistent fouling index reduction to less than 2, which allows for a one- to two-month operation for the NF70 nanofiltration module. Table 10.7 summarizes the results obtained in this study (after Laîné and Anselme, 1995).

TABLE 10.7 Impact of Various Pretreatments to NF/RO on Fouling Indexes and Particle Densities

Waters	SDI	Turbidity, NTU	Particle densities, 0.5–5 μm
Raw water	>25	17	sensor saturated
Conventional treatment	5	0.2	$1 \times 10^3 - 1 \times 10^4$/mL
MF	1–2	<0.1	8.9×10^2/mL
UF	<0.5	<0.1	4.8×10^1/mL

10.3 MEMBRANE AND MODULE CONFIGURATIONS

UF is a low-pressure operation at transmembrane pressures of, typically, 0.5 to 5 bars. This not only allows nonpositive displacement pumps to be used, but also the membrane installation can be constructed from synthetic components, which has cost advantages.

Early in the development of industrial ultrafiltration there was a proliferation of module designs, many of which have stood the test of time. However, through the years, equipment manufacturers steered the development of their membrane module devices toward more energy-efficient and compact designs. Current membrane modules are typically modular with high packing density (ratio of membrane area to equipment volume). Most are suitable for scale-up to larger dimensions.

A broad range of membrane devices, useful for small-scale separation in the laboratory or large industrial-scale operation, is available. As regards flux and retention, the general considerations for a module of good design are those of fluid mechanics and mass transfer. Also very important are membrane packing density (m^2/m^3), energy requirements (kWh/m^3 product), simplicity in design, and cost of membrane replacement (Belfort, 1988).

10.3.1 Membrane Geometry and Type

UF membranes can be fabricated essentially in one of two forms: tubular or flat-sheet. Membranes of these designs are normally produced on a porous substrate material (frequently a spun-bonded nonwoven polyester). The carrier material forms an integral part of the membrane; it imparts mechanical strength to the membrane, protects it during fabrication and subsequent handling, and simplifies module construction. Capillary membranes, which are also of the tubular form, are self-supporting due to their small dimensions.

Pressure-driven membranes must be supported structurally to withstand the pressures under which they are operated, and the single operational unit into which membranes are engineered for use is referred to as a *module*. This operational unit consists of the membranes, pressure support structures, feed inlet ports, fluid distributors, and outlet and permeate draw-off points. A single membrane entity is referred to as an *element,* (tubular element, spiral-wrap element, etc.), and these are used to furnish the module.

Table 10.8 lists various suppliers of UF membranes and equipment; the advantages and disadvantages of the major membrane configurations are summarized in Table 10.9.

TABLE 10.8 Suppliers of UF Membranes and Equipment

Company	Location	Products	Estimated sales, millions US$/yr
Alcoa Separations Tech (Membralox/SCT)	Warrendale, Pa.	Ceramic tubes	—
Amicon (Dorr-Oliver)	Lexington, Mass.	Capillary, flat-sheet, spiral-wrap	11
Asahi Kasei	Tokyo, Japan	Capillary	20
Allied-Signal (Fluid Systems)	San Diego, Calif.	Spiral-wrap	—
AquaSource	Rueil Malmaison, France	Capillary	5
Berghof	Tübingen, Germany	Capillary	—
Brunswick Technetics	Timonium, Md.	Leaf	—
Daicel	Osaka, Japan	Capillary	6
De Danske Sukker-fabriker (DDSS)	Nakskov, Denmark		15
DSI	Escondido, Calif.	Spiral-wrap	—
Kalle		Tubular	—
Koch Membrane Systems (Abcor)	Wilmington, Mass.	Tubular (13mm, 25mm) Spiral-wrap	32
Membratek/Debex Desalination	Paarl, South Africa	Tubular (8mm, 13mm) Capillary	3
Millipore		Flat-sheet	—
Nitto Denko	Osaka, Japan	Tubular (12mm) Capillary Spiral-wrap	4
Osmonics	Minnetonka, Minn.	Spiral-wrap	8
Patterson Candy International	Basingstoke, U.K.	Tubular	—
Schleicher & Schull		Capillary	—
Romicon (merged with Koch)	Woburn, Mass.	Capillary	11
Tech-Sep (Rhône Poulenc)	Paris, France	Plate-and-frame Carbon tube Ceramic tube	16
Wafilin	Wavin, The Netherlands	Tubular	—

Source: Taken in part from Eykamp (1990).

TABLE 10.9 Advantages and Disadvantages of UF Membrane Configurations

	Tubular	Capillary	Plate-and-frame	Spiral-wrap
Cost/area	High	Low	High	Low
Membrane replacement cost (not including labor)	High	Moderate	Low	Moderate/low
Flux, Lmh	Good	Good	Low	Low
Packing density, m^2/m^3	Poor	Excellent	Good/fair	Good
Hold-up volume	High	Low	Medium	Medium
Energy consumption	High	Low	Medium	Medium
Fouling	Excellent	Good/fair	Good/fair	Medium
Cleaning in place	Excellent	Good	Fair/poor	Fair/poor

Source: Adapted from Porter (1990).

10.3.1.1 Configuration

Capillary.   Capillary membranes are thin narrow-bore membranes with outside diameters ranging from 500 to 2000 μm. The inside diameter of these membranes may be 25 to 40 percent of that of the outside diameter. Hollow fibers are geometrically the smallest tubelike membranes available, with inside diameters that range typically from 350 to 1000 μm in ultrafiltration.

Capillary membranes are operated without any pressure support because of their small dimensions and the low operational pressures (<5 bars) normally associated with UF. (Although the membranes appear to be fragile, they can have burst pressures as high as 2.5 MPa, which is substantially higher than their continuous operational pressure ratings.)

The morphologies of the substructures of these membranes differ according to the manufacturer. Capillary membranes can be characterized by their various types of internal structure which can be described as internally skinned with a fingerlike (Nitto Denko) or spongelike (A/G Technology) substructure. It is the thin skin layer on the inside of the asymmetric membrane structure shown in Fig. 10.5 that is permselective. However, these membranes may also be skinned on the outside or may even be double-skinned, as illustrated in Fig. 10.8.

Capillary membranes are normally operated in the cross-flow mode and various methods are used to sweep the membrane surface in this manner. In its simplest form, thin-channel fluid management of the feed stream can be achieved by housing internally skinned membranes in a tube-in-shell arrangement and directing the feed through the membrane lumen. The feed solution may also be directed to flow radially from, or parallel to, the membrane axis when the membranes are skinned on the outside. In this case, the permeate is collected in the membrane lumen (Porter, 1990).

Feed and permeate flow may be either cocurrent or countercurrent, depending on whether the permeate in the fibers flows in the same direction as the feed or against it. The distinction is important since the high frictional losses which occur in lumen-flow modules affect the local transmembrane pressure difference. According to Bansal and Gill (1974), countercurrent flow is always superior to cocurrent flow for axial-flow arrangements. A typical example of a lumen-side-fed cartridge module is shown in Fig. 10.9 (Aquasource). Modules of the type shown in Fig. 10.9 can house membranes of area up to 100 m^2.

Another class of modules in which externally skinned membranes are housed are those that operate in transverse-flow mode. In these, the membranes are arranged perpendicularly to the direction of flow. The advantages claimed are the prevention of areas of stagnant flow or channeling, which commonly occur in shell-side-fed

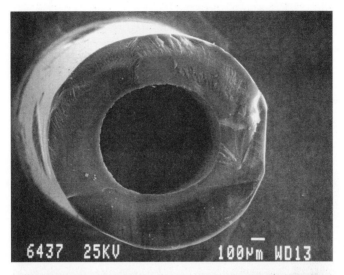

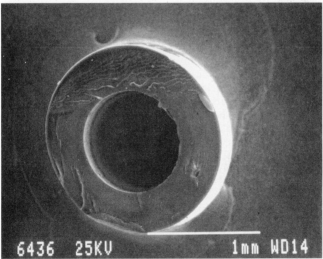

FIGURE 10.8   Micrograph of cross section of double-skinned hollow-fiber Aquasource membrane.

axial flow designs, and higher mass transfer rates and higher shear rates at moderate energy consumption (Roesink and Ràcz, 1989; Knops et al., 1992).

The greatest disadvantage of capillary membranes is the pressure restraint that limits the cross-flow velocity down the lumens of 1.2-m-long membranes. Although shorter membranes will permit higher velocities at 2-bar entrance pressures, the costs per unit length of the shorter modules are higher.

Capillary membranes have the potential of being the most economical of the available membranes, but low manufacturing yields have kept costs on a par with those of spiral-wound modules.

FIGURE 10.9   Industrial-scale hollow-fiber membrane module. *(Aquasource.)*

Tubular Configuration.   The tubular membrane in one form is the simplest configuration in which the membrane is cast on the inside wall of a porous support tube. These tubes have diameters of 6 to 25 mm, with 13 mm being the membrane most commonly used in wastewater treatment. They are less liable to become fouled than membranes of other configurations. Because they provide a simple hydrodynamic flow path, they are most easily cleaned mechanically.

The larger-diameter tubular membranes are self-supporting only at relatively low operating pressures, and the membrane elements are usually fitted on the inside of a porous or perforated pressure support tube. The membranes may be connected either in parallel or in series in a shell-in-tube arrangement. Where high circulation velocities are necessary (e.g., 3 m/s), membranes are arranged in a parallel array inside the vessel to reduce pressure drop across individual membranes and, therefore, modules. Membranes are connected in series when lower circulation velocities (1 m/s) are used.

Porter (1990) noted that there is a trend toward the use of smaller-diameter tubes or of volume displacement rods to reduce the volume of fluid pumped per unit area of membrane.

Some manufacturers have endeavored to reduce costs by using inexpensive materials for making the pressure support tube. With the less-expensive tubes, the membranes are potted in situ and in such a way as to form an integral part of the shroud that acts as permeate collector. In other designs, spent membranes can be replaced, as the membranes are sealed to the tube sheet by means of rubber inserts (see Fig. 10.7).

Disadvantages of the tubular designs are their cost and the low packing density of the membrane modules. A major operational advantage is that tubular membranes can tolerate much higher loads of suspended materials than is possible with any of the other configurations and can therefore be operated with relatively little pretreatment of the feedwater.

Mineral membrane, as illustrated by Fig. 10.10 representing an industrial ceramic Kerasep membrane module from Techsep, are generally available in tubular configuration but new developments are now leading to plate-and-frame mineral membranes.

Spiral Wound.   Spiral-wound modules were originally developed for RO, but sales have expanded increasingly into the UF field. The membrane is essentially a

FIGURE 10.10 Example of a tubular ceramic membrane module. *(Kerasep-Techsep.)*

flat sheet rolled in a prescribed way into a jelly roll or Swiss roll configuration. An envelope of two membranes enclosing a permeate spacer (carrier) is sealed along three edges, and the fourth edge is connected and rolled up onto a perforated tube which carries the product water (see Fig. 10.7). The layers of the rolled-up element are separated by the brine-side or feed-side spacer. The brine-side spacer not only maintains an open channel for feed flow, but also fulfills the very important function of induced mixing to reduce concentration polarization and to improve mass transfer at low energy input. The spacer may be a netting (Da Costa et al., 1993) or a corrugated spacer through which the feed flows parallel to the product tube axis. This spacer improves the solids-handling characteristics of spiral-type modules.

The elements are inserted into pressure vessels; several elements may be fitted into pressure vessels several meters long. Often, a chevron seal is used to seal the outer surface of the module to the inside of the pressure vessel to ensure that the feed is forced to pass between the membrane layers. However, this leads to the formation of stagnant zones of flow on the outside of the element which interferes with in-place cleaning and sterilization. For sanitary reasons, a screen or controlled bypass spacer is used in other designs to permit a small volume of feed to flow in the annular region between the module and the pressure vessel.

The spiral-wound configuration is currently one of the least expensive UF modules as judged by cost per unit of membrane area. Other advantages are concentration-polarization profiles, linear flow velocities (5 to 10 cm/s), and pressure drops (0.7 to 1.05 kg/cm^2), which are low compared with those of other configurations.

However, the elements are more prone to fouling than the supported tubular and some plate-and-frame units (depending on the type of feed channel spacer) but are more resistant to fouling than capillary membranes. However, it is not possible to backwash the supported tubular, spiral-wrap, and plate-and-frame module configurations, an operation generally accepted for capillary membranes.

Plate and Frame. Flat-sheet membranes are used in plate-and-frame modules. The feed solution flows through narrow passages through a membrane stack that can be assembled from individual exchangeable membranes or exchangeable membrane cushions.

The packing density of plate-and-frame units is about 100 to 400 m^2/m^3. The flat-sheet membranes in a plate-and-frame unit offer high versatility compared with other units, but at the highest capital cost.

The configuration has certain advantages. Membrane replacement costs are low, but the replacement labor costs are high. Units can be disassembled to permit access for manual cleaning. In some of the designs, permeate is collected from individual membrane plates which makes the detection of faulty membranes a simple matter.

Millipore Corp. sells a cassette plate-and-frame system. The cassette is a membrane package consisting of two membranes enclosing a filtrate-collection screen sealed around the edges. Alternating holes on the edge of the cassette carry retentate or filtrate to the appropriate manifold. The cassettes are separated by screens or channel plates. When cellular or colloidal suspensions are filtered, the channel plates are preferred to the screens that can collect particles on their cross-members (Porter, 1990).

10.3.1.2 Material

Polymer.   UF membranes are produced from cellulose acetate or noncellulosic engineering plastics (synthetic polymers) such as acrylates or polysulfones. Salt-retention membrane types are generally produced from the relatively more hydrophilic materials (those with hydrogen-bonding properties), whereas the hydrophobic materials are more suitable when mechanical strength and thermal stability are a prime requirement, since water molecules act as plasticizers for hydrophilic materials. For this reason also, membranes fabricated from hydrophobic materials tend to have greater chemical, mechanical, and thermal stability.

The polymer materials from which MF membranes are made are generally crystalline, whereas those used to make UF membranes are amorphous (glassy) polymers. Crystalline polymers show high chemical resistance and thermal stability since the crystalline domains contribute to the effect of cross-linking between amorphous domains and hinder the free rotation of polymer segments. Crystallites inhibit the compaction of polymeric membranes by acting as cross-links.

For UF membranes, which have pores smaller than those of microfiltration membranes, amorphous (glassy) polymers are generally used because the small pore sizes associated with these membranes can be generated, regulated, and controlled with relative ease with these materials during the phase inversion processes used for membrane fabrication.

Polysulfone is one of the most widely used membrane materials principally because of (Cheryan, 1986):

- Wide pH tolerance (continuous exposure range pH 1 to 13)
- High temperature limit (typically 75°C)
- Good resistance to oxidants (chlorine exposure: storage 50 mg/L, short-term sanitation 200 mg/L)
- Wide range of pore sizes (1 to 20 nm) with MWCOs that range from 1000 to 500,000 in commercial-size modules

Polymeric materials with a high glass transition temperature are generally recommended; other materials in this category are polyacrylonitrile, polyethersulphone (slightly more hydrophilic than polysulfone), and polyvinylidene fluoride. The molecular structures of these and various other membrane materials, together with glass transition and melting temperatures, are shown in Table 10.10.

TABLE 10.10   Molecular Structures of Membrane Materials

Polymer	Chemical structure	Tg/°C	Tm/°C
Polyethylene (PE)	$\left(CH_2\right)_n$	$-60 \sim -90$	$137 \sim 143.5$
Polyvinylidene-fluoride (PVDF)	$\left(CH_2-CF_2\right)_n$	-40	$160 \sim 185$
Polypropylene (PP)	$\left(CH_2-\underset{\underset{CH_3}{\mid}}{CH}\right)_n$	-10	$167 \sim 170$
Polycarbonate (PC)		$150 \sim 155$	240
Teflon	$\left(CF_2\right)_n$	-133	327
Cellulose acetate (CA)		—	230
Polyethersulphone (PES)		225	—
Polysulphone (PSf)		190	—
Polyvinylalcohol (PVOH)	$\left(CH_2-\underset{\underset{OH}{\mid}}{CH}\right)_n$	$65 \sim 85$	$228 \sim 256$
Polyacrylonitrile (PAN)	$\left(CH_2-\underset{\underset{CN}{\mid}}{CH}\right)_n$	$80 \sim 104$	319
Polyphenylene-sulphide (PPS)		85	285

Most synthetic UF membranes are hydrophobic and must be stored wet or filled with a wetting agent, since, if they are allowed to dry out without proper treatment, irreversible flux loss might result, the structure of the membrane might collapse, and original wetting will not be achieved.

Due to the nature of colloids and dissolved compounds present in the water, hydrophilic polymers (like cellulosic ones) generally provide membranes which are less subject to fouling than hydrophobic ones (like polysulfone).

Ceramic. Ceramic membranes are made by sintering inorganic materials. Generally, this class of membranes does not exhibit some of the shortcomings of membranes made from synthetic materials, although the resistance of microfiltration ceramic membranes is somewhat better than that of ultrafiltration membranes.

The French company, TechSep, produces a composite tubular membrane which consists of a microporous carbon membrane support, overcoated with a thin zirconium oxide skin layer. SCT, another French company, manufactures an all-ceramic composite tubular membrane. The smallest pore size is 40 nm, and the body of the membrane support is formed from aluminium oxide. The skin layer is formed from aluminium oxide. It was recently shown that unsupported membranes of ferric oxide, with pore sizes as small as 1.7 nm, can be produced (Hackley and Anderson, 1992). TechSep manufactured a titane oxide ceramic membrane (Kerasep) with MWCO of 5 to 50 nm.

Some advantages of ceramic membranes as compared with synthetic membranes are:

- Low maintenance cost
- Wide pH resistance (pH from 0 to 14)
- High temperature resistance (140°C)
- High pressure limitations (2 MPa)
- High flux

TABLE 10.11   Ceramic Membrane Suppliers and Materials

Producer	Configuration	Membrane barrier layer/support	Pore size, μm	Origin
Carre/Du Pont	Tubular	ZrO_2/C	Dynamic membrane	USA
SCT	Multitubular (multiple monolith)	Al_2O_3/Al_2O_3	0.004–5.0 (UF, MF)	France
TechSep Kerasep	Tubular (multiple monolith)	$TiO_2/TiO_2/ZrO_2$	MF (wide range) UF (0.05–0.005)	France
Nihon Gaishi	Tubular (multiple monolith)	Al_2O_3/Al_2O_3	0.2–5.0 (MF)	Japan
Le Carbone Lorraine	Tubular	Carbon composite	0.05–1.4	France
Norton	Tubular	Al_2O_3/Al_2O_3	0.2–1.0 (MF)	USA
TechSep	Tubular (multiple monolith)	ZrO_2/C	0.002–0.15 (UF,MF)	France
Shott Biotech	Capillary	—	0.01–0.09 (UF, MF)	Germany
TDK	Tubular	Al_2O_3/Al_2O_3	Dynamic membrane	Japan

Source: Adapted from Flemming (1987).

- Sharp selectivity
- Resistance to fouling

Table 10.11 gives a summary of some commercially available ceramic membranes.

10.3.2 Compactness

Although UF has been accepted as an industrial filtration operation for more than two decades, existing synthetic membranes are still exhibiting a number of short-comings, the most important of which are fouling, short life expectancy, nonuniform pore-size distribution, low surface-pore density, lack of chemical inertness, low temperature tolerance, and low mechanical integrity. Advantages and disadvantages of UF membranes of different configurations are given in Table 10.9.

Table 10.12 lists names of some commercial suppliers and commercially available membranes with reference to geometric shapes and some performance data.

10.4 SPECIAL CONSIDERATIONS IN PROCESS EVALUATION

10.4.1 Characterization of UF MWCO

UF membrane pore sizes are usually characterized by molecular weight cutoff (MWCO). One of the major discrepancies in the membrane industry is reports concerning membrane cutoff. Depending on the nature and the type of organic molecule and also on operating conditions, an MWCO cutoff characterization may vary by several thousand daltons. Furthermore, membrane cutoffs are often reported using various qualifications such as nominal, apparent, average, and also sometimes absolute cutoffs.

10.4.1.1 Relationship Between MWCO and Targeted Contaminants. UF membranes that have MWCO in the range 1000 to 100,000 dalton and with flux properties suitable for potable water production are available commercially.

In one study it was established that more than 85 percent of the colorant content of a surface water could be attributed to macromolecules with a molecular mass greater than 20 kdalton (Amy et al., 1987a). An analysis performed on Mississippi River water indicated that 40 percent of the organic content in the water was hydrophobic and 60 percent hydrophilic. On the basis of UF fractionation and TOC analysis on the same water, the molecular mass distribution of the macromolecules in this water was found to be: 23 percent <1 kdalton, 15 percent 1–10 kdalton, 50 percent 10–100 kdalton, 17 percent >100 kdalton (Semmens and Staples, 1986).

This illustrates one of the problems associated with the choice of ultrafiltration membranes for a given application: surface water composition changes from site to site, and so will the membranes chosen to treat the waters. For reasons such as these, pilot plant studies are recommended, even though existing plants might be operating satisfactorily.

The average radius of pores in tight UF membranes is less than 5 nm, which would translate, on the basis of PEG, into an MWCO of less than 40 kdalton. Table 10.6 lists membrane MWCO ratings and pore sizes of various commercial membranes. Figure 10.6 shows the size ranges of various materials found in water against

TABLE 10.12 Properties of Various Ultrafiltration Membranes

Polymer	Membrane	MWCO	Water flux, m/d	pH range	Max. temp, °C
Cellulose acetate	PCI tubular, T5/A	20,000	>0.7	2–8	>50
	PCI tubular, T4/A	6,000	1	2–8	30
	Sartorius, flat sheet, SM 14549	20,000	6	4–8	50
	Sartorius, flat sheet, SM 14539	10,000	0.7	4–8	50
	Millipore, flat sheet, PCAC	1,000	0.5	4–8	
	Millipore, flat sheet, PSVP	1,000,000		2–10	
Polysulphone	DDS, flat sheet, GR10PP	500,000	5–10	1–10	80
	DDS, flat sheet, GR40PP	100,000	3–7	1–13	80
	DDS, flat sheet, GR60PP	25,000	5–10	1–13	80
	DDS, flat sheet, GR90PP	2,000	>1	1–13	80
	Abcor, spiral, HFK 30	5,000	3	1–13	85
	Amicon, capillary, H5P1	1,000	~1	1.5–13	75
	Amicon, capillary, H5P10	10,000	2–4	1.5–13	75
	Amicon, capillary, H5P100	100,000	5–10	1.5–13	75
Hydrophilic polysulphone	DDS, flat sheet, GS61PP	20,000	5–10	1–13	80
Polyamide	Abcor, tubular, HFD-300	8,000		2–12	80
Polyvinylidene difluoride	Abcor, spiral, HFM180	18,000		1–13	90
Fluoropolymer	DDS, flat sheet, FS50PP	30,000	5–10	1–12	80
Acrylonitrile	Asahi Chemical Co, capillary, FCV3010	13,000	5	2–10	40
Zirconia/carbon	SFEC/Carbosep, tubular, M4	20,000	4–10	0–14	140
	SFEC/Carbosep, tubular M1	50,000	4–10	0–14	140
Alumina	Ceraver, tubular	40 nm	1		>100
Polyethersulphone	Membratek, tubular, M719 A/G Technologies, capillary	40,000		1–10	70
Cellulosic derivative	AquaSource, capillary	100,000	6	4–8.5	35

Source: In part from Gutman (1987).

various filtration options. Obviously, the pore sizes of most of the membranes are below the sizes required to remove pathogenic forms from water.

However, membrane pore sizes are not absolute. Although a membrane may show sharp cutoff, retention by the sieving mechanism is probabilistic in character in the sense that it depends on the size distribution of species in the feed stream, the pore-size distribution on the membrane surface, and the possibility that species that may pass through pores of a given size may never encounter such pores (Gelman and Williams, 1983).

UF cutoff characterization was probably derived from the food industry, which employs UF technology to concentrate macromolecules. However, for water industry applications, UF technology is employed for clarification/disinfection purposes. Therefore, from a health perspective, the cutoff must or should be defined very precisely. To date, there are no techniques available to accurately determine pore size of

UF membranes (Cheryan, 1986) even with high-resolution technology such as electron microscopy. A French standard method (AFNOR, 1992) is being proposed to characterize membrane molecular weight. Various macromolecules are proposed for the MWCO characterizations:

- Dextrans, ranging in molecular weight from 5000 to 500,000 daltons or more
- Globulin proteins, ranging in molecular weight from 10,000 to 70,000 daltons
- Polyvinylpyrolydones, ranging in molecular weight from 4000 to 400,000 daltons

Removal of these macromolecules by UF membranes is determined using gel permeation chromatography coupled with ultraviolet spectrophotometry. Various operating conditions are also recommended:

- Pressures of 0, 3, 6, and 9 lb/in^2 (0, 0.2, 0.4, and 0.6 bar)
- Cross-flow velocities corresponding to Reynolds numbers of 600 and 1700
- Operating temperature of 68°F (20°C)

Removal profile curves are then generated for each set of compounds at various conditions and an apparent molecular weight is then estimated for a 90 to 95 percent retention of a given MW. It should be noted that if apparent MWCO can be defined using this method, there is no specific and precise characterization of the pore-size distribution. Therefore, it appears that maybe the best way to evaluate a membrane MWCO would be to challenge the membrane with the target water contaminants to be removed (virus, bacteria, etc.) since the European method (AFNOR, 1992) does not provide a disinfection characterization MWCO but only an apparent MWCO, which may or may not be directly related to the membrane disinfection efficiency.

Also, for water disinfection purposes, not only the membrane but also the module must be evaluated in order to avoid misinterpretation of disinfection efficacy due to module imperfections (seal leakages).

10.4.1.2 Mechanisms Involved in Rejection of UF Membranes. An ultrafiltration membrane can be thought of as a screen allowing the retention of particles larger than the largest pore diameter on its surface.

The cutoff threshold of an ultrafiltration membrane can be defined with respect to the molecular mass of macromolecules for which retention is higher than 90 to 95 percent—that is to say, within the ultrafiltration range of several thousand daltons to 500 kdaltons or with respect to an equivalent size of the retained particles by the membrane of 1 to 50 nm. The rejection (or retention) rate for an ultrafiltration membrane for one of the compounds in a solution to be treated can be determined either by porosity (average pore diameter and distribution of pore size on the membrane surface), the membrane's cutoff threshold, or the physicochemical characteristics of the filtrated material.

Therefore, one can distinguish the purely physical mechanism of particulate screening related to the characteristics of the studied membrane from retention mechanisms related to filtration hydrodynamic conditions (tangential velocity, transmembrane pressure), the physicochemical properties of the liquid to be filtered (organic matrix, mineral matrix), and the physicochemical properties of the membrane material (surface charges, polarity) responsible for possible fouling of the membrane.

Rejection mechanisms leading to the definition of the cutoff threshold of an ultrafiltration membrane are very complicated since they depend on the membrane, the fluid to be filtered, and operating conditions, which can be modified with time by fouling and membrane aging. Table 10.13 includes different parameters that can

TABLE 10.13 Physical and Chemical Factors Contributing to the Main Mechanisms of Rejection by an Ultrafiltration Membrane

Membrane-related factors	Rejection mechanisms
Membrane cutoff	Physical straining
Average pore size	by membrane
Pore-size distribution	by cake layer
Membrane surface charge	by gel layer
Membrane roughness	Electrostatic repulsion
Membrane material adsorption	Adsorption
Capacity	Hydrodynamic
Feeding solution–related factors	
pH, mineral content, ionic strength salts and metallic oxide precipitation	
Organic content, fouling, gel polarization, cake formation, adsorption on or in the membrane	
Particle content, cake formation, pore blocking	
Hydraulic regime–related factors	
Cross-flow velocity	
Transmembrane pressure	
Laminar or turbulent flow	

influence the rejection rate determination of an ultrafiltration membrane as well as the main physicochemical mechanisms leading to that rejection.

10.4.2 Membrane Mechanical and Chemical Resistance Required for Reliable Results in Field Application

Optimization of the ultrafiltration process should be considered as a whole, including the definition of mechanical resistance and chemical properties of the membrane. These properties are directly related to on-site aging and thus to membrane life span as well as frequency of replacement of the modules that constitute an ultrafiltration plant. Under actual market conditions, membrane replacement still represents 30 to 60 percent of the cost of producing a cubic meter of ultrafiltered water. Therefore, it is of prime importance to determine the mechanical characteristics required in advance in order to ensure satisfactory module aging.

Starting from the initial properties of the fibers and kinetic laboratory studies of fiber aging with regard to different parameters influencing life span (pH, temperature, free chlorine, the nature of regeneration agents, regeneration frequency, pressure applied during filtration or backwashing, backwashing frequency, etc.), it then becomes possible to limit industrial risks. Of course, theoretical studies carried out in the laboratory should be completed by monitoring on-site aging of working modules in operating production units.

In the following paragraphs, the general case of an ultrafiltration hollow fiber will be discussed. However, it is clear that this kind of rationale can be adapted to flat-sheet, tubular, or spiral-wound membranes.

10.4.2.1 Definition of Initial Mechanical Properties of a Hollow Fiber.
Besides the characteristic hydraulic properties of a hollow fiber (clean water permeability, cutoff threshold), it is possible to define characteristic mechanical properties of the actual resistance of the fibers (Fig. 10.11):

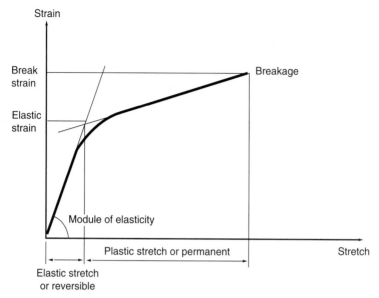

FIGURE 10.11   Strain and stretch diagram of a hollow-fiber membrane.

- Break strain and stretch
- Elastic strain and stretch

The *break strain* is the longitudinal stress that must be applied to a fiber in order for it to break. The *break stretch* is the irreversible deformation of the fiber when subjected to break strain. The *elastic strain* is the maximum longitudinal strain that can be applied in order for the associated deformation (*elastic stretch*) to remain reversible. The relationship between strain and stretch pertaining to elasticity is called the *elasticity modulus* and characterizes the rigidity of the hollow fiber.

Parameters such as hydrostatic bursting pressure and elastic collapsing pressure are the characteristics of radial mechanical resistance of the fibers.

Hydrostatic bursting pressure is the minimum internal pressure applied to hollow fibers that produces bursting. This pressure is measured by applying increasing air pressure to the hollow fiber until bursting occurs. The pressure at which the membrane structure is sufficiently deformed to allow air bubbles to pass through the fiber is called *bubble point pressure*. This characterizes an irreversible fiber deformation that makes it permeable to air and precedes bursting.

Elastic collapsing pressure is the maximum pressure that can be applied to hollow fibers in order for wall deformation to remain reversible. At higher pressures than this, the fibers collapse.

Obviously, these mechanical properties are a function of the characteristics of the fiber, such as:

- Type of polymer
- Internal/external diameter relationship (fiber thickness)
- Fiber structure (homogeneous, dense, asymmetrical)

It is rather difficult to give ideal mechanical resistance values for ultrafiltration hollow fibers, and it is best to remain reserved on the subject because there is a pronounced correlation between the mechanical and hydraulic properties of hollow fibers (for example, between the permeability of a fiber and its mechanical resistance). Therefore, it is necessary to find a compromise. In Table 10.14, maximum values are given for the case of a fiber with an internal diameter of about 1 mm with a diameter ratio of 1.7 and permeability lower than 0.3 $m^3/h \cdot m^2$ and a cutoff of 100 Kdalton.

TABLE 10.14   Mechanical Properties
Characteristic of Ultrafiltration
Hollow Fibers

Break strain: 10 N	
Break stretch:	30 to 35%
Elastic strain:	5 N
Elastic stretch:	2 to 3%
Break pressure:	>20 bars
Bubble point pressure:	15 bars
Elastic collapse pressure:	>4 bars

10.4.2.2   Enhanced Mechanical Aging.   The values given in Table 10.14 should be compared with pressures generally used on this type of hollow fiber. In ultrafiltration with a cutoff threshold of 100 kdalton, applied filtration pressures vary from 0 to 2 bars, whereas backwashing pressures can vary from 0 to 3 bars. In comparison with the values in Table 1, it is noticeable that the operating pressure values are about 10 times lower than the break pressure and 2 times lower than elastic collapse pressure. However, it should be noted that, for UF membranes, burst pressure values may be much higher than the maximum recommended operating pressure because the membrane fiber at high pressure may not break but may lose its operating performance due to a compaction of the membrane fiber material, as presented subsequently. Furthermore, recommended operating pressure (different from maximum operating pressure) may also be provided by a manufacturer in order to minimize fouling.

During filtration/backwashing cycles, the membrane is subjected to alternating internal and external pressure, which can induce dynamic fatigue, leading to modification of mechanical characteristics. To obtain estimates of membrane operation life spans, it is possible to subject the fibers to accelerated pressure/reverse pressure cycles on a laboratory automatic aging bench.

For a fiber like the one described in the preceding example and with pressure/reverse pressure cycles of 2, 3, or 4 bars, the number of cycles tolerated by the fiber prior to breakage is between 180,000 and 200,000 at 2 bars and 100,000 and 150,000 at 4 bars.

Life spans are determined by extrapolation from experimental results, taking into account the fact that the mechanical life span of the fibers is directly related to inversions of pressure applied to them. Therefore, the number of filtration/backwashing cycles based on a backwashing period every 30 min yields a life span of more than 10 years at 2 bars and 6 to 9 years at 4 bars. Of course, these values should be considered as a minimum, since backwashing periods often range from 1 to 3 h, and it is very rare to have a potable water production unit operating 24 h a day.

10.4.2.3   Chemical Aging.   During operation of a water treatment plant, the fibers may be subjected to variations of pH and temperature and the action of different oxidizing or detergent reagents. Therefore, it is indispensable to estimate the effect these products may have on the membrane.

A different sensitivity of the membrane to the aforementioned chemical stresses is a function of the nature of the polymer constituting the membrane. For example, polyethylene or polypropylene membranes are so sensitive to oxidizing agents such as chlorine that the use of chlorine upstream from these membranes is generally forbidden by the manufacturers. At the same time, cellulosic derivative membranes can be hydrolyzed at a certain pH, which limits the use of these materials to fluids with a pH ranging from 4 to 8.5. Generally, this is not a limiting factor in water production applications.

On the other hand, polysulfone membranes are known for their chemical and thermal stability, which makes their use possible for a large range of fluids other than simply for water treatment. Studies of chemical aging of ultrafiltration fibers can be carried out in the laboratory based on the following protocols.

Influence of Temperature and pH.   In the case of a polymer sensitive to hydrolysis, it is possible to measure the decrease in break strain resistance provoked by both different pH and temperatures.

Therefore, it is possible to define a time (T10) corresponding to a 10 percent decrease in break strain resistance. For cellulosic fibers, a decrease by a factor of 10 of the T10 can be noted for each supplementary pH value starting at pH 7 at 25°C. On the other hand, the lower the temperature of the water, the slower the hydrolysis kinetics will be and, thus, the life span of the fibers will be longer for the same pH. Furthermore, there are not necessarily correlations between T10 and the results of an enhanced mechanical aging test such as that previously described.

A T10 value higher than the expected module replacement frequency (three to five years) is a requirement.

Effect of Chlorine.   In potable water treatment, chlorine is one of the most widely used disinfecting agents. Ultrafiltration membranes may be in either continuous contact (prechlorination of raw water) or occasional contact (injection during backwashing, presence of chlorine in chemical regeneration products) with chlorine.

The oxidation of fibers due to process phases in which chlorine is used as a treatment reagent can be quantified in terms of the product (maximum concentration in chlorine) times (contact time of the fibers with chlorinated solutions). This limit value or CT should be determined as a function of pH and temperature, since these two parameters may influence the oxidation kinetics of the membrane due to chlorine. The limit value of CT of the membrane with chlorine can also be defined as the product concentration times contact time leading to a 10 percent lessening of the mechanical properties of the fibers (CT_{10}).

In practice, due to the nature of the polymer used to manufacture the fibers, CT_{10} values as variable as the following can be observed:

- Polysulfone > 10^6 ppm·h
- Cellulosic derivatives > 10^5 ppm·h
- Acrylonitrile < 10^3 ppm·h
- Polypropylene < 10^2 ppm·h

These values forbid the use of chlorine in the case of the two last polymers. For the first two types of fiber, contact time extrapolation for 1 ppm of chlorine leads to life spans greater than 10 years. Obviously, the same reasoning used for chlorine can be applied to the study of other oxidizing agents (hydrogen peroxide, ozone, chlorine dioxide).

Effect of Detergents.   Detergent formulations are often recommended for the cleansing of ultrafiltration membranes. Besides the presence of detergents, one can notice the use of mineral (silicates, phosphates, etc.) and organic (NTA, EDTA, gluconates, citric or oxalic acid) chelatants, and the presence of oxidants (chlorine, peracetic acid, hydrogen peroxide) or hydrolyzing agents (enzymes). The pH of these solutions is either buffered to neutral values to avoid destruction of membranes sensitive to this factor or, on the contrary, set to lower values (pH < 2) if one wishes to dissolve oxides and hydroxides (iron, manganese) or higher values (pH > 10) in the case of an alkaline hydrolysis aimed at breaking organic fouling.

Studies of membrane aging due to these detergent solutions goes back to the study of simultaneous action of pH and oxidizing agents. In practice, the concentrations used in operation amount to several grams per liter for a contact time of several hours at a temperature which is frequently higher than 30°C.

Generally, the CT_{10} determined for a membrane/detergent couple makes it possible to carry out hundreds of chemical washes, which is certainly a much greater number than the membrane will be subjected to during its life span.

10.4.2.4   On-Site Aging.   In actual on-site operations, a membrane is subject to all the various types of stress (mechanical, chemical, and physical) previously discussed. Therefore, it is indispensable to verify that on-site aging of a membrane can be extrapolated from laboratory experiments and monitor the aging process on modules at different stages of their life spans.

An aging study carried out on a module brought back from a site should include at least the following different phases:

- Tests to be carried out on the entire module:
  Integrity test
  Hydraulic permeability measurement
  Circulation head-loss measurement (of the hollow fibers)
- Tests to be carried out on the fibers after module autopsy:
  Verification of the surface condition by microscopy (erosion, bacterial attack, hydrolysis)
  Measurement of mechanical properties
  Determination of mechanical resistance by an enhanced aging test
  Evaluation of variations in chemical resistance (pH, oxidant, detergent)

10.4.3   UF-Specific Recommendations for Pilot Plant Experimentation

Besides its demonstrative nature, a pilot plant experiment is generally carried out in order to be able to design a future industrial-scale plant with precision. The precision level obtained concerning the hydraulic performance of the studied membrane should make it possible to defray the relatively high costs of such experimentation by savings realized on the membrane surface area to be installed.

In order to avoid all interpretation errors and to allow a foolproof extrapolation of the results obtained at the pilot-scale level, it is necessary to comply with a few simple rules in the operation of the pilot experiments. These rules concern the following items:

- Module (size and geometry)
- Pilot plant (hydraulic flowchart)

- Production objectives (flux, energy consumption, overall yield, etc.) for the future plant
- Experiment duration and significance of the raw water quality variations during the experiment.

10.4.3.1 The Module.

The module used for the realization of a test pilot should correspond (in terms of size and geometry) at least to the basic unit of the industrial module that will be used for the full-scale plant. Ideally, if possible, it is also recommended to use the full-scale industrial module itself.

The most frequently encountered errors concern the use of nonindustrial-sized hollow-fiber modules. In this case in particular, it is indispensable to respect the following similarities between pilot modules and industrial modules: diameter, length, and packing density.

These characteristics directly determine the module's performance in terms of production rate, since they influence the filtration and backwashing pressure profiles and, therefore, the actual pressure applied to the membranes during different operation phases. In particular, when using a module of hollow-fiber configuration, a module with a bundle diameter and package density identical to the industrial module must be used in order to assume the same head loss during backwashing as the one that will occur at industrial scale (same backwash efficiency).

10.4.3.2 The Pilot Plant.

The hydraulic design, the process, and the equipment used for regulation phases should be identical on both the pilot and the industrial installation. In particular, the tubing volume/m^2 of installed membrane ratio in the recycle loop must be respected in order to operate the pilot plant membrane(s) under identical shear stress and concentration conditions as those that will be experienced by the industrial unit.

10.4.3.3 Productivity Objectives.

The performance of an ultrafiltration membrane is determined in particular by the hydraulic regime (shear stress, tangential velocity) and concentration (or conversion) rate required. Generally, these parameters are imposed not only by the membrane being used, but also by the economic objectives included in the project being studied. Investment costs are usually of primary importance, which imposes a reduced variation scale for the production flux/conversion rate parameters. The operating cost limits are often estimated, which can limit optimization possibilities, especially the following:

- Energy costs (filtration flow rate or tangential velocity, recycle velocity)
- Water loss (backwashing frequency, module conversion rate, production flux)
- Reagent costs (chemical washing frequency)

Table 10.15 gives an example of the influence operating conditions have on investment and operating costs involved in the use of ceramic ultrafiltration membranes (cutoff threshold 50 nm). In this example, it has been verified that the membrane performances are directly influenced by the frequency of the chemical cleaning used, since the production flux can be doubled with daily chemical washing as opposed to weekly washing. This is the operation method that makes it possible to obtain the best economic compromise when the membrane surface to be installed is restricted.

TABLE 10.15 Comparison of Operating and Capital Costs for a Tubular Ceramic
UF Membrane at a 0.5-mgd (2000-m^3/day) Plant Capacity
Under Two Hypotheses of Operation

Membrane surface to be installed	570 m^2	275 m^2
Flux L/h·m^2 (20°C)	150	300
Cross-flow velocity, m/s	3	3
Backwashing: duration, s/frequency, min/pressure, bar	30/30/5	30/30/5
Chemical cleaning frequency	1/week	1/day
Water loss, %	6	3
Operating costs, F/m^3	2.00	1.60
Energy	0.60	0.30
Coagulant	0.10	0.10
Chemical cleaning	0.30	0.70
Membrane replacement	1.00	0.50

10.4.3.4 Test Pilot Duration. This is also one of the most flexible points with
respect to how representative the obtained results are. A plant design derived from
a test pilot can be only as reliable as the duration of the pilot allows the verification
of performances over a sufficiently long period to include numerous chemical washings of the membrane and to take the variability of resource quality into account.

10.5 DESIGN CRITERIA

10.5.1 Raw Water and Pretreatment Considerations

10.5.1.1 Membrane Fouling: A Brief Review of Mechanisms. Fouling is the limiting phenomenon that is responsible for most of the difficulties encountered in the
generalization of membrane technology for water treatment. Since ultrafiltration is
certainly not exempt from this problem, fouling control and, therefore, membrane
productivity are still important subjects which should be thoroughly researched in
order to have a better understanding of these phenomena and mechanisms.

The term *fouling* includes the totality of phenomena responsible for decrease of
permeate flux over a period of time, except those linked to membrane compaction
and mechanical characteristics modification. By definition, these phenomena are
dynamic and depend on such parameters as physicochemical properties of the membrane, characteristics of the fluid to be filtered, and hydraulic operating conditions
of the system.

Reversibility of these phenomena is therefore one of the critical points to be
resolved. Even though the totality of phenomena leading to membrane fouling are
not clearly understood today, understanding of these mechanisms is still sufficient to
allow construction and operation of UF industrial units for potable water production. Several methods are available that make it possible to cope with, if not totally
control, fouling. These techniques will be discussed later in this section as well as the
following one (pretreatments, chemical washing, hydraulic washing, hydrodynamic
operating conditions: flow rate, shear stress, operating pressure). Technical solutions
making it possible to limit fouling allow industrial implementation of the process,
either by limiting the required membrane surface area or by using less energy per
cubic meter of produced permeate.

Mechanisms leading to fouling of an ultrafiltration membrane can be described by the following physicochemical phenomena.

Concentration Polarization. Strictly speaking, polarization is not a fouling phenomenon: it is only the consequence of the solute being drawn toward the physical barrier of the membrane which stops these solutes (Laîné et al., 1991). This accumulation of solutes on the membrane wall (concentration gradient) has the effect of the occurrence of a maximum permeate flux which can be interpreted as one of the limiting phenomena leading to fouling of the membrane. Theories taking the concentration-polarization phenomenon into account are based on a simple basic equation that provides for equality between the convective flux (proportional to the solute concentration at a distance x from the wall $C(x)$ and the diffusive flux of the solute (proportional to the concentration gradient dc/dx at distance x from the membrane). The equilibrium permeation flux can also be expressed as:

$$v_{\lim} = D \ln \left(\frac{C_w}{C_b} \right) \qquad (10.1)$$

where $v_{\lim}$ = equilibrium permeate flux
         D = solute diffusivity
         C_w = solute concentration on the wall (C_{wall})
         C_b = solute concentration away from the wall (C_{bulk})

The equilibrium permeation flux is directly influenced by the solution concentration and hydraulic filtration conditions (flow rate, pressure). This phenomenon is considered reversible, since stopping the filtration operation provokes the disappearance of the phenomena leading to the existence of the solute concentration gradient. However, if the concentration on the membrane wall is important enough for the solute to gel, reversibility becomes unlikely unless a hydraulic or chemical washing process is used.

Cake Formation (Reversible Fouling). Accumulation of particles on the membrane wall leads to the formation of a cake, which may be considered as a second membrane whose hydraulic resistance adds to the initial resistance of the membrane and eventually leads to a flux reduction. Generally, this type of fouling is partially reversible by hydraulic washing techniques (flushing, backwashing).

Natural Organic Matter Adsorption (Nonreversible Fouling). Natural organic matter (NOM) in waters can lead to membrane fouling, either by surface adsorption, on the particles making up the filtration cake, thereby giving the cake cohesion, or by adsorption in the bulk of the membrane (pores, backup structure). All of these phenomena can be described as fouling due to adsorption and depend mainly on the affinity that the NOM have for the polymer constituting the membrane. This adsorption may be considered a dynamic equilibrium phenomenon with a slow kinetic rate, which leads to progressive saturation of adsorption sites of the membrane material. This type of fouling is difficult to reverse or is slowly reversible because it requires desorption of the organic molecules. This can be effective only when a sudden drop in the NOM concentration occurs in the feedwater. This makes it possible for the adsorption equilibrium to shift towards desorption, thereby cleaning the membrane. The use of oxidizing shocks (chlorine, for example) during membrane backwashing or chemical washing generally makes it possible to limit loss of membrane permeability. However, it has been shown (Mallevialle et al., 1989; Laîné et al., 1989) that hydrophilic polymer membranes (acrylonitrile, cellulosic derivatives, etc.) are less sensitive to this type of cleaning than hydrophobic membranes (polysulfone, polypropylene, etc.). In the case of the second type of membrane, chemical bonds involved in the adsorption process are of higher energy and therefore are more difficult to break, even with chemical cleansing techniques.

Calcium, Iron, and Manganese Precipitation. Unlike nanofiltration (NF) and reverse osmosis (RO), ultrafiltration processes do not generally allow sufficient dissolved salt retention, which can lead to mineral precipitation on the membrane due to ion concentrations higher than solubility products. Nevertheless, precipitation of certain minerals in ultrafiltration processes is a phenomenon that has been identified as one of the causes of fouling. Precipitation phenomena are limited to calcium carbonate precipitation in unbalanced waters (scaling waters) or to precipitation of dissolved metals such as iron and manganese due to oxidation and hydrolysis during the filtration process.

Dissolved metal precipitation, which leads to the formation of an iron oxide and manganese cake on the membrane, is the phenomenon that has been identified most frequently in operating treatment plants. In ultrafiltration, oxidation of iron and manganese is often related to the use of an oxidant during certain process steps (aeration or pretreatment, chlorine injection during backwashing, etc.). Iron and manganese precipitation on the membrane lead to the fouling of the membrane in the concentrate side when operating in filtration. This fouling is reversible by backwashing, but due to an increase of the cake thickness an increase in longitudinal head loss of the module is generally observed, which can induce an increase in energy consumption for concentrate recycling.

In the case of iron/manganese precipitation during the backwashing of the module, a membrane can be fouled on the permeate side, and, consequently, a decrease of the permeability of the membrane during the backwashing phase is observed. This decrease of the backwash flow leads to a lack in efficiency of the backwash, which results in a complete fouling of the membrane that needs a chemical cleaning.

Fouling Reversibility: Model of Hydraulic Resistances in Series. This model gives a relation between permeate flux and hydraulic resistances induced by membrane and filtration cake.

$$\phi = L_p \Delta P \qquad (10.2)$$

where ϕ = flux through membrane, $m^3/h \cdot m^2$
L_p = permeability of the membrane, $m^3/h \cdot m^2$ bar
ΔP = filtration pressure gradient, bar

Inversion of the relation gives the following Darcy's law

$$\Delta P = R_h \frac{\varnothing}{\mu} \qquad (10.3)$$

where R_h = hydraulic resistance to filtration
μ = viscosity of filtrated solution

In this relation, R_h can be seen as a sum of resistant terms, as described after (Laîné et al. 1989; Wetterau et al., 1994):

$$R_h = R_m + R_r + R_{irr} \qquad (10.4)$$

Where R_m is the virgin membrane hydraulic resistance which is transformed with fouling in R_h (apparent resistance of the fouled membrane). Filtration adds to the initial hydraulic resistance of the membrane (R_m) a filtration cake with a specific resistance R_r, as this cake formation is reversible by backwash and, for example, a gel layer or fouling by NOM adsorption with a specific resistance R_{irr} which is nonreversible by hydraulic backwash and reversible by chemical cleaning. Viscosity is a function of temperature; a regression using literature values gives the following explicit relation: μ (T°C) = μ 20°C exp (−0.0239) (T°C − 20). This relation is widely used in the water treatment industry for a temperature between 5 and 50°C.

Material and Fouling Resistance.   Fouling and its effect on product flow are influenced by many factors. The most important of these is the chemical interplay between components in the feed stream and the membrane material at the separation interface. This includes factors such as surface free energy, charge, and composition. If the adsorption of material onto the membrane surface can be prevented or reduced, pore narrowing and pore plugging can be prevented.

Laîné et al., 1989, reported on the importance of membrane chemistry to flux decline in Amicon membranes exposed to untreated surface water. The flux decline and recovery of membranes formed from relatively hydrophobic polyacrylonitrile and polysulfone polymers did not compare well with those of hydrophilic regenerated cellulosic membranes. The regenerated cellulose membranes showed permeabilities of more than 90 percent of their initial flux, although the pure-water permeability of the membranes was low in comparison with that of the two other membranes (Laîné et al., 1989). Regenerated cellulose membranes are made from a polysaccharide which consists of −D(+)-glucose residues. It is these three hydroxyl groups that impart highly hydrophilic properties to the membrane skin (low water-contact angle, high water wetting, sorption and swelling) (Ko and Pellegrino, 1992).

Figure 10.12 shows the effect of an increase in hydrophilicity on the contact angle of a water droplet on a solid surface. Nonspecific binding of protein is normally greater for hydrophobic membranes than for hydrophilic ones. One way to control the rate at which hydrophobic membranes become fouled is to hydrophilize, permanently or semipermanently, the surface of the membrane. Hydrophobic polysulfone membranes may be rendered more hydrophilic by chemical surface modification (Akhtar et al., 1995). One method for achieving a semipermanent modification of the filtration interface is to cause the adsorption of surfactants onto the membrane during cleaning cycles. It is understood that the hydrophobic tail of a surfactant molecule would adsorb onto the membrane with the hydrophilic end facing the aqueous phase. Surfactant-treated membranes have lower fouling rates over the short term (Chen et al., 1992; Bersillon, 1989).

The chemical composition of hydrophobic membrane surfaces may also be modified by adsorption of hydrophilic polymers onto the surface. Methyl cellulose and poly(vinylmethyl ether) were two of the materials tested (Brink et al., 1993).

Fouling Control.   Fouling is associated with steady flux decline that affects the economics of the filtration operation. Fouling is also associated with either a decrease or increase in retention of solute. When a membrane is operated under conditions of high concentration polarization, retention of microsolutes tends to

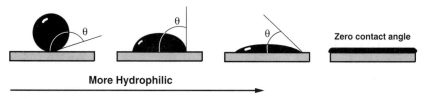

FIGURE 10.12   Evolution of water drop contact angle as a function of membrane surface hydrophobicity.

decrease, whereas that of macrosolutes tends to improve. The increase in retention of macrosolutes may be due to the fouling or osmotic layer acting as a filter aid or partial blocking of pores by adsorbed species. Fouling can be controlled by choice of membrane material, pretreatment, effective cleaning regimes, or process operation (Baudin and Mandra, 1995).

One method by which flux decline can be reduced is to improve the conditions for mass transfer away from the membrane surface. Techniques that could be adopted include operation at reduced pressure and constant draw-off of permeate. These reduce the pressure driving force and thus lower the rate of convective transport toward the membrane surface (Howell, 1995).

Another method is to destabilize the concentration polarization or osmotic boundary layer at the interface; in general, any fluid management procedure that will increase the shear rate at the membrane interface will increase product flow. This may be achieved in a number of ways, such as by use of higher circulation velocities, intermittent foam-ball swabbing, the introduction of turbulence promoters, and by using air sparging, cross-flow reversal, or pulsating flow.

Gupta et al. (1992) reported significant improvement in product flux for a system in which flow and pressure pulsations were superimposed on the inlet flow to ceramic membranes. Pulsations were created by means of a pneumatically operated piston with an approximate sinusoidal motion (fast forward and fast return) at a frequency of 1 Hz; the stroke volume was less than the internal volume of the membranes.

In a study on protein fractionation it was shown that transmembrane pulsing also has a significant effect on the transport properties of membranes, mainly because of reduced pore blocking (Rodgers and Sparks, 1991). Transmembrane pressure pulsing has been proved to be effective in reducing membrane fouling and improving membrane flux in axial-flow devices even in the presence of high polarization resistance. Miller et al. (1993) has reported that permeate fluxes were increased more than tenfold by pulsing only. However, the resistance to transport of protein in the feed solution also decreased, which indicated that steric hindrance caused by adsorption of retained species in the membrane pore matrix was also reduced. This is of significance, since retention is often not controlled by the UF membrane itself, but rather by the polarized and adsorbed layer at the membrane interface. The findings that pulsating flow can affect membrane performance will bear on the choice of feed pumps—for example, whether to use a centrifugal or triplex diaphragm pump in a given application.

10.5.1.2 Impact of Source Water Quality on Design and Membrane Performances. Source water quality directly impacts UF membrane performance, which ultimately affects plant design. To date, few data have been published in the literature (Bersillon et al., 1989; Anselme et al., 1994) regarding raw water quality and UF operational performance. Thus, pilot studies are usually necessary, particularly for plants of significant size (e.g., more than 2000 m³/day), to adjust and optimize operating parameters such as backwash requirements, transmembrane flux rates, operating pressures, and feedwater recovery.

Figure 10.13 shows the effect of different water qualities on membrane fouling. In this study (Jacangelo et al., 1992) one raw water was from the Boise River in Idaho and another from a Ranney collector that draws water from a system of laterals located underneath the Boise River.

Average turbidity for the Ranney collector was 0.5 NTU, and for the Boise River it was 4.9 NTU. Average TOC value for the Ranney collector was 0.9 mg/L as it was 3.2 mg/L for the Boise River.

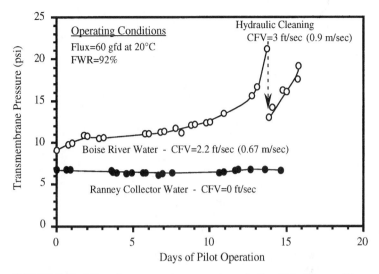

FIGURE 10.13   Effect of two raw waters on membrane fouling (Boise River and Ranney Collector waters). *(Jacangelo et al., 1992.)*

Aquasource cellulosic derivative hollow-fiber membranes were operated at a constant flux of 100 L/h·m² at 20°C (60 gfd) for both source waters. Fouling was monitored by changes in transmembrane pressure. When fouling occurred, an increase of pressure was necessary to overcome the membrane permeability loss. The Boise River pilot plant was initially operated at a cross-flow fiber velocity of 0.73 m/s. However, fouling occurred within two weeks of operation, as demonstrated by the continual increase in transmembrane pressure. Hydraulic cleaning, which consisted of a prolonged chlorinated backwash, and an increase in cross-flow velocity to 1 m/s were not sufficient to control fouling; the transmembrane pressure increased once again within three days. On the contrary, for Ranney collector water and for the same constant production flux, fouling was not observed, even with no cross-flow velocity (dead-end filtration). These results demonstrate that operating parameters and membrane performances were strongly affected by the quality of the raw water treated.

In another example (Bersillon et al., 1989), concerning the first UF plant in operation in France in November 1988 at Amoncourt (240 m³/day), Aquasource membrane performances were also strongly influenced by water quality. This is illustrated by Fig. 10.14, where evolution of membrane permeability is strongly affected by raw water turbidity over a period of six months.

This plant was operated at constant production flux of 100 L/h·m² at 20°C with a cross-flow velocity of 0.7 m/s and a backwash frequency of 45 s/30 mn. Feedwater recovery was 85 percent in this case. Average turbidity of the raw water was 2 NTU as average TOC reaches 2 mg/L. Due to the nature of this groundwater, which is greatly influenced by the surface, numerous turbidity episodes ranging from 50 to 300 NTU were observed during this period.

Increases in turbidity are for this resource, correlated with an increase in TOC up to 7 mg/L. These changes in feedwater quality directly affect the membrane permeability, which decreases from more than 200 L/h·m² bars at 20°C down to 50 L/h·m²

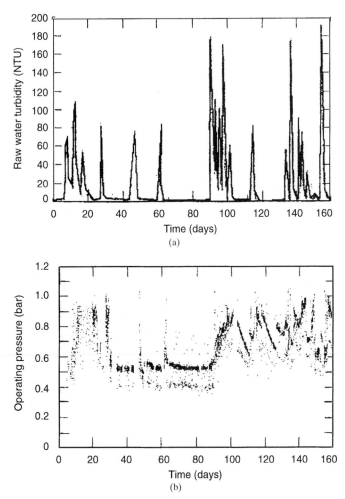

FIGURE 10.14 Turbidity (*a*) and corresponding operating pressure (*b*) profiles of the Amoncourt plant for a constant feed flow of 100 L/h·m² at 20°C (60 gfd).

bars. This corresponds to an increase of transmembrane pressure up to the limit of 1.5 bars in order to maintain the initial production flux. What is really interesting in this case is the self-cleaning effect observed after such an episode. Raw water returns to its characteristic low values of turbidity, and TOC membrane permeability starts to increase up to its initial value within a week, without the need for a chemical cleaning. This can be explained by a slow desorption of organic fouling material when treating a raw water free of these components.

For UF membranes, transmembrane flux as well as backwash requirements and chemical cleaning requirements are strongly related to organic content of the water. The less natural organic matter in the water, the better membranes are operated. Obviously, all membranes are not sensitive to nonreversible fouling due to NOM in the same manner, as illustrated by Fig. 10.15 where cellulosic derivative membrane is less sensitive to fouling than acrylic membrane.

These two polymeric hollow-fiber UF membranes were tested under similar operating conditions for Boise River water (Jacangelo et al., 1989, 1992). For the cellulosic derivative membrane tested, flux rate increased with increasing applied transmembrane pressure. A maximum flux rate of 200 L/h·m² was maintained over a period of five days at a transmembrane pressure of approximately 1.4 bars. For the acrylic membrane tested, flux rates of 80 L/h·m² were maintained at a transmem-

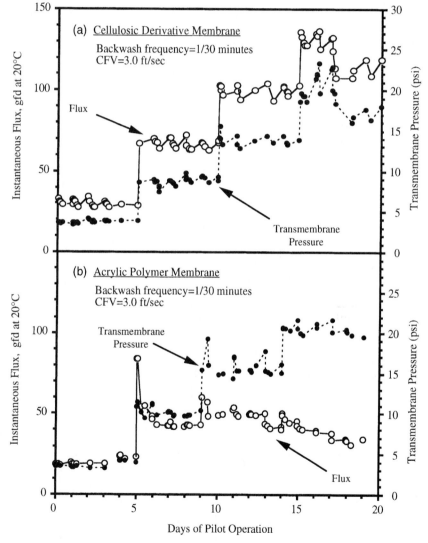

FIGURE 10.15   Effect of membrane material on membrane fouling (Boise River water): (*a*) cellulosic derivative membrane and (*b*) acrylic polymer membrane. *(Adapted from Jacangelo et al., 1992.)*

brane pressure of 0.7 bar. However, fouling occurred when transmembrane pressures greater than 0.7 bar were applied, suggesting a greater fouling potential for this membrane material at the highest flux rates evaluated. Also, the type of macromolecules that composed TOC of water and their potential adsorptive capacity in the membrane material are of great importance when evaluating potential fouling of a UF membrane by a raw water and the resulting operating parameters. UV absorbance at 254 nm is an overall water quality parameter which can be used to characterize aromaticity of TOC in water. The UV 254–to–TOC ratio of a given raw water has been found to be well correlated (Anselme et al., 1994) with membrane fouling and, consequently, membrane performances.

Table 10.16 presents a tentative classification of organic water quality with associated expected membrane performances.

TABLE 10.16   Expected Membrane Performances Associated with Tentative Classification of Water Quality

TOC mg/L	UV/TOC, OD/m/mg/L	Membrane flux, L/h·m² (20°C)	Mode of operation	Feedwater recovery, %
<2	<2	100–150	dead-end	>95
2–4	<2	80–120	dead-end	90–95
	>2	70–100	cross-flow	<90
>4	—	<80	cross-flow	<85

10.5.1.3 Pretreatment.   Ultrafiltration is fairly similar to reverse osmosis, but the requirement for pretreatment of feed is normally much lower. This is because the extent of pretreatment is highly dependent on the quality of the feedwater, and the need for pretreatment will increase as the cross section of the membrane flow channel is decreased. Fortunately, due to the chemical and hydrolytic stability of UF membrane materials, some of the pretreatments essential for RO membranes, such as adjustment of pH or chlorine concentration levels, do not apply. However, it may be necessary to adjust the pH to decrease the solubility of a solute in the feed to improve its filterability, very much as in protein separation.

Ultrafiltration is designed to remove suspended and dissolved macromolecular solids from fluids. The commercially available modules are therefore designed to accept feedwaters that carry high loads of solids. Because of the many uses of UF membranes, pilot studies are normally conducted to test the suitability of a given stream for direct UF. The silt density index (SDI) that was designed for RO membranes does not apply to UF membranes.

The open-flow channels of UF tubular membrane devices make them very suitable for the treatment of industrial wastewater. The amount of pretreatment required with this configuration is negligible, even where a highly contaminated abattoir waste stream is treated (Cowan et al., 1992). This is not so with the narrow-flow-channel flat-sheet or small-diameter capillary membrane devices (below 0.5 mm internal diameter). For spiral-wrap and capillary membranes in particular, minimum treatment standards are specified by manufacturers, as severe blockage of flow paths may occur at the high recoveries necessary to make filtration operations economical.

Depending on their lumen diameters, internally skinned capillary membranes can accommodate particulate matter of diameter less than 50 to 200 μm in the feed.

However, since these membranes can withstand external pressure, fouling matter can be removed from the filtration surface by backwashing with permeate.

Prefiltration. There are few pretreatment requirements for hollow-fiber or tubular systems. Prefilters, ranging between 50 and 200 μm in pore diameter, depending on the internal fiber tube diameter, are necessary to remove large particles which may plug the membrane module inlet (fiber inlet). Various types of prefilters are available, such as disc filters with automatic backwash or, for small plants, filter bags.

Due to a smaller hydraulic channel width, spiral-wound modules are generally difficult to use directly on raw water even with prefiltration systems with sufficiently small pore diameters. Generally, prefilters must have a cutoff corresponding to one-fifth of the hydraulic channel width.

Spiral-wound configuration which cannot be backwashed and exhibits low efficiency of chemical cleaning may necessitate implementation of direct sand filtration pretreatment as required with nanofiltration and reverse-osmosis membrane modules.

pH Adjustment. Adjustment of pH by chemical dosing may be required prior to membrane filtration in order to maintain the pH in the membrane operating range. Hopefully, this adjustment is rarely needed, as most of the commercially available UF membranes have operating pH ranges in agreement with the pH most often encountered in water applications.

Even membranes made of cellulosic derivative polymer, which can be hydrolyzed at specific pH values, have a recommended operating pH of 4 to 8.5 which in most cases does not lead to pH adjustment requirements.

It should be noted that pH adjustment is not required for scaling control, since UF membranes do not remove dissolved ions. Nevertheless, for some naturally unbalanced water which exhibits a calco-carbonic equilibrium favoring calcium carbonate precipitation, it may be necessary to modify pH in order to avoid precipitation of this salt in membrane, pipes, and tanks present in the UF plant.

Preoxidation. As previously mentioned, waters containing dissolved or chelated iron and manganese ions have to be treated by an adequate oxidation process in order to precipitate these ions prior to membrane filtration. This is recommended to avoid precipitation of iron and manganese in the membrane or, even worse, on the permeate side of the membrane (fouling of the membrane during the backwash procedure). Preoxidation processes generally used include aeration, pH adjustment to value greater than 8, or addition of strong oxidants such as chlorine, chlorine dioxide, ozone, and potassium permanganate. In this latest case, for membranes characterized by a low resistance to oxidants, absence of oxidant residuals must be thoroughly controlled in the feedwater prior to membrane filtration.

Coagulation/Adsorption. As previously described, NOM are of great importance in potential fouling of the UF membrane and, consequently, in permeate flux that can be used under normal operating conditions. Thus, it is an interesting design option to pretreat the water to remove natural organic matter and, consequently, decrease the surface of membrane needed, due to the higher flux that is possibly obtained in this case.

Pretreatment in this case must be more accurately defined as a combined treatment, assuming that a UF recycling loop is used as a reactor (CSTR type) to maintain contact between raw water and chemical used. Powdered activated carbon (PAC) and coagulants (alum, ferric chloride, etc.) are generally used for this purpose. Detailed information concerning these combined treatments is given, respectively, in Chaps. 15 and 16.

10.5.2 Typical Values for Design Parameters

This section presents some of the major operating parameters that may directly impact design considerations. It is important to note that substantial savings in construction may be obtained by optimizing the membrane operating parameters for a given water. Depending on the type of membranes and module configurations, several design configurations may be considered. However, since all the UF full-scale plants identified that are currently on-line are hollow-fiber membranes, this section will focus on the design considerations for this type of membrane.

Current designs of hollow-fiber UF systems for municipal applications are composed of membrane modules installed in parallel as opposed to spiral-wound stage-array designs. However, staging of UF systems is being considered in Europe in order to optimize the feedwater recovery, as presented in Fig. 10.16, especially for high-fouling-potential source waters which require fairly intensive membrane cleaning by backwash. The first-stage backwash water can be recovered and treated by a second membrane stage. It should be noted that flux rate and backwash frequency are two parameters which are linked in terms of membrane operating performances, as discussed subsequently. It is expected that by staging UF systems, recoveries as high as 98 percent could be achieved, as illustrated in Fig. 10.16 where the second stage treats the 15 percent backwash water of the first stage. Furthermore, by allowing a lower feedwater recovery for the first stage, chemical cleaning requirements would then be substantially reduced. In summary, a UF staged system could allow a reduction of membrane surface area required by allowing the first stage to be operated at higher flux and at lower feedwater recovery. Such a system, as shown in Table 10.17, could lead to membrane surface area reduction of greater than 15 percent, whereas the feedwater recovery could increase from 90 to 98 percent.

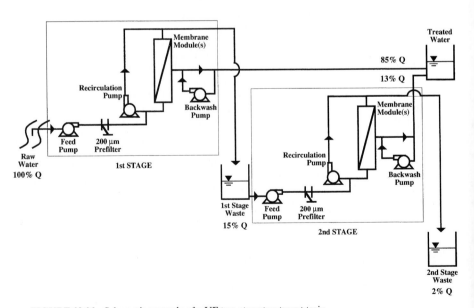

FIGURE 10.16   Schematic example of a UF two-stage treatment train.

TABLE 10.17 Comparison Between Single-Stage and Two-Stage UF Systems for Net Production Capacity of 9.5 mgd (1498 m³/h)

System	Raw water required		Waste generated		UF operating parameters		Feedwater recovery	UF membrane area required	
	mgd	(m³/h)	mgd	(m³/h)	gfd	(L/h·m²)	%	ft²	(m²)
Single-stage	10	(1577)	0.5	(79)	60	(100)	95	169,533	15,771
2-stage system									
Stage 1	9.7	(1528)	1.5*	(229)*	85	(140)	85	117,326	(10,914)
Stage 2	(waste from stage 1)		0.2	(34)	60	(100)	85	24,617	(2290)
Total system	9.7	(1528)	0.2	(34)	—	—	98	141,943	(13,204)

* Waste generated from Stage 1 is recovered as feedwater for Stage 2.

10.45

10.5.2.1 Flux/Pressure. Figure 10.17 illustrates the effect of transmembrane flux on membrane fouling as a function of transmembrane pressure on a surface water (Laîné and Jacangelo, 1992). At a transmembrane flux of 60 gfd (100 L/h·m²) at 20°C, fouling was observed, whereas at a lower flux of 47 gfd 80/h·m²), no fouling occurred. In order to operate this UF system at 92 percent recovery and 60 gfd (100 L/h·m²), a chemical cleaning with detergent would be required every two weeks, whereas at a lower recovery or at a lower operating flux, the chemical cleaning requirement would be reduced.

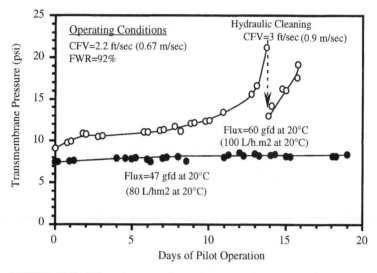

FIGURE 10.17 Effect of transmembrane flux on membrane fouling (Boise River water). *(Jacangelo et al., 1992.)*

10.5.2.2 Cross-Flow Velocity/Dead-End. The recirculation rate of the feed-water across the membrane (cross-flow) is a critical parameter for controlling fouling. The objective of applying a cross-flow fiber velocity (CFV) for hollow-fiber membranes is to maintain solid and organic materials in suspension in the bulk water in order to prevent their deposition at the membrane surface, thereby reducing the potential for fouling. Thus, as the CFV increases, the fouling potential is usually reduced; however, the energy required increases. For low-turbidity, low-fouling-potential source water, as previously discussed, the UF membranes may be operated in dead-end mode, which significantly reduces the energy consumption of the process.

It should also be noted that, as the CFV increases, the pressure head loss through the hollow-fiber increases as well. The fiber diameter may also become a limiting parameter for high-solid-content source waters. The module pressure drop (head loss) in a fiber can be calculated using the following equation, assuming that the membrane fiber roughness is negligible:

$$\Delta P = \frac{32 \times m \times l \times \text{CFV}}{d^2 \times 10^6} \tag{10.5}$$

where   CFV = cross-flow fiber velocity, cm/s
d = internal fiber diameter, cm
ΔP = pressure head loss (pressure drop, inlet pressure minus outlet pressure), bars
l = membrane fiber length, cm
m = water viscosity, g/cm·s
32 = unit conversion factor

It should be noted that most commercially available hollow-fiber membranes have internal fiber diameters ranging from 0.5 to 1.5 mm. These diameters correspond to pressure drop of less than 14.5 lb/in² (1 bar) at 20°C for 1-m-long fibers at velocities as high as 3 ft/s (1 m/s). This appears to be an optimum between the pressure drop and the membrane compactness, since the membrane fiber diameters will dictate the number of fibers which can be fitted in a membrane pressure vessel (module).

10.5.2.3 Backwash Procedure.   Backwash has been found to be one of the most effective methods for controlling fouling for hollow-fiber membranes. Permeate water which is collected in a reservoir is used to backwash the hollow-fiber membranes. Usually, a backwash pump is employed to reverse the permeate flow from the permeate side to the production side of the membrane at an effective backwash pressure ranging from 5 to 50 lb/in² (0.3 to 3.5 bars), depending on the membrane employed. The permeate is used to clean the membrane surface (i.e., membrane cake) and deconcentrate the system piping. The backwash water is then wasted. Figure 10.18 shows the effect of chlorinated backwashing on membrane operation (Jacangelo et al., 1992). Under normal operation, the membrane tested was backwashed for 45 s with permeate water containing 3 mg/L of chlorine. Figure 10.18 illustrates the importance of chlorine on fouling control. The membrane was operated for 21 days at approximately 60 gfd (100 L/h·m²) at 20°C; during this period no

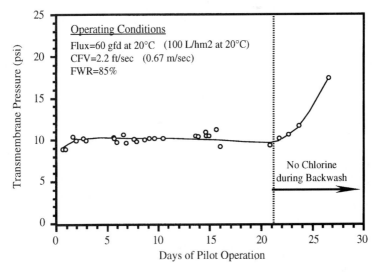

FIGURE 10.18   Effect of backwashing with chlorine on membrane fouling (Boise River water). *(Jacangelo et al., 1992.)*

fouling was observed. However, when the chlorine injection during backwash was discontinued, fouling was observed in less than five days and the transmembrane pressure increased by more than 7 lb/in^2 (0.5 bar). This experiment showed that addition of chlorine to the backwash water was critical for controlling fouling for the UF membrane tested. The mechanisms by which chlorine controls fouling is unclear—i.e., whether the disinfectant oxidizes organics on the membrane, controls microbial growth (biofouling), or both.

10.5.2.4 Temperature Considerations/Peak Flow Demand. Temperature directly affects the design of a UF membrane plant. As shown in Fig. 10.19 (Laîné and Jacangelo, 1992), the transmembrane flux rates achieved at a given pressure varied considerably in relation to the raw water temperature. As much as a 35 percent decrease in flux can be observed for a water approaching 0°C as compared to a water at 20°C. Increases in transmembrane pressure, resulting in increasing transmembrane flux rate, may be required to maintain desired plant production flows at low temperature. This could result in fouling of the membranes. However, since in most cases water demand is lower during winter, overdesign of a UF plant can be minimized if water production is not only paced to water temperature but also to the daily peak demand.

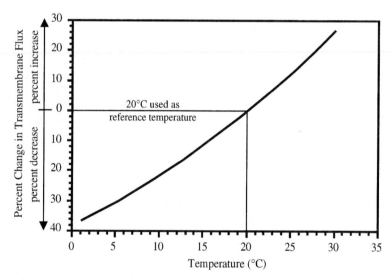

FIGURE 10.19 Effect of temperature on instantaneous flux at constant transmembrane pressure. *(Laîné and Jacangelo, 1992.)*

Table 10.18 summarizes typical predesign parameters for UF systems. Depending on the raw water quality, predesign parameters may be adjusted in terms of flux, cross-flow fiber velocity, or backwash frequency.

TABLE 10.18 Typical Predesign Parameters for UF Plants

Parameters	Levels	
Flux at 20°C	50–100 gfd	80–170 L/h·m²
Cross-flow velocity	0–3 ft/s	0–1 m/s
Feedwater recovery	85–>97%	—
Backwash:		
Duration	10–180 s	—
Frequency	1/30–1/180 min	—
Pressure	5–40 lb/in²	0.35–2.8 bar

Source: Adapted from Laîné and Jacangelo (1993).

10.5.2.5 Chemical Cleaning. Regular sanitation and cleaning are very much part of the operation of a membrane plant. As such, membrane plants are equipped with a second circulation loop that connects the membrane modules to the sanitation equipment, or *clean-in-place* (CIP) equipment, as it is sometimes referred to. Table 10.19 provides a list of commonly used sanitizing agents.

10.5.3 Design Options

10.5.3.1 Package Plants. Two design options can be considered for UF membrane plant systems: package plants and custom design plants. Package plants are usually employed for plant capacities under 1.5 mgd (5800 m³/day). Package plants are skid-mounted standard units which allow significant cost savings. At the time this report was written, only one manufacturer was providing complete package plants. Main skid-mounted system components include auto-cleaning prefilter, raw water pump, recirculation pump, backwash pump, chlorine dosing pump for the backwash water, air compressor (valve actuation), chlorine tank, chemical tank (detergent), programmable logic controller with program and security sensor (high pressure, low level, etc.). Such units comprise 2 to 14 membranes, as illustrated in Fig. 10.20, which

TABLE 10.19 Commonly Used Sanitizing Agents*

Agent	Concentration and duration of sanitation
H_2O_2	1–5% hydrogen peroxide for 30–90 min (preferred in the electronics industry where sodium ion residuals must be avoided)
Cl_2	100–400 mg chlorine gas/L of water for 30–90 min
NaOCl	5–10 mg/L sodium hypochlorite for 60 min
Hot H_2O	180°F for 1–2 h
O_3	Ozone at 0.1–0.2 mg/L for 30–90 min
Citric/oxalic acid	0.5% for 30–60 min in case of iron/manganese oxides fouling
Chelatants/surfactants mixing	10–100 mg/L for each component for 3–6 h at 30°C

* Check with suppliers for chemical compatibility with the filters used in the system before using a sanitizing agent.

Source: Adapted from Porter, 1990.

can be increased from 1084 to 11,137 ft^2 (101 to 1036 m^2) of effective membrane surface area. It should be noted that the level of automation of these systems includes an automatic process switching from dead-end to cross-flow and vice versa, depending on the raw water quality.

10.5.3.2   Custom Design Plants.   For larger plant capacities, the membranes are arranged in racks (blocks) of several modules each. The larger standard rack avail-

(a)

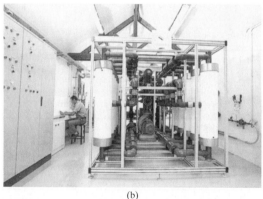

(b)

FIGURE 10.20   Selected examples of package systems: (*a*) 2-membrane module skid; (*b*) 10-membrane module skid. *(Provided by Aquasource, France.)*

able at the time this report was written comprised 28 membranes of 590 ft² (55 m²). An example of rack arrangement is provided in Fig. 10.21, where 8 racks of 28 membrane modules each are presented. Such a design allows significant reduction in terms of ancillary equipment required. For example, only one raw water feed and backwash station are required for the 8 racks. Each rack is controlled with its own PLC, which itself is controlled by a main PLC. Peak flow demand can be adjusted by increasing the transmembrane pressure or by implementing on-line racks which are on standby. The main PLC oversees the plant operation and the operating performance of each rack.

10.5.4 Instrumentation

10.5.4.1 Process Instrumentation. Figure 10.22 represents a typical process and instrumentation drawing (PID) of an ultrafiltration plant. As previously mentioned, due to its physical characteristics, membrane filtration is a highly automated process. Since the number of pumps and valves is quite important, numerous sensors are necessary to automatically operate the process.

Flowmeter and Pressure Sensors. *Flowmeters* provide flow-rate readings for the feed and/or the permeate flows, as well as the recirculation flow. *Pressure sensors* should be located at the inlet, outlet, and permeate side of the membrane banks for transmembrane pressure monitoring. These sensors also serve for monitoring backwashing and chemical cleaning, as well as filtration mode. Other sensors and specific monitoring equipment, as noted subsequently, are employed to protect the membrane from breaks due to high pressure and to monitor process performance (membrane permeability, feedwater recovery, etc.) or determine the filtration mode (dead-end, cross-flow, or feed-and-bleed).

Pressure Release Valves. Hollow-fiber membranes are not designed for high operating pressures, which may cause fibers to rupture. Pressure release valves on the membrane feed and permeate sides protect the system against pressure surges, especially if the plant is operated in a manual mode that may override automated membrane protection devices.

Pressure Switches. Pressure switches are simple sensors that protect the system during normal operation, i.e., rapid membrane fouling, valve failure, or pipe breakage. Pressure switches shut down the pumps and, if remotely controlled, will set off alarms to the main control station.

Turbidimeters. Depending on the water quality, the feedwater flow configuration may be changed from dead-end to cross-flow, especially during high-turbidity episodes. An on-line turbidimeter with a set point can serve as a sensor which automatically changes the configuration of feedwater flow. Other feedwater quality parameters can be used. More information concerning this specific point is given Sec. 10.6.

Temperature Monitoring. A temperature sensor must be set on the unit in order to manually or automatically adjust the feed flow to the observed temperature (correction factor is in the range of a decrease of 2.5 percent of the flow for a decrease of 1°C).

Valve Actuators. For small membrane plants, electric actuators may be used. However, for large UF systems, hydraulically actuated valves may preferentially be employed. It should be noted that, for automation purposes, hydraulic actuators are closed by defaults, whereas electrical actuators may leave the valve halfway open. Thus, the automation of a plant using electric actuators requires more complex programming of the programmable logic controller (PLC) to reset the actuators.

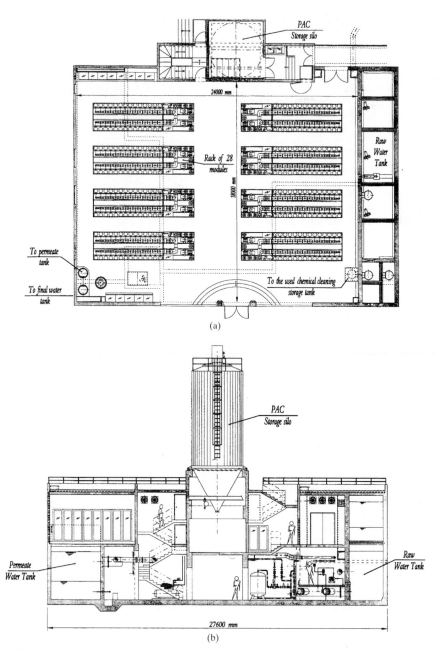

FIGURE 10.21 Example of a 2300-m³/h (15-mgd) full-scale plant lay-out, 8 racks of 28 membrane modules: (*a*) top view; (*b*) and (*c*) side views.

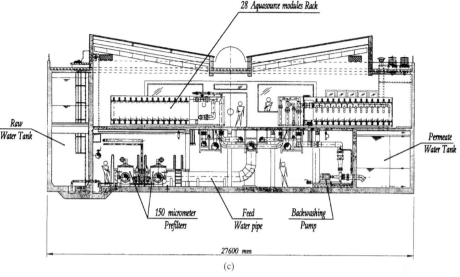

FIGURE 10.21 (*Continued*)   Example of a 2300-m³/h (15-mgd) full-scale plant lay-out, 8 racks of 28 membrane modules: (*a*) top view; (*b*) and (*c*) side views.

Furthermore, cost savings may be expected when using hydraulic actuators. When a UF plant is remotely controlled, all types of sensors can be sent back to the main control station.

10.5.4.2 Process Reliability

Reliability of UF Process.   One of the most critical aspects of employing membrane technology is ensuring that the membranes are intact in order to maintain the initial disinfection efficacy. It should be noted that little information has been published yet regarding long-term evaluation of membranes in terms of reliability. This is mainly due to the fact that membrane technology is a relatively new process in drinking water treatment and that the oldest plant has been on-line only since November 1988 in Amoncourt, France.

Monitoring Methods of Membrane Failure.   There are various methods that can be applied to control the integrity of the UF process, as presented in Table 10.20.

Particle counting.   Particle counting methods have been demonstrated (Jacangelo et al., 1992) to be very sensitive methods for monitoring the integrity of UF membranes, whereas turbidity monitoring was not sensitive enough to detect a fiber break. The high cost of multichannel counters presents the major disadvantages to the widespread use of this technology for membrane-integrity monitoring. However, with further development of this technology, the cost of single-channel counters in particular could be reduced by an order of magnitude. It should be noted that the sensitivity of these measurements will be directly affected by raw water concentration, operating parameters (loop particle concentration, etc.), and also by the number of membrane modules in parallel being monitored by the particle counter. Therefore, the operation of such systems must be carefully assessed to ensure proper membrane integrity monitoring.

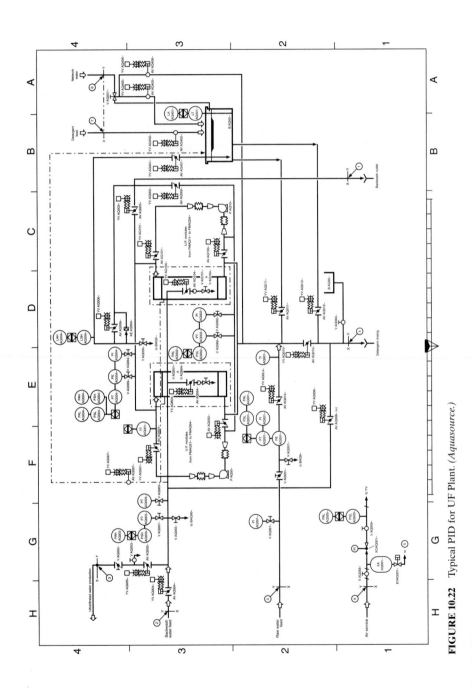

FIGURE 10.22 Typical PID for UF Plant. (*Aquasource.*)

TABLE 10.20 Methods to Evaluate UF Membrane Integrity

Monitoring methods	On-line	Continuous monitoring	Advantages	Disadvantages
Particle monitoring				
Particle counting	Yes	Yes	• May measure several size ranges • One of the most sensitive methods	• High cost • Sensitivity depending on feedwater quality • Requires several sensors for large-scale applications • Indirect measurement of membrane integrity
Particle monitoring	Yes	Yes	• Low cost	• Provide a particle index • Sensitivity depending on feedwater quality • Requires several sensors for large-scale applications • Indirect measurement of membrane integrity
Turbidity monitoring	Yes	Yes	• Extensive water industry experience • Low cost	• Low sensitivity • Indirect measurement of membrane integrity
Other Monitoring Sensors				
Air pressure testing	Yes/No	No	• May be built into membrane system • Direct measurement of membrane integrity • Test applicable to one membrane at a time in manual	• Difficult to operate for large-scale systems • Requires airtight valving • Must drain permeate side • Labor intensive if performed manually
Bubble point testing	No	No	• Direct measurement of membrane integrity	• Must take module off-line to conduct test • Labor intensive
Sonic sensor	Yes	Yes	• Direct measurement of membrane integrity • Sensor installed on individual membrane module • One of the most sensitive methods	• High cost for small-area module (<540 ft^2 / <50 m^2) • Surrounding background noise • System under development
Challenge studies				
Bacteriophage seeding	No	No	• One of the most sensitive methods • Provide direct evaluation of disinfection efficacy • Can be performed on individual module off-line	• Not practical to seed full-scale plants • Labor intensive
Microbial routine analysis	No	No	• Fairly low labor requirement and low cost	• Long response time • Sensitivity depending on feedwater quality • Indirect measurement of membrane integrity

Figure 10.23 presents a comparison of the response sensitivity of particle count and turbidity for the breakage of one fiber over a module composed of more than 2000 fibers (7.2 m²) (Mandra et al., 1994). For an increase of the turbidity of the permeate by a factor of 2 (from 0.07 to 0.14 NTU), which is near the significance limit of the measurement, a 1000-fold increase of particle count is observed (from less than 3 particle/mL to more than 2000) for the total count of particle size ranging 1 to 120 μm. At the same time, this loss of integrity of the membrane reduces the fecal streptococci removal from 4 to 2 log.

Particle monitoring.   Other systems in a similar price range to turbidimeter systems are also available. These systems do not count particles, but they do provide an

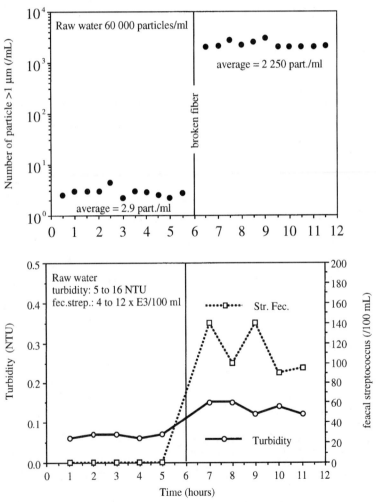

FIGURE 10.23   Comparison of turbidity and particle count sensitivity for the detection of broken fiber in 7.2 m² ultrafiltration module.

index of particles in relation to a particle size range concentration. However, due to the technology employed (LED as compared to particle counters which use laser light), the sensitivity of these systems is lower.

Turbidity monitoring.   Turbidity is the most-used method of evaluating particle removal in drinking water treatment applications. This parameter is also regulated in most countries in the world. However, it has been demonstrated (Jacangelo et al., 1991) that no turbidity increase was discernible for a damaged membrane, whereas other methods such as particle counting showed a substantial increase in particle densities in the UF treated water.

Sonic sensor.   This method is based on hydraulic noise "analysis" of flow perturbation during production, as discussed in detail subsequently. Membrane integrity loss can be monitored efficiently by analyzing the hydraulic noise in a certain frequency range. This method is currently under development and appears to be very promising in terms of integrity monitoring systems. A sensor (i.e., microphone) is located on each of the modules on-line, continuously providing a signal. The sensors are expected to be fairly inexpensive, and this method appears to be cost-efficient for membrane modules with a surface area in the 538-ft^2 (50-m^2) range. It is, in fact, one of the only techniques that can accurately and continuously detect one broken fiber in an industrial module of 28,000 fibers.

Air pressure test/bubble point testing.   These two methods are direct methods to check UF membrane integrity. Bubble point testing is, in fact, a modified method, as discussed subsequently, which is used to identify a broken fiber. This method is fairly labor intensive and necessitates taking the modules to be tested off-line. The air pressure test method can be performed on-line and could be automated. However, industrial applications on a large scale appear to be impractical since UF membranes are not very permeable to air. Such a test may generate air pockets in the system and therefore require a fairly long and complicated procedure to purge the air trapped in the lines.

The bubble point technique has been reported to be the first method to determine pore-size diameters and for testing the integrity of the low-pressure membranes (Cole and Pauli, 1975). If the bubble point requires too high a pressure to determine UF membrane pore size (Cheryan, 1986), this technique may be adapted to verify the membrane integrity during the manufacturing phase and also during the operating phase of the membranes. This test, which is based on the air diffusion in a porous material, can be applied to hollow-fiber membranes to detect potential holes in the membrane structure. The air diffusion flow rate can be expressed by the following equation:

$$Q_J = \frac{66.5DS(P_i - P_o)Ae}{\Delta x P_o T} \tag{10.6}$$

where A = membrane surface area, cm^2
D = diffusion coefficient, cm^2/min
Q_J = air diffusion flow rate at 20°C and 1.013 bars, mL/min
P_i = absolute inlet pressure, bars
P_o = absolute outlet pressure, usually atmospheric pressure, bars
e = membrane porosity (<0.2 for UF membranes)
S = solubility coefficient, mol/bar·mL
T = absolute temperature, °K
Δx = membrane thickness, μm
66.5 = unit correction factor

When the integrity of a membrane is compromised, a flow rate of gas can be measured and expressed by the following equation:

$$Q_g = 0.469\left[\frac{p(P_i - P_o)(P_o + P_{\text{atm}})}{\Delta x \eta_g P_{\text{atm}}}\right]d_h^4 \tag{10.7}$$

where Q_g = gas flow rate due to convective transport, mL/min
         d_h = diameter of membrane fiber hole, μm
         h_g = gas viscosity, μpoise
         0.469 = unit correction factor

For practical reasons, the gas flow rate can be measured by the pressure head loss through the membrane with time. By calculating the gas flow rate from Eq. (10.7) and the pressure drop at a set air pressure, a limit value can be determined for a non-compromised UF membrane. The relationship of the pressure head loss is presented as follows.

$$\Delta P = \frac{V_g P_{\text{atm}}}{V_m} \tag{10.8}$$

where V_g = gas volume passing through membrane, mL
         V_m = permeate side volume of membrane module, mL

Microbial monitoring.   Routine microbial monitoring will provide some indication of the membrane disinfection efficacy depending on the raw water concentration. In order to increase the sensitivity of microbial monitoring, seeding of bacteriophages, which are nonpathogenic viruses similar in size and shape to the smallest enteric virus (i.e., polio), could be performed. Although this method has been successfully applied at the pilot scale, it seems almost impossible to perform such a test for large plant capacities.

Figure 10.24 presents a comparison of the level of sensitivity for various monitoring techniques in the detection of membrane loss of integrity. The different monitoring techniques used are turbidity, particle count (size range 0.5 to 5 μm), and seeding of MS2 bacteriophage. Levels reported in Fig. 10.24 are those of the permeate for an integer module of 7.2 m², the same module with two calibrated holes of 50 μm in a fiber, and another one with one broken fiber. This figure demonstrates that turbidity is not sensitive enough to reliably detect any loss of integrity. On the contrary, particle count gives results which are well correlated with MS2 bacteriophage concentration in the permeate.

10.5.5   Concentrate Disposal

This section presents an overview of the concentrate disposal options for UF treatment plants. Two kinds of concentrates can be generated with UF treatment, depending on whether or not there is a combined treatment, i.e., powdered activated carbon addition.

In conventional drinking water treatment plants, sludge wastes are generated from settling basins and from the filters. The sludge handling usually includes a partial dewatering of wet sludge and recycling of the extracted water. For landfilling, a dewatered solids concentration of 35 percent is usually required. In the United States, the Water Pollution Control Act Amendment (1972) specifies that there shall

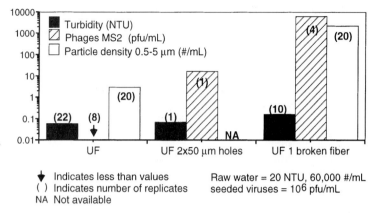

FIGURE 10.24   Comparison of various methods to detect membrane integrity failure.

be zero discharge of wastes from any water treatment plants. In Europe, similar regulations are being proposed, especially for sludges containing metals (i.e., aluminum or iron from coagulant), which will be classified as hazardous waste.

As opposed to conventional treatment, UF wastes can be considered as clean wastes since there is no added chemical residual. It should be noted that, on the contrary, some MF treatment systems currently on-line with coagulation treatment generate sludges comparable to those obtained with a conventional treatment.

10.5.5.1   Regulation Overview.   In France, as in other developed countries, discharge of concentrates to the environment is regulated. Three main parameters are used in order to define the type of discharge authorization required:

- Waste flow rate in m^3/day
- Pollutants daily output in kg/day
- Instantaneous concentration in pollutants in mg/L

The two first parameters are used to define the level of authorization required between a simple declaration and a real discharge permit; these regulated values are variable as a function of the level of quality and protection of the river which will receive the waste. Obviously, the higher the quality of the river, the lower the wasted flux of pollutant that is allowed. Among the regulated quality parameters (e.g., chemical oxygen demand, biological oxygen demand at 5 days, total nitrogen, and total phosphorus), total suspended solids (TSS) concentration is the only parameter which can require the treatment of UF concentrate before discharging. The regulated level of suspended solids is generally 30 ppm for most of the rivers, but the highest level of protection that a river considered as a fish reserve might request is a suspended solids limit of 20 ppm.

10.5.5.2   Characteristics of UF Concentrates.   A backwash procedure which generates a concentrate is currently used to control membrane fouling. Two kinds of concentrates can be generated with UF treatment, depending on whether or not there is a combined treatment, i.e., powdered activated carbon addition (Laîné and Gislette, 1995).

As illustrated in Fig. 10.25, even when operated in one stage, the level of TSS in the concentrate of an ultrafiltration plant is usually higher than the requested limit of 30 ppm for direct discharge without treatment to the environment. This level of TSS is a function of both feedwater concentration and water loss, which is given by the membrane flux selected for operation and the volume of permeate used for the membrane backwashing as well as the backwash frequency.

The composition of the concentrate is mainly particulate matter, since UF does not remove dissolved organic matter to a large extent, and also, as opposed to nanofiltration or reverse osmosis, no salts are removed.

The suspended solids concentration and turbidity profiles of the backwash wasted water are illustrated by Fig. 10.25 in the case of the La Nive (Anglet, France) ultrafiltration plant, which directly treats the River Nive. The quality of the water discharged as waste during the one-minute backwash is characterized by a high TSS concentration—up to 80 ppm—for half of this duration; in the second part of the backwash, the TSS concentration reaches values below the discharge regulation of

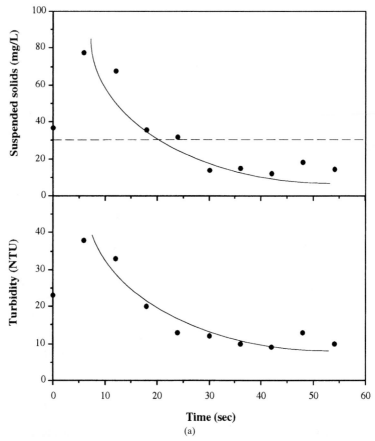

(a)

FIGURE 10.25   Backwash water concentration profiles: (*a*) La Nive UF plant with no PAC; (*b*) Chatel-Gérard UF plant with PAC.

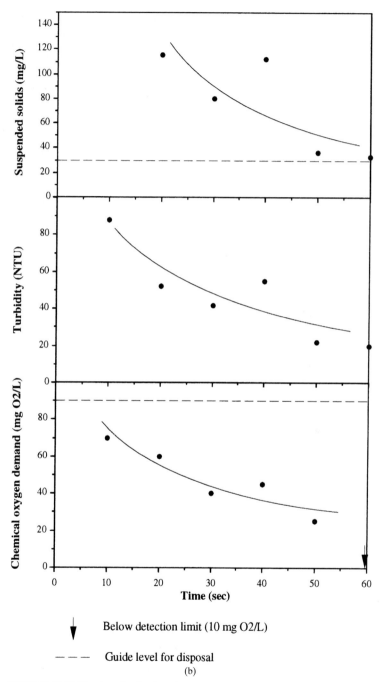

Below detection limit (10 mg O2/L)

— — — Guide level for disposal

(b)

FIGURE 10.25 (*Continued*)   Backwash water concentration profiles: (*a*) La Nive UF plant with no PAC; (*b*) Chatel-Gérard UF plant with PAC.

30 ppm. This is an important point that can be used to reduce by a factor of two the volume of the concentrate to treat as well as increasing its average concentration and treatability.

It should be noted that, since no chemicals are added, the composition of the concentrate water is, in fact, very similar to the raw source water being treated, and in some limited cases, discharge back to the source water can be done. With 85 to 95 percent recovery in one stage, the water to discharge is concentrated 5 to 20 times and characterized by a concentration in TSS generally higher than 30 ppm (see Fig. 10.25).

On the other hand, the combined PAC/UF process is a water treatment system combining UF separation and powdered activated carbon (PAC) adsorption and therefore requires a treatment of the concentrate due to the fairly high solids concentration, including the PAC. Typical concentrate concentration in suspended solids is, in this last case, higher than 200 ppm due to the concentration of the 10 to 20 ppm of PAC added to the feedwater which is discharged as waste during the backwash procedure. It should be noted that, due to its chemical property of being a reducing agent, PAC imparts the measurement of chemical oxygen demand. Therefore, PAC particles which are obviously mineral suspended solids contribute to the COD determination of the PAC/UF concentrate. As the regulated COD value for discharge to the environment is generally low (120 mg O_2/L in France, for example), the problem of concentrate treatment before discharge is enhanced in the case of combined PAC/UF water treatment plants, because both suspended solids and chemical oxygen demand are higher than their requested limits for discharge to the environment. This is illustrated by Fig. 10.26, which gives the relation between PAC concentration and COD. As the slope of the response curve is approximately equal to 2, it is easy to obtain the maximum admissible PAC concentration of 60 ppm, which generates a COD of 120 ppm.

It should be noted that when free chlorine is added to the permeate used for backwash, neutralization of the residual is required using, for example, sulfites addition before discharge.

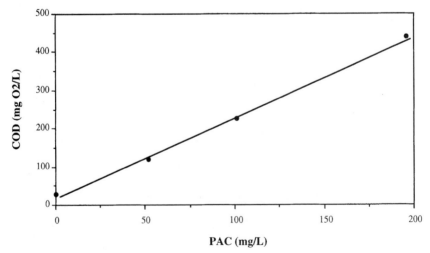

FIGURE 10.26 Relationship between PAC concentration and chemical oxygen demand in the Châtel-Gérard backwash water.

10.5.5.3 Storage or Treatment of UF Concentrates. The main options for concentrate disposal and/or treatment are listed as follows:

- Concentrate returned to the source water
- Discharge to a sewer line
- Lagoon disposal
- Physical filtration
- Coagulant/flocculant pretreatment and filtration
- Complete sludge treatment line

Figure 10.27 illustrates the low settling potential of UF concentrates for both the La Nive UF plant and the combined PAC/UF plant of Châtel-Gérard (France).

Flocculants are generally used in order to increase the settling velocity of UF concentrates. From all types of commercially available polymeric flocculants, anionic acrylamid polymers are the most efficient and are currently used at a dose of 0.5 to 1.0 kg per ton of suspended solids contained in the concentrate. These chemicals are added in the sludge treatment line as a pretreatment before the thickening process.

Most of the concentrate treatments have the objective of reducing the volume of waste as well as increasing the concentration in solids up to a level high enough to allow the transportation of the sludge to the disposal location. Media or multimedia filtration processes are of low interest for this disposal option because of the importance of washing volume needed to clean the filtration media which leads to a dilution of the posttreated concentrate, limiting the overall concentration yield.

Following is a discussion of three plants using the combined PAC/UF process, which requires the treatment of the concentrate before discharge. These three plants illustrate different concentrate treatment and disposal options.

Châtel-Gérard UF Plant. The Châtel-Gérard UF plant is a 700-m³/day plant which treats a groundwater under the influence of surface water. Average raw water quality is characterized by a turbidity of 2 to 5 NTU with peak episodes up to 100 NTU, and a dissolved organic carbon (DOC) concentration of 3 mg/L with frequent increase up to 5 mg/L. A PAC dose of 10 to 20 ppm is added continuously to the feedwater in order to obtain a treated water DOC in the range of 1.5 to 2.0 mg/L. The concentrate flow rate is 4 m³/h and the TSS concentration is usually in the range of 150 to 200 mg/L; the treatment option includes a dosage of anionic flocculant at a dose of 0.8 kg of polymer per ton of suspended solids and the use of a gravity sludge thickener. The design water velocity of the thickener is 2 m/h and leads to the following dimensions of 1.6 m in diameter and 2.75 m in height. The final concentrate, with a solids concentration of 100 g/L, is stored on-site in the thickener and disposed every month by truck transportation to the nearest wastewater treatment plant. The concentrate treated water is characterized by a TSS concentration below 5 mg/L and is discharged in the local river.

L'Apier Saint-Cassien UF Plant. The l'Apier Saint-Cassien UF plant is a 25,000-m³/day plant treating a reservoir water by a combined PAC/UF process. The average quality of the raw water is characterized by a low turbidity of 0.5 to 1 NTU and a low DOC of 1.5 to 2 mg/L, but due to phytoplanktonic activity (up to 3000 algae cells/L), taste and odor are often detected in this water. This has imposed the use of PAC addition in the feed at a dose of 10 ppm, with the objective of complete removal of taste and odor and a DOC concentration below 1 mg/L in the permeate water. Also, the southeast of France is characterized by periods of torrential rain which lead to a deterioration of the resource quality with turbidity up to 150 NTU and DOC up to

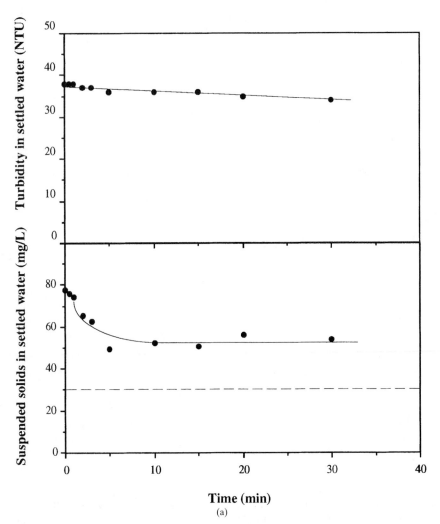

FIGURE 10.27   Backwash water settling profiles: (*a*) La Nive UF plant with no PAC; (*b*) Châtel-Gérard UF plant with 10 mg/L PAC.

3 mg/L. The concentrate treatment includes a complete treatment train, including an anionic polymer dosing setup, a high rate (25 m/h) lamellar settler with sludge recycling system (Densadeg, Degrémont, France), a holding thickener tank, and a dewatering by centrifugation. The concentrate flow rate to treat is 80 m^3/h at an average concentration of 200 mg/L, outlet concentration of the settler is in the range of 30 g/L, allowing for a final solid concentration of the sludge after centrifugation of 350 g/L. The final sludge is disposed by incineration after truck transportation. The direct cost of such a complete concentrate treatment line is limited to 0.05 FF/m^3 of treated drinking water. Because of its treatment with polymeric flocculant addition, which can impart the fouling of the membrane, the water discharged as waste from

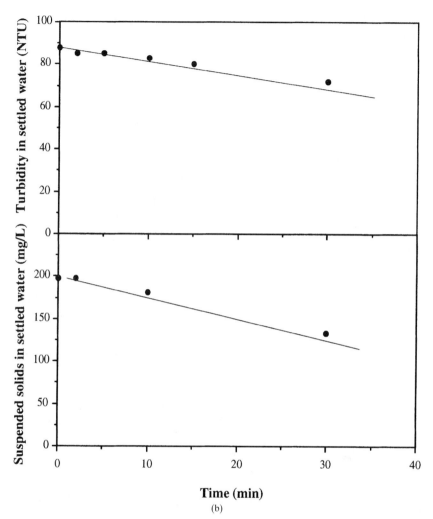

FIGURE 10.27 (*Continued*)   Backwash water settling profiles: (*a*) La Nive UF plant with no PAC; (*b*) Châtel-Gérard UF plant with 10 mg/L PAC.

the rapid gravity settler is not recycled up to the feed of the plant but discharged in the local stream river.

Vigneux sur Seine UF Plant.   The Vigneux sur Seine UF plant is a 55,000-m³/day plant treating the Seine River water upstream from Paris after clarification by a combined PAC/UF process. The feedwater is quite constant in quality with a very low turbidity of 0.2 NTU and a DOC of 2 to 2.5 mg/L; the PAC dosage of 10 ppm allows the production of a permeate with a low 1-mg/L DOC. The concentrate (backwash water) is recycled back to the sludge blanket clarifier. This allows maximum use of the PAC adsorption capacity and minimizes the water loss to less than 2 percent. No additional treatment for the UF concentrate was required since it is recycled to the existing clarifier, where sludge bleed is discharged as waste to the sewer system, which is linked to a wastewater treatment plant serving 1,000,000 inhabitants.

Other Options. Another option being developed is the staging of the UF operation allowing the minimization of the water losses and, therefore, the waste treatment line. Fig. 10.28 presents the isorecovery profiles as a function of membrane flux and backwash frequency in the case of the Aquasource UF membrane. As illustrated in this figure by the slopes of the curves, the recovery of a UF system is more influenced by the operating flux of the membrane than by the backwash frequency. From the various flux and backwash frequency options derived from the 99 percent recovery profile presented in Fig. 10.28, it is clear that either high flux (greater than 150 L/h·m²) or long filtration cycle (>2h) is needed in order to reach high recovery ratio. These performances are possible only in the case of very clean quality water. The other possible option is the use of a two- or more stage system which allows reaching the same water recovery but uses lower and more usual membrane performances. Figure 10.29 illustrates the advantages of such a design arrangement. In order to have the same capacity of production, both one- and two-stage arrangements require approximately the same number of UF modules. However, using a two-stage arrangement, the water losses decrease from 4.3 to 1.4 percent, which translates in terms of waste flow into a reduction from 11.9 to 4.0 m³/h.

Several options of concentrate treatments are available, depending on capacity and the solids concentration in the concentrate. For small plant capacities using combined PAC treatment which require the treatment of backwash water discharged as waste, a conventional gravity sludge thickener can be considered as a simple and low-cost treatment option. However, for larger plants, staging arrangement appears to be a very promising option to reduce the volume of concentrate to treat.

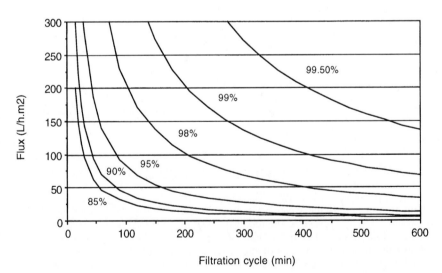

Filtration cycle (min)

FIGURE 10.28   Impact of operating parameters on feedwater recovery (one-stage design).

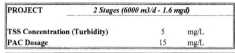

PROJECT	2 Stages (6000 m3/d - 1.6 mgd)	
TSS Concentration (Turbidity)	5	mg/L
PAC Dosage	15	mg/L

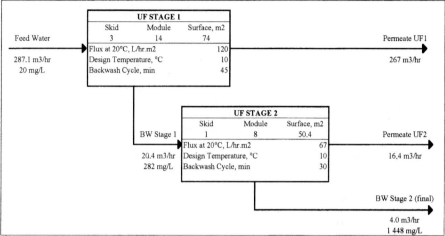

	Water Losses	UF Production		MES	Conc. Ratio
Stage 1	7.1%	267 m3/hr		282 mg/L	14.1
Stage 2	19.5%	16 m3/hr		1 448 mg/L	5.1
Total System	1.4%	283 m3/hr	81 L/hr.m2	1 448 mg/L	72.4

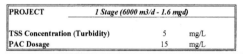

PROJECT	1 Stage (6000 m3/d - 1.6 mgd)	
TSS Concentration (Turbidity)	5	mg/L
PAC Dosage	15	mg/L

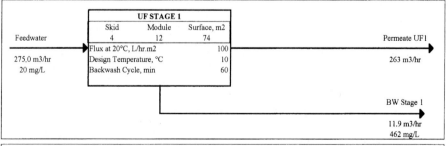

	Water Losses	UF Production		MES	Conc. Ratio
Stage 1	4.3%	263 m3/hr	74 L/hr.m2	462 mg/L	23.1

FIGURE 10.29   Impact of design staging on feedwater recovery.

10.5.5.4 Chemical Cleaning Disposals. It is possible to estimate the timing of chemical washings by monitoring the specific flux or permeability decrease of the membrane down to a specified value. This level varies with the initial value of the specific flux and the feedwater quality as well as the process parameters applied to the UF system (e.g., dead-end or cross-flow filtration, operating flux and pressure, backwashing frequency). Chemical cleanings are generally carried out once every few months and are off-line procedures by definition.

Types of Chemical Agents Used. The types of chemicals to be used for chemical cleaning of fouled membranes depend on the nature of the membrane, its resistance to chemical stress, and the composition of the fouling material. Ceramic-made membranes which are strongly resistant to all types of chemicals are generally cleaned using acid (pH = 2) and base (pH = 10 to 12 at 80 to 100°C); organic membranes which have very variable chemical resistance as a function of the nature of the polymers used in their formulation are generally cleaned using a mixing of chelatants, detergents, and oxidants when this is possible.

Generally, the following chemicals are used, either alone or combined, for specific purposes in order to remove specific components of the fouling matter:

- Strong acids (phosphoric, sulfuric, chlorhydric) and complexing organic acids (citric, oxalic, gluconic) are used to dissolve metallic oxides and hydroxides (iron, manganese, aluminum) and carbonates.

- Sodium hydroxide, either at ambient or hot temperatures and at concentrations allowing pHs in the range of 10 to 12, is used to remove organic matter by an alkalin hydrolysis mechanism.

- Oxidants (sodium hypochlorite, hydrogen peroxide, peracetic acid) are used at concentrations between 50 mg/L and 1 percent to oxidize organic macromolecules.

- Surfactants and detergents, generally combined with chelatants (e.g., ethylene diamine tetracetic acid) are used to lower the superficial tension and to complex metals and calcium.

- Sodium disulfide, which is generally used as a bacteriostatic preservative agent for modules conservation, can also be used for its reductive and acid properties to dissolve metallic oxides.

Characteristics of Chemical Cleaning Disposals. Chemical cleaning disposals include both the volume of chemicals used for the cleaning procedure and the volume of water used to rinse the membrane modules from any trace of chemicals. This rinsing phase is generally realized by restarting the production of the plant to discharge as waste for a duration which is a function of the type of chemicals used for the cleaning, taking into account the fact that formulations containing surfactants require the higher rinsing volume.

Composition of chemical washing concentrate is shown in Table 10.21 in the case of a cellulosic membrane for which two types of surfactant cleaners are needed. The first cleaner is composed of anionic surfactants, chelatants, and 50 mg/L free chlorine when used at 1 percent (m/v) in solution. The second one is composed of anionic surfactants and complexing agents such as gluconates, citric acid, EDTA, and phosphates. It is used at pH 8 for a 0.9 percent (m/v) dose in solution. The operating procedure consists of a rapid 30-min prewash with the chlorinated cleaner followed by a 1-h wash with the same chemical, and ended by a 3-h wash using the second cleaner. During all three washes, cross-flow recycling of the cleaning solution as well as soaking contact time are alternately used. Each washing phase is followed by a backwash for 30 to 60 s. At the end of the procedure, the modules are rinsed using 150 to 200 L of feedwater to discharge as waste per square meter of membrane module. The total volume of chemical washing concentrate is, in this case, as high as three dead-volumes (180 L/module), three backwashings (900 L/module), and the rinsing volume (5 m³ per module). The total volume of concentrate, in the case of an industrial rack of 28 modules of 55 m², averages 240 m³, with pollution as described in Table 10.21, equivalent to 560 inhabitants in chemical or biological oxygen demand (for 90 g COD and 54 g BOD₅ per inhabitant). Due to the high anionic surfactants content (16 mg/L) of such a concentrate, discharging as waste to a sewer collection line should be carefully handled to avoid foaming. The pHs in that case do not required adjustment to acceptable values for sewer diposal. If detectable free-chlorine residual must be removed by addition of a reducing agent, total dissolved solids are generally low in UF chemical washing concentrates and do not require specific treatment. As shown in Table 10.21, suspended solids, chemical and biological oxygen demands, as well as anionic surfactant concentrations are acceptable for sewer disposal and biological or chemical treatments.

Storage or Treatment of Chemical Cleaning Disposals. Chemical cleaning disposal represents one of the most critical issues for the development of ultrafiltration in the water treatment field. On-site treatment is difficult without implementation of a complex treatment line, as off-line treatment in a centralized chemical washing unit is not easy to handle for membrane plants with a large number of modules.

Most of the existing ultrafiltration plants are chemically washed once or twice a year. With this low frequency of chemical washing, specific on-site treatments such as pH correction, lagooning, settling/filtration, and biological processes (activated sludge, biofiltration, membrane bioreactor) are not worth their capital costs when compared to the operating costs derived from concentrate transportation to a municipal wastewater main or treatment plant.

For membrane plants which require the use of chemical cleaning at higher frequency (e.g., once a month), on-site treatment of the cleaning concentrate is the only technical solution when connection to a wastewater main requires too much capital.

Treatment processes to be used are as complex as the chemical cleaning procedures and the composition of the cleaners.

When membranes can be chemically cleaned with formulations that contain no surfactants or chelatants, on-site treatment is reduced to pH correction and oxidant

TABLE 10.21 Analytical Characteristics of Chemical Cleaning Concentrate for a Rack of 28 Aquasource Modules of 55 m²

	COD, g/m³	BOD₅, g/m³	TOC, g/m³	Phosphate, g de P/m³	Anionic surfactants, g/m³	Suspended solids, g/m³	pH
Cleaning concentrate with backwashing, volume = 54 m³	828	217	218	103	56	832	7.8–8.4
Rinsing (90 L/m²), volume = 186 m³	33	9	3.6	0.3	4	0	—
Total waste, volume = 240 m³	212	56	52	23	16	187	7–8
Maximum concentration limit for sewer discharge	2000	800	—	50	—	600	5.5–8.5

residual removal in an equalization tank, before discharge to the environment. This is the case for cleaning procedures such as acid and sodium hydroxide soaking followed by oxidation using sodium hypochlorite or hydrogen peroxide, which are generally used for ceramic or polysulfone membranes.

In the case of formulations containing surfactants and chelatants, implementation of complex treatment lines, including biological processes, are necessary. It should be noted that for the time being, full-scale ultrafiltration plants operated worldwide have avoided the use of on-site treatment of cleaning concentrates when such a complex treatment line is needed. In fact, when sewer discharge is not possible, the solution of truck transportation to a wastewater main or a wastewater treatment plant has always been selected, due to its low cost per treated cubic meter.

10.6 ULTRAFILTRATION OPERATION AND MAINTENANCE

10.6.1 UF as a Single Operation

10.6.1.1 Process Layout and Schematic. Figure 10.30 presents a diagram of a typical UF membrane plant. The plant may be divided in several subcategories:

- Raw water intake and pressure pumps
- Pretreatment, which includes prescreening, prefiltration, and pH adjustment (if required) or any of the needed pretreatments
- Ultrafiltration units (see details in Fig. 10.31)

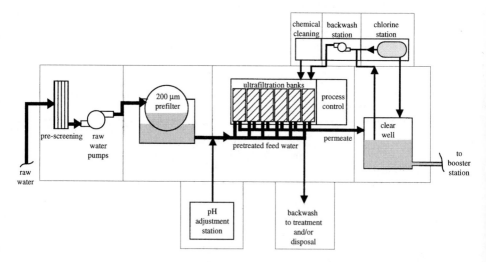

Not to Scale

FIGURE 10.30 Ultrafiltration plant diagram.

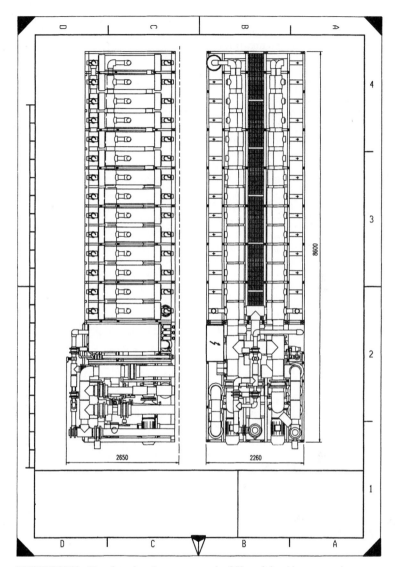

FIGURE 10.31 Drawing of an Aquasource rack of 28 modules. *(Aquasource.)*

- Chemical cleaning station, backwash station (which uses chlorinated product water), chlorine station, conditioner/preservative station
- Line for discharging or treatment of backwash water

Figure 10.22 presents an example of a process flow schematic for an industrial membrane bank, such as that shown in Fig. 10.31. A raw water pump delivers the raw water to a membrane bank unit and provides the required flow and operating pressure (2 to 2.5 bars maximum for polymeric hollow-fiber membranes). A membrane bank is defined as several membrane modules operated as a single unit, as shown in Figs. 10.30 and 10.31. The bank or rack shown in Fig. 10.31 is composed of 28 membrane modules. For small plant capacities, a pump may be dedicated to each membrane bank. However, for large plant capacities, a single pump station can deliver the raw water to the different membrane banks in order to reduce the number of pumps required. In this case, the flow provided by the feed pump is controlled by a flow-monitoring valve dedicated to each membrane rack.

Recirculation of the feedwater is provided by a recirculation pump dedicated to each membrane bank. A recirculation pump, if required, should be able to provide a downflow recirculation–to–raw water flow ratio between 3:1 and 6:1, depending on the raw water quality. For low-turbidity, low-TOC waters, membrane filtration may be operated in dead-end mode, i.e., no feedwater recirculation and associated pumps are required.

Finally, a third pump provides the membrane backwashing. Similar to the raw water pump, a single pump is dedicated for each membrane bank in small plant configurations, and a pump station is used for backwashing all membrane banks for a large plant. Due to the short duration of backwash (less than 1 min) and in order to minimize backwashing pumping requirements, the membrane banks are backwashed in cascade (or series) rather than in parallel; i.e., each membrane bank is backwashed individually. Backwash pressure is generally limited to 1 to 2.5 bars, as a function of radial mechanical resistance of the membrane.

10.6.1.2 Process Operation.

Operation and performance of a UF membrane plant are greatly influenced by raw water quality variations. Turbidity as well as TOC of the raw water are water quality parameters of major importance that drive operation mode and membrane flux for all the UF plants presently in operation worldwide.

Optimization of operating parameters is also of great importance when looking at cost per volume of treated water, particularly for energy savings.

UF plants are generally operated with a high level of automatism in order to be able to minimize energy consumption using external sensors to drive the different process operational modes. This is illustrated in Fig. 10.32, which presents an example of possible operation for a UF plant, at constant production flux.

Three different parameters are used to drive the process:

- Inlet pressure level, which is correlated with membrane-specific flux and fouling. Two *pressure switch high* (PSH) values are preset
- Raw water turbidity level with two preset values
- UV absorbance to TOC ratio level with two preset values

These six different levels (values) or a combination of these values are directly linked to the PLC and can modify the operational configuration of the process.

Inlet pressure and turbidity levels are used to optimize the energy consumption, as those parameters control the operating modes from dead-end to cross-flow, and finally, continuous-bleed filtration. In some cases, PSH can also be used to control backwash frequency of the membrane.

On the other hand, UV-to-TOC ratio levels are used mainly to govern the feed flow of the UF unit, as NOM content of the raw water is a good indicator of the potential fouling of the membrane and, thus, of the possible operating flux that can be treated by the UF plant.

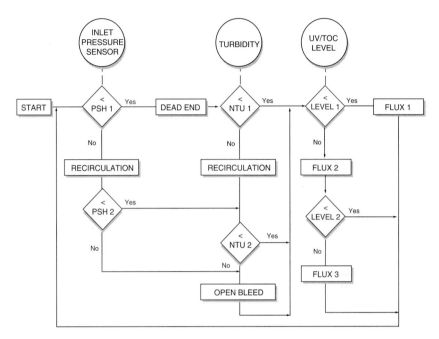

FIGURE 10.32   Example of possible UF plant operation.

This general UF plant operational strategy supposes that the quality of the resource is monitored closely or that sensors can be used continuously to detect any variations in the quality of the resource.

Turbidimeters as well as pressure sensors are considered as reliable sensors. Monitoring systems for continuous UV absorbance at 254 nm are now available to control organic matter variations in the feedwater.

Direct advantages of this operation scheme can be summarized as follows:

• Optimal use of energy, as the process controls its need for cross-flow or continuous-bleed filtration

• Minimum chemical cleaning requirements, as membranes are operated at constant flux levels which decrease with the potential fouling of the membrane

An operational strategy similar to the one presented in Fig. 10.32 is used in full-scale UF plants in France. Since September 1993, the 3500-m^3/day UF plant at Bernay, located in the west of France, has treated a ground resource influenced by the surface. Turbidity levels as well as organic matter (as explained by UV absorbance at 254 nm) contents are subject to a wide range of variations. Figure 10.33 gives an example of turbidity and UV absorbance monitoring of this resource. The highest-quality episodes reported for this resource are a turbidity of 280 NTU, a TOC of 9 mg/L, which corresponds to a UV absorbance of more than 30 (OD/m) in early 1994.

This UF plant is composed of two independent skids of 12 Aquasource 70-m^2 modules (Fig. 10.34); its nominal daily production capacity is 3500 m^3/day at a temperature of 10°C.

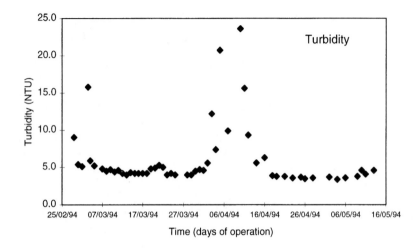

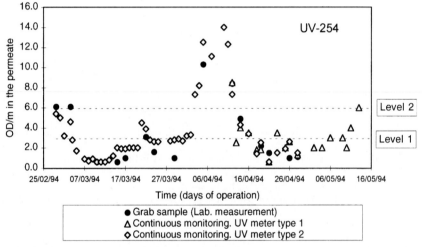

FIGURE 10.33 Evolution of turbidity and optical density at 254 nm in the raw water of the Bernay UF plant.

Table 10.22 summarizes the process parameter levels adopted for the Bernay UF plant. Operation is controlled as explained in Fig. 10.32 except that different back-washing frequencies are associated with filtration modes, from 45 mn in dead-end filtration to 30 mn in cross-flow or continuous-bleed mode.

10.6.1.3 Treated Water Quality

Turbidity and Particles.   The capacity of UF membrane plants in operation world-wide to remove turbidity and particle counts to low levels, regardless of the levels encountered in the raw water, is now well documented (Laîné and Anselme, 1995).

Turbidity levels in UF permeate are consistently reported at values as low as 0.1 NTU even for episodes up to 300 NTU in the feedwater. Particle counts in the size

FIGURE 10.34   Pictures of the Bernay UF full-scale plant (3300 m³/day—0.9 mgd).

range of 0.5 to 5 μm were found by Anselme and co-workers to be lower than 3/mL in the permeate when treating the Seine River raw water, which exhibits counts greater than 60,000/mL. A polymeric hollow fiber with MWCO of 100 kdalton was used in this case.

Microorganisms.   Efficiency of ultrafiltration in the removal of microbial indicators is now well established and documented (Jacangelo et al., 1992, 1995; Mandra et al., 1994).

Log removals as high as summarized in Table 10.23 for various microorganisms have been demonstrated by these authors, both with seeding experiments and naturally occurring organisms.

As previously mentioned, disinfection efficiency of ultrafiltration is theoretically independent of microorganism counts in the feedwater in that the size of the organism is lower than the membrane MWCO. Ultrafiltration must be considered as an absolute physical disinfection technique, with the exception of possible imperfection of membranes or modules. Hollow-fiber configuration is the only module configuration which does not use sealing systems, thus providing a higher reliability to the dis-

TABLE 10.22 Process Parameter Levels Used in Operation at the Bernay UF Plant

	Level 1	Level 2	Level 3
Pressure (PSH), bar	1	1.5	2*
Turbidity, NTU	5	25	—
UV absorbance, DO/m	3	6	—
Flux L/h·m² at 20°C	130	100	70

* Maximum admissible inlet pressure leading to an immediate backwash of the membrane.

infection efficiency of the UF process. In this latest case, the only lack of efficacy which can possibly be suspected is the problem of fiber breakage. Table 10.24 gives a comparison of ultrafiltration and free-chlorine disinfection efficiency in the case of membrane integrity failure. Two types of loss of integrity were tested using an industrial UF module of 50 m² composed of 15,000 fibers:

* Two holes of 50 μm in one fiber
* One fiber broken

Disinfection efficacy of these two modules with integrity failure was compared to a free-chlorine disinfection using a commonly used CT value of 10 to 20 mg × min/L for three types of microorganisms such as *Giardia* cyst, *Cryptosporidium* oocyst, and MS2 bacteriophage (virus).

The results obtained demonstrate that, even with integrity failure, a UF module is a process which gives a higher efficiency for chlororesistant protozoa inactivation than free chlorine and an equivalent or lower efficiency in the case of virus removal.

This demonstrates that, today, UF should be considered as an alternative technique for water disinfection when a system for integrity failure detection is implemented for full-scale application.

NOM and Related Disinfection By-product (DBP) Precursors. UF membranes are commercially available in a wide MWCO range, but data are in good agreement that NOM removal efficiency for the UF process is low, regardless of membrane MWCOs as low as 10,000 daltons.

TABLE 10.23 Efficiency of Ultrafiltration for the Removal of Microbial Indicators

Microorganisms	Size, μm	Reported log removal	Reference
Giardia cysts	7–14	>5.1	JMM East Bay MUD/Contra Costa WD
		>5.0	(Anselme, AGHTM)
Cryptosporidium oocysts	3–7	>4.8	(Anselme, AGHTM)
Escherichia coli	1–3	>8.3	JMM (AWWARF Boise)
Pseudomonas diminuta	0.2–0.5	>7	(Mandra, Hydrotop)
MS2 bacteriophage	0.025	>7.2	JMM—EBMUD CCWD
		>6.5	(Anselme, AGHTM)
Enteric viruses*	0.025	>4	(Mandra, Hydrotop)
Polio virus	0.025	>8.9	(Mandra, Hydrotop)

* Naturally occurring organisms.

TABLE 10.24 Comparison of Ultrafiltration and Chlorine Disinfection Efficiency in the Case of Membrane Integrity Failure

	Concentration (log removal)		
	Giardia cysts, cysts/L	*Cryptosporidium* oocysts, cysts/L	MS2 bacteriophage, pfu/mL
Feedwater concentration	1.1×10^5	7.1×10^4	3×10^6
Intact UF module	<1 (>5 log)	<1 (>4.8 log)	<1 (>6.5 log)
UF module, 50-μm holes	—	—	
1 fiber with 2			6 (5.6 log)
UF module with 1 broken	1.6×10^2	3.6×10^1	6.3×10^3
fiber out of 15,000	(2.8 log)	(3.3 log)	(2.7 log)
Chlorine disinfection,	2 log	<0.5 log	2 log*
CT = 10 to 20 mg/L/mn			>4 log[†]

* Bosch (1989).
[†] Hoff (1989).

Average removals of TOC, UV absorbance at 254 nm, and trihalomethane formation potential (THMFP) for 20 different raw waters when using a hollow-fiber UF membrane with a 100-kdalton MWCO are presented in Fig. 10.35.

Removal of TOC is low and consists of between 0 and 25 percent with an average value of 18 percent. The higher removals are observed when a turbidity episode appears in the resource, since in most cases it is correlated with TOC increases characterized by high molecular weight. UV absorbance as well as THMFP are removed by this UF membrane, with respective average removal of 28 percent and 35 percent.

DOC fractionation of these resources, using a gel-permeation chromatography technique, demonstrates that TOC removal is, in fact, limited to the highest-molecular-weight compounds characterized by an apparent MW greater than 5000 daltons (see Fig. 10.36).

10.6.2 UF Combined with Other Treatments

There are a great number of examples where ultrafiltration membranes have been used in combination with various conventional water treatment processes. This is probably due to either the number of UF membranes available in tubular or hollow-fiber configurations which are favorable to combined-processes implementation and their wide range of molecular weight cutoff. Specific chapters of this book are dedicated to the main combined processes, such as coagulation/flocculation, powdered activated carbon adsorption, and biological process.

A tentative classification of these combined treatments, based on the targeted pollutants, is presented in Table 10.25, in relation to the expected efficiency of these processes.

Inorganic as well as organic pollutants can be efficiently removed by combined UF processes. In most of these cases, in addition to water quality improvement, authors reported an increase in membrane performances due to a decrease of fouling potential of the raw water to be treated. This translated into an increase of flux as high as 30 percent as reported by Moulin et al. (1991) using ozonation with UF

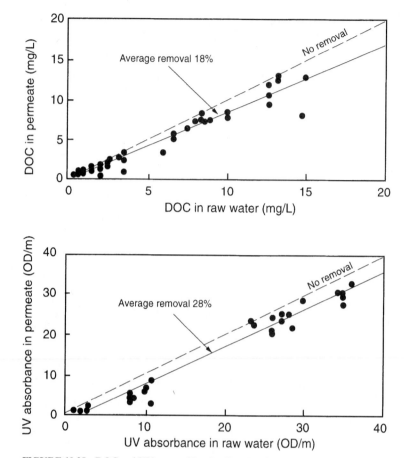

FIGURE 10.35   DOC and UV removal by ultrafiltration (average on 20 sites).

ceramic membrane, or Anselme et al. (1994) using adsorption onto PAC with synthetic UF membrane. The removal of fouling material by these combined processes was also demonstrated by the decrease of chemical cleaning requirement (Anselme, 1995) as well as the improvement in chemical cleaning efficiency (Moulin et al., 1991).

Lahoussine-Turcaud et al. (1990) has used an addition of coagulant (aluminium sulfate or ferric chloride) at a dose equivalent to the optimum of flocculation (50 to 150 mg/L of commercial solution, depending on the type of water to be treated) upstream of the membrane unit. This pretreatment makes it possible to double the flux of polysulfone ultrafiltration membrane as compared to performances obtained with the direct treatment of raw water by the same membrane. At the same time, chemical cleaning frequency was extended and an improvement in the removal of TOC from 15 to 20 percent up to 75 percent was measured. On the contrary, the

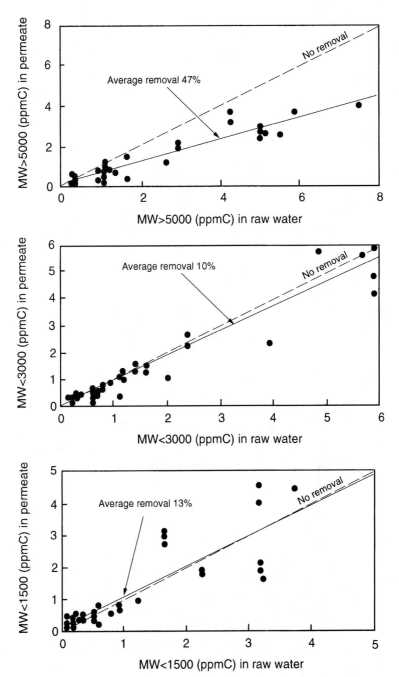

FIGURE 10.36 Removal of GPC/DOC fractions by ultrafiltration (average on 20 sites).

TABLE 10.25 Efficiency of UF Combined Treatments for the Removal of Specific Pollutants

Pollutants	Combined process	UF membrane type	Removal efficiency, %	Reference
Iron, manganese	Aeration + KmnO$_4$ addition	Cellulosic hollow fibers, 100 kdalton	99	James (1992)
Dissolved organic carbon	Ozone + aluminium salt addition	Tubular ceramic, 300 and 30 kdalton	50–60	Moulin (1991)
Dissolved organic carbon	Ozone	Cellulosic hollow fibers, 100 kdalton	<5	Nadeau (1993)
Dissolved organic carbon and atrazine herbicide	Powdered activated carbon addition	Cellulosic hollow fibers, 100 kdalton	70 and 99, respectively	Anselme (1990, 1991, 1992, 1993, 1994)
Dissolved organic carbon and trichlorophenol	Powdered activated carbon	Cellulosic hollow fibers, 100 kdalton	70 and 99, respectively	Adham (1992, 1993)
Nitrates and triazine herbicides	Biological denitrification and PAC addition	Cellulosic hollow fibers, 100 kdalton	80 and 99, respectively	Manem (1993), Urbain (1995)
Total organic carbon	Aluminium sulfate and ferric chloride	Polysulfone hollow fibers, 100 kdalton	75	Lahoussine-Turcaud (1990)

effect of coagulation with either ferric chloride or alum on some cellulosic membrane is not satisfactory, as these coagulants tend to foul the membrane (Jacangelo et al., 1994). Laîné et al. (1990) demonstrates that dissolved organic carbon and trihalomethane formation potential removals are less pronounced in the case of coagulant addition than in the case of powdered activated carbon in combined use with membrane filtration.

Nadeau et al. (1993) combined preozonation with ozone dosage of 2 to 3 mg/L and cellulosic hollow-fiber ultrafiltration and demonstrated that no or adverse hydraulic flux enhancement effect was noticeable, when treating a preozonated surface water, as a function of ozone dose and alkalinity. It should be noted that ozone residual should be avoided in order to protect polymeric membrane from oxidation.

10.7 EXAMPLE FACILITIES AND COSTS

10.7.1 UF Plants in Operation Worldwide

In the water treatment field, ultrafiltration is a process which is still confined to specific applications. Besides the use of ultrafiltration as a pretreatment for reverse-osmosis or ion-exchange systems, generally treating an industrial process water or a secondary treated urban wastewater in a reuse treatment line, the only field of application for ultrafiltration with a growing market is use of this process as the main treatment stage to produce municipal drinking water.

A tentative list of existing UF membrane plants in operation in the world is given in Table 10.26 for the specific application of drinking water production. In this area, the Aquasource system is, today, the only one exhibiting an operating experience. Since 1988 with the first plant in Amoncourt (France), 27 plants have been erected by this company. Installed capacity progressively increased from 250 m³/day to 5000 m³/day. Recent improvements in the technology made it possible to extend the application to nominal capacities up to 28,000 and 55,000 m³/day for the last two projects of l'Apier Saint-Cassien and Vigneux sur Seine in France.

On the other hand, the type of resource treated with UF systems moves from groundwater under the influence of surface water characterized by frequent turbidity episodes to the treatment of surface river or reservoir water using a combined addition of powdered activated carbon in order to remove dissolved organic carbon present in excessive concentration. This combined PAC/UF process can be used to directly treat raw water (like in the l'Apier project) or after a complete clarification of the resource (like in the Vigneux sur Seine project).

10.7.2 Capital Costs

As the UF industrial market is today trusted by only one manufacturer with more than six years of experience, capital cost considerations presented hereafter are mainly based on systems cost and design proposed by the Aquasource company.

As in any other water treatment processes, ultrafiltration capital cost is strongly related to the size of the plant to be designed (Anselme et al., 1993). UF systems can be separated into two types:

- Skid-mounted system for small to medium plant capacity up to 3000 to 5000 m³/day
- Custom-designed plant for large-scale implementation based on a UF multirack design

TABLE 10.26 Identified Full-Scale Ultrafiltration Membrane Plants for Municipal Drinking Water Treatment

On-line date	Plant site location	Raw water source	Installed plant capacity m³/J	mgd	Membrane manufacturer
1988	Amoncourt, Haute Saône, France	Groundwater	256	0.07	Aquasource
1989	Douchy, Loiret, France	Groundwater	1,440	0.38	Aquasource
1990	Graçay, Cher, France	Groundwater	720	0.19	Aquasource
1991	Blomac, Aude, France	Surface water	190	0.05	Aquasource
1991	Macao, Southeast Asia	Surface water	2,600	0.69	Aquasource
1991	Osselle, Doubs, France	Groundwater	220	0.06	Aquasource
1991	Sauve, Gard, France	Groundwater	1,900	0.50	Aquasource
1991	Stonehaugh, England	Groundwater	50	0.01	Aquasource
1992	Charcenne, Haute Saône, France	Groundwater	600	0.16	Aquasource
1992	Hourtin, Gironde, France	Surface water	70	0.02	Aquasource
1992	La Nive, Pyrénées Atlantiques, France	Surface water	5,000	1.32	Aquasource
1992	Le Baizil, Marne, France	Groundwater	220	0.06	Aquasource
1992	Villefrance de Panat, Aveyron, France	Surface water	220	0.06	Aquasource
1993	Avoriaz, Haute Savoie, France	Surface water	3,400	0.90	Aquasource
1993	Bernay, Eure, France	Groundwater	3,300	0.87	Aquasource
1993	Chatel Gérard, Yonne, France	Groundwater	660	0.17	Aquasource
1993	Fontgombault, Indre, France	Groundwater	5,000	1.32	Aquasource
1993	La Filière, Haute Savoie, France	Groundwater	2,000	0.53	Aquasource
1993	Maison Rouge, Haute Saône, France	Groundwater	NA	NA	Koch
1993	New Rochelle, New York, USA	Surface water	320	0.08	Aquasource
1993	Saint Jean d'Arvey, Savoie, France	Groundwater	600	0.16	Aquasource
1993	Urzy, Nièvre, France	Groundwater	1,100	0.29	Aquasource
1994	Jean Prince Hospital, Papeete, Tahiti	Tap water	120	0.03	Aquasource
1994	Chermisey, Vosges, France	Groundwater	200	0.05	Aquasource
1994	Grande Fountaine, Haute Saône, France	Groundwater	800	0.21	Aquasource
1994	Luçay-le-mâle, Indre, France	Groundwater	600	0.16	Aquasource
1995	Jussey, Haute Saône, France	Groundwater	1,000	0.26	Aquasource
1995	East Bay Municipal Utility District, California, USA	Surface water	500	0.13	Aquasource
1995	L'Apier Saint-Cassien, Alpes Maritimes, France	Surface water	28,000	7.40	Aquasource
1995	Tynywaun, UK	Clarified surface water	10,000	2.64	Aquasource
1996	Vigneux sur Seine, Essonne, France	Clarified surface water	55,000	14.53	Aquasource
1996	Yamanashi County, Japan	NA	(300 m²)	(3225 ft²)	Asahi Kasei
32	Total		126,086	33	

NA = not available.

However, in UF systems, as for any membrane processes, the raw water quality and particularly the organic content of the feedwater strongly influence the plant capital cost. This is due to the fouling potential of the raw water for the membrane which can impart performances of the membrane such as design flux of production and water loss.

In a recent study, Jacangelo et al. (1992) have performed Aquasource ultrafiltration system cost evaluation in comparison with various conventional treatment processes for three plant capacities: 400, 4000, and 20,000 m^3/day. This cost study was part of an AWWARF funded project entitled "Evaluation of Low Pressure Membrane Filtration for Particle Removal," realized in Boise, Idaho, where both the Boise River and Boise Ranney Collector were piloted with Aquasource ultrafiltration membranes. In this study, UF-based processes were found to be more expensive than the conventional packaged plants (direct filtration combined or not with ozone) for all simulated production capacities. Capital cost differences increase with increasing production capacity; 10 percent overcost was found for the 400-m^3/day plant, and 30 and 100 percent, respectively, were found for the 4000- and 20,000-m^3/day plants. In another AWWARF-funded study (Jacangelo et al., 1994) and for the same three plant production capacities, total cost of UF combined or not with PAC was compared with the costs of conventional treatment lines (including clarification alone or clarification followed by ozonation and GAC filtration) and other membrane technology treatment lines (including transverse and cross-flow hollow-fiber nanofiltration, and spiral-wound nanofiltration pretreated by ultrafiltration). In this study, the UF process was found to be competitive for capacity up to 4000 m^3/d (1 mgd). A later study (Wiesner et al., 1994) shows that the UF process becomes competitive at higher capacity (5 mgd).

By designing skid-mounted package plants, ultrafiltration technology was found to be cost competitive at a small scale (less than 10,000 m^3/day), as compared to conventional treatment. Today, several projects up to 150,000 m^3/day are under construction or design. At medium and large scale, new design developments and operation procedure optimization were required to make the process cost competitive. Design improvements have generated substantial savings in capital cost—up to 30 percent—as illustrated in Fig. 10.37, for Aquasource systems since the initial design of 1989. The newly developed rack consists of two membrane module lines mounted in parallel on a single unit. The new membrane rack can host up to 28 membrane modules and provide up to 2000 m^2 (20,000 ft^2) of membrane filtration area. The major impacts on cost savings were due to the compactness of the new rack design, which requires less civil work, and to a reduction of equipment—number of valves, pumps, etc. Furthermore, usage optimization of the ancillary equipment, such as the backwash station, has also led to significant savings.

Operating cost savings of these newly designed plants have also been estimated in the neighborhood of 30 percent as compared to the first design generation.

10.7.2.1   Energy Consumption.   The new hydraulic configuration of the rack directly impacts the energy consumption required in cross-flow operation. Furthermore, energy consumption has been reduced by optimizing the automation procedures of the UF process. Based on the water quality and membrane production performance, the programmable logic controller (PLC) can switch from cross-flow to dead-end filtration mode, and vice versa. Furthermore, the operation of the plant is directly assessed to the water demand. The PLC will turn groups of racks on and off in order to maintain the proper water production.

10.7.2.2   Equipment Requirement.   The amount of ancillary equipment has been significantly reduced by optimizing the automation. For example, one backwash sta-

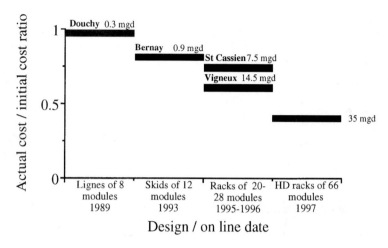

Douchy plant as reference

FIGURE 10.37 Effect of design on the capital cost of Aquasource UF plants as compared to the Douchy plant.

tion is used for several racks of membrane modules. When the backwash of a rack is required, the slave PLC provides a signal to the main PLC to initiate a backwash procedure. The main PLC will provide backwash if the station is not in use; otherwise, the main PLC may either maintain the rack in production or put it in standby until the backwash station is available again.

10.7.2.3 Membrane Replacement. Finally, it should be noted that another saving in operating cost is due to the lower replacement cost of the membranes. This is primarily the result of an optimization of the manufacturing steps and also to an increase in membrane module demand.

In summary, UF membrane technology is becoming more and more attractive from a water quality and a cost perspective as compared to conventional processes. New design and technology developments have made this technology available to medium- and large-scale plants. In the past five years, applications for drinking water treatment have been increasing. Today, a number of plants from 12,000 m³/day to 55,000 m³/day are being constructed.

BIBLIOGRAPHY

Adham, S. S., V. L. Snoeyink, M. M. Clark, C. Anselme, and I. Baudin, 1993, "Predicting and Verifying Organics Removal by PAC in Large Scale Ultrafiltration Systems," *Proceedings of the Membrane Technology Conference AWWA-IWSA,* August 1–4, Baltimore, pp. 601–615.

Adham, S. S., V. L. Snoeyink, M. M. Clark, and C. Anselme, 1993, "Predicting and Verifying TOC Removal by PAC in Pilot Scale UF Systems," *Jour. AWWA,* **12.**

AFNOR (Association Française de Normalisation), 1992, Proposition de Norme suivant le Plan Type Général AFNOR, Groupe "Soluté"/CG/5.0.

Akhtar, S., C. Hawes, L. Dudley, I. Reed, and P. Stratford, 1995, "Coatings Reduce the Fouling of Microfiltration Membranes," *Jour. Membrane Science,* **107**:209–218.

Amirtharajah, A., 1988, "Some Theoretical and Conceptual Views of Filtration," *Jour. AWWA,* pp. 36–46.

Amirtharajah, A., and D. P. Wetstein, 1980, "Initial Degradation of Effluent Quality During Filtration," *Jour. AWWA,* **72**(9):518–524.

Amy, G. L., M. R. Collins, C. J. Kuo, and P. H. King, 1987a, "Comparing GPC and UF for Molecular Weight Characterization of Aquatic Organic Matter," *Jour. AWWA,* **79**(1):43–49.

———, 1987b, "Removing VOCs by Membrane Stripping," *Jour. AWWA,* **79**(1):43–49.

Anselme, C., P. Charles, J. L. Bersillon, and J. Mallevialle, 1991, "The Use of Powdered Activated Carbon for the Removal of Specific Pollutants in Ultrafiltration Processes," *AWWA Seminar Proceedings—Membrane Technologies in the Water Industry,* March 10–13, Orlando, Fla., pp. 571–586.

Anselme, C., V. Mandra, I. Baudin, and J. Mallevialle, 1994, "Optimum Use of Membrane Processes in Drinking Water Treatment," *Proceedings of the 19th Congress and Exhibition IWSA-AIDE: Technical Papers, Water and the Environment, a Common Cause,* Budapest, October 2–8, 1993, **12**(1/2):SS2-1–SS2-11.

Anselme, C., I. Baudin, M. Clark, and S. Adham, 1995, "Polishing Treatment for Optimal Organic Removal from Modelling to the Cristal Process," *Proceedings of the International AIDE/IWSA Workshop,* Membranes in Drinking Water Production, March 27–29, Paris, pp. 75–79.

Bansal, B., and W. N. Gill, 1974, "Theoretical and Experimental Study of Radial Flow Hollow Fibre Reverse Osmosis," *AIChE Symposium Series,* **70**:136–149.

Baudin, I., and V. Mandra, 1995, "Taking into Account Fouling in Ultrafiltration Design and Operation," *Proceedings IWSA-AIDE,* March 27–29, Paris.

Belfort, G., 1988, "Membrane Modules: Comparison of Different Configurations Using Fluid Mechanics," *Jour. Membrane Science,* **35**:245.

Bersillon, J. L., C. Anselme, J. Mallevialle, and F. Fiessinger, 1989, "Ultrafiltration Applied to Drinking Water Treatment: Case of a Small System," *Water Nagoya—ASPAC,* Nagoya, Oct. 29–Nov. 2, Session 6, pp. 209–219.

Bessarabov, D. G., R. D. Sanderson, E. P. Jacobs, C. N. Windt, and V. S. Gladkov, 1994, "Asymmetric Non-Porous Flat Sheet Membranes from Polyvinyltrimethylsilane for Water Oxygenation and Deoxygenation," *S.A. Journal of Chemical Engineering,* **6**(2):26.

Boom, R. M., et al., 1992, "Microstructures in Phase Inversion Membranes: Part 2. The Role of Polymeric Additives," *Jour. Membrane Science,* **73**:277–292.

Brink, L. E. S., S. J. G. Elbers, T. Robbertsen, and P. Both, 1993, "The Anti-fouling Action of Polymers Preadsorbed on Ultrafiltration and Microfiltration Membranes," *Jour. Membrane Science,* **76**:281–291.

Brock, T. D., and K. M. Brock, 1978, *Basic Microbiology with Applications,* Prentice-Hall, Englewood Cliffs, N.J.

Chen, V., A. G. Fane, and C. J. D. Fell, 1992, "Use of Anionic Surfactants for Reducing Fouling of Ultrafiltration Membranes: Their Effects and Optimization," *Journal of Membrane Science,* **67**:249–261.

Cheryan, M., 1986, *Filtration Handbook,* Technomic Publishing, Lancaster, Pa.

Cleasby, J. L., 1990, "Filtration," in F. W. Pontius (ed.), *Water Quality and Treatment: A Handbook of Community Water Supplies (AWWA),* 4th ed., McGraw-Hill, New York, pp. 455–555.

Cole, J., and W. Pauli, 1975, "Field Experience in Testing Membrane Filter Integrity by Forward Flow Test Method," *Parental Bull., Drug Assoc.,* **29**:296.

Cotruvo, J. A., and C. D. Vogt, 1990, "Rationale for Water Quality Standards and Goals," chap. 1 in F. W. Pontius (ed.), *Water Quality and Treatment: A Handbook of Community Water Supplies (AWWA),* 4th ed., McGraw-Hill, New York.

Cowan, J. A. C., T. MacTavish, C. J. Brouckaert, and E. P. Jacobs, 1992, "Membrane Treatment Strategies for Red Meat Abattoir Effluents," *Water Science and Technology,* **25**(10):137–148.

Cui, Z. F., 1993, "Experimental Investigation on Enhancement of Crossflow Ultrafiltration with Air Sparging," in R. Paterson (ed.), *Effective Membrane Processes—New perspectives,* 3d International Conference, Bath.

Da Costa, A. R., A. G. Fane, and D. E. Wiley, 1993, "Ultrafiltration of Whey Protein Solutions in Spacer-Filled Flat Channels," *Jour. Membrane Science,* **76**:245–254.

Degrémont, 1988, Mémento Technique de l'Eau, Lavoisier, Tec et Doc (ed.), Paris.

Ericsson, B., and B. Hallmans, 1991, "Membrane Filtration a Pretreatment Method," *Proceedings 12th International Symposium on Desalination and Water Re-use,* Malta, **2**(11):249–266.

Eykamp, W., 1990, "Ultrafiltration," chap. 11 in *Membrane Separation Systems—A Research and Development Needs Assessment,* Department of Energy Membrane Systems Research Needs Group, Contract No. DE-AC01-88ER30133, vol. 1.

Flemming, H. L., 1987, "Latest Developments in Inorganic Membranes," presented at *The B.C.C. Membrane Planning Conference,* Oct. 20–22, Cambridge, Mass.

Gelman, C., and R. E. Williams, 1983, "Ultrafiltration," chap. 12 in N. P. Cheremisinoff, and D. S. Azbel (eds.), *Liquid Filtration,* Ann Arbor Science Publishers, Woburn, Mass.

Gourgues, C., P. Aimar, and V. Sanchez, 1992, "Ultrafiltration of Bentonite Suspensions with Hollow Fiber Membranes," *Jour. Membrane Science,* **74**:51–69.

Gregor, H. P., and C. D. Gregor, 1978, "Synthetic Membrane Technology," *Scientific American,* **239**(1):112–128.

Grund, G., C. W. Robinson, and B. R. Glick, 1992, "Protein Type Effects on Steady-State Crossflow Membrane Ultrafiltration Fluxes and Protein Transmission," *Jour. Membrane Science,* **70**:177–192.

Gupta, B. B., P. Blanpain, and M. Y. Jaffrin, 1992, "Permeate Flux Enhancement by Pressure and Flow Pulsations in Microfiltration with Mineral Membranes," *Jour. Membrane Science,* **70**:257–266.

Gutman, R. G., 1987, *Membrane Filtration: The Technology of Pressure-Driven Crossflow Processes,* IOP Publishing Ltd., England.

Hackley, V. A., and M. A. Anderson, 1992, "Synthesis and Characterization of Unsupported Ferric Oxide Ceramic Membranes," *Jour. Membrane Science,* **70**:41–51.

Hamann, Jr., C. L., J. B. McEwan, and A. G. Myers, 1990, "Guide to Selection of Water Treatment Processes," chap. 3 in F. W. Pontius (ed.), *Water Quality and Treatment: A Handbook of Community Water Supplies (AWWA),* 4th ed., McGraw-Hill, New York.

Heyden, W., 1985, "Seawater Desalination by RO: Plant Design, Performance Data, Operation and Maintenance (Tanajib, Arabian Gulf Coast)," *Desalination,* **52**:187–199.

Howell, J. A., 1995, "Sub-Critical Flux Operation of Microfiltration," *Jour. Membrane Science,* **107**:165–171.

Hubele, C., 1992, "Automatic Control of the Coagulant Dose in Drinking Water Treatment," chap. 11, in *Influence and Removal of Organics in Drinking Water,* Lewis Publishers, Ann Arbor, Mich., pp. 157–171.

Jacangelo, J. G., E. M. Aieta, K. E. Carns, E. W. Cummings, and J. Mallevialle, 1989, "Assessing Hollow-Fiber Ultrafiltration for Particulate Removal," *Jour. AWWA,* pp. 68–75.

Jacangelo, J., J. M. Laîné, K. E. Carns, E. W. Cummings, and J. Mallevialle, 1991, "Low Pressure Membrane Filtration for Removing Giardia and Microbial Indicators," *Jour. AWWA,* **83**(9):97–106.

Jacangelo, J., N. Patania, J. M. Laîné, W. Booe, and J. Mallevialle, 1992, "Low Pressure Membrane Filtration for Particle Removal," *AWWA Research Foundation and AWWA,* Denver, Colo.

Jacangelo, J., and J. M. Laîné, E. W. Cummings, A. Deutschmann, J. Mallevialle, and M. R. Wiesner, 1994, "Evaluation of Ultrafiltration Membrane Pretreatment and Nanofiltration of Surface Waters," *AWWA Research Foundation and AWWA,* Denver, Colo.

Jacangelo, J., S. Adham, and J. M. Laîné, 1995, "Mechanism of Cryptosporidium, Giardia, and Virus Removal by MF and UF," *Jour. AWWA,* **87**(9):107–121.

Jodellah, A. M., and W. J. Weber, Jr., 1985, "Controlling Trihalomethane Formation Potential by Chemical Treatment and Adsorption," *Jour. AWWA,* **77**(9):95–100.

Knops, F. N. M., H. Futselaar, and I. G. Rácz, 1992, "The Transversal Flow Microfiltration Module: Theory Design Realization and Experiments," *Jour. Membrane Science,* **73**:153–161.

Ko, M. K., and J. J. Pellegrino, 1992, "Determination of Osmotic Pressure and Fouling Resistances and Their Effects on Performance of Ultrafiltration Membranes," *Jour. Membrane Science,* **74**:141–157.

Laîné, J. M., J. P. Hagstrom, M. M. Clark, and J. Mallevialle, 1989, "Effects of Ultrafiltration Membrane Composition," *Jour. AWWA,* **81**(11):61–67.

Laîné, J. M., M. M. Clark, and J. Mallevialle, 1990, "Ultrafiltration of Lake Water: Effect of Pretreatment on the Partitioning of Organics, THMFP, and Flux," *Jour. AWWA,* **82**(12):82–87.

Laîné, J. M., J. G. Jacangelo, N. L. Patania, W. Booe, and J. Mallevialle, 1991, "Evaluation of Ultrafiltration Membrane Fouling and Parameters for Its Control," *Proceeding of the AWWA Membrane Processes Conference,* March 10–13, Orlando, Fla.

Laîné, J. M., and J. Jacangelo, 1992, "System Components and Process Design Considerations for Ultrafiltration of Untreated Drinking Water Supplies," presented at the *Sunday Seminars, AWWA Annual Conference,* June 18–22, Vancouver, B.C.

Laîné, J. M., and J. Jacangelo, 1993, "Low Pressure Membrane Design Considerations," presented at the *Sunday Seminars, AWWA Annual Conference,* June 6–10, San Antonio, Tex.

Laîné, J. M., and C. Anselme, 1995a, "Ultrafiltration Technology Status Overview in Municipal Drinking Water Treatment," Poster presented at the *20th Congress IWSA Conference,* September 9–15, Durban, South Africa.

———, 1995b, "Impact Various Pretreatments on Nanofiltration Performance in Treating a Surface Water," presented at the *AWWA Membrane Technology Conference,* August 13–16, Reno, Nev.

Laîné, J. M., and P. Gislette, 1995, "The Concentrate Issue in Ultrafiltration Operation," *Proceedings IWSA-AIDE,* March 27–29, Paris.

Lahoussine-Turcaud, V., M. R. Wiesner, J. Y. Bottero, and J. Mallevialle, 1990, "Coagulation Pretreatment for Ultrafiltration of a Surface Water," *Jour. AWWA,* **82**(12):76–81.

Lippy, E. C., 1986, "Chlorination to Prevent and Control Water-Borne Disease," *Jour. AWWA,* **78**(1):49–52.

Loeb, S., and S. Sourirajan, 1962, "Sea Water Demineralization by Means of an Osmotic Membrane," *Advan. Chem. Ser.,* **38**:117–132.

Longsdon, G. S., 1987, "Comparison of Some Filtration Processes Appropriate for Giardia Cyst Removal," *Proceedings Calgary Giardia Conference.*

Mallevialle, J., C. Anselme, and O. Marsigny, 1989, "Effects of Humic Substances on Membrane Processes," *193rd National Meeting of the Aquatic Humic Substances: Influence on Fate and Treatment of Pollutants,* April 5–10, 1987, Denver, American Chemical Society, Advances in Chemistry Series: 219, pp. 749–767.

Mallevialle, J., and F. Fiessinger, 1991, "New Guidelines for Drinking Water: The International Trend," *Proceeding of the Australian WWA Annual Conference,* Perth.

Mandra, V., I. Baudin, and C. Anselme, 1994, "Les Techniques Séparatives par Membranes: Procédé de Désinfection," *Techniques et Sciences Municipales, Actes du 74ème Congrès de l'AGHTM,* June 6–10, 1994, Nîmes.

Metcalf & Eddy, 1991, *Wastewater Engineering: Treatment, Disposal and Reuse,* G. Tchobanoglous and F. L. Burton (eds.), 3d ed., McGraw-Hill, New York.

Miller, K. D., S. Weitzel, and V. G. J. Rodgers, 1993, "Reduction of Membrane Fouling in the Presence of High Polarized Resistance," *Jour. Membrane Science,* **76**:77–83.

Moulin, C., M. M. Bourbigot, A. Tazi-Pain, and M. Faivre, 1991, "Potabilization of Surface Waters by Tangential Ultra and Microfiltration on Mineral Membranes: Interest of Ozone," *AWWA, Membrane Processes Conference,* Membrane Technologies in the Water Industry, March 9–13, Orlando, Fla.

Nadeau, I., C. Anselme, O. Wable, I. Baudin, and M. Dore, 1993, "Effect of Ozonation on Organic Membrane Fouling the Microfloculation Phenomena," *11th Ozone World Congress, IOA,* Ozone in Water and Wastewater Treatment, Aug. 29–Sep. 3, San Francisco, **2**:S16-59–S16-79.

Pieterse, M. J., 1989, "Drinking-Water Quality Criteria with Special Reference to the South African Experience," *Water S-A,* **15**(3):169–178.

Porter, M. C., 1990, "Ultrafiltration," chap. 3 in M. C. Porter (ed.), *Handbook of Industrial Membrane Technology,* Noyes Publications, N.J., pp. 136–259.

Rautenbach, R., and R. Albrecht, 1989, *Membrane Processes,* John Wiley & Sons, New York.

Reid, C. E., and E. S. Breton, 1959, "Water and Ion Flows Across Cellulosic Membranes," *Jour. Appl. Polym. Sci.,* **1**:133–143.

Rodgers, V. G. J., and R. E. Sparks, 1991, "Reduction of Membrane Fouling in Protein Ultrafiltration," *AIChE Journal,* **37**(10):1517.

Roesink, H. D. W., and I. G. Rácz, 1989, "Modules with Capillary Microfiltration Membranes: I. Comparison of the Shell-Side Fed Module with the Capillary Bore Fed Module," chap. 7 in "Microfiltration Membrane Development and Module Design," Ph.D. thesis, University of Twente, Enschede, The Netherlands.

Rose, J. B., 1985, "Virus Removal During Conventional Drinking Water Treatment," Dissertation, University of Arizona, Tucson, Ariz.

SABS 241, 1984, "Specifications for Water for Domestic Supply," South African Bureau of Standards.

Sayre, I. M., 1988, "International Standards for Drinking Water," *Jour. AWWA,* pp. 53–60.

Semmens, M. J., R. Qin, and A. Zander, 1989, "Using a Microporous Hollow-Fibre Membrane to Separate VOCs from Water," *Jour. AWWA,* **81**(4):162–167.

Semmens, M. J., and A. B. Staples, 1986, "The Nature of Organics Removed During Treatment of Mississippi River Water," *Jour. AWWA,* **78**(2):76–81.

Sinisgalli, P. D., and J. L. McNutt, 1986, "Industrial Use of RO," *Jour. AWWA,* **78**(5):47–51.

Strohwald, N. K. H., 1992, "Pilot Scale Desalination of Sea Water by Reverse Osmosis," Final report 345/1/92 to the Water Research Commission of South Africa by Membratek (Pty) Ltd.

Strohwald, N. K. H., and E. P. Jacobs, 1992, "An Investigation into UF Systems in the Pretreatment of Seawater for RO Desalination," *Water Science and Technology,* **25**(10):69–78.

Stumm, W., and J. J. Morgan, 1990, "Aquatic Chemistry," 2d ed., John Wiley & Sons, New York, p. 270.

Tate, C. H., and K. F. Arnold, 1990, "Health and Aesthetic Aspects of Water Quality," chap. 2 in F. W. Pontius (ed.), *A Handbook of Community Water Supplies (AWWA),* 4th ed., McGraw-Hill, New York.

Tchobanoglous, G., and E. D. Schroeder, 1985, *Water Quality—Characteristics, Modelling, Modification,* Addison-Wesley, Reading, Mass.

Tweddle, T. A., C. Striez, C. M. Tam, and J. D. Harzlett, 1992, "Polysulfone Membranes: I. Performance Comparison of Commercially Available Ultrafiltration Membranes," *Desalination,* **86**:27–41.

Van Steenderen, R. A., S. J. Theron, and A. C. W. Engelbrecht, 1989, "An Investigation into the Occurrence and Concentration of Trihalomethanes and Their Precursors in South African Drinking Waters," WRC Report 194/89 by the Watertek Division of the CSIR, Pretoria.

Wetterau, G. E., M. M. Clark, and C. Anselme, 1994, "A Dynamic Model for Predicting Fouling During the Ultrafiltration of a Natural Water," *Jour. Membrane Science.*

Wickramanayake, G. B., A. J. Rubin, and O. J. Sproul, 1988, "Effects of Ozone and Storage Temperature on Giardia Lamblia," *Jour. AWWA,* **77**(8):74–77.

Wierenga, J. T., 1985, "Recovery of Coliforms in the Presence of a Free Chlorine Residual," *Jour. AWWA,* **77**(11):83–88.

Wiesner, M. R., J. Hackney, S. Sethi, J. G. Jacangelo, and J. M. Laîné, 1994, "Membrane Processes Might Meet the Needs of Small Systems Better than Conventional Filtration," *Jour. AWWA,* **86**(12):33–41.

CHAPTER 11
MICROFILTRATION

Joseph G. Jacangelo
Montgomery Watson
Herndon, Virginia

Chris A. Buckley
Pollution Research Group
Department of Chemical Engineering
University of Natal
South Africa

11.1 PROCESS APPLICATION

11.1.1 Introduction

Microfiltration (MF) is oldest of the four pressure-driven membrane technologies, which also include *reverse osmosis* (RO), *nanofiltration* (NF), and *ultrafiltration* (UF). The first microfilters were of the depth filter type, used primarily for laboratory and industrial purposes. In depth filtration, particles and microorganisms are entrapped within the internal structure of the microfilter. After the breakthrough of particles or at a specified pressure drop, the microfilters are changed. Because of the need for regular replacement of the filters, depth filtration microfilters have had limited application for primary treatment of drinking water, except in very small scale systems or as pretreatment to reverse osmosis. In contrast to the concept of relative removal by depth filtration, MF membranes provide absolute removal of contaminants from a feed stream by separation based on retention of contaminants on a membrane surface. It is the "loosest" of the membrane processes, having a pore size ranging from 0.05 to 5 μm. As a consequence of its large pore size, it is used primarily for particle and microbial removal and can be operated under ultra-low-pressure conditions. However, it should be noted that MF does not, in all cases, remove contaminants strictly based on pore size of the membrane. A cake layer, consisting of materials present in the feedwater, can form on the membrane surface and provide additional removal capabilities.

11.1.2 Regulations as an Impetus for MF

11.1.2.1 United States. In 1986, the Congress of the United States passed sweeping amendments to the Safe Drinking Water Act (SWDA). The 1986 amendments required the United States Environmental Protection Agency (USEPA) to establish regulations for 83 compounds identified by Congress within three years of enactment. Thereafter, Congress required the USEPA to set 25 additional standards every three years. As a result of these amendments, the Surface Water Treatment Rule (SWTR) was promulgated in June of 1989 and addressed the protozoan cyst, *Giardia,* and viruses through mandating filtration for most surface supplies. The SWTR also included specific disinfection criteria. Surface water systems must now provide a 99.9 percent or 3-log removal/inactivation of *Giardia* and 99.99 percent or 4-log removal/inactivation of viruses. It is currently anticipated that an enhanced SWTR will be promulgated in the future. The rule may specify greater-percentage removals of *Giardia* and/or *Cryptosporidium* based on the concentrations of the protozoan cysts and oocysts in the raw water supply to the utility. The rule will also include a maximum contaminant level goal (MCLG) of zero for *Cryptosporidium.* In addition to the enhanced SWTR, the USEPA is currently developing a groundwater disinfection rule.

While required to provide better disinfection/removal of selected microorganisms, utilities will also be mandated to meet new maximum contaminant levels (MCLs) for disinfection by-products (DBPs). Trihalomethanes, which have an MCL of 0.10 mg/L, are currently the only DBPs regulated; other DBPs such as haloacetic acids will be regulated in the future. Thus, utilities are faced with providing better disinfection while minimizing the concentrations of DBPs in their finished water. As a result of the current and anticipated regulations, there has been an increasing interest in employing MF for the removal of particles and microorganisms from untreated drinking water supplies. By removing microorganisms with this technology, less primary disinfectant would be required, thus lowering the concentrations of DBPs formed through the treatment process. As shown in Fig. 11.1, this membrane process is intended to replace four unit processes in conventional water treatment: rapid mix, coagulation, flocculation, and media filtration. In comparison with conventional treatment, MF is a physical process that removes contaminants primarily by sieving them from the water being treated. Primary disinfection and coagulant addition for *Giardia* and *Cryptosporidium* removal/inactivation is reduced or eliminated. Residual disinfectant is applied in a chlorine contact basin, clearwell, or primary transmission main before the first customer for inactivation of viruses.

11.1.3 World Health Organization Guidelines

The World Health Organization has a set of guidelines for drinking water quality (World Health Organization, 1993). Of particular note is that *Escherichia coli* (or thermotolerant coliform bacteria) and total coliform bacteria must not be detectable in any 100-mL sample collected from water entering the distribution system. While achieving this goal, many European countries are becoming increasingly interested in distributing water with little or no disinfectant residual. Providing a water which is free of pathogens is, therefore, of paramount importance to public health safety. Consequently, membrane filtration is being considered in several cases as a necessary unit process to achieve this goal.

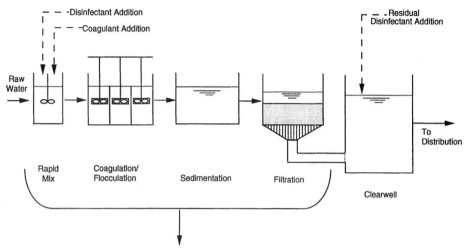

FIGURE 11.1 Conventional water treatment unit processes replaced by MF.

11.1.4 Applications of MF

There are several applications of MF in the drinking water industry (Fig. 11.2). The primary impetus for the more widespread use of MF has been the increasingly more stringent requirements for better removal of particles and microorganisms from drinking water supplies. Additionally, there is a growing emphasis on limiting the concentrations and number of chemicals that are applied during water treatment. In some MF applications for particle and microbial removal, no chemicals are added during the water production cycle of the process. In others, coagulation or adsorption processes are employed as pretreatment to MF. Because of the many potential advantages of MF as compared to conventional water treatment, the technology has been the focus of numerous studies and full-scale applications. A typical treatment train for this application would consist simply of prescreening, MF, and disinfection. For some tubular membrane systems, prescreening is not necessary. An example of how this technology is growing is shown in Fig. 11.3, which is data obtained from a recent questionnaire on low-pressure membrane filtration (Jacangelo et al., 1996). The figure shows that the technology has been growing very rapidly from 1986 to present.

Another application of the technology is for removal of natural or synthetic organic matter. In its normal operation, MF removes little or no organic matter; however, when pretreatment is applied, increased removals of organic material as well as a retardation of membrane fouling can be realized. A treatment train would consist of prescreening, pretreatment, MF, and disinfection. The two most common pretreatments are metal coagulant and powdered activated carbon (PAC) addition. Both have been demonstrated to provide increased removal of natural organic matter (Olivieri et al., 1991a; Wiesner et al., 1992) or synthetic organic chemicals (Pirbazari et al., 1992) as compared to MF alone. In some cases, the cake layer built up on the membrane during the water production cycle can provide some removal of organic materials.

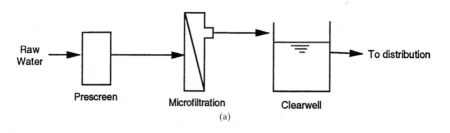

(a)

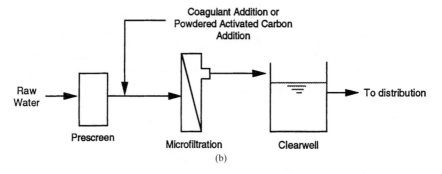

(b)

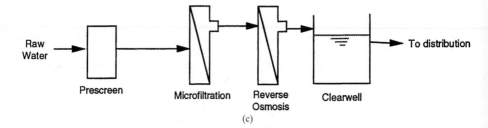

(c)

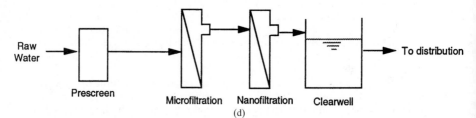

(d)

FIGURE 11.2   Schematic of various applications of MF: (*a*) MF only; (*b*) organics removal by MF pretreatment with coagulant or PAC; (*c*) MF as pretreatment to reverse osmosis; (*d*) MF as pretreatment to nanofiltration.

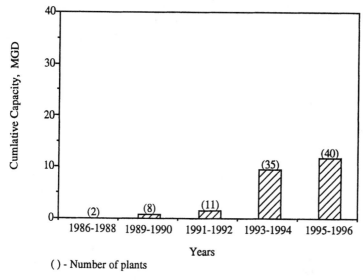

FIGURE 11.3 Cumulative installed capacity of MF. *(From Jacangelo et al., 1996.)*

Two other applications involve using MF as a pretreatment to RO or NF. Both RO and NF have been traditionally employed on groundwaters for desalting or hardness removal. However, with the increasing emphasis on water conservation, RO is being employed by an ever greater number of utilities to treat wastewater for reuse. Consequently, MF is being studied extensively as a pretreatment for RO for water reuse applications. In anticipation of new and more stringent DBP regulations, surface water utilities are considering using NF for removal of precursors to these by-products. As expected, surface waters require more pretreatment than groundwaters when applying NF. Studying a Florida surface water, Reiss and Taylor (1991) showed that pretreatment in the form of scaling control and pre-filtration (5 to 20 μm) was not sufficient to prevent NF membrane fouling. The investigators found that MF was more effective than alum coagulation (followed by settling and sand filtration) or granular activated carbon filtration for maintenance of NF-specific flux.

Other applications of MF involve dewatering of sludge and removal of colloidal materials in waste streams prior to treatment by RO. A woven cloth membrane which contains an array of tubes (as described subsequently) is used for treatment of sewage at various locations in the plant, ranging from primary to tertiary treatment. Additionally, Pillay et al. (1994) showed that a cross-flow MF system using a flexible woven fiber significantly enhanced the performance of anaerobic digesters at a wastewater plant in Durban, South Africa. The investigators reported that the system enabled selective retention of solids, thus decoupling the liquid residence time and solids residence time of the digesters. The process also allowed the digesters to operate at increased biomass concentrations.

11.2 MF MODULE SELECTION

11.2.1 Module Geometries

For drinking water treatment, there are several commercially available membrane geometries:

- Spiral wound
- Tubular
- Hollow fine fiber
- Hollow capillary fiber
- Disc tube
- Plate and frame
- Cassette

Each of these has been described in detail in another chapter. For municipal-scale drinking water applications, spiral wound, tubular, and hollow capillary fiber are the most commonly employed. However, for MF, spiral-wound configurations are not normally employed due to the flat-sheet nature of the membrane, which presents difficulties in keeping the surface of the membrane clean. Unlike the spiral-wound membranes, hollow-fiber and tubular configurations allow the membrane to be backwashed, a mechanism by which fouling due to particulate and organic materials is controlled. The advantages and disadvantages of each are described in Table 11.1.

11.2.2 Hollow-Fiber Capillary Membranes

Hollow-fiber MF membranes usually consist of several hundred to several thousand fibers encased in a module, the fibers bonded at each end with an epoxy or urethane resin (Marcus, 1988). The inner lumens, or internal fiber diameters, which are kept small to avoid fiber collapse when placed under pressure, range from 0.4 to 1.5 mm. The physical strength of these membrane fibers allows them to be backwashed. There are two different flow regimes in hollow-fiber MF: *inside-out* and *outside-in*. Because the water is flowing through a concentric channel or lumen, the inside-out membrane allows good control over module hydrodynamics. On the other hand, it is more difficult to control the flow of an outside-in module (when cross-flow is employed), because it is often difficult to avoid flow channeling and/or dead-end zones. Moreover, with this flow regime it is more difficult to flush the particles from the module when backwashing as compared to an inside-out membrane. However, an advantage of the outside-in membrane as compared to the opposite flow regime is that there is usually lower head loss through the module.

11.2.3 Tubular Membranes

In MF, tubular membranes have relatively large inner diameters, ranging from 1.0 to 2.5 cm (Quinn, 1983). The membranes, which can be composed of polymeric or ceramic materials, are usually placed inside stainless steel or fiberglass-reinforced plastic tubes, which are sealed by means of a gasket and outer ring clamps. They may be single or multiple flow channels. The feedwater, which is under pressure, flows

TABLE 11.1 Advantages and Disadvantages of Hollow-Fiber and Tubular Configurations for MF

Hollow fiber

Advantages
- Linear cross-flow velocities are low, ranging from 0 to 2.5 m/s
- With inside-out flow pattern, shear rates are high due to small inner lumens of the fiber
- High surface area–to–volume ratio or *packing density* of membrane
- Fibers can be backwashed
- Low transmembrane pressures, usually 0.2 to 1.0 bar
- Low pressure drop across the module, ranging from 0.1 to 1 bar

Disadvantages
- Small tube diameter membranes are susceptible to plugging unless prescreening is applied
- Large number of fibers in module present difficulties in detecting loss of membrane integrity in large plants

Tubular

Advantages
- Large diameters of channels (1 to 2.5 cm) allow treatment of waters with high solids content and large particles
- High cross-flow fiber velocities (up to 5 m/s), used to control fouling, are possible
- Large diameters of channels allow easy cleaning; mechanical cleaning can be employed in some cases
- Ceramic tubular membranes exhibit good mechanical strength

Disadvantages
- Low surface area–to–volume ratio or packing density of membrane
- Tubular ceramic membranes higher in cost per m² of filtration area as compared to other membrane configurations

through the inner lumen of the tube, and the permeate is collected in the outer shell of the module.

Another type of membrane system is one which employs flexible, woven polyester tubes that are manufactured as vertical arrays known as *curtains*. In this tube system, the wall is not the main filtration barrier. Instead, rejection of contaminants occurs at the cake layer that forms on the tube wall shortly after the start of the filtration cycle; this is often referred to as *dynamic filtration*. The close packing of the cake enables the retention of particles that are often orders of magnitude smaller than the pores in the tube wall. Hence, a cake or fouling layer, which is undesirable in many conventional MF systems, is effectively employed in the woven fiber system. In instances where the foulants in the feed stream are insufficient to form an effective filtration barrier, a cake is formed by applying another material such as limestone or kaolin prior to the introduction of the feedwater. This is referred to as *precoating* the membrane. When the flux drops to an unacceptably low value, the flexible tubes are mechanically cleaned, destroying or removing the fouling layer and restoring the permeability of the tubes.

11.2.4 Commercially Available Membranes

Table 11.2 shows that there are over 15 MF membranes that are commercially available. The pore sizes range from 0.1 to 5 μm and most have either a tubular or hollow-fiber configuration. The tubular membranes are primarily composed of ceramic materials and the hollow fiber of organic polymers. The modules range considerably in membrane area per single module, from as low as 0.1 to 15 m². Although there are numerous MF membranes commercially available, most have been employed for industrial purposes and have production capacities of less than 30 m³/day. For these applications, long-term fouling and membrane life are not of as great a concern as they are to the drinking water industry because of production-scale differences. It is important to note, however, that only a few membranes have been used in full-scale drinking water plants. The Memtec MF membrane system, which has an integrated system design, is employed in over 75 percent of the MF applications worldwide.

The market for MF membranes is increasing rapidly due to more stringent water quality regulations and scarcer water resources. As a result of market forces, many manufacturers that have focused primarily on industrial use are now placing a greater emphasis on water and wastewater applications. Number bench and pilot studies have been conducted on MF membranes such as those from A/G Technology, TechSep, Renovexx Technology, Akzo, Koch, and US Filter. Some of these membranes hold significant promise for full-scale plant water production.

11.3 PROCESS DESIGN

In the simplest designs, the MF process involves prescreening the raw water and pumping it under pressure onto a membrane in either a direct or cross-flow mode. For the direct filtration mode, the transmembrane pressure can be calculated according to:

$$P_{tm} = P_i - P_p \tag{11.1}$$

where P_{tm} = transmembrane pressure, bars
P_i = pressure at inlet to MF module, bars
P_p = permeate pressure, bars

When the MF system is operated in the cross-flow mode, the average transmembrane pressure is determined by:

$$P_{tm} = \frac{P_i + P_o}{2} - P_p \tag{11.2}$$

where P_{tm} = transmembrane pressure, bars
P_i = pressure at inlet to MF module, bars
P_o = pressure at outlet of MF module, bars
P_p = permeate pressure, bars

When the membranes are operated in a cross-flow mode, there is an accompanying pressure drop across the module, which is defined by:

$$P = P_i - P_o \tag{11.3}$$

TABLE 11.2 MF Membranes Commercially Available

Pore size, μm	Configuration	Material	Membrane area per module, m² (min-max)	Manufacturer
2, 3, 5	T	C	0.02–7.1	US Filters
1.4	T	C	0.005–7.4	US Filters
1	T	C	0.09–10.0	CTI TechSep
0.5, 0.7, 0.8	T	C	0.02–7.1	US Filters
0.5	T	C	0.13–11.5	Ceramem
0.45	T	C	0.09–10.0	CTI TechSep
0.45	HF	PS	0.01–3.7	AG Technology
0.2	T	C	0.02–7.1	US Filters
0.2	T	C	0.13–11.5	Ceramem
0.2	T	C	0.09–10.0	CTI TechSep
0.2	HF	PP	2.0	Akzo
0.2	HF	PP/VF	10.8–15	Memtec
0.2	T	C		Membralox
0.2	HF	VF		Dow
0.2–0.5	T	PWF	4.7–106	Renovexx Technology
0.1	T	C	0.02–7.1	US Filters
0.1	HF	PS	0.01–3.7	AG Technology
0.1	HF	PVA	0.09–12.0	Kuraray
0.1	SW	PTFE	5.0–8.0	Desal
0.05	T	C		Membralox

HF = hollow fiber
SW = spiral wound
T = tubular
C = ceramic
PP = polypropylene
PS = polysulfone
PVA = polyvinylalcohol
VF = fluorinated polymer
PWF = polyester woven fiber

where P = pressure drop across module, bars
         P_i = pressure at inlet to MF module, bars
         P_o = pressure at outlet of MF module, bars

Transmembrane pressures for MF systems usually range from 0.15 to 1 bar and applied pressures from 0.7 to 2 bar. Application of transmembrane pressure produces a filtrate or permeate. The total production flow of a membrane system can be calculated by:

$$Q_p = J_{tm}*S \tag{11.4}$$

where Q_p = system permeate flow rate, L/h
         J_{tm} = transmembrane flux, L/h/m²
         S = total effective membrane surface area, m²

Transmembrane flux is a function of several variables specific to the membrane, quality of the feedwater, and system operating parameters. For most MF systems, transmembrane flux rates range from 80 to 200 L/h/m². The total effective membrane surface area is calculated from the effective membrane area per module and the number of modules. The area for full-scale plant MF modules currently ranges from 1 to 15 m². However, modules with membrane areas as high as 50 m² are currently being considered by the MF manufacturing industry.

11.3.1 Process Configuration

MF can be designed to operate in three different process configurations, as shown in Fig. 11.4. In the first (Fig. 11.4a), the feedwater is pumped with a cross-flow tangential to the membrane. That water which does not pass through the membrane is recirculated as concentrate just ahead of the prefilter and blended with additional feedwater. In this mode, pressure on the concentrate side of the membrane is conserved and deducted from the total head requirements. A bleed stream may be employed to control the concentration of solids in the recirculation loop. At various times, concentrate waste is discharged from the recirculation loop.

The second configuration (Fig. 11.4b) also operates with cross-flow. However, in this mode the water is recirculated back to an open reservoir and blended with additional feedwater. This type of system is usually employed when the pressure of the feedwater to the plant is variable. Water to the reservoir under this scenario is regulated by a level-control switch. It is important to note that, since the concentrate is delivered back to an open reservoir, no energy from the concentrate side of the membrane is recovered; therefore, the total head for the feed system is greater than under the first mode described.

MF may also be operated in a direct-filtration configuration (Fig. 11.4c), during which no cross-flow is applied. This is often termed *dead-end* MF. Prefiltered water is applied directly onto the membrane. Since all of the prescreened feedwater passes through the membrane between backwashings, there is 100 percent recovery of this water. However, some raw water is normally used to flush the system. In the direct-filtration mode, there is considerable energy savings since no recirculation of the concentrate is required and a capital savings because no recirculation pumps or associated piping are necessary.

11.3.2 Pretreatment

11.3.2.1 Prefiltration. In comparison to conventional water clarification processes, where coagulants and other chemicals are added to the water before filtration, there are few pretreatment requirements for hollow-fiber systems when particles and microorganisms are the target contaminants. Prefilters are necessary to remove large particles which may plug the inlet to the fibers within the membrane module. The nominal prescreen or prefilter sizes range between 50 and 200 μm, depending on the membrane fiber inner diameter. Various types of prefilters are available such as disc filters, or for small plants, bag filters. More complex pretreatment strategies are sometimes employed either to reduce fouling or enhance the removal of viruses and dissolved organic matter. In such cases, pretreatment by the addition of coagulants (Olivieri et al., 1991a; Wiesner et al., 1992; Coffey et al., 1993) or powdered activated carbon (Pirbazari et al., 1992) have been employed.

11.3.2.2 pH Adjustment. Adjustment of the feedwater pH by chemical dosing may be required prior to membrane filtration in order to maintain the pH within the

removed by t
achieved by di
the four chem
States Environ
is used for pos
would be requ

Process fo
and P

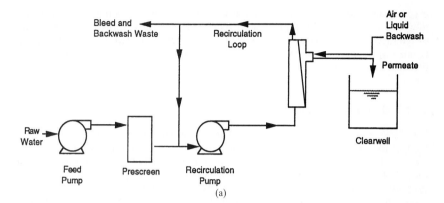

Raw
Water

Prescree

FIGURE 11.5

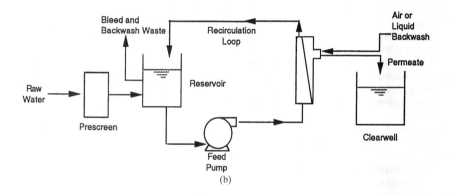

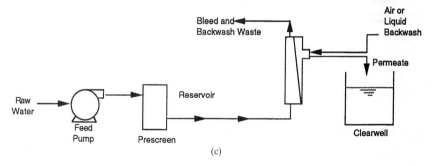

FIGURE 11.4   Various modes of MF operation: (*a*) cross-flow; (*b*) cross-flow with reservoir; (*c*) direct feed.

Ultraviole
against viruse
noted that ne
tion system m
to maintain a

11.3.5 Inst

The instrume
cussed in det

recommer
adjustmer
which usu
is not req
plexed dis

11.3.3   P

MF plants
trifugal fe
bank at th
for polym
progressiv
removed
modules c
individual
plant cap
deliver th
number o
The se
with high
mode of c
provided
tion pump
water flo
good qual
be emplo
plant desi
one manu
use of rec
Finally
ducted us
wash pun
a backwa
MF plant
banks are
i.e., each
more det

11.3.4   I

MF mem
barrier tc
does not
to four-lc
gelo and
MF with
fection m
ramines
virus ina

11.3.6   Concentrate Disposal

The concentrate from an MF plant consists only of that which was removed from the water, which is primarily solids, microorganisms, and chlorine if it is used in the backwash water. If coagulants or powdered activated carbon are employed, then they will also be a component of the concentrate. Concentrate disposal options are very site specific and can include:

- Discharge to a sanitary sewer
- Discharge to a surface water stream
- Discharge to lagoon or holding pond
- Land application
- Recycle back to source water

   UF and MF concentration disposal issues and options are very similar. Figure 11.6 shows disposal options that are currently employed by UF and MF membrane plants (Jacangelo et al., 1995). Interestingly, the largest number of plants either recycle their backwash water to the source water or dispose of it into sanitary sewers. For more details on disposal of wastes, refer to Chap. 10.

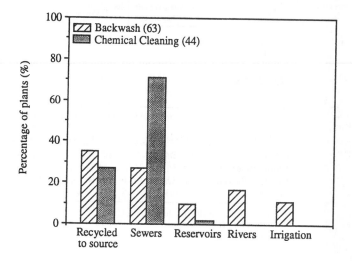

( ) number of responses

FIGURE 11.6   Residual disposal locations of full-scale MF and UF plants. *(From Jacangelo et al., 1995.)*

11.3.7   Design Options

Membrane "packaged" plants are normally employed for plants under 4000-m³/day capacity. The membrane area per packaged plant can range from 4 to 900 m², having production capacities ranging from 10 to 2500 m³/day. Figure 11.7 shows a schematic of a typical packaged system from a large MF manufacturer. The components of the removed by the MF membrane, then 3 logs of virus inactivation will need to be achieved by disinfection. Table 11.3 shows the CT, or concentration × time values, for the four chemical disinfectants to achieve 1, 2, and 3 logs of virus inactivation (United States Environmental Protection Agency, 1990). Thus, in this example, if free chlorine is used for post-disinfection, a CT of 3 mg/L·min at 15°C at pH ranging from 6 to 9 would be required.

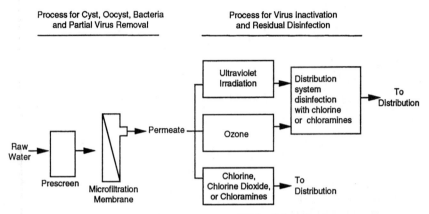

FIGURE 11.5   Various process options to accompany MF for disinfection purposes.

TABLE 11.3   CT Values for Inactivation of
Viruses by Various Disinfectants at 15°C

| | Log inactivation of viruses | | |
Disinfectant	2.0	3.0	4.0
Chlorine*	2	3	4
Chloramines	428	712	994
Chlorine dioxide*	2.8	8.6	16.7
Ozone	0.3	0.5	0.6

Note: Units for all CT values are mg/L·min.
* Data for pH 6 to 9.
Source: USEPA (1990).

Ultraviolet irradiation may also be applied for post-disinfection. It is effective against viruses in wastewater at doses ranging from 20 to 60 mW·s/cm². It should be noted that neither ozone nor ultraviolet irradiation provides a residual for distribution system maintenance; therefore, chlorine or chloramines would need to be added to maintain a disinfectant residual in the distribution system.

11.3.5   Instrumentation and Control

The instrumentation and control for MF is very similar to UF and, as such, is discussed in detail in Chap. 10.

11.3.6 Concentrate Disposal

The concentrate from an MF plant consists only of that which was removed from the water, which is primarily solids, microorganisms, and chlorine if it is used in the backwash water. If coagulants or powdered activated carbon are employed, then they will also be a component of the concentrate. Concentrate disposal options are very site specific and can include:

- Discharge to a sanitary sewer
- Discharge to a surface water stream
- Discharge to lagoon or holding pond
- Land application
- Recycle back to source water

UF and MF concentration disposal issues and options are very similar. Figure 11.6 shows disposal options that are currently employed by UF and MF membrane plants (Jacangelo et al., 1995). Interestingly, the largest number of plants either recycle their backwash water to the source water or dispose of it into sanitary sewers. For more details on disposal of wastes, refer to Chap. 10.

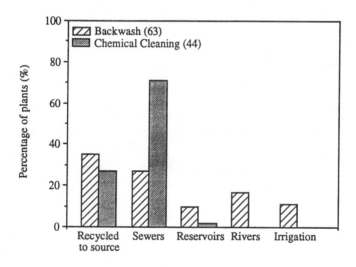

( ) number of responses

FIGURE 11.6   Residual disposal locations of full-scale MF and UF plants. *(From Jacangelo et al., 1995.)*

11.3.7 Design Options

Membrane "packaged" plants are normally employed for plants under 4000-m^3/day capacity. The membrane area per packaged plant can range from 4 to 900 m^2, having production capacities ranging from 10 to 2500 m^3/day. Figure 11.7 shows a schematic of a typical packaged system from a large MF manufacturer. The components of the

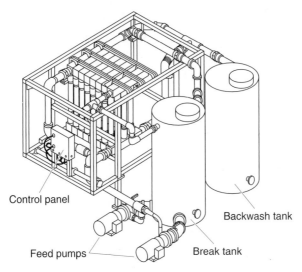

Control panel

Backwash tank

Feed pumps

Break tank

FIGURE 11.7 Schematic of MF membrane package plant. *(From Memcor Technical Bulletin T2093-A.)*

plant may include prescreens, feed pump, cleaning tank, automatic gas backwash system, air compressor, membrane integrity monitor, backwash water transfer tank, pressure break reservoir, air filter for the gas backwash, controls for the programmable logic controller, and a coalescer.

For larger MF plants, the membranes are arranged into blocks of modules (Fig. 11.8), incorporated into banks, which are of a modular array design (Fig. 11.9). The feed and backwash are distributed to the banks by separate manifolds. The main feed pressure is pumped locally at each filter bank, the flow of feedwater being controlled by rate-of-flow control valves. Changes in plant production can be controlled by two methods. Small variations in water demand can be met by adjusting the transmembrane pressure on the modules in respective blocks, thereby adjusting the transmembrane flux and water production. Larger demand variations are addressed by increasing or decreasing the number of blocks of modules on-line at any one time.

11.4 PERMEATE WATER QUALITY

11.4.1 Removal of Microorganisms

One of the primary applications of MF is for the removal of microorganisms. As previously noted, the SWTR requires utilities in the United States to provide a 3-log removal/inactivation of *Giardia* and a 4-log removal/inactivation of viruses. *Cryptosporidium* may also be regulated in the future. The significant resistance of protozoan cysts and oocysts to disinfectants such as chlorine, chloramines, ozone, and chlorine dioxide has increased the interest in employing a physical barrier that would remove high levels of these organisms through the treatment train.

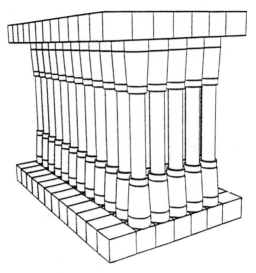

FIGURE 11.8 Schematic of a block of membrane modules. *(From Memcor discussion paper, 1991.)*

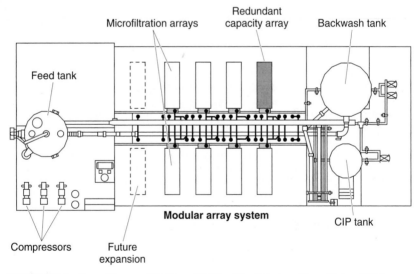

FIGURE 11.9 General layout of 19,000- to 26,000-m³/day MF plant. *(From Vickers, 1992.)*

In order to obtain a more perspicuous understanding of the capabilities of MF in removing microorganisms from water, it is useful to identify the organisms of concern and their sizes. In water treatment, the primary pathogenic targets are viruses, bacteria, and protozoa. Table 11.4 shows the general size ranges of selected microorganisms found in untreated drinking water supplies or those employed as models in microbial challenge studies. Viruses are the smallest organisms ranging from 0.02 to 0.08 µm, followed by bacteria (0.5 to 10 µm) and protozoan cysts and oocysts (3 to 15 µm). As noted previously, the pore sizes for MF range from 0.05 to 5 µm. Thus, through examination of the sizes of the target organisms and the range of pore sizes in MF, it is apparent that removal of these organisms (without pretreatment or particle attachment) is specific to the particular membrane and its pore-size distribution, when considering the membrane as a simple physical barrier. Other physical/chemical mechanisms may also play a role in the removal of microorganisms (Fane, 1994; McGahey, 1994; Jacangelo et al., 1995).

TABLE 11.4 Approximate Sizes of Microorganisms Found in Drinking Water Supplies or Commonly Used in Microbial Challenge Studies

Organism	Model	Approximate size, µm
Enteric virus	MS2 bacteriophage	0.025
Coliform bacterium	*Escherichia coli*	1–3
Protozoan oocyst	*Cryptosporidium parvum*	3–8
Protozoan cyst	*Giardia muris*	7–14

11.4.1.1 Microbial Removal in Laboratory Waters. In order to assess the removal of various organisms by selected MF membranes without the impact of water quality, Jacangelo et al. (1994) exposed MF membranes to viruses, bacteria, and protozoan cysts and oocysts in distilled water buffered at pH 7.0 with 0.001 M phosphate buffer. The use of a laboratory-prepared water eliminated natural water constituents as variables which may have impacted the removal of the seeded microorganisms. (It has been demonstrated that the formation of a cake layer on the membrane surface can increase the removal of virus [Fane, 1994; McGahey, 1994; Jacangelo et al., 1995]). The membranes had nominal pore sizes of 0.1 to 0.2 µm and were composed of various types of materials. A UF membrane with a nominal pore size of 0.01 µm and 100,000 molecular weight cutoff was included for comparison. The purpose of the experiments was to assess the minimum levels of microbial removal by MF under worst-case conditions, i.e., those conditions which favored microbial passage through the membranes. Direct feed flow (no applied cross-flow) and maximum transmembrane operating pressures were applied to each membrane and the laboratory-prepared feedwater was seeded with the unaggregated, selected organism.

Figure 11.10 presents the results of the virus seeding experiments. MS2 bacteriophage, which is a common indicator organism for waterborne enteric viruses, was employed in the experiments. MS2 is similar in size (0.025 µm), shape (icosahedron), and nucleic acid (single-stranded ribonucleic acid) to hepatitis A virus and poliovirus. Since MS2 virus is one of the smallest viruses, seeding MF systems with this organism represents a worst-case microbial challenge to the membranes. The feedwater was seeded with approximately 10^7 viral plaque forming units (pfu)/mL (7 logs of virus/mL). The data show that the average removal of MS2 virus ranged from 0.3

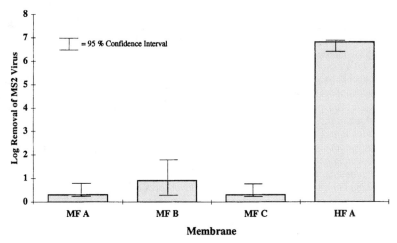

FIGURE 11.10   Removal of MS2 virus by selected MF and UF membranes under direct flow and maximum pressure conditions. *(From Jacangelo and Adham, 1994.)*

to 0.9 logs. None of the removals was statistically different from another; for two of the membranes, the removals were not statistically different than zero. In comparison, the UF membrane employed removed 6.8 logs of virus and no MS2 bacteriophage were detected in the permeate of any of the replicate trials. These data suggested that, under laboratory conditions, in distilled buffered water, MF membranes commonly removed less than 1 log of virus.

Table 11.5 shows the removal of two common water related bacteria, *E. coli* and *Pseudomonas aeruginosa,* by UF and MF membranes under worst-case conditions similar to those previously discussed. Between 7 and 8 logs of bacteria were seeded into the feedwater which was applied to the membrane. The data show that no bacteria were detected in the permeate for all of the MF membranes evaluated and for the UF comparison membrane. Both *E. coli* and *P. aeruginosa* are rod-shaped bacteria with sizes ranging from 1 to 3 μm. Thus, it is not unexpected that all the bacteria

TABLE 11.5   Removal of *P. aeruginosa* and *E. coli* by Various MF Membranes at pH 7 under Direct Flow and Maximum Pressure Conditions

			P. aeruginosa			*E. coli*		
Membrane	Process	Cutoff	Feed, cfu/ 100 mL	Permeate,* cfu/ 100 mL	Log removal	Feed, cfu/ 100 mL	Permeate,* cfu/ 100 mL	Log removal
Memcor	MF	0.2 μm	1.5E + 8	<1	>8.2	1.0E + 8	<1	>8.0
Optimem	MF	0.2 μm	1.5E + 8	<1	>8.2	1.0E + 8	<1	>8.0
Desal	MF	0.1 μm	1.5E + 8	<1	>8.2	1.0E + 8	<1	>8.0
Aquasource	UF	100 kdalton	5.3E + 8 ^	<1	>8.7	1.0E + 8	<1	>8.0

MF = microfiltration.
UF = ultrafiltration.
* Geometric mean of two discrete samples.
Source: Jacangelo and Adham (1994).

were removed under worst-case conditions. Thus, for bacteria, it appears that MF has the same removal efficacy as UF.

Table 11.6 shows the removal of *G. muris* and *C. parvum* by the various MF membranes. The initial protozoan seed densities were approximately 10^4 cysts or oocysts/L. These organisms are large relatively (3 to 15 μm). Thus, it would be anticipated that the MF membranes would act as an absolute barrier to the cysts and oocysts since these organisms are greater than an order of magnitude larger than the pore size of the membrane. The data in Table 11.6 show that all the membranes removed the cysts to less than detection limits (1 cyst or oocyst/L).

TABLE 11.6 Permeate Concentrations of Various Selected Protozoan Cysts and Oocysts after Membrane Filtration at pH 7

Membrane	Process	Pore size or MWCO	Microorganism	
			G. muris, cysts/L	*C. parvum,* oocysts/L
Memcor	MF	0.2 μm	<1	<1
			<1	<1
Optimem	MF	0.2 μm	<1	<1
			<1	<1
Desal	MF	0.1 μm	60*	78*
			<1	<1
Aquasource	UF	100 kdalton	<1	<1
			<1	<1

MWCO = molecular weight cutoff
MF = microfiltration
UF = ultrafiltration
Initial microbial densities: *C. parvum:* 1E4-1E5 oocysts/L
　　　　　　　　　　　　　　G. muris: 1E4-1E5 cysts/L
* Membrane seal was defective.
Source: Jacangelo et al. (1995).

11.4.1.2 Microbial Removal in Natural Waters. For bacteria and protozoa, the removal from natural waters by MF is consistent with that of removal from synthetic waters. *G. muris* as well as total coliforms, *E. coli,* and enterococci have been shown to be removed to detection limits of their respective assays in both water and wastewater (Groves et al., 1990; Kolega et al., 1991; Olivieri et al., 1991a; Olivieri et al., 1991b; Coffey et al., 1993; Moffet, 1994; Jacangelo et al., 1995; Yoo et al., 1995). However, heterotrophic plate count bacteria have been detected in the effluent of MF and other systems (Heneghan and Clark, 1991; Luitweiler et al., 1991; Jacangelo et al., 1991; Coffey et al., 1993; Moffet, 1994). It is probable, however, that the detected organisms were probably due to colonization of a section of the system piping rather than to bacterial penetration of the membrane (Jacangelo et al., 1991).

Although laboratory waters are useful for estimating the removal capabilities of membranes under conservative conditions, the extent of virus removal by MF will vary in natural waters. Olivieri et al. (1991b) showed as much as 4 logs removal of male-specific bacterial virus from an activated sludge effluent. Of 11 samples collected, no virus were detected in the permeate samples. That no viruses were detected in the permeate demonstrates an important difference between virus-seeding studies and natural virus studies. In the former, the viruses are normally prepared so that viruses are as monodispersed as possible (Jacangelo et al., 1991). Thus, the membranes are exposed to many single virions which are not associated with particles or other microorganisms. In natural virus studies, on the contrary, the viruses can be to a great extent associated with particulate matter. By removing the particles, MF membranes can also remove substantial densities of viruses. Hence, viral seeding studies likely represent a conservative estimate of the true virus removal by MF in natural waters.

11.4.2 Removal of Particles

A primary application of MF in water treatment is for the removal of particulate matter. Under the USEPA SWTR, conventional and direct-filtration plants must achieve a filtered water turbidity of ≤0.5 NTU in 95 percent of the samples collected each month with no samples having a turbidity >5 NTU. Several studies have evaluated the removal of turbidity and particles by MF. Olivieri (1991a) showed that a hollow-fiber polypropylene MF system could reduce the turbidity of a surface water (influent ranged from 0.5 to 2.5 NTU) to consistently below 0.2 NTU. With a similar MF system, Letterman et al. (1991) demonstrated greater than 99 percent removal of particles greater than 2 μm (permeate density was generally less than 200 particles/mL) and permeate turbidities consistently below 0.1 NTU. The largest operating MF plant in the United States, which has a capacity of 18,900 m³/day, provides average product water turbidities of 0.05 NTU or less and permeate particle concentrations of less than 1 per mL in the 4- to 10-μm range (Yoo et al., 1995).

Using tubular ceramic membranes, Wiesner et al. (1991) found that permeate turbidity varied with pore size. The turbidity of the raw river water used was 25 NTU. The 0.05-, 0.2-, and 0.8-μm MF membranes employed reduced these turbidities to 0.4, 0.9, and 1.8 NTU, respectively. Pretreatment with coagulants improved the reduction of turbidity over MF alone for the 0.05- and 0.2-μm MF membranes, but not for the 0.8-μm membrane. In the latter case, coagulation increased permeate turbidity, which was attributed to the passage of metal hydroxide particles. In general, particle removal also varied with pore size, the removal decreasing with increasing pore size. With coagulant addition, the 0.05-μm tubular membrane reduced the particle concentration by 99 percent.

11.4.3 Removal of Natural Organic Matter

Because of the large variation of pore sizes (0.05 to 5 μm) and membrane materials associated with MF, the removal of natural organic matter (NOM) is membrane and water specific. Using a polypropylene 0.2-μm MF membrane, Olivieri et al. (1991a) demonstrated a 15 percent removal of the total organic carbon (TOC) and total trihalomethane formation potential (THMFP) from a flowing stream. Wiesner et al. (1991) reported approximately 30 percent removal of TOC using a 0.05-μm ceramic tubular membrane on blended river waters with an average TOC concentration of 8.2 mg/L. The THMFP was reduced by 10 to 20 percent by both the 0.05-μm and a 0.2-μm ceramic tubular membrane.

The use of a coagulant has been employed for the removal of NOM, including DBP precursor materials (Hall and Packham, 1965; Dempsey et al., 1983). Removals of TOC with 50 mg/L of alum, and subsequent reduction in THMFP, have been reported to be as high as 70 percent (Reckhow and Singer, 1990). Conditions under which coagulation is performed are also important. Maximum removals have been achieved at a coagulation pH in the range of 5 to 7 (Reckhow and Singer, 1984; Chadik and Amy, 1983; Dempsey et al., 1984; Kavanaugh, 1978; Hall and Packham, 1965; Semmens and Field, 1980). Based on these studies, it is reasonable to assume that addition of coagulants as a pretreatment to MF would enhance its efficacy in removal of NOM. Olivieri et al. (1991a) demonstrated that THMFP of the surface water could be decreased from 15 to 60 percent by the addition of 10 to 15 mg/L of ferric chloride as a pretreatment to MF. Similar removals were observed by Wiesner et al. (1991) while working with tubular ceramic membranes. Coagulation improved the removal of TOC from 30 to approximately 60 percent using a 0.05-μm ceramic tubular membrane. Reduction in THMFP also improved by approximately 30 percent using this pretreatment.

11.5   MF OPERATION AND MAINTENANCE

MF systems are usually operated in one of two manners: at constant transmembrane water flux with variable pressure to maintain the flux or at constant transmembrane pressure with a variable transmembrane water flux rate. In terms of water treatment plant operation, these operating modes are referred to as *constant rate* and *declining rate filtration,* respectively. The former is more commonly employed in order to assure a constant rate of production.

The major obstacle to maintaining a constant rate of production is the degradation of transmembrane flux due to membrane fouling. This is due to the deposition of materials on the membrane surface and/or in pores. Reduction of flux that can be restored by mechanical or chemical means is termed *reversible fouling.* MF membranes that lose flux, which cannot be restored, are referred to as *irreversibly fouled* membranes.

In MF, there are three methods for maintaining or reestablishing permeate flux after the membranes are reversibly fouled:

- Membrane backwashing
- Membrane pretreatment
- Membrane cleaning

11.5.1   Membrane Backwashing

In order to prevent the continuous accumulation of solids on the membrane surface, backwashing of the membrane is performed. Unlike conventional media filtration, the backwashing cycle takes only a few minutes. Both liquid and gas backwashing is employed with MF technology. Liquid backwashing, which is usually performed for inside-out membranes, can be accomplished by two modes of operation. For most systems, backwashing is fully automatic, being initiated by either a programmable logic controller or a feedback control loop. With the latter, when the transmembrane pressure reaches a certain level, backwash is started in order to reduce the pressure needed to maintain a specified transmembrane water flux rate (constant rate filtration). Backwashing may also be initiated after a preprogrammed period of operation, regardless of the transmembrane pressure by a programmable logic controller. Finally, backwashing may also be programmed to be initiated after a predetermined permeate volume is produced. For most MF systems, backwashing is performed every 30 to 60 min of operation for 1 to 3 min.

11.5.1.1 Internal Backwash Mode. Under this backwash regime, the permeate port of the membrane is closed and cross-flow is applied. The pressures within the membrane fibers and module casing equilibrate and a differential pressure is established across the length of the membrane due to the cross-flow. At the high-pressure end of the membrane, the feedwater flow is from the inside to the outside of the membrane and permeate is produced. At the low-pressure end of the membrane, the permeate produced is reintroduced back into the internal lumen of the hollow fiber, and the membrane is cleaned. Feedwater flow is then reversed to backwash the opposite end of the fibers. An advantage to this method is that no dedicated pump or reservoir for permeate storage is necessary for backwashing. However, the efficacy of backwashing is often limited since backwash pressures and flow rates are not easily adjusted or controlled.

11.5.1.2 External Backwash Mode. Under this scenario, the permeate that is collected in the product water reservoir is employed to backwash the membrane. A liquid stream, which is introduced under pressure (0.5 to 2 bar) from a backwash pump through the permeate port, dislodges solids from the internal surface of the membrane. Chlorine is often employed at concentrations ranging from 3 to 50 mg/L to clean the membrane surface, aid in oxidation of adsorbed organic materials, and control biological growth on the membrane skin or support material. During backwashing, a waste valve is opened in order to discharge the concentrate and backwash water, which may be recovered, treated, recycled, and reused.

11.5.1.3 Gas Backwash. One MF manufacturer employs a patented air backwash system to dislodge solids from the membrane, which is operated in an outside-in mode during normal filtration. Backwashing is initiated every 30 to 60 min for a period of approximately 2 to 3 min. During this time, the membrane is taken off-line. Upon the start of backwashing, the membrane module is drained of permeate and feedwater and the permeate valve is closed. Air is then introduced at relatively high pressure (6 to 7 bar) through the inner lumen of the membrane. When the backwash wastewater valve is opened, the pressurized air penetrates the membrane from inside the membrane lumen to the outside, resulting in a six- to sevenfold expansion of the air atmospheric pressure (Vickers, 1993). Employing compressed air for backwashing and feedwater for solids removal lessens or eliminates the need to design a clearwell or permeate storage tank for backwash purposes. The air pulse is rapid, occurring in 2 to 3 s. As a result of the strong gas-membrane interaction, the solids on the outside of the membrane are dislodged. Feedwater is then reintroduced into the module for approximately 30 s at pressure of 0.8 bar to flush or sweep away the dislodged solids. The flush water is then discharged to a backwash wastewater transfer tank (in large systems). The wastewater from the process is collected and can be directed to either a washwater recovery (treatment) system, to a lagoon or appropriate surface water, or to a local wastewater sewage collection system. Once the backwash cycle is completed, the filtration cycle recommences. The feedwater recovery on total systems using an air backwash is typically greater than 93 percent.

Membrane backwashing is employed at various intervals throughout the filtration cycle to maintain the permeate flux rate or transmembrane pressure. Figure 11.11 shows a schematic of typical flux and pressure profiles of MF membranes. MF membrane systems can be operated at constant pressure with a declining flux or a constant flux with an increasing transmembrane pressure. Backwashing can partially

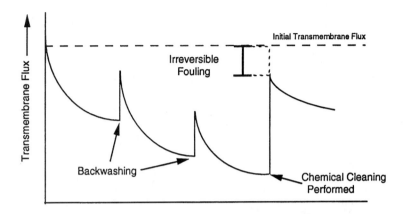

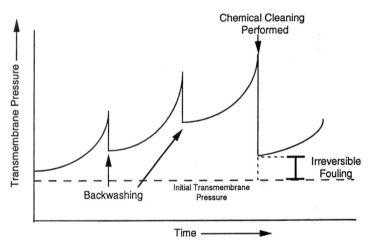

FIGURE 11.11   Schematic of partial restorations of transmembrane flux or pressure by chemical cleaning of MF membranes.

or fully restore the pressure or flux, depending on the extent of fouling. If backwashing is incapable of restoring the flux, then chemical cleaning of the membranes is performed.

11.5.2   MF Pretreatment

As previously noted, feedwater pretreatment can be employed to improve the level of removal of various natural water constituents; it is also used to increase or maintain transmembrane flux rates and/or to retard fouling. The two most common types of pretreatment are coagulant and powdered activated carbon (PAC) addition. Most studies employing coagulant addition have focused on the use of aluminum sulfate or ferric sulfate at concentrations ranging from 5 to 50 mg/L (Wiesner et al., 1992; Coffey et al., 1993). Metal coagulants are normally injected into a raw water through a feed line, after which the water then enters a reservoir for mixing and coagulation. Most retention times in the reservoir range from 5 min to approximately 1 h. Some investigators have injected the coagulant into a mixing pipe loop before addition of the feedwater to the MF membrane (Olivieri et al., 1991a). It appears that the coagulation of smaller particles into larger ones reduces the penetration of various materials, including colloidal and large organic macromolecules, into pores of the membrane. Coagulation may also increase the size of the particles composing the cake layer on the membrane, increasing its porosity, and as a consequence, enhancing the feedwater flux through the membrane (Wiesner, 1992). It is important to note that the effect of coagulants on transmembrane flux may be very sensitive to the conditions under which the coagulant is employed. Figure 11.12 shows the effect of pH on coagulant addition when used for permeate flux enhancement. The addition of alum dramatically improved the transmembrane flux, with the greatest colmatage retardation occurring at pH 6.5.

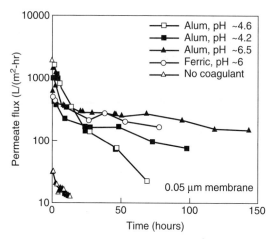

FIGURE 11.12   Impact of coagulant pH on transmembrane flux. *(From Wiesner et al., 1992.)*

The use of PAC as a pretreatment for UF has been fairly well documented in the recent literature. On the other hand, few investigations have focused on PAC pretreatment of feedwaters before MF. Retention times as long as 3 to 5 days of contact time have been employed (Pirbazari et al., 1992). In such cases, the feedwater flows into a reactor to which a PAC slurry is added continuously. The water is then passed across a membrane through which a fraction passes, the concentrate stream being recycled back to the reactor. Therefore the MF system is being operated in a cross-flow mode. The adsorbent is allowed to continually accumulate in subsequently greater concentrations and the full adsorbent capacity of the PAC is exploited. Chapter 15 discusses the use of PAC/membrane reactors in more detail.

11.5.3 Chemical Cleaning

As the concentration of foulant materials accumulates on the membrane surface, the loss of transmembrane flux will continue to increase. Backwashing the membrane is the routine method for removing these materials. However, when foulants can no longer be removed from the membrane surface by backwashing, chemical cleaning is required. An example of this behavior is demonstrated schematically in Fig. 11.13. After chemical cleaning, partial or full restoration of transmembrane flux (or pressure) is achieved. The variables that should be considered in cleaning MF membranes include (Vickers, 1992):

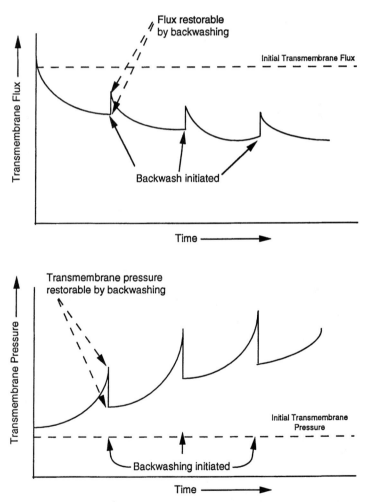

FIGURE 11.13 Impact of backwashing on restoration of transmembrane flux or pressure for MF membranes.

- Frequency of cleaning
- Duration of cleaning
- Chemicals and their concentrations
- Cleaning and rinse volumes
- Temperature of cleaning
- Recovery and reuse of cleaning chemicals
- Neutralization and disposal of cleaning chemicals

There are a variety of different agents that can be employed for the chemical cleaning of MF membranes, including detergents, acids, bases, oxidizing agents, sequestering agents, and enzymes. Chlorine is often employed in doses ranging from 2 to 50 mg/L; it is usually employed with polysulfone or polysulfone-derivative materials, which are resistant to the compound. In addition to acting as an oxidant capable of oxidizing organic material on the membrane, chlorine acts as a disinfectant, inactivating microorganisms which may have a deleterious effect on the membrane surface. Other materials such as polypropylene are sensitive to chlorine; in such cases, alternative cleaning agents are employed. For its polypropylene membrane, the largest manufacturer of MF systems for drinking water recommends a combination of low levels of surfactants (less than 0.1 percent) and sequestering agents such as citrates and gluconates along with sodium hydroxide (0.5 to 2 percent). In some cases, hydrogen peroxide (1 percent) may be employed. Cleaning is usually initiated after the transmembrane pressure exceeds 1 to 1.5 bar. Raising the temperature of the cleaning solution to 35 to 40°C is often employed to enhance the cleaning effectiveness (Wale and Johnson, 1993). The cleaning process is initiated by draining the feedwater from the modules and introducing a cleaning solution from a cleaning tank at a flow of approximately 360 m^3/h and 0.8 bar. The solution is circulated across the membranes for approximately 5 min. The solution may be held inside the modules for a period of 1 h (soak period) and is then purged by air back to the cleaning tank.

Approximately 90 percent of the cleaning solution is recovered by air purging and can be reused for subsequent membrane cleanings. That solution which is not purged from the modules is evacuated by feedwater to a waste drain line. Feedwater flushing continues until the water exiting the module has a pH and conductivity similar to the feedwater. The wastes are usually neutralized before disposal.

11.6 MEMBRANE INTEGRITY TESTING

One of the most critical aspects of employing membrane technology is ensuring that the membranes are intact and continuing to provide a barrier between the feedwater and the permeate or product water. There are several different methods that can be employed to monitor membrane integrity, including:

- Turbidity monitoring
- Particle counting (multichannel counting)
- Particle monitoring (single-channel counting)
- Air pressure testing
- Bubble point testing

- Sonic wave sensing
- Biological monitoring

The first three are indirect methods of measuring membrane integrity; i.e., water quality parameters are measured to detect compromised fibers or modules. The last three are direct measures, detecting changes in the fibers or modules themselves. The potential advantages and disadvantages of each system are presented in Table 11.7.

11.6.1 Indirect Methods of Membrane Monitoring

11.6.1.1 Turbidity Monitoring. Turbidity monitoring is the most commonly used method of measuring performance of filtration systems for conventional drink-

TABLE 11.7 Advantages and Disadvantages of Various Methods to Evaluate Membrane Integrity of MF Membranes

Monitoring method	Advantages	Disadvantages
Particle counting	Continuous on-line measurement Measures several size ranges	High cost Indirect measurement of membrane integrity May require several sensors for large-scale MF applications
Particle monitoring	Continuous on-line measurement Low cost	Does not count particle size ranges May require several sensors for large-scale MF applications
Air pressure testing	Built into membrane system Direct measurement of membrane integrity	Not a continuous monitoring method Available only on one membrane system
Bubble point testing	Direct measurement of membrane integrity	Must take membrane module off-line to conduct test Must be conducted manually Labor intensive for large plants
Sonic leak testing	Direct measurement of membrane integrity Can conduct measurement while module in place; easy to employ Future potential to be an on-line continuous measurement	Must be conducted manually Labor intensive for large plants
Turbidity monitoring	Extensive water industry experience with its use Low cost	Not sensitive to changes in particles at low turbidities Indirect measurement of membrane integrity
Biological monitoring	Most sensitive of indirect methods of monitoring membrane integrity Can conduct measurement while module in place Cost of viral assay is low	Indirect measurement of membrane integrity Often impractical to seed a full-scale plant while serving water to consumers Labor intensive

ing water treatment systems. It is less costly than most other monitoring methods, usually in the range of $1500 to $2000 per on-line turbidimeter. Its long history of use in the water industry makes it a familiar tool to operators and provides a methodology which is more understood than other monitoring methods. However, most membrane systems provide permeate turbidities which are less than 0.1 NTU and recent evidence has shown that turbidity may not be capable of fully detecting changes in particle densities at these levels (Hargesheimer et al., 1991; Jacangelo et al., 1991; Lewis and Manz, 1991). In one study (Jacangelo et al., 1991), when a loss of membrane integrity was observed, there was no readily discernible increase in turbidity. The loss of membrane integrity was detected, however, by particle counting and detection of virus and *Giardia* in the permeate. Adham et al. (1995) demonstrated that artificially compromising a transverse-flow MF membrane with a needle hole did not have any impact on turbidity, which was probably related to the module flow mode and the high number of fibers per module. The large module produced enough permeate to dilute any effect of particles passing through the compromised membrane. Therefore, use of turbidity as the sole method of monitoring for membrane integrity may not provide adequate sensitivity for detection of small pinholes in membrane fibers.

11.6.1.2 Particle Counting. Multichannel particle counting is a standard particle counting method used in many research application systems and in some treatment plants. It has been reported in other studies on conventional media filtration to yield more useful water quality information than turbidity when the turbidity is less than 0.1 NTU (Hargesheimer et al., 1991; Lewis and Manz, 1991). It was also shown to be an adequate method of monitoring fiber breaks of hollow-fiber membranes (Jacangelo et al., 1991; Adham et al., 1995) when permeates from single modules were monitored. The overall efficacy of particle counting for monitoring MF integrity is a function of several factors and can be described by (Adham et al., 1995):

$$\sum_{t=0}^{t=t_b} (P_p * F_p)t = \sum_{t=0}^{t=t_b} \left[\sum_{i=0}^{i=F_c} (P_c * F_c)i \right] t + \sum_{t=0}^{t=t_b} (P_u * F_u)t \tag{11.4}$$

and
$$F_p = N_f * M_t \tag{11.5}$$

where 　P_c = particle concentration in permeate of compromised fiber
　　　　P_u = particle concentration in permeate of uncompromised fiber
　　　　P_p = particle concentration in product line of integral membrane system
　　　　F_c = number of compromised fibers in system
　　　　F_u = number of intact fibers in system
　　　　F_p = number of total fibers connected to product line
　　　　N_f = number of total fibers in membrane module
　　　　M_t = number of membrane modules in integral membrane system
　　　　t_b = backwashing frequency
　　　　t = time of filtration
　　　　i = arbitrary parameter indicating compromised status of fiber

Equation (11.4) assumes that the time required for the particles passing through the compromised fiber to reach the product line is negligible, which is reasonable since the detection sensor would be placed as close as possible to the permeate outlet. The particle concentration parameter of the compromised fiber P_c is an overall collective term which is a function of many variables such as the severity of the compromised status of the fiber, operational conditions, membrane module size and capacity, module flow configuration, feedwater concentration, time of filtration, membrane fouling status, and the nature and extent of the membrane cake layer.

EXAMPLE 11.1 *Calculate the maximum hollow-fiber MF membrane surface area that could be monitored by a particle counting sensor, assuming there is one compromised fiber and it will pass 2.3 × 10⁶ particles per mL. The particle concentration detected in the permeate of the product line is 20 particles/mL. There are 15,000 fibers per module, which has a total membrane surface area of 20 m².*

Solution: Assume that the variables described in the preceding equation are constant; thus, the mass balance on particles at a specified time in the filtration cycle (i.e., at the middle of the filtration cycle) for a particular membrane can be simplified to the following:

$$P_p * F_p = P_c * F_c + P_u * F_u \qquad \text{at } t = \frac{t_b}{2} \qquad (11.6)$$

The particle concentration in the permeate of an uncompromised fiber is usually negligible compared to particle concentration in the permeate of a compromised fiber (i.e., $P_c \gg P_u$); thus, the preceding equation can be further simplified as:

$$P_p * F_p = P_c * F_c \qquad \text{at } t = \frac{t_b}{2} \qquad (11.7)$$

Solving for F_p:

$$F_p = \frac{(2.3 \times 10^6 \text{ particles/mL} * 1 \text{ compromised fiber})}{20 \text{ particles/mL}}$$

$$= 115{,}000 \text{ fibers / sensor}$$

The number of fibers per m² of membrane area = 15,000 fibers / 20 m²

$$= 750 \text{ fibers / m}^2$$

Therefore, membrane area per sensor $= \dfrac{(115{,}000 \text{ fibers / sensor})}{(750 \text{ fibers / m}^2)}$

$$= 153 \text{ m}^2 \text{ of membrane area / sensor}$$

A recently developed instrument for measuring particles, called a *particle monitor,* is based on dynamic light obscuration (Chemtrac Systems, Inc., 1993). The instrument measures fluctuations in intensity of a narrow light beam which is transmitted through the sample. A fluctuating AC signal from a constant DC signal is measured by a detector and amplified. The monitor does not count particle sizes, but rather provides an index (ranging from 0 to 9999) of the water quality. No calibration is required for this instrument since the output is a relative measurement of water quality. The potential advantages of this monitor are its low cost and ease of operation compared to particle counters, while providing high resolution. These units are anticipated to cost 5 to 10 times less than traditional particle counters.

Indirect membrane integrity monitoring through the use of on-line particle monitoring instruments has the advantage of providing continuous evaluation of membrane integrity. Adham et al. (1995) showed that dilution and module flow mode are very important factors in detecting a compromised membrane. As previously noted, the higher the number of fibers/modules connected to a particle monitoring instrument, the lower the probability for detecting the compromised fiber(s). The study found that the rank of sensitivity of four different particle monitoring instruments was:

On-line particle counter > batch particle counter > particle monitor > turbidimeter

11.6.2   Direct Methods of Monitoring

11.6.2.1   Bubble Point Testing.   Bubble point tests, as originally described by Melzer and Meyers (1971), can be employed to measure the integrity of the MF membranes. The membrane is submerged longitudinally in water while air is pumped under pressure through the permeate port (for inside-out membranes). Loss of membrane integrity is immediately detected by the presence of air bubbles rising out of the compromised fiber at the membrane module inlet or outlet. Air rises out of the inner lumen of the membrane at pressures above which the compromised fiber begins to pass air. The pressure required can be theoretically determined by the following:

$$P = 4s \cdot \cos \phi$$

where P = pressure required to produce bubbles from inner lumen of membrane
s = surface tension of permeate water used for bubble point test
ϕ = permeate/membrane wetting angle
d = diameter of hole in compromised membrane

FIGURE 11.14   Photo of a compromised low-pressure, hollow-fiber membrane detected by bubble point testing.

However, since the size and shape of the damage to the fiber are usually unknown, the pressure is difficult to determine. For most applications, approximately 0.1 bar of pressure or less is sufficient to conduct a bubble point test. Figure 11.14 shows a photo of the bubbles protruding from a single compromised membrane fiber (Jacangelo et al., 1991). A major advantage of this method is that it is a direct measurement of membrane integrity and relatively simple to conduct, requiring only an air pump, tubing, and vessel in which to submerge the membranes. The major disadvantage is that each module needs to be taken off-line in order to conduct the test; such a procedure is labor intensive, especially for large plants.

11.6.2.2   *Air Pressure Hold Testing and Sonic Leak Detection.*   Air pressure hold testing is an on-line test used with transverse membrane systems. The lumens of the membranes are pressurized with air to just below the bubble point of the membrane, usually 1.4 to 1.7 bar, and the outer shell of the membrane is kept full of feedwater. The gas within the lumens of the membranes is monitored for approximately 2 to 5 min. If the membrane module is compromised, the decay in the pressure profile within the lumen is observed as the gas passes from the inside to the outside of the lumen. Measurement of the gas flow is automated and can be set to alarms via the MF system control and monitoring system. Figure 11.15 shows an example of the air pressure hold test for four membranes (Adham et al., 1995). The data show that one artificially created needle hole ($\approx$0.7 mm) on only one

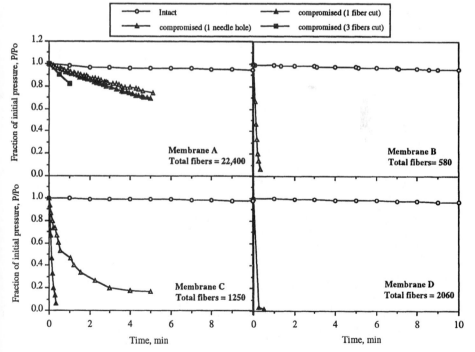

FIGURE 11.15   Air pressure hold test results for compromised and intact membranes. *(From Adham et al., 1995.)*

fiber out of 22,400 fibers resulted in a considerable loss of pressure, indicating a test failure. Similar test results are expected even if more fibers were included, since the test is an integrity check on each fiber in the membrane module. More sensitive pressure transducers will be required as the number of fibers tested increases, however, in order to differentiate between the pressure loss due to natural diffusion through the fiber's pores and minor loss of fiber integrity. For large systems, once a leak or compromised membrane is detected on a block or bank of membranes, the individual faulty module can be identified by use of a sonic analyzer. The analyzers use an accelerometer and signal processing circuitry to detect small bubbles that penetrate the membrane. The sonic sensor needs to be placed at several locations in order to detect the compromised module since the audio sounds at the bottom of the module do not clearly indicate a compromised module.

11.6.2.3 Biological Monitoring. One of the most sensitive methods for evaluating membrane integrity is the use of viral seeding. Using this method, viral densities in the range of 10^6 to 10^7 pfu/mL are seeded into the influent of the plant. At various times, samples are collected from the permeate and assayed for the virus. The most commonly used virus is MS2 bacteriophage, which is a small (0.025-μm) icosahedron virus that uses a bacterium (*E. coli*) as its host (Valegard et al., 1990). A major advantage of this technique is that it is the most sensitive indirect measure of membrane integrity; up to 9 logs of sensitivity can be attained. Additionally, the sizes of virus particles are over an order of magnitude smaller than the smallest size detectable by the most sensitive particle counter (0.3 μm). This technique has been used to detect compromised hollow-fiber membranes when other monitoring techniques were unsuccessful (Jacangelo et al., 1991). The major disadvantage of this monitoring technique is that it is often impractical to seed a full-scale plant while it is on-line producing water to consumers. Because of this consideration, biological monitoring is usually limited to pilot- and bench-scale studies.

11.7 MF COSTS AND CASE STUDY

A city was evaluating alternatives to treat its water supply, which was classified under the influence of surface water. As a result, the supply was subject to the SWTR and 3-log removal/inactivation of *Giardia* and 4-log removal/inactivation of viruses were required. Further, the utility would be subject to any future requirements of an enhanced SWTR. The water quality of the supply is shown in Table 11.8. Given the high quality of the raw water source, the primary water treatment objective was to eliminate waterborne pathogens from the finished water supply. Turbidity removal was also an issue, both as an indicator of filtration efficiency and to prevent interference with the disinfection process.

MF was considered along with various other filtration alternatives. A process schematic for the plant is shown in Figure 11.16. The plant consisted of two stages of membrane filters, the second stage being employed for reclamation of any washwater and to provide feedwater recoveries greater than 99 percent. The system was designed for direct or dead-end filtration. For this application, the waste was discharged to a sanitary sewer.

TABLE 11.8 Water Quality Parameters
for Case Study

Water quality parameter	Value
Turbidity, NTU	0.1–2
pH	7
Total organic carbon, mg/L	<2
Hardness, mg/L as CaCO$_3$	30.0
Color, ACU	<3
Iron, mg/L	<0.5
Manganese, mg/L	<0.5
Temperature, °C	8–22

Table 11.9 shows a selected list of elements for design and an example of some of the criteria used for the 76,000-m^3/day MF plant. Based on these design criteria, capital and operation and maintenance costs were developed (Table 11.10). MF capital costs were approximately $272 per m^3 of installed capacity. For comparison purposes, costs of other filtration options are shown in Table 11.11. The data show that MF was similar in cost to other filtration alternatives. However, MF provides an absolute removal of *Giardia* and *Cryptosporidium* and does not involve chemical optimization requiring operator attention. The technology also requires less land than media filters and expansion is easier than conventional treatment because the system is modular.

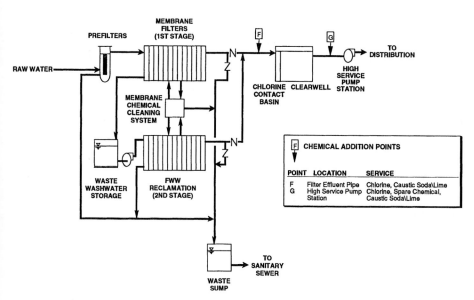

FIGURE 11.16 MF process schematic for case study.

TABLE 11.9 Elements of Process Design Criteria for MF Membrane Plants in Case Study

Design element	Value or description	Units
Plant capacity		
Process design flow	75,700	m³/day
Minimum plant flow	37,850	m³/day
Influent flow metering		
Type	Venturi	
Number	1	
Size	76	cm
Prescreening		
Type	Grooved disc filters	
Number	5	
Size	500	μm
Membranes		
Type	Microfiltration membranes	
Material	Polypropylene	
Pore size	0.2	μm
Membrane banks		
Number	22	
Size	6.3 × 2.5	m
Height	2.8	m
Modules per bank	90	m²
Total number of modules	1,440	
Orientation	Vertical	
Surface area		
per module	15	m²
per bank	1,350	m²
total surface area	21,600	m²
Flow rate		
per module	24.4	L/min
per bank	2,195	L
total flow rate	69,600	m³/day
Transmembrane flux	104	L/h/m²
Feed configuration	Dead end	
Transmembrane pressure differential	0.2–1	bar
Membrane backwash		
Type	Water/air	
Backwash water supply	Raw water	
Rate	5.7	L/min/m²
Flow (per bank)	7,665	L/min
frequency (per bank)	1	per h
duration	90	s
Washwater volume		
per backwash	11.5	m³
per day	6,056	m³
Air backwash		
rate	0.037	m³/min/m²
duration	90	s
compressors	4	
blower capacity	200	scfm
blower motor	50	hp

TABLE 11.9 Elements of Process Design Criteria for MF Membrane Plants
in Case Study (*Continued*)

Design element	Value or description	Units
Filter waste washwater (FWW) handling		
FWW equalization basin		
number of basins	1	
volume	45.4	m^3
water depth	3.0	m
size	3.7×3.7	m
FWW recycle pumps		
number	3	
capacity	114	m^3/h
total head	15.2	m
motor	10	hp
Membranes—second stage		
Type	MF membranes	
Material	Polypropylene	
Pore size	0.2	μm
Membrane banks—second stage		
Size	6.3×2.5	m
Height	2.8	m
Modules per bank	60	m^2
Total number of modules	180	
Surface area		
per module	15	m^2
per bank	900	m^2
total surface area	2,700	m^2
Flow rate		
per module	23.1	L/min
per bank	1,400	L
total flow rate	6,056	m^3/day
Transmembrane flux	111	$L/h/m^2$
Feed configuration	Dead end	
Transmembrane pressure differential	0.2–1	bar
Membrane backwash—second stage		
Type	Water/air	
Backwash water supply	Raw water	
rate	5.7	$L/min/m^2$
frequency (per bank)	3	per h
duration	90	s
Waste stream	1,147	L/min
Waste stream disposal	Sanitary sewer	
Chorine contact basin		
Type	Buried reinforced concrete	
Logs of virus inactivation required	4	
Number of basins	1	
T_{10} required	8	min
Time provided	20	min
Volume	1,400	m^3
Size	4.8×30.5	m
Clearwell		
Type	Buried reinforced concrete	
Number of basins	2	
Volume	5,700	m^3
Size	15×30	m

TABLE 11.10 Capital and Operation and Maintenance Costs for MF in Case Study

Capital cost	
Element	Cost in U.S. dollars*
Sitework and yard piping	930,000
Inlet structure	60,000
Chemical feed facilities	720,000
Flocculation basins	—
Sedimentation basins	—
Roughing filters	—
Filters	—
Membrane prefilters	350,000
Membranes	7,910,000
FWW handling and sludge disposal	2,310,000
CT contact basin/clearwell	1,540,000
Finished water pumping station	820,000
Buildings	2,740,000
Electrical and instrumentation	3,180,000
TOTAL	20,560,000

* Construction costs are based on a 1993 ENR index of 6000. Costs do not include easements, right-of-way, land financing, administration, or engineering costs. All costs are for a 75,700-m³/day facility.

Operation and maintenance[a]	
Element	Cost in U.S. dollars
Personnel[b]	52,000
Misc. operating expenses[c]	10,000
Building & grounds maintenance	5,000
Equipment maintenance	20,000
HVAC[d]	6,000
Electrical	
Lighting & equipment[e]	87,000
Plant head-loss pumping[f]	75,000
Chemicals	65,000
Residuals disposal[g]	22,000
Membrane replacement[h]	250,000
TOTAL	592,000

[a] All costs are for a 75,700-m³/day facility (annual average flow of 47,000 m³/day).
[b] One operator/maintenance person at a rate of $25/h and 2080 h/year/employee.
[c] Includes misc. personnel costs (training, uniforms, etc.), office, and WTP lab expenses.
[d] Based on $10.7/m² heating and cooling costs for operations building areas and $2.69/m² for all other areas.
[e] Based on $0.12/kWh energy cost (includes demand charges); lighting and receptacle allowance at 10.7 kWh/m²/month and equipment itemized individually.
[f] Represents energy cost of pumping for plant head loss only (raw water and finished water pumping not included); based on 7.3-m average head loss through membrane processes, and $0.12/kWh energy costs.
[g] Based on discharge to sanitary sewer.
[h] Based on a 5-year membrane life.

TABLE 11.11   Comparative Costs of MF and Media Filtration Alternatives

Process	Cost of filtration alternatives	
	Capital	Operation and maintenance
MF	$20,560,000	$592,000
Conventional filtration	$21,200,000	$566,000
Direct filtration	$19,550,000	$546,000

Construction costs are based on a 1993 ENR index of 6000. Costs do not include easements, right-of-way, land financing, administration, or engineering costs. All costs are for a 75,700-m^3/day facility.

REFERENCES

Adham, S. L., J. G. Jacangelo, and J.-M. Laîné, 1995, "Low Pressure Membranes: Assessing Integrity," *Journal American Water Works Association,* **87**(3):62–75.

Chadik, P. A., and G. L. Amy, 1983, "Removing Trihalomethane Precursors from Various Natural Waters by Metal Coagulants," *Journal American Water Works Association,* **75**(10):532–536.

Chemtrac Systems, Inc., 1993, "Comparison of Particle Measuring Instruments," *Field Notes,* **93**(1) (product bulletin).

Coffey, B. M., M. H. Stewart, K. L. Wattier, and R. T. Wale, 1993, "Evaluation of Microfiltration for Metropolitan's Small Domestic Water Systems," *Proceedings of the AWWA 1993 Membrane Technology Conference,* Baltimore, Md., p. 373.

Dempsey, B. A., R. M. Ganho, and C. R. O'Melia, 1984, "The Coagulation of Humic Substances by Means of Aluminum Salts," *Journal American Water Works Association,* **76**(4):141–150.

Fane, A. G., 1994, "An Overview of the Use of Microfiltration for Drinking Water and Waste Water Treatment," *Proceedings of Microfiltration for Water Treatment Symposium,* Irvine, Calif., p. 3.

Groves, G. R., K. S. Jackson, and S. L. Lambert, August 1990, "Microfiltration Treatment of Surface Waters," *Proceedings of the 1990 Biennial Conference,* Walt Disney World Village, Fla., **2**:339.

Hargesheimer, E. A., C. M. Lewis, C. M. Yentsch, T. Satchwill, and J. L. Mielke, 1991, "Pilot Scale Evaluation of Filtration Processes Using Particle Counting," *Proceedings of the AWWA 1990 Water Quality Technology Conference,* San Diego, Calif., p. 323.

Heneghan, K. S., and M. M. Clark, 1991, "Surface Water Treatment by Combined Ultrafiltration/PAC Adsorption/Coagulation for Removal of Natural Organics, Turbidity, and Bacteria," *Proceedings of the AWWA 1991 Membrane Processes Conference,* Orlando, Fla., p. 345.

Jacangelo, J. G., and S. A. Adham, 1994, "Comparison of Microfiltration and Ultrafiltration for Microbial Removal," *Proceedings of Microfiltration for Water Treatment Symposium,* Irvine, Calif., p. 57.

Jacangelo, J. G., S. A. Adham, and J.-M. Laîné, 1995, "Mechanism of Cryptosporidium, Giardia, and MS2 Virus Removal by MF and UF," *Journal American Water Works Association,* **87**(9):107.

Jacangelo, J. G., S. A. Adham, and J.-M. Laîné, 1996, "Application of Membrane Filtration Techniques for Compliance with the Surface Water Treatment Rule," Final Report to the American Water Works Association Research Foundation.

Jacangelo, J. G., J.-M. Laîné, K. E. Carns, E. W. Cummings, and J. Mallevialle, 1991, "Low-Pressure Membrane Filtration for Removing *Giardia* and Microbial Indicators," *Journal American Water Works Association,* **83**(9):97.

Kavanaugh, M. C., 1978, "Modified Coagulation for Improved Removal of Trihalomethane Precursors," *Journal American Water Works Association,* **70**(11):163.

Kolega, M., G. S. Grohmann, R. F. Chiew, and A. W. Day, 1991, "Disinfection and Clarification of Treated Sewage by Advanced Microfiltration," *Water Science Technology,* **23**:1609.

Letterman, R. D., S. M. Chiang, D. Herb, X. G. Meng, D. Selger, and Shrodo, June 1991, "Evaluation of Alternative Surface Water Treatment Technologies," Report to New York State Department of Health, Bureau of Public Water Supply Protection.

Lewis, C. M., and D. H. Manz, 1991, "Light-Scatter Particle Counting: Improving Filtered-Water Quality," *Journal Environmental Engineering,* **117**(2):209.

Luitweiler, J. P., T. L. Yohe, E. Crist, and X. Sun, 1991, "Performance Testing of Hollow Fiber Membranes on a Groundwater," *Proceedings of the AWWA 1991 Membrane Processes Conference,* Orlando, Fla., p. 403.

Marcus, D. L., and R. Pastrick, 1988, "Ultrafiltration—Its Role in Today's Water Purification Systems," *Ultrapure Water,* p. 40.

McGahey, C. L., 1994, "Mechanisms of Virus Capture from Aqueous Suspension by a Polypropylene, Microporous, Hollow Fiber Membrane Filter," doctoral dissertation, The Johns Hopkins University, Baltimore, Md.

Melzer, T. H., and T. R. Meyers, 1971, "The Bubble Point in Membrane Characterization," *Bull. Parental Drug Assn.,* **25**(4):165.

Moffet, J. W., 1994, "Membrane Filtration Fits the Bill for Winchester," *Opflow,* **20**:1:1.

Olivieri, V. P., D. Y. Parker, Jr., G. A. Willinghan, and J. C. Vickers, 1991a, "Continuous Microfiltration of Surface Water," *Proceedings of the AWWA 1991 Membrane Processes Conference,* Orlando, Fla., p. 385.

Olivieri, V. P., G. A. Willinghan, J. C. Vickers, and C. McGahey, 1991b, "Continuous Microfiltration of Secondary Wastewater Effluent," *Proceedings of the AWWA 1991 Membrane Processes Conference,* Orlando, Fla., p. 613.

Pillay, V. L., B. Townsend, and C. A. Buckley, 1994, "Improving the Performance of Anaerobic Digestors at Wastewater Treatment Works: The Coupled Cross-Flow Microfiltration/Digestor Process," *Water Science Technology,* **30**(12):329.

Pirbazari, M., B. N. Badriyha, and V. Ravindran, 1992, "MF-PAC for Treating Waters Contaminated with Natural and Synthetic Organics," *Journal American Water Works Association,* **84**(12):95.

Quinn, R. M., 1983, "Ultrafiltration Membrane Processes—The Sleeping Giant," *Desalination,* **46**:113.

Reckhow, D. A., and P. C. Singer, 1984, "The Removal of Organic Halide Precursors by Preozonation and Alum Coagulation," *Journal American Water Works Association,* **76**:4.

Reckhow, D. A., and P. C. Singer, 1990, "Chlorination By-products in Drinking Waters: From Formation Potentials to Finished Water Concentrations," *Journal American Water Works Association,* **82**(4):173.

Reiss, C. R., and J. S. Taylor, 1991, "Membrane Pretreatment of a Surface Water," *Proceedings of the AWWA 1991 Membrane Processes Conference,* Orlando, Fla., p. 317.

Semmens, M. J., and T. K. Field, 1980, "Coagulation: Experiences in Organics Removal," *Journal American Water Works Association,* **72**(8):476.

United States Environmental Protection Agency, 1990, *Guidance Manual for Compliance with the Filtration and Disinfection Requirements for Public Water Systems Using Surface Water Sources,* Science and Technology Branch, Criteria and Standards Division, Office of Drinking Water, Contract No. 68-01-6989.

Valegard, K., L. Liljas, K. Fridborg, and T. Unge, 1990, "The Three-Dimensional Structure of the Bacterial Virus MS2," *Nature,* **345**:36.

Vickers, J. C., 1992, "Engineering Design Considerations for Microfiltration Systems," *Proceedings of the AWWA 1992 Annual Conference,* Vancouver, Canada, p. 81.

Vickers, J. C., 1993, "Meeting the Surface Water Treatment Rule Using Continuous Microfiltration," *Proceedings of the AWWA 1993 Membrane Technology Conference,* Baltimore, Md., p. 213.

Wale, R. T., and R. E. Johnson, 1993, "Microfiltration Principles and Applications," *Proceedings of the AWWA 1993 Membrane Technology Conference*, Baltimore, Md., p. 1.

Wiesner, M. R., D. C. Schroelling, and K. Pickering, 1991, "Permeation Behavior and Filtrate Quality of Tubular Ceramic Membranes Used for Surface Water Treatment," *Proceedings of the AWWA 1991 Membrane Processes Conference*, Orlando, Fla., p. 371.

Wiesner, M. R., S. Veerapaneni, and D. Brejchová, 1992, "Improvements in Membrane Microfiltration Using Coagulation Pretreatment," in R. Klute and H. H. Hahn (eds.), *Chemical Water and Wastewater Treatment II*. Springer-Verlag, New York, p. 281.

World Health Organization, 1993, *Guidelines for Drinking-Water Quality*, vol. 1, recommendation.

Yoo, S. R., D. R. Brown, R. J. Pardini, and G. D. Bentson, 1995, "Microfiltration: A Case Study," *Journal American Water Works Association,* **87**(3):38.

CHAPTER 12
ELECTRODIALYSIS

Japie J. Schoeman
Environmentek, CSIR
Pretoria, South Africa

Mark A. Thompson
Malcolm Pirnie, Inc.
Newport News, Virginia

12.1 PROCESS APPLICATIONS

12.1.1 Introduction

Electrical energy supplies the driving force in *electrodialysis* (ED) for ion-migration through the membranes. Electrolytes are usually transferred from a less concentrated solution through the ion-exchange membranes to a more concentrated solution with the aid of electrical energy. The purification of solvent in ED takes place by the removal of the undesirable solute through the membrane, whereas the purification of solvent in *reverse osmosis* (RO) and *ultrafiltration* (UF) takes place by the selective transport of solvent through the membrane which rejects the solute.

Electrodialysis is an electrochemical separation process in which charged membranes and an electrical potential difference are used to separate ionic species from an aqueous solution and other uncharged components. Electrodialysis today is widely used for the desalination of brackish water, and in some areas of the world it is the main process for the production of potable water. Although of major importance, water desalination and table salt production are by no means the only significant applications (Leitz, 1976; Leitz and Eisenmann, 1981). Encouraged by the development of new ion-exchange membranes with better selectivities, lower electrical resistance, and improved thermal, chemical, and mechanical properties, other uses of ED, especially in the food, drug, and chemical process industries as well as in biotechnology and wastewater treatment, have recently gained a broader interest (Korngold, 1984; Escudier, Cottereau, and Moutounet, 1989).

Electrodialysis in the classical sense can be used to perform several general types of separations, such as the separation and concentration of salts, acids, and bases from aqueous solutions (Nishiwaki, 1972; Korngold, 1978; Schoeman and van

Staden, 1991; Schoeman, 1992), the separation of monovalent ions from multivalent ions and multiple charged components (Nishiwaki, 1972), and the separation of ionic compounds from uncharged molecules (Ahlgren, 1972a, 1972b). Slightly modified ED is also used today to separate mixtures of amino acids or even proteins (Strathmann, 1992). Electrodialysis is also used to produce acids and bases from the corresponding salts when it is combined with electrically forced water dissociation in bipolar membranes (Mani, 1991).

In many applications, ED is in direct competition with other separation processes such as distillation, ion-exchange, reverse osmosis, and various chromatographic procedures. For certain applications there are very few economic alternatives to ED. Although this process has been known in principle for more than 80 years, large-scale industrial utilization began about 25 years ago. Classical or unidirectional standard ED was developed during the 1950s. However, during the past two and a half decades, the main feature has been the development of the polarity reversal process that is known as *electrodialysis reversal* (EDR) (Katz, 1977). This form of ED desalination has virtually displaced unidirectional ED for most brackish water applications and has gained a significant share of this market (Katz, 1977). The latest emphasis in ED development is the application of bipolar ED for resource recovery and pollution control (Mani, 1991; Scott, 1990).

The objectives of this chapter will be to consider ED process applications, module and membrane selection, process design criteria, operation and maintenance of ED applications, a design example, and facilities and costs.

12.1.2 Principles of Operation

The principles of operation of the electrodialysis process will be considered in this section. An ion-exchange membrane is in the form of a sheet while an ion-exchange resin is in granular form. The phenomenon of *ion exchange* is the permeation of ions in the case of an ion-exchange membrane, while it is an adsorptive exchange of ions in the case of an ion-exchange resin (Fig. 12.1a and b). Because of this different phenomenon, an ion-exchange membrane does not require regeneration but can be continuously used for a long period. The mechanism of operation of an ion-exchange membrane under the influence of an electrical potential is shown in Fig. 12.2. The cation-exchange membrane is charged negatively and is permeable to cations such as sodium (Na^+) and calcium (Ca^{2+}), while it is nonpermeable to anions such as chloride (Cl^-), sulfate (SO_4^{2-}), etc. This permselectivity encountered in ion-exchange membranes forms the basis of the ED/EDR process. Anion-exchange membranes are charged positive and behave oppositely.

12.1.2.1 Standard ED Process. In the ED process, water flows between alternately placed cation- and anion-permeable membranes (Fig. 12.3), which are built into a so-called ED *stack*. Direct current (DC) provides the motive force for ion migration through the membranes, and the ions are removed or concentrated in the alternate water passages by means of permselective membranes.

The standard ED process often requires the addition of acid and/or sequesterant to the concentrate stream to inhibit the precipitation of sparingly soluble salts (such as $CaCO_3$ and $CaSO_4$) in the stack. To maintain performance, the membrane stack needs to be cleaned periodically to remove scale and other surface fouling matter. This is done in two ways (USAID Desalination Manual, 1980): by *cleaning in place* (CIP) and *stack disassembly*.

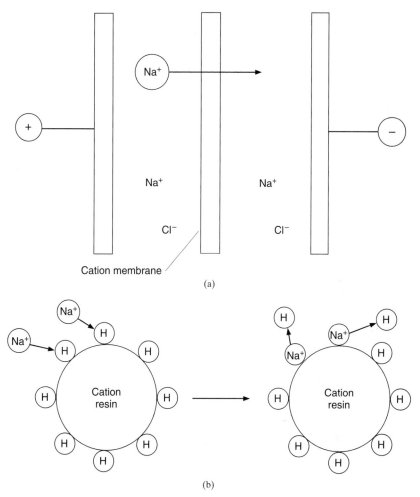

FIGURE 12.1 (*a*) Ion-exchange membrane permeation; (*b*) adsorptive exchange in the case of an ion-exchange resin.

Special cleaning solutions (dilute acids or bases) are circulated through the membrane stack for in-place cleaning but the membrane stack needs to be disassembled and mechanically cleaned at regular intervals to remove scale and other surface fouling matter. Regular stack disassembly is a time-consuming operation and is a disadvantage of the standard ED process.

12.1.2.2 EDR Process. The EDR process operates on the same basic principles as the standard ED process. However, in the EDR process, the polarity of the electrodes is automatically reversed periodically (about three to four times per hour) and, by means of motor-operated valves, the *fresh product water* and *wastewater* outlets from the membrane stack are interchanged. The ions are thus transferred in

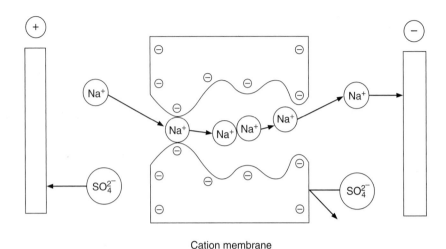

Cation membrane

FIGURE 12.2   Ionic permselectivity of ion-exchange membranes.

opposite directions across the membranes. This is shown in Fig. 12.4. This aids in breaking up and flushing out scale, slime, and other deposits from the cells. The product water emerging from the previous brine cells is usually discharged to waste for a period of one to two minutes until the desired water quality is restored.

The automatic cleaning action of the EDR process usually eliminates the need to dose acid and/or sequesterants, and scale formation in the electrode compartments is minimized due to the continuous change from basic to acidic conditions. Essentially, three methods of removing scale and other surface-fouling matter are used in the EDR process (USAID Desalination Manual, 1980), including clean-in-place (CIP) system, stack disassembly as used in standard ED, and flow and polarity reversal in the stacks. The polarity reversal system greatly extends the intervals between the time-consuming task of stack disassembly and reassembly, providing an overall reduction in maintenance time.

The capability of EDR to control scale precipitation more effectively than standard ED is a major advantage of this process, especially for applications requiring high water recoveries. However, the more complicated operation and maintenance requirements of EDR equipment necessitate more labor and a greater skill level than those of reverse osmosis. This may be a disadvantage of the process (Fraivillig, 1983).

12.1.2.3 Bipolar Electrodialysis Process.   The bipolar ED process uses ion-exchange membranes to separate and concentrate the acid and base constituents from a salt stream. The key element in this electrodialytic process is the bipolar membrane, so called because it is composed of two distinctive layers which are selective to ions of opposite charges (Mani, Chlanda, and Byszewski, 1988). An expanded view of this membrane and its operation is shown in Fig. 12.5. Under the influence of an applied current, water diffuses into the membrane interface where it dissociates to hydrogen and hydroxyl ions. The H^+ and OH^- ions are then transported across the cation and anion selective layers, respectively, to chambers on either side of the bipolar membrane. Acidification/basification of these chambers is the overall result.

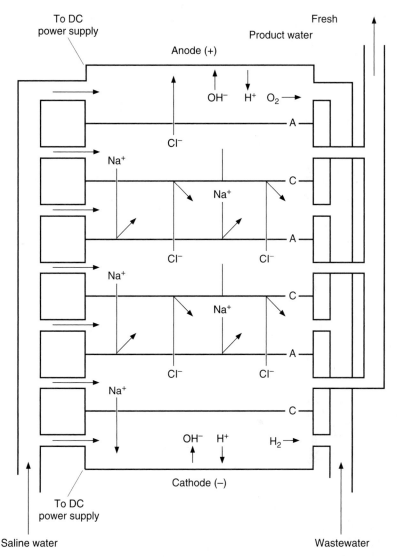

FIGURE 12.3   The ED process. A = anion-permeable membrane; C = cation-permeable membrane. *(Katz, 1979.)*

To achieve net production of acid and base, monopolar (i.e., cation- and anion-exchange) membranes are used in conjunction with the bipolar membrane. A schematic of a generalized three-compartment cell unit is shown in Fig. 12.6. The salt (i.e., sodium sulphate) is fed to a chamber between the cation and anion-selective membranes. The cations (Na^+) and anions (SO_4^{2-}) move across the monopolar membranes and combine with the hydroxide and hydrogen ions, as shown, to form acid and base. In a commercial operation, up to 200 of such cell units are assembled

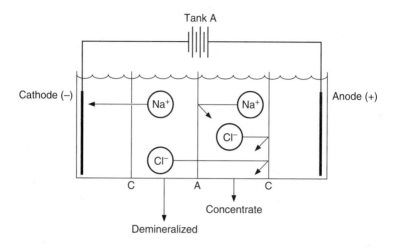

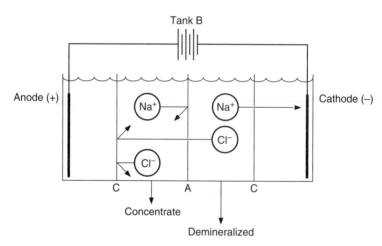

FIGURE 12.4 Controlling of film and scale formation in an EDR stack. *(Meller, 1984.)*

between a single set of electrodes to form a compact water-splitting stack. Feed to the acid, base, and salt chambers is achieved via internal manifolds built into the stack. If only one of the components (NaOH) needs to be obtained in a pure form, the cell can be simplified to a two-compartment unit, as shown in Fig. 12.7. Only the bipolar and cation membranes are used here. The acid product from such a cell using sodium sulphate feed would be a mixture of sulphuric acid and the unconverted salt.

12.1.3 Treatment Objectives

Since the fundamental mechanism of ED/EDR is the transport of electrolytes through membranes, it is apparent that the ED/EDR process offers the opportunity for separation of ionized from nonionized or weakly ionized compounds in solution.

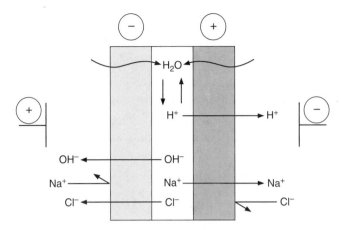

FIGURE 12.5 Operation of bipolar membrane. *(Mani, Chlanda, and Byszewski, 1988.)*

The use of ED/EDR, however, for the removal of contaminants from water and wastewaters is generally restricted to small ions such as sodium, calcium, chloride, sulphate, etc. When large organic anions are present in solution, the electrical conductivity and permselectivity of the membrane decreases with adverse effect on desalination performance (Korngold, 1984).

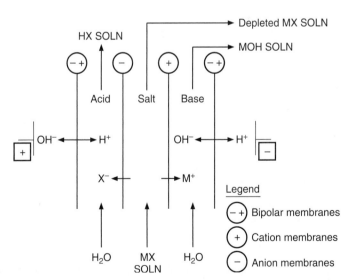

FIGURE 12.6 Three-compartment bipolar ED cell for conversion of salt into acid and base. *(Mani, Chlanda, and Byszewski, 1988.)*

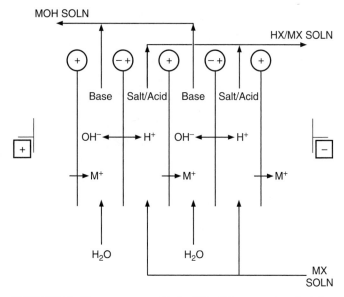

FIGURE 12.7 Two-compartment bipolar ED cell for conversion of salt into base. *(Mani, Chlanda, and Byszewski, 1988.)*

12.1.3.1 ED/EDR Applications. The principal applications of ED/EDR are in the separation of ionic species from neutral species and the concentration and removal of minerals (Scott, 1990). Not surprisingly, ED/EDR has been used widely for desalination where it is a direct competition with RO.

Major and potential ED/EDR applications include (Lacey and Loeb, 1972; Leitz and Eisenmann, 1981; Mani, 1991; Itoi, 1979; Moutounet and Escudier, 1991; Katz, 1982; Hughes, Raubenheimer, and Viljoen, 1992):

1. Desalination of brackish water
2. Concentration of seawater
3. Desalination of seawater
4. Demineralization of whey
5. Recovery of metals and water from electroplating rinse waters
6. Desalination of cooling tower blowdown
7. Recovery of acids and bases from spent pickling acid effluents
8. Demineralization of wine
9. Demineralization of sugar

Substances with a particle size range (ionic to macromolecular) of approximately 0.0004 to 0.1 μm can be removed with ED/EDR from water (Belfort, 1984). The possible size range of permeable species varies from 0.0004 to 0.03 μm. Typical contaminants that can be removed with ED/EDR from water and wastewater are shown in Table 12.1.

TABLE 12.1 Typical Contaminants That Can Be Removed with ED/EDR from Water and Wastewater

Contaminants			
Sodium	Nickel	Chloride	Fluoride
Potassium	Chromium	Sulphate	Chromate
Calcium	Copper	Nitrate	Acetate
Magnesium	Zinc	Phosphate	Hydroxyl
Ammonium	Strontium	Cyanide	Conductivity
Arsenic	Iron	Silver	TDS
	Aluminum		

The salinity level of brackish water can vary from approximately 2000 to 10,000 mg/L. This factor, in particular, affects the relative economics of ED/EDR and RO, as potable water generally must have a salinity level of less than 500 mg/L. Although economics vary from country to country, it appears that RO is superior at salinity levels of around 3000 mg/L and above (Scott, 1990). Strathmann (1992) claims that ED/EDR becomes more economical than ion exchange at about 500 mg/L and higher, while at about 5000 mg/L and higher, RO is the less costly process.

For a given application, there is not always a clear-cut decision as to which type of membrane separation process to select (Scott, 1990). Chemical composition or the size of the material to be removed is then often the key. Reverse osmosis is generally used to remove components of a molecular size near that of water. Electrodialysis can be considered for the removal of ionic constituents and, when greater concentration levels are required above those generally practical with RO, where osmotic pressure limitations restrict the end concentration to around 20% (Scott, 1990).

Recent trends in biotechnology and food processing indicate a growing interest in the electrodialytic separation of organic acids from aqueous solutions (Gudernatsch, Krumbholz, and Strathman, 1989). Electrodialysis has important applications in the pharmaceutical and biochemical industries where gentle processing conditions are required for materials such as human blood plasma and interferon (Scott, 1990). The production of essential amino acids requires various demineralization steps. Certain waste streams in biochemical and pharmaceutical operations contain ammonium sulphate, urea, and quandine hydrochloride which can be recovered by ED/EDR and eliminate certain large BOD concerns (Scott, 1990).

It should be noted that ED/EDR have not proved as attractive as RO for treating effluents with substantial amounts of microbiological contaminants and dissolved organic compounds (Belfort, 1984). Early work on ED treatment of municipal secondary effluents indicated that, for adequate performance, virtually all the dissolved organics had to be removed from the feed prior to treatment (Smith and Eisenmann, 1964, 1967). Consequently, the chief function of the ED/EDR process is the removal of inorganic ions, which leave bacteria, viruses, and neutral organics in the dilute stream. The possible size range of permeable species varies from 0.0004 to 0.03 μm (Belfort, 1984). This could become a serious problem when recycling for potable use.

Concentration of electrolytes by ED/EDR usually accompanies demineralization. The degree of concentration is limited only by the extent to which solvent transport accompanies ion transport (Lacey and Loeb, 1972). In the production of concentrated brine from seawater, for example, it has been difficult to achieve concentrations higher than about 3.5 mol/L because of water transported through the membranes with the ions. Precipitation of insoluble salts may also be a limitation in some concentration processes.

12.1.3.2   Bipolar ED Applications.   Salts like sodium chloride, sodium sulphate, sodium nitrate, potassium fluoride, sodium acetate, etc., occurring in effluents can be converted into their respective acids and bases with bipolar ED (Mani, 1991). Applications of bipolar ED include (Scott, 1990; Mani, 1991; Chiao, Chlanda, and Mani, 1990):

1. Regeneration of spent pickling liquors in stainless steel manufacture (HF, HNO_3, KOH recovery)
2. Flue gas desulphurization to produce sodium sulphite
3. Hydrofluoric acid recovery from fluorosilicic acid, a by-product from the wet process phosphoric acid plant
4. Organic acid and amino acid recovery
5. Ion-exchanger regenerant recovery
6. Acid and base purification
7. Pollution control

The use of bipolar membranes offers outstanding potential for the recovery and reuse of a range of organic and inorganic salts in a wide range of industries.

12.2   MODULE AND MEMBRANE SELECTION

12.2.1   Configuration

Different types of ED/EDR modules (stacks) and module configurations are described in the literature (Nishiwaki, 1972; Huffman and Lacey, 1972; Kedem and Cohen, 1983; Solt and Wen, 1992). The filter press and the unit cell stacks, however, are the most popular.

12.2.1.1   Components of Module.   Components of an ED/EDR stack are shown in Fig. 12.8.

Electrodialysis is carried out in modules with vertically oriented membranes separated from one another by flow spacers. The module or cell stack consists of cell pairs comprising a cation-selective membrane, a diluent flow spacer, an anion-selective membrane, and a concentrate flow spacer. In addition to the cell pairs, each stack contains two electrodes and electrode compartments, plumbing necessary to transport water to and from the stack, and the hardware necessary to hold the stack together.

Membranes.   The ED membranes are flat sheets, usually made of a plastic film formed on a fabric backing of dynel, glass, or other material to provide strength (USAID Desalination Manual, 1980). Ion transfer sites are added to the membranes with the site charge differing between the anion- and cation-permeable membranes to allow each type to selectively pass either anions or cations. Manufacturers vary in how they incorporate the ion transfer sites into the membrane. Two types of membranes currently produced are the homogeneous and heterogeneous types. In the homogeneous membranes, such as those made by Ionics, the sites are uniformly distributed through the membrane. In the heterogeneous membranes, such as those made by Tokuyama Soda, the sites are distributed as discrete points within the membrane.

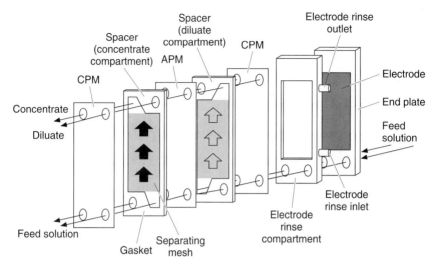

FIGURE 12.8 Components in an electrodialysis stack. APM = anion-permeable membrane; CPM = cation-permeable membrane. *(Chiapello and Bernard, 1992.)*

Membranes are available in a variety of sizes, depending on the manufacturer and the application. A typical membrane size made in Japan is approximately 100 cm by 100 cm and in the United States a rectangle of about 46 cm by 102 cm. The membranes are usually an unbroken sheet except for holes cut out to form flow channels and holes or slots to guide the assembly of the membranes into a stack. Some typical membranes and spacers are shown in Fig. 12.9*a* and *b*. During construction of the stack, the membranes and spacers are aligned in either a vertical or horizontal plane and held together by some type of press or clamping device.

The thickness of the membrane depends on the application, and its selection is a trade-off between membrane properties. Thicker membranes usually have greater strength, increased erosion resistance, and longer life, whereas thinner membranes have lower electrical resistance and, hence, reduced energy requirements. Typically, membranes are about 0.15 to 0.56 mm thick.

Membrane structures can be tailored during fabrication to alter their characteristics by varying components such as their water content. The higher the water content, the "looser" the membrane, i.e., the easier it is for ions to travel through them. With a lower water content, the membrane becomes "tighter." This allows manufacturers to have a family of membranes, enabling them to tailor membranes to various applications and their particular water chemistry characteristics.

Spacers. The spacers separate the membranes and provide a pathway in the cell for the water flow. Three of the major pathway configurations provided by various spacers are shown in Fig. 12.10 (USAID Desalination Manual, 1980). In each of the three spacers, the flow enters via the feed channel at point A and then follows the pathway shown by the arrows until it leaves the cell by way of the discharge channel at point B. Spacers can be formed to provide different types of flow paths as shown in Fig. 12.10. The sheet-flow and tortuous-path flow are two of the most commonly used designs. The slanted strap spacers are a modification of the tortuous-path spacers.

Cells are made up of two membranes with a spacer in between. Cells are stacked with alternating concentrate and dilute cells to form a stage. In each stage the feedwater is exposed only to the electromotive force for the distance of the pathway in one cell, called a *hydraulic stage*. By the use of special spacers, more than one hydraulic stage can be placed between a set of electrodes. The number of stacks, stages, and electrodes is determined at the time of design based on site-specific information.

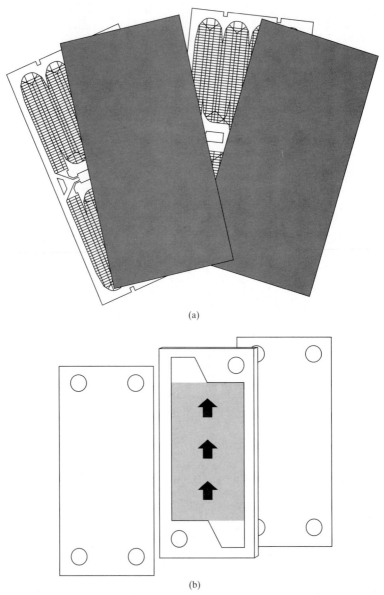

(a)

(b)

FIGURE 12.9   (*a*) Membranes and spacers for an Ionics electrodialysis stack; (*b*) membranes and spacers for an EIVS Corning electrodialysis stack.

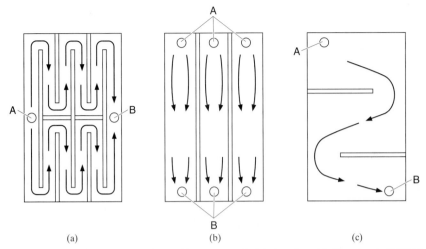

FIGURE 12.10 Three types of spacers: (*a*) tortuous path; (*b*) sheet flow; (*c*) slanted path. *(USAID Desalination Manual, 1980.)*

Electrodes. One pair of electrodes is required for each electrical stage. Normally, no more than two electrical stages are present in a single membrane stack, whereas a pair of electrodes is needed for each electrical stage. The electrodes are generally constructed of niobium or titanium with a platinum coating. Graphite is also sometimes used.

A reaction occurs at each of the electrodes. Hydrogen ions and oxygen and/or chlorine gas are formed at the anode (positive electrode), and hydrogen gas and hydroxyl ions are formed at the cathode (Mason and Kirkham, 1959). Because a separate stream is generally used in the space adjacent to the electrodes in each stack, the by-products of the reactions are confined to these streams. The anode stream is normally acidic due to the hydrogen ions, and the cathode stream is basic. The pH difference is partially neutralized by combining the streams. In most units, the electrode streams are kept apart and then usually discharged to waste, although in some instances this water is treated and recycled. Because the pH increases at the cathode, the formation of a calcium or magnesium precipitate can occur. Acid is often added to the electrode stream to keep the cathode stream acidic and prevent precipitation of the salts.

In the EDR process, the anode and cathode are electrically reversed several times per hour, thus alternating the environment at the electrodes from acidic to basic on a regular basis. This significantly reduces scale formation potential at the eletrodes.

12.2.1.2 Stacks. There are two common membrane stack configurations: the filter-press and the unit-cell stack arrangements. Both will be discussed in this section.

Filter-Press Stack. The filter-press stack has been described by Nishiwaki (1972) and Seko (1962). Alternate cation- and anion-exchange membranes are arranged between compartment frames in an assembly similar to a plate-and-frame filter press. The compartment frames are usually about 1 mm thick. Each frame is provided with gaskets at the ends and edges, screens to support the membranes, and manifolding devices in the gasket areas at each end of the frame. The entire assembly of membranes and compartment frames is clamped between end frames that hold the electrodes. Clamping pressure is applied by bolting or by a hydraulic mechanism at each end. The filter-press stack is widely used for water and effluent desalination applications (USAID Desalination Manual, 1980; Hughes, Raubeheimer, and Viljoen, 1992).

Unit-Cell Stack. The unit-cell-type stack has been described by Nishiwaki (1972); Kedem and Cohen (1983); and Schoeman and van Staden (1991). The stack described by Nishiwaki has a structure especially developed for the concentration of seawater. Each concentrating cell consists of one cation-exchange and one anion-exchange membrane sealed at the edges to form an envelope-like bag. The concentrate cells are separated by screen-like spacers so that seawater can flow between them in sections of multicompartments. A number of such sections are placed into rectangular tanks that have electrodes at each end. Concentrated brine overflows each concentrate unit cell through small tubes attached to the unit cells. More than a thousand cells may be assembled into each unit-cell stack. A central anode is used with cathodes at each end of the stack to keep the total applied voltage to a tolerably low level. The unit-cell stack is used for seawater concentration in Japan.

The cell described by Kedem and Cohen has brine-sealed cells with outlets arrayed in an open vessel, separated by spacers (0.3 mm). The dialysate enters through a suitable port at the bottom of the vessel and runs out through an overflow. Direct current is applied through carbon suspension electrodes (Kedem, Tanny, and Maoz, 1978). This cell design has potential for concentration/desalination of industrial effluents (Schoeman and van Staden, 1991).

12.2.1.3 Staging. The manner in which the membrane stack array is arranged is called *staging* (Meller, 1984). The purpose of staging is to provide sufficient membrane area and retention time to remove a specified fraction of salt from the demineralized stream. Two types of staging are used: *hydraulic staging* and *electrical staging.* In a stack with one hydraulic and one electrical stage, each increment of water makes one pass across the membrane surface between one pair of electrodes and exits. It should be noted that in an Ionics membrane stack, water flows in multiple parallel paths across the membrane surfaces, and a single pass consists of flowing through one water flow spacer between two membranes and exiting through the outlet manifold. In a sheet-flow stack, water enters at the one end of the stack and flows as a sheet across the membrane to exit at the other end in a single pass.

Hydraulic Staging. Examples of hydraulic staging are shown in Figs. 12.11 and 12.12. Additional hydraulic stages must be incorporated to increase the amount of salt removed in an ED/EDR system.

Electrical Staging. Electrical staging is accomplished by inserting additional electrode pairs into a membrane stack. This gives flexibility in system design, providing maximum salt removal rates while avoiding polarization and hydraulic pressure limitations (see Sec. 7.3.4.1). Examples of electrical and hydraulic staging are shown in Figs. 12.13 and 12.14.

12.2.2 Ion-Exchange Membranes

Suppliers of ion-exchange membranes, membrane characteristics, methods to determine membrane characteristics, and guidelines for membrane selection will be presented in this section.

12.2.2.1 Membrane Suppliers and Characteristics. The major suppliers of ion-exchange membranes, membrane characteristics, and methods to determine membrane characteristics are presented in the literature (Lacey, 1972; Korngold, 1984;

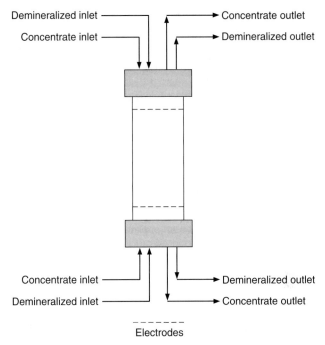

Electrodes

FIGURE 12.11   One stage: one hydraulic stage and one electrical stage.
(Meller, 1984.)

Strathmann, 1992; Simon and Calmon, 1986). Major suppliers of ion-exchange membranes and membrane characteristics are listed in Table 12.2.

12.2.2.2   Ion-Exchange Membranes with Special Characteristics.   Ion-exchange membranes with special characteristics are described in the literature (Korngold, 1984; Strathmann, 1992; Scott, 1990; Nishiwaki, 1972). These membranes are used in the production of table salt, as battery separators, as ion-selective electrodes, for acid concentration, or in the chlor-alkali process. Anion-exchange membranes are prone to fouling by organic matter. This can lead to failure of the ED/EDR process. Consequently, a lot of effort has been put into the development of ion-exchange membranes with low fouling characteristics.

Anion-Exchange Membranes with Antifouling Characteristics.   Significant improvements have been made in the development of anion-exchange membranes with low fouling characteristics (Strathmann, 1992). In conventional ED/EDR plants, the permissible current density in the anion-exchange membrane is smaller than in the cation-exchange membrane, largely due to the risk of precipitates (Korngold et al., 1970). The anion-exchange membrane is more sensitive to fouling. According to some former investigations concerning the permeability of commercial anion-exchange membranes, the upper molecular weight limit for practical electrodialytical separations is in the range of 100 daltons. A molecular weight of 350 daltons is to be considered as a maximum size for any electrotransport through, e.g., the Ionac MA 3475 membrane. The static permselectivity decreases gradually from 98 to 30 percent when the molecular weight of the solute increases from 59 to 171 daltons (Dohno, Azumi, and Takashima, 1975).

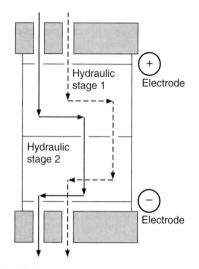

FIGURE 12.12   Two hydraulic stages and one electrical stage. *(Meller, 1984.)*

Fouling of anion-exchange membranes often occurs when the anion is small enough to penetrate into the membrane structure, but its electromobility is so poor that the membrane is virtually blocked. To overcome this problem, different companies have developed membranes such as the Neosepta® AMX, which is characterized by a high permeability for large organic anions. In general, the permselectivity of these membranes is lower than that of regular membranes. Since the pore diameter of ion-exchange membranes is about 0.001 μ (10°A), polyelectrolytes of high molecular weight are not harmful to ion-exchange membranes. However, large ionic compounds with the molecular weight of several hundred daltons can cause membrane fouling. One method of improving the permeability of anion-exchange membranes for large organic acids is based on the adjustment of the degree of cross-linking and the chain length of the cross-linker in the polymer network (Gudernatsch, Krumbholz, and Strathmann, 1989).

Ionics Inc. produces a macroreticular membrane that is less sensitive to traces of detergents. It is produced by dissolving an organic compound in the membrane-forming system. When the material diffuses from the membrane after polymerization, large pores are left behind. Alternatively, passive salts such as potassium iodide are added to the solvent of the binder polymer, mostly dimethylformamide. Through these pores, large anionic molecules can penetrate, thus preventing a steep increase in the electrical resistance.

Another type of antifouling anion-exchange membrane is produced by Tokuyama Soda. The membrane is coated with a thin layer of cation-exchange groups causing electrostatic repulsion of organic molecules. Practically, the coating is done by weak sulfonation of the membrane surface, followed by the ordinary chloromethylation and quaternization steps.

Ion-exchange membranes based on aliphatic polymers show reduced organic fouling in natural waters compared to membranes based on aromatic polymers. The aliphatic membranes also allow operation with solutions containing 0.5 mg/L of chlorine and for shock chlorination up to 20 mg/L of free chlorine.

Ion-Exchange Membranes Permselective Toward Monovalent Ions.   Table salt has been produced in Japan since 1972 by electrodialytic concentration of seawater. For the specific requirements of this process, ion-exchange membranes that can separate monovalent ions from a mixed solution containing monovalent and multivalent ions have been developed (Strathmann, 1992). Tokuyama Soda has commercialized monovalent cation-selective membranes (Neosepta® CMS) prepared by forming a thin cationic charged layer on the membrane surface (Sata, Izuo, and Mizutani, 1984). Monovalent anion-permselective membranes (Neosepta® ACS), which have a thin, highly cross-linked layer on the membrane surface, have also been developed. By such means, the selectivity of sulphate compared to that of chloride can be reduced from about 0.5 to about 0.01 and of magnesium compared to sodium from about 1.2 to about 0.1.

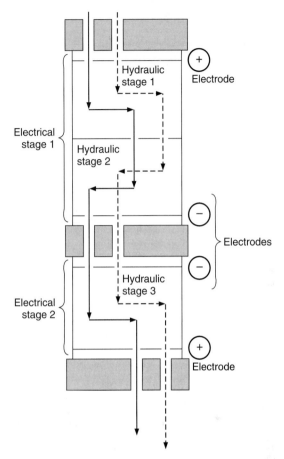

FIGURE 12.13 Two electrical stages and three hydraulic stages. *(Meller, 1984.)*

Anion-Exchange Membranes with a High Retention for Protons. By using traditional membranes, electrodialysis cannot be applied to the recovery of acid in order to reuse the acid because of high proton leakage through the anion-exchange membranes (Strathmann, 1992). In general, protons permeate easily through an anion-exchange membrane. Therefore, acids cannot be concentrated to more than a certain level by electrodialysis with any degree of high efficiency. Recently developed membranes, such as the Neosepta® ACM (Urano, Ase, and Naito, 1984), Selemion AAV, and Morgane ARA (Boudet-Dumy, Lindheimer, and Gavach 1991) exhibit low proton permeabilities and enable efficient acid concentration (Schoeman, 1992).

Cation-Exchange Membranes with a Permselectivity for Protons. Amphoteric-type ion-exchange membranes are preferentially permeable to hydrogen ions (Strathmann, 1992). By coating this membrane with a thin cation charged layer, the selectivity for H^+ versus Na^+ is improved up to $\alpha(H^+/Na^+) = 12$ (Yamane, Izuo, and Mizutani, 1965).

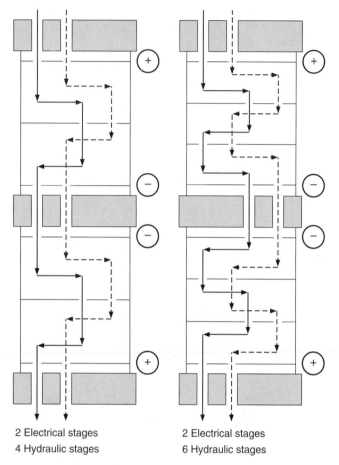

2 Electrical stages
4 Hydraulic stages

2 Electrical stages
6 Hydraulic stages

FIGURE 12.14   Hydraulic and electrical staging. *(Meller, 1984.)*

Fluorocarbon-Type Cation-Exchange Membranes.   Most conventional hydro-carbon ion-exchange membranes are degraded by oxidizing agents, especially at elevated temperatures (Strathmann, 1992). To adapt ion-exchange membranes to an application in the chlor-alkali industry, a fluorocarbon-type membrane with excellent chemical and thermal stability was developed by Du Pont as Nafion® (Connolly and Gresham, 1966; Grot, 1973). The membrane is produced by a several-step procedure, which starts with the synthesis of an ionogenic perfluorovinylether and its copolymerization with tetrafluoroethylene (TFE). The economic lifetimes of chlor-alkali membranes under the aggressive conditions are in the range of three years. The lifetime is determined in part by the morphological changes and in part by loss of carboxylic acid ion-exchange capacity.

Anion-Exchange Membranes Stable in Alkaline Solutions.   In several technically interesting applications, the economics of the process are affected by the limited stability of currently available anion-exchange membranes in strong alkaline solutions (Strathmann, 1992). In the case of cation-exchange membranes, the chemical stability could be improved by perfluorination of the polymer backbone, resulting in membranes such as the Nafion®-, Flemion®-, and Neosepta®-type structures. Comparable attempts with anion-exchange membranes to overcome the poor alkaline stability failed (Matsui et al., 1986). Because the novel fluorocarbon-type anion-exchange membranes cannot overcome the problem of instability under strong basic conditions, it seems reasonable to assume that the alkaline stability of an anion-exchange membrane is determined by the stability of the incorporated positively charged groups against the attack of hydroxyl ions. By determining the disintegration rate of quaternized amines in alkaline solutions, Bauer, Gerner, and Strathmann (1988) developed an anion-exchange membrane with considerably improved alkaline stability. When integrated into a polysulphone (PSU) or polyethersulphone (PES) matrix, the anion-exchange membrane did not show any changes in its mechanical or electrical properties.

Bipolar Membranes. Bipolar membranes have recently gained increasing attention as an efficient tool for the production of acids and bases from their corresponding salts by electrically enforced, accelerated water dissociation (Strathmann, 1992). The process, which has been known for many years, is economically very attractive and has a multitude of interesting technical applications (Liu, Chlanda, and Nagasubramanian, 1977; Mani, 1991). So far, however, large-scale technical use of bipolar membranes has been limited.

Bipolar membranes can be prepared by simply laminating conventional cation- and anion-exchange membranes back to back (Leitz, 1972). The total potential drop depends on the applied current density, the resistance of the two membranes, and the resistance of the solution between them. Since the specific resistance of deionized water is very high, the distance between the membranes of opposite polarity should approach zero. Laminated bipolar membranes exhibit mostly an unsatisfactory chemical stability at high pH values and a rather poor water-splitting capability. Single-film bipolar membranes (Chlanda, Lee, and Liu, 1976) and multilayer bipolar membranes fulfill most of the practical needs. For example, Bauer, Gerner, and Strathmann (1988) prepared a bipolar membrane by casting a cation-selective layer on top of the previously prepared well-defined cross-linked anion-exchange membrane. To minimize the electrical resistance, the thickness of the interphase between the oppositely charged membranes should preferably be less than 5 nm. It could also be shown that in bipolar membranes the chemical stability is determined to a large extent by the properties of the positively charged anion-exchange moieties and by the properties of the matrix polymer. Furthermore, in practical applications, bipolar membranes should not only have good chemical stability but also have adequate water-splitting capabilities. From the observed ion fluxes, calculations show that the water dissociation rate in the bipolar region is much faster than in pure water already at current densities in excess of 1 mA cm^{-2}. The reasons for this acceleration are not completely verified. However, the experimental evidence reported in the literature strongly supports the hypothesis that it is caused by a reversible proton transfer reaction between the charged groups of the membrane and the water molecules at the surface.

12.2.2.3 Characterization of Ion-Exchange Membranes.

Ion-exchange membranes used in electrodialysis can be classified in terms of their mechanical and electrical properties, their permselectivity, and their chemical stability (Strathmann, 1992). A microscopic examination yields information on whether a membrane is reinforced and the type of reinforcement used. The electrical charge of an ion-exchange membrane can be determined qualitatively by using indicator solutions. A drop of 0.05% solution of methylene blue and methyl orange on a sample stains a

TABLE 12.2 Properties of Commercial Ion-Exchange Membranes

Membrane	Type	Structure properties	IEC, meq/g	Backing	Thickness, mm	Gel water, %	Area resistance 0.5 N NaCl, 25°C, $\Omega\cdot cm^2$	Permselectivity 1.0/0.5 N KCl, %
Asahi Chemical Industry Company Ltd., Chiyoda-ku, Tokyo, Japan								
K 101	Cation	Styrene/DVB	1.4	Yes	0.24	24	2.1	91
A 111	Anion	Styrene/DVB	1.2	Yes	0.21	31	2–3	45
Asahi Glass Company Ltd., Chiyoda-ku, Tokyo, Japan								
CMV	Cation	Styrene	2.4	PVC	0.15	25	2.9	95
AMV	Anion	Butadiene	1.9	PVC	0.14	19	2–4.5	92
ASV	Anion	Univalent	2.1		0.15	24	2.1	91
DMV	Cation	Dialysis			0.15		—	—
Flemion®	Cation	Perfluorinated						
Ionac Chemical Company, Sybron Corporation, Birmingham, NJ 08011								
MC 3470	Cation		1.5	Tergal	0.6	35	6–10	68
MA 3475	Anion		1.4	Tergal	0.6	31	5–13	70
MC 3142	Cation		1.1		0.8		5–10	—
MA 3148	Anion		0.8	Tergal	0.8	18	12–70	85
Ionics Inc., Watertown, MA 02172								
61AZL386	Cation		2.3	Modacrylic	0.5	46	~6	—
61AZL389	Cation		2.6	Modacrylic	1.2	48	—	—
61CZL386	Cation		2.7	Modacrylic	0.6	40	~9	—
103QZL386	Anion		2.1	Modacrylic	0.63	36	~6	—
103PZL386	Anion		1.6	Modacrylic	1.4	43	~21	—
204PZL386	Anion		1.9	Modacrylic	0.57	46	~8	—
204SXZL386	Anion		2.2	Modacrylic	0.5	46	~7	—
204U386	Anion		2.8	Modacrylic	0.57	36	~4	—

12.20

			Du Pont Company, Wilmington, DE 19898					
N 117	Cation	Perfluorinated	0.9	No	0.2	16	1.5	—
N 901	Cation	Perfluorinated	1.1	PTFE	0.4	5	3.8	96
			Pall RAI, Inc., Hauppauge, NY 11788					
R-5010-L	Cation	LDPE	1.5	PE	0.24	40	2–4	85
R-5010-H	Cation	LDPE	0.9	PE	0.24	20	8–12	95
R-5030-L	Anion	LDPE	1.0	PE	0.24	30	4–7	83
R-5030-H	Anion	LDPE	0.8	PE	0.24	20	11–16	87
R-1010	Cation	Perfluorinated	1.2	No	0.1	20	0.2–0.4	86
R-1030	Anion	Perfluorinated	1.0	No	0.1	10	0.7–1.5	81
			Rhone-Poulenc Chemie GmbH, Frankfurt, Germany					
CRP	Cation		2.6	Tergal	0.6	40	6.3	65
ARP	Anion		1.8	Tergal	0.5	34	6.9	79
			Tokuyama Soda Company Ltd., Nishi-Shimbashi, Minato-ku, Tokyo 105, Japan					
CL-25T	Cation		2.0	PVC	0.18	31	2.9	81
ACH-45T	Anion		1.4	PVC	0.15	24	2.4	90
ACM	Anion	Low H^+ transport	1.5	PVC	0.12	15	4–5	—
AMH	Anion	Chemical resistant	1.4	—	0.27	19	11–13	—
CMS	Cation	Univalent	>2.0	PVC	0.15	38	1.5–2.5	—
ACS	Anion	Univalent	>1.4	PVC	0.18	25	2–2.5	—
AFN	Anion	Antifouling	<3.5	PVC	0.15	45	0.4–1.5	—
AFX	Anion	Dialysis	1.5	PVC	0.14	25	1–1.5	—
Neosepta®-F		Perfluorinated						—

Source: Strathmann, 1992.

golden yellow on top of an anion-exchange membrane and a deep blue on top of a cation-exchange membrane.

Mechanical Examination and Swelling Behavior of Membranes. Detailed mechanical characterization involves the determination of thickness, dimensional stability, tensile strength, and hydraulic permeability (Strathmann, 1992). All mechanical measurements should be conducted with pretreated and well-equilibrated membranes. Information related to storage and handling characteristics, the durability, and type of reinforcing material is obtained from determining the dimensional changes between the wet and dry states of the membrane.

The presence of pinholes in ion-exchange membranes will not only obscure the hydraulic permeability test but also invalidate any application. Pinholes can be determined by placing a wet membrane sheet on a sheet of white absorbent paper. A 0.2% solution of methylene blue for an anion-exchange membrane or a 0.2% solution of Erythrocein-B for a cation-exchange membrane is spread over the entire surface. If no spots of the dye can be observed on the paper, the membrane is free of pinholes and can be tested for its hydraulic permeability. Hydraulic permeability measurements provide information on the diffusive transport of components through a membrane under a hydrostatic pressure driving force. The test is carried out at room temperature using deionized water and a hydrostatic pressure driving force. The permeability can then be calculated from the volumetric flow rate in $L/m^2 \cdot h$ (Strathmann, 1992).

The gel water content or swelling capacity not only determines the dimensional stability of the membrane but also affects its selectivity, electrical resistance, and hydraulic permeability (Strathmann, 1992). The swelling of a membrane depends on the nature of the polymeric material, the ion-exchange capacity, the cross-linking density, and the homogeneity of the membrane. Usually, the gel water content is expressed by the weight difference between the wet and dry membranes. In a test for the gel water content, a sample is equilibrated for two days in deionized water. After removing the surface water from the sample, the wet weight of the swollen membrane W_{wet} is determined. The sample is then dried at 75°C over phosphorus pentoxide under reduced pressure until a constant weight W_{dry} is obtained. The gel water weight percent is obtained from

$$\% \text{ gel water wt.} = 100 \, \frac{(W_{wet} - W_{dry})}{W_{dry}}$$

Long-Term Chemical Stability of Membranes. The economics of ion-exchange membranes in different applications are determined to a large extent by their chemical stability under process conditions (Strathmann, 1992). Membrane deterioration after exposure for certain time periods to various test solutions containing acids, bases, or oxidizing agents is estimated by visual comparison with new, unexposed samples and by determining changes in their mechanical and electrical properties (Kneifel and Huttenbach, 1980).

Determination of Membrane Ion-Exchange Capacities. The ion-exchange capacity of charged membranes is determined by titrating the fixed ions such as $-SO_3^-$ or $-R_4N^+$ groups with 1 N NaOH or HCl, respectively (Strathmann, 1992). For these tests, cation- and anion-exchange membranes are equilibrated for about 24 h in 1 N HCl or NaOH, respectively, and then rinsed free from chloride or sodium for 24 h with deionized water. The ion-exchange capacity of the samples is then determined by back titration with 1 N NaOH or HCl, respectively. Weak base anion-exchange membranes are characterized by equilibration in 1 N sodium chloride and titration with standardized 0.1 N silver nitrate solution. The samples are then dried and the ion-exchange capacity is calculated for the dry membrane.

Permselectivity of Ion-Exchange Membranes. The permselectivity of an ion-exchange membrane relates the transport of electric charges by specific counter-ions to the total transport of electrical charge through the membrane (Strathmann, 1992). An ideal selective cation-exchange membrane would, for example, transmit positively charged ions only. The permselectivity approaches zero when the transport numbers within the membrane $\bar{t}_\pm$ are the same as in the electrolytic solution $t_\pm$. The degree of permselectivity depends on the concentration of electrolytes in the membrane and therefore on the ion-exchange capacity and the cross-linking density. When a membrane separates dilute and concentrate solutions, there will be a concentration gradient across the membrane. In this case, the permselectivity $S_\pm$ can be calculated from the transfer of counter-ions (dynamic method) according to the following equation (Tuwiner, 1962):

$$S_\pm = \frac{\bar{t}_\pm - t_\pm}{1 - t_\pm} \tag{12.1}$$

A faster method for the determination of the apparent permselectivity $S_{\pm,obs}$ is found from a potential measurement (static method). The experimental setup of the test procedure is illustrated in Fig. 12.15. In this special case, the transport of water through the membrane is not taken into account. The apparent transport number $\bar{t}_{\pm,obs}$ of an ion is obtained as

$$\bar{t}_{\pm,obs} = t_\pm - 0.018 t_w c_\pm \tag{12.2}$$

where 0.018 mL·mmol^{-1} means the molar volume of water and t_w, the water transfer rate. As an advantage, the determination of the potential between two solutions of different concentrations is considered independent of any polarization effects at the membrane surface.

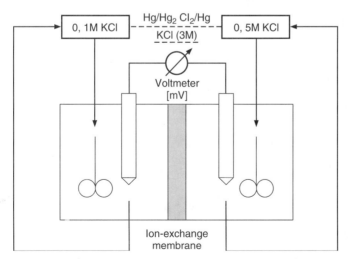

FIGURE 12.15 Experimental setup for determining membrane permselectivities. *(Strathmann, 1992.)*

The actual test system consists of two cells separated by the membrane sample. The potential difference across the membrane is measured using a set of calomel electrodes. The selectivity is then calculated from the ratio of the experimentally determined to the theoretically calculated potential difference for a 100 percent permselective membrane.

For a system consisting of standardized aqueous solutions of 0.1 N and 0.5 N KCl at 25°C, for example, this theoretical potential difference amounts to exactly 36.94 mV. It is calculated using the Nernst equation:

$$\Delta\Psi_{theo} = \frac{RT}{zF} \ln\left(\frac{c_1'\gamma_1'}{c_2'\gamma_2'}\right) \tag{12.3}$$

where R = gas constant
             T = absolute temperature
             z = electrochemical valence of ions in solutions
             F = Faraday's constant
             c_1' and c_2' = the concentrations
             γ_1' and γ_2' = the activity coefficients of the two solutions separated by the membrane. The apparent permselectivity $S_{\pm,obs}$ of the membrane is given by

$$S_{\pm,obs} = \frac{\Delta\Psi_{exp}}{\Delta\Psi_{theo}} \cdot 100\% \tag{12.4}$$

where $\Delta\psi_{exp}$ is the measured potential difference between the two electrolyte solutions. The absolute value of $\Delta\psi_{exp}$ is positive for a cation-exchange membrane and negative for an anion-exchange membrane (Korngold, 1984).

Electrical Resistance of Ion-Exchange Membranes. The electrical resistance of ion-exchange membranes is one of the factors that determines the energy requirements of electrodialysis processes (Strathmann, 1992). It is, however, generally considerably lower than the resistance of the dilute solutions surrounding the membrane since the ion concentration in the membrane is very high. The specific membrane resistance is usually reported in $\Omega\cdot cm$. From the engineering point of view, the membrane area resistance in units of $\Omega\cdot cm^2$ is more useful. The area resistance of ion-exchange membranes is determined by conductivity measurements in a cell that consists of two well-stirred chambers separated by the membrane as shown in Fig. 12.16.

The cell is filled with a 0.5-N solution of NaCl. The electrical resistance of the cell is measured with and without a membrane separating the two chambers. The membrane area resistance is then calculated from the difference in the conductivity of the two measurements using the equation

$$R_{mA} = K\left(\frac{1}{k_T} - \frac{1}{k'}\right) \tag{12.5}$$

where R_A = area resistance of membrane per unit area
        k_T = conductivity of cell and membrane
        k' = mean conductivity of cell
        K = cell constant

12.2.2.4   Guidelines for Membrane Selection.   The following characteristics represent features of an ideal ED/EDR membrane (USAID Desalination Manual, 1980):

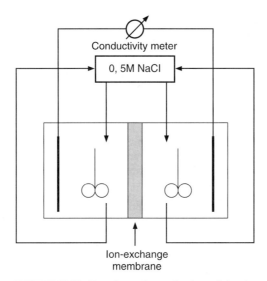

FIGURE 12.16 Experimental setup for determining the electrical resistance of ion-exchange membranes. *(Strathmann, 1992.)*

1. High thermal stability
2. High mechanical strength
3. High resistance to chemical and biological degradation
4. High dimensional stability
5. High ionic selectivity
6. Low cost
7. Low electrical resistance
8. Low diffusion of water

Since some of these properties are mutually exclusive, the commercial membranes produced are necessarily a compromise of these characteristics.

The degree of concentration that can be achieved in ED/EDR is limited by the amount of water that is transferred through the membranes along with the ions by osmosis and electroosmosis (Huffman and Lacey, 1972). The flux of water that occurs with a flux of ions is highly dependent on the nature of the membranes. The concurrent fluxes of water and ions have not been studied extensively, but Lakshminarayanaiah (1965) and Lacey (1967) have reported some data on the subject. In general, the number of milliliters of water transported per faraday decreases with increases in current density, decreases with increases in solution concentration, and decreases as the water content of the membrane decreases.

Another limitation on the degree of concentration that can be accomplished in some application of ED/EDR processing is that some compounds present in the feed solutions may exceed their maximum solubility if the feed is concentrated too much (Huffman and Lacey, 1972). For example, in the electrodialytic concentration of seawater to furnish brines for the chlor-alkali industry, the formation of precipitates of calcium sulphate in the concentrating compartments limits the degree of concentration that can be achieved. The developers of this process solved the problem by developing ion-exchange membranes that allowed the transfer of Na^+ and Cl^- ions in preference to Ca^{2+} and SO_4^{2-} ions. Consequently, monoselective membranes must be used when the degree of concentration that can be achieved is limited.

Although some membranes can be used in either sheet-flow or tortuous-path stacks, the optimum physical or mechanical properties of membranes differ for the two types of stacks (Huffman and Lacey, 1972). The membranes used in tortuous-flow stacks must be sufficiently rigid to support themselves over the width of the solution-flow channel. Membranes that are very flexible or that stretch or undergo creep are usually not suitable for use in tortuous-path stacks, because such membranes can collapse or creep into the solution-flow channels and thus prevent the attainment of equal flow rates through each compartment.

Membranes for sheet-flow stacks may be more flexible, and consequently thinner, than those for tortuous-flow stacks, because of the support provided by the spacer screens. The more rigid membranes that are normally used in tortuous-path stacks can be used in some sheet-flow stacks, but their suitability depends on the design of the solution entrance and exit ports. The entrance and exit ports of some sheet-flow stacks are not in the same place as the rest of the membrane and require the use of flexible membranes in order to obtain satisfactory sealing at the solution manifolds (Huffman and Lacey, 1972).

If the highest degree of electrolytic concentration is to be effected with low energy requirements, the electrolyte must be transferred through the membranes with high coulomb efficiencies, and the amount of water that permeates the membranes and transfers to the concentrated solutions must be small (Nishiwaki, 1972). It was shown that high coulomb efficiencies are obtained when the transference numbers are high, the diffusion coefficients are low, and thick membranes are used (Nishiwaki, 1972). Because of the decrease in Donnan exclusion in ion-exchange membranes with an increase in electrolyte concentration in the solution in contact with membranes, the transference number of ion-exchange membranes decreases as the degree of concentration increases. In addition, when the degree of concentration is high, the rate of backdiffusion is relatively high because of large concentration differences between the dilute and concentrated solutions. In general, higher-quality ion-exchange membranes are needed for concentrating electrolyte solutions than are needed for desalting saline water (Nishiwaki, 1972).

It was shown (Nishiwaki, 1972) that highly concentrated solutions can be obtained if the transference numbers are high, the diffusion coefficients of salt and water are low, the electroosmotic water transport numbers are low, and thick membranes are used. Thick membranes with high densities of ion-exchange groups and low water contents best achieve the foregoing desirable properties of high transference numbers, low diffusion coefficients, and low electroosmotic transport numbers.

To minimize electrical energy requirements for electrodialytic concentrations, the membranes should have the lowest possible electrical resistances. Thin membranes with high water contents are desirable to achieve low electrical resistances. Those properties needed for low electrical resistance are opposite to those needed to achieve high coulomb efficiencies and high degree of concentration. Therefore, in the development of ion-exchange membranes for the electrodialytic concentration process, compromises are necessary.

Before choosing the best membranes for electrodialytic concentration, the relationships between the costs for electrodialytic concentration and the costs for further evaporation should be studied so that the degree of electrodialytic concentration that results in lowest overall production costs can be selected. Once the desirable degree of electrodialytic concentration has been selected, the concentrations of feed solution to the process and brine from the process may be established. Then, a choice of the best ion-exchange membranes may be made by measuring transference numbers, electroosmotic water transport numbers, and diffusion coefficients of salt and water at the two solution concentrations of interest for the membranes of interest (Nishiwaki, 1972).

Strathmann (1992) (see Table 12.2) and Korngold (1984) described ion-exchange membranes with special properties. This information can also serve as guidelines to select membranes for certain applications.

The Ionac Company, for example, produces five types of heterogeneous reinforced membranes, two cation-exchange membranes, and three anion-exchange membranes (Korngold, 1984). The MC-3470 cation-exchange membrane and the MA-3475 anion-exchange membrane have excellent mechanical properties and display high stability in various chemicals, including chlorine. These membranes may still be classified as among the best on the market, although their electrical resistance is relatively high (especially in dilute solutions). The IM-12 anion-exchange membrane is produced in such a way that it is less sensitive to organic materials. In addition, it can be used, according to the manufacturer's specifications, at higher current densities than other membranes.

The American Machine and Foundry Co. (AMF) membranes have good chemical, electrical, and mechanical properties, but they undergo shape changes while in use because they are not reinforced, causing many technical problems (Korngold, 1984). This is especially true of the cation-exchange membrane.

Ionics membranes are homogeneous, reinforced, and have very good mechanical, chemical, and physical properties (Korngold, 1984). They are not sold separately but as part of a complete EDR installation. Ionics membranes have been supplied for about 25 years to some 500 installations the world over, establishing Ionics as a leader in this field.

The permselective membranes produced and marketed by Asahi Glass Co., Asahi Chemical Industry Co., and Tokuyama Soda Co. are fairly similar (Korngold, 1984). All are thinly reinforced and relatively low priced when sold in large quantities. These membranes have lower electrical resistance than the American counterparts but are mechanically weaker and vulnerable to damage under dry conditions. They may also be used for purposes such as separation by diffusion (especially between acid and salt) or ion exchange.

Neginst polyethylene membranes from Research and Development Authority, Ben-Gurion University of Negev, Beer-Sheva, Israel, are aliphatic and are, therefore, less sensitive to fouling (Korngold, 1984; Korngold et al., 1970). The form, stability, and mechanical properties of the reinforced membranes are very good. The electrical resistance is higher than that of the Japanese membranes and similar to that of the American membranes.

Membranes for acid concentration must have low proton leakage. Such membranes, for example, are supplied by Tokuyama Soda Co. and Asahi Glass Co. (Table 12.2) (Boudet-Dumy, Lindheimer, and Gavach, 1991).

12.3 DESIGN CRITERIA

Design criteria for ED/EDR applications are described by many authors in the literature (Huffman and Lacey, 1972; Rogers, 1984; Strathmann, 1992; Meller, 1984; Wilson, 1960). Raw water, pretreatment considerations, process limitations, process configuration, posttreatment, controls and instrumentation, and concentrate disposal will be considered in this section.

12.3.1 Raw Water

Fouling of ED/EDR membranes by dissolved organic and inorganic compounds in the raw water may be a serious problem in electrodialysis (Korngold et al., 1970; Lacey and Loeb, 1972; Van Duin, 1973) unless the necessary pretreatment is provided. Organic fouling is caused by the precipitation of large negatively charged anions on the anion-permeable membranes in the dialysate compartments, while inorganic fouling is caused by the precipitation (scaling) of slightly soluble inorganic compounds (such as $CaSO_4$ and $CaCO_3$) in the brine compartments and the fixation of multivalent cations (such as Fe and Mn) on the cation-permeable membranes. Organic anions or multivalent cations can neutralize or even reverse the fixed charge of the membranes, causing significant reduction in efficiency. Fouling also causes an increase in membrane stack resistance which, in turn, increases electrical consumption and adversely effects the economics of the process.

The following constituents occurring in raw water are, to a greater or lesser extent, responsible for membrane fouling (Katz, 1982):

1. Traces of heavy metals such as Fe, Mn, and Cu
2. Dissolved gases such as O_2, CO_2, and H_2S
3. Silica in diverse polymeric and chemical forms
4. Organic and inorganic colloids
5. Fine particulates of a wide range of sizes and composition
6. Alkaline earths such as Ca, Ba, and Sr
7. Dissolved organic materials of both natural and man-made origin in a wide variety of molecular weights and compositions
8. Biological materials—viruses, fungi, algae, bacteria—all in varying stages of reproduction and life cycles

Many of these foulants present in raw water may be controlled by pretreatment steps which usually stabilize the ED/EDR process. However, according to Katz (1982), the development of the EDR process has helped to solve the pretreatment problem more readily in that it provides self-cleaning of the vital membrane surfaces as an integral part of the desalting process.

12.3.2 Pretreatment

Pretreatment techniques for ED/EDR are similar to those used for RO (USAID Desalination Manual, 1980). Suspended solids are removed by sand and cartridge filters ahead of the membranes. Suspended solids, however, must be reduced to a much lower level for RO than for ED/EDR. The precipitation of slightly soluble salts in the standard ED process may be minimized by ion-exchange softening and/or reducing the pH of the brine through acid addition and/or the addition of a precipitation-inhibiting agent.

The filtration requirements depend on the size and shape of the particles (Lacey and Loeb, 1972). Colloidal and gelatinous particles and "stringy" particles give particular trouble because they adhere more tenaciously to the meshlike spacers within the solution compartments than do granular particles. It is usually good practice to remove particulate matter as completely as possible. Five-micron cartridge filters can remove particulate material effectively prior to ED/EDR treatment.

Organic matter is removed by carbon filters. Hydrogen sulphide is removed by aeration or oxidation and filtration. Biological growths are prevented by a chlorination-dechlorination step. The dechlorination step is necessary to protect the membranes from oxidation. Iron and manganese are removed by green sand filters, aeration, or other standard water treatment methods. The concentration levels of these ions in feed solutions should not be greater than 0.05 ppm (Lacey and Loeb, 1972). It may be desirable to remove traces of some heavy metals by treatment with H_2S or Na_2S and filtration. It has been suggested that multivalent metal and organic ions, and hydrogen sulphide, however, must be reduced to a lower level for EDR than for RO (Fraivillig, 1983).

For applications in which the feed solutions contain small amounts of tri- or tetravalent cations (such as iron, manganese, or thorium) in addition to the main ions that are to be transferred through the membranes, the multivalent ions should be removed (Lacey and Loeb, 1972). It has been found that multivalent cations can attach themselves to the fixed-negative charges within cation-exchange membranes in such a way as to partially, or completely, neutralize the negative charges. This neutralization of charge results in reduced values of transport numbers for other cations with an attendant decrease in coulomb efficiencies. DeKörösy (1968) found that not only charge neutralization but even charge reversal could occur with tetravalent ions. Apparently, the tetravalent positive ions attached themselves to the fixed negative charges in a manner that left at least one of the four positive charges unneutralized by fixed-negative charges so that the membrane matrix eventually acquired a net positive charge instead of a negative charge. The formerly cation-exchange membrane then becomes an anion-exchange membrane.

The overall requirements for pretreatment in ED/EDR may be somewhat less rigorous than for RO due to the nature of the salt separation and the larger passages provided (USAID Desalination Manual, 1980). In ED/EDR, the ions (impurities) move through the membranes, while in RO the water moves under a high pressure through the membranes while the salts are rejected. Salts with a low solubility can, therefore, more readily precipitate on spiral and hollow, fine-fiber RO membranes to cause fouling and to block the small water passages. Suspended solids can also more readily form a deposit. However, this might not be the case with tubular RO membranes. With the EDR process, precipitated salts in the brine compartments can be more readily dissolved and flushed out of the system using polarity reversal without the need for chemical pretreatment.

However, high removals of suspended solids, iron, manganese, organic matter and hydrogen sulphide are still critical to pretreatment of the feedwater (USAID Desalination Manual, 1980) if it contains the following ions: Fe > 0.3 mg/L, Mn > 0.1 mg/L, H_2S > 0.3 mg/L, free chlorine and turbidity > 2 NTU. In every case, of course, careful examination of the prospective water would be necessary to determine suitability and pretreatment.

A certain degree of fouling is, however, unavoidable. Membranes should be regularly washed or cleaned in place with dilute acid and alkali solutions to restore performance when required.

12.3.3 Process Limitations

There are a number of considerations that are important when the ED/EDR process is considered for water and effluent treatment. These considerations include:

1. Membrane polarization
2. Current efficiency

3. Scaling potential of the effluent for the membranes
4. Fouling potential of the effluent for the membranes
5. Energy consumption
6. Process operating variables

These considerations and accompanying process limitation will be discussed in this section.

12.3.3.1 Polarization. Polarization in electrodialysis is widely described in the literature (Korngold, 1984; Davis and Brockman, 1972; Hodgkiess, 1987; Meller, 1984; Rubenstein, 1990). Current density in ED/EDR can be increased until the current to transfer ions exceeds the number of ions available to be transferred (Meller, 1984). This point is called the *limiting current density*. Limiting current density is usually expressed as $(CD/N_d)_{lim}$, where CD is current density (the amount of current carried by a unit area of membrane surface) and N_d is normality of the demineralized outlet stream. This limit is a function of fluid velocity in the flow path, stream temperature, and types of ions present. While practically all the ions are transported through the membranes in ED/EDR by electrical transport, only about half of the ions arriving at the membrane surfaces from the bulk of the solution are carried by electrical transport. The remaining ions arrive at the membrane surfaces from the flowing stream as a result of diffusion and convection processes. As ions are electrically transferred from the demineralizing cell through the membranes, the concentration of ions in the demineralizing cell in the thin layer immediately adjacent to the membrane surfaces becomes depleted. As the current density is increased, the solution layer next to the membrane becomes so depleted in ions that the electrical resistance rises sharply. The increased resistance results in increased voltage, which eventually exceeds the breakdown voltage for water molecules causing them to dissociate, forming hydrogen (H^+) and hydroxyl (OH^-) ions. When such dissociation of water molecules occurs, the *polarization* point is reached. Transfer of hydrogen ions in the case of cation-exchange membranes and hydroxyl ions in the case of anion-exchange membranes becomes appreciable. The extent of transfer of hydrogen and hydroxyl ions depends on the ratio of the concentration of hydrogen ions to other cations at the surface of a cation-permeable membrane and the ratio of hydroxyl ions to other anions at the surface of anion-permeable membranes. Polarization thus occurs gradually as the voltage applied to a membrane cell (and, hence, the current density) is increased.

Polarization, as discussed here, occurs only in the demineralizing compartments, since it is in these cells that depletion is taking place. Polarization does not usually become significant at both membranes at the same time. When polarization becomes pronounced at an anion transfer membrane, hydroxyl ions are transferred into the concentrate stream, making it alkaline; the hydrogen ions remaining in the demineralizing cell from the dissociated water cause a decrease in the pH of the demineralized stream. Polarization of cation transfer membranes results in transfer of hydrogen ions into the concentrate stream, decreasing its pH and increasing the pH of the demineralized stream. Therefore, pH changes of the process streams can indicate polarization.

Operation at current densities sufficient to cause significant polarization can cause several inefficiencies (Meller, 1984):

• Electrical resistance per cell pair increases so that the energy consumption increases.

- Current efficiency decreases, since the transfer of hydrogen and/or hydroxyl ions is usually not the desired object of the operation.
- In the polarization region, each additional increment of current is less efficient in transferring the desired ions.

It has been observed that the limiting current density can be increased if the fluid temperature, fluid velocity, and/or solution concentration is increased.

In the design of commercial ED/EDR systems, 70 to 80 percent of the limiting current density is used as the maximum allowable operating current density. This provides a reasonable level of safety in system design. When dealing with very low TDS waters, higher operating currents relative to the limiting current densities are permissible.

Determination of Limiting Current Density.   Evaluation of operating conditions for a new product is done by determining the polarization curve. The product tank of the ED/EDR unit is filled with the product to be treated, and the brine and electrode tanks are filled with a 10 g/L sodium chloride solution. The water in the feed tank is passed through the membrane stack at an appropriate flow rate while the voltage is progressively increased and current measured at the steady state. A typical polarization curve is shown in Fig. 12.17.

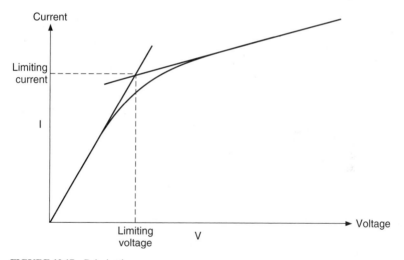

FIGURE 12.17   Polarization curve.

The operating point is the voltage corresponding to 70 to 80 percent of the limiting current.

12.3.3.2   *Current Efficiency.*   At the salt concentrations encountered in the practical application of the ED/EDR process, membranes do not exhibit ideal ion selectivity and the resulting deviation from ideal performance is expressed by the *coulomb* or *current efficiency.* This is the ratio of equivalents of electrolyte displaced to faradays of electricity passed (96,500 A-s) in the process. Coulomb efficiency expresses the performance of the ED/EDR process with respect to current. A low coulomb effi-

ciency implies poor selectivity, lower brine concentration, impurer product, and higher electrical energy consumption. Therefore, depending on the application, membranes with as high as possible current efficiency should be selected.

Determination of Current Efficiency. Current efficiency is measured as follows: fill the brine and feed tanks of an ED/EDR unit with 3 g/L each sodium chloride and the electrode rinse tank with 10 g/L sodium chloride. Measure the sodium chloride feed concentration before voltage is applied across the stack. Circulate feed, product, and brine through the stack at appropriate flow rates, apply constant voltage across the stack, measure current, and measure the sodium chloride concentration level of the desalinated product. The current efficiency η_c can then be calculated from the following formula:

$$\eta_c = \frac{26.8\,QN}{n_{pr}I} \tag{12.6}$$

where Q = product flow rate, m^3/h
        N = salt quantity removed, g eq/m^3
        I = applied current, A
        n_{pr} = number of membrane pairs

12.3.3.3 Scaling Potential of Water for ED/EDR Membranes. The scaling potential of a water for ED/EDR membranes is determined by evaluation of the *Langelier index* (LI) for calcium carbonate saturation and the calcium sulphate saturation of the water (Meller, 1984).

Langelier Index. The accumulation of Ca^{++}, TDS, and HCO_3^- in the concentrate stream resulting from purification of the demineralized stream can cause $CaCO_3$ scaling. The Langelier index indicates the scaling tendency for a specific water. This index is the difference between the actual pH and the pHs at which calcium carbonate scaling can occur with the particular composition of ions present. Thus, a positive index indicates a scaling tendency, while a negative index indicates a nonscaling tendency. Ionics ED systems use polarity reversal to control membrane scaling. A Langelier index is acceptable up to a value of +2.2 in an Ionics ED reversal system. Adjustments can be made in the system to decrease the Langelier index, which include decreasing the percent salt removal, decreasing the pH, or decreasing the percent water recovery.

Calcium Sulphate Saturation. Since the solubility of $CaSO_4$ is only slightly pH sensitive, control and prevention of $CaSO_4$ precipitation is brought about by limiting the waste-stream concentrations of calcium and sulphate. As with the Langelier index, this can be achieved by decreasing the percent salt removal or decreasing the percent water recovery. Chemical alternates include the injection of a sequestering agent in the feedwater or concentrate stream that inhibits precipitation of $CaSO_4$. Removal of Ca^{++} from the feed stream can also be done.

Fortunately, even high concentrations of $CaSO_4$ or the removal of Ca^{++} above recognized saturation levels can be handled satisfactorily in the EDR system. This is due to a time lag in $CaSO_4$ precipitation combined with the effect of polarity reversal. The upper design limit for the EDR system, without chemical addition is 175 percent $CaSO_4$ saturation. With chemical addition to the concentrate stream, this limit can be extended to the 300 percent saturation range in commercial water treatment plants. Pilot plant operations at over 400 percent saturation have been successfully demonstrated.

12.3.3.4   Oxidation of Membranes.   If substances are present in water that can cause oxidation of the membranes, such as dissolved halogens or nitric acid, they must be removed or rendered inactive (Lacey and Loeb, 1972). Most ion-exchange membranes are attacked by strong oxidizing agents, as indicated by the manufacturers' literature.

Anion-exchange membranes used in electrodialysis equipment are based mainly on styrene divinyl-benzene or acrylic chemistry. Membranes based on styrene divinyl-benzene have two major problems when used in equipment treating surface and wastewaters (Elyanow et al., 1994). The first problem is poor resistance to chlorine. Second, they become irreversibly fouled by organic materials. Acrylic-based anion-exchange membranes which were introduced in 1981 in EDR service overcome these problems and have superior performance.

Membrane desalination systems, with their small passages, are likely to have problems with bacterial growths when treating these waters if a disinfectant residual is not maintained. The acrylic-based anion-exchange membranes have substantial resistance to chlorine. Ion-exchange capacity data was shown by Elyanow et al. (1994) where both membrane types were operated with 10 mg/L of free residual chlorine in the feedwater. After 1000 h, the styrene divinyl-benzene-based membrane was seriously degraded. Systems using these membranes need residuals below 0.1 mg/L for long life. The acrylic-based membrane shows only relatively minor capacity loss after 4000 h of operation. With this resistance, EDR systems are operated with an average of 0.3 to 0.5 mg/L of free chlorine residual in the feedwater, which has proven to be very effective for controlling bacterial growth. Much higher residuals are used for shorter periods for sterilization and other purposes.

Alternate disinfectants that have been and can be used in ED applications include:

1. Choramines
2. Ozone
3. Hydrogen peroxide
4. Chlorine dioxide (Elyanow et al., 1994)

12.3.3.5   Particulate Matter.   If insoluble particulate matter is present in the feed solution, it must be removed by filtration or other means prior to ED/EDR. The removal of particulate matter is discussed under pretreatment (sec. 12.3.2).

Membrane processes, by their nature, have driving forces that tend to deposit colloidal material on the membrane surface (Elyanow et al., 1994). In RO, this is the flow of water perpendicular to the membrane. When the particles reach the surface, water flows through the membrane and other forces tend to hold the deposits in place (Elyanow et al., 1994).

Colloidal particulates interact with water to form an effective charge at the surface of their bound water layer. This charge is negative in natural and most wastewaters. The DC electric power field applied to electrodialysis stacks is a driving force that moves colloids toward the anion membrane. When the particles reach the membrane surface, the electric field and electrostatic attraction to ion-exchange sites in the anion-exchange membrane tend to hold the deposit in place (Elyanow et al., 1994).

EDR employs periodic reversal of the DC electric field. Typical field reversal frequencies range from 15 to 30 min. When the field is reversed, the electric driving force is reversed, which tends to remove deposited colloids into the brine stream.

Because the mechanisms for depositing and holding colloids on the membrane surface in EDR equipment depend on the degree of charge on the particles, precise limits for turbidity or SDI cannot be defined for EDR as a general limit (Elyanow et al., 1994). Colloidal fouling has never been experienced below a feedwater 5-min SDI of 12, while fouling is likely if the 5-min SDI is above 16. These values generally correspond to turbidities of 0.25 to 0.5 NTU. Exceptions do occur, as a few EDR plants have operated for sustained periods of several months with feedwater 5-min SDI values in excess of 19 without problems.

Spacer design helps to reduce colloidal fouling. The tortuous flow path spacers have cross straps which are needed to give the spacer physical integrity and are positioned to optimally promote turbulence, which helps minimize colloidal deposition.

Colloidal material in feedwater to an ED/EDR plant can best be removed with coagulation-flocculation and media filtration prior to ED/EDR treatment (Elyanow et al., 1994).

12.3.3.6 Fouling Potential of Effluent for ED/EDR Membranes. Organic material in effluents can foul ion-exchange membranes and membrane fouling can lead to process failure. Therefore, it is necessary to evaluate the fouling potential of an unknown effluent for ED/EDR membranes when ED/EDR is considered for treatment of such an effluent. Large organic anions are usually responsible for membrane fouling in ED/EDR and should be considered as a potential fouling mechanism (Korngold, 1984).

Electrotransport of Large Ions Through ED/EDR Membranes. The use of ED/EDR is generally restricted to small ions, such as chloride, sulphate, sodium, and calcium, because the electrochemical properties of permselective membranes are changed on their association with large ions. When these ions are present in the solution, the electrical conductivity and permselectivity of the membrane decreases. If large ions are driven by an electrical current into permselective membranes, they may become blocked there, poisoning the membrane.

Influence of the Molecular Mass of Ions on the Electrochemical Properties of the Membrane. Investigation of the permeability of anion-exchange membranes for carboxyalate anions showed (Dohno et al., 1975; Lightfoot and Friedman, 1954) that the permselectivity decreases gradually from 98 to 30 percent when the number of carbon atoms in the molecule is increased from two (MW = 59) to nine (MW = 171). The electrical resistance of the membrane increases with the rise in MW. The ED/EDR process can be used with solutes ranging in MW up to about 100 without significant changes in the electrochemical properties of the membrane.

Fouling and Poisoning by Large Ions. Materials that cause fouling and poisoning can be classified into three categories (Kobias and Heertjes, 1972; Korngold, 1973, 1976; Korngold et al., 1970; Kusomoto et al., 1973; Tamamushi and Tamaki, 1959; Van Duin, 1970, 1973).

1. *Organic anions that are too large to penetrate the membrane and accumulate on its surface:* Mechanical cleaning can restore the original electrical resistance of the membrane. Anions such as humates and algenates can precipitate on the anion-exchange membrane under polarization conditions in the form of humic acid and alginic acid, causing a sharp increase in electrical resistance. These precipitates can be dissolved with a dilute base, 0.1 N NaOH, and the original values of electrical resistance can be restored (Korngold et al., 1970).

2. *Organic anions that are small enough to penetrate the membranes but whose electromobility is so low that they remain inside the membrane, causing considerable* *increase of resistance:* Different kinds of detergents can cause this type of poisoning, and it is difficult to restore the original electrical resistance of membranes poisoned in this way (Korngold, 1976).

3. *Organic anions that are smaller than those of category 2, but still cause a certain increase in electrical resistance of the membrane:* These anions can be eluted by electroelution with sodium chloride, and the original properties of the membrane can be obtained.

Experiments conducted with dodecylbenzenesulfonate (DBS, molecular weight [MW] = 347.45), methyl orange (MW = 327.34), and congo red (MW = 696.67) with different commercial anion-exchange membranes illustrate the various types of behavior of permselective membranes with large ions (Korngold, 1976). DBS passed through all membranes. Methyl orange did not cross the Ionac and Neosepta membranes at all, while a small amount was transported through the Neginst HD-(I) membrane and a much larger amount through the highly swollen Neginst HD-(II) membrane. The electrical efficiency was low (less than 4 percent), and most of the current caused co-ion passage and water splitting rather than electrotransport of counter-ions. The congo red indicator did not pass through any of the membranes tested and did not cause significant changes in electrical resistance. At the end of the experiment with this indicator, slight precipitation was observed on the surface of the membrane and the precipitate was easily removed by mechanical cleaning (Korngold, 1976).

Rapid Method for Determination of the Fouling Potential of Industrial Effluents for ED/EDR Membranes. A simplified diagram of a fouling cell that can be used to evaluate the fouling potential of industrial effluents for ED/EDR membranes is shown in Fig. 12.18 (Schoeman, 1986).

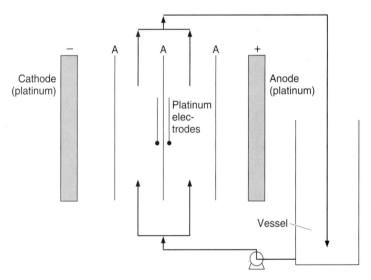

FIGURE 12.18 Simplified diagram of flow-through fouling cell. A = anion-permeable membrane.

The fouling apparatus consists of four perspex cells which can be clamped together to hold two membranes (7.1 cm² exposed area) in position. Approximately 100 L of test solution is circulated through the two middle compartments and a 0.1-M sodium chloride solution through the electrode compartments (not shown in Fig. 12.18). Where necessary, the conductivity of the test solution is increased by addition of a solution containing about 5 or 10 g/L NaCl. A DC current density of 5 to 20 mA/cm² membrane area is used to supply the motive force for ion migration. The voltage drop across the central membrane is measured with platinum electrodes connected to a voltmeter. An increase in potential drop across the membrane indicates fouling. The AC membrane resistance before and after fouling is measured in a 0.5 N NaCl solution. Foulant DBS concentrations are measured with a spectrophotometer at 223 nm.

Membrane Fouling with Different DBS Concentrations. The fouling of Selemion AMV (anion-permeable) membrane with different DBS concentrations is shown in Fig. 12.19. The AC membrane resistance ($\Omega \cdot cm^2$) before and after fouling is shown in Table 12.3.

The results (Fig. 12.19 and Table 12.3) demonstrate that, in this case, membrane fouling is highly sensitive to DBS concentration. Severe fouling was obtained with 100 and 50 mg/L DBS, while very little or no fouling took place with 10 mg/L DBS. A high fouling potential can therefore be determined quickly with the simple fouling test in the laboratory.

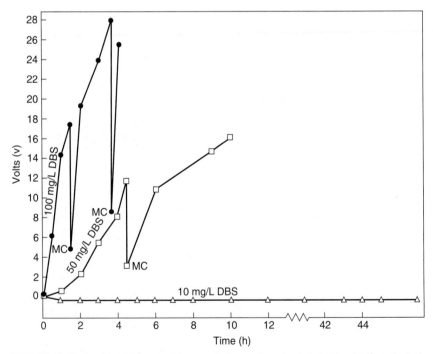

FIGURE 12.19 Effect of different DBS concentrations and mechanical cleaning (MC) on the fouling of Selemion AMV membranes. *(Schoeman, 1986.)*

TABLE 12.3 Membrane Resistance $\Omega \cdot cm^{-2}$, Before and After Fouling with Different DBS Concentrations

	DBS concentration in feed		
Fouling time, h	10 mg·L^{-1} DBS	50 mg·L^{-1} DBS	100 mg·L^{-1} DBS
0	58.0	58.0	58.0
1.5			95.5 (90.0*)
3.5			163.0 (138.0)
4.5		94.6 (93.6)	210.0 (191.0)
10		163.0 (162)	536.0 (376.0)
44	59.2		

* Values between parentheses show resistance values after membranes have been mechanically cleaned.
Source: Schoeman, 1986.

Mechanical cleaning (removal of surface fouling) had only a temporary effect on membrane resistance (Fig. 12.19) and the membrane resistance could not be reduced to what it originally was (Table 12.3). At the elevated concentration, both surface fouling and internal fouling took place. As fouling proceeded, membrane fouling occurred at approximately the same rate as before cleaning. DBS passed through the anion-permeable membrane to a substantial degree only at the lowest DBS concentration of 10 mg/L (Schoeman, 1986).

In this case, a situation in between categories **1** and **2** (mentioned before) occurs with the addition of surface accumulation at the higher concentrations. At the lowest DBS concentration, the DBS ions are relatively few compared to the inorganic ions. They may move through the membrane at a lower speed than the chloride ions, but there are too few DBS ions to impede overall ionic transfer. At an elevated DBS concentration, the fraction of the ionic paths containing the slow-moving DBS ions is so high that it has a marked effect on membrane resistance. The ionic flow through unobstructed paths increases but so does the entry of DBS ions into these "open" paths. The obstruction is thus accelerated. Furthermore, as more ions try to enter fewer "open" paths at the surface, a surface accumulation of the relatively large DBS ions occurs. In addition, it is likely that the presence of closely spaced DBS ions in a membrane pore may cause local structural (swelling) changes which decrease the mobility of ions in the area. The real fouling situation thus appears to be considerably more complicated than envisaged in Korngold's categorization.

12.3.3.7 Energy Requirements in Electrodialysis. The two main energy inputs in ED/EDR plants are (Hodgkiess, 1987):

1. The DC power supplied to operate the ED/EDR stack
2. The energy required to pump the process water through the plant

Although the latter is far from insignificant, it is the former which primarily dictates the circumstances in which electrodialysis is more or less competitive with other desalination processes and will be discussed in some detail here (Hodgkiess, 1987).

The DC Energy Requirement in an Electrodialysis Stack. The theoretical minimum energy required to operate an ED/EDR cell is obtained when the applied voltage is big enough to overcome potential differences which exist across the electrodes and membranes. Because of the large number of membranes in commercial units, the electrode potentials are very small in comparison to the sum of the membrane potentials; hence, emphasis is given to the impact of the latter on the energy consumption (Hodgkiess, 1987).

1. *Minimum energy for electrodialysis.* Consider the situation when no voltage is applied from the external DC source. In this case, a naturally occurring potential difference develops across an ion-selective membrane when the solutions on either side are of different concentration. This is because there is a diffusion-driven, spontaneous transport of ions through the membrane from the side containing salts at higher concentration to the compartment containing more-dilute solution. Since this ion movement will consist preferentially of either cations or anions, it will lead to a charge separation (i.e., a potential difference) across the membrane whose buildup will gradually oppose further charge transfer, and eventually an equilibrium will be established at a specific value of this potential difference which can be denoted by $E_{m(e)}$. The (e) in the subscript indicates that the membrane potential is the value obtained when the system is at equilibrium, i.e., with no net charge transfer occurring across the membrane.

As an indication of the parameters which govern the value of $E_{m(e)}$, for solutions consisting of a single salt behaving thermodynamically ideally and consisting of singly charged ions, we can express $E_{m(e)}$ as follows:

$$E_{m(e)} = -(t_{m+} - t_{m-}) \frac{RT}{F} \ln\left(c_{wc} / c_{wd}\right) \tag{12.7}$$

where t_{m+} and t_{m-} are the ion-transport numbers of cations and anions in the membrane and c_{wc} and c_{wd} are the salt concentrations in the bulk solutions on either side of the membrane.

The important point is that $E_{m(e)}$ arises from spontaneous transport of ions across the membrane from the concentrated to the more dilute solutions (i.e., in the wrong direction for desalination), and to reverse this natural direction of flow requires the application of a voltage of magnitude slightly greater than $E_{m(e)}$. In other words, $E_{m(e)}$ represents a potential drop which must be overcome by the external applied voltage in order for the ion flows to occur in the required direction for water purification. Moreover, the sum of the values of $E_{m(e)}$ for a cation-permeable membrane (CPM) and an anion-permeable membrane (APM), when multiplied by the total number of CPM/APM cell pairs in a particular ED/EDR stack, can be thought of as a measure of the minimum energy required for operation of the electrodialysis cell.

2. *Practical energy requirements.* In order to operate an ED/EDR stack at finite currents and, hence, finite desalinated water production rates, the required energy is considerably higher than the minimum values previously discussed. This is essentially because of two factors.

- The occurrence of IR-drops through the solution and membranes
- Increases in the values of the membrane potentials from the equilibrium (zero-current) values $E_{m(e)}$ previously defined

This is due to the phenomenon of concentration polarization which occurs for all finite cell currents and involves salt depletions and enrichments in the membrane boundary layers. The value of the membrane potential is determined partly by the salt concentrations at the membrane/solution interface. Concentration gradients in the boundary layers steepen with rising current density, and, vice versa, at zero current (equilibrium) the salt concentration at the membrane/solution interface is equal to the bulk solution value (hence, the use of the latter in Eq. 12.7). However, the use of finite currents necessitates the substitution of (c_{cmb}/c_{dmb}) inside the logarithmic term in Eq. 12.7 (where c_{cmb} and c_{dmb} represent the salt concentrations at the membrane/boundary layer interfaces on the concentrate side and diluate sides of the membrane, respectively, once concentration polarization is occurring. Since the value of (c_{cmb}/c_{dmb}) is clearly greater than (c_{wc}/c_{wd}), the values of the membrane potentials E_m in an electrodialysis cell operating at finite current will be greater than the equilibrium values $E_{m(e)}$.

Concentration polarization also causes a rise in the cell resistance because of the increased resistances in the depleted diluate boundary layers. As the cell current rises, the progressive steepening of the salt gradients in the boundary layers causes continuous increases in cell resistance and membrane potential. The associated increases in energy requirement become particularly severe as the limiting current density is approached, and the problems are accentuated by a decline in current efficiency as increments in current become increasingly used for moving H^+ and OH^- ions through the membranes. Consequently, the operating current densities in electrodialysis are usually kept significantly less than the limiting current density (say, $0.7\ i_{lim}$; but on occasions higher than this).

Taking these factors into account but ignoring the relatively small contribution to the stack energy associated with the electrode reactions, the following equation can be developed relating the DC electrical power consumption P to a number of design and operational factors (Hodgkiess, 1987).

$$P = \left[\frac{nFQ\Delta c}{\eta_c} \right]^2 \frac{R_{mT}}{n_{pr}A} + \left[\frac{nF\Delta c E_{mT}}{\eta_c} \right] \tag{12.8}$$

A is the membrane area, R_{mT} is the total resistance of one *cell-pair* (namely, one dilute and one concentrate stream plus one pair of membranes), and E_{mT} is the sum of the (membrane) potential differences across one CPM and one APM.

It is evident from Eq. 12.8 that P is a strong function of the amount of salt required to be removed (Δc). To give a rule-of-thumb indication of the energy used per kg of salt removed, data gleaned from the literature and obtained in the laboratory (Hodgkiess, 1987) suggest a figure of around 0.4 to 1 kWh/kg. This value will vary depending on the detailed mode of operation of the plant and also on the salinity of the feed. In another view, namely in terms of the DC energy per m³ of product water, the data in Table 12.4 show the trend towards increasing energy with degree of desalination carried out (Hodgkiess, 1987).

TABLE 12.4 Energy Consumption

Feed type/temp	Feed ppm	Product ppm	DC energy kWh/m³
River (12°C)	395	46	0.15
Na₂SO₄ (18°C)	1025	180	0.54
Well (15°C)	1170	89	0.55
Canal (19°C)	1609	130	0.78
Na₂SO₄ (18°C)	1900	625	1.0
Na₂SO₄ (18°C)	2750	1025	1.6
Na₂SO₄ (18°C)	3500	1425	2.3

Source: Hodgkiess, 1987.

Equation 12.8 also demonstrates that P increases with R_{mT}, with production rate Q and with decreasing current efficiency η_c.

It is therefore concluded that the desalination of high-salinity water such as seawater, to say potable water standards (around 500 ppm), is likely to be relatively expensive in energy costs. This is mainly because the Δc term in Eq. 12.8 trends toward lower current efficiencies in high-salinity water due to increased water transport membranes, and lower membrane permselectivities increase system energy requirements.

The use of electrodialysis alone to produce high-purity water is unlikely to be economically attractive because of its high resistivity. Inefficiencies in the DC-rectification equipment also impose an increment to the total DC power requirement of around 5 percent.

Pumping Energy. An electrodialysis plant requires a supply of pressurized feedwater generally at about 4 bar for pumping the feed/product water, concentrate stream, and electrode rinse solution through and around the plant. This varies with plant size and feedwater TDS from around 0.5 to 1 kWh/m³ of product water when operating on brackish feedwater of around 1000 ppm TDS to about 2 to 3 kWh/m³ for seawater.

Total Energy Consumption. Total energy consumption depends on the salinity of the feedwater and the total amount of desalination required. A very rough estimate of the energy requirement, for a given application, can be obtained from the following expression:

Total energy = 0.7 kWh/m³ of product water + 0.7 kWh/m³ per 1000 ppm salt removed

Under ambient-temperature conditions and assuming product water of 500 ppm TDS, the typical total energy consumption would be around 1.5 and 3.5 to 4 kWh/m³ for feedwaters of 1500 and 3500 ppm, respectively. For higher-salinity water, the energy consumption rises significantly to perhaps 7 to 10 kWh/m³ with 11,000-ppm TDS feedwater. However, these figures are indicators only, for the energy requirement also depends on the detailed design, operation, and production rates required from the plant. Some specified performance data for one small EDR plant are presented in Table 12.5 to illustrate this point of the trade-off of production rate and energy consumption.

TABLE 12.5 Performance Data for Small EDR Plant

Feed TDS (ppm)	3500		10,000	
Number of stages	4	6	8	10
Product m³/day	5.7	2.8	1.5	0.85
Energy consumption kWh/m³	3.7	2.7	9.0	6.3

Source: Hodgkiess, 1987.

At very high salinities, the increases in energy consumption do not derive only from the concentration term in Eq. 12.8. The phenomenon of water transport through membranes can mean that, when desalting seawater, 15 to 20 percent of the seawater fed to the product channels is transferred to the concentrate. This and other factors such as increased backdiffusion will lead to much lower current efficiencies (perhaps down to 70 percent) than those (around 90 percent) when desalting brackish water.

The energy consumption can be reduced by increasing the membrane area. One source in the literature quotes a reduction from 3.5 to 1.3 kWh/m³ accompanying an increase in membrane area of 1 to 2.6 m² when desalting a feed of 3000 ppm down to 500 ppm in a large, 5000-m³/day plant.

Elevated-Temperature Electrodialysis. The electrical conductivity of saline water is very dependent on temperature, approximately 2 to 3 percent per degree. This is because both the degree of ionization and the mobility of the ions increase, resulting in a decrease in both solution and membrane resistance. Thus, the voltage (hence, energy) requirement for a given current (i.e., given desalting rate) decreases with increasing temperature (Hodgkiess, 1987).

Other possible benefits from elevated-temperature operation arise from the associated decreased viscosity, which implies lower pumping costs and thinner membrane boundary layers. This latter factor, taken together with the increase in salt diffusion coefficients as the temperature rises, points to increased limiting current densities at higher temperatures (Hodgkiess, 1987). For instance, some research work has indicated that the limiting current density increases by 1 to 4 percent for every degree of temperature increase.

An illustration of some of the aforementioned benefits for the simple case of desalination of 2000 ppm NaCl solution is presented in Fig. 12.20. The increased limiting current at 35°C (95°F), as against 20°C (68°F), is clearly evident, and comparison of the voltages for a given current indicates a power reduction of about 25 percent secured by the 15°C (59°F) increase in temperature. These results were obtained in much simpler (laboratory) circumstances than will prevail in commercial practice and, about a 1 percent decrease in power consumption might be expected with each one-degree rise in operating temperature. Other research work has indicated that the energy requirement can be reduced by 60 to 70 percent when

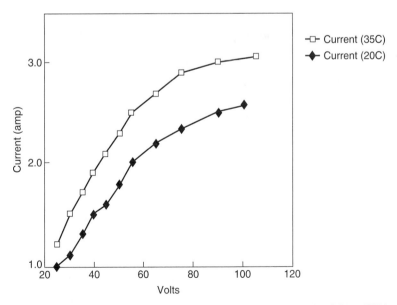

FIGURE 12.20   Effect of temperature on electrodialysis performance. *(Hodgkiess, 1987.)*

the operating temperature is raised from ambient to 70°C (158°F). An alternative way of utilizing the benefits of higher temperature operation is to secure a reduction in capital cost by decreasing the specific membrane area.

However, there are counteracting disadvantages associated with elevated temperature operation (Hodgkiess, 1987):

1. Deterioration of the spacers and the membrane: these are polymeric materials which are particularly susceptible to thermal degradation at rather moderate temperatures. With respect to the membrane, both the active groups and the matrix itself have to be stable.

2. Increased scale tendencies because of the inverse solubility/temperature characteristics exhibited by the important low-solubility compounds.

The maximum operating temperatures in projected "high-temperature" electrodialysis units are limited to about 50 to 70°C (122 to 158°F) largely due to factor 1. Advances in polymer technology might enable this temperature limitation to be raised in the future. Current commercial ED/EDR units are restricted to an upper temperature of about 45°C (113°F) due to loss of rigidity at higher temperatures of the presently used, low-density polyethylene spacer materials.

12.3.3.8 Operating Variables. Operating variables that should be taken into account when evaluating ED/EDR (small-scale units) for applications will be described in this section (Ionics Inc). Independent, semi-independent, and dependent operating variables will be considered.

Independent Operating Variables

1. *Feed concentration of diluting streams.* Sometimes, the stream to be demineralized contains nonionized solutes as well as ionized solutes. For a given ratio of nonionized to ionized solutes, the electrical conductivity of such solutions usually reaches a maximum value as the concentration (the amount of total solids, ionized and nonionized) increases, after which the conductivity decreases because of viscosity effects. It is usually desirable to electrodialyze at the concentration of maximum conductivity because less DC energy is required, and the limiting current density is higher.

However, if the optimum concentration must be achieved by preconcentrating, such preconcentrating may involve added energy costs. If the optimum concentration must be achieved by prediluting, subsequent reconcentrating may be required at added energy costs, and more volume must be handled at added pumping costs. Therefore, the feed concentration should be optimized if possible.

2. *Feed concentration of concentrating stream.* The concentrating stream should be as concentrated as possible without impairing membrane efficiency due to backdiffusion, without causing an electrical short circuit through the manifolds, and without giving rise to precipitation of the transferred ionic species. Precipitation sometimes can be circumvented by pH control and by adding supersaturation-promoting chemicals.

3. *Electrode stream feed composition.* An electrode stream consisting of 0.1 to 0.2 mol/L Na_2SO_4, acidified to pH 2 to 3 to prevent scaling at the electrodes is suggested. The composition of this stream is of no immediate significance in laboratory evaluation of standard diluting-concentrating applications of ED/EDR.

4. *Temperature.* The higher the operating temperature, the lower will be the electrical resistance of the system and the viscosities of the fluids and, therefore, the

lower will be the power and pumping requirements. However, membrane life probably decreases and membrane replacement cost therefore probably increases with increasing temperature. In some cases, preheating the fluids will require an additional expense.

In general, operation of Ionics' standard Stackpacks at temperatures above 50 to 60°C (122 to 140°F) is to be avoided.

5. *Flow rate.* In relatively dilute solutions, when current density is a critical factor, the higher the flow rate, the higher will be the permissible current density. The following approximate relation holds:

$$i_{\lim} = aQ^{0.5-0.6} \tag{12.9}$$

where a is a constant and Q is flow rate.

For a given overall demineralization, area is inversely proportional to current density. Therefore,

$$A = \frac{b}{Q^{0.5-0.6}} \tag{12.10}$$

where b is a constant and A is area.

The higher the flow rate, the lower will be the total required area. The maximum permissible flow rate is limited only by the pressure drop through the stack. However, it sometimes is more economical to operate below the maximum allowable pressure drop because of pumping energy costs.

6. *Membranes.* An ED/EDR stack is shipped with standard cation and anion membranes. Modifications of these standard membranes are available which will give lower or higher osmotic liquid transfer or loss. However, this is coupled with higher or lower electrical resistance, respectively. Membrane backing can also be selected to meet special chemical resistance requirements in some cases.

Semi-Independent Variables

1. *Current density.* Current density is defined as current per unit area of available membrane through which the current passes. In a 9-in × 10-in (228.6-mm × 254-mm) stack, the available area is 220 cm^2, so that $i = l/220$, where i = current density (m/A/cm^2) and l = current (mA) (note that 1 mA/cm^2 = 0.929 A/ft^2).

The higher the applied current density, the lower the amount of cell area that will be required to achieve a specified degree of ion transfer; therefore, the lower the capital cost and membrane replacement cost. On the other hand, the energy cost will be higher because the voltage varies in proportion to the current density. For each system, there is an optimum current density where the inversely varying costs balance to a minimum. For every system there also is a maximum allowable or limiting current density, and it sometimes happens that this limiting current density is lower than the optimum. Consequently, limiting current density is the first operating variable that should be established experimentally.

For a given system of fixed concentration, limiting current density increases with increasing flow rate about to the 0.5 to 0.6 power, i.e.,

$$\left[\frac{i\ 1}{i\ 2} = \left(\frac{Q_1}{Q_2} \right)^{0.5-0.6} \right] \tag{12.11}$$

and increases about one percent for every degree Fahrenheit increase in temperature (*Note:* these are rules of thumb which should be verified for every system). At any given flow rate and temperature, the limiting current density depends strongly on the nature of the solutions being treated and on their concentration. In general, the ratio of limiting current density to average conductivity of the concentrating stream and the diluting stream is a constant.

2. *Current density in concentrated solutions.* Since the ratio of limiting current density to conductivity is fairly constant, it follows that the more concentrated a solution is in ions, the greater the permissible current density. In solutions having a concentration above 0.1 to 0.5 normal, too high current densities may cause membrane overheating and excessive power consumption. In these cases, the maximum current density may be limited by a limiting voltage. Current densities above 50 to 150 mA/cm^2, depending on the solutions, are not recommended except under special conditions. Note that the standard rectifier supplied by Ionics with its 9-in × 10-in (228.6-mm × 254-mm) laboratory stack has a maximum rating of 30 A. This allows operation up to $i = 135$ mA/cm^2.

3. *Current density in dilute solutions.* Since the electrical current is carried by the ion, it follows that the more dilute the solution, the lower the current transmittal capacity of the solution will be. Application of too much voltage will result in ion starvation at the membrane solution interface and resultant decomposition of solvent and/or other materials that may be present. This phenomenon is referred to as *polarization* (see sec. 12.3.3.1) and manifests itself through pH disturbances in the solution, loss of current efficiency, and increase in electrical resistance. If pH-sensitive substances are present in the solution, decomposition, precipitation, or coagulation may result in damage to the product and/or to the membranes.

In solutions having a concentration below 0.1 to 0.5 normal, polarization usually occurs at ratios of current density to conductivity between 2 and 12 (A-Ω/cm), with the lower end of the range holding for viscous solutions and/or solutions containing large, slow-moving ions.

Dependent Variables

1. *Product-stream concentration.* The concentration of the streams after one pass through the demineralizer is a function of the flow rate, current density, and current efficiency.

As demineralization progresses, any nonelectrolyte present will become concentrated due to electro-endosmotic transfer of water. The amount of water transferred is a function of the total current passed and of the "tightness" of the membranes. A typical range for water transfer is 200 to 400 mL per equivalent of electrical current, depending on the ionic concentration.

2. *Membrane-area requirement (plant size).* From the foregoing, it follows that the total area required to achieve a specified degree of ion-transfer depends on the concentration of the product stream after one pass. If one pass does not yield the required transfer or *cut,* multiple passes are needed. On laboratory as well as on production scale, this can be accomplished by batch operation or series staging.

3. *Voltage.* Required voltage depends on the electrical resistance of the stack and on the current density. In actual practice, the voltage applied to the stack is varied manually, and the resultant current is shown on an ammeter. Calculations then show what voltage to apply in order to obtain the desired current density.

There exists a maximum permissible voltage in design considerations for larger stacks. This limiting voltage varies depending on the solution to be processed. As a first approximation, it is recommended that the following voltage not be exceeded:

- For solutions having an electrical resistance of 25 Ω-cm or greater:

$$E_{\max,64} = (1.4 + 0.02R_s)n_{\text{pr}} \tag{12.12}$$

- For solutions having an electrical resistance below 25 Ω-cm:

$$E_{max,64} = (0.9 + 0.04R_s)n_{pr} \qquad (12.13)$$

where R_s = resistance of most conductive stream (excluding electrode stream), Ω-cm

n_{pr} = number of cell pairs

E_{64} = voltage across n cell pairs (excluding electrodes) at 64°F

At temperatures greater than 64°F, the allowable voltage becomes:

$$E_{max} = \frac{E_{max,64}}{1 + (0.01)(t - 64)} \qquad (12.14)$$

and at temperatures below 64°F, the allowable voltage becomes:

$$E_{max} = \frac{E_{max,64}}{1 + (0.01)(64 - t)} \qquad (12.15)$$

where t = operating temperature, °F.

4. *Pumping power.*   Pumping power requirements depend on the viscosity and the flow rate of the solutions and on the number of required passes.

5. *Membrane life.*   The life of the membranes in commercial electrodialysis plants varies anywhere from several weeks to many years. A relatively short life provides a high membrane-replacement rate that can be tolerated when treating valuable chemicals. A longer life and a lower membrane replacement rate is desirable when treating lower-value chemicals.

An important factor in membrane life is the chemical aggressiveness of the solution being processed. Since membrane deterioration due to chemical attack is a chemical reaction phenomenon, it is possible that membrane life is reduced at elevated temperatures.

Sometimes the electrical resistance of the membranes will increase rapidly. This is evidenced by a faster decrease in current after the start of a run that could be explained by the increase in resistance of the stream being demineralized. This usually is due to *fouling* by materials present in the solution being demineralized and results in a buildup of these materials on the membranes. Such materials must be preeliminated from the solution.

Sometimes membranes will show a gradual increase in their electrical resistance without visual signs of deterioration or material buildup. In those instances, original membrane properties can sometimes be restored.

12.3.4   Process Configuration

Different process configurations can be used in the design of ED/EDR systems. These configurations will be presented in this section.

12.3.4.1   Stack Design.   A typical flow diagram for a two-stage ED/EDR plant for concentrating or depleting electrolytes in solution is shown in Fig. 12.21 (Huffman and Lacey, 1972). Solution to be fed to the concentrating compartments of the ED/EDR stacks (assumed to be free of gross particulate matter) is stored in the concentrate feed tank. It is pumped from the feed tank through a filter and into the concentrating compartments of the first-stage ED/EDR stack. From there it goes through the concentrating compartments of the second-stage stack. The flow path

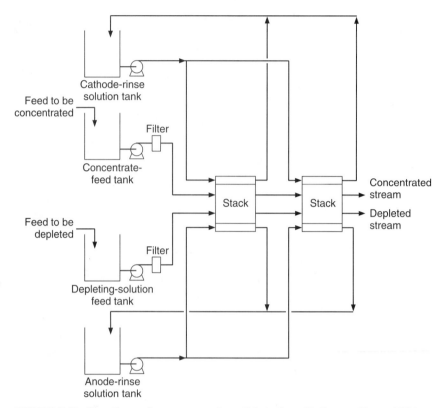

FIGURE 12.21   Flow diagram for a two-stage electrodialysis plant. *(Huffman and Lacey, 1972.)*

for the solution to be depleted of electrolytes is similar except that the depleting stream is pumped through the depleting compartments of the first- and second-stage ED/EDR stacks. In the ED/EDR stacks, ions are transferred through the cation- and anion-exchange membranes, as previously described to effect depletion of the electrolytes in the solution in one set of compartments and concentration of the electrolytes in the alternate compartments.

Anode and cathode rinse solutions, which may be the solutions being treated or specially prepared solutions of various kinds, are pumped from the electrode rinse solution tanks through the electrode compartments of the stacks and returned to the proper tank. The pH of each of these solutions is usually maintained at a value that will counteract the bases produced by the electrochemical action at the cathodes or the acids produced at the anodes so that H^+ or OH^- ions do not enter the concentrating and depleting compartments of the ED/EDR stacks. The concentrated and depleted streams leaving the second-stage stack may be pumped to product storage tanks if the desired amount of concentration or depletion is achieved, or to further stages of ED/EDR if further concentration or depletion is desired.

In addition to the conventional ED/EDR process, there are several other electromembrane processes that make use of various combinations of nonselective membranes, bipolar electrodes, and cation- and anion-exchange membranes. The flow schemes used with some of these processes are described by Ahlgren (1972a, 1972b).

Each ED/EDR system is designed for the particular needs of the application. The capacity of the system determines the size of the ED/EDR unit, pumps, piping, and stack size. The fraction of salt to be removed determines the configuration of the membrane stack array. The purpose of staging is to provide sufficient membrane area and retention time to remove a specified fraction of salt from the demineralized stream. Two types of staging are used: *hydraulic staging* and *electrical staging.*

Hydraulic Staging.   Typically, maximum salt removal for any hydraulic stage is 55 to 60 percent with normal design values at 40 to 50 percent (Meller, 1984). To increase the amount of salt removed in an ED/EDR system, additional hydraulic stages must be incorporated. In systems where high capacities are required, additional hydraulic stages are made by simply adding more stacks in series to achieve the desired water purity (Fig. 12.22). Note that in this arrangement each stack has only one electrical stage (one anode and one cathode).

In systems where additional hydraulic stages are incorporated within a single membrane stack, one or more interstage membranes are used. This membrane is a heavy cation membrane with all of the properties of the regular cation membrane. However, it is made twice as thick (1.0 mm) as the normal cation membrane (0.5 mm) to withstand a greater hydraulic pressure differential than that of a normal membrane. The heavy cation membrane has only two manifold cutouts, as shown in Fig. 12.23, in contrast to the four manifold cutouts in a regular membrane (dashed lines).

The heavy cation membrane is included as one of the components which makes up a cell pair since it performs the same ion transfer functions as the regular cation membrane. A hydraulic stage is formed by placing the heavy cation, or interstage membrane, at an appropriate place in the membrane stack. The two manifold openings are placed over the stack outlet manifolds to achieve the effect shown in Fig. 12.24.

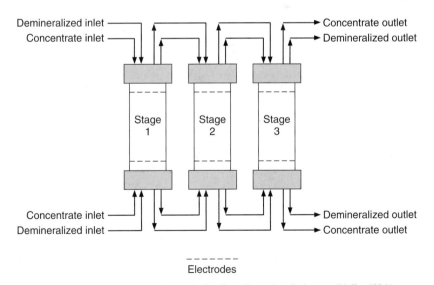

FIGURE 12.22   Each stage contains one hydraulic and one electrical stage. *(Meller, 1984.)*

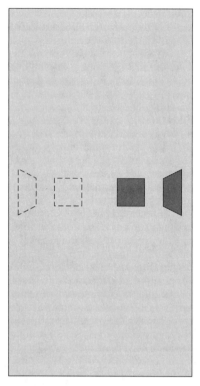

FIGURE 12.23   Heavy cation membrane with two manifold cutouts. *(Meller, 1984.)*

An example will further illustrate hydraulic staging (Meller, 1984). If the inlet water is 2000 ppm and the desired product water salinity is 250 ppm, three hydraulic stages would be required, assuming 50 percent salt removal per stage. The salt levels reduced to each stage would be 1000 ppm (Stage 1), 500 ppm (Stage 2), and 250 ppm (Stage 3).

The salt removal from a given volume of water is directly proportional to the current and inversely proportional to the flow rate through each cell pair. Higher currents will transfer greater amounts of salt. Higher flow rates will cause lesser amounts of salt to be transferred from a given amount of water due to shorter retention time in the membrane stack.

If the three hydraulic stages were contained within one electrical stage, whereby the current and, hence, the salt removed remained constant for every cell pair, the flow rate per cell pair would have to be doubled for each successive stage to obtain the decreasing ppm exhibited in the example. That is, the number of cell pairs in Stage 2 would have to be half the number in Stage 1, etc. This requires the flow rate in Stage 2 cell pairs to be twice that of Stage 1 cell pairs, and the flow rate through Stage 3 cell pairs to be twice that of Stage 2 cell pairs. Since the total flow into each hydraulic stage is identical, the only way to increase the flow rate per cell pair in successive stages is to decrease the number of cell pairs in those stages.

However, a hydraulic problem may prevent this arrangement. By decreasing the number of cell pairs in a hydraulic stage and increasing the flow rate per cell pair, the pressure drop through that stage is increased. At some point, the pressure drop through the entire stack will exceed the stack inlet hydraulic pressure limit of 50 lb/in^2 (3.4 bar). If, in this example, the first hydraulic stage had 200 cell pairs and required 5 lb/in^2 (0.34 bar) to force a given volume of water through the stage, stages 2 and 3 would require the following:

Stage 2:   Only one-half the amount of salt is removed in the second stage as is removed in the first stage; the flow rate (velocity) must be increased by a factor of two, thus doubling flow of water per unit of time and increasing the required pressure to 15.6 lb/in^2 (1.06 bar). This is accomplished by decreasing the number of cell pairs by a factor of 2 to 100.

Stage 3:   In Stage 3 the number of cell pairs would be 50 to again double the flow rate. However, 49.4 lb/in^2 (3.36 bar) would now be required to force the same volume of water through Stage 3.

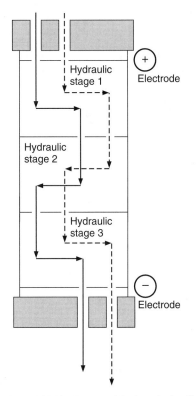

FIGURE 12.24 System with three hydraulic and one electrical stage. *(Meller, 1984.)*

The total pressure drop in the stack, therefore, would be 70 lb/in^2 (5 + 15.6 + 49.4) or 4.76 bar. Having exceeded the limit of a 50 lb/in^2 (3.4 bar) stack inlet pressure, an additional type of staging, known as *electrical staging,* must be incorporated.

Electrical Staging. Electrical staging is accomplished by inserting additional electrode pairs into a membrane stack. This gives flexibility in system design, providing maximum salt removal rates while avoiding polarization and hydraulic pressure limitations. Each electrical stage allows the use of an independently controlled current (I) to the cell pairs within that stage. In the previous example, if two electrical stages are used, there would now be two independent currents, I_1 and I_2. Suppose the stack is arranged as shown in Fig. 12.25.

Hydraulic Stages 1 and 2 are now in electrical stage 1, and hydraulic stage 3 is in electrical stage 2. With the ability now to introduce a lower current (I_2) to hydraulic Stage 3, additional cell pairs can be added to reduce the pressure drop (Δp) to an acceptable level. By using 100 cell pairs in hydraulic stage 3, the pressure drop would be 15.6 lb/in^2 (1.06 bar) in Stage 3 and would give a total pressure drop for the stack of 36.2 lb/in^2 (5 + 15.6 + 15.6) or 2.46 bar, which is well within the acceptable limit of 50 lb/in^2 (3.4 bar).

The concept of staging leads to great flexibility in system design with standard components. If 75 percent demineralization is required, then sufficient stages are installed to give the required 75 percent. If more stages should ever be needed, then additional stages can be added. This may be desired, for example, if it is found that a brackish aquifer becomes more saline with time, if it is desired to use the system with a more saline aquifer at the same or a different location, or if it is desired to produce a higher-purity product.

Pressure Drop/Water Flow Spacers. The pressure drop through a membrane stack is the sum of the pressure drop through each hydraulic stage, which is dependent on the spacer type, the flow rate per stage, and the number of cell pairs in each stage. Each of the Ionics tortuous-path spacers has differing pressure velocity profiles. To calculate the water velocity in the stack, the conversion equations from Table 12.6 are utilized (Meller, 1984).

The pressure vs. velocity curves (Fig. 12.26) predict the pressure drops for the various spacer models through a single hydraulic stage once the velocity in that stage has been calculated.

Differential Pressure. Differential pressure is the difference in hydraulic pressure between the demineralized stream and the concentrate stream. During normal

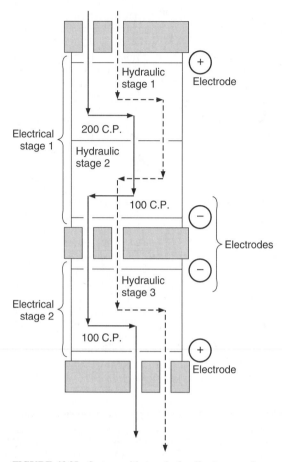

FIGURE 12.25 System with two hydraulic stages and one electrical stage (first electrodes) and one hydraulic stage and one electrical stage (second electrodes) or three hydraulic stages and two electrical stages. *(Meller, 1984.)*

membrane stack operation, the pressure of the demineralized stream is maintained at 0.5 to 1 lb/in^2 (10 to 30 in water; 254 to 762 mm water) higher than the pressure of the concentrate stream (Meller, 1984). The purpose of maintaining a higher demineralized stream pressure is to ensure that if internal stack leakage (cross-leakage) occurs between the demineralized and concentrate manifolds, the demineralized stream will leak into the concentrate stream. If the reverse were to occur, the result would be contamination of the demineralized stream. Measurement of the differential pressure is done at the inlet manifold piping.

Water Transfer. Some water is electrically (but not hydraulically) transferred through the membranes along with the ions. The amount of water transferred varies with the membrane type and solution concentration. Typically 0.5 percent of the demineralized stream flow is transferred per 1000 ppm salt removed. This represents the limiting water loss from an ED/EDR system (Meller, 1984).

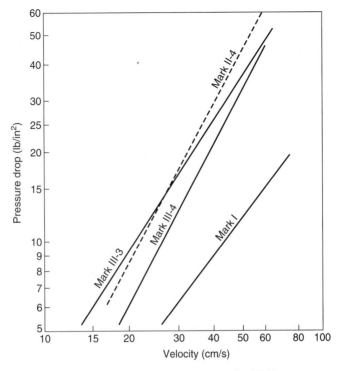

FIGURE 12.26 Pressure-versus-velocity curve. *(Meller, 1984.)*

Temperature Limits. Ionics membrane stacks have an upper operating limit of 46°C (115°F) (Meller, 1984). This limit is determined by the low-density polyethylene spacer material which tends to lose its rigidity at elevated temperatures. The anion and cation transfer membranes can withstand higher temperatures, generally in the range of 60 to 70°C (140 to 158°F). Higher-temperature spacers have been developed but are not now in standard commercial use.

Distance Between Membranes. The distance between the membrane sheets (i.e., the cell thickness) should be as small as possible since water has a relatively high electrical resistance (Strathmann, 1992).

TABLE 12.6 Conversion Equations
for Different Spacer Modules

Spacer model	Equation
Mk III-4	$V = 76.5 \times \text{gpm/cp}$
Mk III-3	$V = 100.5 \times \text{gpm/cp}$
Mk II-4	$V = 153 \times \text{gpm/cp}$
Mk I	$V = 1035 \times \text{gpm/cp}$

Where V = velocity, cm/s; gpm/cp = demineralized inlet flow rate in US gal/min per cell pair.
Source: Meller, 1984.

In industrial-size electrodialysis stacks, membrane distances are typically between 0.5 and 2 mm (Mintz, 1963). A spacer is introduced between the individual membrane sheets both to support the membrane and to help control the feed solution flow distribution. The most serious design problem for an ED/EDR stack is that of assuring uniform flow distribution in the various compartments. In a practical electrodialysis system, 200 to 1000 cation- and anion-exchange membranes are installed in parallel to form an electrodialysis stack with 100 to 500 cell pairs (Lacey, 1966).

Solution Flow Velocities. Solution flow velocities in sheet-flow stacks lie typically between 5 and 10 cm/s, whereas in tortuous-path stacks solution flow velocities of 15 to 50 cm/s are used (Strathmann, 1992). Because of higher flow velocities and longer flow paths, higher pressure drops of the order of 2 to 3 bar are obtained in tortuous-path stacks than in sheet-flow systems where pressure drops of 1 to 2 bar occur. However, higher velocities help to reduce the deposition of suspended solids and biological materials. Several other stack concepts are described in the literature; most of them provide three or four independent solution flow cells, which are used, for example, in combination with bipolar membranes.

Membrane Area, Membrane Life, and Cost. For typical brackish water of approximately 3000 ppm total dissolved solids (TDS) and an average current density of 12 mA/cm^2, the required membrane area for a plant capacity of 1 m^3 of product per day is about 0.4 m^2 of cation-exchange membranes and 0.4 m^2 of anion-exchange membranes (Strathmann, 1992). Other items such as pumps, piping, and tanks, do not depend on feedwater salinity but on plant size. For desalination of brackish water with a salinity of approximately 3000 ppm, the total capital cost for a plant with a capacity of 1000 m^3/day will be in the range of US$200 to $300/m^3 per day capacity. The cost of the actual membrane is less than 30 percent of the total capital costs. Assuming a useful life of 5 years for the membranes (up to 7 years is common for many brackish water applications) and 10 years for the rest of the equipment, a feedwater salinity of 3000 ppm, and a 24-h operating day, the total amortization of the investment is about US$0.10 to $0.15/m^3 of potable water with a salinity of less than 500 ppm. The operating costs are determined mainly by the required energy. For some applications that have high brine concentrations, the current density may be limited by the manifold shorting, resulting in a requirement for more membrane area to achieve equivalent desalination.

12.3.4.2 Electrodialysis Stacks and Criteria for Design of Stacks.

Most of the equipment used in electromembrane processes are standard items commonly used in the process industries, such as pumps, tanks, or filters. However, the membrane stack is unique. An electromembrane stack is essentially a device to hold an array of membranes between electrodes in such a way (Fig. 12.27) that the streams being processed are kept separated (Huffman and Lacey, 1972).

Figure 12.27 is an exploded view of part of an electromembrane stack. Component 1 in Fig. 12.27 is one of the two end frames, each of which has provisions for holding an electrode and introducing and withdrawing the depleting, the concentrating, and the electrode-rinse solutions. The end frames are usually made relatively thick and rigid so that pressure can be applied easily to hold the stack components together. The inside surfaces of the electrodes are recessed so that an electrode-rinse compartment is formed when an ion-exchange membrane, component 2, is clamped in place. Components 3 and 5 are spacer frames. Spacer frames have gaskets at the edges and ends so that solution compartments are formed when ion-exchange membranes and spacer frames are clamped together.

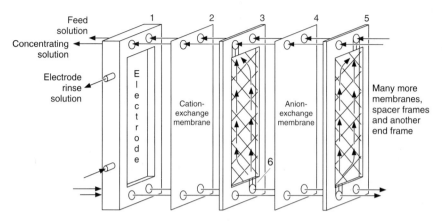

FIGURE 12.27   Exploded view of components in an electromembrane stack. *(Huffman and Lacey, 1972.)*

Usually, the supply ducts for the various solutions are formed by matching holes in the spacer frames, membranes, gaskets, and end frames. Each spacer frame is provided with solution channels (six in Fig. 12.27) that connect the solution-supply ducts with the solution compartments. The spacer frames have mesh spacers, or some other device, in the compartment space to support the ion-exchange membranes so that they cannot collapse when there is a differential pressure between two compartments. In newer gasket-spacer design, the spacer is melt into the gasket and very thin solution compartments are formed.

An electromembrane stack usually has many repeated sections, each consisting of components 2, 3, 4, and 5 with a second end frame at the end.

The final design of a stack is almost always a compromise among a number of conflicting criteria and considerations. Some criteria for the design or selection of electromembrane stacks have been given by Wilson (1960); Lacey and Lang (1969); and Schaffer and Mintz (1966). Some of the general criteria to be considered in the selection or design of an electromembrane stack are discussed under three general classifications: mechanical, hydrodynamic, and electrical (Huffman and Lacey, 1972).

Mechanical Criteria.   The maximum amount of the area of the membranes should be utilized for demineralization, because the cost of membranes constitutes a percentage of the total cost of an electromembrane stack. Maximum utilization of membrane area requires that the minimum possible membrane area be obscured in the edge seals and solution-distribution devices.

The stacks should permit quick and easy assembly and disassembly. The dimensions of the membranes should be limited to a size that can be handled easily by one or two men. There should be as few separate components as possible and the components should be designed so that the chance for incorrect assembly is minimized.

The gasket material should have negligible creep to prevent distortion of the gaskets and the resulting misalignment of solution entrance and exit channels. The gaskets should be capable of providing positive sealing with low compression force so that only a simple, low-cost means of clamping is required. The combined requirements for gaskets permit several alternatives:

1. Hard gaskets, with no creep, if the variations in thickness are small

2. Hard gaskets with no creep that have resilient faces

3. Hard gaskets with no creep, with provision for a resilient *line-seal* type of seal similar to O-rings

The solution-distribution design for each stream should ensure equal solution flow through each compartment. Some electromembrane stacks have performed well in this respect in initial operations, but later failed because of compression and creep of the solution-distribution devices. The solution-distribution channels should not be easily blocked by small particles in the solution being treated.

The spacer screens should provide closely spaced support points for the membranes so that creep of the membranes into the solution compartments is minimized. The spacer screens should have a high degree of open area in the direction perpendicular to the membrane faces so that utilization of membrane area is high and electrical resistance is low. The hydraulic resistance of the spacer screen should be low in the direction of solution flow to minimize the energy needed to pump the solutions through the membrane compartments.

The gaskets and spacer screens should be in one plane so that the membranes are not bent or distorted. This permits the use of stiff or brittle membranes, as well as more flexible membranes.

The end plates of the stack, which hold the membranes and spacer frames between them and transmit the force to seal the compartments against leakage of solution, should be strong and rigid. The rigidity of the end plates is especially important for wide stacks, where beam deflection or bowing of the end plates may cause insufficient clamping pressure to be applied to some parts of the assembly of membranes, spacers, and gaskets. The required rigidity may be obtained by designing the end plates to have low beam deflection or by the use of auxiliary clamping devices on the end plates.

Hydrodynamic Criteria. It is necessary to have essentially equal distribution of the solutions being processed over the width of each compartment. In narrow compartments, the solution may be introduced at one point, but in wider compartments, introduction of solutions at a single point results in stagnation in some regions. Except in narrow compartments, the solutions should be introduced at multiple points across the widths of the compartments.

The solution velocity should be essentially equal at all points within a compartment. To meet this requirement, no channels or spaces between spacer screens and edge gaskets can be allowed, the compartment thickness must be essentially uniform throughout a compartment, and the hydraulic resistance of the spacer screen must be essentially uniform.

The solution velocities should be essentially equal in all solution compartments. Good alignment of the holes that form the supply ducts at the entrance and exit ends of the compartments must be provided so that the hydraulic pressure drops in the supply ducts are low. All the solution channels between the supply ducts and the individual compartments should have identical hydraulic resistances and the hydraulic resistances of the spacer screens in the compartments should be identical, or the resistances of the solution channels should be high relative to the resistances in the compartments so that the flows through the compartments are controlled by the resistances of the solution channels.

The spacer screens used to support the membranes should cause mixing of the solution from each membrane face with the bulk solution when solutions are pumped through the membrane compartments. Effective mixing of the solutions is desirable to maintain thin interfacial boundary layers at the surfaces of the membranes.

Low total pressure drop across the stack is desirable to minimize the energy required for pumping and to minimize internal pressure within the stacks so that bowing of the side gaskets and bulging of the compartments is minimized.

Electrical Criteria.   The electrical leakage through the solutions in the supply ducts should be small. To achieve low electrical leakage, the resistance of the solution in the supply ducts should be high and the cross-sectional area of the supply ducts should be small. This requirement is in conflict with the need for large cross-sectional areas to provide the low hydraulic resistances.

The solution channels between the supply ducts and the individual compartments also should have high electrical resistances. This requirement means that the cross-sectional areas of the solution channels should be small and coincide with the desirability of relatively high hydraulic resistances for the solution channels.

Basic Types of Stacks.   Most commercial stacks may be considered to be one of two basic types: tortuous path or sheet flow. These designations refer to the type of solution flow in the compartments of the stack.

In the tortuous-path type of electromembrane stack, the solution flow path is a long narrow channel, as illustrated in Fig. 12.28, which makes several 180° bends between the entrance and exit ports of a compartment. The bottom half of the spacer gasket in Fig. 12.28 shows the individual narrow solution channels and the cross-straps used to promote turbulence, whereas the individual channels have been

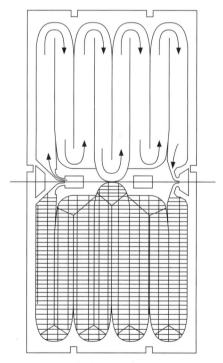

FIGURE 12.28   Diagram of a tortuous-path spacer for an electrodialysis stack. *(Huffman and Lacey, 1972.)*

omitted in the top half of the figure so the flow path could be better depicted. The ratio of the solution channel length to its width is high, usually greater than 100:1.

Solution flow in a sheet-flow type of stack is approximately in a straight path from one or more entrance ports to an equal number of exit ports of a compartment, as illustrated in Fig. 12.29. Therefore the solution flows through the compartment as a sheet of liquid. The ratio of the solution channel length to width in this sheet-flow type of stack is much lower (about 2:1) than in the tortuous-flow type. An example of a combined spacer-gasket sheet-flow type of stack is shown in Fig. 12.30.

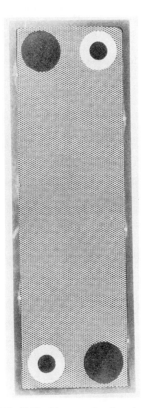

Solution velocities in sheet-flow and tortuous-path stacks are discussed under sec. 12.3.4.1. Membranes with suitable characteristics for use in sheet-flow and tortuous-path stacks are described in sec. 12.3.2.4.

Spacer screens are usually not needed in the solution compartments of tortuous-path stacks to support the membranes because of the narrow width of the channels and the rigid nature of the membranes used. Small straps across the solution channels provide some support for the membranes and also cause turbulence in the flowing solution. This turbulence causes mixing of the depleted or enriched solution near the membrane surfaces with the bulk of the solution and reduces the thickness of the boundary layers at the membrane surfaces.

Spacer screens are generally used in the solution compartments of sheet-flow stacks, both to support the membranes and to produce turbulence in the flowing solution.

FIGURE 12.29 Photograph of a sheet-flow spacer for an electrodialysis stack. *(Huffman and Lacey, 1972.)*

Arrangements of Electromembrane Stacks for Various Purposes. For applications in which the desired throughput or degree of demineralization is greater than is practical with one stack, combinations of stacks in parallel or in series, or both, can be used to meet the requirements. Throughput is increased in proportion to the number of stacks in parallel hydraulically. The degree of demineralization is increased progressively by adding stacks in series.

1. *Batch ED.* The first ED system to be commercially developed was the batch-type ED system. Figure 12.31 illustrates the flow diagram of this system (Meller, 1984).

In a batch-type ED system, the saline or brackish feedwater is recirculated from a holding tank through the demineralizing spacers of a single stack until the desired final purity is obtained. The production rate is dependent on the concentration of

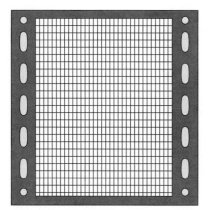

FIGURE 12.30 Diagram of a combined sheet-flow spacer gasket for an electrodialysis stack.

dissolved minerals in the raw feedwater and on the degree of demineralization desired.

The flow diagram shows two main streams flowing in parallel through the membrane stack. One of these streams is progressively demineralized as it is recirculated through the system. This stream is referred to as the *dilute stream* or the *product stream*.

The other main stream recirculated through the system is the *concentrate stream*. Its function is to collect the salt that is transferred from the dilute stream. The concentrate stream is recirculated to reduce the quantity of wastewater. Recirculation, however, increases the stream concentration to a level much higher than that of the feedwater. To control mineral concentrations and prevent membrane stack scaling, a fraction of the concentrate stream is sent to waste and a continuous addition of feedwater, acid, and conditioning chemicals is required.

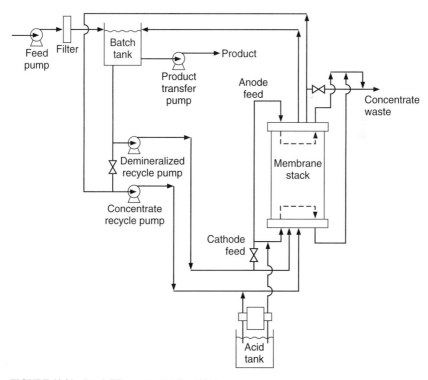

FIGURE 12.31 Batch ED system. *(Meller, 1984.)*

Also shown in Fig. 12.31 are two low-flow electrode streams. Acid is continuously added to the cathode stream to prevent scaling in the electrode compartments. A batch system is normally used when the TDS of the feed is greater than 7000 mg/L (Rogers, 1984a).

2. *Feed-and-bleed system.* A feed-and-bleed system is sometimes used when large variations in the concentration of the feed solution are encountered and a continuous flow of product is desired. This system is also useful when the desired degree of demineralization is low. In the feed-and-bleed system a portion of the product solution is recirculated and blended with the raw feed solution. This blended solution then becomes the actual feed to the electromembrane stack. The production rate is the part of the product stream that "bleeds" out of the system and is not recirculated to the stack. A feed-and-bleed system is shown in Fig. 12.32.

3. *Continuous ED (unidirectional).* The third commercial system is unidirectional continuous–type ED. Figure 12.33 illustrates the flow diagram. In this system, the membrane stack contains two stages in series, internally manifolded so that the two streams pass first through the bottom half of the stack or first stage and then through the top half or second stage. More than two stages can be arranged in a single stack and/or two or more stacks can be arranged in series.

The dilute stream makes a single journey through the stack and exits as product water. The concentrate stream is partially recycled to reduce waste and injected with acid and other chemicals to avoid scaling in the stack.

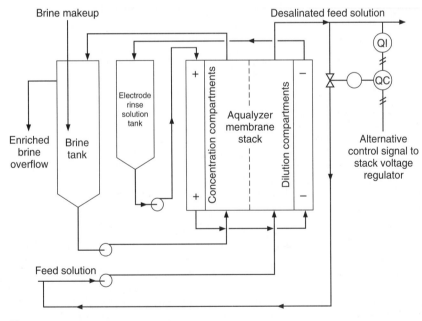

Key:
    QI – Conductivity meter
    QC – Conductivity controller

FIGURE 12.32   Feed-and-bleed system.

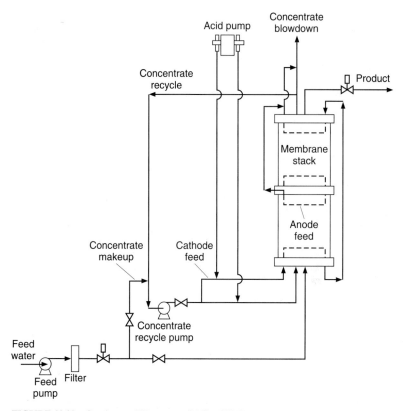

FIGURE 12.33 Continuous ED system. *(Meller, 1984.)*

There is a metal electrode at the top and bottom of each of the two stages. The two cathodes occur at the very top and bottom of the stack, whereas the two anodes occur in the center of the stack on opposite sides of the center plastic block.

The two cathodes are made of stainless steel (or similar inert metal) and are fed in series, with the bottom cathode being fed externally by a plastic line which is piped off the concentrate inlet stream and injected with acid.

The two anodes in the center of the stack are made of a platinum-plated metal and are also fed in series. The bottom, or first-stage anode, is fed from the concentrate stream inside the stack and connected hydraulically to the top, or second-stage anode, through the center plastic block.

The types of ED systems previously discussed are unidirectional. That is, ionic movement is in one direction only with cations moving towards a fixed cathode, anions moving towards a fixed anode. In such a system, chemical dosing is usually required to prevent the formation of scale caused by the precipitation of calcium carbonate or calcium sulphate on the membrane surfaces (Meller, 1984). Calcium carbonate scale is controlled by adding acid to the recirculating concentrate stream and calcium sulphate scale is controlled by adding chemical inhibitors such as SHMP (sodium hexametaphosphate) to the concentrate stream. In addition to scale formation, colloidal particles or slimes may accumulate on the surface of the anion transfer membranes and cause membrane fouling. These fouling materials require removal by flushing with cleaning solutions, the frequency depending on the concentration of such materials in the feedwater.

Control of scale and fouling materials is critical to the operation of unidirectional desalting systems of any type. The costs to install, operate, and maintain chemical feeding systems for strong acids and chemicals must be provided for. The ability to transport acids and other chemicals to water plant locations, particularly in remote areas, can present many practical operational problems. These problems can result in higher operating and maintenance costs.

4. *Electrodialysis reversal (EDR).* The EDR reversal system is designed to continuously produce demineralized water without constant chemical addition during normal operation, thus eliminating the major problems encountered in unidirectional systems (Meller, 1984). The EDR system uses electrical polarity reversal to continually control membrane scaling and fouling.

In this system the polarity of the electrodes is reversed three to four times each hour. This reverses the direction of ion movement within the membrane stack, thus controlling film and scale formations (see Fig. 12.4).

Upon reversal, the streams which formerly occupied demineralizing compartments become concentrate streams, and the streams which formerly occupied concentrate compartments become demineralized streams. Therefore, at reversal, automatically operated valves switch the two inlet and outlet streams so that the incoming feedwater flows into the new demineralizing compartments and the recycled concentrate stream flows into the new concentrating compartments. The effect of this reversal is that the concentrate stream remaining in the stack whose salinity is higher than the feedwater must now be desalted. This creates a brief period of time in which the demineralized stream (product water) salinity is higher than the specified level. This slug of water is known as *off-spec product.*

Because of reversal, no flow compartment in the stack is exposed to high solution concentrations for more than 15 to 20 min at a time. Any buildup of precipitated salts is quickly dissolved and carried away when the cycle reverses. Figure 12.34 is a typical EDR flow diagram.

5. *Electrode compartments.* When reversal takes place, the chemical reactions within the electrode compartments are also reversed since the electrodes change polarity (Meller, 1984). The alkaline environment at the new cathode will now occur at the anode of the previous polarity. Similarly, the acidic environment at the new anode will now occur at the cathode of the previous polarity. The acidic environment generated at an anode which helps prevent scale formation can now be taken advantage of at both electrodes since each electrode is alternately an anode. To obtain maximum benefit from anodic acid generation, the anode stream may be operated in a nonflow condition which allows the concentration of H^+ ions to increase and lower the pH to about 2 to 3. However, chlorine and oxygen gases are also being generated at the anode. These gases tend to accumulate on the electrode to form a layer of gas which greatly increases the resistance at the electrode. This is known as *gas blanketing.* To reduce this problem, the anode stream is allowed to flow for a brief period to flush out the gases. This action is known as a *bump.* While the anode is in the bumping mode, the cathode flows continuously to minimize OH^- concentration since an alkaline environment will increase scaling tendency. Solenoid valves in the electrode streams automatically control the flows in the anode and cathode compartments. Since both electrodes are at some time in the anode mode, they must be constructed of a platinum-coated metal.

To summarize, the EDR process has five major effects on the operation of a membrane system:

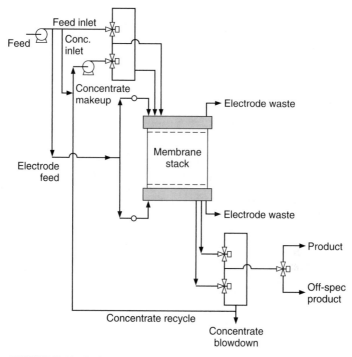

FIGURE 12.34 Typical EDR flow diagram. *(Meller, 1984.)*

- Breaks up polarization films three to four times each hour, thus preventing polarization scale
- Breaks up freshly precipitated scale or seeds of scale and flushes them to waste before they can grow or cause damage
- Reduces slime or similar formations on membrane surfaces by electrically reversing the direction of colloidal particle movement
- Eliminates the entire complex of practical problems associated with the need for continuous feed of acids or complexing chemicals
- Automatically cleans the electrodes with acid formed during anodic operation

12.3.4.3 Typical Values for Design Parameters. There are several parameters which must be considered for the successful design of an ED/EDR plant (Meller, 1984; Huffman and Lacey, 1972; Wilson, 1960; Rogers, 1984a; USAID Desalination Manual, 1980). These parameters will be considered in this section and include:

- Characteristics of ED/EDR feed, product, and brine
- Desired process flow diagram
- Materials of construction
- Pretreatment of feed
- Pumps and pressure drops
- Storage tanks

- Power supply
- Controls
- Limiting current density
- Current leakage
- Backdiffusion
- Langelier index
- Calcium sulphate saturation
- Membrane stack design

Feedwater Characteristics and Product Water Requirements. The design of an ED/EDR plant is based on the product water requirement and the characteristics of the raw water to be treated. An example of the characteristics of raw water to be treated and product water requirements is shown in Table 12.7 (USAID Desalination Manual, 1980).

TABLE 12.7   Characteristics of ED/EDR Feed, Product, and Brine
(70% water recovery)

Constituent	Feed, mg/L	Product, mg/L	Brine, mg/L
Sodium	19	7	44
Calcium	295	105	873
Magnesium	58	21	153
Chloride	36	13	89
Bicarbonate	294	105	757
Sulphate	699	249	2082
TDS	1401	500	3998

Source: USAID Desalination Manual, 1980.

Materials of Construction. The materials used in electromembrane stacks, pumps, piping, tanks, and other equipment associated with electromembrane processes are usually chosen on the basis of the requirements for corrosion resistance, physical strength, and resistance to electrochemical attack by the solutions to be treated (Huffman and Lacey, 1972). For applications in which conventional ED/EDR stacks are used, satisfactory materials for construction are usually chosen by the manufacturers of ED/EDR equipment. For applications in which special electromembrane stacks are used, such as those with bipolar electrodes or special arrangements of membranes, the stack components may have to be developed to meet specific requirements. In such instances, data on the materials used in conventional electrochemical equipment, such as chlor-alkali cells, hydrogen-oxygen cells, and hydrogen peroxide cells have been found to be helpful. Such data can be found in the chemical literature. Information from manufacturers of reinforced plastic materials and electrode materials has also been found to be helpful.

Piping that is not electrically conducting, such as polyvinyl chloride and reinforced epoxy-resin pipe, is often used to transfer solutions to and from the electromembrane stacks. Such piping also can offer advantages in corrosion resistance.

Plastics, both reinforced and unreinforced, are often the material of choice for tanks, although plastic and fiberglass-lined steel tanks can also be used.

Pumps and Typical Hydraulic Pressure Drops. The primary considerations in choosing pumps for use in electromembrane processing are corrosion resistance, introduction of undesirable metallic contaminants, and the hydraulic pressure drops through the equipment (Huffman and Lacey, 1972). The resistance of materials to corrosion in use depends on the nature of the solutions to be treated in the particular application. The corrosion-resistance requirements for pumps are no different from those for other chemical processes. If pumps are chosen to be satisfactory on the basis of corrosion resistance, they will usually not introduce undesirable amounts of heavy metals into the feed solutions.

Storage Tanks. Minimum sizes of interstage tanks (if used) depend on the time needed to shut the entire system down when a tank level is too low, or on the time needed to start a pump when a tank level is too high (Huffman and Lacey, 1972). Feed tanks and interstage tanks should be chosen to be large enough to ensure a supply of feed solutions equal to at least a few minutes of pumping capacity for the pumps involved. The sizes of feed tanks may also depend on surges, shutdowns, and other features of the process furnishing the feed. Sizes of product storage tanks depend on the inventory of stored product desired, which is dependent on factors other than the operation of the electromembrane process itself.

Power Supply. DC energy for electromembrane stacks is usually supplied with transformers and rectification equipment, although motor generator sets have been used on occasion (Huffman and Lacey, 1972). Power supplies suitable for use may be made with single-phase or three-phase transformers and silicon diodes arranged as a full-wave rectifier. Capacitors are generally used with single-phase circuits to give a low-ripple DC voltage, but with three-phase transformers and full-wave rectifiers the output voltage has acceptably small ripple without the use of capacitors.

Limiting Current Density. Seventy to eighty percent of the limiting current density is used as the maximum allowable operating current density (see sec. 12.3.3.1).

Current Leakage. For a given ED/EDR system, there is a limiting voltage which can be applied (Meller, 1984). This limit is determined by water temperature, stream concentrations, membrane stack size, and internal manifold area. If this limit is exceeded, excessive electric current will travel from the electrode laterally through an adjacent membrane to the concentrate stream manifold, generating enough heat to damage some membranes and spacers in the vicinity of the electrode.

This limiting voltage is defined as the voltage at which only the cell pair adjacent to the electrodes is affected. Normal design practice limits the voltage applied to a membrane stack to 80 percent of this voltage (Meller, 1984).

Backdiffusion. Backdiffusion occurs when the ion concentration in the concentrate stream is substantially higher than the ion concentration in the demineralized stream. The result is that some of the ions from the concentrate stream diffuse back through the membrane, against the force of the DC potential, into the demineralized stream. Backdiffusion must be taken into account during the design of an ED/EDR plant. Typically, membrane stack efficiency is derated when the concentration ratio exceeds 150:1 (Meller, 1984).

Langelier Index. A positive Langelier index (LI) indicates that there is a potential for calcium carbonate to precipitate. An LI value up to +2.2 is acceptable in an Ionics EDR system (see sec. 12.3.3.3).

Calcium Sulphate ($CaSO_4$) Saturation. The upper design limit for the EDR system, without chemical addition, is 175 percent $CaSO_4$ saturation. With chemical addition to the concentrate stream this limit can be extended to the 300 percent saturation range in commercial water treatment plants (see sec. 12.3.3.3).

12.3.5 Posttreatment

The ED/EDR product water is usually less aggressive than the RO product. This is because acid is usually not added in ED/EDR for scale control (Fraivillig, 1983) and ED/EDR removes only the level of salts desired as opposed to greater than 98 percent with most RO processes. Post-pH adjustment may, therefore, not be required as with RO. Nonionic matter in the feed, such as silica, particulates, bacteria, viruses, pyrogens, and organics, will not be removed by the ED/EDR process and must, if necessary, be dealt with during pre- or posttreatment.

12.3.6 Controls and Instrumentation

The following variables are usually measured or controlled, or both (Huffman and Lacey, 1972):

1. DC voltage and current supplied to each electrodialysis unit
2. Flow rates and pressures of the depleting and concentrating streams, and of the electrode rinse streams
3. Electrolyte concentrations of the depleting and concentrating streams at the inlets and outlets to the ED stacks
4. pH of the depleting stream and the electrode rinse streams

All of these variables are interrelated. Automatic control of the flows of the depleting and the concentrating streams can be achieved by the use of flow-type conductivity cells in the effluent depleting and concentrating streams along with a controller that compares the conductivities of the depleting and concentrating streams with that of a preset resistance and actuates flow control valves in the liquid supply lines.

To prevent damage to the membranes or other components of an electromembrane stack in the event of stoppage of liquid flows to the stacks, the equipment should be provided with fail-safe devices that will turn off the power to the stacks and pumps. This can be accomplished by placing flow-measuring controllers in the feed streams which will turn off the power to the stacks and pumps if the flow rate of any stream drops below a preset value.

12.3.7 Concentrate Disposal

Any desalination process produces a freshwater product and a waste concentrate containing the dissolved solids in a concentrated form. This concentrate must be disposed of in an environmentally acceptable way so as not to cause harm to the water environment or to pollute freshwater sources. The quantity of concentrate produced depends on the quality of the feedwater and the characteristics of the desalination process. Concentrate disposal is described by Schutte (1983) and Andrews and Witt (1993).

12.3.7.1 Water Recovery. The quantity of product water produced from a unit volume of feedwater is termed *water recovery* and is usually expressed as a percentage. Water recoveries can be higher than 95 percent or lower than 25 percent, resulting in concentrate volumes of less than 5 percent or more than 75 percent of the feedwater, depending on the process employed and characteristics of the feedwater.

The most important factor which determines water recovery during desalination is the presence of scale-forming substances, such as calcium carbonate, magnesium hydroxide, and calcium sulphate, in the feedwater. When the concentrations of these substances exceed their solubility limits in the concentrate, precipitation and scale formation take place. This seriously affects the performance of the process. For this reason, feedwater is pretreated to remove these ions or to prevent them from precipitating by means of acid treatment and dosing of scale inhibitors. However, in many cases, especially with calcium sulphate, the only solution is to operate within the solubility limits of the compound.

12.3.7.2 Methods of Disposal. Methods for the disposal of ED/EDR concentrate are considered to be the same as for RO concentrate. The concentration of dissolved salts in ED/EDR concentrate, however, can be significantly higher than in RO concentrate.

Sea Disposal. The disposal of ED/EDR concentrate into the ocean is the most economical method of disposal. However, care must be taken to ensure that concentrate discharge meets regulatory requirements and that the water environment is not adversely affected.

Evaporation. In inland areas where there is a high net rate of evaporation, such as in arid areas, disposal of ED/EDR concentrate into evaporative ponds is an effective method. However, where land costs are high, other disposal methods may prove to be more cost-effective. Care, however, must be taken to prevent seepage of concentrate to groundwater sources or overflow of runoff to surface sources. Properly lined ponds of sufficiently large capacity should prevent problems in this respect.

Saltwater Sinks. In areas where large volumes of ED/EDR concentrate must be disposed of, the creation of a saltwater environment in natural or artificial salt ponds will greatly facilitate disposal.

Deep Well Injection. In this method, ED/EDR concentrate can be injected in wells much deeper than any water source in the area to prevent any possibility of pollution. Deep defunct mines may be used for this purpose provided that no contamination of underground waters takes place or that the concentrate does not flow to active mines from which water is being pumped.

Further Processing. Further processing of ED/EDR concentrate in crystallizer-evaporators which produce an essentially dry product is the recommended way of disposing of ED/EDR concentrate derived from desalination of water and industrial effluents. Concentrate from evaporators can be absorbed on fire ash to result in zero effluent discharge.

12.4 OPERATION AND MAINTENANCE

Operation and maintenance of ED/EDR plants and operational examples will be considered in this section.

12.4.1 Operation

The ED/EDR process can be operated continuously, in the batch or feed-and-bleed modes (see sec. 12.2.4.2). The simplest continuous flow sheet contains a single stage of ED/EDR. For greater purity of product, a multistage plant is used. It should be noted that the purpose of staging is not the extraction of greater water fractions from the feed. Instead, each stage upgrades the purity of the dilute from the preceding stage. A single ED/EDR stage may be expected to remove 45 to 55 percent of the dissolved salts in the feed. That is, its purified effluent will have a salt content of 55 to 45 percent of the original feed salinity. A product water containing 410 mg/L of total dissolved solids, for example, can be produced from a 4500-mg/L feed by means of a three-stage ED/EDR plant (Rogers, 1984a).

When the dilute is not purified sufficiently in one pass through an ED/EDR stage, either a batch or a continuous system with a number of stages may be employed (Rogers, 1984). The dilute is returned to the feed tank and recirculated to the ED/EDR stack in a batch operation. Operation is continued until the contents of the feed tank attain the required purity. This is known as batch-type operation (see sec. 12.3.4.2). It may be operated by manual control, thus requiring almost no instrumentation. If desired, it may be automated by modulating the voltage across the stack as the salinity changes. In addition, the change in conductivity of the liquid may actuate a switch to start the pump, transferring the product to the storage tank when the required purity is attained. Immediately thereafter, a fresh batch is admitted and the cycle restarted. The batch system is normally used when the salinity of the feed is greater than 7000 mg/L. The batch system requires a minimum of membrane area. Its disadvantage is that (Rogers, 1984a):

1. The supply tank must be large enough to hold a complete batch of feed.
2. The prolonged holdup and exposure of feed encourages bacterial growth.
3. Current efficiencies are low.

A feed-and-bleed system is sometimes used when large variations in the concentration of the feed solution are encountered and a continuous flow of product is desired (see sec. 12.3.4.2). This system is also useful when the desired degree of demineralization is low. In the feed-and-bleed system, a portion of the product solution is recirculated and blended with the raw feed solution. This blended solution becomes the actual feed to the membrane stack. The production rate is the part of the product stream that "bleeds" out of the system and is not recirculated to the stack (Huffman and Lacey, 1972).

It is common practice to operate an ED/EDR plant (see sec. 12.4.4.2) with partial blowdown of concentrate while recycling a portion of the concentrate stream in order to decrease the volume of reject brine and improve the overall conductivity of the cell (Rogers, 1984a). The resulting decrease in voltage drop reduces power consumption. The total salinity of the recycle stream may be allowed to rise to as much as 0.4 mol/L. There are several reasons it seldom exceeds 0.2 mol/L, including (Rogers, 1984a):

1. The rate of backdiffusion of salts from the concentrate into the diluate increases as the concentration of the concentrate stream increases.
2. At high brine concentrations, the solubility of scale-forming compounds such as $CaSO_4$ and $CaCO_3$ may be exceeded.
3. Internal short circuits of the current by highly concentrated brine will begin to introduce operational difficulties.
4. Osmotic pressure drives an appreciable percentage of purified water back into the brine at high brine concentrations.

The flow rates of concentrates and dilute are generally maintained approximately equal to ensure comparable pressures in both compartments and to minimize cross-leakage and membrane distortion.

The splitting of water and the passage of hydroxyl ions at higher current densities may deposit alkaline scale on the anion-selective membrane (see sec. 12.3.3.1). This condition is controlled by adding acid to the feed (Rogers, 1984a). The problem is particularly severe in the cathode compartment. The tendency for the cathode compartment to become alkaline is counteracted by circulating the acidic effluent from the anode compartment along with some additional acid through it. A number of ED manufacturers combat these problems by periodically reversing the applied voltage while simultaneously rerouting the dilute and concentrate flows. Several manufacturers recommended a reversal once a day, others every 15 min on an automatic cycle. As a result, electrode compartment scaling is avoided, and the acid requirement of the feed is reduced or eliminated completely. Polarity reversal has the further advantage of purging the membrane surfaces of any deposited fouling materials. The only apparent disadvantage of this system is the loss of productivity while sweeping the fluid from the cells following each current reversal. This loss is reported to amount to about 15 percent of the plant output for plants on a 15-min reversal cycle. The loss of off-specification water produced during the voltage reversal can be minimized by routing it to the recirculating brine stream or by returning it to the feedwater tank.

One of the most important operating variables is the feed temperature. As operating temperature increases, the resistance of aqueous solutions decreases, resulting in a reduction in power consumption. A further benefit resulting from operation at elevated temperatures is the improvement in salt transport through the membranes. Each 1°F (0.55°C) increase in temperature improves the *cut* (salt removal per stage) by approximately 1 percent (Katz, 1971). One of the commercially available ED/EDR plants is claimed to operate successfully up to 110°F (43°C) on a continuous basis and up to 120°F (49°C) for a short period of time.

12.4.1.1 Cleaning of the Membranes.
Cleaning of an ED/EDR plant is an infrequent procedure, provided the pretreatment of the feed is adequate, 5- or 10-μm filters are used ahead of the stack, and conservative operating conditions are maintained (Rogers, 1984a). Generally, a decrease in purity of the product water indicates the need for cleaning. For cleaning in place, a 5 percent by volume solution of commercial hydrochloric acid is circulated through the stack for 30 to 60 min to remove scale and inorganic slime. This is followed by a water rinse. Organic contaminants are removed by a subsequent flushing with a 5 percent by weight solution of NaCl adjusted to pH 10 to 13 with NaOH. The operation is simple if flushing connections are properly located on the stack. Cleaning in place may be required at intervals of twice a week or one to four weeks, depending on the quality of the feedwater. The waste can be stored in a tank or pond and adjusted to a neutral pH prior to disposal. The spent cleaning solution can also be dumped with the brine.

If the membranes have become severely fouled, it may be necessary to resort to the more time-consuming method of disassembling the stack and cleaning the membranes individually. Manufacturers recommend sponging, brushing, or even scouring the membrane surfaces with steel wool. Skilled labor is not required for these operations. The frequency of stack disassembly and manual cleaning will depend on feed quality and operating conditions, varying from a monthly to a semiannual operation.

12.4.1.2 Maintenance.
Little maintenance is required by the feed and circulating pumps if suitable construction materials are used. If it is necessary to inject additives into the feed stream, experience indicates the likelihood of replacing valves, seats, springs, and diaphragms on the metering pumps at least every six months (Rogers, 1984a).

Approximately 10 percent of the membranes will require yearly replacement for general brackish water desalination applications. The replacement rate may be less with a well-water feed stream, somewhat greater with surface waters or aqueous water streams. However, membrane replacement rate for industrial effluent treatment, such as whey demineralization, can be 100 percent per year. The membranes most likely to fail are those nearest the electrodes where more severe temperature and chemical environments are encountered. When the stack is disassembled, it may be necessary to replace spacers and gaskets (when separate gaskets are used) if they have distorted in service or have been damaged as a result of stack disassembly (Rogers, 1984a).

If pressure is maintained on the assembled stack by tie bolts rather than a hydraulic ram, the manufacturer's directions must be carefully followed as to the torque and the sequence of bolt tightening during reassembly of the stack. Failure to do so may result in excessive liquid leakage from the assembled stack or irreparable damage to the stack components (Rogers, 1984a).

12.4.2 Examples

12.4.2.1 Seawater Desalination. There is limited application of ED/EDR for seawater desalination because of high costs (USAID Desalination Manual, 1980). A small batch system (120 m^3/d) has been in operation in Japan since 1974 to produce water of potable quality at a power consumption of 16.2 kWh/m^3 product water (Miva, 1977). A 200-m^3/d seawater EDR unit was evaluated in China (Shi and Chen, 1983). This unit operated at 31°C and had stable performance with a total electric power consumption of 18.1 kWh/m^3 product water. The product water quality of 500 mg/L TDS met all the requirements for potable water. When the stacks were disassembled for inspection, there were no signs of scale formation.

With the commercial ED/EDR units currently available, the energy usage for seawater desalination is relatively high compared with that of RO. However, work under the Office of Water Research and Technology (OWRT) programs has indicated that high-temperature ED/EDR may possibly be competitive with RO (Parsi, Prato, and O'Donoghue, 1980). Results have shown that the power consumption can be reduced to the levels required for seawater RO (8 kWh/m^3) and that a 50 percent water recovery can probably be attained.

12.4.2.2 Production of Table Salt. The production of table salt from seawater by the use of electrodialysis to concentrate sodium chloride up to 200 g/L prior to evaporation is a technique developed and used nearly exclusively in Japan (Strathmann, 1992). More than 350,000 tons of table salt are produced annually by this technique, requiring more than 500,000 m^2 of installed ion-exchange membranes (Tokuyama Soda, 1988). The key to the success of this technology has been the low-cost, highly conductive membranes with a preferred permeability for monovalent ions. However, note that in Japan this salt production procedure is highly subsidized.

12.4.2.3 Brackish Water Desalination. In terms of the number of installations, the most important large-scale application of electrodialysis is the production of potable water from brackish water. Here, electrodialysis competes directly with reverse osmosis and multistage flash evaporation (Strathmann, 1992).

A considerable number of standard ED/EDR plants for the production of potable water from brackish water are in operation (Urano, 1977; Kusakari et al., 1977; USAID Desalination Manual, 1980). These plants are operating successfully.

The major application of the EDR process manufactured by Ionics is for the desalination of brackish water. The power consumption and, to some degree, the cost of equipment required is directly proportional to the TDS to be removed from the feedwater (USAID Desalination Manual, 1980). Thus, as the feedwater TDS increases, the desalination costs also increase. In the case of the RO process, costs are also related to TDS removal, but it is not as pronounced. Often, the variation in the scaling potential of the feedwater and its effect on the percentage of product water recovery can be more important than the cost to TDS relationship.

For applications requiring low TDS removals, ED/EDR is often the most energy-efficient method, whereas with highly saline feedwaters RO may be expected to use less energy and is preferred. The economic crossover point between ED/EDR and RO based on operating costs is, however, difficult to define precisely and needs to be determined on a site-specific basis. One significant feature of ED/EDR is that salts can be concentrated to comparatively high values, in excess of 18 to 20 percent by weight without affecting the economics of the process severely.

Apart from local power costs, other factors must also be considered in determining the overall economics. Among those to the advantage of ED/EDR, are the high recoveries possible (up to 90 percent), the elimination of chemical dosing with EDR, and the reliability of performance that is characteristic of the ED/EDR process.

The energy consumption of a typical EDR plant is as follows (USAID Desalination Manual, 1980):

Pump:	0.5 to 1.1 kWh/m^3 product water
Membrane stack:	0.7 kWh/m^3 product water/1000 mg of TDS removed
Power losses:	5 percent of total energy usage

The major energy requirement, therefore, is for pumping the water through the ED/EDR unit and for the transport of the ions through the membranes.

Most modern electrodialysis units operate with so-called electrodialysis reversal (EDR), i.e., the anode and cathode, and with that, the dilute and concentrate cell systems are exchanged periodically, preventing scaling due to concentration-polarization effects (Ionics, 1988). In brackish water desalination, more than 2000 plants with a total capacity of more than 1,000,000 m^3 of product water per day are installed, requiring a membrane area in excess of 1.5 million m^2 (Strathmann, 1992). Installations in Russia and China for the production of potable water are estimated as being of the same order of magnitude.

In many cases now, EDR is used in combination with RO and/or UF. UF/EDR provides an excellent pretreatment for RO. A large number of trailer-mounted units with these three membrane systems are used in the power and semiconductor industries.

Case Study: Corfu, Greece, 14,800-m^3/d EDR Plant.   This EDR plant treating groundwater was constructed for the Municipality of Corfu to provide drinking water for the island and began operation in 1977 (USAID Desalination Manual, 1980). The 14,800-m^3/d EDR plant is divided into six parallel modules, each containing ten membrane stacks arranged in four parallel banks. Two of the banks have two stages and the other two have three stages. Each group can be operated independently with its own flow controls, rectifiers, and instrumentation.

With a feedwater flow (treated through 10-μm cartridge filters) of about 21,200 m^3/d, there is a recovery of about 70 percent, with 14,800 m^3/d of product water and a concentrate discharge of approximately 6430 m^3/d. The characteristics of the feedwater, product, and concentrate streams are shown in Table 12.8. The salt rejection is about 64 percent.

TABLE 12.8 Characteristics of the Feed, Product, and Concentrate Blowdown at the Corfu Facility

Constituent	Feedwater, mg/L	Product water, mg/L	Concentrate blowdown, mg/L
Sodium	19	7	44
Calcium	295	105	873
Magnesium	58	21	153
Chloride	36	13	89
Bicarbonate	294	105	757
Sulphate	699	249	2082
TDS	1401	500	3998

Source: Arnold, 1979; Andreadis and Arnold, 1978.

Some operating problems were encountered during the startup and initial operation, primarily with the stacks, hardware, and external factors (Arnold, 1979). In the membrane stacks there were some initial problems with minor variations in the spacers between the membranes, causing a poor hydraulic flow distribution and, hence, some blockage of cells. This was alleviated by the installation of spacers of proper and more uniform thickness. Some hardware problems were encountered with pneumatically operated valves for the reversal system. These were corrected by resizing some of the pneumatic operators and adding equipment to improve the air drying in the pneumatic system. Outside the system reduced production and disruptions have been caused by a lack of sufficient water and frequent breakages of old lines in the collector system for the feedwater. This latter has resulted in an increase in suspended solids and iron slimes in the feedwater after the repairs were performed and the pipelines flowed again. This caused plant upsets and premature clogging of the cartridge filters.

Case Study: Suffolk, Virginia, EDR Facility. The initial evaluation of groundwater desalination for Suffolk, Virginia, was completed in the spring of 1986 (Thompson, 1991). It provided preliminary operating and capital cost estimates which were within the projected range for surface water supply expansion. Reverse osmosis appeared to be the best membrane technology for Suffolk. EDR also appeared to be feasible.

The city constructed a 280-m (918-ft) -deep well capable of producing up to 15,100 m^3 per day (4.0 mgd) that became operational in 1987 and was designed to discharge into the local surface water reservoir feedwater directly into an existing surface water treatment facility's finished water reservoir. Flexibility was also included in the well design, which would allow groundwater to be fed into a membrane desalination process.

Water quality in the well is characterized by moderate levels of total dissolved solids (563 mg/L), fluoride (4.77 mg/L), and sodium (185 mg/L). Without treatment, this water would not meet the current fluoride maximum contaminant level of 4.0 mg/L and would exceed Suffolk's water quality guideline for sodium of 50 mg/L.

Results from a feasibility study showed that both RO and EDR were technically feasible treatment concepts, with the new deep well being the best groundwater source for desalination. These conclusions were based on the following preliminary findings:

- RO would operate effectively at an 85 percent and EDR at a 95 percent recovery rate with silica being the limiting element.

- On the basis of 11,355 m³ per day (3.0 mgd) annual average production from the well, the EDR system would produce 10,785 m³ (2.85 mgd) per day of product. Since a portion of the well water could bypass the RO process, annual average RO product would be 9955 m³ per day (2.63 mgd).

Both RO and EDR pilot units were obtained on a lease basis and installed in 6.1-m (20-ft) shipping containers which had previously been purchased by the city. The containers were modified by city personnel to provide for electrical service, lighting, ventilation, and feedwater service.

The RO pilot unit was configured to replicate the proposed RO treatment strategy recommended in the feasibility study. The array was 2:1:1 with 10.16-cm (4-in) elements in the first two stages, with the third stage containing 6.35-cm (2½-in) elements. Each pressure vessel held three 101.6-cm (40-in) -long elements, utilizing Fluid Systems TFCL membranes. The estimated feed pressure was 16.3 bar (240 lb/in²), recovery of 85 percent, and production of 83.3 m³ per day (22,000 gpd). Acid and scale inhibitor were also provided. This system operated as anticipated and was able to reduce fluoride to acceptable levels.

The EDR pilot unit was a modified Aquamite V unit, containing a single stack of 267 cell pairs. The stack had three hydraulic stages and three electrical stages. The unit was designed to produce approximately 113.6 m³ per day (30,000 gpd) of product at a recovery rate of 94 percent. To achieve this with a single stack, the off-spec product resulting from the reversal cycle was recycled by placing it in the feedwater tank. No pretreatment chemicals were used with the EDR pilot unit.

The EDR unit was started up at its design recovery rate of 94 percent, and with a stack voltage of 75 V on each stage. The predicted voltages were 69, 70, and 76 for first, second and third stages, respectively. The amperages were:

Stage	Predicted	Actual
1st	6.0	7.4
2d	3.8	5.0
3d	2.5	3.1

After about 750 hours of operation, it was decided to eliminate the off-spec product recycle. Since the single-stack configuration did not provide for staggered stage reversal, it was skewing some of the operating conditions. All off-spec product was being dumped into the feed tank at one time, substantially altering the feedwater characteristics. Having demonstrated a high-recovery operation, the piping was changed to divert the off-spec product to waste, reducing recovery to 90 percent. The rest of the performance parameters now more closely resembled full-scale operation. The unit was run in this mode for an additional 650 h. The actual voltage and reduced amperage requirements for this modified condition were recorded as follows:

Stage	Voltage	Actual amperage
1st	75	5.3
2d	69	3.7
3d	69	2.3

Concentrate disposal is a major issue when evaluating desalination systems. Not only should concentrations be evaluated, but also the loading factors, which take into consideration the volume of concentrate as well as the water quality concentrations. This is evident when evaluating the phosphate levels from the pilot testing. Even though the EDR pilot unit had a high phosphate concentration, the EDR system had a lower phosphate loading due to its lower volume of concentrate.

Due to lower nutrient loading in the concentrate, higher recovery rates, and lower operational costs, the EDR process was selected for full-scale application in Suffolk.

As the desalination facility was to be located at the existing surface water treatment facility site, design considerations were given to compatible operation of both membrane and conventional treatment systems. These considerations included providing standby power, providing parallel transfer pump stations, an additional 11,355m^3 (3.0 million gal) finished water reservoir, consolidation of the controls for the two plants to one location, and physically locating the desalination facility to allow expansion of either facility without conflict. Capital costs for these improvements and the 14,232-m^3 per day (3.76-mgd) rated EDR facility including engineering was approximately US$6,920,000. Of this cost, US$2,900,000 was for the delivery of the EDR equipment.

The only pretreatment required for the feedwater prior to entering the EDR process units is filtering. Two cartridge filters were provided with each filter sized to handle 7570 m^3 per day (2.0 mgd) of feedwater flow. The filter cartridges were constructed of wound polypropylene and can remove 90 percent of all particles greater than 10 μm in size and will be 101.6 cm (40 in) long. Each cartridge filter contains seventy 6.35-cm (2.5 in) nominal outside diameter (OD) filter cartridges. In addition to the cartridge filters, provision for additional chemical pretreatment was provided should chemical pretreatment with acid be necessary at a future date.

A total of 73 (1 uninstalled spare) membrane stacks were provided. Each stack contains 500 membrane cell pairs configured into three hydraulic and three electrical stages in series. The EDR units are arranged and piped so that the feedwater will pass through one of three parallel EDR (Auamite 120) units. Each EDR unit contains 24 Mark III 45.72-cm × 101.60-cm (18-in × 40-in) membrane stacks. The piping within the individual EDR units was arranged to pass the feedwater through a series of three membrane stacks with eight parallel trains of these three-series membrane stacks.

A membrane-cleaning system was included to allow periodic cleaning of the EDR membrane stack equipment. The system, referred to as a *clean-in-place* (CIP) system, will allow cleaning of one Aquamite 120 unit at a time. The system consists of a 5.68 m^3 (1500-gal) FRP tank and a recirculation pump which can circulate acid solutions (down to pH 1) or caustic solutions (up to pH 10) through the membrane stacks and associated piping for cleaning. Use of this system is anticipated at once per month. An electrode cleaning-in-place (ECIP) system is also included to allow acid feed to the electrode compartments. The system includes a 114-L (30-gal) tank and three pumps, each capable of pumping 45.42 L/min (12 gpm) of an acid solution.

The EDR process monitoring and control system consists of a local panel at each Aquamite 120 unit with conductivity controllers and local indicators and a remote *programmable logic controller* (PLC) system located in the control room. The PLC system consists of three central processing units (CPUs), with digital and analog input/output modules, I/Q racks, and power supplies for control of the EDR process equipment. Each CPU communicates input/output status and register data to every other CPU. The PLC system contains timing, sequencing, and logic functions for control of the Aquamite 120 units, including concentrate pump, electrode recycle pump, cleaning-in-place systems, and various electric actuated valves (i.e., inlet reversal valves, outlet reversal valves, electrode valves, product valves, and off-spec product valves). The PLC system also collects data from the EDR process and transmits it via a communications link to a personal computer (PC) in the main control room, where data from the city's distribution and SWTP is collected.

After a 13-month construction period, the EDR plant began serving Suffolk on August 19, 1990. Power usage to operate the EDR stacks and brine recirculation pumps is the most significant operating cost at 1.4 kWh per 3.785 m^3 (1000 gal) produced. Other examples include cartridge filter replacement, cleaning chemicals, replacement mechanical parts, and replacement membranes. As of July 1994, no membranes have been replaced. A summary of operational and maintenance costs is presented in Table 12.9.

TABLE 12.9   EDR Operating Costs

Item	Predicted	Actual
Stack and pump power	0.11 (1.5 kWh)	0.11 (1.4 kWh)
Well pump	0.11	0.11
Cartridge filters	0.01	0.01
Cleaning chemicals	0.01	0.01
Stack parts	0.03	0.02
Nonstack parts	0.01	0.00
Membrane replacement cost	*0.06*	*0.00*
Total cost	$0.33	$0.26

Treatment of a High-Scaling, High-TDS Water with EDR.   The successful performance of EDR on high–calcium sulphate waters has been reported (Katz, 1977). Brown (1981) has described the performance of an EDR plant treating 300 m^3/d of a high–calcium sulphate water with a TDS of 9700 mg/L. The only pretreatment applied was iron removal with green sand filters. The quality of the feed, product, and brine is shown in Table 12.10.

TABLE 12.10   Water Quality Before and After EDR Treatment

Constituent	Feed, mg/L	Product, mg/L	Brine, mg/L
Na^+	2090	79	3694
Ca^{++}	652	4	1390
Mg^{++}	464	4	964
Cl^-	3687	111	7084
HCO_{3-}	134	25	175
SO_4^-	2672	19	5000
TDS	9727	242	18,307
pH	7.0	6.8	7.2

Source: Brown 1981.

The main developments in EDR during the past number of years have been the following:

1. EDR has achieved $CaSO_4$ saturation in the brine stream of up to 440% without performance decline on tests of several hundred hours duration (Elyanow, Sieveka, and Mahoney, 1981).

2. EDR has desalted a hard (Ca^{2+} approximately 150 mg/L) brackish water of 4000 mg/L TDS at water recoveries of up to 93 percent without cumbersome and expensive presoftening (Elyanow, Parent, and Mahoney, 1980).

3. An EDR unit has achieved 95 percent or greater recovery of a limited 4000 mg/L TDS brackish water resource by substituting a more abundant 14,000 mg/L saline water in the brine stream (Elyanow, Parent, and Mahoney, 1980). The substitution of seawater in the brine stream would be freely available in coastal or island regions with limited high-quality brackish water resources.

4. A new family of thick (0.5 mm), rugged antifouling anion-permeable membranes have been developed, extensively field-tested, and put to large-scale commercial use in the United States. They have much higher current efficiencies and chlorine resistance than those formerly available (Elyanow, Parent, and Mahoney, 1980).

12.4.2.4 Brackish Water Desalination for Industrial Purposes. In the past, most ED/EDR plants treated brackish waters of 1000 to 10,000 mg/L TDS and produced general-purpose industrial product water of 200 to 500 mg/L TDS. However, ED/EDR capital and construction costs have declined during recent years to the point where it is already feasible to treat water containing 200 to 1000 mg/L TDS and produce product water containing as little as 3 to 5 mg/L TDS (Katz, 1971). These low TDS levels are achieved by multistaging. The systems, which often employ ion-exchange (IX) units as *polishers,* are usually referred to as *ED/IX systems.*

ED/EDR/IX System. New and existing ion-exchange facilities can be converted to ED/EDR/IX systems by addition of ED/EDR units upstream of the ion-exchange units. The ED/EDR unit reduces chemical consumption, waste service interruptions, and resin replacement of the ion exchanger in proportion to the degree of prior mineral removal achieved (Katz, 1971). For small-capacity systems (2 to 200 m³/d), the optimum ED/EDR demineralization will usually be 90 percent or greater; for larger installations, and particularly those where adequate ion-exchange capacity is already provided, the optimum demineralization via ED/EDR is more likely to be in the 60 to 80 percent range.

It must be stressed that RO may also be used for the aforementioned application. However, the choice of the treatment method (ED/EDR or RO) would be determined by the specific requirements and costs for a specific situation.

Honeywell in the United States, which manufactures printed circuit boards and does zinc plating and anodizing, used IX for the treatment of their process waters before they changed over to an ED/IX system (Highfield, 1980). ED was chosen instead of RO because of lower membrane replacement costs. Process waters of varying degrees of purity are required, dissolved solids being the primary concern. Water with a TDS of about 50 mg/L is suitable for zinc plating and anodizing, and water with a TDS with a minimum specific resistance of 100,000 Ω is satisfactory for circuit-board fabrication operations (Highfield, 1980). The purity of the treated water (raw water TDS—350 to 500 mg/L) after treatment with the ED/IX system was better than expected. Service runs have been up to 10 times longer than before.

Costs. A comprehensive cost study was carried out to compare the costs of owning and operating the old IX and new ED/IX systems (Highfield, 1980). The cost of removing the TDS from the water, the prime purpose of the process water plant, was shown to be reduced by about 85 percent (US$1.94 to $0.30 per 3.785 m^3) for the ED/IX system. The total operating cost (water cost, electrical power, chemicals, labor) was reduced from US$2.25 to US$0.65 per 3.785 m^3, a reduction of slightly more than 70 percent.

The economic crossover point between an ED/IX system and IX system only is not stated for this application. However, this economic crossover point depends on the specific circumstances for each situation. The ED/IX system becomes more economical than the IX system alone at a TDS of approximately 300 mg/L and higher.

More recently, electrodialysis has been used for the production of ultrapure water for the semiconductor industry (Strathmann, 1992). A combination with mixed-bed ion-exchange resins seems attractive because completely deionized water is obtained without the chemical regeneration of the ion-exchange resin (Kedem and Maoz, 1976). This process has been commercialized recently.

12.4.2.5 Industrial Wastewater Treatment. Large volumes of water containing varying amounts of salt, which are generated by washing and regenerating processes, blowdown from cooling towers, disposal of dilute chemical effluents, and other processes present significant problems. This is particularly true when zero effluent discharge is required. The problem is one of too much water carrying comparatively little salt but still having a TDS content too great for acceptance to a receiving stream. Many industries face this problem today and have to consider the application of processes for concentrating salts or desalting water. The ED/EDR system for water recovery and brine concentration may be one of those best suited to alleviate the problem.

The main application of ED/EDR in wastewater treatment systems is in processing rinse waters from the electroplating industry. Here, the complete recycle of the water and metal ions is achieved by ED/EDR. Compared to RO, ED/EDR has the advantage of being able to utilize more thermally and chemically stable membranes so that processes can be run at elevated temperatures and in solutions of very low or high pH values. Furthermore, the concentrations that can be achieved in the brine can be significantly higher. The disadvantage of ED/EDR is that only ionic components can be removed and additives usually present in a galvanic bath cannot be recovered.

Several other potential applications of ED/EDR in wastewater treatment systems have been studied on a laboratory scale and are reported in the literature (Strathmann, 1992). While, in most of these applications, the average plant capacity is considerably lower than that for brackish water desalination or table salt production, there are also a significant number of large plants installed for the treatment of refinery effluents and cooling tower waste streams.

Electrodialysis of Nickel-Plating Solutions. During many plating operations, a substantial amount of bath solution adheres to plated work pieces as they leave the plating tank. In this manner, valuable materials are lost as drag-out into the subsequent rinse tank. This contaminated rinse solution can be passed through an ED/EDR system where these valuable materials can be recovered and returned to the plating tank.

One such opportunity of significant industrial importance is provided by nickel electroplating operations (Itoi, 1979). Earlier work by Trivedi and Prober (1972) demonstrated the successful application of ED to nickel solutions. Later, Eisenmann (1977) and Itoi (1979) reported the use of ED to recover nickel from electroplating rinse waters. The wash water from a nickel galvanizing line is treated by ED as shown in Fig. 12.35 (Itoi, 1979). The results achieved in an existing facility are given in Table 12.11.

TABLE 12.11   Electrodialysis of a Nickel Galvanization Effluent

Constituent	Effluent, g/L	Concentrate, g/L	Diluant, g/L
$NiSO_4$	12.47	133.4	1.27
$NiCl_2$	1.81	29.1	0.039

Source: Itoi, 1979.

The concentration ratio of the concentrated solution to the dilute solution is greater than 100. The concentrated solution is reused in the plating bath while the dilute solution is reused as wash water. The recovery of nickel discharged from the wash tank is approximately 90 percent or greater.

If organic electrolytes are present in the additives used in the galvanization bath, they must be removed prior to ED treatment to prevent organic fouling of the ED membranes.

For a plant recovering about 3500 kg nickel sulphate per month, the net profit was shown to be over US$4000 (Itoi, 1979).

Fromonot (1992) did studies on nickel recovery with ED and found that the nickel concentration in the rinse tank could be maintained between 0.5 to 2 g/L, while a nickel concentration of 50 g/L could be obtained in the ED brine. Drag-out recovery was 97 percent and plant payback period was determined at approximately 1.5 years.

Electroplating Company in Bridgeport, Connecticut, uses an EDR system to recover nickel-plating salts from rinse water (Elyanow and Batchelder, 1994). In operation since 1986, the typical feed has 3600 ppm of nickel, and the EDR produces a concentrate of 50,000 ppm nickel. The concentrated nickel salts are recycled back to the plating bath.

Electrodialysis of Copper-Plating Solutions.   An application that has been studied in a pilot plant stage is the regeneration of chemical copper-plating baths (Strathmann, 1992). In the production of printed circuits, a chemical process is often

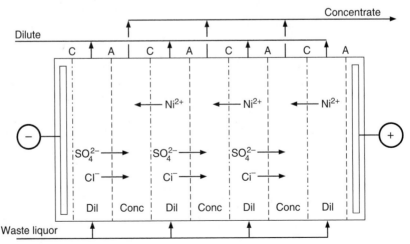

FIGURE 12.35   Electrodialysis of the wash water from a nickel galvanizing operation. *(Itoi, 1979.)*

used for copper plating. The components to be plated are immersed into a bath containing, besides the copper ions, a strong complexing agent, for example, ethylenediaminetetraacetic acid (EDTA), and a reducing agent such as formaldehyde. Since all constituents are used in relatively low concentrations, the copper content of the bath is soon exhausted, and $CuSO_4$ has to be added. During the plating process, formaldehyde is oxidized to formate. After prolonged use, the bath becomes enriched with Na_2SO_4 and formate and, consequently, loses its useful properties. By applying ED in a continuous mode, the Na_2SO_4 and formate can be selectively removed from the solution, without affecting the concentrations of formaldehyde and the EDTA complex. Therefore, the useful life of the plating solution is significantly extended (Korngold, Kock, and Strathmann, 1978).

Treatment of Copper- and Silver Cyanide–Plating Rinse Waters with ED. Fromonot (1992) and Chiapello and Gal (1992) have described treatment of copper- and silver cyanide–plating rinse waters with ED. Copper concentration could be maintained between 0.5 and 1 g/L Cu^{2+} in the rinse tank while a brine concentration of 60 g/L could be obtained. Approximately 92 percent drag-out could be recovered and plant payback was determined at approximately one year. Silver could be maintained between 0.3 to 1 g/L in the rinse tank, while a silver concentration of 27 g/L could be obtained in the brine. Drag-out recovery was approximately 95 percent and plant payback was determined at approximately 2 years.

Treatment of Cooling Tower Blowdown. The range of TDS levels encountered in cooling tower blowdown waters usually varies from about 1500 to 4000 mg/L and higher levels at about 4000 to 12,000 mg/L have also been reported (Jordan, Bearden, and McIlhenny, 1975). The disposal of large volumes of this saline effluent can be a serious problem. The application of ED/EDR for the treatment of blowdown streams to recover good-quality water for reuse and produce a small volume of concentrate promises to be the best prospective system available (Jordan, McIlhenny, and Westbrook, 1976; Westbrook and Wirth, 1976).

Blowdown waters from cooling towers can be concentrated tenfold or more using ED/EDR, while recovering and recycling the desalted water to the cooling tower at one-half its original concentration (Wirth and Westbrook, 1977). To accomplish this, blowdown is pretreated, filtered, and passed through the ED/EDR system. By recirculation of the brine, it is possible to concentrate the salts into a small stream, while allowing for a water recovery of about 90 percent.

Detailed design studies and cost estimates for ED/EDR and several other alternative blowdown recovery/concentration systems have been reported (Wirth and Westbrook, 1977). The side-stream process design which utilizes ED/EDR results in the lowest capital costs for the conditions specified. According to Wirth and Westbrook (1977), it is expected that if the cost comparison were made on overall annual operating costs, the same results would occur.

The concentration of cooling blowdown waters in an EDR pilot plant at one of Escom's power stations in South Africa was evaluated (Melzer and van Deventer, 1983). Pretreatment of the blowdown water with lime softening, clarification, pH reduction, filtration, and chlorination was found to be a basic precondition for successful operation. The operating experience on the EDR pilot plant was sufficiently positive to warrant full-scale application.

Hughes, Raubenheimer, and Viljoen (1992) have reported on seven years of operating and design experience of an ED reversal plant treating cooling tower blowdown at Tutuka Power Station in South Africa. Tutuka Power Station is operated on Eskom's zero-effluent discharge philosophy; i.e., no effluents are discharged to the environment. Effluents are absorbed in an ashing system. Zero-effluent discharge is achieved by concentration of effluents, recovery of good-quality water, and absorption of the concentrated effluents in the ashing system.

The power station employs an EDR plant and evaporators to concentrate effluents. The capacity of the EDR plant has recently been increased from 5400 to 13,200 m³/day. The function of the EDR plant is to reduce the volume of cooling tower blowdown while rejecting as much salt as possible. The plant consists of six Ionics Aquamite 50 ED trains with polarity reversal (EDR) as well as the necessary pretreatment system, feed and product storage, and transfer system.

Some of the trains are designed with an off-spec brine system, where low-conductivity brine is returned to the feed tank. This ensures that the brine is not diluted during polarity reversal.

The raw water to the filters is dosed with acid to improve chlorine disinfection effectiveness and to prevent $CaCO_3$ scaling on concentration in the EDR trains. The feedwater is dosed with chlorine to control organic growth in the filter plant as well as to oxidize organic matter to some extent. Shock dosing with chlorine is carried out periodically.

Typical composition of the EDR feed, product, and brine is shown in Table 12.12. The average stack component replacement over the past seven years has been as indicated in Table 12.13.

TABLE 12.12 Chemical Composition of EDR Feed, Product, and Brine

Constituent	Feed	Product	Brine
Conductivity, mS/m	271	65	996
TDS, mg/L	1725	431	6818
Na, mg/L	448	130	2056
Ca, mg/L as $CaCO_3$	100	16	256
Cl, mg/L	388	240	1800
SO_4, mg/L	545	9	2443
Rejection, %	75		
Recovery, %	82		
Current efficiency	57		

Note: The calcium sulphate saturation levels in brine varied between 40 and 150 percent with a single excursion of 300 percent during the test period.
Source: Hughes, Raubenheimer, and Viljoen, 1992.

TABLE 12.13 Stack Component Lifetime

Component	% replacement/annum
Anion	11
Cation	1.7
Spacers	1.3
Heavy anion	22
Heavy cation	22
Electrodes	22

Source: Hughes, Raubenheimer, and Viljoen, 1992.

The consumption of electrodes and heavy membranes has been acceptable while the consumption of other components has been very low. The consumption of electrodes and heavy membranes has been lower in later years due to the improved electrode feedwater. In addition, Ionics has developed a new platinizing technique for electrodes which, based on real plant trials, has resulted in a fourfold life extension. Thus, electrode consumption is expected to decrease in the future as new electrodes progressively replace the old.

The membrane life has proven to be exceptionally good, particularly in light of the arduous duty. The membranes are subject to organic fouling and chemical cleaning every 10 days and have also been mechanically cleaned on a number of occasions. Some of the stacks have also been scaled up on occasion when the train has been left standing without flushing after an emergency stop.

Operating and maintenance costs are given per megaliter treated (Table 12.14). The power costs are based on a rate of $0.047/kWh. The maintenance cost is based on seven years of operation.

TABLE 12.14   Operating and Maintenance Cost

Operating	Cost, $/ML
Power	12.18
Feed treatment	26.67
Chemical cleaning	4.98
	43.83
Maintenance	
Stack components	25.35
Other (mechanicals)	0.33
Labor	3.88
	29.56

Total production cost = $73.39/ML.
Source: Hughes, Raubenheimer, and Viljoen, 1992.

The EDR plant has adapted well to changing needs and provided a key tool in achieving the power station's aim of zero liquid discharge. Satisfactory salt rejection and water recovery have been achieved. Operating and maintenance costs have been contained, given attractive costs per megaliter produced for such a duty. Membrane life has proven to be very good and well in excess of guaranteed figures.

Treatment of a Fertilizer Manufacturing Process Effluent with EDR.   Schoeman et al. (1988) described treatment of a fertilizer manufacturing process effluent with an EDR pilot plant for water and chemical recovery and effluent volume reduction. A typical example of the chemical composition of the EDR feed, product, and brine is shown in Table 12.15.

Very good ion removals were obtained. Ammonium and nitrate showed rejections of 96.1 and 97.8 percent, respectively. However, phosphate, chloride, and COD rejections were not as good. It appeared that the product water, with the exception of TDS and phosphate, complied with the quality requirements for cooling tower makeup. Ten-stage EDR should reduce TDS and phosphate to within the limits for cooling tower makeup.

TABLE 12.15 Chemical Composition of EDR Feed, Product, and Brine after 512 h of Operation

Constituent	Feed	Product	Brine	% rejection
pH	6.9	8.9	2.5	
Conductivity, mS/m	1,777	64	6,501	96.4
Na^+, mg/L	67	4	269	94.0
K^+, mg/L	94	7	426	92.6
Ca^{2+}, mg/L	141	9	1,000	93.6
Mg^{2+}, mg/L	33	2	146	93.9
NH_4^+, mg/L	2,575	100	11,839	96.1
NO_3^-, mg/L	8,292	177	36,326	97.9
Si^{4+}, mg/L	13.2	12.1	14.0	
SO_4^{2-}, mg/L	1,000	20	4,200	98.0
PO_4^{3-}, mg/L	32	8	107	75.0
Cl^-, mg/L	135	31	575	77.0
Alkali (as $CaCO_3$), mg/L	27	68		
COD, mg/L	23	16	37	30.4
TDS (calculated), mg/L	12,415	452	54,905	
Current efficiency was 78.8 percent.				

Source: Schoeman et al., 1988.

The membrane stack was opened at the end of the 1000-h test run and a membrane inspection showed that there was a slight whitish scale on the anionic membrane surfaces of both the fourth hydraulic stages of the two electrical stages. Membranes from the other hydraulic stages showed no visual scale formation and the membranes appeared to be in very good condition. Membrane properties of the membrane edges (ME) and membrane flow paths (MFP) are shown in Table 12.16.

TABLE 12.16 Anion Membrane Properties of Membrane Edges (ME) and Membrane Flow Paths (MFP)

Stage[a]	Resistance[b], $\Omega \cdot cm^2$ Before		After		Capacity, me/dg		% H_2O		% weight change
E:H	ME	MFP	ME	MFP	ME	MFP	ME	MFP	MFP
1:1[c]	9.8	9.6	9.6	9.6	2.69	2.69	37.6	37.8	+0.8
1:3[c]	9.6	9.8	9.7	9.8	—	—	—	—	—
2:1[c]	10.1	9.8	9.8	9.5	—	—	—	—	—
2:2[c]	9.6	9.7	9.3	9.5	—	—	—	—	—
1:4[c]	10.0	20.9	9.8	16.2	2.64	2.10	38.2	34.6	+6.6
2:4[c]	9.8	20.9	9.3	14.8	2.66	2.15	38.5	36.3	+5.4
1:4[d]	10.0	20.9	9.8	17.2	2.64	1.95	38.2	32.5	—
1:4[e]	10.0	20.9	9.8	11.1	2.64	2.40	38.2	37.0	+2.8

[a] 1E.1E: 1st electrical.1st hydraulic, etc.
[b] Resistance before and after conditioning.
[c] Soaked in 0.1 N NaCl.
[d] Soaked for 1 h in 1000 mg/L NaOCl.
[e] Soaked for 1 h in 4 N HCl.
Source: Schoeman et al., 1988.

The anionic membranes from the first three hydraulic stages were as good as new except for the membranes from the fourth hydraulic stages. These membranes were scaled internally, causing an increase in weight and resistance and a decrease in ion-exchange capacity and gel water content. Strong acid (4 N HCl) effected removal of the scale and improved capacity with a close return to the membrane's original properties. (*Note:* No changes in cationic membrane properties were noticed.)

The same current density was applied across all hydraulic stages of each electrical stage during the tests. Therefore, a too-high current density could have caused polarization in the fourth hydraulic stages. However, in a full-scale application, each hydraulic stage will have its own electrical stage which can be controlled independently, thus preventing polarization. It is expected that a full-scale plant should run well with electrical adjustments and/or frequent acid cleanings.

The scale on the anion membrane surfaces (1E.4H) consisted mainly of calcium phosphate. Traces of sulphate, manganese, iron, and nickel were also present. Acid treatment (5% HCl) removed most of the scale.

Process design criteria for a full-scale EDR plant can be derived from the EDR pilot results. Preliminary estimates had indicated that a 30 m^3/h EDR plant and clariflocculator for phosphate removal would cost approximately US$750,000. It was concluded that a full-scale EDR plant should run well with electrical adjustments and/or frequent acid cleaning.

Ammonium Nitrate Recovery. The ED system from Explosive Company, Louisiana, was used to concentrate ammonium nitrate solution for recycle to the process while reducing a nitrate effluent problem beginning in mid-1992 (Elyanow and Batchelder, 1994). This company manufactures nitrate chemicals for the explosives industry.

Pickle Liquor Recovery with Bipolar Electrodialysis. A pickle liquor recovery process with bipolar ED is described by McArdle, Piccari, and Thornburg (1991). The initial step in the Acid Recovery System is neutralization of the spent pickle acid with potassium hydroxide (KOH). Neutralization of the free nitric (HNO_3) and hydrofluoric (HF) acid components in the spent material forms the associated potassium salts shown as follows:

$$HF + KOH \rightarrow KF + HOH$$

$$HNO_3 + KOH \rightarrow KNO_3 + HOH$$

Studies have shown that the majority of fluoride present in spent pickle acid is in the form of complexed metal fluorides rather than as free acid. Unlike other recovery systems, such as acid purification, the Acid Recovery System recycles both free and metal complexed metal fluoride components. The conversion of complexed metal fluorides to form the associated potassium salts and metal hydroxides is shown as follows:

$$FeF_3 + KOH \rightarrow 3 KF + Fe(OH)_3$$

$$CrF_3 + KOH \rightarrow 3 KF + Cr(OH)_3$$

$$NiF_2 + KOH \rightarrow 2 KF + Ni(OH)_2$$

Metal values are recovered as metal hydroxides in a relatively dry filter cake suitable for recycle to the steel-processing facility after treatment using readily available commercial techniques.

After filtration, the clean potassium salts are processed in the ED water-splitting step. For this application, the ion-exchange membranes are arranged in a three-compartment cell stack configuration. Electric current applied across the cell stack results in the conversion of potassium salts to the mixed HF and HNO_3 acid and KOH base as follows:

$$KF + HOH \rightarrow HF + KOH$$

$$KNO_3 + HOH \rightarrow HNO_3 + KOH$$

The general flow configuration of the Acid Recovery System (Fig. 12.36) shows its closed-loop operation for recovery of the spent steel pickle acids. Both free and complexed fluoride and nitrate components are recycled to the pickle lines as mixed acids at the required volume and concentration. The potassium components are internally recycled to the neutralization step. Metal values are recovered as a metal hydroxide filter cake suitable for recycle after treatment to the steel-processing facility.

Maintaining the closed-loop operation of the acid recovery system requires an appropriate water balance around the integrated pickle line and acid recovery operations. This is achieved by incorporating a standard ED unit (Fig. 12.37) into the acid recovery system. Proper operation of the integrated system relies on the ability of the system to adjust the concentration of the spent pickle acid. Water is removed by the ED unit into temporary storage if the spent acid concentration delivered to it is lower than desired, and added from temporary storage if the spent acid concentration is higher than desired.

The Acid Recovery System recovers both the free acid and metal complex forms of the fluoride component, as well as the free nitric acid. The economic analysis for a system to recycle 9084 m^3/year (2,400,000 gal/year) of spent pickle acid shows a 2.2-year payback. Operating credits include the value of the recycled acid and the avoidance of disposal costs for the spent acid. It is apparent that the investment in the acid recovery system would become increasingly attractive at higher acid disposal costs. For example, the payback at US$0.75/3.785 L for acid disposal would be approximately 1.5 years.

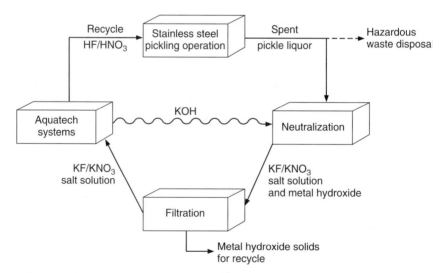

FIGURE 12.36 Acid Recovery System general flow configuration. *(McArdle, Piccari, and Thornburg, 1991.)*

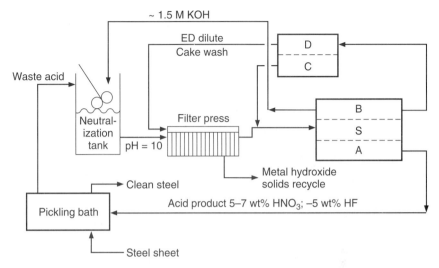

FIGURE 12.37   Acid Recovery System process flow configuration. *(McArdle, Piccari, and Thornburg, 1991.)*

The acid recovery unit has processed an average of approximately 5678 m³ (1.5 million gal) of waste acid each year from Washington Steel's pickle lines during 1989 and 1990. This has led to a 75 percent reduction in waste acid disposal from the pickle lines with a reduction of 86 percent during the last quarter of 1990. These improvements, in large part, resulted from improved integration of the acid recovery unit with the pickle line.

A significant reduction in purchase requirements for HF and HNO₃ pickling acids has been realized after installing the acid recovery unit (Table 12.17). The acid purchases indicated on Table 12.17 include those purchases required for makeup of acid losses in the pickle line. Acid recovery has been greater than 90 percent in the acid recovery unit itself.

Sludge generation in the pickle lines has decreased significantly as a result of operating the pickle tubs at a lower average metals concentration. The reduction in sludge generation is difficult to quantify. However, approximately 75 percent less maintenance downtime has been required to remove sludge from the pickle tubs since startup of the acid recovery unit.

TABLE 12.17   Effect of Acid Recycle on Overall Pickle Line Operations at Washington Steel

	Quantity		
Material	Before acid recovery	After acid recovery	Reduction, %
HF purchase, tons/year (100%)	695	290	58
HNO₃ purchase, tons/year (100%)	1,385	740	47
Spent acid disposal, gal/year	1,750,000	250,000	86

Source: McArdle, Piccari, and Thornburg, 1991.

Current operation of the acid recovery unit is saving Washington Steel approximately US$870,000/year for the pickle line operation. This savings represents a better than 3.0 year payback (Table 12.18) on the initial investment in the unit.

TABLE 12.18 Economics of Acid Recovery System at Washington Steel
(Annual rate)

Operating credits, $	
HF purchase, $1,500/ton (100%)	610,000
HNO$_3$ purchase, $250/ton (100%)	160,000
Spent acid disposal, $0.50/gal	750,000
Sludge disposal, $500/ton	100,000
Total	1,620,000
Operating costs, $	
Membrane maintenance	300,000
Electricity	150,000
Equipment maintenance	100,000
Miscellaneous	200,000
Total	750,000
Net operating credit, $	870,000

Source: McArdle, Piccari, and Thornburg, 1991.

These economics are sensitive to the disposal cost of waste acid. For example, at a disposal cost of US$1.00/3.785 L, payback for the acid recovery unit is better than 2.0 years.

Chemical Applications.   In chemical applications, ED is used to recover valuable metal salts from plating bath rinse water and to purify for reuse such chemicals as diethylene glycol, pentaerythritol, and glycerin (Elyanow and Batchelder, 1994).

Glycol desalting.   One area of promise is desalting of glycols and other polyalcohol solutions. These applications can be put in two general categories:

* Operations, such as oil wells and natural gas pipelines, where glycol is used as a desiccant, with gradual buildup of salt contamination
* Formaldehyde-based chemical reactions which result in the presence of sodium formate as impurity in a polyalcohol product

An ED plant in Italy is used to remove sodium formate from pentaerythritol solution. This facility has been in operation since 1986 with a typical feed of 20 percent pentaerythritol and 15 percent sodium formate. Typical ED product is 25 percent pentaerythritol and 0.5 percent sodium formate. The product is then crystallized for further purification. Pentaerythritol is used in manufacture of resins.

In a petrochemical plant in Colorado, EDR units are used to desalt diethylene glycol solution. In operation since 1986, the typical feed includes 40 percent DEG, 2000 mS/cm conductivity, and 170 ppm Ca. Typical EDR product is 40 percent DEG, 150 µS/cm conductivity, 5 ppm Ca. The product is then evaporated back to strength and used to desiccate CO_2 in oil field application.

An ED plant in Japan is used to desalt polyethylene glycol (PEG) solution. In operation since 1990 the typical feed quantity is 25 percent PEG, 6400 ppm NaCl. Typical ED product is 25 percent PEG and conductivity levels are less than 25 µmho/cm. Desalted PEG is then disposed of by burning.

A pilot plant in Canada is used to remove sodium formate from an ethylene glycol solution. Feed to the EDR unit will come from an electrochemical process in which formaldehyde is reacted to ethylene glycol/sodium formate. The desalted EG from the EDR unit will go to distillation for further purification. The concentrated stream (sodium formate) will be recycled to the reactor. The commercial potential for this application appears to be quite large.

Surface and Wastewater Applications. The first major cooling tower blowdown recovery EDR installation is installed at Eskom's Tutuka Power Station in South Africa (see sec. 12.4.2.5).

Reclamation of biologically treated wastewaters is a more recent application of EDR (Elyanow et al., 1994). Several pilot studies showed EDR works on these waters. In Italy, 2400 h of pilot testing was performed on biologically treated refinery wastewater. Based on the results, a 4088-m³/d (1.08-mgd) plant was installed to recover this water for use as boiler feed. The plant has operated successfully for over five years.

In the Middle East, a 1120-m³/d (317,000-gpd) plant recovers a blend of 60 percent secondary treated municipal wastewater and 40 percent cement manufacturing wastewater. On-site pretreatment consists of chlorination, clarification, and multimedia filtration. The plant is in its ninth year of operation. Newer plants are now in operation on similar applications. A 3785-m³/d (1-mgd) facility in the Far East and a 1136-m³/d (300,000-gpd) system in Texas are now in operation, recovering municipal wastewaters for industrial use.

To date, over 40 EDR systems are treating surface and wastewaters around the world. Five surface and three wastewater facilities have production capacities in excess of 3785 m³/d (1 mgd). Pretreatments range from simple contact filtration with chlorination to clarification and filtration with lime softening. Not one of these plants has had to replace cation or acrylic anion membranes as a result of colloidal, bacterial, organic, metals, or polymer fouling. In addition, no plant has required membrane replacement due to attack by oxidizing disinfectants. While a few plants have required disassembly and washing to remove deposits when serious pretreatment problems have occurred, many plants have never been cleaned in this manner. These are the reasons EDR has a reputation as a membrane process that works on surface and wastewaters.

Food Applications. In food applications, electrodialysis is used for demineralization of whey, demineralization of soya sauce, demineralization/deacidification of molasses and other sugar solutions, and for deacidification of fruit juices (Elyanow and Batchelder, 1994).

Biological Applications. In biological applications, ED is used to demineralize human blood plasmas in the preparation of AHF (antihemophilic factor) and as a purification step in the manufacture of interferon (Elyanow and Batchelder, 1994).

12.4.2.6 Other Potential Applications

Concentration of a Sodium Sulphate Wastewater and Its Conversion into Acid and Alkali. Smirnova et al. (1983) did pilot studies to examine the concentration of a sodium sulphate–type industrial wastewater and the subsequent conversion of the concentrated sodium sulphate solution into sulphuric acid and caustic soda. The sodium sulphate solution was concentrated from 20 to 40 g/L to 260 to 320 g/L in a multicompartment electrodialyzer. The dialysate concentration was about 2 g/L, comprising 90 percent of the feed solution. Solutions of sulphuric acid and caustic soda with concentrations of 17 to 19 percent by weight were manufactured at a second stage in a three-compartment electrodialysis stack from the brine. The highly selective membranes used guaranteed the content of sodium sulphate in both products to be less than 1 percent. The current efficiency for the base was 77 percent and for the acid 65 percent at a current density of 70 mA/cm^2. The power consumption was 3.1 to 3.3 kWh/kg sodium sulphate decomposed in the electrodialysis process.

Concentration of a Sodium Nitrate– and a Carbonate-Sulphate–type Wastewater. The concentration of a sodium nitrate–type wastewater was described by Smirnova and Laskorin (1980). The electrodialysis of the wastewater which contains mainly sodium nitrate (c_o = 1.7 g/L, pH 4.3) and small amounts (up to 1 mg/L) of iron, aluminum, chromium, copper, nickel, and manganese was carried out under constant voltage. The process was completed when the concentration of the feedwater reached about 0.5 g/L. The circulation solution velocity in the dialysate cells was 3 to 4 cm/s. The two membrane pairs MAK-1, MKK-1P and MAK-2, MKK-1P were tested. The results at various current densities are shown in Table 12.19.

No circulation of liquid was applied in the brine compartments to obtain the highest possible concentration. The transfer of water decreases with rise in current density and this promotes concentration of the brine. The highest brine concentration was achieved with the membrane pair MAK-2, MKK-1P. The reason for this is the lower rate of water transported across membrane MAK-2 in comparison with membrane MAK-1.

A maximum brine concentration of 225 g/L was achieved using the membrane pair MAK-2, MKK-1P. The average current efficiency was 70 percent, and the power consumption was about 2 kWh/m^3 of the feed solution. The volume of brine varied between 0.5 and 5 percent of the original feed volume. The water recovery varied between 95 and 99.5 percent.

It was also shown that the treatment of carbonate-sulphate wastewater with a salt content of about 15 g/L yielded a brine concentration of 150 to 180 g/L (10 percent of feed volume) and a dialysate with a salt content of 1 to 2 g/L (90 percent of feed volume). The current efficiency varied between 60 and 80 percent and the energy consumption was about 2 kWh/m^3 of feed solution at a current density of 25 mA/cm^2. No change in membrane resistance was observed over a five-year period.

This work shows that electrodialysis units without circulation of brine in the concentrate compartments operate quite successfully. However, scale-forming chemicals must be absent for successful operation.

Concentration of Sodium Sulphate, Potassium Chloride, Sodium Chloride, and Calcium Chloride. The concentration of sodium sulphate, potassium chloride, sodium chloride, and calcium chloride solutions by electrodialysis was described by Smagin et al. (1983). The concentrations and current efficiencies obtained at different current densities are shown in Table 12.20.

TABLE 12.19   The Electrodialysis of Sodium Nitrate Solution

	Current density, A/m^2		Brine concentration, g/L	Current efficiency, %	Consumption of energy		Water transport, L/m^2·h
Membrane	Initial	End			kWh/kg	kWh/m^3	
MAK-1	8.3	4.2	77.2	73.0	0.79	0.43	3.52
MKK-1P	16.7	7.5	125.3	73.3	1.16	1.44	1.68
	33.3	15.0	170.0	75.1	2.38	2.95	0.84
	50.0	19.2	180.0	59.1	3.91	4.73	0.71
	80.0	32.5	206.0	36.5	7.98	9.02	—
MAK-2	16.7	9.1	186.0	68.5	1.53	1.81	0.95
MKK-1P	50.0	22.5	225.9	64.1	4.67	5.19	0.63

Source: Smirnova and Laskorin, 1980.

TABLE 12.20 Concentration of Sodium Sulphate, Potassium Chloride, Sodium Chloride, and Calcium Chloride at Different Current Densities

Current density, mA/cm^2	Concentrate concentration, ge/L	Current efficiency, %
	Sodium sulphate, $c_o = 43, 8$ me/L	
1.25	1.51	89
2.5	1.91	72
5.0	2.67	72
	Potassium chloride, $c_o = 100$ me/L	
2.5	1.39	77
5.0	1.97	83
7.5	2.21	79
10.0	2.25	79
15.0	2.93	81
	Sodium chloride, $c_o = 50$ me/L	
4.3	2.3	90
10.0	3.4	88
15.0	3.33	85
20.0	3.54	84
30.0	3.96	89
	Calcium chloride, $c_o = 100$ me/L	
5.0	2.16	81
10.0	2.90	87
20.0	3.48	90
30.0	3.60	92

$t = 18°C$, MA-40 and MK-40 membranes.
Source: Smagnin et al., 1983.

Concentration of Calcium Chloride, Sodium Acetate, Hydrochloric Acid, Aluminum Chloride, and Sodium Sulphide. The results obtained from studies of the concentration of calcium chloride, sodium acetate, hydrochloric acid, aluminum chloride, and sodium sulphide are shown in Table 12.21 (Nishiwaki, 1972).

In the calcium chloride studies, the feed solution contained dilute calcium chloride (about 2.9%) contaminated by organic impurities. The calcium chloride was not only concentrated to about 17% but also separated from the organic impurities. In the sodium acetate studies, the effect of current density on performance was studied. With increase in current density, the degree of concentration increased, the current efficiency decreased, and the energy requirements increased. The hydrochloric acid studies show that concentration and separation of hydrochloric acid from a dilute impure feed solution are possible but that the current efficiencies are low. In the sodium sulphide studies, neither the degree of concentration nor the current efficiency were changed much by a 30 percent increase in current density. The energy requirements increased approximately in proportion to the increase in current density. It has been found in other studies that salts of organic acids that have relatively high degrees of dissociation, such as sodium or potassium formates or acetates, could be effectively concentrated by electrodialysis.

Concentration of Electroplating Wastewaters. The concentration of electroplating wastewaters is shown in Table 12.22 (Nishiwaki, 1972). Table 12.21 shows that nickel in electroplating waste liquors can be concentrated more than tenfold so that

TABLE 12.21   Concentration of Dilute Aqueous Solutions by Means of Ion-Exchange Membrane Electrodialysis

	Calcium chloride	Sodium acetate		
Salts				
Current density, mA/cm^2	20	20	40	60
Concentration of dialysate, g/L	29.6	76	76	68
Concentration of concentrate, g/L	170	188	208	222
Current efficiency, %	77.0	83.5	81.3	79.3
Energy consumption, kWh/ton	1320	243	496	759
Hydrochloric acid				
Current density, mA/cm^2		19	50	
Concentration of dialysate				
HCl, g/L		20.0	4.8	
AlCl$_3$, g/L		40.0	3.8	
Concentration of concentrate				
HCl, g/L		65.0	75.2	
AlCl$_3$, g/L		trace	trace	
Current efficiency (hydrochloric acid), %		—	14	
Energy consumption, kWh/ton of HCl		650	—	
Sodium sulphide				
Current density, mA/cm^2		33	42	
Concentration of dialysate, g/L		27	27	
Concentration of concentrate, g/L		125	122	
Current efficiency, %		77.3	78.5	
Energy consumption, kWh/ton		590	790	

Source: Nishiwaki, 1972.

it can be discharged back into the plating baths. Copper in waste liquors can be similarly recovered. The treatment of nickel wash waters by ED for nickel and water recovery is a commercialized process in Japan (Itoi, 1979).

Treatment of Sodium Acetate by Water-Splitting Electrodialysis.   Jeffries et al. (1979) described the treatment of sodium acetate and the recovery of acetic acid by water-splitting electrodialysis. The acetate solution was batch-fed into the water-splitting electrodialysis unit. In this process, anionic, cationic, and bipolar membranes are used to carry out the separation of salts into their respective acids and bases. For example, sodium acetate is converted into sodium hydroxide and acetic acid. The feed stream is introduced into the center chamber and the concentrated acidic and basic streams are removed from the side chambers. The streams are circulated through the chambers in order to decrease the effects of concentration polarization at the membrane surfaces. The three compartments form a repeating cell. Generally, the voltage drop was about 3 V for each unit-cell. Operation was carried out at a current density of 20 to 50 mA/cm^2 and current efficiencies approached 90 percent.

This technology has potential in a number of areas, including resource recovery, pollution control, and chemical processing (Krieger, 1985). In pollution control, for example, it turns liabilities into assets by converting waste salts into reusable or soluble raw materials, sharply reducing waste generation. In stainless steel manufacture, hydrofluoric acid and nitric acids are used as a pickling liquor to remove surface oxides (see Sec. 12.4.2.5). In the process, they pick up metals and become

TABLE 12.22 Concentration of Electroplating Waste Liquors

Nickel plating		
Current density, mA/cm^2	10	15

Waste liquor		
NiSO$_4$, g/L	13.63	12.47
NiCl$_2$, g/L	1.94	1.81
H$_3$BO$_3$, g/L	3.15	3.18

Concentrate		
NiSO$_4$, g/L	115.0	133.4
NiCl$_2$, g/L	29.9	29.7
Current efficiency*, %	35.8	32.9

Copper plating		
Current density, mA/cm^2	15	20

Waste liquor		
K$_2$Cu(CN)$_3$, g/L	8.47	8.41
KCN, g/L	1.07	1.17
KOH, g/L	11.67	11.90

Concentrate		
K$_2$Cu(CN)$_3$, g/L	91.1	85.6
KCN, g/L	8.64	9.4
KOH, g/L	89.1	94.7
Current efficiency[†], %	24.7	17.4

* Based on nickel ions.
[†] Based on copper ions.
Source: Nishiwaki, 1972.

inactive or spent. Treatment of the spent pickling liquor comprises neutralization with potassium hydroxide, forming potassium fluoride/potassium nitrate solution. Metal hydroxides precipitate out and are removed by filtration. It has been shown that a bipolar cell can convert potassium fluoride/potassium nitrate solution into hydrofluoric and nitric acids (typically 3N) and potassium hydroxide (typically 2N) (Krieger, 1985).

The fresh acid mixture can be recycled to the pickling plant and potassium hydroxide to the neutralization plant. Thus, chemicals can be recovered in the process.

Concentration of Calcium. The effect of the concentration of calcium ions in a simulated chloride-sulphate mine water (NaCl 0.29 g/L, Na$_2$SO$_4$ 0.71 g/L) on electrodialysis process performance was studied by Pisaruk et al. (1982). The study was conducted in a five-chamber electrodialyzer with a working surface area of each membrane of 10 cm^2 and a distance of 6 mm between the membranes. The calcium concentration of the feed was changed by calcium chloride addition.

At a calcium concentration level of 25 mg/L in the feed, the calcium in the brine reached a concentration of 610 mg/L. This exceeds the solubility of the calcium ion in the chloride-sulphate brine, as obtained by a solubility method. Consequently, the brine became supersaturated with calcium sulphate. During storage of the brine in a glass container, a calcium sulphate precipitate formed only after 30 days. At a calcium concentration of 41 mg/L, the maximum concentration in the brine reached 1200 mg/L. Calcium sulphate precipitated from this solution after a 12-day storage period. In both these cases there was no precipitate formation on the surface of the ion-exchange membranes. On further increase of the calcium concentration of the feed (>50 mg/L), a precipitate was visually noted in the concentration chambers. A change in the pH of the dialysate from 3 to 10 had no significant effect on the maximum allowable concentration of calcium in the brine.

In order to prevent the formation of a calcium sulphate precipitate on the membranes in the concentration compartments and to obtain a brine of higher concentration, the concentrate was passed through a column filled with a cation-exchanger in the calcium form. It was thought that particles of calcium sulphate on the surface of the cation-exchanger would serve as centers for crystallization of calcium sulphate. However, it was not possible to significantly decrease the calcium concentration in the brine by this means.

The use of an external filter with seed crystals of freshly precipitated calcium sulphate was also investigated. It was found that the maximum allowable concentration of calcium in the original feed solution could be increased from 40 to 50 mg/L by passing the brine through the filter. The allowable concentration of calcium in the dialysate may be increased by approximately 20 percent by using external filters with calcium sulphate seed crystals, connected to the brine circulation system.

Recovery of Gold from Gold-Plating Operations.   Some of the most successful applications of electrodialysis according to Millman and Heller (1982), have been on gold plating baths. Gold recoveries in excess of 99 percent have been obtained. It was demonstrated that electrodialysis played a major role in gold recovery due to low initial costs, low energy consumption, and the ability to produce a product suitable for direct recycle to the plating tank.

The electrodialysis system was operated on the drag-out rinse following the plating bath. A small ion-exchange column was installed on the second rinse to recover the last gram of gold. The concentrate recovered by electrodialysis was returned directly to the plating tank. Approximately 1 k of gold was recovered in the first operation and, with the price of gold at that time, the electrodialysis system had paid for itself.

The successful recovery of cobalt, nickel, chrome, and silver has also been described. Electrodialysis systems are also operating on palladium chloride and cyanide-cadmium wastes. It is expected that the number of electrodialysis plants that treat these wastes will be increased in the future. The wide range of applications and economical cost of electrodialysis has established this process as the preferred method for the recovery of many chemicals, according to Millman and Heller.

Concentration of Radioactive Strontium and Cesium.   Golutivina et al. (1964) described the concentration of radioactive strontium and cesium from water by electrodialysis. This method makes it possible to concentrate radioactive strontium and cesium by a factor of 60 to 70 in 1 h, which is a considerable saving in time as compared with the concentration of these isotopes by evaporation, ion-exchange, and coprecipitation.

Concentration of Sulphuric and Hydrochloric Acids.   Korngold (1978) described the recovery of sulphuric acid from rinsing waters from a pickling process. Sulphuric acid was concentrated from 9100 mg/L to 34,300 mg/L while the diluate contained 3700 mg/L sulphuric acid. About 70 percent of the sulphuric acid in the rinsing water could be recovered by electrodialysis treatment.

The possibility was also demonstrated of concentrating a hydrochloric acid solution containing 4.4 g/L hydrochloric acid, 59 g/L sulpholene-3, and 20 g/L 3-chloro-4-oxysulpholane (chlorhydrin) (Kononov et al., 1984). At a current density of 10 mA/cm^2, a brine was obtained containing 51 ± 1 g/L of hydrochloric acid with a current efficiency of 35 percent. The low current efficiency is explained by the diffusion of acid from the brine into the dialysate and the decrease of the selectivity of the membranes in contact with the concentrated hydrochloric acid solution (50 g/L). The presence of chlorhydrin and sulpholene in the hydrochloric acid solution did not impair the electrochemical properties of the membranes. The concentration of hydrochloric acid is accompanied by transfer of chlorhydrin and sulpholene into the brine. No fouling was experienced. Type MK-40 and MA-40 membranes were used.

By using newly developed anion-exchange membranes (Selemion AAV) whose transport number for hydrogen ions is very small, the acids can be more effectively concentrated by electrodialysis (Urano et al., 1984; Schoeman, 1992).

Concentration of Calcium Chloride. The neutralization of industrial effluents by the precipitation of heavy metal ions with lime leads to a high degree of mineralization of the effluent and makes it unsuitable for reuse. Therefore, desalination of such effluents is necessary. An investigation was therefore carried out to study the desalination of a model solution of calcium chloride ($c_o = 6$ g/L) and to elucidate the role of the ratio of the volumes of solution in the brine and dialysate chambers (Bobrinskaya et al., 1981). The solution was desalinated by a batch scheme in an 11-chambered laboratory electrodialyzer until the dialysate concentration reached 0.5 g/L. Type MK-40 and MA-40 membranes were used. It was shown that it would be desirable to perform the electrodialysis treatment with a volume of diluate 25 to 50 times greater than the amount of brine at a voltage of 1.4 to 1.8 V per cell pair. The highest concentration in the brine was obtained with the smallest volume of brine.

Desalination of a Hard Water. The possibility was demonstrated of desalting a hard water of the chloride-sulphate type (NaCl, 2.9 g/L; Na_2SO_4, 2 g/L; $CaCl_2$, 0.5 g/L; $MgCl_2$ and $NaHCO_3$, 1.3 g/L each) in an electrodialysis unit containing cation-exchange membranes coated with a deposit of disperse anion-exchange resin (Ponomarev et al., 1982). The deposition of the anion-exchange resin on the surface of the cation-exchange membrane decreases the passage of calcium ions into the concentrate and practically eliminates the passage of magnesium ions. Electrodialysis is accompanied by a sharp increase in the pH of the dialysate and a fall in the pH of the concentrate. This is caused by the transfer of H^+ and OH^- ions formed during the dissociation of water in the regions of contact of particles of anion-exchange resin with the surface of the cation-exchange membrane. The increase in the pH of the dialysate causes the formation of suspended particles of calcium carbonate and magnesium hydroxide which, in time, deposit on the bottom of the intermediate vessel. At a current density of 20 mA/cm^2 and an initial concentration of calcium in the model water of 0.19 g/L, a brine with a total salt content up to 120 g/L was obtained.

Concentration of Carbonate Solutions. The concentration of a carbonate-type solution was investigated in the laboratory (Lasskorin et al., 1973). The working surface area of a membrane was 100 cm^2. The feed solution had the following composition: Na_2CO_3, 4 to 7 g/L; $NaHCO_3$, 4 to 7 g/L; Na_2SO_4, 2 to 3 g/L. The total salt content of the solution did not exceed 15 g/L.

The first series of experiments was carried out with liquid circulation in both the diluting and concentrating compartments. A linear liquid velocity of 5 to 6 cm/s was used at a current density of 20 mA/cm^2. The duration of the desalting cycle was 1.5 to 2 h. A fresh portion of feed was introduced after each desalting cycle. The portion of concentrate remained unchanged for 10 cycles. MKK cation– and MAK-V anion–selective membranes were used. The concentrate concentration was increased from 22.9 g/L at the end of the first cycle to 87.8 g/L at the end of the 10th cycle at a current efficiency of about 81 percent. The diluate concentration at the end of the cycles varied between 0.16 and 0.47 g/L.

A second series of experiments was conducted without circulation of liquid through the concentrating compartments. The solvent entered the concentrating compartments as a result of electroosmotic transport through the membranes. The concentrate salt content reached a value of 182.8 g/L after three cycles. The current efficiency varied between 70 and 75 percent, and the energy consumption was about 2.7 kWh/kg salt. A higher brine concentration was obtained without circulation of brine through the concentrate compartments.

Concentration of Caustic Soda and Sodium Chloride. The maximum possible concentration of caustic soda and sodium chloride obtainable by electrodialysis was determined (Smagin and Chukhin, 1975). Caustic soda concentration of 0.07 N and sodium chloride concentration of 1.07 N were chosen as the feed solutions. No circulation of the brine was used. The change of the brine concentration in relation to the current density was determined. MA-40 and MK-40 membranes were used. Maximum concentrations of 346 g/L caustic soda and 305.1 g/L sodium chloride were achieved at current densities of 249 mA/cm^2 and 116.6 mA/cm^2, respectively.

Concentration of Glycerine. The concentration of heat-sensitive nonelectrodialyzable solutes can be carried out by adding dry salt to dilute solutions of these solutes and subsequently electrodialyzing the solution (Harkare et al., 1978). Harkare et al. showed that a partial concentration of glycerine could be obtained while desalting crude glycerine containing 8 percent salts.

Concentration of Waste Seawater from a Flash Evaporator. Asahi Glass developed new membranes (Selemion CMR and ASR) for the concentration of seawater (Kawahara and Suzuki, 1981). As a result of this achievement, it was possible to increase the brine concentration from 170 to 230 g/L. Electric power consumption was reduced by about 35 percent.

Waste seawater from a multistage flash evaporating seawater desalination plant was quite successfully concentrated by the new membranes. The concentration of the waste seawater was approximately twice as high as that of seawater. A concentrated brine of 261 g/L was produced with a power consumption of 225 to 240 kWh/ton salt.

Concentration of Wastewaters from Demineralization Plants. Vysotskii, Parykin, and Ulasova (1983) described the concentration of wastewaters from demineralization plants. The composition of the wastewater feed after sodium cation exchange was: sulphate = 465 mg/L; chloride = 171 mg/L; bicarbonate = 317 mg/L; calcium and magnesium = 0.1 me/L; sodium = 460 mg/L; pH = 8.6; Fe(tot) = 120 − 150 μg/L. The TDS of this water was about 1420 mg/L.

The desalting and concentrating loops of the electrodialysis modules operated in circulating mode. The experiments were conducted at a constant voltage of 2 V per cell pair. When the treated water reached a TDS of 450 to 550 mg/L, it was replaced by raw water. The TDS of the brine was in one case increased from 2145 to 35,700 mg/L and the power consumption was 0.052 kWh/ge salt transported. It is claimed that concentrate concentrations of 100 to 150 g/L can be obtained and that large reduction in effluent volume can be achieved.

Concentration of Reverse Osmosis Brine. The concentration of reverse osmosis brine was described by Jordan, Bearden, and McIlhenny (1975). In this example, the electrodialyzer feed was the brine from a reverse-osmosis unit with a TDS of about 10,000 mg/L. The stream compositions are shown in Table 12.23.

The electrodialyzer produces a brine concentrate (about 13,500 mg/L) and a partially desalted stream that can be returned to the desalting circuit. The volume of brine from the reverse-osmosis unit was reduced from 119 m^3/d (31,400 gpd) to 0.19 m^3/d (50 gpd) after electrodialysis treatment.

Treatment of Industrial Effluents with Sealed-Cell Electrodialysis. Sealed-cell electrodialysis has been described by Kedem, Tanny, and Moaz (1978); Kedem and Cohen (1983); and Kedem and Bar-On (1986); and Schoeman and van Staden (1991). Schoeman and van Staden evaluated SCED for treatment of industrial effluents.

TABLE 12.23 Reverse-Osmosis Brine, Electrodialysis Product, and Electrodialysis Brine Compositions

	Na	Ca	Mg	Cl	SO₄	HCO₃
RO brine	3,820	33	12	3,960	2,610	176
ED product	2,980	27	8	2,840	2,200	—
ED brine	49,500	600	151	62,600	22,200	40

Source: Jordan, Bearden, and McIlhenny, 1975.

Sealed-cell electrodialysis (SCED) was evaluated for desalination/concentration of industrial effluents. It was found that relatively dilute (500 to 3000 mg/L) salt solutions ($NaCl$, NH_4NO_3, Na_2SO_4, $NaNO_3$, $CaCl_2$) could be effectively desalinated (<300 mg/L)/concentrated (up to 15%). Electrical energy consumption of 0.34 to 5.9 kW·h/m³ product water was obtained. Brine volume comprised approximately 2 percent only of the treated feed. However, SCED became less efficient in the 5000 to 10,000 mg/L feed concentration range due to high electrical energy consumption (3.86 to 13.06 kW·h/m³).

It was demonstrated that a relatively dilute ammonium nitrate–type effluent (3600 mg/L TDS) could be treated with SCED for water and chemical recovery. Brine volume comprised 2.8 percent of the treated water, and the electrical energy consumption was determined at 2.67 kW·h/m³. It was shown that it would be difficult to treat a concentrated "ammonium nitrate" effluent (123,700 mg/L TDS) with SCED. However, it would be much easier to desalinate/concentrate a much more dilute effluent (16,557 mg/L TDS). This effluent could be desalinated to 88 mg/L TDS (17.9 kW·h/m³). Brine volume comprised 8.4 percent of the treated feed. Nitrate and ammonium removals of greater than 99 percent were obtained. It was also shown that it would not be possible to desalinate/concentrate an effluent saturated with calcium sulphate with SCED. However, the effluent could be desalinated/concentrated after sulphate removal by chemical precipitation. The effluent was desalinated from 4401 mg/L TDS to 299 mg/L (5 kW·h/m³). Brine volume comprised 5.3 percent of the treated feed.

Scale-forming ions (Ca^{2+}, SO_4^{2-}) affect the SCED process adversely due to membrane scaling and these ions should be removed by nanofiltration, chemical precipitation, or ion exchange prior to SCED treatment. Sealed-cell ED has potential for treatment of relatively low-TDS (<3000 mg/L) nonscaling waters for chemical and water recovery. However, higher-TDS (up to 16 000 mg/L) waters should also be treated effectively depending on the value of the products that can be recovered.

Sodium Silicate Purification. Chemical Company, Illinois (pilot plant), uses the EDR system to remove sodium sulphate from a sodium silicate solution (Elyanow and Batchelder, 1994). It began operation in 1990. This company is looking at EDR to enhance or replace ion exchange, which is the present method for purifying this chemical stream.

12.5 EXAMPLE FACILITIES AND COSTS

The economics of the ED/EDR process for brackish and seawater desalination are described by many authors in the literature (Gleuckstern and Arad, 1984; Rogers, 1984b; Larson and Leitner, 1979; Urano, 1977; and others).

12.5.1 Brackish Water Desalting by Electrodialysis

Costs to desalinate typical brackish waters (Table 12.24) occurring in the United States are shown in Figs. 12.38, 12.39, and 12.40.

Three figures are presented here for ED: one showing capital equipment costs (both inclusive and exclusive of site-related costs and indirect capital costs), one showing operating costs (exclusive of capital charges), and one showing overall water costs, including all inputs. Each one of the figures includes a curve for systems operating on each of the four waters.

Figure 12.38 shows that the capital cost for ED increases as the number of stages increases. It also shows that the costs vary from a maximum of US$274/m³/day (US$1.04/gpd) for a small four-stage system, to a minimum of US$148/m³/day (US$0.56/gpd) for a large two-stage system.

Figure 12.39 shows the operating cost for systems operating on these four waters. It is seen that the costs range from US$0.14/m³ (US$0.52/1000 gal) for a small four-stage system on a difficult water to US$0.08/m³ (US$0.30/1000 gal) for a large two-stage system on a relatively easy water.

Finally, Fig. 12.40 shows the water costs for brackish water desalting by ED. These include not only the pure operating costs, but also a capital charge based on all other inputs. Water costs are seen to vary from a low of US$0.19/m³ (US$0.71/1000 gal) to a high of US$0.35/m³ (US$1.32/1000 gal).

Capital and total costs for brackish water desalination in Japan have also been determined (Urano, 1977). Costs were determined for the following conditions:

1. TDS of raw water 1000, 3000, and 10,000 mg/L.

2. TDS of product water lower than 500 mg/L.

3. Water temperature 35°C.

4. Water intake work and civil engineering work are excluded. The results are shown in Figs. 12.41 and 12.42. The ED process appears to be the most economical process when the TDS of the raw water is in the range of 1000 to 3000 mg/L.

TABLE 12.24 Chemical Composition of Typical Brackish Waters

	Brackish waters			
Chemical composition, ppm	No. 1	No. 2	No. 3	No. 4
Sodium (Na)	886	125	630	900
Calcium (Ca)	118	316	116	250
Magnesium (Mg)	72	69	15	70
Chloride (Cl)	131	67	1054	1450
Sulfate (SO₄)	1943	900	115	590
Bicarbonate (HCO₃)	473	357	78	210
Hardness as CaCO₃	590	1073	354	912
Manganese (Mn)	1	0.10	NIL	0.1
Fluoride (F)			2	
Iron (Fe)	2	1.0	0	0.4
Potassium (K)	16	13	0	5
Nitrate (NO₃)	6.3	19	9	1
Silicate (SiO₃)			17	
Total dissolved solids	3648	1800	2076	3475

Note: All systems operate on water recoveries from 78 to 87 percent and each system is designed to produce a product water with a salinity less than 500 mg/L.
Source: Larson and Leitner, 1979.

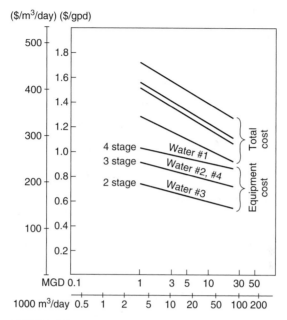

FIGURE 12.38   Capital cost—brackish water desalination by electrodialysis.

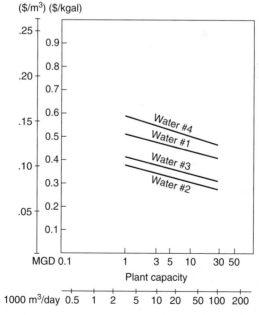

FIGURE 12.39   Operating cost—brackish water desalting by electrodialysis.

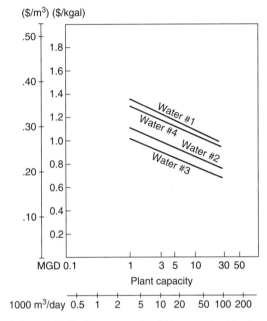

FIGURE 12.40 Water cost—brackish water desalting by electrodialysis.

Typical capital and unit water costs for operating ED plants are shown in Table 12.25 (Glueckstern and Arad, 1984). The capital charges of these plants were based on a 20-year plant life, a 6 percent annual interest rate, and the actual operating-plant factors experienced during the corresponding operating period when the costing was made. It should, however, be noted that the rather low operating plant factors for some of the plants are not necessary because of plant inoperativeness, but are because of other factors, especially low water demand.

12.5.2 Seawater Desalination

The first ED seawater desalting plant in the world began operation in September 1974 in Japan (Glueckstern and Arad, 1984). The capital and operating costs of this small-capacity (120-m^3/day) plant have been reported (Seto et al., 1976).

The construction cost of the plant, excluding site development, buildings, and seawater intake, was 47.7 million Japanese yen. At an exchange rate of approximately 250 yen per U.S. dollar, this amounts to approximately $190,000 or about $6/gpd (1974 prices).

The specific power consumption of the plant was reported to be 16.2 kW h/m^3, and the reported operation and maintenance cost, including membrane replacement, based on a 10-year membrane life, was 80 Japanese yen. By assuming an electrical power cost of $0.04/kWh, a capital fixed-charge rate of 10 percent per year, a 300-day-per-year operation at full capacity, and an exchange rate of 250 Japanese yen per U.S. dollar, the unit water costs were obtained as listed in Table 12.26 (Glueckstern and Arad, 1984). It should be noted that this plant operates automatically and has, therefore, a relatively low labor cost (25 yen/m^3).

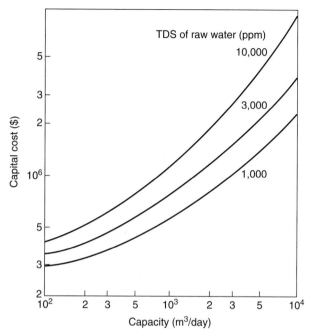

FIGURE 12.41   Capital cost of electrodialysis process.

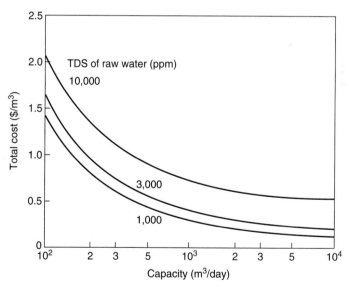

FIGURE 12.42   Total cost of produced water.

TABLE 12.25 Typical Cost Data of Operating ED Plants

Plant location	Year	Type	Feed TDS, ppm	Capacity, 1000 gpd	Capital cost $1000	Capital cost $/gpd	Plant factor, %	Unit water cost, $/m³ Capital charges	Unit water cost, $/m³ Operating cost	Unit water cost, $/m³ Total	Operating period
Siesta Key, Florida	1969	ED	1307	1830	1004	0.55	60.7	0.055	0.082	0.137	1/74–9/74 (9 mo)
Gilette, Wyoming	1972	ED	1840	1500	666	0.44	21.6	0.195	0.134	0.329	7/73–6/74 (12 mo)
Sanibel Island, Florida	1973	ED	2930	1200	456	0.38	62.1	0.107	0.150	0.267	1/74–6/74 (6 mo)
Sorrento Shores, Florida	1973	ED	2786	70	88	1.26	87.0	0.089	0.235	0.324	5/74–2/75 (10 mo)

Source: Glueckstern and Arad, 1984.

TABLE 12.26   Unit Water Costs

Capital cost (neglecting site costs)	$0.53/m^3
Power cost	0.65/m^3
Operation and maintenance cost	0.32/m^3
Total	**$1.50/m^3**

Source: Glueckstern and Arad, 1984.

As already discussed, conventional ED is a relatively high energy consumer when used to convert high-saline waters. Therefore, it cannot compete with the RO process to desalt seawater. The current development work relating to high-temperature ED may, however, make this process much more attractive, primarily because of the potential of a significant reduction in energy consumption (Glueckstern and Arad, 1984).

An economic evaluation of this process was made by the U.S. Bureau of Reclamation by assuming a reasonable range of membrane costs and lifetimes. Based on this data an estimate of ED equipment and operating cost of a 165-m^3/h (1.05-mgd) plant was published at the Fifth International Symposium on Fresh Water from the Sea (Leitz, 1976).

Based on the previously mentioned basic data, an estimate of the total capital investment and the corresponding unit water cost, in accordance with the economic ground rules were made (Glueckstern and Arad, 1984). The reported ED equipment cost was escalated to mid-1977 prices and all other site-related costs were estimated with the relevant costing factors used for RO plants. The resulting capital investment and unit water cost, along with sensitivity analysis of the main economic parameters, are summarized in Table 12.27.

TABLE 12.27   Capital Investment and Unit Water Cost of High-Temperature ED Seawater Desalting Plant

General design data	
Plant capacity	165 m^3/h (1.05 mgd)
Current density	350 A/m^2
Operating temperature	65°C
Power consumption	11 kWh/m^3
Capital cost, $10^6	
Desalting equipment (excluding membranes)	1.97
Membranes at $40/m^2	0.68
Site development, buildings, seawater intake, and pretreatment system	1.10
Total direct cost	3.75
Indirect cost (30%)	1.13
Total capital cost	4.88
Unit water cost, c/m^3	
Capital cost	42.6
Power cost	44.0
Membrane replacement (5-yr membrane life)	10.4
Operation and maintenance including chemicals	14.4
Total	111.4
Effect of economic parameters (percent of base cost)	
25% change in capital cost or fixed charges	9.6
1-c/kWh change in power cost	10.1
50% change in membrane replacement cost	4.8
50% in operation and maintenance cost	6.6

Source: Glueckstern and Arad, 1984.

Other cost studies (Parsi et al., 1980) reported an estimated cost of $4.69 per 3.785 L/day day (gpd) for a 3785-m³/day (1-mgd) plant. The total cost, including indirect cost, was estimated at $6.18 per 3.785 L/day (gpd), with the costs being based on the second quarter of 1980. The resulting unit water cost, evaluated for a 2.5-cent/kWh power cost and a capital recovery factor of 0.1174, was $1.24/m³. After adjustment to the higher power cost and somewhat lower capital recovery factor (Glueckstern and Arad, 1984), the resulting unit water cost amounts to approximately $1.4/m³. This cost is close to the one obtained by escalating the cost from the 1977 study to 1980 prices.

ACKNOWLEDGMENT

The authors would like to thank René Cassidy for typing the manuscript and André Steyn for drawing the graphs.

REFERENCES

Ahlgren, R. M., 1972a, "Electromembrane Processes for Recovery of Constituents from Pulping Liquors," in R. E. Lacey and S. Loeb (eds.), *Industrial Processing with Membranes,* Wiley Interscience, New York.

Ahlgren, R. M., 1972b, "Electromembrane Processing of Cheese Whey," in R. E. Lacey and S. Loeb (eds.), *Industrial Processing with Membranes,* Wiley-Intersience, New York.

American Water Works Association Research Foundation, 1973, *Desalting Techniques for Water Supply Quality Improvement.*

Andreadis, G., and J. W. Arnold, 1978, "Development of a Quality Water Supply for the Island of Corfu, Greece," in *Proceedings of the Sixth Annual Conference of NWSIA,* July 16–20, 1978, Sarasota, Fla.

Andrews, L. S., and G. M. Witt, 1993, "An Overview of RO Concentrate Disposal Methods," in Zahid Amjad (ed.), Reverse Osmosis—Membrane Technology, Water Chemistry and Industrial Applications, Van Nostrand-Reinhold, New York.

Arnold, J. W., 1979, "Operating Experience of a 15,000 Cubic Meter per Day Municipal Desalting Plant at Corfu, Greece," in *Proceedings of the International Congress on Desalination and Water Reuse,* October 21–27, 1979, Nice, France.

Bauer, B., F. J. Gerner, and H. Strathmann, 1988, "Development of Bipolar Membranes," *Desalination,* **68**(2–3):279–292.

Belfort, G. (ed.), 1984, *Synthetic Membrane Processes—Fundamentals and Water Applications,* Academic Press, New York.

Bobrinskaya, G. A., G. A. Lebedinskaya, and A. Yasha Tolov, 1981, "Electrolytic Desalting of Water after Lime Coagulation Purification," *Khimiya i Technologiya Vody,* **3**(4):349–351.

Boudet-Dumy, M., A. Lindheimer, and C. Gavach, 1991, "Transport Properties of Anion-Exchange Membranes in Contact with Hydrochloric Acid Solutions: Membranes for Acid Recovery by Electrodialysis," *Journal of Membrane Science,* **57**(1):57–68.

Brown, D. R., 1981, "Desalting High Salinity Water Using 10-Stage EDR," *Proceedings of the Ninth Annual Conference and International Trade Fair of the National Water Supply Association 2.*

Chiao, Y. C., F. P. Chlanda, and K. N. Mani, 1990, *Bipolar Membranes for Purification of Acids and Bases,* Aquatech Systems, Allied-Signal Inc., Warren, N.J.

Chiapello, J. M., and J. G. Gal, 1992, "Recovery by Electrodialysis of Cyanide Electroplating Rinse Waters," *Journal of Membrane Science,* **68**(3):283–291.

Chlanda, F. P., L. T. C. Lee, and K. L. Liu, 1976, Bipolar Membranes and Methods of Making, U.S. Patent 4,116,889.

Conolly, D. J., and W. F. Gresham, 1966, Fluorocarbon Vinyl Ether Polymers, U.S. Patent 3,282,875.

Davis, T. A., and G. F. Brockman, 1972, "Physiochemical Aspects of Electromembrane Processes," in R. E. Lacey and S. Loeb (eds.), *Industrial Processing with Membranes,* Wiley Interscience, New York.

De Körösy, F., 1968, "Influence of Strongly Bound Counter-Ions on Permselective Membranes," U.S. Off. Saline Water Res. Dev. Rep. 380.

Dohno, R., T. Azumi, and S. Takashima, 1975, "Permeability of Monocarboxylate Ions Across an Anion-Exchange Membrane," *Desalination* **16**(1):55–64.

Eisenmann, J. T., 1977, "Recovery of Nickel from Plating Bath Rinse Water by ED," *Plat. Surf. Finish,* **64**(11):34.

Elyanow, D., and B. Batchelder, 1994, "Electrodialysis for the Recovery of Value Added Products," presented at *African Water Technology Conference,* June 6–9, 1994, Johannesburg, South Africa.

Elyanow, D., R. P. Allison, B. P. Hernon, and the Staff of Ionics Inc., 1994, "Electromembrane Technology in Water Treatment: ED, EDR, and EDI," presented at *African Water Technology Conference,* June 6–9, 1994, Johannesburg, South Africa.

Elyanow, D., A. Sieveka, and J. Mahoney, 1981, "The Determination of Super-Saturated Limits in an EDR Unit with Aliphatic Anion Membranes," *Technical Proceedings, Ninth Annual Conference, National Water Supply Improvement Association,* Washington, D.C., published by NWSIA, Ipswich, Mass.

Elyanow, D., R. G. Parent, and J. Mahoney, 1980, "Parametric Tests of an Electrodialysis Reversal System with Aliphatic Anion Membranes," Report to OWERT U.S. Department of the Interior (Contract No. 14-34-0001-9510), Washington, D.C.

Escudier, J. L., P. Cottereau, and M. Moutounet, 1989, "Electrodialysis Applications in the Treatment of Grape Musts," *Bull. O.I.V.* **62**(695–696), 20–33, Fr.

Fraivillig, J. B., 1983, *Reverse Osmosis/Electrodialysis Reversal Comparison,* Permasep Products (Du Pont), Hemel Hampstead, Herts. HP2 7DP.

Fromonot, G., 1992, E.I.V.S. S.A., Filiale de Corning France, 11, Chemin de Ronde, B.P. 36, 78110 Le Vesinet, France.

Glueckstern, P., and N. Arad, 1984, "Economics of the Application of Membrane Processes, Part 1: Desalting Brackish and Seawater," in G. Belfort (ed.), *Synthetic Membrane Processes—Fundamentals and Water Applications,* Academic Press, New York.

Golutivina, M. M., G. M. Nikow, G. A. Kuznetsova, and T. A. Kazakova, 1964, "Concentration of Radio-Active Strontium and Cesium from Water by Electrodialysis," *Radiokhimuya,* **6**:738–742.

Grot, W. G., 1973, Laminates of Support Material and Fluorinated Polymer Containing Pendant Side Chains Containing Sulfonyl Groups, U.S. Patent 3,770,567.

Gudernatsch, W., Ch. Krumbholz, and H. Strathmann, 1989, "Development of an Anion-Exchange Membrane with Improved Permeability for Organic Acid of High Molecular Weight," in *Proc. 6th Intl. Symp. on Synthetic Membranes in Science and Technology,* September 4–8, 1989, Tübingen, Germany, pp. 223–226.

Harkare, W. P., B. S. Joski, V. K. Indusekhar, G. T. Gadre, and N. Krishnasnamy, 1978, "Metathesis by Electrodialysis Technique," *Proc. Ion-Exchange Symp.,* February 1978, Bhavnaagar, India.

Highfield, W. H., 1980, "Electrodialysis in Industrial Water Treatment," presented at the *41st Annual Meeting of the International Water Conference,* October 1980, Pittsburgh, Pennsylvania.

Hodgkiess, T., 1987, "Electrodialysis," *European Desalination Association Seminar on Small Plant Applications for Desalination Technology,* June 19, 1987, London.

Huffman, E. L., and E. Lacey, 1972, "Engineering and Economic Considerations in Electromembrane Processing," in E. R. Lacey and S. Loeb (eds.), *Industrial Processing with Membranes,* Wiley Interscience, New York.

Hughes, M., A. E. Raubenheimer, and A. J. Viljoen, 1992, *Electrodialysis Reversal at Tutuka Power Station, RSA: Seven Years Design and Operating Experience,* Eskom Chemical Engineering and Chemistry Division, Johannesburg, South Africa.

Ionics Inc., 1988, Product Bulletin, Watertown, Mass.

Ionics Inc., 65 Grove Street, Watertown, Massachusetts.

Itoi, S., 1979, "Electrodialysis of Effluents from Treatment of Metallic Surfaces," *Desalination,* **28**(3):193–205.

Jeffries, T. W., D. R. Omstead, R. R. Cardenas, and H. P. Gregor, 1979, "Membrane-Controlled Digestion: Effect of Ultrafiltration and Anaerobic Digestion of Glucose," *Biotechnology and Bio-Engineering Symp.* (USA), No. 8, 1978, 37–49, John Wiley and Sons, 1979.

Jordan, D. R., M. D. Bearden, and W. F. McIlhenny, 1975, "Blowdown Concentration by Electrodialysis," *Chemical Engineering Progress,* **71**(7):77–82.

Jordan, D. R., W. F. McIlhenny, and G. T. Westbrook, 1976, "Cooling Tower Effluent Reduction by Electrodialysis," *American Power Conference.*

Katz, W. E., 1979, "The Electrodialysis Reversal (EDR) Process," *Desalination,* **28**:31–40.

Katz, W. E., 1977, "The Electrodialysis Reversal (EDR) Process," presented at the *International Congress on Desalination and Water Reuse,* November/December 1977, Tokyo, published by Ionics Inc., Watertown, Mass. (Bulletin TP. 307).

Katz, W. E., 1971, "Electrodialysis for Low TDS Waters," *Ind. Water Engng.,* June/July (Bulletin TP.301).

Katz, W. E., 1982, "Desalination by ED and EDR—State-of-the-Art in 1981," *Desalination,* **42**:129.

Kawahara, T., and K. Suzuki, 1981, "Utilization of the Waste Concentrated Seawater in Desalination Plants," *Desalination,* **38**(1–3):499–507.

Kedem, O., and Y. Maoz, 1976, "Ion Conducting Spacer for Improved Electrodialysis," *Desalination,* **19**(1–3):465–470.

Kedem, O., and J. Cohen, 1983, "EDS-Sealed-Cell Electrodialysis," *Desalination,* **46**(3):291–299.

Kedem, O., G. Tanny, and Y. Maoz, 1978, "A Simple Electrodialysis Stack," *Desalination,* **24**(1–3):313–319.

Kedem, O., and Z. Bar-On, 1986, "Electro-Osmotic Pumping in a Sealed-Cell ED Stack," *AIChE Symposium Series., Ind. Membrane Processes,* **82**(248):19–27.

Kneifel, K., and K. Huttenbach, 1980, "Properties and Long-Term Behaviour of Ion-Exchange Membranes," *Desalination,* **34**(1–2):77–95.

Kobias, E. J. M., and D. M. Heertjes, 1972, "The Poisoning of Anion-Selective Membranes by Sodium Dodecylsulphate," *Desalination,* **10**:383.

Kononov, A. V., M. I. Ponomarev, L. V. Shkaraputa, V. D. Gebenyuk, and V. T. Sklyar, 1984, "Removal of Hydrochloric Acid from Wastewaters Containing Organic Synthesis Products," *Khimiya i Tekhnolo Giya Vody,* **6**(1):66–68.

Korngold, E., 1970, "Present State of Technological Development of Permselective Polyethylene Membranes at the Negev Institute for Arid Zone Research," *Water Desalination Symposium,* Beersheva, Israel.

Korngold, E., 1973, "The Development of New Membranes for Use in Desalination," *Proc. Fourth Int. Symp. Fresh Water Sea,* **3**:99–109.

Korngold, E., 1976, "Electro-Transport of Large Organic Anions Through Anion-Exchange Membranes," *Proc. Fifth Int. Symp. Fresh Water Sea,* **3**:33–42, May 16–20.

Korngold, E., 1978, "Electrodialysis in Advanced Waste Water Treatment," *Desalination,* **24**(1–3):129–139.

Korngold, E., 1984, "Electrodialysis—Membranes and Mass Transport," in G. Belfort (ed.), *Synthetic Membrane Processes,* Academic Press, New York.

Korngold, E., F. Dekörösy, R. Rahav, and M. F. Taboch, 1970, "Fouling of Anion Selective Membranes in Electrodialysis," *Desalination,* **8**:195–220.

Korngold, E., K. Kock, and H. Strathmann, 1978, "Electrodialysis in Advanced Waste Water Treatment," *Desalination,* **24**(1–3):129–139.

Krieger, J. H., 1985, "Process Converts Wastes, Pollutants into Valuable Raw Materials," *Chemical and Engineering Technology,* Dec. 23, p. 23.

Kusakari, K., F. Kawamata, N. Matsumoto, H. Sakeiki, and A. Terada, 1977, "Electrodialysis Plant at Hatsushima," *Desalination*, **21**(1):45–50.

Kusomoto, E., T. Sata, and Y. Mitzutani, 1973, "New Anion-Exchange Membrane Resistant to Organic Fouling," *Proc. Fourth Int. Symp. Fresh Water Sea*, **3**:111–118.

Lacey, R. E., 1966, *Membrane Processes for Industry.* Southern Research Institute, Birmingham, Ala.

Lacey, R. E., 1967, "Transport of Electrolytes Through Membrane Systems," U.S. Off. Saline Water Res. Dev. Rep. 343.

Lacey, R. E., and E. W. Lang, 1969, U.S. Off. Saline Water Res. Dev. Rep. 398.

Lacey, R. E., 1972, "Basis of Electromembrane Processes," in R. E. Lacey and S. Loeb (eds.) *Industrial Processing with Membranes,* Wiley-Interscience, New York.

Lacey, R. E., and S. Loeb (eds.), 1972, *Industrial Processing with Membranes,* Wiley-Interscience, New York.

Lakshminarayanaiah, N., 1965, "Transport Phenomena in Artificial Membranes," *Chem. Revs.,* **65**(5):526.

Larson, T. J., and G. Leitner, 1979, "Desalting Seawater and Brackish Water: A Cost Update," *Desalination,* **30**(1–3):525–539.

Lasskorin, B. M., N. M. Smirnova, Y. I. Tisov, G. N. Gorina, and A. V. Borisov, 1973, "Use of Electrodialysis with Ion-Selective Membranes for Concentrating Carbonate Solutions," *Zhurnal Priklad No. 1 Khimii,* **46**(9):2117–2119.

Leitz, F. B., 1972, Apparatus for Electrodialysis of Electrolytes Employing Bilaminar Ion-Exchange Membranes, U.S. Patent 3,654,125.

Leitz, F. B., 1976, "Electrodialysis for Industrial Water Cleanup," *Environmental Science and Technology,* **10**(2):136–139.

Leitz, F. B., 1976, "Desalination of Seawater by Electrodialysis," *5th International Symposium on Fresh Water from the Sea,* Alghero, Sardinie, Italy, **3**:105–114.

Leitz, F. B., and J. L. Eisenmann, 1981, "Electrodialysis as a Separation Process," *AIChE Symposium Series,* **77**(204):204–212.

Lightfoot, E. N., and J. Y. Friedman, 1954, "Ion-Exchange Membrane Purification of Organic Electrolytes," *Ind. Eng. Chem.,* **46**:1579.

Liu, K. J., F. P. Chlanda, and K. J. Nagasubramanian, 1977, "Use of Bipolar Membranes for Generation of Acid and Base: An Engineering and Economic Analysis," *J. Membr. Sci.,* **2**:109.

Mani, K. N., 1991, "Electrodialysis Water Splitting Technology," *Journal of Membrane Science,* **58**:117–138.

Mani, K. N., F. P. Chlanda, and C. H. Byszewski, 1988, *Aquatech Membrane Technology for Recovery of Acid/Base Values from Salt Streams,* Aquatech Systems, Allied-Signal Inc., Warren, N.J. (Desal, **68**:149–166).

Mason, E. A., and T. A. Kirkham, 1959, "Design of Electrodialysis Equipment," in *Chemical Engineering Progress Symposium Series:Adsorption, Dialysis and Ion-Exchange,* **55**(24):173–189.

Matsui, K., E. Tobita, K. Sugimoto, K. Kondo, T. Seita, and A. Akimoto, 1986, "Novel Anion-Exchange Membranes Having Fluorocarbon Backbone: Preparation and Stability," *J. Appl. Polym. Sci.,* **32**(3):4137–4143.

McArdle, J. C., J. A. Piccari, and G. G. Thornburg, 1991, "Aquatech Systems Pickle Liquor Recovery Process—Washington Steel Reduces Waste Disposal Costs and Liability," *Iron and Steel Engineer,* May 1991, pp. 39–43.

Meller, F. H., 1984, *Electrodialysis—Electrodialysis Reversal Technology,* Ionics Inc., 65 Watertown, Mass.

Melzer, J., and J. van Deventer, 1983, "Electrodialysis Used for Desalination of Waste Water in an Eskom Power Station," presented at a *Symposium on Desalination, New Developments and Industrial Application,* October 27, CSIR Conference Centre, Pretoria.

Milman, W. G., and R. J. Heller, 1982, "Some Successful Applications of Electrodialysis," *4th Conference on Advanced Pollution Control for the Metal Finishing Industry,* January 18–20, 1982, Lake Buena Vista, Spain, EPA-600/9-82-0225, pp. 70–74.

Mintz, M. S., 1963, "Electrodialysis: Principles of Process Design," *Ind. Eng. Chem.,* **55**(6):18–28.

Miva, T., 1977, "Desalting by Electrodialysis," *Desalination,* **20**:375.

Moutounet, M., and J. L. Escudier, 1991, "Tartaric Stabilization of Wines and Electrodialysis: New Prospects," *Filtration and Clarification '91,* No. 332, August 1991.

Nishiwaki, T. N., 1972, "Concentration of Electrolytes Prior to Evaporation with an Electromembrane Process," in R. E. Lacey and S. Loeb (eds.), *Industrial Processing with Membranes,* Wiley-Interscience, New York.

Parsi, E. J., T. A. Prato, and K. O'Donoghue, 1980, "Status of High Temperature Electrodialysis," presented at the *Eighth Annual Conference of the National Water Supply Improvement Association,* San Francisco, July 6–10.

Pisaruk, V. I., O. R. Shendruk, R. L. Voloshiva, and V. D. Grebenyuk, 1982, "Producing Brine by Electrodialysis from Water of the Chloride-Sulphate Type with a Varying Content of Calcium," *Khimiya i Tekhnologiya Vody,* **4**(1):62–64.

Ponomarev, M. I., O. R. Shendrik, V. T. Pisaruk, and V. D. Grebenyuk, 1982, "The Desalination of Hard Waters by Electrodialysis," *Khimiya i Technologiya Vody,* **4**(2):159–161.

Rogers, A. N., 1984a, "Design and Operation of Desalting Systems Based on Membrane Processes," in G. Belfort (ed.), *Synthetic Membrane Processes—Fundamentals and Water Applications,* Academic Press, New York.

Rogers, A. N., 1984b, "Economics of the Application of Membrane Processes, Part 2: Wastewater Treatment," in G. Belfort (ed.), *Synthetic Membrane Processes—Fundamentals and Water Applications,* Academic Press, New York.

Rubenstein, I., 1990, "Theory of Concentration Polarization Effects in Electrodialysis on Counter-Ion Selectivity of Ion-Exchange Membranes with Differing Counter-Ion Distribution Coefficients," *J. Chem. Soc. Faraday Trans.,* **86**(10):1857–1861.

Sata, T., R. Izuo, and Y. Mizutani, 1984, "Study of Membrane for Selective Permeation of Specific Ions," *Soda to Enso,* **35**(425):313–336.

Schoeman, J. J., 1986, "Rapid Determination of the Fouling of Electrodialysis Membranes by Industrial Effluents," *Water SA,* **12**(2):103–116.

Schoeman, J. J., 1992, "Electrodialysis of Salts, Acids and Bases by Electro-Osmotic Pumping," Ph.D. diss., University of Pretoria, Pretoria, South Africa.

Schoeman, J. J., I. J. M. Buys, I. B. Schutte, and H. MacLeod, 1988, "Pilot Investigation on the Treatment of Fertilizer Manufacturing Process Effluent Using Lime and Electrodialysis Reversal," *Desalination,* **70**(1–3):407–429.

Schoeman, J. J., and J. F. van Staden, 1991, "Evaluation of Sealed-Cell Electrodialysis for Industrial Effluent Treatment," *Water SA,* **17**(4):307–320.

Schutte, C. F., 1983, *Desalination: A South African Perspective,* Water Research Commission, Pretoria, South Africa.

Scott, K., 1990, *Membrane Separation Technology—Industrial Applications and Markets,* Scientific and Technical Information, Oxford, England.

Seko, M., 1962, *Dechema Monograph,* **47**:575.

Seto, T., L. Ehara, R. Komoni, A. Yamaguchi, and T. Miva, 1976, "Seawater Desalination by Electrodialysis," *5th International Symposium of Fresh Water from the Sea,* Alghero, Sardinie, Italy, **3**:131–138.

Shaffer, L. H., and M. S. Mintz, 1966, Chap. 6 in K. S. Spiegler (ed.), *Principles of Desalination,* Academic Press, New York.

Shi, S., and P. Chen, 1983, "Seawater Desalination by Electrodialysis," *Desalination,* **46**:191–196.

Simon, G. P., and C. Calmon, 1986, "Experimental Methods for the Determination of Non-Transport Properties of Membranes," *Desalination,* **59**(106):103.

Smagin, V. N., and V. A. Chukhin, 1975, "Concentration of Brines of Desalination Plants in Electrodialysis," *5th International Symposium on Fresh Water from the Sea,* **3**:139–148.

Smagin, V. N., V. A. Chukhin, and V. A. Kharchuch, 1983, "Technological Account of Electrodialysis Apparatus for Concentration," *Desalination,* **46**(1–3):283–290.

Smirnova, N. M., B. N. Laskorin, J. S. Mishukova, and A. V. Borisov, 1983, "The Application of Electrodialysis with Ion-Exchange Membranes for Treatment of Sodium Sulphate Solutions," *Desalination,* **46**:197–201.

Smirnova, N. B., and B. N. Laskorin, 1980, "Concentration of Electrolytes in Seawater by Electrodialysis," *Proceedings of the 7th International Symposium on Fresh Water from the Sea,* Amsterdam, 1980, **2**:75–80.

Smith, J. D., and J. L. Eisenmann, 1964, "Electrodialysis in Waste-Water Treatment," *Eng. Bull. Purdue Univ. Eng. Ext.,* **Ser. 117**:738–760.

Smith, J. D., and J. L. Eisenmann, 1967, "Electrodialysis in Waste-Water Recycle," Report WP-20-AWTR-18, Federal Water Pollution Control.Administration.

Solt, G. S., and T. Wen, 1992, "Characterising the Spirally Wound Electrodialysis Modules," in *Engineering of Membrane Processes,* May 13–15, 1992, Garmisch-Partenkirchen, Bavaria, Germany.

Strathmann, H., 1992, "Electrodialysis," in W. S. Winston, Ho and Kamalesk K. Sirkar (eds.), *Membrane Handbook,* Van Nostrand-Reinhold, New York.

Tamamushi, B., and K. Tamaki, 1959, "Adsorption of Long-Chain Electrolytes at the Solid-Liquid Interface, Part 3—The Adsorption of Ion-Exchange Resins," *Trans. Farad Soc.,* **55**:1013–1016.

Thompson, M. A., and M. P. Robinson, Jr., 1991, "Suffolk Introduces EDR to Virginia," *American Water Works Association Membrane Conference Proceedings,* pp. 129–148.

Tokuyama Soda, 1988, Product Bulletin, Tokyo, Japan.

Trivedi, D. S., and R. Prober, 1972, "On the Feasibility of Recovering Nickel from Plating Wastes by Electrodialysis," *Ion Exch. Memb.,* **1**:37.

Tuwiner, S. B., 1962, *Diffusion and Membrane Technology,* Reinhold Publishing Co., New York.

Urano, K., 1977, "Present Status of Electrodialysis Processes in Japan," *Desalination,* **20**:365–374.

Urano, K., T. Ase, and Y. Naito, 1984, "Recovery of Acid from Wastewater by Electrodialysis," *Desalination,* **51**(2):213–226.

USAID Desalination Manual, 1980, CH2M, International Corporation, Gainesville, Fla.

Van Duin, P. J., 1970, *Proc. Third Int. Symp. Fresh Water Sea,* **2**:141.

Van Duin, P. J., 1973, "Poisoning of Electrodialysis Membranes," *Proc. Fourth Int. Symp. Fresh Water Sea,* **3**:253–259.

Vysotskii, S. P., V. S. Parykin, and S. A. Ulasova, 1983, "Use of the Series-Manufactures UEO-50-4/12-5 Electrodialysis Plants for Concentrating the Waste Waters from Demineralization Plants," *Thermal Engineering,* **30**(9):540–542.

Westbrook, G. T., and L. F. Wirth, Jr., 1976, "Water Reuse by Electrodialysis," *Ind. Water Engng.,* April/May, p. 8.

Wilson, J. R. (ed.), 1960, *Demineralization by Electrodialysis,* Butterworths Scientific Publications, London.

Wirth, J. R., and G. Westbrook, 1977, "Cooling Water Salinity and Brine Disposal Optimized with Electrodialysis Water Recovery/Brine Concentration System," *Comp,* May, pp. 33–37.

Yamane, R., T. Izuo, and Y. Mizutani, 1965, "Ion-Exchange Membranes XXIV: Permeability of the Amphoteric Ion-Exchange Membranes," *Denki Kagaku,* **33**(8):589–593.

CHAPTER 13
PHASE CONTACT PROCESSES AND APPLICATIONS

P. Aptel
Laboratoire de Génie Chimique, CNRS
Université Paul Sabatier
Toulouse, France

M. J. Semmens
Department of Civil Engineering
University of Minnesota
Minneapolis, Minnesota

13.1 ANALYTICAL APPLICATIONS

A number of studies have been conducted to use the selective transport properties of membranes as a means of separating gases and specific organics from liquid matrices for analytical purposes (e.g., Dheandhanoo, 1989; Melcher and Morabito, 1990; Fogelquist et al., 1986). Teflon, silicone, and microporous polypropylene membranes have been evaluated and used in a variety of different applications including studies of reaction kinetics (Calvo et al., 1981), direct determination of organics in fermenters (Calvo et al., 1981; Bier et al., 1987), measurement of blood-gas concentrations (Meyer et al., 1988), and automated on-line measurements of organics in aqueous solutions (Dheandhanoo, 1989; Melcher and Morabito, 1990).

To illustrate the way in which these authors exploit the selective transport properties of different membranes, we focus on an example application: that of measuring the concentration of volatile organic compounds in water. A common method of analyzing the concentration of low concentrations of volatile organics in water is the use of head-space analysis. This method effectively separates the volatile contaminants from the nonvolatile (organic and inorganic) components present in the sample matrix. In this approach, a water sample is transferred to a sealed vial, leaving a small head space. The vial is shaken and the contaminant is allowed to equilibrate and partition between the phases, as depicted in Fig. 13.1. If the initial concentration in the water is C_{wo} and the volume of the water V_w, then the total mass of the compound in the water sample is $V_w C_{wo}$. When the compound equilibrates between the phases, the distribution of the compound between the phases is determined by the

Henry's constant H_u and the relative volumes of the two phases. The mass balance and the equilibrium relationship enable the initial water concentration to be calculated from the measured gas-phase concentration as shown:

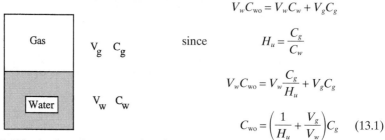

$$V_w C_{wo} = V_w C_w + V_g C_g$$

since

$$H_u = \frac{C_g}{C_w}$$

$$V_w C_{wo} = V_w \frac{C_g}{H_u} + V_g C_g$$

$$C_{wo} = \left(\frac{1}{H_u} + \frac{V_g}{V_w} \right) C_g \qquad (13.1)$$

FIGURE 13.1 A schematic representation of the equilibrium distribution of a volatile specie between the water and gas phases in a sealed vial.

From an examination of Eq. (13.1), when the head space is very small, the ratio of the gas volume to the water volume becomes negligible and the gas-phase concentration may be assumed to be in equilibrium with the initial water-phase concentration.

Large water volumes may be needed for this analytical approach, and it is clumsy and prone to errors. Organic compounds may be lost from the water in the process of transferring the water to the container. Organics may adsorb to the walls of the container or leak from the system during the equilibration period, and, finally, manual sampling techniques for the gas phase are also prone to error. An automated method for the more rapid and routine measurement of dissolved contaminant concentrations is desirable.

Dissolved gases and organic compounds in water can diffuse across gas-permeable hollow-fiber membranes and partition into the gas phase on the other side of the membrane, and, using an approach akin to head-space analysis, membranes can be designed to provide a sensor for the dissolved concentrations of such species.

An example of a hollow-fiber contactor that could be used to measure the concentration of volatile organic compounds (VOCs) in water is illustrated in Fig. 13.2. In this configuration, the water containing the VOCs is drawn past hollow-fiber membranes in a cross-flow mode while a carrier gas is pumped through the inside of the hollow fibers. As the gas is pumped through the fibers, the dissolved gases and VOCs in the water permeate through the membrane wall and partition into the flowing gas. If the fibers are long and the gas flow rate is slow enough, then the gas can approach equilibrium with the water before the gas exits the hollow fibers. The gas can then be sent directly to a gas chromatograph (GC) or, if the concentrations are too low, to a tenax trap which can later be purged to a GC. The measured gas-phase concentration can then be used to estimate the concentration in solution.

In designing such a membrane contactor, it is necessary to consider the rate of mass transfer across the membrane and to determine the length of membrane, water velocity, and gas flow rate required to ensure that equilibrium is approached.

From Table 8.3 we obtain the mass transfer correlation for the water flow over the outside of the fibers. Since in this case we do not wish the concentration in solution to change significantly across the module, we use the correlation of Yang and Cussler (1986) for a low fiber-packing density.

$$\text{Sh} = 1.38 \text{Re}^{0.34} \text{Sc}^{0.33} \qquad (13.2)$$

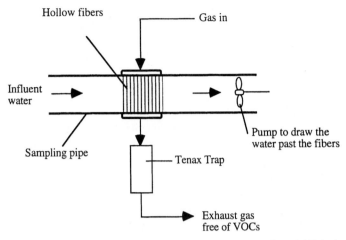

FIGURE 13.2   An apparatus for measuring the concentration of VOCs in water.

where $\mathrm{Re} = v d_e / v$, $\mathrm{Sh} = k_L d_e / D$ and $\mathrm{Sc} = v/D$ and the value of d_e is based on the outside diameter of the fibers.

For the gas flow inside the fibers, we use the Gratz-Leveque correlation for low Reynolds numbers and, in this case, the inside fiber diameter is used:

$$\mathrm{Sh} = 1.85\left(\mathrm{ReSc}\frac{d_e}{L}\right)^{0.33} \tag{13.3}$$

The membrane resistance is given by:

$$k_m = \frac{\varepsilon D_m}{\delta\tau} \tag{13.4}$$

For this problem, it is best to describe the overall mass transfer on the basis of the gas-phase concentrations, and it is calculated as follows:

$$\frac{1}{K_G} = \frac{H_u}{k_l} + \frac{1}{k_m} + \frac{1}{k_g} \tag{13.5}$$

Next, we examine the mass transfer to the inside of the fiber and the increase in concentration in the gas phase along the fiber length. A mass balance across a differential segment of the fiber under steady-state conditions yields the following equation:

$$v\,\frac{dc_g}{dx} = 4\,\frac{K_G}{d_f}(c_g^* - c_g) \tag{13.6}$$

The concentration of the solute along the length of the fiber is obtained by assuming that the concentration of the solute in the aqueous phase remains constant and by integrating this expression with the boundary condition that $c_g = 0$ at $x = 0$:

$$c_g = \frac{c_g^* e^{4K_G L/d_e v} - c_g^*}{e^{4K_G L/d_e v}} \qquad (13.7)$$

$$\frac{c_g}{c_g^*} = \frac{e^{4K_{OL} L/d_e v} - 1}{e^{4K_{OL} L/d_e v}} \qquad (13.8)$$

This equation suggests that an infinite length of membrane is required to achieve an equilibrium between the internal carrier gas and the external liquid. However, for a specified approach to equilibrium, the fiber length is readily calculated as a function of the fiber diameter and the gas velocity inside the fiber.

EXAMPLE 13.1 *A hollow-fiber membrane is to be used to measure the concentration of carbon tetrachloride in water. The membrane is supplied with a nitrogen carrier gas to provide a velocity of 1 cm/s inside the fiber and, under the conditions of operation, the overall mass transfer coefficient is calculated to be 0.003 cm/s. Calculate the length of fiber needed to obtain a gas concentration of carbon tetrachloride within 1 percent of equilibrium.*

In this example, $c_g/c_g^* = 0.99$ and Eq. (13.8) may be simplified and rearranged as follows:

$$0.99 = \frac{e^{4K_G L/d_e v} - 1}{e^{4K_G L/d_e v}}$$

$$\therefore e^{4K_G L/d_e v} = 100$$

taking logs and rearranging:

$$L = \frac{d_e v}{4K_G} \ln 100$$

$$L = \frac{(0.0280)1}{4(0.003)} \ln 100 = 10.7 \text{ cm}$$

Note that the Henry's constant is not needed to solve this equation since it is already incorporated into the overall mass transfer coefficient, as shown in Eq. (13.5).

This example demonstrates that small fiber lengths are needed to approach an equilibrium between the internal gas phase and the external water.

13.2 PERVAPORATION

In the pervaporation process, a membrane is used to selectively separate solvents by reducing the vapor pressure on the permeate side of the membrane. The term *pervaporation* is derived from the two major steps involved in the separation process—namely, permeation and evaporation. Components in the stream being treated must permeate through the membrane by first dissolving into and then diffusing through the membrane. The evaporation process occurs on the permeate side of the membrane and is assisted by providing a vacuum. The driving force for each component is determined by the difference in chemical potential (or partial vapor pressure in first approximation) across the membrane, and its rate of transfer across the membrane is determined largely by its solubility and its diffusity in the membrane.

The permeate is generally removed as a vapor and is recovered by condensation. A number of applications exist for pervaporation processes, including dehydration of solvents, organic-organic separations, and recovery of volatile solvent from wastewater. Pervaporation is mainly used for the dehydration of alcoholic azeotropes (Aptel et al., 1976; Néel, 1991; Néel, 1992). The largest commercial industrial applications of the process are in the dehydration of ethanol and isopropanol, and 55 units have been built by GFT of Germany for these applications (Zhang and Drioli, 1995). Pilot plants already exist for organic-organic separations and the first full-scale plants are in project.

Other recent applications include: the extraction of solvents from industrial wastewaters first developed by Membrane Technology & Research Inc. (MTR), United States (Wijmans, Baker, and Athayde, 1994); the recovery of volatile by-products such as alcohols or aroma compounds from fermentation broths (Böddeker, 1994); and the elimination of volatile organic compounds (VOCs) from drinking water (Castro and Zander, 1995). No commercial systems have yet been developed for these applications.

13.2.1 Mass Transfer Considerations

13.2.1.1 Membrane and Liquid Resistances. The removal of organics from water by pervaporation is accomplished using dense or composite organophilic polymer membranes which exhibit high fluxes for organic compounds and a relatively low flux of water. Since organics are preferentially extracted from dilute aqueous solutions, the concentration polarization at the liquid-membrane interface can control the mass transfer rate of the organics (Psaume et al., 1988; Côté, 1988) and, as noted in Chap. 8, the selectivity of the membrane may not be fully exploited.

From Table 8.2, we find the expression for the overall mass transfer coefficient for dense membranes.

$$\frac{1}{K_{OL}} = \frac{1}{k_l} + \frac{1}{K_d k_m} + \frac{1}{H_u k_g} \tag{13.9}$$

In this equation, the values of K_d and k_m will vary with the organics to be removed and, for convenience, we may lump these parameters and call the product a permeability coefficient, P. Since the membrane is selective for the organics P_i is much greater than P_w, where P_i and P_w are the permeability coefficients for the organic species and water, respectively. Since the preceding equation is based on the liquid-phase concentrations, we can calculate the flux of organics and water across the membrane, assuming that the vapor concentrations on the permeate side of the membrane are small (high vacuum) and that the mass transfer limitations in the gas phase are negligible. With these assumptions, the flux of VOCs reduces to:

$$J_i = \left(\frac{k_l P_i}{k_l + P_i}\right) C_i \tag{13.10}$$

where C_i is the bulk liquid concentration of the organic species i. The flux of water is even simpler. Since the activity of the water is 1.0 at the surface of the membrane, the flux of water across the membrane is given by

$$J_i = P_w \tag{13.11}$$

Several general trends, which have been experimentally verified, can be deduced from consideration of Eqs. (13.10) and (13.11):

- For a given membrane and the same hydrodynamic conditions, the organic compound flux is proportional to its concentration, while the water flux is constant.
- For volatile compounds that have a high affinity for the membrane (i.e., K_d is large), the resistance of the membrane is far lower than the liquid film resistance (i.e., $1/K_d k_m \ll 1/k_l$).

As a first consequence, the organic compound flux can be increased only by improving the hydraulic conditions. A second and unusual consequence is that, if a thicker membrane were used, there would be no decrease in the organic compound flux, but the flux of water across the membrane would decrease (i.e., the membrane would appear to be more selective in separating the organics, and a more organic-rich permeate would be obtained). A third and also unusual consequence is that the organic compound flux is independent of the nature of the membrane (i.e., the membrane can be dense or porous, isotropic or composite, etc.).

In other words, an optimal choice for the membrane should be a compromise depending on the objective of the process. If the objective is simply to remove the organic contaminants from the water, the least expensive membrane should be used. However, if the objective is to recover the organic compound, a dense, thick membrane which is less permeable to water will produce a more concentrated permeate.

13.2.1.2　Permeate Side Head Losses and Heat Transfer Considerations. The pressure drop associated with permeate flow and the heat loss from the feed must be considered carefully in pervaporation (Rautenbach, Herion, and Meyer-Blumenroth, 1991). For example, under certain circumstances it is possible for the combined effects of temperature drop along the feed channel and pressure rise in the permeate compartment to cause the more permeable species to be transferred back from the vapor phase to the liquid feed (Aptel, 1988). However, in applications related to water and wastewater treatment, the solvent concentrations and the corresponding fluxes are normally low so that these issues are less important. As such, the design of pervaporation modules is typically controlled by the liquid film and membrane resistances.

13.2.2　Membranes and Modules

The performance of pervaporation is dependent on the choice of membrane material and the design of the module.

13.2.2.1　Membranes. Membrane permeation is governed by both the chemical nature and physical structure of the membrane, as well as the chemical properties of the organics to be removed. Polymers are usually divided into two groups: glassy polymers and rubbery polymers (elastomers). The distinction is based on the state of the polymer at room temperature. Polymers with a glass transition temperature below the room temperature are classified as rubbery, and those with a glass transition temperature above room temperature are classified as glassy. The glassy polymers are further distinguished as being crystalline, semicrystalline, and amorphous polymers. The degree of crystallinity of the polymer influences several of the polymer properties such as elasticity, solubility, and diffusivity.

Glassy polymers are normally used for dehydration processes, while the elastomers are used for separating organics from water. The first commercial membrane used for dehydration was developed by GFT (Bruschke, 1983). It is a composite membrane prepared by coating a mesoporous polyacrylonitrile (PAN) support with a thin, dense layer of cross-linked polyvinylalcohol (PVA).

The elastomeric polymers are very flexible with main chain bonds like C—C, Si—C, or C—O and they do not contain polar groups or large side groups. These properties give them a hydrophobic character and preferential absorption of organics (Koops and Smolders, 1991; Favre et al., 1994). These polymers do not have strong intermolecular forces, such as H-bonding or dipole-dipole interactions, so they are flexible. This flexibility gives the polymers a relatively high permeability for the organics and results in a higher selectivity for the organic compound.

The separation factor can be quantified by dividing the permeability of an organic component by the permeability of water under the same operating conditions. Typical separation factors from Losin and Miller (1990) are presented in Table 13.1. Large separation factors are obtained for solvents which are immiscible in water and have a high activity in dilute aqueous solution.

The most widely used elastomeric polymers are silicone rubber and copolymers such as poly(dimethylsiloxane), or PDMS. Two different flat-sheet silicone-based membranes are commercialized (Néel, 1992). The standard membrane is obtained by deposition of a liquid functionalized oligodimethylsiloxane onto a porous polyacrylonitrile film, followed by curing under an electron beam. A variant is made more selective but less permeable by introducing, in the silicone layer, a zeolite-type filler such as silicalite (Dotremont et al., 1995). Other commercial composite membranes are manufactured by plasma polymerization of siloxane monomers or by dip-coating polymers on mesoporous flat-sheet or hollow-fiber membranes (Zhang and Drioli, 1995).

13.2.2.2 Modules.   The first commercial module was developed by GFT for dehydration or organic/organic separations. It is a plate-and-frame module manufactured with stainless steel plates and spacers. Gaskets are made of special grades of polymers, polytetrafluoroethylene (PTFE), or flexible graphite, depending on the applications. Up to 50 m^2 of membranes can be fitted in a single module. For other applications, attempts have been made to reduce module cost. Low-cost versions of plate-and-frame modules, using plastic parts such as the GKSS "GS mod-

TABLE 13.1   Typical Pervaporation Separation Factors for Organics Removal from Water

Relative separation factor	Organic compounds
100+	Benzene, toluene, ethyl benzene, xylenes, TCE, chloroform, vinyl chloride, ethylene dichloride, methylene chloride, perchlorofluorocarbons, hexane
10–100	Ethyl acetate, propanols, butanols, MEK, acetone, aniline, amyl alcohol
5–10	Methanol, ethanol, phenol, acetaldehyde
0–5	Acetic acid, ethylene glycol, DMF and DMAC

Source: From Losin and Miller, 1993.

ule" have been developed for the separation of organic vapors from gas stream (Peinemann and Ohlroge, 1994). Spiral-wound modules are also now used for wastewater treatment (GFT and MTR). Bore-side and shell-side feed hollow-fiber modules have also been developed. The Zenon cross-flow pervaporation system (USEPA, 1995) uses silicone hollow fibers aligned in parallel to form one membrane layer. Separate membrane layers are aligned in series, with the interior of the capillary fibers exposed to vacuum. Wastewater flow is normal to the fiber axes to minimize the liquid mass transfer resistance.

13.2.3   System Configurations

The vapor pressure gradient across the membrane can be created in a number of ways using either an applied or induced vacuum, or a sweep fluid, as illustrated in Fig. 13.3. The simplest, conceptually, creates a low vapor pressure on the permeate side using a mechanical vacuum pump as in Fig. 13.3a. However, on a commercial scale, this configuration may be impractical due to the size of the vacuum pump required.

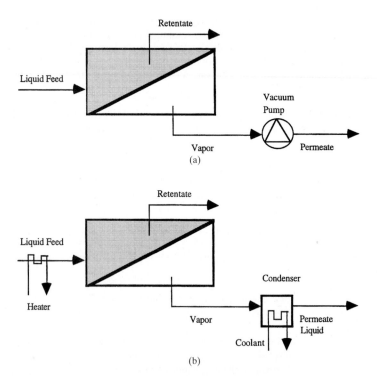

FIGURE 13.3   Different system configurations for pervaporation: (a) vacuum driven; (b) temperature gradient driven; (c) vacuum and temperature driven; (d) carrier gas driven; (e) condensable, permeate/immiscible liquid carrier driven.

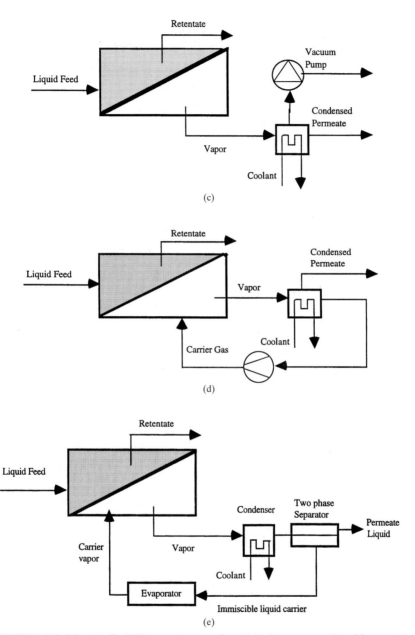

FIGURE 13.3 (*Continued*)   Different system configurations for pervaporation: (*a*) vacuum driven; (*b*) temperature gradient driven; (*c*) vacuum and temperature driven; (*d*) carrier gas driven; (*e*) condensable, permeate/immiscible liquid carrier driven.

A vapor pressure gradient may also be obtained by a temperature gradient, as shown in Fig. 13.3b. This has been referred as *thermopervaporation* by Aptel et al. (1976) to underline the fact that the energy which is supplied to the system is then reduced to thermal energy. A heater is required to raise the temperature of the feed stream to encourage vaporization. On the permeate side of the membrane the vapor is condensed using a chiller, and the condensation sustains a vacuum which encourages transfer across the membrane.

Most commercial units use the configuration illustrated in Fig. 13.3c. where noncondensable gases can be continuously removed by the vacuum pump. A vapor pressure gradient can also be generated without a vacuum by using an inert sweep gas on the permeate side, as shown in Fig. 13.3d. However, this configuration is very similar to air stripping and suffers from some of the same disadvantages. The driving force, especially for semivolatile organics, may be reduced and the economic feasibility of the process is questionable (Fleming and Slater, 1992). As an alternative to the inert carrier gas, a condensable vapor can be used, as shown in Fig. 13.3e. The vapor sweeps the permeate from the pervaporation module to a condenser where it forms an immiscible liquid and the permeate is separated from the carrier in a multiple-phase separator. This approach has not been commercialized, since the driving force is also low and the recovery system is expensive and complex.

An inert sweep liquid has been considered as an alternative to the inert sweep gas. This operation initially proposed by Cabasso (1974) is now known as *pertraction*. Among its major drawbacks are an additional liquid resistance on the permeate side of the membrane, possible contamination of the water by the "inert" sweep liquid, and the production of a diluted "permeate" which necessitates an additional separation step to regenerate the sweeping solvent.

A somewhat more complex system, called the *hollow-fiber contained liquid membrane* (HFCLM) pervaporation system, was developed by Yang et al. (1995). In this system, a set of hydrophobic microporous hollow fibers and a second set of nonporous silicone hollow fibers are packed into the same tubular module. Wastewater is fed to the inside of the hydrophobic membranes and a vacuum is applied to the bore of the silicone membranes. A high-boiling organic liquid, in which water has very low solubility, fills the shell side of the module and separates the two sets of fibers. This organic carrier liquid is recirculated on the shell side of the fibers. The VOCs to be removed from water diffuse through the organic liquid immobilized in the porous walls of the first set of hollow fibers and into the recirculated liquid. The VOCs are continuously stripped from the recirculating organic phase by the evacuated silicone fibers. Basically, this system solves the separation step of the pertraction process by incorporating a pervaporation step within the same module. However, this process has two drawbacks: it has two liquid resistances in series, and there is still a potential for contamination of the treated water by the organic carrier.

Another configuration of a liquid-liquid membrane contactor which can be related to pervaporation is the *extractive membrane bioreactor* (EMB) developed by Brookes and Livingston (1995). The purpose of this novel aqueous-aqueous membrane contactor is to operate a continuous selective biodegradation and/or to avoid biological inhibition by using a membrane that is selective for the substrate of interest. For example, consider the treatment of an acidic wastewater that contains toxic concentrations of heavy metals and low-molecular-weight aromatic organic compounds. This type of mixed waste is difficult to treat by normal biological treatment processes. The conventional approach would include a physicochemical pretreatment to neutralize the pH and precipitate the toxic metals, followed by biological treatment for oxidation of the organics. However, this sequence of processes poses problems for engineers. For example, the metal sludge generated in the pretreatment process is likely to be contaminated with the organics from the wastewater. In addition, nutrients (nitrogen and phosphorus) may need to be added to the wastewater to encourage biological growth. The EBM process overcomes these problems effectively and provides a means of selectively removing the organics first. This is accomplished by using a silicone membrane to separate the wastewater to be treated from an active biological culture that has been acclimated to the specific organic compounds in the wastewater. The biological culture is supplied with ample nutrients, and it is kept at a pH that is optimal for the effective biodegradation of the organics of interest. This is easy to do since the volume of the recirculating culture is small. By continuously degrading the organics on one side of the membrane, the microorganisms maintain a concentration driving force for the effective extraction of the organic pollutants from the wastewater.

13.2.4   Applications for VOC Removal

13.2.4.1   Importance of VOC Emissions.   According to the U.S. Environmental Protection Agency (USEPA), VOCs are defined as stable products having a vapor pressure greater than 0.1 mm Hg under normal temperature and pressure conditions. Organic molecules that fit this general definition are found, therefore, in almost every branch of the chemical industry and differ widely in chemical character (e.g. alcohols, ketones, aromatics, chlorinated hydrocarbons, etc.). Large quantities of these compounds are released to the environment annually. The Centre Interprofessionel Technique d'Etude de la Pollution Atmospherique (CITEPA) has estimated the total European Economic Community (EEC) VOC emission in 1985 to be 20 million tons per year. The main sources include transportation, solvent uses, industrial processes, refineries, combustion, cleaning, and miscellaneous industrial operations.

In recent years, the environmental significance of these VOC releases has been a focus of numerous studies. VOCs are sometimes toxic, carcinogenic, irritating, and/or flammable compounds and the presence of VOCs in groundwaters and wastewaters may pose a risk to consumers or plant personnel. In addition, the emission of VOCs to the atmosphere is being restricted by increasingly more stringent air quality standards. As a result, a great deal of research is being devoted to find means to reduce or suppress VOC emissions. These efforts include designing new and cleaner processes that reduce emissions, finding ways to recover and reuse the VOCs, and developing processes to oxidize the VOCs.

13.2.4.2   Conventional Treatment.   The removal of trace organic contaminants from water sources has traditionally been accomplished by packed tower aeration. This low-cost approach of air stripping has become more expensive in recent years since increasingly stringent air pollution standards have required the installation of granular activated carbon (GAC) canisters for the treatment of the exhaust air. In order for the air-stripping process to remove a high percentage of semivolatile organics, very high air/water ratios are required, and the high air flow rates make the total cost of treatment (air stripping plus GAC) prohibitive.

13.2.4.3   Case Studies.   In water and wastewater treatment applications, volatile compounds are generally present in low concentration. If the feed stream contains more than 1 to 2 percent by weight of solvent(s), the economic benefits of solvent recovery generally drive the selection of the pervaporation process. However, if the solvent concentration in the aqueous stream is very low, the process could be selected on the basis of pollution control considerations.

13.2.4.4   Industrial Effluents.   The first commercial applications in water treatment were for the extraction of solvents from industrial wastewater. For instance, isopropylether contaminates the wastewaters from chemical plants in which oligomeric cyclosiloxanes are synthesized. The concentration ranges from 0.5 to 0.9%. By using pervaporation with an air sweep on the permeate side of the membrane, one obtains a gas stream sufficiently rich in organics that it can be incinerated, and the treated water contains a reduced chemical oxygen demand (COD) that can be handled economically by biological treatment (Néel, 1992).

The recovery of ethyl acetate for reuse is becoming economically attractive (Wijmans, Baker, and Athayde, 1994). For example, it appears feasible to treat and recover 90 percent of the solvent in an effluent that contains 2% ethyl acetate. Based on a separation factor of 100, a pervaporation system can produce a permeate containing 45.7 percent organic, which spontaneously separates into an organic phase (96.7 percent organic) which can be purified and reused, and an aqueous phase which is recycled to the feed stream. The capital cost of a 20,000-gal/day plant is estimated at $140,000 and the annual operating costs are $100,000, or $0.75/gal of solvent recovered. The current cost of technical-grade ethyl acetate is $5/gal, so the capital cost of the pervaporation system can be recovered in a short period of time.

13.2.4.5   Groundwater.   Pervaporation also appears to be a viable option in the treatment of groundwater containing low concentrations of organics such as trichloroethane (TCE), toluene, benzene, and phenols. Vacuum-assisted pervaporation has several major advantages over air stripping:

1. The volume of the exhaust gas phase is many orders of magnitude smaller since only the dissolved gases, VOCs, and a little water vapor are extracted from the water.

2. The concentration of VOCs in the exhaust gas phase is very much higher and alternative methods for solvent removal and/or recovery may be feasible. If, however, GAC is used, its capacity for the VOCs will be correspondingly greater and much smaller volumes of GAC will be required.

3. Since the pervaporation process is diffusion limited, semivolatile organic compounds are removed as effectively as their more volatile counterparts.

4. Problems associated with iron fouling of the packed tower aerators may be avoided since no oxygen is introduced to the water.

5. The modules required for pervaporation are compact, and no unsightly towers or tall structures are required.

When very high levels of VOC removal are required, both the air-stripping and pervaporation processes become very expensive. This is because the organic compound flux is proportional to its concentration and the driving force for transfer approaches zero as the VOC concentration is reduced to low concentrations. When a high degree of removal is required, pervaporation may be used in conjunction with other processes, such as GAC.

A recent paper by Losin and Miller (1993) considered a case study in which pervaporation was used to treat a groundwater contaminated with 0.3 percent by weight of benzene, toluene, and xylene (BTX). A 10-gpm system produced an effluent containing <1 ppm of BTX and the condensed permeate phase separated to produce a 0.03-gpm permeate stream that contained in excess of 99 percent BTX. The small volume of condensed water that was separated from the BTX permeate was recycled to the feed stream. The authors noted that the permeate stream was of sufficiently high purity to justify recovery.

Côté and Lipski (1990) investigated the use of pervaporation for the removal of organic contaminants from water. They completed an economic evaluation of several pervaporation modules in systems designed to remove 99 percent of 10 mg/L of TCE from water at a feed flow rate of 10 m³/h. Four module configurations were considered: a spiral-wound module (SWM), two inside flow modules (IFM), and a transverse flow module (TFM). The spiral-wound module was a FilmTec FT30 4-in × 40-in. The inside flow modules were 1 m long with fiber diameters of 500 µm and 1000 µm, respectively. The transverse flow module was constructed with 500-µm fibers packed into a 10-cm shell on a center-to-center spacing of 1.5 mm. All modules were made from a composite membrane in which the dense layer of silicone rubber varied in thickness from 2 to 150 µm. A range of Reynolds numbers from 100 to 5000 was investigated and an optimal condition was sought for each module within the specified ranges. The cost of treated water (1990 dollars) for the respective modules was given as follows:

IFM	500-µm fibers	Re = 500	75-µm thickness	$3.80/m³
	1000-µm fibers	Re = 4000	10-µm thickness	$1.41/m³
SWM		Re = 700	30-µm thickness	$1.10/m³
TFM		Re = 250	30-µm thickness	$0.56/m³

A comparative cost evaluation for the same initial conditions was also conducted for carbon adsorption and air stripping. The costs were based on various configurations combining air stripping with granular activated carbon (GAC) to liquid-phase GAC by itself. The cost for a complete system varied from $0.40/m³ for air stripping to $0.80/m³ for liquid-phase GAC. The authors concluded that pervaporation is technically feasible and economically competitive with conventional technologies for treating VOC-contaminated water. The process is also compact, continuous, has a low overall system capital cost, and has reduced energy costs compared to the conventional alternatives. In addition, pervaporation offers the possibility of direct recovery of the organic compound for reuse.

The transverse flow module was recently successfully evaluated during several site demonstrations (USEPA, 1995). Periodic cleaning of the membranes is necessary to maintain the treatment efficiency, which tends to decrease due to mineral-scaling problems.

13.3 VACUUM DEGASSING

Dissolved gases or volatile and semivolatile compounds can be removed from an aqueous stream by vacuum degassing using microporous hollow-fiber membranes. The membrane modules provide a high surface area-to-volume ratio and create a large liquid-gas interface. By applying a vacuum to one side of the hollow fibers, a partial pressure gradient is created and dissolved gases or volatile compounds transfer from the water across the membrane and into the vacuum phase. Hydrophobic microporous membranes such as polypropylene and polyethylene, with pore sizes around 0.1 µm and porosities of approximately 70 percent, are generally used for vacuum degassing.

The hydrophobic and organophilic characteristics of the hollow-fiber membranes ensure that the pores remain dry and gas filled and, as a result, the transfer through the membrane is by gaseous diffusion and the membrane resistance is negligible compared with the liquid film diffusion resistance. As a result, for relatively volatile organics and dissolved gases (e.g., $H_u > 0.01$), the rate of mass transfer in a gas-stripping module is limited by the liquid boundary mass transfer coefficient and is very dependent on the module design and hydraulics. In selecting a membrane module, it is therefore important to ensure that the liquid film mass transfer coefficient is maximized within the constraints of the hydraulic design.

If the pores are wetted and the aqueous solution enters the pores, the conditions for gas stripping are no longer fulfilled and the system ceases to function. Wetting of the pores occurs when the feed pressure exceeds the liquid entry or penetration pressure. This pressure is a function of the pore size and the organic concentration of the aqueous stream. To ensure that the feed pressure does not exceed this critical value and cause spontaneous wetting of the membrane pores, the critical entry pressure can be estimated by the Washburn equation (Néel, 1992).

EXAMPLE 13.2 *A groundwater flow of 0.06 L/s is contaminated with 100 μg/L of 1,1,2 trichloroethane (TCA). A point-of-use treatment system is needed to provide at least 90 percent TCA removal. Vacuum degassing with hollow-fiber membranes is one of the alternatives considered. Design a recirculating system that could meet this objective.*

Assume that the flow of groundwater is pumped to a holding tank which is assumed to be well mixed. Hollow-fiber modules are available from a local supplier in a standard size; the specifications are listed in Table 13.2. The module(s) is to be used in a recirculating system of the type depicted in Fig. 13.4. A separate recirculation pump is activated when the well pump is activated and water from the pressurized holding tank is recirculated through the membrane module. The flow rates and concentrations of gases in the system are also shown in the figure.

TABLE 13.2 Fiber Module Specifications

Shell size (ID):	5 cm	Number of fibers:	6000
Fiber length:	100 cm	Fiber diameter:	0.04 cm
Fiber area:	75,360 cm²	Flow area:	12.1 cm²
Effective diameter:	0.0157 cm	a (fiber area/vol)	62.3 cm⁻¹

TABLE 13.3 Physical Properties Information

Water temperature	20°C
Kinematic viscosity	0.010068 cm²/s
H_u for TCA (Cg/C)	0.032
Diffusivity of TCA in water	8.13×10^{-10} m²/s
Diffusivity of TCA in gas	7.97×10^{-6} m²/s

The operating system will arrive at a steady-state condition, and we develop the equations here to represent this condition. A mass balance on the holding tank provides the following equation:

$$V\frac{dC}{dt} = QC_o - QC_e - Q_m C_e + Q_m C = 0 \tag{13.12}$$

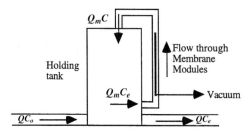

FIGURE 13.4 System configuration for gas stripping of TCA-contaminated groundwater.

The dissolved gas concentration C leaving the membrane module must be calculated by integrating a separate differential equation for the mass transfer in membrane module. Figure 13.5 shows the detail of the membrane module used in Fig. 13.4.

Treating the membrane module as a plug flow reactor at steady state, we obtain the following equation that describes the change in concentration along the length of the module:

$$-v\frac{dC}{dx} = K_{OL}a(C - C^*) \tag{13.13}$$

where   *v = water velocity within module*
a = interfacial area of membrane per unit volume in module
C* = concentration of dissolved gas in equilibrium with vacuum phase within fibers
If C* is assumed to be zero, Eq. (13.13) simplifies to the following form:

$$-v\frac{dC}{dx} = K_{OL}aC \tag{13.14}$$

The validity of the assumption of C* = 0 is dependent on the capacity of the vacuum pump and the operating conditions of the system. We are assuming in this case that the vacuum pump is sized well and has the capacity to maintain a high vacuum on the fibers. Equation (13.14) may be integrated with the boundary conditions C = C_e at x = 0, and C = C_o at x = L where L is the length of the fibers. The integrated form of the equation is as follows

$$C = C_e e^{-K_{OL}a(L/v)} \tag{13.15}$$

Equation (13.15) is substituted back into Eq. (13.12) to provide an equation for the overall system:

$$QC_o - QC_e - Q_mC_e + Q_mC_e e^{-K_{OL}a(L/v)} = 0 \tag{13.16}$$

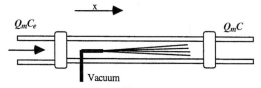

FIGURE 13.5 A schematic of the hollow-fiber membrane module used in the recirculating system.

Equation (13.16) can be rearranged:

$$K_{OL}a\frac{L}{v} = -\ln\left[1 - \frac{QC_o - QC_e}{Q_m C_e}\right] \qquad (13.17)$$

For 90 percent TCA removal (i.e., $C_e = 0.1C_o$), Eq. (13.17) becomes:

$$K_{OL}a\frac{L}{v} = -\ln\left[1 - \frac{Q(0.9C_o)}{Q_m(0.1C_o)}\right]$$

$$K_{OL}a\frac{L}{v} = -\ln\left[1 - \frac{9Q}{Q_m}\right] \qquad (13.18)$$

Note that this equation is independent of the influent concentration. This is typical for a first-order removal process; the same membrane area will remove 90 percent of the VOCs regardless of their initial concentration. The equation is also independent of the volume of the mixed tank since the volume does not influence the rate at which TCA has to be removed.

Assume that we choose Q_m to be 1 L/s; then, from the manufacturer's information previously provided, we know the following:

$$a = 62.3 \text{ cm}^{-1}$$

$$v = \frac{4000Q_m}{(5)^2\pi - 6000(0.04)^2\pi} = 82.7 \text{ cm/s}$$

and substituting all the known parameters into Eq. 13.18, we obtain:

$$K_{OL} \times 62.3 \times \frac{L}{82.7} = -\ln\left[1 - \frac{9 \times 0.06}{1}\right]$$

$$K_{OL}L = 1.03$$

This result means that the product of the overall mass transfer coefficient and the fiber length must have a value of 1.03 in order to meet the 90 percent removal requirement. These are independent variables; the value of K is unaffected by the value of L. We also know that the value of L must be 100 cm or multiples of 100 if we are to use the commercially available modules.

The overall mass transfer coefficient K_{OL} can be calculated using the correlations given in Chap. 8. In this example, we assume that the membrane modules are designed for water flow on the shell side of the fibers and the fibers are unconfined as shown in Fig. 13.5. The correlation developed by Ahmed (1991) is the most appropriate for this configuration; $Sh = 0.018Re^{0.83}Sc^{0.33}$. Substituting the information provided in the preceding tables, we use this correlation to find the following values: $Re = 129$, $Sc = 1238$, and $Sh = 10.67$. From the Sherwood number, we find the value of the overall mass transfer coefficient to be 0.00552 cm/s.

The required length of membrane for the desired removal is then determined:

$$L = \frac{1.03}{0.00552} = 187 \text{ cm}$$

This result means that it will be necessary to install two 100-cm modules in series in order to meet the 90 percent removal level.

13.4 MEMBRANE DISTILLATION

Membrane distillation is, as the name implies, essentially distillation with a membrane. The process is based on evaporation of the permeate (usually water) through a microporous hydrophobic membrane, and the permeate is condensed or removed from the other side of the membrane. Due to the hydrophobic characteristics of the membrane used and the small size of the membrane pores, the membrane pores remain dry and air-filled. Vapors transfer through the membrane by gaseous diffusion. The membrane creates a barrier between the liquid and the vapor phases, does not alter the vapor-liquid equilibrium of the different components, but provides a large interfacial area for mass transfer. Like distillation, this process is effective in separating water from nonvolatile species, but if volatile contaminants are present they will cross the membrane. Potential applications of membrane distillation include the production of desalted or demineralized water from seawater, brackish water, or wastewater; concentration of aqueous solutions of salts or inorganic acids to higher concentrations; and in the concentration of fruit juices.

13.4.1 System Configurations

Following the recommendations of the European Society of Membrane Science and Technology (Koops, 1995), membrane distillation processes have been designed using four different configurations, as illustrated in Fig. 13.6.

The first is defined as *direct-contact membrane distillation* (Fig. 13.6a). The liquid on both sides of the membrane is in direct contact with the membrane, and the liquid on the permeate side is used as the condensing medium. By heating the feed stream and cooling the permeate stream, a driving force is generated based on the difference in water vapor pressure between the hot and cold solutions. As a result, water vapor diffuses through the gas-filled membrane pores and the rate of transfer is determined by the temperature difference (vapor pressure gradient).

The second is gas-gap membrane distillation: a difference in the water vapor pressure can also be generated by making a cold surface on the permeate side. In this configuration (Fig. 13.6b), an air gap—sometimes referred to as a *diffusion gap*—is made between the cold surface and the membrane where the water vapor diffuses through the air and condenses on the cold surface.

Vapor can also be extracted from the feed stream by creating a difference in the partial pressure between the feed stream and the permeate side. This can be done using either a vacuum (low-pressure membrane distillation, Fig. 13.6c) or an inert sweep gas (sweeping gas membrane distillation, Fig. 13.6d). By generating a vacuum on the permeate side, a difference in partial pressure is created such that the vapor is drawn across the membrane. Similarly, a partial pressure difference is generated when the vapor is carried away from the membrane distillation unit in an inert sweep gas. For the last two configurations, no condensation takes place within the membrane distillation module. The permeate is transported away from the membrane by convection, and this results in higher fluxes compared to the first two configurations (Sarti, Gostoli, and Cavuoti, 1990). The water vapor is subsequently recovered in a condenser.

13.4.2 Membranes

Membrane distillation membranes are made from hydrophobic polymers such as polypropylene, polytetrafluoroethylene (PTFE), and polyvinylidenefluoride (PVDF).

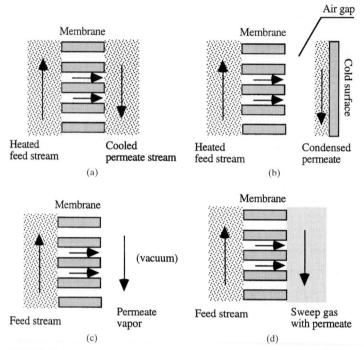

FIGURE 13.6 Membrane distillation configurations: (*a*) direct-contact membrane distillation; (*b*) air gap membrane distillation; (*c*) vacuum membrane distillation; (*d*) sweep gas membrane distillation.

The membranes have pore sizes ranging from 0.1 to 3 μm and a porosity of around 70 percent. More commonly, a pore size range of 0.1 to 0.45 μm is applied to membrane distillation applications (Sirkar, 1992). A fundamental requirement of the membrane is that none of the solutions in contact with the membrane spontaneously wet the pores. This condition must be fulfilled during the operation of the unit. To prevent wetting of a membrane, it is important that the feed stream pressure be kept below the value required to force the solution into the pores of the membrane. The pressure required for liquid entry into the pores is a function of the surface tension of the aqueous solution, which is, in turn, a function of the organic concentration in the solution. At a critical organic concentration, the liquid entry pressure is zero and spontaneous wetting of the membrane pore occurs. The entry pressure or penetration pressure can be calculated using the Washburn equation (Néel, 1992).

$$P = \frac{4\sigma}{d_{max}} \cos \varphi$$

where σ = surface tension of contacting liquid
    d_{max} = maximum pore diameter of membrane
    φ = contact angle between liquid and membrane

The preceding must be taken into consideration when designing a membrane distillation system, in choices of membrane type, pore size distribution, and operating conditions for a particular application.

13.4.3 Applications

The following discussion is limited to *direct contact* and *gas-gap* membrane distillation. The latter process was suggested for water desalination in the early '50s (Henderyckz, 1970) and was referred to successively as *air film doublet, vapor gap reverse osmosis,* and *thermoosmosis in vapor phases.* More recently, the application of membrane distillation in the desalination of seawater was investigated by Kubota et al. (1988). Two plate-and-frame membrane distillation modules were tested, with a capacity of 100 to 200 L/d. Both modules were constructed with a parallel stack for diffusion gap (gas-gap membrane distillation). The modules were fitted with hydrophobic microporous membranes, a 0.1-mm-thick polytetrafluoroethylene (PTFE) membrane with a 75 percent porosity and an area of 1.92 m², and a 0.15-mm-thick polyethylene membrane with a 60 percent porosity and an area of 1.86 m², respectively. The temperature of the feed stream was 60°C and the cooling stream was held at 20°C. Although the permeability of the membrane depended mainly on the operating temperature and the temperature difference across the membrane, the effect of varying flow rates of both the brine and the cooling water was investigated. The water permeability for the range of flow rates studied was in the range of 2 to 10 kg/h. As the brine and cooling water flow rates were increased, an optimum permeability was attained. The investigators also observed different declining patterns in the permeate flux; it was thought to be related to heat loss. The electric conductivity of the product water was measured at about 10 μS/cm, indicating a very good quality. Theoretically, the product quality should have been better; however, the result was attributed to a minor leak at the membrane seal. The overall conclusions from the experiment point out that the heat loss from the test modules was large, a consideration that needed to be taken into account with future designs of membrane modules. Although the flux mainly depends on the operating temperatures, it also varies with the brine and cooling water flow rates where an optimum flow rate for a system should be determined. The flow rate of the two solutions also appears to affect the heat loss of the membrane modules.

13.5 AIR STRIPPING

The use of gas-permeable membranes for removing organic contaminants has been discussed in all of the preceding sections. In this section we focus on the use of an air sweep on the permeate side of the membrane to maintain the driving force for rapid mass transfer. This process is essentially an air-stripping operation.

The specific surface area in membrane contactors can be about an order of magnitude higher than that in conventional packed towers. As a result, at a given air-to-water ratio, the membrane will generally be more effective in removing the organics or volatile contaminants than conventional packed tower aeration (Zander et al., 1983; Semmens et al., 1989). Membrane contactors are compact and modular, and they can be used in some applications where tall stripping towers cannot be installed. In addition, the use of hollow-fiber membrane contactors allows the two phases to be maintained at different pressures and so, for example, volatile contaminants can be stripped from pressurized water by air at atmospheric pressure.

Membranes have not found a wide application for air stripping to date for several reasons:

1. The high specific surface area of the membrane contactors causes them to have higher head losses for fluid flow than conventional packed towers.

2. The large surface area of membranes can become fouled with organics and inorganic contaminants in the water. Fouling can cause a dramatic decline in performance and an increase in the head losses across the contactor. Iron fouling is particularly troublesome in groundwater applications (Schwarz et al., 1991). The ferrous iron content of a groundwater is oxidized to ferric oxide at the surface of the membrane. This oxide effectively coats the membrane and, in time, seals the pores of the microporous membrane, preventing the transport of volatile contaminants across the membrane.

3. The cost of the membrane is much higher than the cost of dumped packings on a per-unit-area basis.

4. The modular character of the membrane contactors does not provide economies of scale. Large-scale applications require a larger number of the small-scale modules.

Nevertheless, small-scale point-of-use applications appear well suited to the use of membrane modules. The membrane modules are compact and easy to install in a basement, and they can be used on a pressurized water supply. One such application is for radon removal.

13.5.1 Radon Removal

Radon is a naturally occurring element that is formed by the radioactive decay of radium 226. The radon is generated in three isotopic forms: Rn-222, Rn-220, and Rn-219. The latter two isotopes decay very quickly with half-lives on the order of seconds, while the Rn-222 form has a half-life of 3.82 days and is the isotopic form that represents a health hazard. Most inhaled radon is exhaled, but some decay will occur within the lungs. The products of decay (polonium-218 and polonium-214) are solids that can deposit in the lungs and as these deposited solids continue to decay they emit short-range but powerful alpha particles that can damage the surrounding cells. The EPA has estimated that on the order of 5000 to 20,000 lung cancer deaths per year are attributable to exposure to natural radon levels from all sources (Hanson, 1989).

As a result of the toxicity of radon, its release from radon-laden groundwater is currently being scrutinized by the EPA. This source of radon appears to be significant with an estimated 100 to 1800 lung cancer deaths per year being attributable to waterborne radon emissions (USEPA, 1987).

The EPA estimates that some eight million citizens may be at risk from high radon levels in small community water systems and private wells. Wells with waterborne radon levels in excess of 100,000 pCi/L have been found in seven states, while private wells with levels in excess of 1 million pCi/L have also been reported (Lowry, 1987). The American Water Works Association has estimated that if the EPA adopts a standard for radon in the 100 to 1000 pCi/L range, approximately 75 percent of its supplies would require radon removal to meet the standard (Dixon and Lee, 1988).

Radon is quite volatile with a Henry's constant ($H = C_g/C_w$) of 4.0, and, as a result, it is readily removed by air stripping. The air-stripping process is quite economical for a large-scale water treatment facility and, even if GAC is needed for treatment of the exhaust gas, the decay of the adsorbed radon will provide for effective regeneration of the GAC if the system can be engineered to provide the residence time required for substantial decay. However, air stripping is not very readily incorporated into a small-scale point-of-use system such as the treatment of a homeowner's well.

Typically, a homeowner's water supply system consists of well pump and a pressurized supply tank equipped with a pressure control system. When the pressure in the tank falls below approximately 40 lb/in², the well pump is automatically activated to repressurize the supply. In order to air-strip the water supply, air would need to be pressurized to the same level as the water and some mechanism for releasing the exhaust from the system would need to be engineered into the system. Alternatively, additional tankage, pumps, and air supply would need to be added and this would take up more space and represent a costly approach for radon removal.

Granular activated carbon is an alternative approach that is sometimes used. The radon is effectively adsorbed by the GAC, and the decay to nonvolatile products effectively regenerates the carbon so that an installed canister of GAC may be effective for a long period of time. There are several problems with GAC for small-scale applications: (1) several cubic feet of carbon may be needed for effective removal of radon and the normal under-the-sink type of canisters, which are smaller in volume, are not likely to provide sufficient residence time for effective treatment; and (2) the carbon accumulates the other radionuclides that are formed by the decay of radon and some of these are long-lived (e.g., Pb-210 and Po-210) and may emit gamma radiation at levels that require special shielding or installation of the canister outside of the house.

Matson (1992) describes a process that employs membrane-assisted air stripping to remove radon from a domestic water supply. In this process, a groundwater is pumped through a filter cartridge, if necessary, to remove solids from the water and then through a membrane module to remove the radon. The system is depicted in Fig. 13.7. The membrane performance is unaffected by the pressure of the water, and the water is typically maintained at pressures between 20 to 80 lb/in^2 by the normal pressure controllers on the well pump. The design employs a simple air blower to blow air at atmospheric pressure across the shell side of the fibers as illustrated. The exhaust air is vented outside of the house. The treated water flows to the water storage tank which provides the water supply for the home. The storage tank may be fitted with a recirculation pump to control water flow rates and achieve better radon removal in the system.

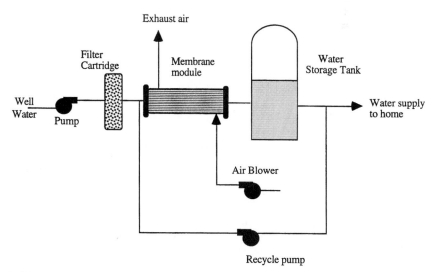

FIGURE 13.7   Household radon removal system as described by Matson, 1992.

REFERENCES

Aptel, P., 1988, "Pervaporation in Hollow Fibers Bundles: Some Aspect of Heat and Mass Transfer," in N. N. Li and H. Strathmann (eds.), *Separation Technology,* Engineering Foundation, New York.

Aptel, P., N. Challard, J. Cuny, and J. Néel, 1976, "Application of Pervaporation Process to Separate Azeotropic Mixture," *J. Membrane Sci.,* **1:**271–287.

Bier, M. E., and R. G. Cooks, 1987, *Analytical Chemistry,* **59:**597.

Böddeker, K. W., 1994, "Recovery of Volatile Byproducts by Pervaporation," in J. G. Crespo and K. W. Böddeker (eds.), *Membrane Processes in Separation and Purification,* NATO ASI Series, vol. 272. Kluwer Academic Publishers, Dortrecht.

Brookes, P. R., and A. G. Livingston, 1995, "Aqueous-Aqueous Extraction of Organic Pollutants Through Tubular Silicone Rubber Membrane," *J. Membrane Sci.,* **104:**119–137.

Bruschke, H. E. A., 1983, "Multilayer Membrane and Its Use in the Separation of Liquids by Pervaporation," E.P. 0.096.339.

Cabasso, I., J. Jaguar-Grodzinski, and D. Vofsi, 1974, "A Study of Permeation of Organic Solvents through Polymeric Membranes based on Polymeric Alloy of Polyphosphonates and Acetylcellulose," *J. Appl. Polym. Sci.,* **18:**2137–2147.

Calvo, K. C., C. R. Westerberger, L. B. Anderson, and M. H. Klapper, 1981, *Analytical Chemistry,* **53:**981.

Castro, K., and A. K. Zander, 1995, "Membrane Air-Stripping: Effects of Pretreatment," *Journal AWWA,* **87**(3):50–61.

Côté, P., and C. Lipski, 1988, "Mass Transfer Limitations in Pervaporation for Water and Wastewater Treatment," in *Proc. of the Third International Conference on Pervaporation Processes in the Chemical Industry,* Nancy, France.

Côté, P., and C. Lipski, 1990, "The Use of Pervaporation for the Removal of Organic Contaminants from Water," *Environmental Progress,* **9**(4):254–261.

Dheandhanoo, S., 1989, "The Application of a Membrane Interface to On-Line Mass Spectrometric Measurements of Volatile Compounds in Aqueous Solutions," *Rapid Communications in Mass Spectrometry,* **3**(6):177–179.

Dixon, K. L., and R. G. Lee, 1988, *Journal AWWA,* **80**(7):65–70.

Dotremont, C., B. Brabants, K. Geeroms, J. Mewis, and C. Vandecasteele, 1995, "Sorption and Diffusion of Chlorinated Hydrocarbons in Silicalite-Filled PDMS Membranes," *Journal of Membrane Science,* **104:**109–117.

Favre, E., P. Schaetzel, Q. T. Nguyen, R. Clément, and J. Néel, 1994, "Sorption, Diffusion and Vapor Permeation of Various Penetrants through Dense Poly(dimethylsiloxane) Membranes: A Transport Analysis," *Journal of Membrane Science,* **92:**169–181.

Fleming, H. L., and C. S. Slater, 1992, "Pervaporation: Design," in W. S. W. Ho and K. K. Sirkar (eds.), *Membrane Handbook,* Van Nostrand-Reinhold, New York, pp. 123–131.

Fogelquist, E., M. Krysell, and L. G. Danielson, 1986, *Analytical Chemistry,* **58:**1516.

Hanson, D. J., 1989, *C&E News,* February 6, pp. 7–13.

Henderyckz, Y., 1970, "Utilisation de la Thermoosmose en Phase Vapeur dans un Appareil de Dessalement," *Proceedings of 3d International Symposium on Fresh Water from the Sea,* **1:**291–305.

Koops, G. H., and C. A. Smolders, 1991, "Estimation and Evaluation of Polymeric Materials for Pervaporation Membranes," in R. Y. M. Huang (ed.), *Pervaporation Membrane Separation Processes,* Elsevier Science Publishers B. V., Amsterdam, pp. 253–278.

Kubota, S., K. Ohta, I. Hayano, M. Hirai, K. Kikuchi, and Y. Murayama, 1988, "Experiments on Seawater Desalination by Membrane Distillation," *Desalination,* **69:**19–26.

Losin, M., and B. D. Miller, 1993, "Applications of Pervaporation Technology in Wastewater Processing," *AIChE Summer National Meeting,* August 1993, Seattle, Wash.

Lowry, J. D., et al., 1987, *Journal AWWA,* **79**(4):162–169.

Matson, S. L., 1992, "Method and System for Removing Radon from Radon Containing Water," U.S. Patent Number 5,100,555.

Melcher, R. G., and P. L. Morabito, 1990, "Membrane/Gas Chromatography System for Automated Extraction and Determination of Trace Organics in Aqueous Samples," *Analytical Chemistry,* **62:**2183–2186.

Meyer, B. U., W. Schroter, Z. Zuchner, and G. Hellige, 1988, *Biomedical Technology,* **33:**66.

Néel, J., 1992, "Current Trends in Pervaporation," in *Proc. of the CEE-Brazil Workshop on Membranes Separation Processes,* May 3–8, 1992, Rio de Janeiro, Presso La Tipolitographia Teti, Napoli, pp. 182–198.

Néel, J., 1991, "Introduction to Pervaporation," in R. Y. M. Huang (ed.), *Pervaporation Membrane Separation Processes,* Elsevier Science Publishers B. V., Amsterdam, pp. 1–109.

Peinemann, K. V., and K. Ohlroge, 1994, "Separation of Organic Vapor from Air with Membranes," in J. G. Crespo and K. W. Böddeker (eds.), *Membrane Processes in Separation and Purification,* NATO ASI Series, vol. 272, Kluwer Academic Publishers, Dortrecht, pp. 357–372.

Psaume, R., P. Aptel, Y. Aurelle, J. C. Mora, and J. L. Bersillon, 1988, "Pervaporation: Importance of Concentration Polarization in the Extraction of Trace Organic from Water," *Journal of Membrane Science,* **36:**373–384.

Rautenbach, R., C. Herion, and U. Meyer-Blumenroth, 1991, "Engineering Aspect of Pervaporation: Calculation of Transport Resistances, Module Ptimization and Plant Design," in R. Y. M. Huang (ed.), *Pervaporation Membrane Separation Processes,* Elsevier Science Publishers B. V., Amsterdam, pp. 181–223.

Sarti, G. C., C. Gostoli, and G. Cavuoti, 1990, "Vacuum Membrane Distillation through Capillary Polypropylene Membranes," in *Proc. International Congress of Membranes,* August 20–24, 1990, Chicago, Ill., vol. 1, pp. 65–67.

Schwarz, S. E., M. J. Semmens, and K. Froehlich, 1991, "Membrane Air Stripping: Operating Problems and Cost Analysis," *45th Purdue Industrial Waste Conference Proceedings,* Lewis Publishing, Chelsea, Mich.

Semmens, M. J., R. Qin, and A. K. Zander, 1989, "Using Microporous Membranes to Separate VOCs from Water," *Journal AWWA,* **81:**162–167.

Sirkar, K. K., 1992, "Other New Membrane Processes," in W. S. W. Ho and K. K. Sirkar (eds.), *Membrane Handbook,* Van Nostrand-Reinhold, New York, pp. 885–912.

USEPA, 1987, "Removal of Radon from Household Water," EPA OPA-87-011, September 1987.

USEPA Demonstration Bulletin, 1995, "Zenon Cross-Flow Pervaporation Technology," EPA/540/MR-95/511, USEPA, Cincinnati, OH.

Wijmans, J. G., R. W. Baker, and A. L. Athayde, 1994, "Pervaporation: Removal of Organics from Water and Organic/Organic Separation," in J. G. Crespo and K. W. Böddeker (eds.), *Membrane Processes in Separation and Purification,* NATO ASI Series, vol. 272. Kluwer Academic Publishers, Dortrecht, pp. 283–316.

Yang, D., S. Majunmdar, S. Kovenklioglu, and K. K. Sirkar, 1995, "Hollow Fiber Contained Liquid Membrane Pervaporation System for the Removal of Toxic Volatile Organics from Waste Water," *J. Membrane Sci.,* **103:**195–210.

Yang, M. C., and E. L. Cussler, 1986, "Designing Hollow-Fiber Contactors," *AIChE Journal,* **32**(11):1910–1916.

Zander, A. K., M. J. Semmens, and R. M. Narbaitz, 1989, "Removing VOCs by Membrane Stripping," *Journal AWWA,* **81**(4):76.

Zhang, S., and E. Drioli, 1995, "Pervaporation Membranes," *Separation Science Technology,* **30:**1–31.

CHAPTER 14
FIELD EVALUATION
AND PILOTING

Jean-Luc Bersillon
Mark A. Thompson
Malcolm Pirnie, Inc.
Newport News, Virginia

14.1 INTRODUCTION

Performing pilot plant studies and field evaluations represents a considerable investment in equipment and labor for a consultant, manufacturer, developer, or operator. It is driven by the need for information required to help in the decision-making process involved at any stage of a membrane plant life. If examined thoroughly, all the needed information revolves around one "simple" issue: one must have the best estimates about the membrane-water interactions in order to make or suggest the best decisions in either installing or operating a membrane plant.

14.1.1 Why Pilot?

14.1.1.1 New Process and Situation. If all membrane processes were understood and consistent enough, an adequate estimate of the operational performance and product water quality provided from a thorough set of design specifications could be developed. Through the use of the appropriate set of models, a first choice of membrane nature and the associated process designs could be determined without pilot testing. Unfortunately, with the possible exception of reverse osmosis and electrodialysis, membrane processes are too new as water treatment processes to have reached this maturity.

Therefore, membrane piloting and field evaluations are needed to make it possible to establish the relationships between the preexisting information constituted by:

- Raw water quality
- Product water quality and quantity specifications
- Waste-stream specification

and the required knowledge to make a choice on the following issues:

- Membrane nature and associated process characteristics
- Pretreatment requirements
- Operating conditions

14.1.1.2   Competitive Comparison.   When several membranes or several processes qualify for a given set of treatment objectives, or when a new product comes
on the market, competitive comparison of the membrane system may be needed. In
these cases, it is important that one of the treatment lines be used as a reference. If
this is not possible, data collected from the pilot testing should be normalized to a
standard set of operating and cost conditions to allow proper comparison.

14.1.1.3   Owner Request.   Even though a selected process is very well known, the
owner may request that a demonstration test be performed. This may happen when
the extent of the new work is considered important or new to a particular owner. In
this case, hands-on operation of the pilot equipment by the owner should be done to
provide experience in the membrane system's operation.

14.1.1.4   Regulatory Needs.   Regulatory agencies will sometimes require pilot
testing to either develop a level of comfort with the process or generate reject. Concentrate disposal requirements may require toxicity testing or analysis to allow permitting prior to final design and construction of a membrane facility. Identification
and permitting disposal of the reject stream can sometimes be the most complex
component of designing and building a membrane facility.

For the resulting data to be useful, they must be generated under conditions that
ensure their reliability. This issue is related to the availability of information about
the variability of raw water characteristics and the possibility to extrapolate the
results to raw water conditions not necessarily met during the experiments. These
considerations have bearing on all issues of the piloting process, from pilot design to
data generation, storage, and processing. Ultimately, pilot experiments and field
evaluations should be less and less necessary for the design and operation of membrane plants. In some cases, pilot testing may still be very useful when demonstration
is needed or marginal raw water characteristics are encountered.

14.2   RAW WATER QUALITY

The analysis of the stream to be treated by a membrane process should be performed in order to answer the following questions:

- What are the target pollutants to be removed in order to comply with the treated
  water specifications and what are the corresponding raw water levels?
- Which membrane process(es) should be used that would most efficiently produce
  the specified treated water?
- What are the secondary contaminants to be removed?
- What pretreatment process(es) are needed to remove contaminants that inhibit
  performance or to alleviate their effect?
- According to the preselected membrane process, what other water quality parameters need to be adjusted in the product water?

- What are the disposal concerns for the respective waste stream from the selected membrane process(es)?

The first two items are relevant to the specifications of the plant. However, the evaluation of the raw water quality should not be assessed on only a few grab samples. A careful examination of the water quality history over an extended period of several years is preferred in order to minimize the risks of process failure.

Each major class of pollutant can typically be associated to a membrane process, according to pollutant size or physical chemical properties. This is illustrated in Table 14.1.

TABLE 14.1 Membrane Process Selection for Target Pollutants

Type of target pollutant*	Corresponding processes
Volatile compounds	Pervaporation
Salinity, nitrates	Reverse osmosis (RO)
	Electrodialysis
Hardness, heavy metals	Reverse osmosis
	Electrodialysis
	Nanofiltration (NF)
Specific organic compounds, natural organic matter	Nanofiltration
	Activated carbon-UF
	Coagulation-MF
	Membrane bioreactor
Microbial contaminants, viruses	Ultrafiltration (UF)
	Coagulation-MF†
Microbial contaminants, suspended solids	Ultrafiltration
	Microfiltration (MF)

* The relationship between pollutant size and membrane cutoff is presented in Chap. 3.
† The combination coagulation-MF is discussed in detail in Chap. 16. As far as pollution removal is concerned, the coagulation applied prior to MF makes it possible to affect pollutants smaller than the membrane cutoff. The limitations are then those of the coagulation process.

Since a membrane retains any particle or molecule it rejects, an accumulation or concentration of these rejected substances exists at the membrane–raw water interface. These substances may be target pollutants or other compounds. The concentration of these species at the membrane surface may determine interactions between themselves or with the membrane material. When dealing with a truly porous membrane, these interactions may occur within the membrane and between the membrane material and species contained in the filtered water.

These interactions are most often at the origin of detrimental phenomena responsible for flux decay due to fouling. When known, these phenomena can be controlled through pretreatment of the raw water in order to remove or mitigate fouling material. Of course, the fouling material should be identified and quantified whenever possible, as these data make it possible to design the proper pretreatment process.

Typical known fouling materials are listed in Table 14.2.

Membrane fouling is discussed in detail in Chap. 6 and others. For the purpose of this chapter, it should be kept in mind that, depending on the nature and the intensity of the interaction between the fouling material and the membrane material, different remedies may be applied. Pretreatment processes are used where additional

TABLE 14.2   Membrane Processes and Their Most
Frequent Foulants

Membrane process	Common fouling materials
Electrodialysis	Dissolved organic material
	Calcium carbonate or sulfate particles
Reverse osmosis	Dissolved organic compounds
	Microbes (biofouling)
	Calcium carbonate or sulfate silica particles
Nanofiltration	Dissolved organic material compounds
	Microbes (biofouling)
	Calcium carbonate or sulfate particles
Ultrafiltration	Colloidal material
	Organic matter
Microfiltration	Colloidal material
	Organic matter

processes are needed prior to membrane treatment in order to avoid any contact between the fouling material and the membrane material. Reverse osmosis is often very demanding in that respect and most of the need for piloting testing is done to optimize the design of the pretreatment steps. On the other end of the spectrum of remedies, the application or modifications of tangential velocity may suffice when the flux decay is caused by the accumulation of mineral, well-crystallized solids.

In addition to prefiltration, which is required for almost all membrane systems, pretreatment may involve any of the processes commonly used in water treatment. The fouling character of natural organic matter for any of the membrane processes explains the interest of the membrane operators for carbon adsorption or preoxidation. The adsorption-membrane filtration combination is presented in detail in Chap. 15. As removal of smaller ions or particles is required, the pretreatment requirements also increase. Whereas microfiltration and ultrafiltration systems typically require a coarse prefiltration, reverse osmosis and nanofiltration may require a complete drinking water treatment train for pretreatment of a surface water source.

Posttreatment is necessary when either the membrane process or the pretreatment changes the water quality to a point where degassing or corrosion problems may occur in the distribution system. The resulting treated water for some desalination processes may be "too pure" and contain no minerals to provide stability in the water. Posttreatment processes are designed to strip excessive levels of dissolved gasses or correct the pH and chemistry of the water so that it is no longer corrosive. In addition, disinfection of the product water is also provided as an additional microbial barrier and to maintain a residual throughout the distribution system.

Summarizing, four classes of measurements that must be measured prior to or during pilot plant experiments include the following:

- Particulate material (including microbial species)
- Dissolved or colloidal organic material
- Dissolved minerals
- Physical parameters

These four classes of measurements are reviewed in the following sections.

14.2.1 Particulate Contents

Particles found in natural waters can be chemical precipitates, inert silts, or biological organisms. Identification of the type and size of the particles present will affect the selection of the membrane process, membrane material, and membrane geometry. Some level of particle removal prior to the membrane process is typically provided. The level of particle removal required can range from using a mechanical screen to remove large debris to using 5-µm cartridge filters for a reverse-osmosis membrane process.

During pilot studies and system operation, the particle content in the feed and treated water must be closely monitored. Common methods for this include turbidity measurement, silt density index, and particle counting.

Turbidity is a measurement of the sample's optical property that results from the scattering and absorbing of light by the particles in the water. The amount of turbidity measured is dependent on the size, shape, and refractive properties of the particles present. However, it can be used to measure relative levels of particles present in the water and indicate removal efficiencies obtained.

Membrane filtration processes such as microfiltration and ultrafiltration can operate with turbidity levels in excess of 100 ntu when the proper membrane and operating configurations are used. However, turbidity levels of less than 2 ntu are needed for desalting membrane processes. For these membrane processes, the *silt density index* (SDI) is also used to characterize the nature of the feedwater. A discussion of how to measure the SDI and other similar indices is presented in Chap. 9.

A typical SDI limit for desalting membranes is 2 for hollow-fiber membranes and 3 for spiral-wound membranes based on recommended levels by various membrane manufacturers. Above these levels, treatment of the feedwater may be required to reduce particles or other foulants in the feedwater. Operational changes, such as reducing flux rates, can also be used to maintain operability with higher-than-desired SDI levels. Reducing flux rates can reduce fouling tendencies indicated by high SDI levels. In some cases, lower SDI limits are used, depending on site-specific and membrane characteristics.

Because of the relative nature of turbidity and SDI measurements, particle counters and monitors have become very common particle measurement tools for pilot studies. These are most commonly used for determining membrane filtration removal efficiencies using microfiltration or ultrafiltration membranes. Because the smallest particle that most process particle counters and monitors can detect is approximately 1 µm, any filtered water particles detected with these instruments indicate a failure of the membrane. However, air entrapped in filtered water, electronic noise, or fouling of the sensor can cause false particle counts, and this needs to be considered when evaluating particle counter results.

The particles found in water can be in several different forms. The general classification of particles found in natural water include chemical colloids, silts, and microbial particles.

14.2.1.1 Chemical Colloids. Colloids are very small particles that can remain suspended in water for long periods of time. Relative to membrane processes, colloids are a class of compounds that can cause fouling in membrane processes due to their size and charge. These characteristics allow colloids to plug membrane pores or flow spacers by either attachment to the membrane material or physical entrapment. Pretreatment to remove or chemically condition may be required, depending on the membrane process being evaluated.

An example of a chemical colloid is silica. Silica is an amorphous, inorganic compound that can foul desalting membranes if its solubility is exceeded. Because silica is a polymer rather than a crystalline ionic lattice, the use of sequestering chemicals to allow supersaturation is difficult. Silica solubility, like most other chemical colloids, is related to water temperature and pH. In addition to these parameters, silica solubility is also affected by the level of particles in the water. Water with higher levels of suspended solids provides additional sites for the silica to attach and will precipitate silica at lower concentrations than low-particle-level water.

14.2.1.2 Silts. As with colloids, silts can also plug membranes. These typically comprise inert clays and due to their small size and charge, are difficult to settle out of the water. Filtration membranes are ideally suited to remove these contaminants; however, they must be removed prior to RO or NF membrane processes.

14.2.1.3 Microbial Particles. The interior of a membrane module is an ideal location for microbiological growth due to its damp environment. Once established on the membrane's surface, microbiological organisms can adhere to the organic polymer materials used for most membranes. In addition to the microbes themselves, biofilm secretions that they release can uniformly cover a membrane's surface and decrease flux and possibly salt rejection. In most brackish water ground sources, the number of organisms is low, limiting the microbial effects. Ultrafiltration and MF processes typically treat surface waters that have high levels of biological activity and require treatment to control biological fouling of the membrane.

Microbial fouling can be controlled by addition of a disinfectant either continuously or intermittently to the feedwater. For some membranes this is possible, while others will degrade when exposed to a typical disinfectant such as chlorine. Other types of disinfection such as a caustic flush at pH 10 or the use of an organic disinfectant can be considered.

14.2.2 Organic Contents

As stated earlier in this chapter, organic material affects two fields of interest:

• Treated water quality (target pollutants and secondary quality standards)
• Process efficiency and reliability

14.2.2.1 Organic Water Quality Parameters. Organic compounds are detrimental to water quality for many reasons. Some of them are pollutants for which maximum concentration in treated waters is regulated. Others are responsible for tastes, odors, or color either by themselves or after final chlorination. Therefore, they are undesirable as well. These organic pollutants are often target compounds for water treatment operations. These compounds are measured according to well-established methods that will not be discussed here. These methods all require off-line measurements. It is therefore necessary to properly choose the sampling method.

In addition to specific organic material, it is recommended to follow surrogate or lumped parameters such as:

• Color
• Total organic carbon/dissolved organic carbon (TOC/DOC)
• UV absorbance (@ 254 nm).

These measurements can be performed on raw water and treated water, most of them on-line. Their evolution is a good indication of changes in raw water quality. Since organic material is most often involved in flux decay with most membrane materials, their frequent monitoring is advisable. Also, a degradation in their removal by a membrane unit may indicate membrane failure.

Disinfection by-product formation potential, in particular, simulated distribution system trihalomethane (SDS-THM) and haloacetic acids (HAA), should be monitored on treated waters for drinking water supplies. This measurement is performed off-line as it involves contact with chlorine. Several methods are applicable as no standard exists for this measurement. SDS-THM is assessed in the presence of excess chlorine (2.0, 5.0, 10.0 mg/L) for a preset period of time (48 h, 168 h) so that less than 2.0 mg/L but not less than 0.5 mg/L free chlorine residual remains. Residual chlorine, THMs, HAAs, and organic halides (TOX) are then measured. Since this measurement is not standardized, other methods can be used. They differ by the chlorine dosage and the exposure time. Chlorine dosage may be calculated as the sum of the chlorine demand due to ammonium and nitrite plus a dosage based on the TOC. A dose similar to common final chlorination dose can also be chosen. Exposure time can be preset to a constant value (72 h) or to the expected residence time of the water in a distribution system.

14.2.2.2 Fouling Material. The nature of the fouling material may vary with the membrane material. Recent work on this topic showed that for ultrafiltration, operational (averaged) fluxes can be related to rather simple raw water quality parameters (turbidity, DOC, and UV). These parameters are, of course, indexes of the amount of material likely to foul the membranes, but they do not require sophisticated methods to be quantified.

Reverse-osmosis module manufacturers and system engineers recommend that feedwater be checked for suspended particles and fouling indices such as SDIs. Particle counters or turbidimeters can be installed on-line. Fouling index is a method based on refiltration time, as already discussed. There are standards that differ from one manufacturer to the other.

It is also useful to analyze the nature of the fouling material as sampled from fouled membranes. The nature (and texture) of the fouling material may then help in the choice of cleaning techniques or even in the choice of improved pretreatment steps.

14.2.3 Dissolved Mineral Contents

Dissolved mineral content is commonly referred to as the *total dissolved solids* (TDS) level of a water. This is a measurement of the total mineral content weight in the water once the water is evaporated away.

Membrane desalting processes such as reverse osmosis, nanofiltration, electrodialysis, and electrodialysis reversal are used to remove dissolved ions from contaminated water. In doing so, dissolved salts are concentrated in the reject stream and limit the processes operation. Concentrating salts beyond their saturation can cause precipitation of salts concentration past their saturation in the membranes. This will then either form a scale on the membrane's surface or plug the flow spacers between the membranes. In either case, increased operating pressures and lost performance will occur.

Solubility calculations can be used to predict the tendency of precipitates to scale the membrane. While the information provided by the solubility calculations is useful, it is not absolute and should be used only as a guideline. Solubility calculations have inherent limitations, as they typically include only the primary water quality characteristics in their calculation. Other salts or water quality constituents can also affect the potential for precipitation. For example, silica can be saturated beyond 100 percent in the absence of colloidal particles; however, it will precipitate quickly in the presence of colloidal particles, sometimes even below 100 percent saturation in some wastewater applications. Recently developed silica scale inhibitors have been successfully used for preventing silica fouling and may allow higher recoveries in the future in high-silica waters.

There are only a few common scales that form when concentrating brackish or seawater with membrane processes. These include calcium carbonate, calcium sulfate, barium sulfate, strontium sulfate, magnesium carbonate, magnesium hydroxide, silica, and magnesium silicate. Solubility of these compounds is typically dependent on the ionic strength, pH, and temperature of the water, in addition to their concentration. The impact of the other water quality parameters on a compound's tendency to precipitate varies and must be evaluated for each compound. For example, lowering the pH of the feedwater is a common method to increase the solubility of calcium carbonate. However, the pH adjustment does little to inhibit the precipitation of calcium sulfate. In these cases, a scale inhibitor is added to the feedwater to allow concentration of the ions beyond solubility levels.

14.2.4 Physical Parameters

Physical parameters play an important role in selecting the appropriate membrane material and operating conditions. Variations in feedwater pH, temperature, and other physical parameters will affect membrane process performance and need to be monitored and considered when evaluating membrane processes.

The pH of the feedwater will affect salt rejections with typically higher rejections at lower pH values. For some membrane materials, the pH must be maintained within a narrow range to prevent degradation of the membrane. An example is the hydrolysis of RO and NF membranes made from cellulose acetate. The minimum rate of hydrolysis occurs at a pH of 5.5 and increases rapidly as the pH is increased or decreased from that point. Hydrolysis in a cellulose acetate membrane breaks down the acetyl bond in the membrane matrix and *loosens* the membrane, allowing salt passage to increase. Changes to the salt permeation coefficient due to hydrolysis as a function of time can be described by Eq. 14.1.

$$B_t = B_o \times e^{kt} \qquad (14.1)$$

where B_t = salt permeation coefficient at time t, cm^2/s
B_o = salt permeation coefficient at day 1, cm^2/s
k = hydrolysis rate constant, day^{-1}
t = time, in days

This is not a concern with UF and MF membranes made from cellulose acetate because of the already high molecular weight cutoff and/or pore size.

The water temperature can have a significant effect on membrane performance. A typical temperature correction curve is presented in Fig. 14.1. Lower temperatures will typically decrease membrane flux rates. To compensate for this phenomenon, additional membrane area and/or higher feed pressures must be provided to maintain the same production capability.

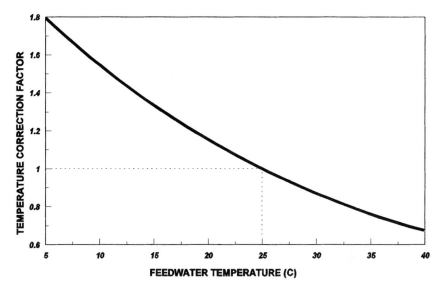

FIGURE 14.1   Temperature effects on RO feed pressure.

14.3   BENCH-SCALE TESTS

Bench tests performed on membranes allow characterization of the membrane itself and in some cases to qualify some of the interactions that the membrane may have with the raw water. Unlike activated carbon adsorption or coagulation, membrane filtration performance cannot be entirely determined from bench-scale tests. Models that exist to predict membrane performance from bench test results enable the user to determine performance in orders of magnitude at best. Pilot systems that hydraulically mimic a full-scale system remain necessary to set operating conditions, as extrapolation from one module to another is not yet possible, let alone the extrapolation from one membrane to another.

14.3.1   Membrane Characteristics

The following sections describe the membrane characteristics that should be evaluated during a bench- or pilot-scale study.

14.3.1.1   Permeability.   Membrane permeability is generally given by membrane manufacturers. In the case of microfiltration and ultrafiltration membranes, this permeability is generally measured as the ratio between membrane flux and operating pressure. For desalting membrane processes, the permeability is given for a standard feed pressure and salt concentration. The salt concentration and pressure depend on the pressure rating of the membrane. Lower pressure membranes will have lower standard pressures and TDS levels, whereas seawater membranes will have high standard pressures and salt concentrations.

Membrane permeability is, in fact, a function of several physical and chemical parameters that should be checked on the raw water. For UF and MF membranes, *clean water permeability* constitutes a membrane characteristic as it is measured using "clean" water. This calls for a standard definition of "clean" water so that membranes can be compared independently of site contingencies—i.e., of physical or chemical water properties. Manufacturers generally use ultrapure water for the permeability measurements. For practical purposes, it is more convenient to use treated water as clean water. This measurement is useful to evaluate the efficiency of chemical cleaning.

Temperature strongly influences water viscosity and volumetric mass. Viscosity is tabulated in most fluid dynamics handbooks. The use of these tables makes it possible to apply temperature correction and to calculate a membrane flux at a standard temperature. For 20°C, the following relationship has been used:

$$J_{20°C} = J_T \, e^{-0.0239(T - 20)} \tag{14.2}$$

where T is the field temperature.

For tighter membranes, such as reverse osmosis and nanofiltration, temperature effects can change the performance characteristics of the membrane material. Typically, higher feed pressures or lower flux rates occur as the temperature decreases.

14.3.1.2 Pore Diameter Distribution Function. This characteristic is, of course, relevant only to truly porous membranes, i.e., microfiltration and ultrafiltration membranes. The pore size distribution function can be measured according to different techniques:

- *Bubble point technique* (ASTM Method F 316-70 (1976)). According to Cheryan (1986), for a microfiltration membrane, discrepancies between actual pore size and calculated (Cantor equation) pore size come from nonzero contact angles. Therefore, it is difficult to rely on such a method, as the contact angle is a function of the membrane material. Other liquids with low surface tension can be used, provided that they are compatible with the membrane material. This is a common method of testing the integrity of microfiltration membranes.

- *Direct microscopic observation.* This is very useful to provide pore statistics. *Scanning electron microscopy* (SEM) is limited to microfiltration membranes. *Transmission electron microscopy* (TEM) can be used on replicas, as suggested by Cheryan (1986). Most recently, new electron microscopic techniques have been tentatively used. They enable identification of many foulants and changes to the physical characteristics on the membrane surface.

- *The solute passage technique.* This constitutes the most reliable way of evaluating a pore size distribution function. The membranes are exposed to solutions of well-defined compounds whose molecular weights are known. These compounds are polysaccharides (dextrins, dextrans), aminoacids, or proteins.

For water treatment purposes, membranes should be challenged with suspensions or solutions containing relevant pollutants. For truly porous membranes, bacterium and virus challenges can be used at either bench scale or pilot scale.

14.3.1.3 Artificial/Accelerated Aging. Artificial aging of UF and MF membranes involves the exposure of the membrane and the module to extreme physical and chemical conditions likely to occur sometime during the operation of the module. These tests should be performed when operation experience is not available. This is a case of new membranes that were not necessarily designed for water treatment. Chemical agents and physical stresses affect the mechanical resistance of the membrane material. Therefore, *membrane integrity* and *breakage risk* must be evaluated during these tests. Membrane integrity can be evaluated easily, as presented in Sec. 14.2.1.2, whereas breakage risk requires specific equipment rarely available in water treatment research laboratories.

Chemical Aging. Exposure of the membrane material to chemical agents used in the process is advisable. These chemicals include the washing mixtures used in the regeneration of the membranes. *Chlorine exposure* and *pH* are the two most common chemical aggressions likely to occur to a membrane material.

The *pH resistance* is related to the hydrolysis of the membrane material. Figure 14.2 presents an example of the time required to observe a 10 percent modification

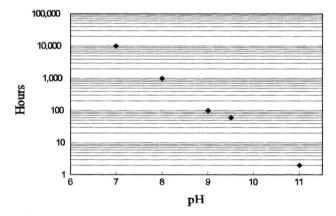

FIGURE 14.2   Membrane life versus pH.

of the mechanical properties of a membrane when exposed to water at different pH values. It can be observed that, in this case, the evaluated membrane is very sensitive to pH.

Chlorine resistance and, more generally, resistance to chemicals is expressed as mg/L·h or g/L·h. Since it is the product of a concentration by time, it is possible to accelerate the aging process by exposure of the membrane material to higher concentrations. *Detergent resistance* is also important, as washing mixtures may contain oxidizing agents and pH buffers. Furthermore, surfactants may modify surface properties of the material, thereby potentially modifying water permeability and fouling susceptibility.

Physical Aging. Membranes and modules that can be backwashed experience repeated pressure cycles that are responsible for material fatigue. Since time is not involved, accelerated aging can be performed. Results are expressed as a number of cycles before a set percentage risk of breakage occurs. This can then be converted into lifetime expectancy, as illustrated in Table 14.3.

Combination. Temperature always interferes as chemical kinetic rates increase with increasing temperature. Therefore, for a given property or set of properties, maximum exposure to chemicals (expressed as CT) should either specify a temperature or integrate the temperature in an expression for CT.

TABLE 14.3 Membrane Mechanical Resistance to Backwashing

Pressure in bar (filtration/backwash)	Cycles before breakage	Lifetime in years (1 backwash/30 min)
2/2	101,000–180,000	5.9–10.5
3/3	133,000–200,000	7.8–11.7
4/4	100,500–155,000	5.9–9.0

14.4 PILOT EVALUATIONS

14.4.1 Time Period

Conventional water treatment processes are dependent on the retention time in the rapid mix, flocculation, sedimentation, and finished water tanks for proper treatment. Hydraulic residence times in conventional systems are often several hours. In membrane treatment units, residence times are much shorter. Because water crosses a relatively thin layer of material, cumulative effects on membrane permeability, mechanical resistance, and integrity take years to show up. By then, each gram of membrane material may have contacted tons of water.

Flux decay, which is the most important concern of the membrane system operator will show several time scales for UF and MF membranes:

- Mechanically reversible fouling may occur within an hour of operation.
- Chemically reversible fouling may occur within hours to days or months.
- Irreversible flux decay due to physical deterioration of the membrane material occurs over months or years.

This is illustrated in Fig. 14.3.

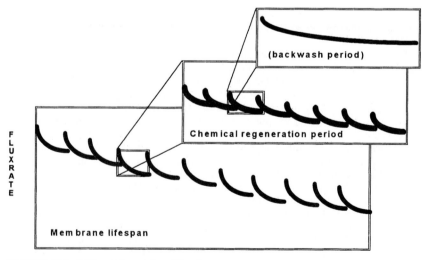

FIGURE 14.3 Different time scales of flux decay.

Desalting membrane processes typically take longer periods of time for mechanical and chemical reversible fouling to be seen. Typically, at least 2000 h of operating time is required. Pretreatment is provided for these membrane processes to reduce mechanical and chemical fouling because of the less aggressive cleaning procedures that can be used.

14.4.1.1   Short-Term, Mechanically Reversible Fouling.   This type of fouling is due to the accumulation of particles, precipitates, and/or organic matter at the water-membrane interface. In MF and UF processes, this material is removed from the membrane by either flushing or backwashing. It is important to evaluate the period at which the mechanical cleaning should be performed in order to know down periods and water losses. The choice of the membrane geometry (plate and frame, tubular, hollow fiber) is also important, as most plate-and-frame or spiral-wound modules cannot be backwashed, whereas most tubular and hollow-fiber membranes can be backwashed. Downtime is therefore important in operating plate-and-frame modules, as they must be periodically dismantled for cleaning.

The criteria used to determine when to mechanically clean the modules must be carefully set, as cleaning too may often lower the average production. This is illustrated by Fig. 14.4.

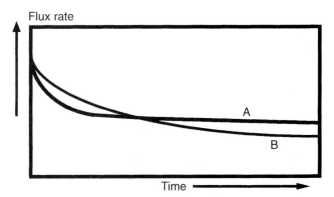

FIGURE 14.4   Different fouling rates as determining the mechanical cleaning period: (*a*) rapidly stabilizing process; (*b*) slowly degrading process.

A compromise must be found between instantaneous flux and average production. Also, when working on-site, it has been shown that average production stabilizes after at least eight days with unchanged operation conditions. This time limit was found by modeling the permeate flow rate during a filtration cycle between two backwashes, using the following equation:

$$Q(t) = Q_{eq} + (Q_0 - Q_{eq})e^{-k\alpha^\beta} \tag{14.3}$$

where Q = flowrates at time t, initial (0) and at equilibrium (eq)
             α and β = constants

and to calculate the time required for the values of the constants α and β to stabilize. This is illustrated in Fig. 14.5 where flux decay was processed on a specific site during a period of relatively stable feedwater quality.

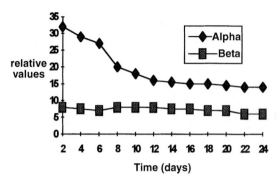

FIGURE 14.5   Evolution of the parameters α and β as a function of time.

It is very difficult to reverse fouling in spiral-wound membranes used for most RO and NF systems mechanically. Only fast-flushing the feed stream can be done to attempt to remove the trapped particles from the flow spacers. For this reason, pretreatment to remove particles from the feed stream is very important for these processes.

14.4.1.2   Long-Term Fouling.   Long-term or *chemical* fouling of the membranes may occur very rapidly if the membrane and the process are not adequate. However, in drinking water treatment by membrane processes, chemical cleaning frequency varies widely from once a week to once a year. It is impossible to precisely evaluate this frequency from pilot plant experiments, as long-term cumulative effects must be taken into consideration.

On the other hand, specific challenges may be performed if the nature of the potential foulants is known. Operational process conditions, such as higher-than-desired flux rates, may be used to enhance the fouling process. These experiments will provide fouled membranes on which chemical regeneration methods can be investigated.

14.4.1.3   Irreversible Fouling or Membrane Deterioration.   Long-term flux decay and deterioration of the mechanical and chemical properties of the membrane that are not remediated by either mechanical or chemical regeneration may be due to changes in the membrane structure. Unless specially equipped, this type of deterioration is not accessible to the water treatment operator. This characteristic is necessary to estimate membrane replacement time (hopefully several years) and should be available from membrane manufacturers.

14.4.2   Module Scale and Process Parameters

The size and arrangement of the membrane modules used during pilot testing should be similar to the proposed full-scale treatment system. This includes using the same membrane module, array or staging, and operating conditions.

14.4.2.1   Membrane Module.   It is important to use the same membrane material, geometry, size, and packing density in pilot testing to simulate performance of the process being evaluated.

Different membrane materials can have different affinities for contaminants in the feedwater that may promote fouling. This requires the selected membrane material to be exposed to the feedwater to allow identification of any incompatabilities present. Use of the same membrane geometry and fiber diameter or membrane thickness are used to identify hydraulic or physical fouling that may occur. This also requires that the membrane modules being proposed for the full-scale facility also have the same packing densities for hollow-fiber membrane modules or flow spacer thickness for spiral-wound membrane elements. Simply, the hydraulic, material, and geometric characteristics of the membrane modules being piloted should be the same as the proposed modules, even though they are typically smaller in size and contain less membrane area per module.

14.4.2.2 Module Array. Piloting desalting processes such as RO, NF, and ED requires staging of the membrane modules. Pressure vessels used for RO and NF desalting systems typically contain six membrane elements and are configured into an array to minimize operational and capital costs while producing the needed volume and quality of product water. It is necessary to maintain adequate cross-flow velocities in the concentrate stream and permeation rate throughout the membrane array.

Of the process configurations, the most common for UF and MF are either a single module or parallel modules. For pilot purposes, the use of at least three modules operated in parallel is desired to obtain statistical verification that the module performance criteria are obtained.

RO, NF, and ED processes use either reject staging or product staging to obtain the desired recoveries of salt removals. Reject staging is done to increase product recovery by using the concentrate from the first stage as feedwater into the second, then using the concentrate from the second into the third, etc., until the supersaturation of specific salts or permeate quality prohibits further staging. This process is also referred to as a multiple-stage, cascade, pyramidal, or tapered array configuration. No additional pumping is typically used between stages but may be required when different membranes are used in the later stages or if higher rejections are desired. The most common pilot testing array is 2:1 where two pressure vessels containing six 4-in-diameter membrane elements are used in the first stage and six 4-in-diameter elements are in the second stage. To minimize the size of the pilot unit for three stage systems, 2.5-in-diameter elements are used in the third stage to obtain hydraulic similarity with full-scale design conditions.

Product staging is used with RO processes for high-TDS feedwater such as seawater or with ED processes, to provide higher salt rejects than can be obtained with a single stage. Unlike reject staging, permeate or the dilute stream in ED processes where the salts are removed is fed to the next stage for further salt removal. In RO seawater applications, repumping of the permeate is required but not chemical pretreatment. This is because most of the limiting elements in the feedwater have been removed in the first stage, allowing the second stage to operate at very high recoveries and increasing the overall system product water recovery and quality. The ED process removes approximately 50 percent of the salts in the feedwater per stage. Therefore, two stages will remove approximately 75 percent of the salts and three stages will remove approximately 87 percent of the salts in the raw water.

Selection of the staging and array configuration for the pilot system should match the full-scale treatment process. This is done by using smaller membrane elements for the pressure-driven membrane processes, as discussed earlier, and by using smaller ED membranes but arranged in the same number of hydraulic and electrical stages as the proposed facility.

14.4.2.3 Operating Conditions. If the pilot system has the appropriate module and staging needed to simulate the full-scale facility, it should be possible to operate at design pressures and hydraulic conditions. Parameters used for evaluating the performance differ between membrane filtration and membrane desalting processes. Common concerns with all pressure-driven membrane processes are the changes in pressure differential across the membrane and decreasing flux rates. Desalting membrane processes also include monitoring changes in salt passage, also referred to as *salt flux*. ED processes also monitor salt flux but use electrical power consumption for evaluating driving force changes instead of pressure.

14.4.3 Data Acquisition

14.4.3.1 Required Data. Besides its demonstrative character, piloting enables the evaluation or refinement of *design* parameters influencing the *capital cost* of the final plant. Also, elements of the *operating cost* can be evaluated or refined during pilot plant experiments. Three types of data are worth measuring in a pilot plant run:

- Pressure and flow rates
- Raw water quality parameters
- Treated water quality parameters

   Pressure and Flow Rates. Truly porous membranes can be operated either at constant flux (variable operating or transmembrane pressure) or at constant operating pressure (variable operating flux). This choice determines the type of pump and the kind of process control transducers and actuators to be installed. Desalting membrane systems such as RO and NF operate at constant flow. In any case, the pilot unit should be equipped with monitoring systems, allowing the measurement of pressures and flow rates in order to evaluate the process response to variations in the operating conditions, pretreatment strategies, or raw water quality. In addition to their value as design criteria (recording membrane surface area, temperature, conductivities, pH), pressures and flow rates allow for the monitoring of the following characteristics:

- Membrane permeability
- Module physical fouling rates
- Process energy requirements

   Pressure and flows are affected by operating conditions such as pH and temperature and should be normalized to obtain the true performance during pilot plant operation. This is especially true for RO, NF, and ED processes. In addition to the preceding, applied voltage and amperage for each stage of an ED or EDR pilot system should be monitored to evaluate process energy requirements for these stages.
   Raw Water and Treated Water Quality Parameters. Raw water quality should be monitored during field evaluation for its physical and chemical quality. Parameters to be monitored are the main pollutants to be removed as well as the quality parameters likely to have an influence on the treated water flux. Therefore, the water analysis performed prior to the piloting itself as presented in the paragraph on bench-scale tests should also be performed during the field evaluation. Of course, treated waters should be analyzed as well in order to monitor the process

efficiency. Membrane integrity can be indirectly monitored through the analysis of naturally occurring markers (particle counts or conductivity) or challenging markers (bacterial viruses).

14.4.3.2   Parameter Measurement.   Process parameters are monitored on-line either by the use of ordinary pressure gauges and flow meters or by digital devices allowing automation. It is now more and more frequent that pilot units are computerized. This presents many advantages as continuous recording of on-line monitored parameters is possible. Proper calibration of all instrumentation is critical to obtaining useful information.

Alarms and process failures can be managed automatically to a point. Significant or exceptional events can be identified and analyzed with a substantially lower risk of missing them. Also, digital recording eases the data transfer to off-site computers for data analysis. When possible, automatic process control and optimization can be installed. In this latter case, pilot plants are demonstration units. Automatic process optimization should be minimized when dealing with rather new processes or with poorly known feedwater quality.

Water quality monitoring should also be performed on a continuous basis, at least for parameters such as particle counts, turbidity, temperature, pH, and conductivity. UV absorbance is also an important parameter where organic fouling is a concern. Most of these parameters may heavily influence flux decay or, more generally, water production. Some of them may have alarm levels, as they indicate pretreatment failure and, if not remediated, process failure. For quality parameters that are not crucial to the operation of the system, off-line analysis is acceptable.

14.4.4   Data Processing

Physical data from pilot plant experiments may take many forms, depending on the type of process and the degree of automation of the pilot plants. To illustrate what can be accomplished by thoroughly processing the data, the case of tangential ultrafiltration is described in this section.

Raw physical data from the measurements taken on the ultrafiltration pilot plant were the following:

1. Tangential and transmembrane flow rates
2. Module inlet and outlet pressures
3. Temperature
4. Module characteristics (geometrical description)
5. Pilot run operating conditions

The run performance can be characterized by the evolution of the following derived data:

1. Volumetric flux and permeability
2. Tangential Reynolds number, hydraulic diameter (in the tangential mode), and effective tangential velocity
3. Filtration cake characteristics
4. Energy consumption

The performance characteristics are accessible from the set of raw physical data by applying the equations of fluid mechanics, using temperature corrections where needed, and performing mass balances. Temperature corrections should be performed at least for the permeate flow rate and for all the terms involved in the energy requirement calculation.

Volumetric flux and permeability are calculated from the permeate volumetric flow rate, the total membrane surface area, and the transmembrane pressure. This latter derived parameter is set to the half-sum of the module inlet and outlet pressures:

$$P_{transmembrane} = \frac{(P_{inlet} + P_{outlet})}{2} \tag{14.4}$$

under the assumption that the tangential pressure gradient is linear.

Tangential Reynolds number, hydraulic diameter (in the tangential mode) and effective tangential velocity are derived characteristics that may be useful to verify that the flux decline rate proceeds normally as the filtration cake is formed. They are derived from the flow-rate data and the geometrical description of the module, using the Hagen-Poiseuille and the Blasius equations. In the case of hollow fibers, a module contains n_f fibers, each with a diameter d_f. The units should all be consistent, i.e., from the SI system. The Reynolds number is accessible from raw physical data through the following process:

$$Q_f = \frac{Q}{n_f} \tag{14.5}$$

$$v_f = \frac{4Q_f}{\pi d_f^2} \tag{14.6}$$

$$Re = \frac{v_f d_f}{v} \tag{14.7}$$

The effective diameter d_{eff} is accessible through an iteration on the values of d and v:

Effective diameter $\qquad d_{eff} = \left(\frac{(128\mu L Q_{inlet})}{n_f \pi (P_{outlet} - P_{inlet})}\right)^{1/4} \tag{14.8}$

where L is the length of the fibers:

Effective tangential velocity $\qquad v_{eff} = \frac{4Q_{inlet}}{\pi d_{eff}^2} \tag{14.9}$

The Reynolds number Re can then be calculated from d_{eff} and v_{eff} using Eq. (14.7).

If the hydraulic regime is found turbulent (Re > 4000), the effective diameter must then be calculated using the Blasius equation instead of the Hagen-Poiseuille equation:

$$d_{eff} = 1.5337 \left(\frac{L}{P_{outlet} - P_{inlet}}\right)^{4/19} \left(\frac{Q_{inlet}}{n_f}\right)^{7/19} \tag{14.10}$$

This new value should then be used in the calculation of v_{eff} and Re until the Reynolds number value converges.

Cake characteristics such as its thickness and its specific permeability are accessible from the fiber permeability, the fiber diameter, and the effective diameter. The resistance in series model is applied to calculate the permeability of the cake:

$$R = \frac{P_{transmembrane}}{permeate\ flux \cdot \mu} = R_{membrane} + R_{cake} \qquad (14.11)$$

where the hydraulic resistance R_i is reciprocal of permeability.

The *energy requirement* is calculated from the energy balance between the inlet and outlet of the process. The energy requirement should be reduced to the unit of produced water. It should include the energy requirements for backwashing when relevant. The calculation involves masses of water brought to a certain pressure. Since flow rates and pressure are time dependent, the work involves a time summation:

$$W = \int_0^t [Q_{inlet}\ P_{inlet} + Q_{permeate}\ P_{permeate} - Q_{outlet}\ P_{outlet}]\ dt \qquad (14.12)$$

This piece of information is very important for the evaluation of the operating cost, as membrane processes are generally considered as requiring less labor and more energy than conventional clarification treatments.

The energy dissipated along the membrane by recirculation in the tangential mode constitutes a very good standard when comparing several membranes. In this type of piloting, it is preferable to make the different membranes work at the same shear stress value at the membrane surface.

14.4.4.1 Design Information.
Full-scale plant design can be extrapolated from pilot plant experimental results. However, if pilot plant experimental results were always obtained, plant costs could be quite high. Some design information is normally derived from analytical data gathered or generated on the candidate feed stream using flux prediction tools. When concentration polarization is the limiting factor, numerous computerized design aids are available and used by membrane manufacturers or by consultants to obtain design information. When dealing with truly porous membranes for which flux decay is a consequence of the formation of a cake, deterministic models are often limited in their predictive ability.

If numerous pilot plant experiments have been run and the data are formatted properly, one can derive design information directly from the statistical analysis of the data. This has been performed for a UF membrane with the results obtained on 36 different sites (worldwide) using five different sizes of pilot plants. A total of 362 full events, i.e., containing flux, process and analytical data were used as a database for a typology. It was found that the data could be satisfactorily parted into seven groups associating values of UV absorbance, total organic carbon, and turbidity of the candidate resource to stable permeate fluxes using optimum operating pressure and backwash frequency.

It was found that a given resource may change from one group to another with time or season. This stresses the great interest in the knowledge of the feedwater quality variability for design purposes.

14.4.4.2 Operating Cost Information.
Operating cost is made of several elements that can be summarized as follows:

- Energy consumption
- Process chemicals

- Labor requirements
- Module regeneration rate
- Module replacement rate

Setting operating conditions is certainly part of the pilot plant experimentation when the system to be used is flexible. Once installed, however, none of the membrane processes are very flexible. Generally, operating pressure and chemical addition rates are selected during design and cannot be easily modified during operation.

Pilot plant experiments may then yield operating cost elements such as labor and energy requirements (as already discussed). Module regeneration rate can be either determined from extensive piloting or extrapolated from slow flux decay rate. Module replacement rate can be evaluated only over long-term operation, as it depends on the operating conditions themselves. If operated close to its limits, a module would certainly age faster than if "treated gently." Precessions on these running cost elements generally come only from actual plant operation or known characteristics of the membrane being used.

14.4.4.3 Process Selection.   Process selection involves many kinds of choices that should be done. Membranes are generally preselected and field validation will involve only a limited number of modules or membranes.

Operating conditions that should be chosen through piloting, such as operating pressures, pretreatment chemicals, recycle rate, or process conversion yield, have adverse effects on the investment cost that increases with increasing membrane surface area and running cost. This calls then for a compromise solution, as illustrated in Fig. 14.6.

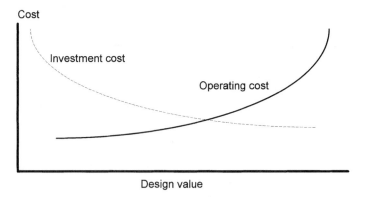

FIGURE 14.6   Costs as a function of design criteria.

14.5 *MEMBRANE PILOT CONSTRUCTION*

The methods for evaluating and installing membrane process systems vary as much as the potential for membrane process applications. Selection of the appropriate membrane process, evaluation methodology, and construction procedure are all based on the cost and size of the proposed system. Other variables such as specific project requirements, schedule, and regulatory concerns will also impact the selected methodology; however, cost is typically the primary criterion—especially when being compared to non-membrane-process options.

Once a potential membrane process application is identified and raw and product water quality is determined, evaluation of the proposed membrane system must be done. Projections for the pre- and posttreatment needs, system performance, and residual characterization to identify permitting needs are done to determine capital and operational costs. Two types of membrane process test units are commonly used:

- Single-element test stand
- Pilot test unit

14.5.1 Single-Element Test Stand

A single-element test stand can be used to determine what the rejection characteristics are for a given parameter using a given membrane element. Ultrafiltration and microfiltration membrane processes typically do not stage membranes and would use only a single membrane module on a pilot unit. This section discusses the evaluation of a single reverse-osmosis or nanofiltration membrane element.

When using a single element, it is important to remember that it only simulates what a single membrane element's rejection properties are and cannot provide system information for RO or NF systems. The feedwater quality to the second, third, etc., membrane element changes as permeate water is removed from the initial feedwater containing salts, and organic matter are concentrated in the remaining flow that feeds subsequent membrane elements. However, the percentage of removal for each contaminant will remain approximately the same throughout the membrane array, even though the feedwater quality is constantly changing.

By recycling concentrate from the test stand into the feedwater stream, higher levels of contaminants can be obtained. This will simulate feedwater to one of the later membrane elements based on the concentration factor used. For example, doubling the feedwater concentrations will simulate a concentration factor of two or simulate the water quality feeding the last element in a membrane system operating at 50 percent recovery. This procedure can also be used to determine what precipitates may be formed in the tail elements at elevated recoveries and the impacts of different pretreatment conditions. An example of single-element test stand process design is presented in Fig. 14.7.

Using this and other operating procedures, certain information can be obtained for reverse-osmosis and nanofiltration membranes using a single-element test stand. This information can include:

- Membrane rejections for specific contaminants
- Production of a small amount of permeate and/or concentrate volume for further evaluation
- Precipitation of salts in a membrane element at simulated high recovery for identification
- Determination of compatibility of membrane materials with feedwater or pretreatment chemical characteristics

14.5.1.1 Microfiltration and Ultrafiltration Pilot Plant Design and Operation. A single pilot unit may accommodate different types of modules to be tested. However, it may require the construction of several pilot units that can be run in parallel or, if the allocated time is long enough and the water resource quality stable enough,

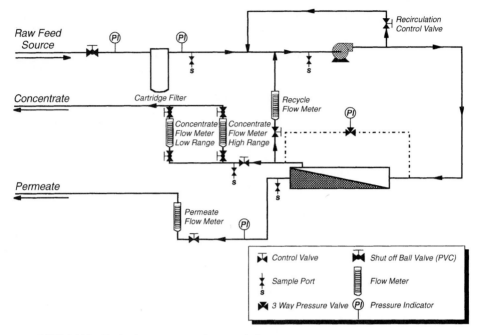

FIGURE 14.7 Single-element test stand process design.

a versatile pilot unit with adaptation systems. Scale-up issues should be examined at this stage so that generated data can be relevant to a full-scale plant. As already stated in Chap. 10, a module representative of the full-scale modules that will be used should be used for piloting. A process diagram for an ultrafiltration or microfiltration pilot unit is presented in Fig. 14.8.

The main constituents of the pilot plant are the following:

- Prefiltration system
- Equalization and cleaning tanks
- Pressure pumps: feed, recirculation, and backwash when relevant
- Frequency drive on pumps to allow easy setup
- Backwash and cleaning system

Membrane and Module Selection. This type of pilot experimentation is often meant to assess permeate production, fouling resistance, and particulate and organics removal. Generally speaking, since microfiltration and ultrafiltration are used in water application to remove particulate and biological material, it is preferable to preselect modules and systems that can be cleaned effectively.

Tubular and hollow-fiber modules can be backwashed, which constitutes an advantage over other module geometries (spiral wound or plate and frame).

Long-Term Evaluation of Membrane Performance. As in any pilot experiment or demonstration, long-term testing is used to generate a large enough database on the treatment efficiency and operating characteristics to derive reliable design or operation parameters. Therefore, the following monitoring is recommended.

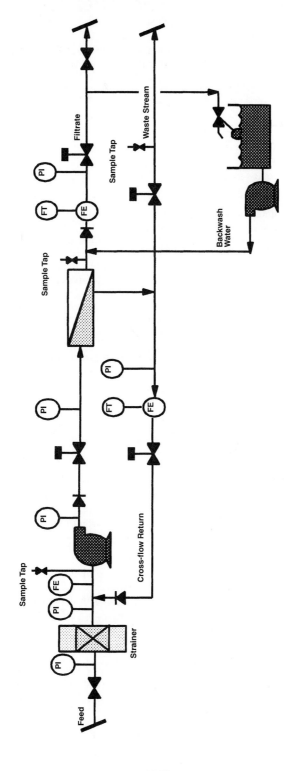

FIGURE 14.8 UF/MF pilot process diagram.

Treatment efficiency: These data should include the performance characteristics of the system, with reference to the treatment objectives. They include:

- pH
- Turbidity
- Suspended solids
- Particle counts
- Total organic carbon
- UV absorbance (@ 254 nm)
- Coliforms
- Heterotrophic plate counts (HPC)

This list may be completed by viruses, protozoa, or their determination whenever relevant, for example, *Giardia* cyst and *Cryptosporidium* oocyst.

An analytical scheme specifying the frequency for each measurement should be designed and followed during experiments. Table 14.4 gives an example of such a program.

Operating Characteristics. This part of the performance evaluation should gather data related to the design of a possible full-scale plant such as the size of the filtration system as well as the size of the related equipment (prefiltration and cleaning and backwashing system). This data subset should also include the elements necessary to estimate energy and chemical costs. Therefore, they include:

- Transmembrane flux
- Transmembrane pressure
- Washwater requirements
- Recirculation ratio
- Cleaning frequency

Feasibility and Cost Analysis. This part of the work is the data processing necessary to estimate capital and operation and maintenance costs. The costs are typically dependent on the feedwater characteristics, whether the membrane system is operating in dead-end or cross-flow, power costs, and other site-specific issues.

TABLE 14.4 Example of Analytical Scheme Followed on a UF Piloting Experiment

		Sampling location and frequency	
Parameter	Sampling method	Plant influent	Plant effluent
pH	On-line	Continuous—1/day	Continuous—1/day
Turbidity	On-line	Continuous—1/day	Continuous—1/day
Particle counts	On-line	10/week	10/week
Total organic carbon	Grab	1/week	1/week
UV absorbance (254 nm)	Grab	1/week	1/week
Temperature	On-line	Continuous—2/day	Continuous—2/day
HPC bacteria	Grab	3/week	3/week
Total coliforms	Grab	3/week	3/week

14.5.1.2 Reverse Osmosis and Nanofiltration Pilot Design and Operation.
Pilot testing for reverse osmosis and nanofiltration is used to obtain full-scale system performance information. A pilot unit consists of a miniature full-scale membrane treatment system. This is done by using smaller membrane elements, typically 4- or 2.5-in elements. The same hydraulic conditions and membrane operating conditions including feed pressure, staging, flux rates, and pretreatment conditions on the pilot unit are used as proposed for the full-scale installation. A typical reverse-osmosis and nanofiltration two-stage pilot unit process diagram is presented in Fig. 14.9.

Membrane and Array Selection. Selection of the appropriate RO or NF membrane depends on the raw water characteristics and the desired level of salt removals. To meet the range of requirements, RO membranes can be offered that reject over 99 percent salts and operate at pressures over 1000 lb/in^2 for seawater desalination. Low-pressure NF membranes that reject only 30 percent salts but provide excellent organic matter removal from some waters can operate at pressures less than 100 lb/in^2.

Once the rejection characteristics are identified, membrane materials must be considered. Cellulose acetate (CA) membranes have been historically used when high fouling potential from organic contaminants is present in the raw water, such as wastewater. This is due to the hydrophilic characteristic and smooth surface that can reduce potential fouling. However, the feedwater pH must be lowered to approximately 5.5 to minimize hydrolysis of the CA membranes.

The use of improved pretreatment methods are currently being evaluated to allow the use of thin film composite (TFC) membranes for water containing high levels of organic matter as well as the low-organic-content water. These membranes are typically constructed using polyamides (PA) and operate at lower driving pressures than the CA membranes for the same or higher levels of salt rejection. Because of the operational cost savings and a greater flexibility of operating pH levels, PA membranes are preferred for most low-pressure desalting applications.

Most pilots operate in a 2:1 array or a 2:1:1 array for high recovery systems. This means that two parallel pressure vessels containing, typically, six membrane elements each constitute the first stage with the concentrate stream off this stage entering a second stage consisting of a single pressure vessel containing six membrane elements. Most pilot units use 4-in-diameter pressure vessels for the first two stages. As water permeates from the feed stream, a lower volume of water is available to feed the following pressure vessels or stages. The required cross-flow velocity and flows are maintained by either reducing the number of pressure vessels operating in parallel or by going to a smaller-diameter pressure vessel. Therefore, if a third stage is required, a 2.5-in-diameter vessel is commonly used to obtain the needed hydraulic characteristics.

Long-Term Evaluation of Membrane Performance. Most long-term performance characteristics that are evaluated in a pilot study can be obtained after approximately 2000 h of operation. As with the MF and UF membrane processes, several parameters should be monitored including:

- pH
  Raw water
  Feedwater
  Concentrate

- Pressure
  Feed to first stage
  Interstage pressures
  Concentrate
  Permeate

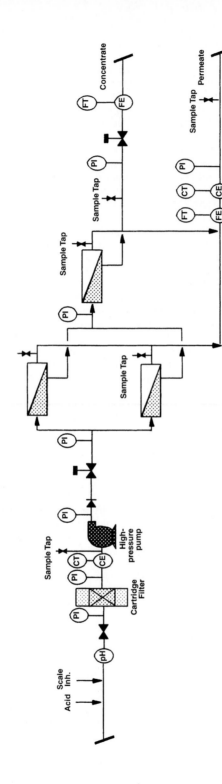

FIGURE 14.9 RO/NF pilot process diagram.

- Flows
  Feed
  Concentrate and/or permeate
  Bypass (if used)
- Temperature
  Feedwater
- Conductivity
  Feed
  Permeate
  Blended product (if bypass used)
  Concentrate
- Other Specific Water Quality Parameters
  Feed
  Permeate and/or concentrate

A written protocol specifying the frequency and location of each sample should be developed and followed during testing. Data collected should then be entered into a spreadsheet or software program for normalization. Without normalization, the raw data can provide misleading conclusions, especially when feed pH and temperature values are not consistent during the test program. Three parameters are monitored for a membrane desalting process to monitor performance: salt rejection or passage, normalized permeate flow, and pressure drop.

Salt rejections are presented as the percent of either *salt rejection* or *salt passage*. Either is acceptable and one can be easily converted to the other by the relationship

$$\% \text{ salt rejection} = 100 - \% \text{ salt passage}$$

The percent of salt rejection can be calculated by using Eq. (14.13) as follows:

$$\%\text{SR} = \frac{C_f - C_p}{C_f} \times 100 \qquad (14.13)$$

Pressure, temperature, and salt concentration determine the flux, or permeate flow, that a membrane will operate at. If these parameters remain constant, the loss in permeability through the membrane over time is due to fouling. However, without normalization, even slight variations in one of these parameters could mislead the operator by showing no decrease in flux when one occurs or vice versa. The normalized permeate flow can be calculated using the following equation:

$$\text{NPF} = \frac{\text{NDP}_i}{\text{NDP}_t} \times \frac{\text{TCF}_i}{\text{TCF}_t} \times Q_p \qquad (14.14)$$

where NPF = normalized permeate flow
  NDP_i = initial net driving pressure
  NDP_t = net driving pressure at time t
  TCF_i = initial temperature correction factor
  NDP_t = temperature correction factor at time t
  Q_p = permeate flow

The *net driving pressure* (NDP) used in this equation is the available force to drive the water through the membrane minus the permeate and osmotic backpressures. These are based on average pressures where:

$$\text{NDP} = \overline{P}_f - \overline{P}_o - \overline{P}_p \qquad (14.15)$$

The *average feed pressure* is the average of the feed and concentrate pressure. Average osmotic pressure can be *approximated* by using total dissolved solids (TDS) levels in the feed and concentrate streams and Eq. (14.16). This is only an approximation, as osmotic pressure is a function of the type and amounts of salts in the water.

$$P_o = \frac{TDS_f + TDS_c}{2} \times 0.01 \tag{14.16}$$

Temperature correction factors were discussed earlier in this chapter and can be calculated by using Eq. (14.2). Typically, a 10 to 15 percent loss in normalized permeate flow indicates unacceptable fouling levels and chemical cleaning is recommended.

When comparing different membranes that provide the desired salt rejections, it is useful to calculate the unit flux or amount of water that passes through a unit area of membrane per unit of feed driving pressure. This is done by dividing the normalized flux by the net driving pressure for the system. The higher the unit flux, the lower the amount of energy required to produce a unit of permeate flow and the more cost-effective the membrane process will be.

Pressure drop through a pressure vessel or a membrane system array is from the resistance of flow through the concentrate flow spacers in the membranes and associated piping. An increase in the pressure drop (Δp) across a pressure vessel or array can be caused by membrane compaction during initial operation or may indicate fouling by particulate matter or precipitated saturated salts that increase flow resistance through the membrane spacers. Other factors such as changes in feed flow into the membrane system and water temperature can also affect the Δp. Increases in feed flows will increase the Δp where increases in water temperature will decrease the Δp.

Operating Characteristics.   Design of a full-scale plant will require the pilot processes' operating characteristics to be monitored to determine the amount of membrane area, array, staging, and operating conditions needed for proper performance.

Information needed to estimate energy, chemical, and other operating costs include:

- Pretreatment chemical and filtering requirements
- Operating pressures
- Flux rates
- Posttreatment requirements
- Cleaning frequency
- Concentrate toxicity (as required by disposal permit)

Feasibility and Costs.   Pilot testing RO and NF processes can provide the largest savings with identification and confirmation of pretreatment needs to provide an effective treatment process. Maximizing flux and system recovery rates that can be maintained without adverse impacts on the process are also factors that pilot testing should provide. A cleaning should also be performed to determine the proper procedure to recover lost performance based on the specific fouling characteristics of the water being evaluated. Once determined, capital and operational cost estimates can be developed with a high degree of certainty.

Cost estimates are divided into three different categories and consist of the following major components:

- Capital costs
  Membrane process system (typically includes pretreatment system)
  Posttreatment and treated water storage and distribution facilities
  Building and other support facilities
- Fixed operation and maintenance costs
  Membrane replacement
  Labor
- Process operation and maintenance costs
  Pump energy requirements
  Chemical usage
  Cartridge filter replacement

When developing annual operating costs for budget purposes, the annualized capital and fixed operational and maintenance costs are estimated based on the design capacity of the treatment facility. This is because these costs occur independently of the volume of water treated. If six persons are required to operate and maintain the treatment facility, they will need to be paid the same whether the plant is operating at full capacity or half capacity. Membrane replacement is also based on an estimated life of the membranes and, once put into operation, will age more quickly whether they are being used or not. In fact, most membranes will age quicker when not in operation than when being operated. Therefore, operating at a reduced flux rate but keeping water flowing through all of the membranes is better than operating one train and leaving one off-line. This is limited only by the need to maintain a minimum flow through the membrane trains.

Process operational and maintenance costs are based on the average annual flow, as the costs are directly related to the volume of water treated. Chemical feed systems are based on dosage per unit of water being treated. Therefore, lower flows will require fewer chemicals and lower operational costs.

14.5.1.3 Electrodialysis Pilot Design and Operation. As with the RO and NF pilot systems, the proposed electrodialysis process system should be replicated on a smaller scale for pilot testing. This is done by using smaller-sized membranes but retaining the same number of hydraulic and electrical stages anticipated in the full-scale installation. One major difference between electrodialysis and pressure-driven membrane desalting processes is that additional stages are added in electrodialysis to increase salt removal, whereas pressure-driven membrane desalting processes add stages to improve system water recoveries. Production water capacity needs are used to size pumps, pipes, valves with the number of stages, or stacks placed in series, determining the amount of salts that will be removed.

Membrane and Array Selection. Pilot units are available from electrodialysis system manufacturers and range from approximately 8 to 114 L/min. The number and types of stages are incorporated into the pilot depending on the treatment objectives that the electrodialysis system is being evaluated for. Most pilot units contain only one membrane stack containing multiple stages. There are two types of stages: hydraulic and electrical.

A typical electrodialysis stage removes from 40 to 50 percent of the salt in the feedwater. Additional stages are added to remove additional salts, approximately 40 to 50 percent of the remaining salts each stage, until the product water contains the desired salt levels. For example, an electrodialysis system with two hydraulic stages would remove 65 to 75 percent, and a three-stage would remove from 82 to 99 percent of the salts in the feedwater.

Electrical stages are used to control the electrical current used to remove the salts. These stages are determined from Faraday's law, which determines the amount of current needed to transfer a specific quantity of salt through the electrodialysis membrane, and Ohm's law that provided the voltage requirements for each stack. These laws and how they apply to the selection of the appropriate number of electrical stages for a system are presented in Chap. 12.

Because most electrodialysis pilot units use only one membrane stack, recycling and phased reversals will not provide actual, complete, full-scale operational information. Recycling of the concentrate stream to simulate high recovery systems will produce higher-than-expected power usage than the full-scale system. This is due to the high concentration of salts in a single stack where the true dilution cannot be simulated. Phased reversals cannot be done, as only one stack is being operated. All of the simulated hydraulic stages must be reversed at the same time when piloting electrodialysis reversal systems.

Long-Term Evaluation of Membrane Performance.   Pilot testing of electrodialysis systems is used to generate a large enough database on the treatment efficiency and operating characteristics to develop reliable design or operation parameters. Most electrodialysis pilot studies are done over a period of 1500 to 2000 h of continuous operation. Many of the same parameters monitored for reverse-osmosis systems are also monitored for electrodialysis. However, monitoring for the electrical voltage and amperage applied to stages becomes the primary performance-monitoring tool for electrodialysis. Increases in amperage to obtain the same salt removals indicate an increase in electrical resistance from fouling of the membranes.

Treatment efficiency data include operational and performance characteristics such as:

- pH
    Feedwater
    Dilute stream
    Recycle
    Concentrate
    Electrode flush

- Electrical
    Voltage applied to each electrical stage
    Amperage applied to each electrical stage

- Pressure
    Feed
    Dilute stream
    Concentrate stream

- Flows
    Feed
    Dilute
    Recycle
    Concentrate
    Electrode flush

- Temperature
    Feedwater

- Conductivity
    Feed
    Dilute
    Recycle
    Concentrate

- Other specific water quality parameters
  Feed
  Dilute and/or concentrate

As with other pilot studies, a written protocol specifying the frequency and location of each sample should be developed and followed during testing. The protocol should incorporate the specific requirements of the testing being done and allow for possible limitations of the pilot unit discussed earlier.

Operating Characteristics. This part of the performance evaluation should gather data related to the design of a possible full-scale plant such as the number of stages required as well as the size of the related pretreatment and cleaning equipment. Information to determine required energy and chemical costs should also be verified.

Feasibility and Costs. Capital costs for an electrodialysis system are related to the capacity and the levels and types of salts being removed. Electrical energy applied to the membrane stacks is typically the largest operational cost and is related to the amount and type of salts being removed. Other site-specific issues can also affect the feasibility and cost of a proposed electrodialysis or electrical dialysis reversal system.

14.5.1.4 Combined Treatments: Membrane Bioreactors and PAC Membrane Combinations

PAC-Membrane Combination (CRISTAL Process). Recommendations for pilot experiments with a powdered reagent such as powdered activated carbon are similar to those related to UF and MF. Recommendations about module geometry are even more relevant since particulate material is added to the feed stream and concentrated during operation, until the system is operated on a feed-and-bleed mode or blown down.

Precautions should be taken in the choice of the PAC granulometry, as fiber plugging is possible if particle diameters and fiber diameters are close enough. Typically, particle diameters should not exceed a tenth of the fiber diameter.

The sizing of the pilot plant can be approached using models such as those presented in Chap. 15.

In addition to the standard set of parameters, treatment performance assessment should include

- Color
- DBP formation potential
- Organic micropollutant analysis (pesticides)
- Taste- and odor-causing material

Operating characteristics should be monitored especially regarding module head loss, as this parameter is the index of possible fiber plugging.

Membrane Bioreactor Pilot Plant Design and Operation. Membrane bioreactors (MBRs) are biological systems that can be entirely and rigorously controlled through the proper choice of the *hydraulic residence time* (HRT) and the *sludge residence time* (SRT). MBRs can be operated within the following range:

0 h < HRT < 300 h

0 days < SRT < 300 days

The choice of the HRT and SRT values determines the biomass concentration $[X]$ in the reactor. Two operation strategies are possible: (1) operating MBRs at very low biomass concentration ($[X] < 1$ g/L) or (2) operating MBRs at high biomass concentration ($[X] > 8$ g/L). The first strategy is very efficient for the treatment of slightly polluted waters, such as in the case of groundwater denitrification ($50 < NO_3 < 150$ mg/L). The second strategy should be preferred in the case of high-strength or very polluted waters (landfill leachates or industrial wastewaters). These waters have the following characteristics:

NH4*1500 mg/L

$4000 < COD < 100,000$ mg/L

The second strategy is also recommended if compactness of the treatment system is the issue. This is the case of wastewater treatment and reuse within buildings.

MBR pilot plants should be operated on a continuous mode. Pretreatment (pre-screening) should also be automated.

Membrane Selection. The membrane, as well as the module configuration, are chosen according to the desired biomass concentration and the raw water quality. Raw waters with low particle content treated with a diluted biomass, hollow-fiber, or spiral-wound modules can be used. On the other hand, tubular or plate-and-frame modules should be preferred in the case of a highly charged raw water or the use of a concentrated biomass treatment operation.

Membrane cutoff threshold should be chosen according to the solid retention and the disinfection objectives of the treatment. In any case, biomass retention constrains limits the cutoff threshold to 0.15-m diameter.

Long-Term Evaluation of Membrane Performance. In addition to operational parameters that must be monitored during a UF or MF pilot experiment, the following should be measured:

- Module head loss
- Fouling kinetics
- Dissolved oxygen
- pH
- Temperature

Furthermore, the efficiency of the biological system should be assessed through the following routine parameters, depending on the type of treatment:

- COD
- BOD5
- Ammonia
- Nitrate

These parameters are added to those associated to the membrane treatment efficiency, as already listed earlier.

14.6 CONCLUSIONS

For processes that are not fully controlled by polarization concentration, the future of pilot field testing may disappear after having provided data relating raw water quality and desired treated water quality to the required type of membrane and the most economical process and operation procedures to run them. This is already the case with reverse osmosis and nanofiltration, for which there are computerized decision aids. When doubtful raw water qualities are encountered, piloting may be necessary to provide information on a limited number of design and process parameters, hence, involving simpler pilot equipment. Pilot plant experiments should be run by membrane or system manufacturers when proposing new products. Pilot demonstrations are still required for new membrane processes as long as operation references are scarce.

BIBLIOGRAPHY

"Solar Pilot Plant Feeds Hydroponics," 1987, *World Water,* **10**(8):42.

"Application of Microfiltration for Water and Wastewater Treatment," 1991, *Environ. Sanit. Rev.,* **31**:184.

Argo, D. R., 1984, "Use of Lime Clarification and Reverse Osmosis in Water Reclamation," *Jour. WPCF,* **56**(12):1238.

Bersillon, J. L., et al., 1989, "L'ultrafiltration appliquée au traitement de l'eau potable: le cas d'un petit systeme," *L'eau, l'ind. les nuisances,* **130**:61.

Bersillon, J. L., et al., 1990, "Ultrafiltration Applied to Drinking Water Treatment: Case of a Small System," *Water Supply,* **8**(3/4):209.

Blau, T. J., et al., 1992, "DBP Control by Nanofiltration: Cost and Performance," *Jour. AWWA,* **84**(12):104.

Bourdon, F., et al., 1988, "Microfiltration tangentielle des eaux d'origine karstiques," *L'Eau, l'ind les nuisances,* **121**:35.

Cheryan, M., 1986, *Ultrafiltration Handbook,* Technomic Publishing, Lancaster, Pa.

Clark, R. M., et al., 1988, "Removing Contaminants from Groundwater," *Environ. Sci. and Tech.,* **22**(10):1126.

Conn, W. M., et al., 1985, "Practical Operation of a Small Scale Aquaculture," *3d Water Reuse Symp. Proc.,* San Diego, Calif., **2**:703.

Demers, J. F., et al., 1990, "Use of Nanofiltration for the Removal of Colour and Organic Matter of Natural Origin During the Production of Drinking Water," *Proceedings 13th International Symp. on Wastewater Treatment and 2d Workshop on Drinking Water,* Montreal, PQ, Canada. p. 247.

Duguet, J. P., et al., 1992, "Elimination des pesticides par de nouvelles techniques de traitement," *L'eau, l'ind les nuisances,* **153**:41.

Duranceau, S. J., et al., 1992, "SOC Removal in a Membrane Softening Process," *Desalination,* **84**:1.

Finken, H., 1984, "Flux Stabilized Cellulose Acetate Membranes and Their Applications to Reverse Osmosis for Water Desalination and Purification," *Ind. and Eng. Chem. Prod. Res. and Dev.,* **23**(1):112.

Fox, K. R., 1989, "Field Experience with Point of Use Treatment Systems for Arsenic Removal," *Jour. AWWA,* **81**(2):94.

Gandhilon, A., et al., 1992, "An Experimental Study on a Membrane Ultrafiltration Process for Drinking Water Production," *Aqua,* **41**(4):203.

Goodrich, J. A., et al., 1991, "Drinking Water from Agriculturally Contaminated Groundwater," *Jour. of Environ. Quality,* **20**(4):707.

Hrubec, J., et al., 1984, "Reuse of Municipal Wastewater," *H₂O,* **17**(11):228.

Jacangelo, J. G., et al., 1991, "Low Pressure Membrane Filtration for Removing Giardia and Microbial Indicators," *Jour. AWWA,* **83**(9):97.

Jacangelo, J. G., et al., 1989, "Assessing Hollow Fiber Ultrafiltration for Particulate Removal," *Jour. AWWA,* **81**(11):68.

Jacangelo, J. G., et al., 1992, "Low Pressure Membrane Filtration for Particle Removal," *AWWARF Report.*

Kopfler, L. C., et al., 1987, "A Comparison of Seven Methods for Concentrating Organic Chemicals from Environmental Water Samples," I. H. Suffet (ed.), *ACS,* **214:**425.

Lange, P. H., et al., 1989, "THM Precursor Removal and Softening—Ft Myers 12 MGD RO Membrane Plant, Florida, USA," *Desalination,* **76**(1&3):39.

Lauer, W. C., et al., 1991, "Process Selection from Potable Reuse Health Effect Studies," *Jour. AWWA,* **83**(11):52.

Lewis, J. W., et al., 1992, "Desalting Oil Field By-Product Water," *Desalination,* **87**(1):229.

Loebb, S., 1984, "Circumstances Leading to the First Municipal Reverse Osmosis Desalination Plant," *Desalination,* **50:**53.

Lozier, J. C., et al., 1992, "Selection, Design and Procurement of a Demineralization System for a Surface Water Treatment Plant," *Desalination,* **88**(1&3):3.

Lozier, J. C., et al., 1992, "Nanofiltration Treatment of a Highly Organic Surface Water," *AWWA Annual Conf. Proceedings,* p. 715.

Lykins, B. W., 1988, "Removal of Organics from Drinking Water," USEPA EPA/600/D688/008.

Lykins, B. W., et al., 1986, "Chemical Products and Toxicological Effects of Disinfection," *Jour. AWWA,* **78**(11):66.

Marquardt, K., 1985, "Demonstration Plant for Seawater Desalination by Reverse Osmosis," *Marine Technology/Meerestechnics,* **16**(4):155.

Marquardt, K., et al., 1986, "Demonstration Plant for Seawater Desalination by Reverse Osmosis (RO-2 Hollow Fiber Module Line)," *Desalination,* **60**(1):25.

Moulin, C., et al., 1991, "Design and Performance of Membrane Filtration Installations: Capacity and Product Quality for Drinking Water Applications," *Environmental Technology,* **12**(10):841.

Pieterse, T., et al., 1991, "Plant Quartet Proves Potable Water Reuse," *Water Quality International,* **4:**31.

Rogers, S. E., et al., 1986, "Disinfection for Potable Reuse," *Jour. WPCF,* **58**(3):193.

Schippers, J. C., et al., 1985, "Reverse Osmosis for Treatment of Surface Water," *Desalination,* **56:**109.

Schneider, L., 1992, "Drinking Water for San Francisco," *Gas Wasser Abwasser,* **72**(3):174.

Sorour, M. H., et al., 1992, "Desalination of Agricultural Drainage Water," *Desalination,* **86**(1):63.

Taylor, J. S., et al., 1987, "Applying Membrane Processes to Groundwater Sources for Trihalomethane Precursor Control," *Jour. AWWA,* **79**(8):72.

Taylor, J. S., et al., 1986, "Cost and Performance Evaluation of In-Plant Trihalomethane Control Techniques," USEPA Report EPA/600/S2/85/138.

Thompson, M. A., and M. P. Robinson, Jr., 1991, "Suffolk Introduces EDR to Virginia," *AWWA Membrane Technologies in the Water Industry Proceedings,* p. 129.

Thompson, M. A., 1992, "The Roles of Membrane Filtration in Drinking Water Treatment," *AWWA Proceedings, Engineering and Operations,* p. 5.

Vail, J. W., et al., 1986, "Reclamation of Secondary Sewage Effluent by Reverse Osmosis: A Pilot Plant Study," *Water SA,* **12**(1):37.

Vanopbergen, G., et al., 1983, "Reduction of Nitrate Concentration in Drinking Water by Hybrid Process with Zero Discharge Based on Reverse Osmosis," *Desalination,* **47:**267.

Wong, J. M., 1990, "Treatment Options for Water Supplies Contaminated with DBCP and Other Pesticides," *Public Works,* **121**(7):78.

CHAPTER 15
MEMBRANE-POWDERED ACTIVATED CARBON REACTORS

M. M. Clark
Environmental Engineering and Sciences Program
Department of Civil Engineering
University of Illinois
Urbana, Illinois

I. Baudin and C. Anselme
Centre International de Recherche sur l'Eau et l'Environment
Lyonnaise des Eaux
Le Pecq, France

15.1 INTRODUCTION AND MOTIVATION FOR MEMBRANE-POWDERED ACTIVATED CARBON REACTORS

Previous chapters have demonstrated the excellent particle-removing capabilities of membrane processes. The term *particles* includes here clays, algae, and other natural suspended particulate matter. However, only the "tighter" higher-pressure membrane processes such as nanofiltration (NF) and reverse osmosis (RO) can remove dissolved organic matter to a significant degree. By dissolved organic matter, we mean natural organic matter such as humic substances and taste- and odor-producing compounds, as well as synthetic organic compounds such as the trihalomethanes (THMs) and pesticides. But what if these organic materials could be associated with a particulate phase, for example, with powdered activated carbon (PAC)? It should then be possible to use the lower-pressure (and lower-operating-cost) membrane processes such as microfiltration (MF) and ultrafiltration (UF) to remove both natural particulate matter, powdered activated carbon, as well as the organics adsorbed on the powdered activated carbon (Laîné et al., 1989, 1990; Clark and Heneghan, 1991; Heneghan and Clark, 1991; Clark, 1992; Anselme et al., 1991, 1992; Pirbazari et al., 1992; Baudin et al., 1993).

The purpose of this chapter is to demonstrate that the membrane-powdered activated carbon reactor (called the *PAC/UF process* here—see Figs. 15.1, 15.3, and 15.5) is effective for removal of both particles and organics from drinking water supplies such as surface waters and organic-containing groundwaters. Far from harming or fouling the membrane system, we show that the addition of powdered activated carbon to an ultrafiltration system can improve flux through limitation of membrane fouling. We review the theory and modeling of powdered activated carbon adsorption in the context of low-pressure membrane systems. The testing of this model on the pilot scale is demonstrated. Operational and water quality data for several existing PAC/UF installations in France are reviewed, and the role of the PAC/UF model in plant design is discussed. The chapter ends with a discussion of some recent developments in the treatment of PAC/UF concentrate.

15.2 ROLE OF POWDERED ACTIVATED CARBON IN HYBRID PROCESS

15.2.1 Advantages of PAC in Comparison to Granular Activated Carbon Adsorption Systems

As was stated in the introduction, UF by itself cannot remove a very significant amount of natural or synthetic organic compounds; this was the motivation for combining PAC adsorption with UF. Occasionally, the use of UF in conjunction with granular activated carbon (GAC) is suggested. However, on reflection, this does not seem to be an efficient use of either the membranes or GAC. For example, if the GAC columns preceded the membrane and raw water was directly applied to the GAC, the GAC would become fouled by most surface and groundwaters. We have also learned in experiences with full-scale plants and simulations that removal of micropollutants by GAC beds depends significantly on the portion of the cycle operated in. For example, seasonal spikes of atrazine may be effectively removed by GAC early in the adsorption cycle, but will not be effectively removed near the end of the adsorption cycle when bed exhaustion approaches. On the other hand, if the GAC columns followed the membrane, the superior disinfection capabilities of the membrane process could very likely be compromised, due to biological growth on the carbon. By enumeration of phytoplankton and zooplankton, and particle counting, we have verified that full-scale GAC columns release a great number of biologically active particles in their effluents.

PAC dose, on the other hand, can be adjusted rather quickly to meet changing treatment goals. Especially for seasonally varying water quality problems such as micropollutants (e.g., pesticides) and taste and odors, it will be an easy matter to increase PAC doses during critical periods. In current PAC/UF designs, PAC is concentrated in the recirculation loop, helping to make efficient use of the carbon. Finally, we have verified that direct addition of PAC to UF processes tends to minimize membrane fouling and prolong filtration cycles.

15.2.2 PAC Adsorption and Water Quality

The dose of PAC is, of course, determined by the quality of the water resource and desired final water quality. For groundwater resources with a total organic carbon (TOC) content less than 2 mg/L, we have found that UF can generally be used alone for removal of particulate contaminants and for disinfection. However, if these waters have seasonal problems with micropollutants (such as atrazine) or taste and odor problems, then PAC addition at small concentrations can be made (see Secs. 15.5, 15.6, and 15.7).

However, the majority of natural water sources are surface waters with relatively high TOC values. In this case, the PAC addition has to solve a lot of problems, such as high TOC, UV absorbance, and THM formation potential, as well as color, taste, and odor problems. In practice, we have found that the dosage is often related to the required TOC removal, since TOC is often correlated with other problems, such as color and THM formation potential. The models described in Sec. 15.4 demonstrate the connection of PAC dose to the removal of TOC and/or micropollutants. Some case histories are examined in Secs. 15.6 and 15.7.

15.2.3 Effect of PAC on Ultrafiltration Membrane Fouling

PAC addition to UF can increase the efficiency of the membrane process. Laîné et al. (1990) found that addition of 250 mg/L of PAC to a lake water during batch UF could decrease irreversible fouling of several Amicon UF membranes and decrease the hydraulic resistance of the cake formed on the membrane surface. Adham et al. (1991) continuously added 25 mg/L of PAC to a continuous UF pilot system (Fig. 15.1), treating groundwater over long periods of time, resulting in a steady-state buildup within the system of 200 mg/L PAC. The buildup of PAC in a continuous waste system is set by the rate of wastage of PAC and the volume of the recycle system (see Secs. 15.4.2 and 15.6). Although the addition of PAC did not

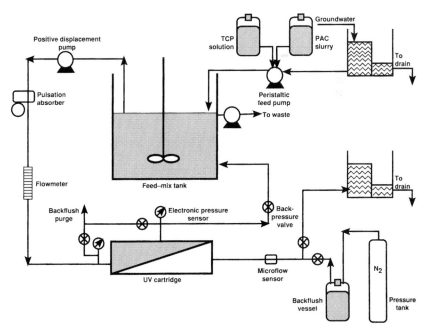

FIGURE 15.1 Laboratory-scale PAC/UF pilot used for testing of perfect mixing model. *(Adham et al., 1991.)*

increase the average water production in the system, the PAC did result in significantly longer filtration runs prior to backflush (about 22 h without PAC and 39 h with PAC). The membrane used by Adham et al. (1991) (and in the following studies) was an Aquasource hollow-fiber UF membrane made of a cellulosic derivative (molecular weight cutoff [MWCO] = 100,000).

In a number of longer pilot studies in France, PAC has been found to improve UF performance. The system was a CSTR + UF with a mean hydraulic contact time in the CSTR of 5 min. Table 15.1 shows results for the performance of ultrafiltration of relatively polluted raw Seine River water near Paris, with and without a 40-mg/L PAC dose (Chemviron TL 9003). This PAC has a high affinity for the TOC in this water, $K_f = 80$ (mg/g)(L/mg)$^{1/n}$. The Seine River water quality is characterized by turbidities as high as 100 NTU, TOC as high as 5 mg/L, and UV absorbance (254 nm) as high as 15 m^{-1}. Note from the table that addition of PAC is mainly responsible for less frequent chemical cleaning and a higher flux recovery after cleaning.

To show the effect of PAC addition on UF efficiency with a less-polluted water, Table 15.2 shows results for the filtration of a clarified (presettled) Seine River water with an initial (postsedimentation) TOC of 2.5 mg/L and UV = 3.5 m^{-1}. The PAC was Norit W35 ($K_f = 60$ (mg/g)(L/mg)$^{1/n}$) added at a concentration of 20 mg/L. The main effect of PAC addition in the case of this cleaner water was to increase sustainable water production, increase length of filtration cycle, and decrease the frequency of chemical cleaning.

Finally, addition of PAC can improve UF performance in water recycling applications. In this case, UF was used to disinfect secondary treated wastewater. This effluent has a high TOC of 20 mg/L. A dose of 10 mg/L of Norit W20 PAC was applied in an effort to increase production only. Table 15.3 shows that the PAC addition did increase production by 50 percent.

TABLE 15.1 Comparison of Performance of UF of Seine River Water with and without PAC

		PAC/UF	
	UF alone	Lower frequency chemical washing	Higher frequency chemical washing
Production flux, L/m²·h at 20°C	60	60	80
Filtration cycle (time between system backflushes), min.	30	30	30
Chemical washing frequency, days	15 (winter) 30 (summer)	45 80	20 60
Washing efficiency, % permeability recovery	80	100	100

TABLE 15.2 Comparison of Performance of UF of Clarified Seine River Water with and without PAC

	UF alone	PAC/UF
Production flux, L/m²·h at 20°C	140	180
Filtration cycle (time between system backflushes), min.	60	180
Chemical washing frequency, days	60	90
Washing efficiency	100	100

TABLE 15.3 Comparison of Performance of UF for Treatment of Secondary Effluent with and without PAC

	UF alone	PAC/UF
Production flux, L/m²·h at 20°C	40	60
Filtration cycle (time between system backflushes), min.	30	30
Chemical washing frequency, days	15	15
Washing efficiency	50	50

15.3 CRITERIA FOR THE SELECTION OF PAC

15.3.1 Effect of Particle Size on Physical Plugging, Adsorption Kinetics, and Adsorption Isotherms

Selection of PAC particle size is based on two main considerations when the PAC is used in hollow-fiber systems with inside-out flow: physical plugging of the module fibers (not the pores) and efficiency of adsorption.

PAC particles should not be so large as to physically plug the hollow fibers. A practical guideline for selecting PAC particle size is that particle diameter be less than one-fifth of the fiber inside diameter. For example, for 0.93-mm inside diameter hollow-fiber UF membranes filtering relatively clean water, we have verified that when more than 5 percent of the total particle number are greater than 200 μm, individual fibers in the module begin to plug. When PAC is added to water resources containing a high level of turbidity, or when floc from previous clarification processes enters the fibers, then even the one-fifth guideline may not be restrictive enough. In this case, specially sieved PAC may have to be requested from the supplier. This, in turn, can affect the overall operating costs of a PAC/UF system (see Sec. 15.3.3).

There is also a maximum PAC concentration which should not be exceeded in inside-out hollow-fiber applications. For example, in filtration of groundwater in the PAC/UF process, Adham et al. (1993a) found that PAC concentration in the recirculation loop should not exceed 600 mg/L, although other studies in France have shown that higher concentrations can be used.

As shown in Sec. 15.4, the kinetics of adsorption are closely tied to the characteristic diffusion time scale τ:

$$\tau = \frac{r_p^2}{D_{ef}} \tag{15.1}$$

Here, D_{ef} is the effective diffusion coefficient for the particle, and r_p is the particle radius. D_{ef} depends on the type of carbon, the type of adsorbate, the temperature, and, in some cases, even the particle size. Equation (15.1) is very helpful in understanding the effects of particle radius and diffusion coefficient on adsorption kinetics: as particle size decreases and/or diffusion coefficient increases, the characteristic time scale decreases; hence, adsorption kinetics are enhanced. On the other hand, adsorption kinetics in larger particles and/or in systems with smaller diffusion coefficients is slower. The moral here is that, if possible, smaller-diameter PAC particles should be considered in PAC adsorption: faster kinetics generally translates into smaller unit processes (i.e., smaller capital costs). However, the PAC particle size may also affect the porosity of cakes formed on the membrane surface and smaller PAC fractions are more expensive (see Sec. 15.3.3); hence, adsorption kinetics are not the only factor to consider in particle size selection.

The effective diffusion coefficient in Eq. (15.1) is not the diffusion coefficient of the adsorbate in water, so D_{ef} cannot be looked up in a reference book. In fact, the reader will find that effective diffusion coefficients in PAC adsorption are 4 to 5 orders of magnitude less than pure-water diffusion coefficients. The diffusion path in the particle is very complicated, because of the extremely tortuous internal structure of the PAC. The tortuous route the adsorbate takes as it moves in the radial direction slows the average radial diffusion. So when we speak of "radial" diffusion in the spherical PAC particle, we are talking about the *effective* radial diffusion in the particle. The other thing which slows the average radial diffusion in the particle is the adsorption itself. It is well known that adsorption retards or hinders radial diffusion in particles.

The second important aspect of PAC selection is the capacity of the carbon for the target adsorbate and the adsorption isotherm. In all the experiences recounted in this chapter, the Freundlich isotherm has proven most useful:

$$q_e = K_f c_e^{1/n} \tag{15.2}$$

Here, q_e is the equilibrium mass of adsorbate adsorbed per mass of PAC, c_e is the equilibrium adsorbate concentration in solution, K_f is the Freundlich constant, and $1/n$ is the Freundlich exponent. Activated carbons with larger K_f values have larger capacity for adsorption. K_f and $1/n$ values are not only related to the type of activated carbon, but also the particle size, temperature, the nature of the adsorbate, and, in some cases, the presence of oxygen. Table 15.4 gives some indication of the variability of the Freundlich parameters for one type of PAC with various average particle sizes and three types of natural water.

15.3.2 Laboratory Data Required for Establishing PAC Dose

As shown in Sec. 15.4, the PAC dose is a fairly complicated function of the system hydraulics (i.e., how the PAC is contacted with the adsorbate), the adsorption isotherm, and the adsorption kinetics. For example, for the same flow rate and reactor size, less PAC is usually required in plug flow reactors than in completely mixed reactors to achieve the same treatment objective. Further, PAC dose will be lowest with carbons with the greatest K_f; the characteristic diffusion time scale (Eq. [15.1]) can also impact carbon dose in systems with short detention times.

TABLE 15.4   Selected Freundlich Isotherm Parameters for Adsorption of Total Organic Carbon (TOC) at Room Temperature

Units of K_f are $(mg/g)(L/mg)^{1/n}$. Carbon was Calgon WPH.

Water source	$\overline{d}_p$ (geometric), μm	K_f	$1/n$
Raw Seine River (6/10/91)	10	63.2	1.2
Raw Seine River (3/18/91)	10	42.6	1.2
Clarified Seine River (6/23/91)	10	68.6	1.5
Illinois groundwater	5	43.0	1.13
Illinois groundwater	20	20.8	0.80

Source: Adham et al. (1993a).

For determination of K_f and $1/n$ in the Freundlich isotherm, the approach of Adham (1993) and Adham et al. (1991, 1993a, 1993b) is followed. Weighed samples of dried powdered activated carbon are added to water samples containing the target compound in acid-washed, amber glass bottles. The bottles are sealed with aluminum crimp caps and Teflon-lined silicon septa. The bottles are maintained in the dark in a laboratory agitator until equilibrium is achieved. Samples are removed from the glass bottles with a gas-tight glass syringe; the syringe head contains a 0.22-μm nylon filter for separation of the PAC from the water. The target compound is then analyzed with the appropriate instrument, such as a total organic carbon analyzer, a gas chromatograph, or a liquid scintillation counter.

Batch kinetics tests are required to fit the D_{ef} value to the diffusion model. In these tests, the PAC and water sample are contacted in glass beakers under good mixing conditions. At timed intervals, samples are withdrawn with the glass syringe/filter assembly previously described and the relevant organic carbon measurement is made.

Adham et al. (1991) described the problem of attempting to fit kinetic data with the K_f value determined from the equilibrium isotherm. They often found that a better correspondence of experimental and modeled results could be achieved if an apparent K_f value was used in fitting the D_{ef} value in the diffusion model. In this approach, the solution-phase adsorbate concentration at the end of the batch test (around 4 h) is assumed to be the equilibrium adsorbate concentration for the particular carbon dose. The capacity parameter is then called the *apparent equilibrium capacity constant*, K_a. Adham et al. (1991) believe that the K_a approach works better than the conventional procedure because of the heterogeneous nature of natural organic matter and the occasional importance of dissolved oxygen in the oxidation of some micropollutants in longer-term isotherm experiments.

Another problem arises in measuring adsorption isotherms for micropollutants in the presence of natural organic matter. It has been fairly conclusively demonstrated that the adsorption isotherm for a compound like atrazine in pure water can be much different than the same isotherm measured in a natural water. This difference is believed to be caused by competition for adsorption sites between the micropollutant and the background organic matter. For example, Najm et al. (1991) and Qi et al. (1994) have shown that, as the initial concentration of atrazine in an adsorption experiment decreases, the capacity of the carbon for atrazine adsorption decreases [i.e., a smaller K_f in Eq. (15.2)]. In effect, it appears that the atrazine becomes less "competitive" as its concentration decreases. When the plant inlet concentration of both the background organic matter and the micropollutant are fairly constant, then the competitive effect can be assumed to be constant. The isotherm parameters determined from a single experiment using the plant inlet concentration can be sufficient. However, if the concentration of the micropollutant or background organic matter changes, then the competitive effect will change and a different isotherm is generally required. Najm et al. (1991) developed a practical method for dealing with this phenomenon, called the *equivalent background compound* (EBC) method. The method utilizes the ideal adsorbed solution theory (IAST) and a straightforward parameter fitting procedure. The results of the EBC technique allow one to predict the adsorption isotherm for a micropollutant such as atrazine for different initial atrazine concentrations. The reader is referred to Najm et al. (1991) and Qi et al. (1994) for more details. In the remainder of this chapter, when adsorption of micropollutants is discussed, we will assume the correct adsorption isotherm parameters are available, either because of constant water quality or through evaluation using something like the EBC technique.

15.3.3 Costs

PAC costs vary greatly among manufacturers and appear to be somewhat correlated with adsorption capacity (higher-adsorption-capacity carbon usually costs more) and definitely correlated with particle size, the smaller size fraction carbons being more expensive. Table 15.5 pulls together the costs, size, and performance data for a number of PAC products commercially available in France. Note that for the four manufacturers shown, PAC price increases significantly as the product is sieved into smaller-size fractions.

15.4 MODELING OF ADSORPTION OF TOC AND MICROPOLLUTANTS IN UF-PAC PROCESS

15.4.1 Modeling PAC/UF System as a Plug Flow Reactor

When there is no limitation to mass transfer at the spherical particle surface (called an *external mass transfer limitation*), the modeling of adsorption of TOC (i.e., background natural organic matter) and micropollutants or synthetic organics involves solution of the diffusion equation in spherical coordinates. If there is only variation of concentration in the radial direction r, Fick's second law becomes

$$\frac{\partial q}{\partial t} = D_{ef}\left(\frac{\partial^2 q}{\partial r^2} + \frac{2}{r}\frac{\partial q}{\partial r}\right) \tag{15.3}$$

In Adham (1993) and Adham et al. (1991, 1993a, 1993b), D_{ef} is interpreted as the PAC surface diffusion coefficient; however, there is no mathematical difference between this conception of diffusion and radial pore diffusion (Crank, 1956). Two relevant boundary and initial conditions are:

$$\text{At } t = 0 \quad \text{and} \quad 0 \le r \le r_p, q = 0 \tag{15.4}$$

$$\text{For } t \ge 0 \quad \text{and} \quad r = 0, \frac{\partial q}{\partial r} = 0 \tag{15.5}$$

TABLE 15.5 Price, Size, and Performance Data for Several Commercially Available PAC Products

Water was raw Seine River water (20°C).

Manufacturer/product	$\bar{d}_p$, μm	% particles > 200 μm	K_f (mg/g)(L/mg)	c_n, mg/L	PAC Price, FF/kg
Chemviron RB	40	8	98	0.4	35–40
Sieved Chemviron RB	15	0	100	0.4	40–45
Norit W35	35	8	60	0.3	8–10
Sieved Norit W35	26	0	65	0.3	10–15
CECA Acticarbone 3S	46	6	48	0.35	18–24
Sieved Acticarbone 3S	15	0	59	0.35	20–28
PICA CNB 200 15/30 μm	35	3	34	1	5–7
PICA CNB 200 8/15 μm	11	0	30	0.5	12–15

where r_p is the PAC particle radius. By considering a mass balance on adsorbate in a plug flow reactor at steady state, a second differential equation relating liquid-phase adsorbate concentration C and q results for the case of no external mass transfer limitation

$$\text{For } t \geq 0, \frac{dc}{dt} = \frac{3c_{PAC}}{R^3} \frac{\partial}{\partial t} \int_0^{r_p} qr^2 \, dr \tag{15.6}$$

where c_{PAC} is the inlet PAC concentration and t is interpreted as the travel time in the plug flow reactor. The boundary and initial conditions for this differential equation are

$$\text{For } t \geq 0 \quad \text{and} \quad r = r_p, c = c_s \tag{15.7}$$

and $$\text{At } t = 0, c = c_{in} \tag{15.8}$$

where c_s is the liquid-phase adsorbate concentration at the particle surface and c_{in} is the adsorbate concentration at the head of the plug flow reactor. One final equation is needed to relate c_s to q at the particle surface (i.e., at $r = r_p$). This is accomplished with the aid of the Freundlich isotherm Eq. (15.2),

$$q_s = K_f c_s^{1/n} \tag{15.9}$$

Knowing K_f, $1/n$, and D_{ef}, Eqs. (15.3) and (15.6) can be solved simultaneously with the appropriate initial and boundary conditions provided (see Adham, 1993).

15.4.2 Modeling of PAC/UF System as a Perfect Mixer at Steady State

In certain continuous PAC/UF configurations, permeate is produced at a roughly constant rate and PAC is wasted at a constant rate. If the mixing within the system is intense, the reactor can be assumed to be a perfect mixer; hence, both water and PAC can be assumed to have exponential residence time densities (although with different mean detention times). In a steady-state perfect mixer, every PAC particle entering the system sees the same adsorbate concentration throughout its time in the reactor; this concentration is also the effluent concentration from the reactor C_{out}. Hence, the boundary condition in this case is the constant adsorbate concentration in the reactor. A solution of the diffusion equation for this boundary condition is given by Crank (1956). Adham (1993) and Adham et al. (1991, 1993a, 1993b) performed a mass balance using Crank's solution and the Freundlich isotherm to solve for the steady-state adsorbate concentration exiting the perfect mixer,

$$c_{in} - c_{out} - c_{PAC} K_f C_{out}^{1/n} \left[1 - \frac{6}{\pi^2} \sum_{n=1}^{\infty} \frac{1}{n^2 \{1 + [(\pi^2 n^2 \bar{t}_{PAC})/\tau]\}} \right] = 0 \tag{15.10}$$

where c_{in} is the influent adsorbate concentration entering the reactor. $\bar{t}_{PAC}$ is the mean detention time of the PAC,

$$\bar{t}_{PAC} = \frac{V}{Q_{PAC}} \tag{15.11}$$

Here, V is the total system recycle volume, and Q_{PAC} is the PAC waste flow rate during a filtration cycle. The only unknown in Eq. (15.10) is c_{out}, which is easily solved for by a number of methods, including the one-dimensional secant method (Adham, 1993; Adham et al., 1993a) or by a number of equation-solving software packages (see Sec. 15.6).

15.4.3   Modeling of PAC/UF System for Uniform PAC Age Distribution

A common UF/PAC operational scheme discussed in Sec. 15.5.2 is more akin to a semibatch operation; PAC is added continuously to the system but is not purged until the filtration cycle is ended. Quite often, the recirculation + UF module volume is approximately completely mixed; hence, the residence time distribution of the fluid can be assumed to be exponential (i.e., a CSTR). However, PAC is added continuously during the filtration cycle, and its concentration builds approximately linearly over time. Therefore, the residence time distribution of the PAC is uniform (Adham et al., 1993b). This is a true nonsteady system with regard to adsorption: at the beginning of the filtration cycle there is very little PAC in the system, while at the end of the filtration cycle, the concentration of PAC in the system is a maximum. However, as shown in Sec. 15.5.2, even though the PAC concentration at the beginning of the cycle is low, the adsorbate concentration is also low because the membrane system is flushed with permeate during backflushing. Hence, the changes in adsorbate concentration at the system exit are not as great as might be anticipated. This led to a pseudo steady-state modeling procedure outlined by Adham et al. (1993b). Using an approach similar to the steady-state perfect mixer, the average effluent concentration of the adsorbate exiting the system is determined by solution of the following equation

$$c_{in} - c_{out,ave} - c_{PAC} K_f c_{out,ave}^{1/n} \left\{ 1 + \frac{6\tau}{\pi^4 t_b} \sum_{n=1}^{\infty} \frac{\{\exp[-(\pi^2 n^2 t_b)/\tau] - 1\}}{n^4} \right\} = 0 \quad (15.12)$$

$c_{out,ave}$ is the unknown average adsorbate concentration in the effluent, and t_b is the time to backflush, which is the maximum age of PAC in the system. Equation (15.12) is solved like Eq. (15.10).

15.4.4   Parameter Fitting

When there is no external phase film mass transfer limitation at the PAC particle surface (see Adham (1993) and Adham et al., 1991, 1993a, 1993b) then the parameters required in the models are $1/n$, K_f (or K_a), and the effective diffusion coefficient D_{ef}. Following standard procedures, $1/n$ and K_f are fit to a linearized version of the Freundlich isotherm. When the apparent capacity constant (K_a) is required, the $1/n$ value from the standard Freundlich fit is assumed to apply here as well. However, the K_a is calculated from the ending concentration in the batch kinetics test, i.e., the ending batch concentration is considered the isotherm equilibrium concentration; hence,

$$K_a = \frac{q_e(\text{batch})}{c_e(\text{batch})^{1/n}} \quad (15.13)$$

The D_{ef} value was fit using the same solution technique discussed in Sec. 15.4.1, but with an optimization program which searches for the optimal D_{ef} value. The programs were based on the work of Traegner-Duhr and Suidan (1989).

When two adsorption reactors are used in series (for example, when a plug flow or stirred reactor is followed by the PAC-UF process—see Secs. 15.5.1 and 15.5.2), a modification of the modeling procedure is required. Although parameter fitting for the first reactor can be accomplished as outlined in Sec. 15.3.2, the PAC entering the second reactor in series can be considered to be already partially loaded with organic carbon; hence, a modified parameter fitting is required. In this procedure, the batch kinetic data for times greater than the mean detention time of the first reactor are considered as a separate data set for fitting a new K_f and D_{ef} for the second reactor in series. In effect, we assume the PAC is uniformly loaded at the start of the second period.

15.5 APPLICATION OF PAC/UF MODEL AT LABORATORY AND PILOT SCALES

The models described in the previous section can be used to predict the removal of total organic carbon and micropollutants in a number of different treatment configurations. In this section, application of those models to real treatment systems is outlined, and experimental measurements are compared with model predictions. Additional information on expected organic removal by the PAC/UF process can be found in Chap. 10.

15.5.1 Removal of TOC and TCP from Groundwater Using the PAC/UF Process

Adham et al. (1991) modified a commercially available UF pilot* for operation with PAC feed (Fig. 15.1). The target compounds for removal in the system were TOC and 2,4,6-trichlorophenol (TCP) in a groundwater supply. The PAC was Calgon WPH, and the PAC dose was 15 mg/L. Representative results of that work are presented here. The pilot shown in Fig. 15.1 was operated in the steady-state perfect mixer mode previously described (continuous wastage of PAC). Figure 15.2 shows the predictions of the model for the removal of TCP from groundwater compared with actual data. For the two runs with 13- and 5-μm-diameter PAC particles,[†] model prediction and experimental data agree well. Two other curves are shown, which are predictions for hypothetical 20- and 10-μm samples. Note two important effects of PAC particle size: faster kinetics (hence, potentially smaller recycle systems) as particle size decreases and greater adsorption capacity (hence, potentially lower carbon doses) as particle size decreases.

Later experiments and modeling were done to simulate larger-scale pilots evaluated in France, for the purpose of predicting TOC removal. This modification, shown in Fig. 15.3, is composed of a stirred reactor followed by a second PAC/UF system with continuous recycle but with no waste during the filtration cycle. Hence, in the modeling, the first reactor is modeled with the steady-state perfect mixing model (Sec. 15.4.2), and the second reactor is modeled with the uniform age distribution model (see Sec. 15.4.3 and discussion of parameter fitting in Sec. 15.4.4). The results of these measurements and modeling are shown in Fig. 15.4. Good correspondence between model prediction and measurements is found here as well for a PAC dose of 42 mg/L.

---

* Nautilus Pilot, Lyonnaise des Eaux-Dumez, Le Pecq, France.
[†] In these experiments, PAC was air-classified into the various fractions. Average particle size was determined with an electrical sensing zone particle counter (Adham et al., 1991).

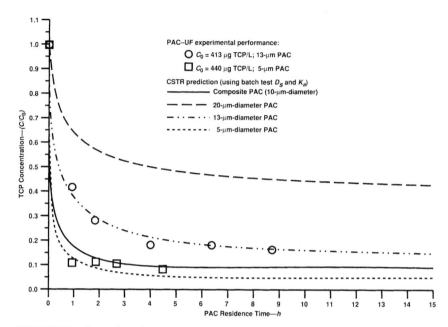

FIGURE 15.2 Comparison of model predictions and experiments for adsorption of TCP from groundwater in PAC/UF system. $C_0 = c_{in}$ and $D_s = D_{ef}$ in nomenclature of Chap. 15. *(Adham et al., 1991.)*

15.5.2 Removal of TOC from Surface Waters Using the PAC/UF Process

The remainder of the PAC/UF pilot plants reviewed in this chapter are one or the other of the two generic types shown in Fig. 15.5. The CEB plant at Mont Valerien, France, consists of a hydrocyclone followed by PAC feed, a long feed pipe, and a UF module with continuous solids accumulation; hence, the system is modeled with the generic reactor configuration in Fig. 15.5*b*. This pilot is operated with variable pressure in order to maintain a constant membrane flux during the filtration cycle. Solids are wasted from the system at the end of each filtration cycle. The PAC used in this pilot was Chemviron TL 9003 with a geometric mean diameter of 18 μm. Generally good correspondence between the model predictions and measurements is shown in Fig. 15.6. It is very interesting that the six months' worth of predictions shown in Fig. 15.6 were based on parameters determined from just two isotherm and kinetics tests at the start of the test period. This suggests that, in real applications of the model, seasonal variations in organic carbon may not be significant enough to affect the fidelity of the model.

15.6 SIZING PAC/UF PLANTS WITH THE SIMULATION MODEL

The previous sections have shown that the PAC/UF model allows us to accurately predict the removal of TOC and micropollutants in real systems. The purpose of this section is to demonstrate that the PAC/UF model can be used to simulate plant

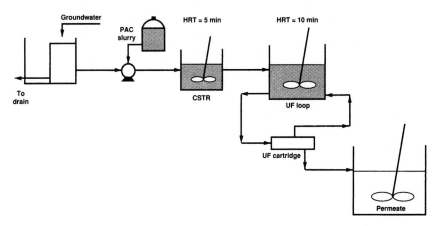

FIGURE 15.3 Laboratory-scale PAC/UF pilot used for testing reactor configuration consisting of perfect mixer followed by UF with solids accumulation in recycle loop. HRT = mean hydraulic detention time. *(Adham et al., 1993c.)*

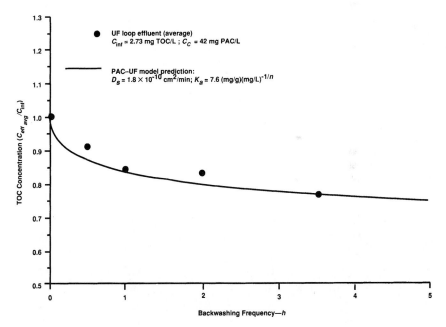

FIGURE 15.4 Model predictions and experiments for TOC adsorption from groundwater in PAC/UF system. $C_{inf} = c_{in}$, $C_c = c_{PAC}$, and $D_s = D_{ef}$ in Chap. 15 nomenclature. *(Adham et al., 1993c.)*

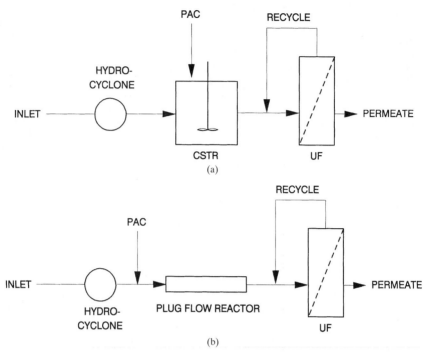

FIGURE 15.5 General reactor configuration used in simulations of PAC/UF performance of French pilot plants: (*a*) CSTR plus UF, and (*b*) plug flow plus UF.

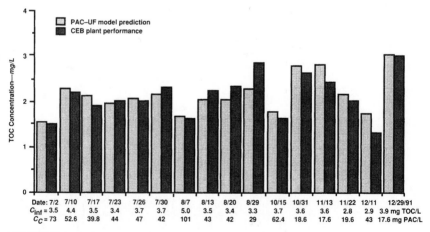

FIGURE 15.6 Performance of CEB plant compared with PAC/UF model predictions. $C_{inf} = c_{in}$ and $C_c = c_{PAC}$ in nomenclature of Chap. 15. *(Adham et al., 1993c.)*

performance for different-size contact units and treatment scenarios. This is an extremely useful capability in the sizing, design, and marketing of ultrafiltration facilities.

The performance of the PAC/UF process was simulated for a groundwater source from central France (St. Thibault). The TOC was 2.2 mg/L and the UV absorbance (254 nm) was 3.5 m^{-1}, and the source suffers from occasional atrazine pollution (up to 2 µg/L). The treatment goal is to reduce TOC concentration to 1.1 mg/L. This value was selected because independent adsorption testing showed that for TOC values below 1.1 mg/L, the atrazine concentration would generally fall below the European standard of 0.1 µg/L. The proposed reactor configuration is shown in Fig. 15.5a. The carbon was Norit W35, and for this source water, $K_f = 70$ (mg/g)(L/mg)$^{1/n}$, $1/n = 1.07$, and $D_{ef} = 2.15 \times 10^{-10}$ cm^2/min. The flux was set at 120 L/m^2·h, and the backflushing frequency was set at 60 min. In the first design scenario, the carbon dose was set at 20 mg/L, and different CSTR mean detention times were investigated. The following results were found:

1. For a contact time of 15 min, the predicted permeate TOC was 1.3 mg/L.
2. For a contact time of 30 min, the predicted permeate TOC was 1.2 mg/L.
3. For a contact time of 60 min, the predicted permeate TOC was 1.13 mg/L.

Because these mean contact times were considered unrealistic for the proposed site, the PAC/UF model quickly showed that the 20-mg/L dose would not be sufficient to meet the treatment goal. The next treatment strategy was to fix the CSTR mean detention time at 10 min and determine what PAC concentration would be required to reach the target TOC concentration of 1.1 mg/L. Through several iterations of the model, it was shown that the treatment goal could be reached with a 30-mg/L carbon dose. The last treatment strategy considered was to compute the piston flow reactor size required to achieve the treatment goal using the 30-mg/L carbon dose (i.e., the configuration in Fig. 15.5b). Through several iterations, the model showed that the required plug flow unit would need a mean detention time of 3 min. Since a plug flow unit of size sufficient for a 3-min contact time was considered unrealistic due to practical site limitations, the final design selected for actual implementation at this site was the 10-min mean detention time CSTR with 30-mg/L PAC dose.

The following design scenario concerns another groundwater source in France. It has a TOC of 3.5 mg/L, but due to connection with the surface, the TOC can occasionally rise to 13 mg/L. The water can also occasionally have a color problem (30 °Hazen) and tastes and odors (threshold = 70). Two PAC samples were used. The first carbon was Norit W35 with $K_f = 18$, $1/n = 0.99$, and a nonadsorbable organic matter concentration of $c_n = 1.5$ mg/L for this water source. The second carbon used was higher-capacity Chemviron TL 9003, with $K_f = 50$, $1/n = 1$, and nonadsorbable concentration $c_n = 1$ mg/L for this water source. The filtration cycle is 60 min, and the membrane flux is maintained at 55 L/m^2·h (20°C). In the simulations, the inlet TOC of the raw water was set at 10 mg/L. Table 15.6 shows the results of a number of simulations for the reactor configuration of Fig. 15.5a. The treatment goal in this case is TTO < 3, which was independently related to the permeate TOC. The data in Table 15.6 show that, by successive simulations with different carbons, carbon doses, and mean detention times, we can fairly quickly arrive at feasible PAC/UF design (last row Table 15.6).

EXAMPLE 15.1 *This example concerns treatment of the St. Thibault water with Norit PAC, as discussed in this section. For the reactor configuration shown in Fig. 15.1, develop a relationship between the PAC mean residence time and the effluent TOC*

TABLE 15.6 Simulation of Permeate TOC Concentration in a Groundwater Source for Different Reactor Mean Detention Times, Carbon Doses, and Carbon Types in PAC/UF Reactor

Carbon	PAC concentration, mg/L	Mean detention time, min	Permeate TOC, mg/L	Color (°Hazen)	TTO
Norit W35	20	15	8.8	8	NA
Norit W35	20	30	8.2	NA	NA
Norit W35	40	15	7.3	NA	NA
Norit W35	40	30	6.9	6	NA
Norit W35	60	15	6.3	NA	NA
Norit W35	60	30	5.9	NA	NA
Norit W35	60	60	5.4	4	NA
Chem. TL9003	20	30	4.5	5	NA
Chem. TL9003	40	30	4.2	4	3
Chem. TL9003	50	30	3.8	3	<2

value. Assume the following parameter values: mean hydraulic detention time $\bar{t} = 10$ min, $c_{PAC} = 30$ mg/L, $r_p = 17.5$ μm, as well as the Norit W35 PAC parameters, $K_f = 70$ $(mg/g)(L/mg)^{1/n}$, $1/n = 1/07$, and $D_{ef} = 2.15 \times 10^{-10}$ cm²/s. Also, assume that the PAC in the recirculation loop is not to exceed a concentration of 2 g/L.

SOLUTION. *The reactor configuration shown in Fig. 15.1 is treated as a perfect mixer at steady state, described in Sec. 15.4.2. The performance equation is*

$$c_{in} - c_{out} - c_{PAC}K_f C_{out}^{1/n}\left[1 - \frac{6}{\pi^2}\sum_{n=1}^{\infty}\frac{1}{n^2[(1 + \pi^2 n^2 \bar{t}_{PAC})/\tau]}\right] = 0 \qquad (15.10)$$

Before solving Eq. (15.10), it is useful to have a better understanding of the trade-offs between the hydraulic and PAC mean residence times, and the influent and effluent TOC values. A mass balance on PAC written around the recirculation tank ("feed mix tank" in Fig. 15.1) yields the following equation

$$Q_{in}c_{PAC} = Q_{PAC}c_{sys}$$

Here, Q_{in} is the influent flow to the recirculation tank, c_{PAC} is the PAC feed concentration, Q_{PAC} is the waste flow rate from the recirculation system, and c_{sys} is the PAC concentration in the recirculation system (which is also the waste concentration in a perfect mixer). The mean hydraulic detention time in the system is defined as usual as

$$\bar{t} = \frac{V}{Q_{in}}$$

Now, from Eq. (15.11), $Q_{PAC} = V/\bar{t}_{PAC}$. Combining this and the last two equations, we find that

$$c_{sys} = c_{PAC}\frac{\bar{t}_{PAC}}{\bar{t}}$$

This equation compactly expresses the concentration of PAC in the recirculation loop in terms of the PAC feed concentration and the two relevant mean detention times. Since the PAC feed concentration ($c_{PAC} = 30$ mg/L) and hydraulic detention time ($\bar{t} = 10$ min) are fixed in this problem, and since the PAC concentration in the recirculation loop is not allowed to exceed $c_{sys} = 2$ g/L, the maximum $\bar{t}_{PAC}$ is therefore 666 min. In the *simulations to follow, we will explore $\bar{t}_{PAC}$ values between 10 and 666 min. The only remaining unknown parameter in Eq. (15.10) is τ, the characteristic diffusion time in the PAC particle. But this can be calculated from Eq. (15.1)*

$$\tau = \frac{r_p^2}{D_{ef}} = \frac{(17.5 \times 10^{-4}\ \text{cm})^2}{2.15 \times 10^{-10}\ \text{cm}^2/\text{min}} = 1.42 \times 10^4\ \text{min} \tag{15.1}$$

Equation (15.10) is easily solved by trial and error or with a number of equation solvers. We solved the equation using a commercial software package called Mathcad, for values of $\bar{t}_{PAC}$ between 10 and 666 min, and the other parameter values previously discussed. The results for c_{out} versus $\bar{t}_{PAC}$ are shown in Fig. 15.7. The x axis is the PAC mean residence time, and the y axis is the effluent TOC concentration. For reference, the equilibrium effluent TOC concentration (0.721 mg/L) is also shown; this corresponds to a reactor with a very long mean detention time (i.e., a very large reactor), or a batch experiment with the same c_{PAC} and c_{in} as above. The equilibrium value was calculated using Eq. (15.2) and the equilibrium parameters previously discussed. Note from the figure that, as $\bar{t}_{PAC}$ is increased, the effluent TOC value decreases. The last point in the curve corresponds to the maximum allowable c_{sys} value in this problem, 666 mg/L. Recall that we were able to achieve an effluent TOC of 1.1 mg/L with the treatment scheme described in Sec. 15.6 (a CSTR followed by nonsteady PAC buildup in the UF unit only) and the parameter values used in this problem. For the scheme used in this example problem, Fig. 15.7 indicates we would need to maintain a PAC mean residence time near 550 min in order to achieve an effluent concentration of 1.1 mg/L. From the equation for c_{sys} previously given, this would correspond to a PAC concentration in the recirculation loop of 1.65 gm/L.*

15.7 FULL-SCALE IMPLEMENTATIONS OF THE PAC/UF PROCESS

A number of full-scale PAC/UF installations are already operational in France, and more will be on-line in the very near future. In this section, we review their special water quality problems, and how the PAC/UF process was implemented in order to meet various treatment goals.

15.7.1 Saint-Cassien (near Cannes)

In this situation, the capacity of an existing conventional plant needed to be upgraded, especially because of summer demands in this popular tourist area. The source water is a lake, and occasional taste, odor, color, and microbiological problems are experienced. The design capacity for the PAC/UF plant is 25,000 m^3/day, which for a planned flux of 140 L/(h·m^2), is met with 160 Aquasource hollow-fiber UF modules, each with 50 m^2 of surface area. Planned backflushing frequency is once each 60 min, with one chemical washing per year. The generic reactor configuration is as shown in Fig. 15.5b, with a plug flow contact time of 3 min. The planned PAC dose is between 10 to 15 mg/L. Capital costs are estimated at 3000 FF/(m^3/day),

---

* Mathsoft Inc., Cambridge, Mass.

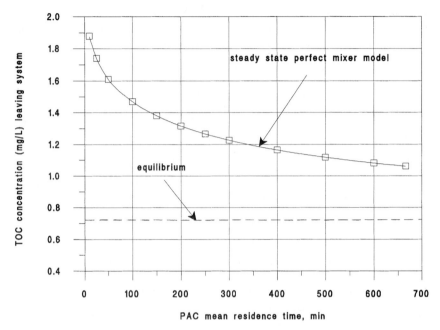

FIGURE 15.7 Simulated effect of PAC mean residence time on TOC removal in the PAC/UF system treating the St. Thibault water.

and operational costs are estimated to be 0.6 FF/m³. These costs are comparable to an equivalent direct filtration plant upgrade. The plant will be put into service in early 1996.

15.7.2 Chatel-Gerard

In this location, a new plant was needed to replace the existing direct filtration plant. The raw water source was a groundwater, which, because of surface water contact, suffered occasional (and significant) increases in turbidity, color, TOC, coliforms, and atrazine. Consideration of equivalent plug flow and CSTR units and space limitations showed the capital cost of the CSTR + PAC/UF system (Fig. 15.5a) was the lowest. The plant capacity is 600 m³/d, which, with an average flux of 100 L/(m²·h), translates to six Aquasource modules each with a 70-m² surface area. The PAC dose varies between 10 and 15 mg/L, backflushing is once per hour, and chemical cleaning is once per year. Plant data show that the average atrazine is kept below the European standard of 0.1 μg/L, with excellent removal of turbidity, coliforms, and odors. Production at Chatel-Gerard began in 1993.

15.7.3 Fontgombault

This plant, which went into production in 1993, has a capacity of 5000 m³/d. The source is a surface water in an agricultural area, and the plant was designed to replace an existing pump and chlorinate system. The plant was originally designed for just UF, since turbidity spikes as high as 50 mg/L had been recorded. The production goal was met with 28 Aquasource modules each with 70 m^2 of surface area. The membranes are backflushed each hour, and there are two chemical washings per year. After construction, an occasional atrazine contamination problem was noted, with spikes as high as 0.2 µg/L. The existing UF plant was easily upgraded to the system in Fig. 15.5a by addition of a stirred tank before the UF. The stirred tank variation was preferred in this case because a sufficiently large stirred tank was already present on-site. The PAC dose varies between 0 and 20 mg/L (depending on season). Average atrazine concentrations are now maintained below 0.02 µg/L.

15.7.4 Vigneux (Seine River)

In this case, a conventional surface water treatment plant needed to be upgraded. Two alternatives were considered for tertiary treatment at Vigneux: either oxidation (O$_3$) plus PAC/UF and postchlorination, or oxidation (O$_3$) plus GAC and postchlorination. In the end, the PAC/UF plant was chosen for the following reasons:

1. The O$_3$/PAC/UF system has superior removal of mineral, organic, and biological particles compared to the O$_3$/GAC.

2. The O$_3$/PAC/UF system can be more easily adapted (through the type and dosage of carbon) to specific pollutants and variable pollutant concentrations.

3. There was a desire to demonstrate an "innovative" process.

The plant will make use of the plug flow + UF configuration of Fig. 15.5b; this configuration was chosen over that in Fig. 15.5a because an extensive channel (plug flow) system already existed at the plant. The plant is scheduled to go into production in mid-1996, and will have a production of 55,000 m^3/d, which is met using 224 Aquasource modules, each with 70 m^2 of surface area. The flux is projected to average 160 L/(m^2·h), with a PAC dose of 15 mg/L, one backflushing per hour, and six chemical cleanings per year. On-site piloting of the PAC/UF process has demonstrated excellent reductions of tastes, odors, and THM precursors (reduced from 50 to 10 µg/L average). Average atrazine concentrations have been reduced from 0.06 to less than 0.02 µg/L.

15.8 FUTURE DEVELOPMENTS AND RESEARCH NEEDS

One of the most common questions about the PAC/UF process (and most membrane processes) concerns the processing of reject. From the environmental and cost perspective, the best possible situation in the PAC/UF process would be if all the PAC could be regenerated. Unfortunately, there are almost no data on PAC regeneration. Adham et al. (1993a) studied the regeneration of PAC loaded with methylene blue in a fluidized bed regeneration furnace (Waer et al., 1992). For the conditions studied, mass loss was between 18 and 22 percent; more than 90 percent of iodine number and adsorption capacity were recovered. Much more work needs to be done in this area, especially studies of the regeneration of carbon samples loaded in real natural waters. Another important aspect of the regeneration question is the process configuration. For example, successful regeneration would definitely be more likely if the natural solids loading on the PAC/UF reactor could be minimized. This would suggest that regeneration would tend to be more efficient with groundwater sources or surface water sources with some kind of presedimentation or other particle removal.

Another possibility for treatment of PAC/UF reject is dewatering. Typically, the reject and backflush from the PAC/UF consists of PAC and other particulate matter of natural origin. There are actually two goals in dewatering: to be able to dispose the liquid directly to a receiving body and to make the solids as dry as possible. In France, for the reject water to be directly disposed in a surface water, the suspended solids concentration must be less than 30 mg/L. Four technologies are currently being evaluated in France for treatment of reject. The first is the Howden-Wakeman filter (Wakeman et al., 1994). This is a cylindrical depth fiber filter, which works in basic filter and cleaning cycles. In the filtering cycle, a piston compresses the filter; during cleaning, the piston is retracted to mechanically expand the filter. The second reject filtration system is composed of a synthetic sponge or foam material. The third system is a tubular microfiltration cartridge. Finally, a sedimentation clarifier is also being evaluated.

Finally, when the PAC/UF process is used as a polishing process following conventional coagulation-flocculation and clarification, the PAC can be returned to the settler. In this case, the PAC is processed with the normal sludge dewatering and processing equipment. This variation also ensures that nearly the full capacity of the PAC is used, since there can be considerable additional adsorption in the settler.

Further details on handling of UF reject are provided in Chap. 10.

15.9 CONCLUSIONS AND FUTURE PERSPECTIVES

An innovative drinking water treatment process combining powdered activated carbon and ultrafiltration has been successfully demonstrated in France and has already entered the production stage. The PAC/UF process has been shown to be effective for enhanced removal of natural and synthetic organics. It is a "supple" process, meaning it can be easily varied to meet new or changing treatment goals. It is an efficient process in terms of space requirements since it is based around compact membrane modules. Since the PAC is added before the membrane, the final product water quality is excellent in terms of removal of mineral, organic, and biological particles. This leads to lower postchlorination requirements.

The use of mathematical modeling in design of the PAC/UF process has matured quickly. As shown in this chapter, the mathematical model can be used to investigate various treatment variations, both physicochemical (e.g., PAC type and dose variations) and hydraulic.

In terms of future perspectives, work continues on application of new and different adsorbents, e.g., zeolites, clays, and/or catalysts (Bottero et al., 1994). These processes are being developed for possible future enhancement of the removal of NH_4^+, NO_3^-, iron, manganese, and heavy metals.

ACKNOWLEDGMENTS

The authors want to acknowledge the profound contributions of Samer Adham (Montgomery-Watson Engineers) and Vernon Snoeyink (University of Illinois) to the development of the modeling and experimental protocol reviewed in this chapter.

SYMBOLS

The following symbols from Chap. 15 were agreed to at the Martha's Vineyard meeting. During the revision process, five new symbols were added. They are indicated in the following list.

c_e	equilibrium adsorbate concentration, M/L^3
c_{in}	inlet adsorbate concentration, M/L^3
c_n	nonadsorbable organic fraction, M/L^3
c_{out}	outlet adsorbate concentration, M/L^3
$c_{out,ave}$	average outlet adsorbate concentration, M/L^3
c_{PAC}	PAC feed concentration, M/L^3
c_s	adsorbate concentration at particle surface, M/L^3 (new)
c_{sys}	PAC concentration in recirculation system, M/L^3
d_p	particle diameter, L
$\overline{d}_p$	average particle diameter, L
D_{ef}	effective diffusion coefficient, L^2/T
K_a	apparent Freundlich isotherm capacity parameter, $(M/M)(L^3/M)^{1/n}$
K_f	Freundlich isotherm capacity parameter, $(M/M)(L^3/M)^{1/n}$
MWCO	molecular weight cutoff in daltons
n	Freundlich isotherm exponent (dimensionless)
q	local adsorbed mass per mass adsorbent, M/M
q_e	equilibrium adsorbed mass per mass adsorbent, M/M
q_s	adsorbed mass per mass adsorbent at particle surface, M/M (new)
Q_{PAC}	PAC waste flow, L^3/T
r	radial position, L
r_p	particle radius, L
t	time, T
$\overline{t}$	mean hydraulic detention time, T
t_b	time to backflush (or length of filtration run), T
$\overline{t}_{PAC}$	PAC mean detention time, T
V	volume, L^3
τ	diffusion time scale for spherical particle (T)

REFERENCES

Adham, S., 1993, "Evaluation of the Performance of Ultrafiltration with Powdered Activated Carbon Pretreatment for Organics Removal," Ph.D. thesis, University of Illinois, Dept. of Civil Engineering, Urbana, Ill.

Adham, S. S., M. M. Clark, and V. L. Snoeyink, 1993a, "Evaluation of the Performance of Ultrafiltration with Powdered Activated Carbon Pretreatment for Organics Removal," final project report submitted to Lyonnaise des Eaux-Dumez, July.

Adham, S. S., V. L. Snoeyink, M. M. Clark, and C. Anselme, 1993b, "Predicting and Verifying TOC Removal by PAC in Pilot-Scale UF Systems," *Jour. AWWA,* **85**(12):58–68.

Adham, S. S., V. L. Snoeyink, M. M. Clark, and J. L. Bersillon, 1991, "Prediction and Verification of the Performance of Powdered Activated Carbon for Removal of Organic Compounds in the PAC/UF Process," *Jour. AWWA,* **83**:10.

Anselme, C., I. Baudin, P. Mazounie, and J. Mallevialle, 1992, "Production of Drinking Water by Combination of Treatments: Ultrafiltration and Adsorption of PAC," presented at the *AWWA Annual Conference,* June 1992, Vancouver, B.C.

Anselme, C., J-L. Bersillon, and J. Mallevialle, 1991, "The Use of Powdered Activated Carbon for the Removal of Specific Pollutants in Ultrafiltration Processes," presented at the *AWWA Membrane Processes Conference,* March 10–13, 1991, Orlando, Fla.

Baudin, I., C. Anselme, and M. R. Chevalier, 1993, "The Removal of Turbidity and Taste and Odor Problems in Drinking Water—Advantages of the Ultrafiltration Process," *J. Water SRT—Aqua,* **42**(5):295–300.

Bottero, J. Y., K. Khatib, F. Thomas, K. Jucker, J. L. Bersillon, and J. Mallevialle, 1994, "Adsorption of Atrazine onto Zeolites and Organoclays in the Presence of Background Organics," *Water Research,* **28**(2):483–490.

Clark, M. M., 1992, "Ultrafiltration of Lake Water: Optimization of TOC Removal and Flux," chap. 23 in *Influence and Removal of Organics in Drinking Water,* Lewis Publishers.

Clark, M. M., and K. S. Heneghan, 1991, "Ultrafiltration of Lake Water for Potable Water Production," *Desalination,* **80**:243–249.

Crank, J., 1956, *The Mathematics of Diffusion,* Oxford at the Clarendon Press, London.

Heneghan, K., and M. M. Clark, 1991, "Surface Water Treatment by Combined Ultrafiltration/PAC Adsorption/Coagulation for Removal of Natural Organics, Turbidity, and Bacteria," *Proceedings of the AWWA Membrane Processes Conference,* March 10–13, 1991, Orlando, Fla.

Laîné, J.-M., M. Clark, and J. Mallevialle, 1989, "Optimization of Organic Removal in the Ultrafiltration of Natural Waters," *AWWA Annual Conference,* June 19–22, 1989, Los Angeles, Calif.

Laîné, J.-M., M. Clark, and J. Mallevialle, 1990, "Ultrafiltration of Lake Water: Effect of Pretreatment on Organic Partitioning, THMFP, and Flux," *Jour. AWWA,* **82**(12):82–87.

Najm, I., V. Snoeyink, and Y. Richard, 1991, "Effect of Initial Concentration of a SOC in Natural Water on its Adsorption by Activated Carbon," *Jour. AWWA,* **83**:57.

Pirbazari, M., B. N. Badriyah, and V. Ravindran, 1992, "MF-PAC for Treating Waters Contaminated with Natural and Synthetic Organics," *Jour. AWWA,* **84**, 12:95.

Qi, S., S. Adham, V. Snoeyink, and B. Lykins, 1994, "Prediction and Verification of Atrazine Adsorption by PAC," *J. Environmental Engineering,* **120**:202–218.

Traegner-Duhr, U. K., and M. T. Suidan, 1989, "Parameter Evaluation for Carbon Adsorption," *J. Environmental Engineering,* **115**(1):109.

Waer, M. A., V. L. Snoeyink, and K. L. Mallon, 1992, "Thermal Regeneration of Activated Carbon for the Treatment of Drinking Water: Time and Temperature Dependance," *Jour. AWWA,* **84**(3):82.

Wakeman, R. J., D. R. Burgess, and R. J. Stark, March-April 1994, "The Howden-Wakeman Filter in Wastewater Treatment," *Filtration and Separation,* pp. 183–187.

CHAPTER 16
COAGULATION AND MEMBRANE SEPARATION

Mark R. Wiesner
Department of Environmental Science and Engineering
Rice University
Houston, Texas

Jean-Michel Laîné
CIRSEE
Lyonnaise des Eaux
Le Pecq, France

16.1 INTRODUCTION

The addition of coagulants to the feedwater of membrane units may be done to improve permeate flux or permeate quality. Coagulation pretreatment appears to yield the greatest benefits for subsequent microfiltration (MF) of the pretreated water. Research and practice have shown that, in surface water treatment, coagulation pretreatment is essential for maintaining higher permeate flux using ceramic microfiltration membranes[1-4] (Fig. 16.1). Under certain conditions the performance of polymeric microfiltration membranes[5-7] and ultrafiltration (UF) membranes[8] may also be enhanced by coagulation pretreatment. However, coagulation may also be detrimental to some membrane applications. While coagulant addition to the membrane feed improved filtrate quality, it also appeared to decrease permeate flux in a pilot study of surface water treatment using an outside-in polypropylene membrane.[7] Similar incompatibilities between alum coagulation and an inside-out cellulosic membrane have been observed.[9]

There appear to be no advantages and, in fact, there may be disadvantages of coagulation pretreatment for nanofiltration (NF)[9] or reverse-osmosis (RO) membranes. However, there is less motivation for considering coagulation of NF or RO feedwaters since these membranes are capable of rejecting many dissolved contaminants. Thus, membrane compatibility and treatment goals are key factors to be considered in applying coagulants to membrane feed streams.

In this chapter, we consider the rationale for coagulation pretreatment of membrane feedwaters, review laboratory and field experience with the process, and outline operational considerations in implementing this treatment scheme.

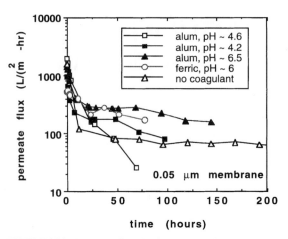

FIGURE 16.1 Permeate flux of a zirconia ceramic membrane in surface water filtration with various coagulation pretreatment conditions. *(Data from Wiesner, Veerapaneni, and Brejchová.[3])*

16.2 COAGULATION PRETREATMENT TO IMPROVE PERMEATE FLUX

Coagulation pretreatment may, under certain conditions, increase the permeate flux of an operating membrane. In other instances, coagulation pretreatment may reduce the frequency of hydrodynamic or chemical cleaning procedures used to maintain permeate flux. The potential for enhanced membrane performance as a result of coagulation pretreatment depends on the nature of foulants in the raw water and their potential for interaction with the membrane.[1,2,8,10–12] Coagulation pretreatment may enhance permeate flux by: (1) reducing foulant penetration into membrane pores, (2) conditioning the layer of materials deposited on the membrane, and (3) improving particle transport characteristics (Fig. 16.2).

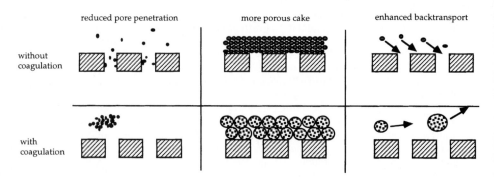

FIGURE 16.2 Possible mechanisms by which coagulation pretreatment may enhance permeate flux across membranes.

16.2.1 Reduced Foulant Penetration

Materials that might otherwise enter membrane pores and deposit or adsorb within the pores and constrict flow may be complexed, aggregated, or sorbed onto flocs of precipitated metal hydroxide. These flocs are rejected at the membrane surface. Small colloidal particles may be destabilized and form larger aggregates that are rejected at the membrane surface. Coagulants that form precipitates, such as metal salts, may envelop small colloidal materials in the larger precipitate. Such precipitates of iron, aluminum, or calcium may also be effective in adsorbing foulants such as natural organic matter (NOM). The adsorption of NOM to membranes is recognized as a frequent cause of long-term permeate flux decline in membranes (Chap. 4). Complexation of these materials before they encounter the membrane may therefore yield some benefit to membrane performance. Coagulation pretreatment has been found to decrease the rate of permeate flux decline in the treatment of laboratory solutions of humic or tannic acids treated using polysulfone membranes.[1] However, a related study showed that the principal benefit of coagulating a surface water before filtering with these membranes was to increase the amount of permeate flux recovered by backflushing, thereby decreasing the need for chemical cleaning.[8] The rates of permeate flux decline with and without coagulation were similar. Unfortunately, the fractions of NOM that appear to be implicated most strongly with membrane fouling are poorly complexed by metal coagulants.[8] Therefore, the benefits of coagulation pretreatment in reducing membrane fouling by NOM may be limited.

16.2.2 Conditioning of Materials Deposited on the Membrane

The specific resistance (resistance per depth of cake deposited) of a cake of particles on a membrane decreases as particle size increases (Eq. [4.12]). Aggregation of small colloids in the feed stream may lead to a larger effective particle size that, when deposited in the cake, results in less specific resistance. Aggregation may also lead to a cake with different properties of compressibility. For example, a cake composed of unaggregated colloidal clay particles may be largely incompressible. Although incompressible, the small size of the particles in the cake creates a large specific resistance. Aggregates of these particles may present less specific resistance but may also be more compactible. At the limit of compactibility, a cake composed of aggregated particles may have a specific resistance greater than that of a cake of unaggregated particles if the coagulant itself fills in void space between the clay particles.

Specific resistance and compactibility of the cake is closely related to the mechanisms by which colloids and dissolved organic materials are destabilized by the coagulant. These are, in turn, determined by the conditions of pH, coagulant dose, and coagulant aids applied to the feed stream. Cakes formed from humic acid destabilized with an aluminum coagulant have been found to present minimal specific resistance when humic acid is coagulated under conditions of precipitation/charge neutralization.[12] This appears to be due in part to the formation of larger particles of the precipitated/aggregated aluminum-humate. However, it is possible that cakes formed from these particles are also less compressible than are cakes formed predominantly from amorphous metal hydroxides.

16.2.3 Improved Particle Transport

The deterministic component of the trajectories of particles less than approximately $0.1\ \mu m$ is virtually identical to the movement of the water near the particles. In laminar flow, this means that particles follow paths that are nearly indistinguishable from the fluid streamlines. Wall effects, such as those resulting when particle-bearing water flows over a membrane, may lead to inertial forces on particles that cause them to deviate from streamlines (see Chap. 4). These inertial lift forces increase with particle size and may act to reduce particle deposition on the membrane. Also, transport of particles along or away from the membrane due to shear forces is also predicted to increase with particle diameter. Thus, coagulant-induced aggregation of small particles in the membrane feed stream may result in larger particles that are deposited to a smaller extent on the membrane. If deposited, these aggregates may be more efficiently removed from the cake. This latter consideration may also be of importance in backflushing, or fastflushing membranes to remove the deposited layer. In addition to these considerations, coagulation pretreatment may alter the surface chemistry or speciation of membranes and foulants to reduce adhesion, improve removal with backflushing or otherwise increase permeate flux.

16.3 COAGULATION PRETREATMENT TO IMPROVE PERMEATE QUALITY

16.3.1 Enhanced Removal of Microbial Contaminants

In principle, MF membranes do not serve as an absolute barrier to the passage of all microorganisms. The lower limit of pore diameter typically considered to be within the range of microfiltration is approximately 0.1 μm. While this is sufficiently small to remove virtually all bacteria and protozoa, the sizes of viruses of concern in water are small enough to pass through the pores of MF membranes. Nonetheless, significant removals of viruses by MF membranes have been observed, apparently due to the tendency of some viruses to form larger aggregates or to stick to the surfaces of larger particles which are retained by membranes. Coagulation pretreatment takes advantage of and enhances this tendency for viruses to associate with other particles. Excellent removals of viruses by MF with coagulation pretreatment have been reported.[13]

16.3.2 Natural Organic Matter

Coagulation and subsequent liquid-solid separation are widely utilized to reduce the concentrations of numerous particulate and dissolved contaminants in water. Aluminum and iron salts are added regularly in water and wastewater treatment to reduce the concentrations of dissolved organic materials, nutrients, heavy metals, and hydrophobic micropollutants. For example, the ability of coagulants to remove organic materials from the dissolved phase has recently been institutionalized in U.S. regulatory practice under the name *enhanced-coagulation,* the goal of which is to reduce the concentrations of natural organic materials (NOM) that may serve as precursors to disinfection by-products.

The ability of metal salts and cationic polymers to coagulate natural organic materials has been recognized for some time. Cationic polymers may form polymer-humate precipitates. Metal salts may also destabilize NOM by charge neutralization as well as by adsorption to the precipitated flocs. In either case, coagulant addition under the correct conditions may convert the problem of removing dissolved NOM to one of conventional liquid-solid separation. The conditions of pH and coagulant dose determine the speciation of the metal coagulant and the mechanisms by which particles and dissolved organic materials are coagulated. Closely related to these mechanisms are the properties of the precipitated phase such as particle size, number, and charge that also vary with pH and coagulant dose.[14] These variables impact on the operational efficiency of conventional unit processes such as packed-bed filtration, sedimentation, and flotation as well in membrane processes.

The efficacy of membranes as processes for liquid-solid separation suggests that membranes may be particularly useful in achieving the maximum benefits to be derived from coagulating a water to reduce NOM. Liquid-solid separation using membranes is not dependent on the formation of a settleable floc. Also, removal does not require that particle surface chemistry be properly adjusted to promote adherence of the particles to a surface as in packed-bed filtration. Small particles that might otherwise pass through settling basins, and perhaps even packed-bed filters, can be removed by membrane filters. This allows for greater flexibility and economy in the use of coagulant.

It is useful to visual coagulant dosing strategies for membrane pretreatment to use a stability diagram for the coagulant. Such a diagram is constructed from equilibrium relations for speciation under the assumption that the system is at equilibrium with a solid phase as described by a solubility product (Fig. 16.3). The example shown in Fig. 16.3 essentially serves as a map of the combinations of aluminum concentration and pH that result in the precipitation of $AL(OH)_3$ at equilibrium. In this example, aluminum is predicted to be at its minimum solubility near a pH of 7. A coagulant dose and pH that minimize the solubility of the metal (e.g., Al) should also reduce the concentration of the metal in the permeate as well as the possibility of reprecipitation and ensuing high particle concentrations.[3] At values of pH less than 7, positively charged species of dissolved aluminum are dominant. Also, the charge of precipitated $AL(OH)_3$ tends to be positive. At higher pH values, dissolved aluminum is present predominantly as negatively charged $Al(OH)_{4-}$ and the surface charge of precipitated aluminum is negative.

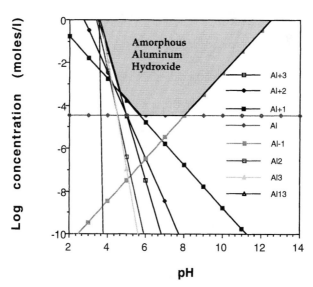

FIGURE 16.3 Stability diagram for aluminum in equilibrium with amorphous aluminum hydroxide.

Conditions of aluminum dose (alum) and pH comprising the left-hand and bottom borders of the amorphous aluminum hydroxide region in Fig. 16.3 are thought to favor the removal of particles and NOM by charge neutralization. The interior of the hydroxide region corresponds to conditions favoring *sweep-floc* formation, a condition in which contaminants are enveloped in the hydroxide precipitate.

In practice, stability diagrams must be constructed from jar tests in which regions of contaminant removal (by sedimentation or filtration) are noted. Different coagulants produce different stability diagrams. The aluminum concentrations and pH values that result in significant removal of NOM using a polyaluminum coagulant without sulfate differ from those observed using a polyaluminum coagulant with sulfate, or an aluminum chloride, or alum. Stability diagrams can also be constructed for iron coagulants, cationic polymers, and other coagulants. Raw water quality also affects the stability diagram. For example, higher concentrations of NOM tend to shift the diagram upward. Stability diagrams derived from jar test data typically show an optimal pH for NOM removal between pH 5 and 6. On a molar basis, iron coagulants are usually observed to remove more NOM than do aluminum coagulants.

There is some experimental evidence that waters with high concentrations of NOM produce cakes with minimum resistance to permeation when coagulant is dosed to produce particles with a zeta-potential near zero.[12] Excellent removal of NOM is also produced at these doses. The coagulant dose required for charge neutralization can be determined by a jar test if the raw water contains significant concentrations of NOM and particles. Such waters are excellent candidates for coagulation and membrane filtration. Coagulant dose for charge neutralization can also be controlled automatically using streaming current detectors. This offers some advantage in that the conditions for charge neutralization may be limited and accurate control of coagulant dose may be required to assure optimal operation. In a water containing small amounts of NOM and particles, charge neutralization may not produce the desired degree of aggregation due to insufficient opportunities for particle contact. In this case, liquid-solid separation by MF or UF without coagulation pretreatment may be preferable.

16.3.3 BOD

In domestic wastewater treatment, coagulants may be added to remove dissolved or particulate materials that exert a *biological oxygen demand* (BOD). These materials are predominantly organic compounds. As such, the criteria for their removal are similar to those for removing natural organic materials such as fulvic acid from surface waters. Based on comparative removals of TOC and BOD, it appears that most of the BOD reductions obtained from coagulation appear to be associated with the removal of particulate matter.[6,15]

16.3.4 Phosphorus

Dissolved phosphorus in wastewater can be removed by adding lime, aluminum, and iron salts. When lime is added at pH values greater than approximately 10, hydroxylapatite ($Ca_{10}(PO_4)_6(OH)_2$) is precipitated. Phosphate ions compete with carbonate ions for the precipitation of calcium. It is for this reason that the reaction typically occurs at higher pHs and that the dose of lime is dependent primarily on the alkalinity of the wastewater rather than on the concentration of phosphate. Theoretically, phosphate removal by reaction with iron or aluminum to form a precipitate of the form $MePO_4$ requires one mole of iron or aluminum (Me) for each mole of phosphate. (Phosphate may also be precipitated as ferrous phosphate, $Fe_3(PO_4)_2$.) However, there are also competing reactions for the metal, principally through its reaction with organic materials, that may alter the stoichiometry of precipitation. In addition to direct precipitation, phosphorus may be removed by adsorption onto the metal precipitate. Phosphate may also form complexes with positively charged hydrolyis species, a factor that appears to be limited largely to the case where prepolymerized iron or aluminum coagulants are added. However, prepolymerized (partially neutralized) coagulants have been found to be less effective in removing phosphorus.[16] This may be due to both the decreased availability of the polymerized metal for direct precipitation and the effect of preneutralization on pH. The minimum solubility of aluminum phosphate is near pH 6. The optimum pH range for phosphate precipitation with Fe(III) is between 4 and 5 and between 7 and 8 for Fe(II).

As with NOM removal, jar tests are required to select chemicals for phosphate removal and to determine chemical doses. Once precipitated, phosphorus is removed by subsequent liquid-solid separation. UF or MF membrane can be used in the liquid-solid separation step to remove precipitated phosphorus. However, optimum conditions for phosphate removal may not correspond to those for permeate flux maintenance. For example, polyaluminum coagulants appear to offer some advantages over alum in maintaining permeate flux, while they are not the coagulant of preference for the removal of phosphate.

16.3.5 Nitrogen

Ammonia nitrogen constitutes one of the principal sources of nitrogen in the effluent from conventional biological treatment facilities. This form of nitrogen is virtually untouched by coagulation and therefore passes through coagulation/membrane reactors using UF or MF membranes. Measurements of total Kjeldahl nitrogen (TKN) include organic nitrogen that may include particulate species. Coagulation pretreatment has a small effect in increasing the removal of TKN. However, pilot studies performed to date indicate that coagulation pretreatment does not substantially enhance the removal of nitrogen beyond the removal of particulate nitrogen retained by membranes without coagulation pretreatment.

16.3.6 Heavy Metals

Industrial wastes containing Ba, Pb, Hg, arsenic, and other inorganic pollutants can be treated by adding aluminum or iron salts and subsequently filtering the coagulated water using membranes. These contaminants are removed by direct or coprecipitation. While there are differences in the optimal pH range for the removal of a given contaminant using a given coagulant, a range of 6 to 8 is typically required. Because ferric hydroxide is insoluble over a wider pH range than is aluminum hydroxide, iron coagulation appears to be more effective than aluminum in removing dissolved inorganic contaminants. In some cases, attention to the redox state of the contaminant is required to ensure its removal in coagulation. For example, arsenic as As^{3+} must be oxidized to As^{5+} before coagulation, while Cr^{6+} should be reduced to Cr^{3+} before coagulation with either Fe(III) or aluminum salts.

16.4 OPERATIONAL CONSIDERATIONS

The manner in which coagulated materials are wasted from the membrane system will significantly affect the efficacy of the coagulation pretreatment. Coagulant added to the feed stream concentrates within the membrane system. As metal concentrates, the pH of the concentrate and permeate decreases to the point that metal residuals in the permeate, the operating pH, or both may become unacceptable. The drop in pH may be compensated in part by adding base to the feed stream. However, the key to successful operation of the pretreatment is to remove the coagulated materials from the system.

In a feed-and-bleed operation, the concentration of the retentate is a function of the system recovery as determined by the waste or bleed flow Q_w.

For a constant permeation rate, the time required for a feed-and-bleed system to reach steady state t_{ss} with respect to materials rejected by the membrane can be estimated as

$$t_{ss} = \frac{3V_{sys}}{Q_f(1 - Rr)} \tag{16.1}$$

where Q_f is the flow of raw water to the membrane pilot; V_{sys} is the volume of the water within the membrane system; R is the recovery of the system (permeate flow rate Q_p divided by raw water flow rate Q_f or, equivalently, $(1 - Q_w/Q_f)$; and r is the rejection factor calculated as $(1 - C_p/C_r)$. C_p and C_r are the concentrations of the material of interest in the permeate and rejectate, respectively. The time to steady state in Eq. (16.1) is defined as the time required for C_r to be equal to 95 percent of the concentration that would be measured if the system were operated at constant conditions for an infinite period of time. In pilot studies of coagulation pretreatment with membrane filtration utilizing a feed-and-bleed wastage strategy, accurate estimates of system performance can be obtained only after a period of at least t_{ss}. In this mode, higher recoveries require that the membrane operate with a higher concentration in the retentate. This is likely to correspond to a lower permeate flux, a greater potential for membrane fouling, and poorer permeate quality.

These problems can be partially avoided by wasting a greater quantity of coagulant from the system. However, in the feed-and-bleed mode, this entails operating the system at a lower recovery; either the waste flow must be increased or the system must be drained periodically. An alternative wastage strategy is to remove precipitates of coagulant and associated contaminants that accumulate on the membrane in a concentrated *flush*. In this case, the waste flow Q_w is composed of water used to carry away materials dislodged during periodic backflushing of the membrane. If no other wastage occurs, the system recovery is determined by the frequency of these backwashing operations. Because the concentration of coagulant is higher in the backflush water, lower concentrations of coagulant are achieved in the retentate flow without reducing recovery.

As the preceding general discussion and some of the case studies subsequently presented indicate, careful control of pH and coagulant dose in the pretreatment is required. There may also be trade-off in the pretreatment with respect to optimizing permeate quality versus permeate flux. Consideration must be given to monitoring coagulant residuals in the permeate, the effectiveness of each backflush cycle, the pH and alkalinity of the feed and permeate, and the service time between chemical cleanings of membranes.

16.5 OPERATIONAL EXPERIENCE

16.5.1 Hollow-Fiber MF

Olivieri and co-workers[7] evaluated the addition of coagulant to the feed stream of a hollow-fiber MF unit used to treat a low-turbidity surface water and noted improved removal of THM precursor material and improved permeate flux. The performance of a similar microfiltration membrane for the case of no pretreatment was compared with that of coagulation pretreatment using two polyaluminum coagulants in an application to secondary wastewater filtration.[15] A skid-mounted microfiltration membrane pilot* with a nominal capacity of 8 L/min was used to treat water drawn from a secondary clarifier at the Houston Southwest Wastewater Treatment Plant. The pilot used four polypropylene membrane modules, each with 1 m^2 of outside-in hollow-fiber membrane area with an effective pore size of 0.2 μm. The membrane unit was operated in a constant permeate flux mode in which the transmembrane pressure (TMP) increases over time as materials deposit on and in the membrane. Permeate flux was set at 340 L/h and readjusted as needed to maintain this flow. The pilot was automatically air-backflushed every 20 min, resulting in a recovery (permeate flow divided by feed flow) for the unit of 82 percent. Samples from the feed and permeate flows were analyzed for turbidity, total organic carbon (TOC), phosphorus concentration, and particle size distribution. Samples were also analyzed for total and fecal coliforms. Similar analyses were performed on samples of influent and effluent water from the packed-bed filters at the treatment plant during the same period of operation.

Coagulation pretreatment significantly improved process operation by maintaining lower transmembrane pressures during the pilot runs (Fig. 16.4). The two coagulants investigated reduced fouling to different degrees. The polyaluminum coagulant containing sulfate performed better than the sulfate-free product in maintaining low transmembrane pressures (less fouling). In all cases, chemical cleaning of the membrane was required weekly.

Coagulation pretreatment was found to have a small impact on permeate quality. Particle concentrations were reduced by 2 to 3 logs with and without pretreatment. Most samples indicated no coliforms detected, the occasional exceptions being attributed to contamination of the samples during sampling. Turbidities were consistently near 0.1 ntu or less in both cases. Orthophosphate concentrations in the clarified water were low and little additional removal (with or without coagulation) was observed. Similarly, ammonia was not removed by the membrane. Coagulation did produce a slight improvement in the removal of carbonaceous oxygen demand. However, the most significant improvement in permeate quality produced by coagulation pretreatment was an approximately 50 percent reduction in TOC concentrations.

Coagulation-microfiltration was investigated as a pretreatment for reverse osmosis in an evaluation of wastewater recovery in Florida.[5,6] The pilot* was operated at a flow rate of approximately 5.7 m^3/h at transmembrane pressures less than 100 kPa. Characteristics of the membrane were similar to those reported in the previous study. Chlorinated and filtered effluent from the Reedy Creek wastewater treatment facility was used as feed to the membrane pilot. The membranes were operated in a dead-end filtration mode with air-backwash every 20 to 25 min. In contrast with the Houston study, chemical cleaning of the membranes was required every three to four

---

* Memtec America Corporation, Timonium, Md.

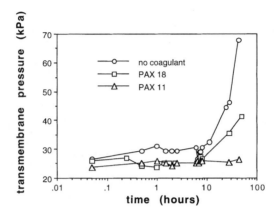

FIGURE 16.4 Transmembrane pressure versus time for three pilot runs. The polyaluminum coagulant PAX 11 contains sulfate; PAX 18 is sulfate-free.

weeks. The use of filtered wastewater rather than water drawn directly from the secondary clarifiers appears to have substantially extended the period between chemical cleanings. Alum dosed at approximately 2 to 4 mg Al/L and, in some cases, ferric sulfate were used as coagulants. Coagulation reduced the phosphorus concentration from an average of 0.32 mg P/L in the raw water to 0.02 mg/L in the MF permeate. TOC was reduced an average of 22 percent while total nitrogen concentrations were virtually unchanged across the MF membrane.

16.5.2 Ceramic Membranes

The effect of alum and ferric sulfate addition on the performance of ceramic membranes* used to treat a high-turbidity/high-TOC surface water in Houston, Texas, was evaluated at pilot scale and compared with the performance of a conventional potable water treatment plant.[3] The effect of pH on coagulation pretreatment using alum and ferric sulfate was evaluated. Pilot work was performed at the Houston water purification facility located east of central Houston. Waters from two nearby rivers, the Trinity and the San Jacinto, were pumped to the Houston facility and initially discharged into presedimentation basins where feed to the membrane pilot was drawn.

The pilot consisted of two skid-mounted units. The first unit was a chemical pretreatment reactor used to add coagulant, adjust pH, and mix raw water with recycled concentrate from the membrane skid. The membrane unit on the second skid was operated using a single Alcoa P19-40 module, with a membrane area of 0.2 m². A P19-40 module consists of a single ceramic rod, 100 cm in length, with 19 tubular membrane channels or *elements* running through the length of the rod. The diameter of each tubular element is 0.4 cm. Bonded to the interior of these tubes is a sintered layer of α-alumina or zirconia that defines the nominal pore size of the membrane. Membranes with nominal pore sizes of 0.8, 0.2, 0.5, and 0.05 μm were used. However, most work was performed with the 0.05-μm zirconia membrane, as it was found to produce

---

* Alcoa Separations Technology Division, now US Filter, Warrendale, Pa.

the best permeate quality and flux. The unit was operated at a cross-flow velocity of approximately 167 cm/s and a pressure drop across the membrane of 50 to 340 kPa, depending on the period of operation, the pretreatment conditions, and the nominal membrane pore diameter. Backflushing was automatically initiated every 96 s. The pilot was operated in a feed-and-bleed mode with recovery ratios (permeate flow/raw water feed) of 90 to 95 percent. The turbidities of the raw, feed-, and permeate waters were measured, as well as the pH and absorbance of UV light at 254 nm. Particle size distributions and the concentrations of total and dissolved organic carbon (TOC and DOC, respectively) in the raw, feed-, and permeate waters were measured.

Without coagulation, backflushing the membrane every 96 s did not increase permeate flux. This indicates that backflushing was either ineffective in removing deposited materials or that the deposition rate of materials was rapid in comparison with this backflush frequency. When these waters were coagulated with alum or ferric sulfate, frequent backflushing substantially enhanced permeate flux (Fig. 16.1). Although coagulation with alum or ferric sulfate improved permeation rates, membrane performance was found to be sensitive to the conditions of pH and coagulant dose. The highest permeation rates were obtained when pH was maintained near neutral values that minimize the solubility of aluminum or iron. The accumulation of aluminum or iron in the recycle loop of the pilot over time results in a decrease in pH. Without compensating for this drop in pH by adding base, permeate flux decreases and the concentration of soluble aluminum or iron in the permeate increases. In addition to improving permeate flux, coagulation pretreatment also significantly improved the quality of the membrane permeates. Without coagulant, the turbidity of the average value of the membrane permeate was 1.29 ntu. With coagulation pretreatment, permeate turbidities were less than 0.5 ntu. Alum reduced the TOC in the permeate by approximately half, while ferric sulfate pretreatment reduced TOC to approximately one-third the concentration observed in the permeate when pretreatment was not applied. In summary, while pretreatment was found to enhance most aspects of membrane performance, satisfactory performance was sensitive to the conditions of coagulant dose and pH. Near-neutral pH values that minimized the solubility of aluminum or iron produced the highest permeate fluxes and the lowest concentrations of TOC in the permeate. However, in comparison with lower pH values, turbidities and particle counts were higher when the pH for coagulation was maintained near values of 6 to 7.

Coagulants have been found to be indispensable in the operation of full-scale facilities for potable water treatment using ceramic membranes. In one instance, an 850-m³/day plant was put into operation in the city of Marques, France. The 0.2-μm α-alumina membranes experienced massive fouling and the system was removed from service. A subsequent study[2] demonstrated that coagulant addition before the membranes avoided the fouling problems experienced earlier without coagulation.

REFERENCES

1. V. Lahoussine-Turcaud, M. R. Wiesner, and J. Y. Bottero, "Fouling in Tangential-Flow Ultrafiltration: The Effect of Colloid Size and Coagulation Pretreatment," *Journal of Membrane Science,* **52**:173–190 (1990).

2. V. Lahoussine-Turcaud, et al., "La Microfiltration couplée à la floculation: application à la potabilisation des eaux souterraines en terrain karstique," *L'Eau, L'Industrie, Les Nuisance,* **146**(5):1–4 (1991).

3. M. R. Wiesner, S. Veerapaneni, and D. Brejchová, "Improvements in Membrane Microfiltration Using Coagulation Pretreatment," in H. Hahn and R. Klute (eds.), *Chemical Water and Wastewater Treatment II,* Springer-Verlag, Berlin (1992).

4. M. R. Wiesner, D. C. Schmelling, and K. Pickering, "Permeation Behavior and Filtrate Quality of Tubular Ceramic Membranes Used for Surface Water Treatment," *Proceedings of the AWWA Specialty Conference on Membranes,* Orlando, Fla. (1991).

5. H. R. Kohl, et al., "Reclamation of Secondary Effluent with Membrane Processes," *Proceedings of the Annual Conference of the American Water Works Association,* San Antonio, Tex. (1993).

6. H. R. Kohl, et al., "Evaluating Reverse Osmosis Membrane Performance on Secondary Effluent Treated by Membrane Microfiltration," *Proceedings of the Annual Conference of the American Water Works Association,* San Antonio, Tex. (1993).

7. V. P. Olivieri, et al., "Continuous Microfiltration of Surface Water," *Proceedings of the Membrane Technologies in the Water Industry,* Orlando, Fla. (1991).

8. V. Lahoussine-Turcaud, et al., "Coagulation Pretreatment for Ultrafiltration of a Surface Water," *Journal American Water Works Association,* 82(12):76–81 (1990).

9. J. G. Jacangelo, et al., *Evaluation of Ultrafiltration Membrane Pretreatment and Nanofiltration of Surface Waters,* American Water Works Association Research Foundation (1994).

10. M. R. Wiesner, and S. Chellam, "Mass Transport Considerations for Pressure-Driven Membrane Processes," *Journal American Water Works Association,* 84(1):88–95 (1992).

11. J.-M. Laîné et al., "Effects of Ultrafiltration Membrane Composition," *Journal American Water Works Association,* 81(11):61–67 (1989).

12. M. R. Wiesner, M. M. Clark, and J. Mallevialle, "Membrane Filtration of Coagulated Suspensions," *Journal of Environmental Engineering, American Society of Civil Engineers,* 115(1):20–40 (1989).

13. V. P. Olivieri, et al., "Continuous Microfiltration of Secondary Wastewater Effluent," *Proceedings of the Membrane Technologies in the Water Industry,* Orlando, Fla. (1991).

14. M. R. Wiesner, V. Lahoussine-Turcaud, and F. Fiessinger, "Organic Removal and Particle Formation Using a Partially Neutralized AlC_{13}," *Proceedings of the Annual Conference of the American Water Works Association,* Denver, Colo. (1986).

15. R. Patel, et al. "Membrane Microfiltration of Secondary Wastewater Effluent," in R. Klute and H. H. Hahn (eds.), *Chemical Water and Wastewater Treatment III,* Springer-Verlag, Berlin (1994).

16. H. Ratnaweera, J. Fettig, and H. Ødegaard, "Particle and Phosphate Removal Mechanisms with Prepolymerized Coagulants," in R. Klute and H. H. Hahn (eds.), *Chemical Water and Wastewater Treatment II,* Springer-Verlag, Berlin (1992).

CHAPTER 17
MEMBRANE BIOREACTORS

Jacques Manem
CIRSEE-Lyonnaise des Eaux
Le Pecq, France

Ron Sanderson
Institute for Polymer Science
University of Stellenbosch
Stellenbosch, South Africa

17.1 INTRODUCTION

Biological treatment is an important aspect of industrial and municipal wastewater treatment and reuse processes (Metcalf and Eddy, 1991) and is rapidly assuming an important role in drinking water treatment (Rittmann and Snoeyink, 1984).

Biological processes can be defined as engineered systems, designed to accumulate microorganisms which oxidize organic (chemical oxygen demand, or COD) and mineral (NH_3, Fe^{2+}, etc.) pollutants that are electron donors and reduce O_2, NO_3, SO_4, or CO_2 that are electron acceptors (Rittmann, 1987).

The first recorded use of biological processes in the production of drinking water was by slow sand filtration, designed by the English engineer, Simpson, for the city of London in 1829 (Imbeaux, 1935). The English also developed the first engineered biological wastewater treatment system (trickling filters) in the 1880s and then, at the beginning of the 20th century, the *activated sludge* process, which is still used in most wastewater treatment plants throughout the world.

The efficiency of biological processes depends on two main factors: the biomass concentration in the reactor and the specific conversion rate of the microorganisms. Efforts to improve biological processes over the past hundred years have generally aimed at increasing the concentration of microorganisms in the bioreactor, either by separating the solids and liquids and then recirculating the biomass (activated sludge) or by developing fixed culture reactors in which the microorganisms are fixed on a support. This support may itself be static (fixed bed) or mobile (fluidized or turbulent bed) (Lazarova and Manem, 1994).

The recent development of a new generation of more productive and less expensive ultra- (UF) and microfiltration (MF) membranes has prompted the emergence of a new concept in biological treatment: the *membrane bioreactor* (MBR) (Cheyran and Mehala, 1986). This new technology offers several advantages over conventional processes used to date; these include reliability, compactness, and, above all, excellent treated water quality. The resulting high-quality and perfectly disinfected effluent means that MBR processes can be used for many purposes—e.g., drinking water, industrial and municipal wastewater treatment and reuse, recycling in buildings, and landfill leachate treatment.

This chapter describes the principles, advantages, and also the limitations of MBR processes. Since the investment and operating costs of this new process are attributable mainly to the filtration unit, the first part of this chapter will focus on the various parameters which influence the membrane performances. The second part presents an overview of the existing industrial applications in the fields of water and wastewater treatment and reuse. The current major research trends will also be discussed.

17.2 FUNDAMENTALS OF MEMBRANE BIOREACTORS

17.2.1 Membrane Bioreactor Process Description

Membrane bioreactors can be defined as the combination of two basic processes—biological degradation and membrane separation—into a single process where suspended solids and microorganisms responsible for biodegradation are separated from the treated water by a membrane filtration unit (Fig. 17.1a). The entire biomass is confined within the system, providing both perfect control of the residence time for the microorganisms in the reactor (sludge age) and the disinfection of the effluent.

According to this definition, the MBR process should be distinguished from other treatment processes where membrane filtration is installed downstream of the biological processes, such as activated sludge or fixed film processes, as a refining stage or tertiary treatment, as shown in Fig. 17.1b. The description and performances of such systems are described elsewhere in the following sections.

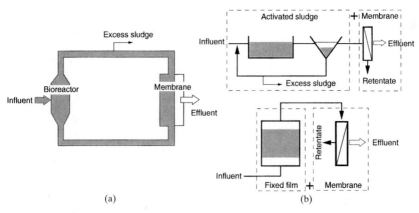

FIGURE 17.1   Principle of membrane bioreactor (a) MBR and (b) tertiary treatment.

The general operation of the MBR is illustrated in Fig. 17.1*a*. The influent enters the bioreactor, where it is brought into contact with the biomass. The mixture is pumped from the bioreactor and then, under pressure, filtered through the membrane. The permeate is discharged from the system while the entire biomass is returned to the bioreactor. Excess sludge is pumped out in order to maintain a constant sludge age and the membrane is regularly cleaned by backwashing, chemical washing, or both.

The biological reactor and membrane units of the MBR can be combined externally (in this case, the biomass must circulate between the bioreactor and the membrane, as illustrated by Fig. 17.2*a*, or by integrating the membranes inside the bioreactor (Fig. 17.2*b*). The second configuration, which can be defined as an *integrated* MBR, requires outer skin membranes (OSM). Configuration 17.2*a*, currently the most common MBR, corresponds to a *recirculated* MBR and can be operated with either inner (ISM) or outer skinned membranes. Apart from the differences in the types of membranes, these two configurations can be distinguished by the technology used to create the pressure gradient between the two sides of the membrane (driving force). The pressure across the membrane in an *integrated* MBR can be applied only by suction through the membrane (Kayawake, Narukami, and Yamagata, 1991; Ishida et al., 1993; Chiemchaisri, Yamamoto, and Vigneswaran, 1993) or by pressurizing the bioreactor. In a recirculated MBR, pressure across the membrane can also be created by recirculating flow through the membrane (most common).

17.2.2   Advantages of the MBR Process

One of the major advantages of the MBR lies in the quality of the treated water, since this system is capable of simultaneously biologically treating and disinfecting the effluent. As shown in Tables 17.1 and 17.2, typical COD, biological oxygen demand (BOD), and suspended solids (SS) removal percentages are often greater than 95, 98, and 99 percent, respectively.

The complete separation between the hydraulic retention time (HRT) and the suspended solids retention time (SRT) provides optimum control of biological reactions and greater reliability and flexibility in use. A key element to MBR technology is its ability to absorb variations and fluctuations in the applied hydraulic and organic load to the system. Complete control of the sludge age is particularly important to allow the development of slow-growing microorganisms such as nitrifying bacteria.

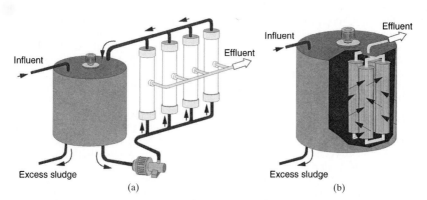

FIGURE 17.2   The two main MBR configurations: (*a*) recirculated MBR; (*b*) integrated MBR.

TABLE 17.1 Examples of MBR Process Performances
for Municipal Wastewater Treatment

Influent, mg/L			Effluent, mg/L				
SS	COD	BOD	SS	COD	BOD	Membrane	Reference
80–460	100–365*	200–1000	<5	<40	<10	UF	Aya et al., 1981
236	96*	134	<5	6	1.2	24.000	Arika et al., 1977
96	89*	349	<5	12	3.7	UF	Roullet, 1989
280	620	230	<5	11	<5	MF	Manem et al., 1993
153	79*	176	<1	6	1.5	MF	Ishida et al., 1993
110–164	292–411	—	<5	15–19	—	UF	Fan et al., 1996
1315[†]	—	1130[†]	5	—	5	UF	Irwin, 1990

* COD measurements were made in accordance with Japanese methods, using potassium permanganate; therefore, COD levels are lower than those measured in Europe and U.S.
[†] Urban effluent from a building.

UF and MF membranes used in MBR function under greater biomass concentrations and, therefore, greater loads than traditional clarified activated sludge processes based on gravity separation. Consequently, the systems can be very compact. The membrane can also retain soluble material with a high molecular weight, increasing its retention time and, therefore, improving the opportunity for its biodegradation in the biological reactor.

Several studies have shown that MBRs produce less sludge than conventional activated sludge systems (Krauth and Staab, 1994; Ishida et al., 1993; Bouillot, 1988; Chaize and Huyard, 1991). Although the high sludge age frequently used in this process partially explains this advantage, the theory concerning biological reactions in a perfectly confined space, such as in an MBR, is still not well understood. The high shearing forces, used in the membrane unit to prevent clogging, probably stimulate hydrolysis and mass transfer, which play key roles in degrading suspended substances in wastewater.

17.2.3 Main Design Parameters

The optimum design of an MBR process is very complex since many dependent factors have to be considered, including membrane performance and cost, energy consumption, and sludge treatment and disposal. Furthermore, the majority of these are interrelated and can influence capital and operation costs in adverse ways.

An example of how these various factors can influence and depend on each other is the effect the biomass concentration has on the design of the MBR. As stated previously, the most frequently quoted advantage of the MBR is the possibility of using high biomass concentrations and thereby increasing the volumetric load and lowering the civil works investment. However, this advantage should be viewed in relation to the risks presented by the action of high biomass concentrations on membrane performance, which in turn can lead to increased membrane surface area requirements, thus increasing investment costs. High concentrations of suspended matter also influence oxygen transfer and sludge viscosity and, subsequently, the energy cost.

The interdependence of the main design parameters shown by this example illustrates the necessity for a global and systemic approach in order to optimize the MBR from an economic point of view. A general analysis of the main relationships that exist between the design parameters is given here, along with a review of the principal bibliographical data available.

TABLE 17.2 Examples of MBR Process Performances for Industrial Wastewater Treatment

Industry	Process A/ANA*	Influent				Effluent				Reference
		COD mg/L	BOD₅ mg/L	SS g/L	N-NTK mg/L	COD mg/L	BOD₅ mg/L	SS mg/L	N-NTK mg/L	
Cosmetic	A	6,500	2,400	1,900	40	<100	20	<5	0.4	Manem et al., 1993
Milk processing	A	4,200	2,600	650	110	40	<10	<5	4.2	Manem, 1995
Textile	A	10,000				600				Krauth et al., 1994
Fruit juice	A	2,250				24				
Tannery	A	7,600				190				
Oily WW	A	4,300–6,900	919–1,360	253–889	—	180–660	3–34	1–11	—	Knoblock et al., 1994
Sludge heat-treated liquor	ANA	9,200–10,600	4,300–5,000	180–520	160–310	1500–2200	150–230	<5	250	Kayawake et al., 1991
Sweet whey	ANA	58,000	34,000	5,200	—	700	300	<10	—	Li et al., 1985
Wheat starch	ANA	35,000	15,000	13,000	—	270	70	<10	—	Li et al., 1985
Starch synthetic wastewater	ANA	9,700	—	—	—	300	—	—	—	Cadi et al., 1994

* A Aerobic/ANA anaerobic

17.2.3.1 Membrane Selection. Technical and economical analyses of the first industrial-scale MBR projects reveal that the filtration installation (membrane block) and the associated energy necessary to operate it are the major investment and operating costs.

Filtration of biological solutions has been the object of numerous experiments. Most types of membranes have been tested; these include ultra- and microfiltration, inner and outer skin, hollow fiber, tubular and flat, organic, metallic, and ceramic. Depending on the membranes used, filtration performances vary from 0.05 to 10 m/d as reported in Table 17.3. This table shows that typical fluxes for internal skin membranes are between 0.5 and 2 m/d, whereas outer skin membrane fluxes are often between 0.2 and 0.6 m/d.

Besides the flux, the main factors that affect filtration costs are the price of the membrane itself and its installation, including all required appurtenances, operating costs (energy, membrane replacement, regeneration cost, etc.), and the desired effluent quality. A technical and economic analysis including all these parameters tends to level the actual cost of filtration when considered over a period of several years for the different membranes. This conclusion partially explains the large diversity of membranes being used in present industrial applications (Table 17.4).

With the current technology, outer skin membranes (Fig. 17.2*b*) appear to be the most promising. Indeed, outer skin membranes offer many advantages, including low energy consumption and important reduction in skid manufacturing costs. Current research tends toward the optimization of fluxes that are three times lower than for internal skin membrane.

Dynamic membrane technology is another innovative concept which can offer a good compromise between effluent quality and capital and operation costs, if irreversible fouling can be controlled (Pillay, Townsend, and Buckley, 1994).

At the present time, the use of MBR is reserved for specific situations that require excellent effluent quality (COD, disinfection) and mainly for water reuse (see sec. 17.3) or for high-strength industrial effluents. Large-scale use of MBR in urban wastewater treatment will require newer technological developments or a significant decrease in the price of the membrane. It can offer a solution in highly populated areas, or in areas where land used to disperse bioreactor effluents can be better used for other purposes (Ross et al., 1993).

Whatever system is adopted, the membrane must satisfy a certain number of basic criteria. First of all, to limit fouling, the size distribution of membrane pores should have as little interference as possible with the size distribution of the particles or molecules to be filtered. Highly porous membranes with evenly distributed pores enhance filtration efficiency (Fane et al., 1989). The membrane should preferably be hydrophilic (Fane et al., 1989) and be either negatively charged or neutral in order to limit biomass adsorption (Shimizu et al., 1989). In addition, it should have an internal diameter (ISM) or minimum spacing (OSM) to limit clogging by the biomass; however, clogging depends on sludge concentration and circulation flow rate (for more details, see Kiat, Yamamoto, and Ohgak, 1992; Krauth and Staab, 1989, 1993). The membrane should be nonbiodegradable by the microorganisms present in the solution and easy to clean and regenerate if fouled. The latter criterion is important because biological failure or sudden effluent modification (as in industrial wastewaters) could lead to severe membrane fouling.

17.2.3.2 Parameters Influencing Membrane Performances Filtration Model. Filtration of cellular suspensions containing both dissolved and suspended matter is a complex process which has been the subject of numerous modeling projects (see details in Sec. 17.4). One of the most frequently used models is the resistance-in-

TABLE 17.3 Characteristics and Performances of Some Membrane Filtration Units in MBR Process

Application	MLSS concentrate, g/L	Membrane material	Type of membrane	Porosity or MW cutoff, μm or kdalton*	Filtration area, m^2	Velocity, m/s^{-1}	Flux at 20°C, m/d	Transmembrane pressure, bar	Running period regeneration, d	Reference
					Internal skin membrane					
MWW/IWW	2.5–3.5	PS	T	25*		8.70	10.00	5.00		Krauth and Staab, 1994
Human excreta	10–15	PAN / POF	HF	13* / 20*			1.80 / 1.2–1.4	1–3 / 2–3.5		Magara and Itoh, 1991
IWW	28.00	Organic	T	UF	195		1.50			Knoblock et al., 1994
Synthetic MWW	10–32	Ceramic	T	0.05–0.2		3.80	0.5–0.7	1.1–1.4	3–7	Lacoste et al., 1993
MWW	2.50	Ceramic	T	0.10	1.10	3.00	1.4–2	0.5–1.5	15.00	Trouve et al., 1994
DW	0.50	Organic	HF	0.01	7.20	0.90	1.4–1.7	0.5–0.8	60	Urbain and Manem, 1996
SD	25–60	WPT	T	20–80*		2.00	0.48–1.0	2.00		Pillay et al., 1994
IWW	9–21	PES	T	MF	668.00	1.60	0.14–0.6	2–3	>100	Ross et al., 1994
MWW	1–5	WPT†	T	UF		2.2–3.6	0.5	2–2.5		Bailey et al., 1994a, b
MWW	20.00	PAN	PF			2.50	2.4–2.9		45.00	Roullet, 1989
MWW	4–27		UF	24*	25.00	1.80	0.60	0.14	150.00	Arika et al., 1977
					External skin membrane					
Synthetic MWW	5–15	PE	HF	0.10			0.2–0.6	–0.2–0.8	120.00	Yamamoto et al. 1989
MWW	16.00	PE	PF	0.40			0.2–0.9	–0.1–0.5	2.00	Ishida et al. 1993
MWW	13–15	PS	RD	750*	0.2–1.74	2.6–3.9*	0.6–0.8	0.80	240.00	Ohkuma et al. 1994
Synthetic MWW	0.2–6	PE	HF	0.1	0.30		0.2–0.5	–0.4–0.8	150–250	Chiemchaisri et al., 1993
Sludge liquor‡	0.10–20	Ceramic	T	0.1	1.06	0.2–0.3	0.05–0.3	–0.20	30.00	Kayawake et al., 1991

PF = plate and frame
PE = polyethylene
POF = polyolefin
DW = drinking water
IWW = industrial wastewater
MWW = municipal wastewater
HF = hollow fibers
PS = polysulfone
SD = sludge digestion

T = tubular
PES = polyethersulphone
RD = rotary disk
WPT = woven polyester tube
PAN = polyacrylonitrile
* Disk peripherical velocity.
† Precoated with diatomaceous.
‡ Liquor from sludge heat treatment.

TABLE 17.4 Principal Industrial-Scale MBR Process

Company	Trade name	Country	Compound removed	Plant, Nbr	Capacity, m^3/d	Reference
			Drinking water			
Aquasource Lyonnaise des Eaux	Biocristal	F	NO_3, pesticides	1	400	Urbain et al., 1996
			Domestic wastewater			
Rhone Poulenc—TechSep	UBIS	F	C, NH_4, NO_3	>40	<400	Roullet, 1989
Dorr Oliver	MSTS	USA	C, NH_4	1	10	Smith et al., 1967
Thetfort Syst	Cycle-LET	USA	C, NH_4	>30	<200	Irwin, 1989
Kubota		J	C, NH_4	8	10–110	Ishida et al., 1993
Mitsui Petroc	Asmex	J	C, NH_4, NO_3	>6		Lambert, 1983
Degrémont/Lyonnaise des Eaux	BRM	F	C, NH_4	2	10, 150	Manem, 1995
			Industrial wastewater			
Zenon Env Inc.	Zenogem	CDN	C, NH_4	1	116	Knoblock et al., 1994
Degrémont/Lyonnaise des Eaux	BRM	F	C, NH_4	1	500	Trouvé et al., 1994b
					160	
Dorr Oliver	MARS	USA	C	1	38	Sutton et al., 1983
University of Stuttgart		G	C, AOX	1	162	Krauth and Staab, 1994
Membratek	ADUF	RSA	C	2	500	Ross et al., 1994
					100	
			Landfill leachate			
SITA/Lyonnaise des Eaux		F	C, NH_4, NO_3	3	10.50	Trouvé et al., 1994a
Thames W/PC/PWT	Biomembrat	GB	C, NH_4, NO_3	1	65	
Whrle Werk		G	C, NH_4	1	160	
Zenon Env Inc.	Zenogem	CDN	C, NH_4	1	10	
Enviro/ICL		D	C, NH_4	1	136	

series model, which provides a simple means of describing the permeate flux–transmembrane pressure relationship over the entire domain of pressure. Based on this model, the dependence of permeate flux on the applied transmembrane pressure is expressed by the following equation (Cheryan, 1986):

$$J = \frac{\Delta P}{\mu(R_m + R_n + R_c)} \qquad (17.1)$$

where J = permeate flux, $m^3/m^2/h$
         μ = permeate viscosity, Pa·s
         P = operating pressure, Pa
         R_m = hydraulic resistance, L/m
         R_n = irreversible fouling, L/m
         R_c = resistance due to cake layer, L/m

This equation makes it possible to show the different parameters that can influence filtration performance. Hydraulic resistance R_m is characteristic of the membrane being studied and corresponds to the intrinsic resistance of this membrane. Irreversible fouling R_n, which results in supplementary resistance to filtration and can have many different causes related to membrane porosity, is often due to adsorption of soluble organic molecules. Resistance due to the cake that forms on the membrane surface R_c is a function of the concentration and composition of suspended matter as well as the applied hydraulic conditions. According to some authors, the resistance of this polarization layer R_c can be written as $R_c = \Theta \Delta P$, where Θ is a function of the system's mass transfer properties. Equation (17.1) then becomes:

$$J = \frac{\Delta P}{\mu(R_m + R_n + \Theta \Delta P)} \qquad (17.2)$$

From Eq. (17.2), two regions of interest can be determined: a low-pressure zone where the hydraulic resistance of the membrane is the governing factor $[(R_m + R_n) \gg \Theta \Delta P]$ and where the flux is proportional to applied pressure (below a limiting value) and a high-pressure zone where the cake resistance is predominant $[(R_m + R_n) \ll \Theta \Delta P]$. Equation (17.2) also shows that the flux is inversely proportional to the viscosity. The latter parameter, which corresponds to viscosity of the permeate (treated water), is often close to that of pure water. It is important not to confuse the permeate viscosity with sludge viscosity. Sludge viscosity is higher and will influence the hydraulic regime inside above the membrane surface (laminar or turbulent regime).

Although the concentration and composition of the sludges to be filtered do not appear in Eq. (17.2), these factors strongly influence several parameters: R_m, R_n, and Θ. Sludge concentration is probably one of the key parameters in process design. It influences both permeate flux for OSM and ISM and head loss for ISM (changes in viscosity and, therefore, capital and operating costs).

The relationship between the membrane flux and the biomass concentration has been the subject of numerous, and sometimes contradictory, research efforts. For example, Magara and Itoh (1991) found that a semilogarithmic relationship conveniently described the influence of biomass concentration on the limiting permeate flux. This relationship, described by Eq. (17.3), was developed for biomass concentrations between 5 and 15 g/L:

$$J\left(\frac{m}{d}\right) = -1.571 \log (MLSS) + 7.84 \qquad (17.3)$$

where MLSS = mixed liquor suspended solid concentration, in mg/L.

On the other hand, Ross et al. (1990) have shown that, in the case of anaerobic sludge, using a UF membrane, the flux is relatively stable up to a concentration of 40 g/L, but then decreases sharply and stabilizes at about 60 g/L. Several experiments performed in the author's laboratory demonstrated that for aerobic sludge very little flux decline was observed for sludge concentration between 5 and 12 g/L. Other experiments by Li, Kothari, and Corrado (1985) on anaerobic sludges seem to indicate that a semilogarithmic relationship would best describe the influence of suspended matter concentration on the permeate flux obtained with polysulfone UF membranes for concentrations between 10 and 50 g/L.

Two observations can be drawn from these results. First, the operating zone (i.e., low or high transmembrane pressure) must always be properly identified for a good assessment of the influence of operating parameters on the filtration performance of the process. Second, increasing the biomass concentration generally decreases the membrane performance. According to the data reported here, two main types of relationship seem to predominate: semilogarithmic, which is the most frequent, and linear. This apparent diversity probably stems from the fact that the biomass concentration parameter camouflages situations that, in reality, are very different, and this parameter is not sufficient to predict how the membrane will respond. Two other variables, (1) the structure and composition of the polarization layer, and (2) the nature and concentration of molecules susceptible to be adsorbed on the membrane surface, are more useful to describe sludge/membrane interactions and to estimate process performance. Certain operating conditions can influence these two parameters: hydrodynamic conditions, the physiological state of the biomass, and the sludge concentration through its viscosity.

From these observations, it is clear that the design and operation of a membrane bioreactor is not an easy task, and it requires a careful matching of filtration and biological operating parameters. In the following sections, the influence of several of these parameters on the overall process performance will be examined.

Effects of Flow Rate and Transmembrane Pressure (TMP) on Membrane Performance.   Equation (17.2) predicts that the permeate flux will remain proportional to the transmembrane pressure for a cake resistance much lower than the membrane hydraulic resistance. As the transmembrane pressure increases, the resistance of the filtration cake becomes predominant and the flux becomes independent of the applied pressure. This behavior has been verified by several authors (Magara and Itoh, 1991; Jeannot et al., 1992) who have also shown that cross-flow velocity plays an important, although indirect, role in this relation. As illustrated in Fig. 17.3, increasing cross-flow velocity can increase the permeation flux by widening the range over which transmembrane pressure is linearly related to permeate flux. However, in order to be able to take full advantage of this extended range, it is important to know as precisely as possible what the response of the system will be, since, for instance, increasing transmembrane pressure when in the pressure-independent zone would promote only membrane fouling without any significant increase in the membrane performance. One parameter that could help in the characterization of a membrane bioreactor is the *critical* or *optimal* transmembrane pressure (TMP_c), which can be defined as the transmembrane pressure at which the membrane hydraulic resistance ($R_m + R_n$) and the resistance due to the formation of the cake are equal, such as

$$\text{TMP}_c = \frac{(R_m + R_n)}{\Theta} \tag{17.4}$$

where TMP_c is the optimal transmembrane pressure with regard to permeation flux and membrane fouling. As with other operating parameters, the optimal pressure

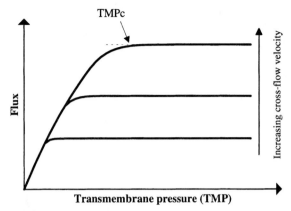

FIGURE 17.3 Theoretical flux evolution as a function of the transmembrane pressure.

depends on the structure and concentration of the polarization layer, the shear forces applied to it, and the biological component of the system. This is illustrated in Fig. 17.4 where the variation of the optimal pressure with cross-flow velocity is shown for two different biological systems.

Optimal performance in terms of cross-flow and transmembrane pressure should thus be expected from operating at, or near, optimal transmembrane pressures, provided that no significant variations are observed in other important parameters such as suspended solids concentration, temperature, and suspension viscosity.

Physiological State of the Biomass. The biological community present in a membrane bioreactor is very complex and, as in all other biological processes, is mostly composed of several species of bacteria and higher grazing microorganisms.

Bacteria are either agglutinated in a structure called a *floc* or are individual free-swimming cells. They are principally gram-negative microorganisms; hence, their membrane structure contains an outer layer of exopolymeric substances (EPS).

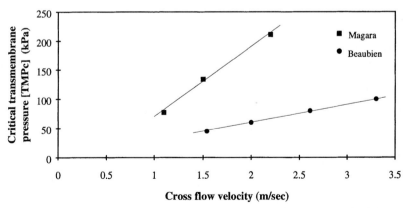

FIGURE 17.4 Experimental evolution of the critical transmembrane pressure (TMP$_c$) as a function of the cross-flow velocity for two different biological systems. *(After Jeannot, 1992, and Magara and Itoh, 1991.)*

These substances, mainly proteins and polysaccharides, play an important role in floc formation.

Several authors have shown the importance of the biochemical structure of the bacteria outer layer in the cake formation process when filtering bacterial suspensions with membranes. For instance, Leslie et al. (1992) have shown that the specific resistance of a filtration cake composed of gram-negative marine bacteria can be significantly lowered by enzymatic (pronase) or physical-chemical (EDTA) modification of the EPS. These treatments lead to a collapse of the *outer layer* structure, thereby modifying the surface of the bacteria and, therefore, membrane bacteria-bacteria and bacteria-surface interactions.

The role played by EPS in fouling is confirmed by results obtained by Chaize (1990), who studied the influence of growth rates of a solution of *E. coli* on the performance of a ceramic membrane. This work showed that an increase in the growth rate is related to a higher membrane fouling rate and to a modification of the microorganism surface which became more hydrophobic.

Lacoste, Drakidés, and Rumeau (1993) showed that, by specific enzymatic hydrolysis of the adsorbed "slime" on a ceramic membrane surface, which had been previously fouled with a bacterial suspension, the initial permeability could be partially restored. As the enzymes used were specific to cell-wall debris and bacterial capsules (lysozyme and pronase), these authors concluded that these biological materials played an important role in the bacterial fouling by their adsorption onto the membrane.

These studies illustrate the importance of the physiological state of the biomass in the fouling phenomenon, and notably on the following aspects:

1. The specific hydraulic resistance of the filtration cake, mainly composed of microorganisms, partially depends on the surface state of the bacterial EPS, therefore depending on the environmental parameters (sludge age, mass load, dissolved oxygen, etc.) (Urbain, Block, and Manem, 1993).

2. There is adsorption of cell debris and of the bacterial soluble microbial by-products (SMPs) at the membrane surface. Therefore, overaged sludge and starving conditions that lead to an aging of the culture and to an increase in the concentration of cell debris and SMPs should, therefore, be avoided. High hydraulic strains and stresses may also enhance cell and floc breakage, therefore favoring cell debris and bacterial metabolite release (Wisniewski and Grasmick, 1994; Choi et al., 1992).

From a practical point of view, based on the author's experience, MBR plants operated in aerobic conditions provide least fouling effect when the biomass is well aerated and the soluble COD concentration in the bioreactor is minimum. These two criteria are, generally, satisfied when redox potentials are consistently maintained above 300 mV. This conclusion is confirmed by the work of Sato and Ishii (1991), who demonstrated that filtration resistance can be described by:

$$R = 842.7 \Delta P(\text{MLSS})^{0.926}(\text{COD})^{1.368}(\mu)^{0.326} \tag{17.5}$$

where R = filtration resistance, m^{-1}
ΔP = transmembrane pressure, Pa
MLSS = mixed liquor suspended solids, mg/L
COD = soluble chemical oxygen demand, mg/L
μ = viscosity of activated sludge suspension

In Eq. (17.5) it can be seen that soluble COD concentration plays a major role in filtration resistance. Chang and Lee (1994) also reported that activated sludge with a low SVI index had a lower fouling tendency than sludge with a high SVI index.

Unfortunately, experiments on the relationship between the physiological state of the biomass and the fouling mechanism remain sparse and further research in this field is necessary in order to optimize MBR operation.

Fouling can also be due to biofilm growth on the membrane. Refer to Chap. 6 for more information on biofilm formation and its consequences on membrane filtration performance.

Influence of Sludge Viscosity on Membrane Performance. The viscosity of the suspension to be filtered is relevant at various stages in the MBR design and its relationship with the sludge concentration is therefore very important. The study of sludge rheology is a very difficult field because of the plastic or pseudo plastic behavior of the sludges and because viscosity measurement values often depend on the type of equipment used (Campbell and Crescuolo, 1982).

Krauth and Staab (1994) suggest an exponential relationship between suspended solids concentrations and sludge viscosity for aerobic sludges, as presented in Fig. 17.5. Measurements performed in the author's lab (CIRSEE) on sludges originating from an aerobic MBR operating on dairy wastewaters (SLVO) confirm this relationship (Fig. 17.5). Ross et al. (1990) found similar results for an anaerobic sludge; they found an exponential dependence between the sludge concentration and the yield stress which is related to viscosity.

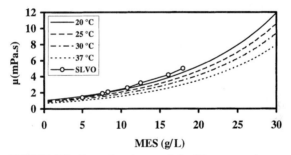

FIGURE 17.5 Examples of suspended solid concentrations on sludge viscosity. *(After Krauth and Staab, 1994, and Manem, 1995 (SLVO), see text.)*

Sludge viscosity influences inner skin membrane process performance at two levels: longitudinal head loss (PCL) and the hydraulic regime in the membrane. The calculation of the values of these two parameters, involving the Reynolds's number (Re), make it possible to estimate the energy cost related to the sludge circulation in the membrane and the shear stress at the filtration cake surface.

Figure 17.6 illustrates Re values as a function of viscosity for several velocities in tubular membranes with 4-mm-diameter channels. It shows that for viscosity higher than 6 centipoises (Biomass > 20g/L at 20°C, according to Krauth and Staab, 1994) it is necessary to recycle at velocities higher than 6 m/s to reach a turbulent regime that is favorable to filtration performance (Ross et al., 1990; Bailey, Hansford, and Dold, 1994b). For a 2-m/s velocity and for a 9-mm-diameter tubular membrane, Ross et al. (1990) showed experimentally that the transition from turbulent to a laminar regime

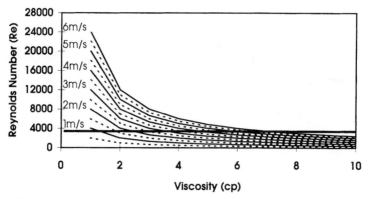

FIGURE 17.6   Reynolds number (Re) as a function of viscosity for a 4-mm tubular membrane.

will occur at an anaerobic sludge concentration of about 40 g/L (at 35°C), which should be accompanied by an important decrease in the permeate flux.

A limiting factor to the commercialization of inner skinned MBRs is the high energy cost related to sludge circulation in the membrane. Head loss, which is a function of membrane characteristics (L, d), and circulation velocity both depend on the viscosity of the suspension. Using Colebrook's formula (Degrémont, 1989), it can be shown that the impact of viscosity on the head loss becomes more important at higher circulation velocities (Fig. 17.7). The slope breaks observed in Fig. 17.7 correspond to changes in the hydraulic regime (turbulent to laminar).

It should be kept in mind that viscosity is a function of temperature, and an increase of about 2 percent of membrane flux per degree Celsius (Magara and Itoh, 1991; Ross et al., 1990) can be expected. Consequently, thermophilic biological processes are favored.

In conclusion, an increase in sludge concentration in the biological reactor can, due to its effect on viscosity, have a detrimental effect on the permeate flux rate and

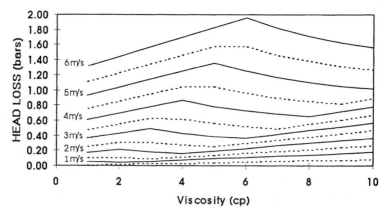

FIGURE 17.7   Head loss in a 4-mm tubular membrane as a function of viscosity (see text).

thus increase investment costs. Furthermore, head loss in the membrane can lead to extra energy costs. Systematic sludge rheology studies should make it possible to estimate the optimal economic use of a membrane in many cases.

Effects of Physical-Chemical Parameters on Membrane Performance. The structure of the filtration cake depends on biomass physiology, TMP, cross-flow velocity, and physical-chemical conditions. Fane et al. (1989) showed that, for a low biomass concentration, lower ionic forces favor the filtration flux by increasing the cake's porosity, thus decreasing its specific hydraulic resistance. Besides their effect on the filtration cake, the ionic strength and pH of the solution can, in some cases, lead to a modification in the membrane's surface properties, especially for some types of highly charged ceramic membranes. In order to limit adsorption of the negatively charged biomass, it is preferable to adjust the pH to obtain a negatively charged surface (Shimizu et al., 1989).

17.2.3.3 Biological Performance of MBRs.

Conventional biological processes, either with suspended or attached biomass, exert a selection pressure on the biomass tending to favor the formation of bacterial flocs or biofilms. In both cases, microorganisms are held together by a complex mixture of exopolymers (Jorand et al., 1995) that hinder the substrate diffusion and, therefore, the degradation rate.

The use of membrane bioreactors makes it possible to counteract this selection pressure and the typically high tangential velocities limit the floc size (Wisniewski and Grasmick, 1995; Bailey, Hansford, and Dold, 1994a, 1994b), leading to increased mass transfer rates toward the microorganisms. An example of floc size distribution present in samples taken from an activated sludge and an MBR reactor fed with the same influent (Fig. 17.8) confirmed the smaller size of floc units in MBR sludge (~40 μm) compared to activated sludge (~60 μm). The decrease in the selection pressure for floc-forming microorganisms and the higher mass transfer rates may be responsible for the higher specific yields reported by several authors. They may also be involved, to a certain extent, in the MBR low sludge production reported in the literature.

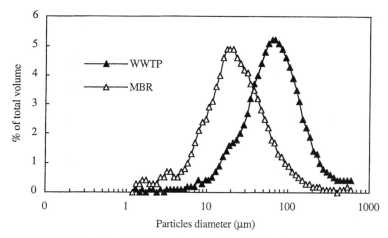

FIGURE 17.8 Floc size distributions in the activated sludge from the conventional WWTP and MBR (cross-flow velocity is 3 m/s) fed with the same municipal wastewater.

Effluent Quality. One of the main advantages of the membrane bioreactor is the quality of the effluents produced. Tables 17.1 and 17.2, which summarize results obtained from treating municipal and industrial wastewaters, respectively, show COD and BOD removals of more than 95 and 98 percent, respectively, and the absence of all suspended matter.

The excellent quality of the treated water can be attributed to better biological degradation (see below), and to the membrane, which removes suspended matter and some soluble macromolecules due to its retentive fractionation (inherent membrane- coupled to cake-retentivity).

Specific Rates. The high mass loads applied to aerobic membrane bioreactors can be explained by high sludge concentrations and apparent specific substrate removal rates. For example, Manem (1995) has shown that the mass loads for a dairy effluent are sixfold compared to conventional treatments, although the biomass is only twice as concentrated; this indicates a specific rate multiplied by 3. Using a complex synthetic substrate, Lacoste et al. (1993) obtained excellent COD removal (more than 95 percent) for mass loads higher than 5 kg $COD/m^3/d$. Similar results have been published by Wisniewski and Grasmick (1994) and Bouillot (1988), who found an average increase in the mass loads by a factor of between 3 and 5.

Unlike their action in aerobic processes, the use of anaerobic MBRs does not seem to favor specific degradation rates in the case of carbonaceous substrate degradation. Whether in the laboratory (Cadi, 1994) or on a semi-industrial scale (Jeannot et al., 1992; Ross et al., 1994) or on a full scale (Ross et al., 1994), the massic loads are similar to conventional processes and often lower than 1 kg COD/kg VSS/d for complex substrates.

Results obtained with denitrifying microorganisms for the production of drinking water (Urbain and Manem, 1996) also show similar specific substrate removal rates between MBR and conventional processes.

Observed differences in behavior between aerobic, anoxic, and anaerobic ecosystems can be explained by the following.

1. There is a selection of the more efficient strains, as they would normally be washed out in the activated sludge process because of a relative inability to flocculate. The higher specificity of the microorganisms involved in anaerobic processes, especially in the methanization step, and in the denitrification process would alleviate the selection pressure in this case.

2. There is a better mass transfer rate (O_2, BOD_5) toward and within smaller aerobic bacterial flocs (Bailey, Hansford, and Dold, 1994b).

3. There are changes in the metabolism due to biological stress provoked by high shear stress (Wisniewski and Grasmick, 1994).

4. There is an apparent reduction of the biomass production due to the removal of part of the exopolymeric matrix involved in the flocculation process and which is very important in aerobic processes.

Ongoing microbiological studies on the MBR bacteria should make it possible to discriminate among the preceding hypotheses.

Sludge Production. Although reduction of biological sludge production in aerobic MBRs is generally accepted (Bailey, Hansford, and Dold, 1994b; Bouillot, 1988), the mechanisms responsible are still under debate. Two possible explanations for this phenomenon have been suggested. The first possibility, which is not specific to MBRs, is the high sludge ages maintained in the bioreactor. The other is the modification of the floc composition and structure and bacterial metabolism provoked by the high shear forces in inner skin membranes that lead to smaller floc units which (1) favor the mass transfer and hydrolysis reactions and (2) reduce the synthesis of EPS (see Sec. 17.2.3.3).

It is interesting to note that anaerobic and anoxic MBR experiments (Urbain and Manem, 1996) show biomass production that is equivalent to conventional processes. These results, which can be compared to those described for specific rates (see Sec. 17.2.4.3), seem to confirm the hypothesis that the use of membrane bioreactors modifies the aerobic ecosystem considerably but has less influence on anaerobic or anoxic biomass.

Pretreatment and Excess Sludge Treatment.   Filtration of high sludge concentrations by inner skin membranes often requires a prescreening to avoid membrane clogging, especially when low-diameter tubes are used (between 2- and 6-mm inner diameter). A one-to-five ratio between the membrane inner diameter and the screen mesh size should limit membrane clogging.

Very little research has been done on the problem of MBR excess sludge dewatering. However, it seems that the quality of sludge produced with inner skin membrane requires the use of centrifuge or filterpress-type dewatering processes. This conclusion can be explained by the MBR sludge structure which is less flocculated than traditional activated sludge and, therefore, cannot be dewatered with a traditional beltpress. This problem deserves to be thoroughly researched in the near future.

On the other hand, outer skin membrane MBR sludge structure is very similar to activated sludge and, therefore, can be dewatered by conventional processes.

17.2.3.4   Global Approach for MBR Design.

For given wastewater characteristics, MBR design brings together three main groups of independent parameters: (1) biological factors, (2) hydrodynamic factors, and (3) membranes. Their influence on investment and operating costs is complex and often antagonistic, as described in previous sections. Figure 17.9 describes the main relationships between the independent parameters and the different items which compose investment and operating costs. The optimum MBR design should integrate all these parameters.

Even though the influence of some of these parameters seems clear and easy to quantify—e.g., investments relative to bioreactor size and operating costs for excess sludge treatment—the consequences of others, such as biological and hydrodynamic parameters, on filtration performances are subject to discussion and will require a considerable amount of research.

The relatively high investment and operating costs of actual industrial-scale MBR processes are mainly due to the filtration unit using inner skin membranes. A better understanding of the relationships between the different parameters listed in Fig. 17.9 and the development of new MBR processes with, for example, outer skin membranes, should considerably lower these costs and extend MBR applications in the near future.

17.3   INDUSTRIAL APPLICATIONS

17.3.1   Introduction

The first research efforts on membrane bioreactors began in the 1960s and the first industrial plant was built by the American company, Dorr Oliver, in 1967. This plant

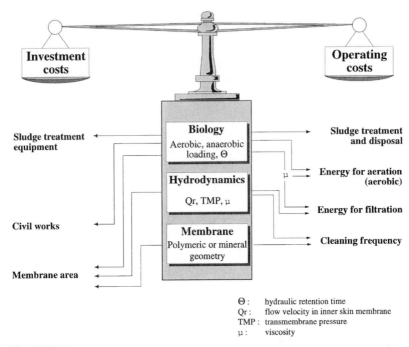

Θ :   hydraulic retention time
Qr :   flow velocity in inner skin membrane
TMP :   transmembrane pressure
μ :   viscosity

FIGURE 17.9   Main relationships between the three independent design parameters and the principal items of the investment and operating cost.

could treat 14 m³/day of "domestic" wastewater from a factory in Connecticut (Smith, DiGregorio, and Talcott, 1967). It was not until 1977, however—10 years after the Dorr Oliver project—that the first water recycling schemes were installed in a Japanese building. In Japan, the process used was UBIS, developed by Rhône Poulenc (Roullet, 1989). More recently (during the 1980s), buildings in the United States have been designed using the Cycle-let process (Irwin, 1989, 1990). Currently, most MBR installations are still used for water recycling in buildings (Table 17.3), but landfill leachate and industrial wastewater treatment applications are becoming more and more important.

This current limited number of MBR applications around the world is primarily due to the high cost of the filtration units, developed over the last 20 years, which are based mainly on the inner skin membrane design. Using this technology, MBR processes are competitive to conventional treatment when at least one of the following conditions is fulfilled: (1) high-strength effluent (some industrial wastewaters, sludge, etc.) in which the MBR system can take advantage of high biomass concentration and, therefore, smaller biological basin counterbalances the cost of filtration; (2) stringent disinfection requirements for treated water (markets include water reuse or drinking water production); and (3) biomass composed of very low growth rate microorganisms (xenobiotic degradation, soil or groundwater remediation). Most of the applications described in the following satisfied at least one of these conditions.

17.3.2 Municipal Wastewater Treatment and Reuse

17.3.2.1 Water Reuse in Buildings. Over the last 15 years the shortage of water in many Japanese regions such as Tokyo and Kyushu has led some municipalities to legislate regulations to encourage water recycling in large buildings over 15 stories high. Government incentives in Japan are also being studied in the Chiba and Osaka regions.

The membrane bioreactor process is probably the best-adapted biological technology for water recycling inside buildings requiring compact systems and excellent water quality (Fig. 17.10). Two different systems are experiencing large commercial success: UBIS in Japan (Lambert, 1983; Roullet, 1989) and Cycle-let in the United States (Irwin, 1989, 1990).

In 1980, the Japanese company MPC introduced the Ultra Biological System (UBIS), developed by Rhône Poulenc (France), in Japan and applied it to wastewater recycling in the Marunouchi building in Tokyo. Wastewaters originating from toilets, kitchens, and sinks and floor washing are collected and sent to an aerobic-activated sludge reactor. Plate-and-frame ultrafiltration membranes (Pleiade) are used to treat the water and maintain sludge separation. The hydraulic retention time is about 1 h and the sludge concentration is 20 g/L. Chemical cleaning should be performed every 45 days to ensure and maintain membrane fluxes of between 100 and 120 L/h·m². The MBR effluent, which contains less than 5 mg/L of BOD and no suspended solids, is recycled as toilet flush water. Today, the UBIS system is installed in more than 40 buildings and produces more than 5000 m³/day.

In the United States, the Thetford Corporation in Ann Arbor, Michigan, has developed another process: the Cycle-let Wastewater Recycling System, which is based on the combination of a two-phase biological treatment system (anoxic and oxic) and tubular organic membranes (Irwin, 1990). Activated carbon for color removal and ozone for disinfection are used to finish and improve the water quality. Used in conjunction with conserving fixtures, such as 1.5-gal toilets, this process cuts external water use by 70 to 90 percent. Investment and running cost were estimated in 1991, around US\$1.25 and US\$0.09 for a 300,000-ft² building housing 1500 people. Today, the Thetford Corporation can claim the existence of more than 30 plants across United States (Irwin, 1990).

Wastewater treatment and recycling in apartment buildings is also performed in Europe, where the Lyonnaise des Eaux group developed an MBR process using a ceramic tubular-type membrane.

17.3.2.2 Collective Night-Soil Treatment Plants. In Japan, the population relies on three major domestic wastewater treatment categories: public sewage, on-site treatment tanks, and collected human excreta treatment systems. Each of them directly serves around one-third of the population, but the last system, which is also called the "night-soil" treatment system, is also used for treating the sludge generated by on-site treatment (Magara and Itoh, 1991).

Magara et al. (1994) reported that there are about 1200 night-soil treatment systems across Japan that treat more than 42 million population equivalents of night soil and 30 million population equivalents of on-site treatment tank sludge.

Typical night-soil BOD, SS, total-N, and phosphorus concentrations are 12,000, 12,000–20,000, 3000, and 400 mg/L, respectively, whereas effluent concentrations should be less than 10 mg/L for BOD, 30 mg/L for nitrogen, and 5 mg/L for phosphorus (Lambert, 1983; Roullet 1989; Magara and Itoh, 1991). The more recent night-soil treatment plants typically incorporate biological denitrification processes

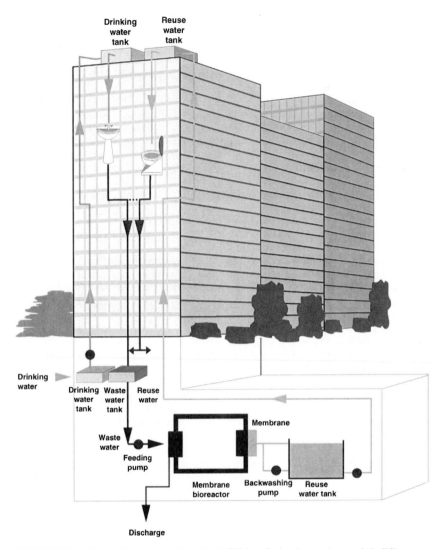

FIGURE 17.10 Schematic representation of an MBR installation for water reuse in buildings.

followed by tertiary treatments, including coagulation, filtration, and granular acti-vated carbon (GAC) adsorption (Magara et al., 1994).

The MBR technology is particularly adapted to this type of concentrated effluent. Several pilot studies have been reported and at least six full-scale plants are under operation with the Activated Sludge and Membrane CompleX System (AMEX) technology (Roullet, 1989). This process, which was adapted from the UBIS system by the Mitsui Petrochemical Industry, comprises a pretreatment unit followed by a multiphase bioreactor to remove nitrogen and carbonaceous pollution and a Pleiade filtration unit. Posttreatment is performed by an activated carbon filter and dephosphatation resin column. The average hydraulic retention time (HRT) ranges from six to nine days for influent and effluent BODs of 13,000 and 20, respectively.

Recently, Magara et al. (1994) developed a high-rate biological denitrification system based on the MBR technology followed by a tertiary treatment composed of UF filtration after coagulation, GAC filtration, and disinfection. Denitrification is controlled in the MBR by oxydo-reduction potential (ORP) and dissolved oxygen (DO) measurements.

17.3.2.3 Urban Wastewater Treatment. Several processes using outer skin membranes (Ishida et al., 1993; Turner, 1994) or inner ceramic tubular membranes (Manem et al., 1993; Fan et al., 1996) were recently developed for urban wastewater treatment. Due to the high investment and operating costs of the actual membrane units, the first applications of MBR processes are limited to small treatment plants. Nevertheless, one can predict from the large research effort undertaken around the world and, particularly, on outer skin membrane that this application should experience a high development in the next five years.

17.3.2.4 Sludge Treatment. Anaerobic digestion is widely used in large wastewater treatment plants to stabilize primary and secondary sludges. A conventional anaerobic digester is a single-stage, single-pass process (Metcalf and Eddy, 1991) with identical hydraulic (HRT) and solids (SRT) retention times, ranging from 20 to 30 days. Pillay, Townsend, and Buckley (1994) suggest the application of cross-flow microfiltration (CFMF) to decouple the HRT from the SRT. A pilot-scale experiment using an anaerobic MBR process demonstrated a twofold increase in both the volumetric loading and sludge concentration, which led to a significant reduction in the digester volume, the downstream sludge concentration, and dewatering equipment.

Using flexible woven fiber tubes precoated with a limestone suspension, Pillay, Townsend, and Buckley (1994) suggested that the sludge filtration cake acts as a formed-in-place membrane which prevents irreversible fouling. Fluxes between 0.48 and 1 m/d were obtained for sludge concentrations ranging from 25 to 60 g/L.

An economic evaluation of this MBR process using 25-mm-diameter flexible woven fiber tubes was carried out in order to upgrade an existing sludge digester at the Northern Wastewater Treatment Plant in Durban, South Africa. According to Pillay, Townsend, and Buckley (1994), the MBR process reduces the sum of capital and operating cost as compared to the conventional process. Application of low-priced, high-cutoff microfiltration membranes can be a good compromise between effluent quality and capital and operation costs if irreversible fouling can be controlled.

17.3.3 Industrial Wastewaters

17.3.3.1 Anaerobic Membrane Bioreactors. Anaerobic wastewater treatments have become increasingly popular for the treatment of industrial effluents, especially those containing a high level of easily fermented products. Despite the obvious advantages—i.e., the production of usable energy and low sludge production—the development and use of anaerobic processes have been neglected to some extent, mainly due to the long biomass retention time required to maintain slow-growing methanogenic bacteria. Due to their ability to withhold methanogens and the resulting high-quality effluent, anaerobic membrane bioreactors have generated a growing interest.

Dorr-Oliver's patented Membrane Anaerobic Reactor System, better known by the name MARS, was the first process developed using the advantage of membrane separation in an anaerobic process (Sutton and Evans, 1983). The MARS system is composed of Dorr-Oliver organic ultrafiltration membranes (polyethersulfone) coupled with a single-phase anaerobic reactor. Sutton et al. (1983) point out that, for the anaerobic MBR process, the treatment cost per kilogram of COD is highly dependent on the effluent strength. This conclusion is very important and can generally be applied to all types of MBR, whether they are aerobic or anaerobic. In fact, the main expense of this process, which is related to the cost of the membranes themselves, is more dependent on the wastewater flow rate and, therefore, the surface membrane rather than the COD (reactor volume) to be treated.

Several types of industrial wastewater including those from grain processing and the dairy industry were tested at the pilot-scale level, and a full-scale demonstration plant consisting of a 38-m^3 suspended-growth anaerobic reactor coupled with UF membranes was operated at a Midwestern United States cheese plant (Sutton et al., 1983). More than 99 percent COD removal was achieved for a volumetric load of 83.5 $kg/m^3/d$ and a solids retention time of 50 days. The authors noticed that full-scale performances were superior to pilot-scale results, which can be explained by membrane overdesigning in the pilot study, which may imply higher shear stresses of the biomass.

More recently, the new combination of an anaerobic digester and an external ultrafiltration step (the ADUF process) was proposed and successfully applied for the treatment of corn and egg processing effluents (Ross et al., 1994). An average COD removal rate of over 90 percent was obtained for organic massic loads of 0.24 and 0.33 COD/kg VSS d, respectively. The membrane permeate fluxes obtained with the 9- and 12.5-mm tubular polyethersulfone UF membranes (Membratek, S.A.) operated at low pressure (500 kPa) varied from 10 to 70 L/h·m^2 at 30 to 35°C (Ross et al., 1994). Sludge concentrations below 40 g/L were maintained to ensure turbulent flow inside the membrane (Ross et al., 1990).

Capital and operating costs for a 1500-m^3/d ADUF plant treating a brewery effluent were estimated at 0.1 and 0.045 US$/$m^3$, respectively (Ross et al., 1994).

17.3.3.2 Aerobic/Anoxic MBR Processes *Landfill Leachate.* Landfill leachates are produced by (1) the biological activity in the waste dumps, and (2) the percolation of rainwater (Coulomb et al., 1993). These effluents contain high levels of organic and inorganic compounds and must be treated before reuse or discharge into the environment occurs. The quality and quantity of these effluents are highly variable and depend on the waste composition and the site's geoclimatic and operating conditions (Millot, 1986).

Conventional biological and physical-chemical processes are typically used to remove landfill leachate suspended solids (SS), organics (COD), and minerals (NH_3, NO_3) (McArdle, Arozarena, and Gallagher, 1987). Recently, reverse osmosis (RO) was suggested as a process to comply with new regulations requiring low concentrations of salt and heavy metals. Such technology should be coupled with a pretreatment unit to remove suspended solids and organic matter in order to avoid rapid fouling and/or clogging of the RO membrane.

The combination of the MBR process for removing SS, organic and nitrogen removal with a reverse osmosis (RO) process step for salt and heavy metal removal has been successfully applied and produces excellent quality water (Coulomb et al., 1993). Several industrial-scale treatment plants are in operation at the present time (Table 17.3). Figure 17.11 shows a general view of the plant operated by Dectra

FIGURE 17.11   General view of industrial-scale treatment plant operated by Dectra.

(SITA group) that treats around 50 m³/d of landfill leachate by an MBR-RO process (see Coulomb et al., 1993; Trouvé et al., 1993; Urbain et al., 1993a, for details).

Industrial Effluents.   Aerobic membrane bioreactor processes for industrial wastewater treatment have experienced a relatively fast growing market over the last two to three years. At least three different companies have developed their own process with ceramic (Degrémont) or organic membranes (Zenon and University of Stuttgart) (Knoblock et al., 1994; Krauth and Staab, 1994; Manem et al., 1993; Trouvé et al., 1994a). Four industrial-scale plants, ranging from 110 to 500 m³/d, have recently been built to treat cosmetic, milk, oil, and textile processing effluents. Most of these plants produce high-quality effluents which can then be reused as industrial waters (Table 17.1).

The process developed by the Degrémont company is based on the coupling of an aerobic bioreactor (6 to 8 m deep) equipped with floor aeration (small bubbles) and a ceramic membrane. This MBR process was successfully implemented at the Lancôme plant in the north of France (L'Oreal group). This MBR wastewater treatment plant (see Fig. 17.12) produces a high-quality effluent in terms of organic carbon and nitrogen (see Table 17.2, Cosmetic) which is perfectly disinfected and can be reused for different applications, including watering, floor cleaning, and toilet flushing.

The process, designed by University of Stuttgart, allows operation under pressure (bioreactor + membrane), which makes it possible to optimize oxygen transfer (Krauth and Staab, 1989). The optimal pressure is approximately 3 bars, leading to high energy consumption of about 10 kW/m³ (Krauth and Staab, 1994). The first industrial realization also showed a very low excess sludge production (0.077 kg MLSS/kg COD removal) which could be partially explained by a very high sludge age of up to 250 days.

In conclusion, the aerobic membrane bioreactor is a very attractive process for concentrated biodegradable industrial wastewaters, since the high-quality effluent can often be reused inside the industrial facilities.

FIGURE 17.12   Lancôme wastewater treatment plant.

17.3.4   Drinking Water Production (Denitrification)

Nitrates and pesticides are widely used as fertilizers and as protection substances, respectively, in agriculture and are, therefore, often found at high concentrations in underground water resources (Spalding and Exner, 1993; Mirgain et al., 1993). Thirteen years ago, European Economic Community guidelines set a maximum contaminant level at 50 mg/L for nitrates, 0.1 µg/L for individual pesticides, and 0.5 µg/L for total pesticide (80/778/EEC). The USEPA has set up an MCL of 44 mg/L for nitrates and 3 and 4 µg/L for atrazine and simazine, respectively (USEPA, 1992). Guidelines for nitrites are much more severe than for nitrates: respectively, 0.1 mg NO_2/L in Europe and 3.3 mg NO_2/L in the United States.

Nitrates can be removed either by physicochemical processes (electrodialysis, reverse osmosis, ion exchange), or by biological fixed film culture processes (Liessens et al., 1993; for a review see Hiscock, Lloyd, and Lerner, 1991, and vol. 10, no. 3, of *Water Supply*, 1992). The advantage of the latter is that nitrates are completely removed in the form of nitrogen gas and the only waste product of the process consists of excess biological sludge.

Pesticides are usually removed by adsorption on granular or powdered activated carbon (PAC) or by oxidation with ozone or ozone plus hydrogen peroxide (Hultman, 1982; Duguet et al., 1992).

The Lyonnaise des Eaux group developed an industrial MBR process combining biological denitrification, pesticide adsorption, and turbidity removal with organic hollow-fiber UF membranes, manufactured by Aquasource (France) (Chang, Manem, and Beaubien, 1993; Urbain and Manem, 1996). Denitrification takes place in a completely mixed bioreactor where PAC adsorbs pesticides and a fraction of the organic matter from raw water. Both the biological and filtration performances have been optimized during the pilot scale, and a 400-m^3/d plant was built at Douchy (France).

The results reported in Figs. 17.13 and 17.14 outline the well-known general advantages of MBRs. These advantages include the excellent and constant quality of the treated water disinfection when ultrafiltration membranes are used, flexibility in response to turbidity, nitrate, and organic carbon-loading variations, and automation (Urbain and Manem, 1996).

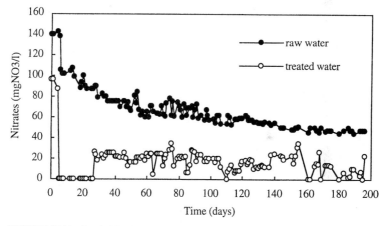

FIGURE 17.13 Denitrification performances of the Biocristal-DN process (nitrites are always lower than 0.1 mg NO$_2$/L).

As PAC is continuously added to the MBR and when an ultrafiltration membrane is used (cutoff = 0.01 μm), the dissolved organic carbon (DOC) concentration in the treated water was always lower than in the raw water. Moreover, pesticides, mainly the triazine compounds (atrazine, simazine, desethylatrazine), were removed up to the detection limit which is 0.02 μg/L (Fig. 17.15).

The pilot-scale study has shown that a flux of 60 L/h·m^2 (at 15°C) can be maintained in the MBR system with a total suspended solids (TSS) concentration of about 1 g/L when PAC is added in order to reach a concentration of 0.5 g/L (50 percent of the TSS content) in the system. These conditions can be maintained for a period of at least two months without chemical cleaning of the membranes. Mild conditioning of the membranes with chlorine facilitates the stabilization of these performances.

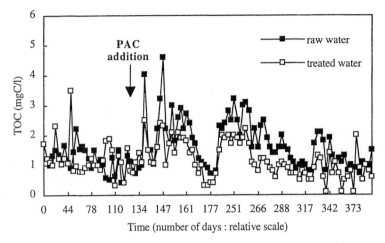

FIGURE 17.14 Organic carbon removal in the Biocristal-DN process (dissolved organic carbon in the sludge is between 10 and 30 mgC/L).

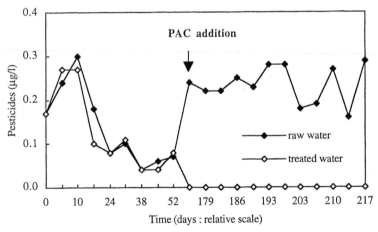

FIGURE 17.15 Pesticides removal in the Biocristal-DN process.

Denitrification can be perfectly controlled with this process, assuming that the appropriate amount of ethanol is fed.

17.4 CONCLUSIONS

The recent emergence of a new concept, the membrane bioreactors, should greatly enhance the performance and the reliability of biological treatment in the water and wastewater areas. This new process can be defined as the combination of two basic processes: biological oxidation and membrane separation. Its main advantages are the production of a high-quality effluent in terms of C, N, P and a perfect disinfection, meeting world health standards, and the maintenance of higher biomass concentrations that lead to very compact systems.

MBR processes have already experienced rapid development in areas where global water management becomes a major issue: (1) industrial wastewater treatment—lower pollution discharge and water reuse, and (2) water treatment and reuse in buildings located in areas experiencing water shortages.

In terms of drinking water, the membrane bioreactor has, to date, been used only for nitrate and pesticide removal. There remains, therefore, much scope for meaningful advances such as ammonia or organic biodegradable removal. The use of ultrafiltration membranes offers further water quality advantages (Anselme, Baudin, and Chevalier, 1992).

The membrane cost, which is the major capital input, deserves the most attention as regards module design and filtration operation (cleaning, regimes, etc.). Actual filtration units, mostly based on inner skin membranes, limits applications to high-strength industrial wastewater or wastewater reuse in building. Future development of outer skin membranes should lead to the rapid development for the municipal wastewater treatment.

A great deal can also be learned about membrane bioreactors from fields other than water. One can refer to, for instance, excellent review articles (Cheryan, 1986; Belfort, 1988; Engasser, 1988), or see in news items the varied applications to which MBRs can be put (Chemtech, 1989; Rogers, 1988; Meldrum, 1987).

As seen in the foregoing sections, fundamental concepts have been established which provide an integrated basis for process design and control. Continuing research will confidently be expected to lead to more favorable economics and improved reliability.

ACKNOWLEDGMENTS

The authors are very grateful to E. Trouvé, V. Urbain, A. Beaubien, A. Huyard, and E. Tardieu for their willing help and collaboration.

The technical review and various suggestions by Doctor M. Suidan and Professor Magara were very helpful in completing this chapter and are gratefully acknowledged.

REFERENCES

Anselme, C., I. Baudin, and M. R. Chevalier, 1992, "The Removal of Turbidity and Taste and Odour Problems in Drinking Water: Advantages of the Ultrafilration Process," *Water Supply,* **10:**91–97.

Arika, M., H. Kobayashi, and H. Kihara, 1977, "Pilot Plant Test of an Activated Sludge Ultrafiltration Combined Process for Domestic Wastewater Reclamation," *Desalination,* **23:**77–86.

Aya, H., G. Inoue, T. Okabe, and Y. Murayama, 1981, "Development of Compact Wastewater Treatment Plants for Non-potable Water Re-use System," *Water Reuse Symposium,* Washington, D.C. pp. 456–475.

Bailey, A. D., G. S. Hansford, and P. L. Dold, 1994a, "The Enhancement of Upflow Anaerobic Sludge Bed Reactor Performance Using Crossflow Microfiltration," *Water Research,* **28:** 291–295.

Bailey, A. D., G. S. Hansford, and P. L. Dold, 1994b, "The Use of Crossflow Microfiltration to Enhance the Performance of an Activated Sludge Reactor," *Water Research,* **28:**297–301.

Beaubien, A. M., M. Baty, H. Harduin, J. Manem, A. Mechinot, and F. Ehlinger, 1993, "Traitement des condensat de distillerie par bioréacteur anaérobie," *16éme Symposium International sur le traitement de eaux usées,* Montreal, Canada, pp. 151–165.

Belfort, G., 1988, "Membranes and Bioreactors: A Technical Challenge in Biotechnology," *Biotechnology and Bioengineering,* **33:**1047–1066.

Bouillot, P., 1988, "Bioréacteurs à recyclage des cellules par procédés membranaires: application à la dépollution des eaux en aérobiose," Ph.D. dissertation, INSA, Toulouse, France.

Cadi, Z., A. Huyard, J. Manem, and R. Moletta, 1994, "Anaerobic Digestion of a Synthetic Wastewater Containing Starch by a Membrane Reactor," *Environmental Technology,* **15:**1029–1039.

Campbell, H. W., and P. J. Crescuolo, 1982, "The Use of Rheology for Sludge Characterization," *Water Sciences Technology,* **14:**475–489.

Chaize, S., 1990, "les Bioréacteur à Membrane: Ulilisation en Traitement d'Eau Résiduaire Urbaine. Etude du Colmatage Bactérien," Ph.D. dissertation, University of Montpellier II, Montpellier, France.

Chaize, S., and A. Huyard, 1991, "Membrane Bioreactor on Domestic Wastewater Treatment: Sludge Production and Modeling Approach," *Water Sciences Technology,* **23:**1591–1600.

Chang, I. S., and C. H. Lee, 1994, cited in Lee, C. H., 1994.

Chang, J., J. Manem, and A. Beaubien, 1993, "Membrane Bioprocesses for the Denitrification of Drinking Water Supplies," *Journal of Membrane Science,* **80:**233–239.

Chemtech, 1989, *Chemical Engineering and Biotechnology Briefings,* pp. 299–302.

Cheryan, M., 1986, *Ultrafiltration Handbook,* Technomic Publishing, Lancaster, Pa., p. 375.

Cheyran, M., and M. A. Mehala, 1986, "Membrane Bioreactors," *Chemtech,* p. 676.

Chiemchaisri, C., K. Yamamoto, and S. Vigneswaran, 1993, "Household Membrane Bioreactor in Domestic Wastewater Treatment," *Water Sciences Technology,* **27:**171–178.

Choi, Y. S., et al., 1992, "Microfiltration of MLSS in an Activated Sludge System Treating Organic Wastewater," in *Proc. IMSTEC '92,* pp. 397–402.

Coulomb, I., P. Courant, J. Manem, V. Mandra, P. Renaud, and E. Trouvé, 1993, "Le Traitement des Lixiviats de Décharge," *Biofutur,* Déc, p. 3236.

Degrémont, 1989, commercial communication.

Duguet, J. P., C. Anselme, F. Bernazeau, I. Baudin, and J. Mallevialle, 1992, "L'Elimination des Pesticides par de Nouvelles Techniques de Traitement," *L'eau, l'industie et nuissances,* **153:** 41–44.

Engasser, J-M., 1988, "Bioreactor Engineering: The Design and Optimisation of Reactors With Living Cells," *Chemical Engineering Science,* **43**(8):1739–1748.

Fan, X-J., V. Urbain, Y. Qian, and J. Manem, 1996, "Nitrification and Mass Balance with MBR for Municipal Wastewater Treatment," submitted to the *1996 IAWQ Conference,* Singapore.

Fane, A. G., C. J. D. Fell, P. H. Hodgson, G. Leslie, and K. C. Marshall, 1989, "Microfiltration of Biomass and Biofluids: Effects of Membrane Morphology and Operating Conditions," *Proc. of the Vth World Filtration Congress,* pp. 320–329.

Hiscock, K. M., J. W. Lloyd, and D. N. Lerner, 1991, "Review of Natural and Artificial Denitrification of Groundwater," *Water Research,* **25:**1099–1111.

Hultman, B., 1982, "Elimination of Organic Micropollutants," *Water Sciences Technology,* **14:**73–86.

Imbeaux, I. D., 1935, "Qualités de l'Eau et Moyens de Correction," *Ed DUNOD,* Paris, France.

Irwin, J., 1989, "Wastewater Recycling in Commercial Building," *Plumbing Engineer Magazine,* September 1989.

Irwin, J., 1990, "On-site Wastewater Reclamation and Recycling," *Water Environment & Technology,* pp. 90–91.

Ishida, H., Y. Yamada, M. Tsuboi, and S. Matsumura, 1993, "Submerged Membrane Activated Sludge Process (KSMASP)—Its Application into Activated Sludge Process with High Concentration of MLSS," *Proceedings of the 2nd International Conference on Advances in Water and Effluent Treatment,* pp. 321–330.

Jeannot, F., E. Francoeur, J. Manem, and A. Beaubien, 1992, "Treatment of Industrial Effluents by Anaerobic Membrane Bioreactors," *Proceedings of the Enforced Techniques for Water and Wastewater Treatment Symposium,* pp. 37–47.

Jorand, F., F. Zartarian, F. Thomas, J. C. Block, J. Y. Bottero, G. Villemin, V. Urbain, and J. Manem, 1995, "Chemical and Structural (2D) Linkage between Bacteria within Activated Sludge Flocs," *Water Research,* **27**(7):1639–1647.

Kayawake, E., Y. Narukami, and M. Yamagata, 1991, "Anaerobic Digestion by a Ceramic Membrane Enclosed Reactor," *Journal of Fermentation and Bioengineering,* **71**:122–125.

Kiat, W. Y., K. Yamamoto, and S. Ohgaki, 1992, "Optimal Fiber Spacing in Externally Pressurized Hollow Fiber Module for Solid Liquid Separation," *Water Sciences Technology,* **26**:5–6.

Knoblock, M. D., P. M. Sutton, P. N. Mishra, K. Gupta, and A. Janson, 1994, "Membrane Biological Reactor System for Treatment of Oily Wastewaters," *Water Environment Research,* **66**:133–139.

Krauth, K. H., and K. F. Staab, 1989, "Substitution of the Final Clarifier by Membrane Filtration within the Activated Sludge Process with Increased Pressure; Initial Findings," *Desalination,* **68**:179–189.

Krauth, K. H., and K. F. Staab, 1993, "Pressurized Bioreactor with Membrane Filtration for Wastewater Treatment," *Water Research,* **27**:405–411.

Krauth, K. H., and K. F. Staab, 1994, "Pressurized Biomembrane Reactor for Wastewater Treatment," *Hydrotop,* **94**:555–562.

Lacoste, B., C. Drakidès, and M. Rumeau, 1993, "Study of an Aerobic Concentrated Culture Reactor Coupled to Separation by Crossflow Micro- or Ultra-filtration through Inorganic Membranes. Initial Approach to a Depollution Application," *Revue des Sciences de l'Eau,* **6**:363–380.

Lambert, S., 1983, "L'Ultrafiltration: Application aux Eaux Résiduaires Industrielles et au Recyclage des Eaux d'Immeubles," *l'Eau, l'Industrie, les Nuisances,* **74**:35–38.

Lazarova, V., and J. Manem, 1994, "Advance in Biofilm Aerobic Reactors Ensuring Effective Biofilm Activity Control," *Water Sciences Technology,* **29**(10–11):319–327.

Lee, C. H., 1994, "Membrane Fouling Mechanisms in Membrane-Coupled Bioreactor," in *Korea-Australia Joint Symposium on Sensor and Membrane Tech.,* Seoul, Korea, pp. 229–245.

Leslie, G. L., P. H. Hodgson, A. G. Fane, R. P. Schneider, and K. C. Marshall, 1992, "The Role of Bacteria Surface Characteristics in the Microfiltration of Bacterial Suspensions," in *Proceedings Euromembrane 92,* Paris, **6**:215–220.

Li, A., D. Kothari, and J. J. Corrado, 1985, "Application of Membrane Anaerobic Reactor System for the Treatment of Industrial Wastewaters," *Proceedings of the 39th Purdue Industrial Waste Conference,* May, Lafayette, Ind., pp. 627–636.

Liessens, J., R. Germonpré, S. Beernaert, and W. Verstraete, 1993, "Removing Nitrate with a Methylotrophic Fluidized Bed: Technology and Operating Performance," *J.A.W.W.A.,* **85**:144–154.

Magara, Y., and M. Itoh, 1991, "The Effect of Operational Factors on Solid/Liquid Separation by Ultra-Membrane Filtration in a Biological Denitrification System for Collected Human Excreta Treatment Plants," *Water Sciences Technology,* **23**:1583–1590.

Magara, Y., K. Nishimura, M. Itoh, and M. Tanaka, 1994, "Biological Denitrification System with Membrane Separation for Collective Human Excreta Treatment Plant," *Proceedings of the 7th International Symposium on Anaerobic Digestion,* Cape Town, South Africa.

Manem, J., E. Trouvé, A. Beaubien, A. Huyard, and V. Urbain, 1993, "Membrane Bioreactor for Urban and Industrial Wastewater Treatment: Recent Advances," *66th Annual Conference,* Anaheim, Calif., pp. 51–59.

Manem, J., 1995, personal communication.

McArdle, J. L., M. M. Arozarena, and W. E. Gallagher, 1987, *A Handbook on Treatment of Hazardous Waste Leachate,* EPA 600/8-87/006, U.S. Environmental Protection Agency, Washington, D.C.

Meldrum, A., 1987, "Hollow Fibre Membrane Bioreactors," *Chemical Engineer,* p. 28.

Metcalf & Eddy, 1991, *Wastewater Engineering Treatment Disposal Reuse,* 3d ed., McGraw-Hill, New York.

Millot, N., 1986, "Les Lixiviats de Décharge Controlée, Caractérisation Analytique, Etudes des Filiéres de Traitement," Ph.D. dissertation, INSA Lyon, Lyon, France.

Mirgain, I., C. Schenk, and H. Monteil, 1993, "Atrazine Contamination of Groundwaters in Eastern France in Relation to the Hydrogeological Properties of the Agricultural Land," *Environmental Technology,* **14:**741–750.

Ohkuma, N., T. Shnoda, T. Aoi, Y. Okaniwa, and Y. Magara, 1994, "Performance of Rotary Disk Modules in Collected Human Excreta Treatment Plant," 1994, 17th Biennial Conference of the International Association on Water Quality, Budapest, Hungary, pp. 141–150.

Pillay, V. L., B. Townsend, and C. A. Buckley, 1994, "Improving the Performance of Anaerobic Digestors at Waste Water Treatment Works: The Coupled Cross-Flow Microfiltration," *Digestor Process Proceedings of the 17th International Symposium on Anaerobic Digestion,* Cape Town, South Africa, pp. 577–586.

Rittmann, B. E., 1987, Aerobic Biological Treatment. *Environmental Sciences Technology,* **21**(2):128–136.

Rittmann, B. E., and V. L. Snoeyink, 1984, "Achieving Biologically Stable Drinking Water," *Journal American Water Works Association,* **76**(10):106.

Rogers, P., 1988, "New Directions with Membrane Bioreactors," *Membrane News,* **1**(2):12.

Ross, W. R., J. P. Barnard, J. Le Roux, and H. A. de Villiers, 1990, "Application of Ultrafiltration Membranes for Solids-Liquid Separation in Anaerobic Digestion Systems: The ADUF Process," *Water SA,* **16:**85–91.

Ross, W. R., N. K. H. Strohwald, C. Grobbelaar, and E. P. Jacobs, 1993, "Anaerobic Digestion in Ultrafiltration (ADUF) Applications to the Treatment of Industrial Effluents," *Proceedings of the International Congress on Membranes and Membrane Processes,* Heidelberg, Germany, pp. 6–19.

Ross, W. R., N. K. H. Strohwald, C. J. Grobler, and J. Sanetra, 1994, "Membrane-Assisted Anaerobic Treatment of Industrial Effluents: The South African ADUF Process," *Seventh International Symposium on Anaerobic Digestion,* Cape Town, South Africa, pp. 550–560.

Roullet, R., 1989, "The Treatment of Wastewater Using an Activated Sludge Bioreactor Coupled with an Ultrafiltration Module," *Proceedings of Workshop on Selected Topics on Clean Technology,* Asian Institute of Technology, Bangkok, Thailand, pp. 171–179.

Sato, T., and Y. Ishii, 1991, "Effects of Activated Sludge Properties on Water Flux of Ultrafiltration Membrane Used for Human Excrement Treatment," *Water Sciences Technology,* **23:**1601–1606.

Shimizu, Y., M. Rokudai, S. Tohya, E. Kayaware, T. Yazawa, H. Tanaka, and K. Eguchi, 1989, "Filtration Characteristics of Charged Alumina Membranes for Methanogenic Waste," *Journal of Chemical Engineering of Japan,* **22:**635–641.

Smith, C. V., D. DiGregorio, and R. M. Talcott, 1967, "The Use of Ultrafiltration Membranes for Activated Sludge Separation," *Proceedings of the 24th Annual Purdue Industrial Waste Conference,* pp. 1300–1310.

Spalding, R. F., and M. E. Exner, 1993, "Occurrence of Nitrate in Groundwater, a Review," *Journal Environmental Quality,* **22:**392–402.

Sutton, P. M., and R. R. Evans, 1983, "Anaerobic System Designs for Efficient Treatment of Industrial Wastewater," *Proceedings of the Third Symposium on Anaerobic Digestion,* Boston, Mass.

Sutton, P. M., A. Li, R. R. Evans, and S. R. Korchin, 1983, "Dorr-Oliver's Fixed-Film, Suspended-Growth Anaerobic Systems for Industrial Wastewater Treatment and Energy Recovery," *Proceedings of the 37th Waste Conference,* Purdue, Ind., pp. 667–675.

Thetford Systems, Inc., 1990, Commercial documentation, Ann Arbor, Mich.

Trouvé, E., V. Urbain, V. Mandra, D. Amar, and I. Coulomb, 1993, "Coupling of Membrane and Reverse Osmosis for Landfill Leachate Treatment," *Proceedings of the WEF Annual Conference,* October 3–6, Anaheim, Calif.

Trouvé, E., C. Dupont, A. Huyard, and J. Manem, 1994a, "Cost of Anaerobic Treatment of Wastewaters on Membrane Bioreactors: Optimal Filtration Device," *Proceedings of the 7th International Symposium on Anaerobic Digestion,* Cape Town, South Africa, pp. 567–576.

Trouvé, E., V. Urbain, and J. Manem, 1994b, "Treatment of Municipal Wastewater by a Membrane Bioreactor: Results of a Semi-Industrial Pilot-Scale Study," *Proceedings of the IAWQ Biennal Conference,* Budapest, Hungary.

Turner, A., 1994, in *Water Bulletin,* October 1994, pp. 10–11.

Urbain, V., J. C. Block, and J. Manem, 1993a, "Bioflocculation in Activated Sludge: An Analytical Approach," *Water Research,* **27**(5):829–838.

Urbain, V., A. Attal, J. Manem, P. Courant, and D. Amar, 1993b, "Membrane Bioreactor: An Innovative Process for Leachate Treatment," *Proceedings IV International Landfill Symposium,* Sardenia 93, Italy.

Urbain, V., and J. Manem, 1996, "Membrane BioReactor, a Powerful Process for Nitrate, Natural Organic Matter, and Pesticides Removal for Drinking Water Production," submitted for publication in *Journal American Water Works Association.*

USEPA, 1992, Drinking Water Regulations and Health Advisories, Office of Water, USEPA, Washington, D.C.

Wisniewski, C., and A. Grasmick, 1994, "Mieux gérer l'eau," *Actes du colloque scientifique et technique Hydrotop, Marseille,* pp. 535–542.

Wisniewski, C., and A. Grasmick, 1995, "Elimination d'Ethanol par Cultures Mixtes en Bioréacteur à Membrane," *Actes du colloque sur les Procédes d'épuration des effluents et déchets des Industries agroalimentaires, Gergy-Pontoise,* June 1995.

Yamamoto, K., M. Hiasa, T. Mahmood, and T. Matsuo, 1989, "Direct Solid-Liquid Separation Using Hollow Fiber Membrane in an Activated Sludge Aeration Tank," *Water Sciences Techniques,* **21**:43–54.

CHAPTER 18

ION-EXCHANGE MEMBRANE REACTORS

Susan Wadley and Chris A. Buckley

Pollution Research Group
Department of Chemical Engineering
University of Natal
Durban, South Africa

18.1 INTRODUCTION

Many bulk industrial chemicals are based on sodium hydroxide, ammonia, hydrochloric acid, sulfuric acid, and nitric acid and salts thereof. Waste streams containing high concentrations of dissolved salts are produced in many large-scale chemical processes such as metal pickling, rayon manufacture, fermentation, flue-gas scrubbing, and ion-exchange resin regeneration. Conventional disposal techniques for brines include evaporation (by evaporative crystallization or evaporation ponds) and dilution by disposal into waterways. These methods of disposal are becoming increasingly expensive and environmentally unacceptable and represent the loss of a potentially recoverable resource.

Electrochemical techniques for wastewater treatment are becoming increasingly important, since they often allow regeneration, reuse and recycle of valuable or hazardous chemicals as well as water recycling. Hence, electrochemical processing is considered to be a *clean technology*. The processes can often form part of a *closed-loop recycling* approach in which the potential pollutant is recycled to the production process and, hence, does not enter the aquatic environment.

This chapter gives a description of various types of ion-exchange membrane reactors and their application in wastewater treatment and the generation of water treatment chemicals. Some theory on electrochemical processes and reactor design considerations is included. The theory of ion-exchange membrane transport and selectivity has been covered in Chap. 7. Electrodialysis applications are presented in Chap. 12.

18.2 TYPES OF ION-EXCHANGE MEMBRANE REACTORS

Ion-exchange membrane reactors can be divided into two groups, according to the main driving force for the process:

1. *Diffusion dialysis cells,* where the process is driven purely by concentration gradients across the membranes
2. *Electrolytic cells,* where the process is driven primarily by electrical potential gradients

18.2.1 Diffusion Dialysis Cells

Diffusion dialysis equipment is usually constructed in the parallel plate (plate-and-frame) configuration, consisting of alternating dialysate and diffusate compartments separated by ion-exchange membranes (all of one type), which are installed vertically. The *dialysate* is the feed liquor containing the ions to be separated, while the *diffusate* is initially pure water into which the recovered ions diffuse, forming the recovered solution. The feed liquor (dialysate) flows upward, while the recovered solution stream (diffusate) flows downward (see Fig. 18.1). Hence, the densities of the feed liquor and the recovered solution stream are higher at the bottom of their respective compartments, which prevents backmixing. The countercurrent flow arrangement maximizes the concentration gradients, which leads to a high transfer rate of the diffusing components. The concentration of the recovered component can be almost as high as in the original feed liquor.

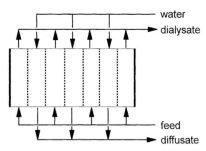

FIGURE 18.1 Diagram of flows involved in a typical diffusion dialysis stack.

18.2.2 Electrolytic Cells

Electrolytic cells are usually constructed in the parallel plate configuration, with the cell consisting of an anode (positively charged electrode) and a cathode (negatively charged electrode), between which are placed one or more ion-exchange membranes, which divide the cell into compartments. There are four main types of electrolytic cells (see Fig. 18.2):

1. Two-compartment electrolysis cells for synthesis, solution regeneration, and contaminant removal (with a single cation- or anion-exchange membrane)
2. Three-compartment electrolysis cells for salt splitting (with cation- and anion-exchange membranes)
3. Multiple-compartment electrodialysis stacks (with several alternating cation- and anion-exchange membranes)
4. Multiple-compartment bipolar membrane electrodialysis stacks for salt splitting (with repeating ion-exchange membrane sequence: anion, bipolar, cation)

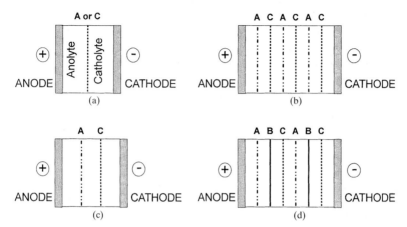

FIGURE 18.2   Main types of electrolytic cells: (*a*) two-compartment electrolysis cell; (*b*) three-compartment electrolysis cell for salt splitting; (*c*) multiple-compartment electrodialysis stack; (*d*) multiple-compartment bipolar membrane electrodialysis stack. (Key for ion-exchange membranes: A = anion, B = bipolar, C = cation.)

For each main type of cell, various membrane arrangements and flow patterns are possible, some of which will be described in Sec. 18.4. Electrodialysis and bipolar membrane electrodialysis have already been dealt with in Chaps. 7 and 12; hence, they will not be discussed in this chapter.

18.3 THEORY OF ELECTROCHEMISTRY

The main features of a typical two-compartment electrochemical cell with cation-selective membrane separation (i.e., type 1) are shown in Fig. 18.3. The various charge transport mechanisms that are present in the cell and external circuit are illustrated in Fig. 18.3. In the *anolyte* (compartment adjacent to anode) and *catholyte* (compartment adjacent to cathode), the charge is carried by the anions and cations which move toward the anode and cathode, respectively. The cation-exchange membrane allows selective passage only to cations; hence, positive charge is transported from the anolyte to the catholyte. The charge is transported by electrons in the external circuit and across the electrode/electrolyte interfaces. These mechanisms ensure that electroneutrality is maintained throughout the circuit. Also, due to the electroneutrality requirement, the electric current through the external circuit I is a measure of the rate of reaction at the surfaces of the electrodes. The total reaction rate can be determined using Faraday's laws of electrolysis, i.e.,

$$q = \int I \, dt = n n_e F \eta_p \tag{18.1}$$

where q = charge passed, C
       I = electric current, A
       t = time for which electrolysis has occurred, s

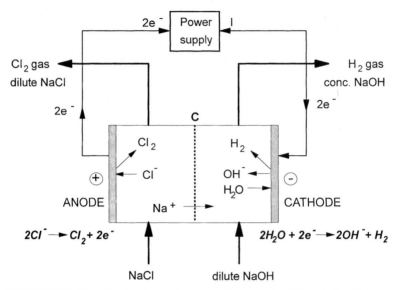

FIGURE 18.3 Example of an ion-exchange membrane reactor (chlor-alkali cell) showing ion transport, electrode reactions, and charge transport. (C = cation-exchange membrane.)

n = number of moles of species which undergo oxidation or reduction, mol
n_e = number of electrons transferred for the reaction (as written)
F = Faraday's constant (96,485 C/mol)
η_p = fractional current efficiency (dimensionless) for the reaction of interest, defined in Eq. (18.18)

18.3.1 Electrode Reactions

Electrode reactions occur via a sequence of fundamental steps (Pletcher, 1992):

1. *Mass transport of reactants* toward the electrode surface by diffusion, convection, and/or migration
2. *Electron transfer* at the electrode surface
3. *Mass transport of products* away from the electrode surface by diffusion, convection, and/or migration

In addition, there are usually other, more complex reactions such as:

4. Adsorption of reactants, intermediates, and/or products on the electrode surface
5. Coupled (competing) chemical reactions in the solution or at the membrane surface
6. Formation of new phases (solid or gaseous)
7. Multiple electron transfer in a sequence of reactions

18.3.1.1 Thermodynamics of Cell Reactions. For the cell shown in Fig. 18.3, the electrode reactions are:

Anode reaction: $\qquad\qquad 2Cl^- \rightarrow Cl_2\uparrow + 2e^-$ (18.2)

Cathode reaction: $\qquad\quad 2H_2O + 2e^- \rightarrow 2OH^- + H_2\uparrow$ (18.3)

Hence, the overall chemical change for the cell is:

$$2Cl^- + 2H_2O \rightarrow 2OH^- + Cl_2\uparrow + H_2\uparrow \qquad (18.4)$$

The cell potential at equilibrium, E^*_{cell} (V), is related to the free-energy change of this reaction, ΔG (J/mol), via the equation,

$$E^*_{cell} = -\frac{\Delta G}{n_e F} \qquad (18.5)$$

The equilibrium cell potential is also given by,

$$E^*_{cell} = E^*_C - E^*_A \qquad (18.6)$$

where the equilibrium potentials for the anode E^*_A and for the cathode E^*_C can be calculated from the Nernst equation,

$$E^* = E^{*o} + \frac{RT}{n_e F} \ln\left(\frac{a_{ox}}{a_{red}}\right) \qquad (18.7)$$

E^{*o} is the standard (formal) equilibrium potential of the redox couple, V, T is the absolute temperature, K, R is the gas constant (8314 J/K mol), and a_{ox} and a_{red}, mol/m^3, are the activities of the oxidized form and the reduced form in the solution, respectively. For dilute solutions, the activities can be approximated by the concentrations c_{ox} and c_{red} (mol/m^3), i.e.

$$E^* = E^{*o} + \frac{RT}{n_e F} \ln\left(\frac{c_{ox}}{c_{red}}\right) \qquad (18.8)$$

Hence, for the example in Eqs. (18.2) to (18.4), Eq. (18.6) becomes (at 25°C)

$$E^*_{cell} = \left[-0.83 + \frac{8314 \times 298}{2 \times 96,485} \ln\left(\frac{1}{[OH^-]^2 \times p_{H_2}}\right)\right]$$
$$- \left[-1.36 + \frac{8314 \times 298}{2 \times 96,485} \ln\left(\frac{p_{Cl_2}}{[Cl^-]^2}\right)\right] \quad (18.9)$$

18.3.1.2 Electron Transfer. To design an electrolytic cell, it is necessary to know the electrical potential difference required to drive the electron transfer reactions at any particular electric current density i (ratio of current passing to effective electrode surface area). The electrode potential determines which electrode reactions are thermodynamically favorable, as well as affecting the kinetics of electron transfer. Hence, a suitable electrode potential must be determined experimentally by obtaining a current density versus potential curve (called the $i = E$ characteristic).

At equilibrium, the potential of an inert electrode is given by the Nernst equation (Eq. [18.7]) and, since no chemical change would be occurring, the measured electric current density i would be zero. However, both oxidation and reduction would be occurring at the electrode surface, but at equal rates. Hence, the current densities of the cathodic i_C and anodic reactions i_A would be related by:

$$i = i_A + i_C = 0 \tag{18.10}$$

i.e., $|i_A| = |i_C| = i_0$, the exchange current density.

When an electrode is at a potential which is different from the equilibrium potential, current will pass until the ratio c_{ox}/c_{red} is such that the Nernst equation is satisfied. If the imposed potential is greater than the equilibrium potential (E^*), oxidation will occur and an anodic current will be observed, while if the imposed potential is lower than the equilibrium potential, reduction will occur and a cathodic current will be observed. The rate of the oxidation or reduction reaction depends on the kinetics of the electron transfer couple and these reactions are considered to be first order; hence,

$$i_A = n_e F k_A c_{red} \quad \text{and} \quad -i_C = n_e F k_C c_{red} \tag{18.11}$$

The rate constants for the oxidation and reduction reactions k_A and k_C depend on the potential field driving the transport of electrons across the electrode/electrolyte interface and are proportional to the potential E applied to the electrode versus a reference electrode. It is generally assumed that these relationships are of the form

$$k_A = k_A^o \exp\left\{\frac{(1 - \alpha)n_e FE}{RT}\right\} \quad \text{and} \quad k_C = k_C^o \exp\left\{\frac{(-\alpha)n_e FE}{RT}\right\} \tag{18.12}$$

where α is the transfer coefficient (typically around 0.5) and k_A^o and k_C^o are the standard equilibrium rate constants (m/s) for the anodic and cathodic reactions, respectively.

The electrode overpotential is defined as

$$E_{ov} = E - E^* \tag{18.13}$$

hence, the measured current density is given by the sum of the partial anodic and cathodic current densities,

$$i = i_0\left[\exp\left\{\frac{(1 - \alpha)n_e FE_{ov}}{RT}\right\} - \exp\left\{\frac{(-\alpha)n_e FE_{ov}}{RT}\right\}\right] \tag{18.14}$$

In practical applications, the potential is usually not close to the equilibrium value; hence, a large overpotential exists and one of the terms in Eq. (18.14) will dominate. For example, if the overpotential is positive, then

$$i = i_A \approx i_0 \exp\left[\frac{(1 - \alpha)n_e FE_{ov}}{RT}\right] \tag{18.15}$$

18.3.1.3 Mass Transport. The rate of an electrode reaction is limited by the rate at which the reactants reach the electrode surface. Hence, consideration of the mass transport conditions within the cell is required to design the cell correctly. Three forms of mass transport are important in electrolytic cells (Pletcher, 1992):

1. *Diffusion:* The movement of species due to a concentration gradient. The electrode reaction leads to the depletion of reactant and accumulation of product at the electrode surface; hence, concentration gradients are set up in the layer of

electrolyte next to the electrode surface. Reactants will diffuse toward the electrode surface and product away from it.

2. *Convection:* The movement of species due to external mechanical forces, such as electrolyte flow or electrode rotation.

3. *Migration:* The movement of charged species due to an electrical potential gradient.

Convection is usually the dominant mode of mass transport, except at the electrode/electrolyte interface where diffusion is dominant.

18.3.1.4 Interaction of Electron Transfer and Mass Transport.

The rate of conversion of reactant to product is determined by the slowest step in the sequence; for example, for a simple reduction reaction the following sequence occurs:

$$O_{\text{solution}} \xrightarrow{\text{mass transport}} O_{\text{surface}} \xrightarrow{\text{electron transfer}} R_{\text{surface}} \xrightarrow{\text{mass transport}} R_{\text{solution}} \qquad (18.16)$$

where O is the oxidized form and R is the reduced form. The rate of electron transfer increases exponentially with overpotential. Four different situations can occur, depending on the potential applied.

1. At the equilibrium potential, the current density is zero and no net chemical change occurs.

2. At sufficiently low overpotentials (positive or negative), the rate-determining step is the electron transfer, so the i-E characteristic can be calculated using Eqs. (18.13) and (18.14). The current density is completely independent of the mass transport conditions, and the difference between the concentration at the electrode surface compared with that of the bulk solution will be negligible.

3. At intermediate overpotentials, the rate of electron transfer and mass transport become similar and mixed control occurs. The overpotential range in which this occurs depends on the kinetic parameters of the redox couple, i_o and α. In this region the concentration of reactant at the electrode surface will be significantly lower than in the bulk solution and the current density will depend on the rate of convection. Hence, there are no simple expressions for the i-E characteristic.

4. At large overpotentials, the rate of electron transfer will become very high, causing mass transport to become the rate-limiting step. The current density is strongly dependent on the mass transport conditions but becomes independent of the potential. The concentration of reactant at the electrode surface decreases to zero. The current density at which this occurs is referred to as the *limiting current density.*

18.3.2 Cell Design

18.3.2.1 Electrodes.

The electrode may serve merely as a source or sink of electrons, or it could also be a catalytic surface or a surface on which deposition of metals occurs. In the first case, the reaction kinetics are independent of the electrode material used, but if the electrode surface is electrocatalytic, then the type and rate of the electrode reaction depends on the specific interactions between the electrode surface and the electroactive species.

In the chlor-alkali cell, the anode must evolve chlorine gas rather than oxygen gas, which is the thermodynamically preferred product. This selectivity is achieved by using an electrode surface material with an exchange current density which is high for chlorine but low for oxygen—for example, a ruthenium oxide coating on titanium.

In the reduction of hydrogen ions to hydrogen gas, the adsorption of the hydrogen ions onto electrode surfaces made from platinum, ruthenium, palladium, or nickel makes the pathways for the evolution of hydrogen gas more favorable than if the metal is, for example, mercury, lead, or cadmium. The extent of adsorption of a particular species depends on competition of this species with other species and with the solution phase and, hence, depends on the applied potential, the electrode material, the solvent, the pH, the temperature, and other dissolved species (Pletcher, 1992). Some typical electrode materials are given in Table 18.1. All these electrode surfaces are electrocatalytic except for steel and stainless steel.

18.3.2.2 Ion-Exchange Membranes.

18.3.2.2 Ion-Exchange Membranes.   Ion-exchange membranes are made from organic polymers containing cation- or anion-exchange groups. Cation-exchange membranes have sulfonate (strong acid membranes) or carboxylate groups (weak acid membranes), while anion-exchange membranes have quaternary ammonium groups. Perfluorinated ion-exchange membranes are often used due to their high chemical stability.

Ion-exchange membranes generally distinguish well between cations and anions; however, they do not usually exhibit good selectivity between ions of the same charge. In addition, they tend to allow passage to solvent (mostly in the form of the solvation shell of the migrating ions) and neutral molecules (by diffusion under the concentration gradient).

The ability of cation-exchange membranes to restrict the passage of OH^- ions and the ability of anion-exchange membranes to restrict the passage of H^+ ions is generally poor, especially when the solution is concentrated. This limits the concentration of acids and bases produced. In the chlor-alkali industry, the cation-exchange membranes used have a special barrier layer on the catholyte side of the membrane to prevent backmigration of OH^- ions, hence, allowing the production of high concentration (35%) sodium hydroxide.

Ion-exchange membranes prevent mixing of the gases produced at each electrode (provided the solubility of the gases in the electrolytes is low).

TABLE 18.1   Some Typical Electrode Materials

Electrode reaction	Medium	Electrode material
H_2 evolution at cathode	Neutral or basic	Steel, stainless steel; high area Ni on steel or Ni
	Acidic or basic	Pt and other precious metals (e.g., Ru, Pd, Ni)
	Strongly basic	Ni alloys (e.g., Ni-Mo) on Ni or steel
Cl_2 evolution at anode	Neutral or acidic (conc. NaCl)	RuO_2-based coatings on Ti (known as DSA,* also contains TiO_2 and other additives), spinels, e.g., PdO_2 and Co_3O_4 (avoids DSA patents)
O_2 evolution at anode	Various	IrO_2 on Ti (known as DSA O_2)
	Acidic	Pt on Ti; PbO_2 on lead alloys, Ti, or C
	Basic	Ni oxides; Ni- and Co-based spinels, e.g., $NiCo_2O_4$ on Ti
	Neutral or basic	Steel, stainless steel

* DSA: Dimensionally stable anode.
Source: Pletcher (1992), Walsh (1993).

If the current density through the cell is higher than can be maintained by the flux of ions to the membrane surface, other processes (such as water splitting) will occur at the membrane surface, which causes the membrane resistance to rise sharply and results in large potentials drops across the cell.

18.3.2.3 Separators and Turbulence Promoters. Plastic mesh separators are often placed in the electrolyte compartments to prevent physical contact between the electodes and the membrane or to provide dimensional support for fragile electrodes and membranes. In addition, they promote turbulence in the electrolyte compartments, which enhances mass transport.

18.3.2.4 Mass Transport. The design of the cell should be such that the mass transport is efficient and uniform throughout. Efficient mass transport conditions are required because the limiting (maximum) current density is proportional to the mass transfer coefficient. Uniform mass transport ensures that the reaction proceeds at nearly the same rate in all parts of the reactor. For the parallel plate design, the following factors affect the mass transport conditions:

1. Flow rate of the various electrolyte streams
2. Design of the inlet and outlet flow distributors on each compartment
3. Design of the turbulence promoters in the compartments
4. Extent of gas bubble evolution

18.3.2.5 Factors Affecting Success of Processes. Some important considerations in the determination of the success of an effluent treatment process which involves ion-exchange membrane reactors are:

1. Achievement of desired levels of ion removal or conversion
2. Value of recovered chemicals
3. Energy consumption
4. Rate of effluent treatment
5. Cost and lifetime of various parts of the system
6. Methods of dealing with by-products
7. Stability of performance and sensitivity to changes in feed composition

18.3.2.6 Cell Performance Indicators. There are several ways of expressing cell performance. Some of the most commonly used terms are defined as follows (Pletcher and Walsh, 1990).

Material Yield. The material yield (or overall operational yield) Y_{mat} is the maximum amount of desired product obtained per mole of reactant, taking the stoichiometry of the reaction into account,

$$Y_{mat} = \frac{\text{moles of reactant converted to product}}{\text{total moles of reactant consumed}} \tag{18.17}$$

Current Efficiency. The current efficiency η_c is the yield based on the electrical charge passed:

$$\eta_c = \frac{\text{charge used in forming product}}{\text{total charge}} \tag{18.18}$$

Cell Voltage. The cell voltage E_{cell} depends on several process parameters such as temperature, composition of electrolytes, flow of electrolytes, and the material, form, and surface condition of the electrodes. The cell voltage is given by:

$$E_{\text{cell}} = E_C^* - E_A^* - |E_{ov,A}| - |E_{ov,C}| - IR_e^{\text{circuit}} - IR_e^{\text{electrodes}} - IR_e^{\text{electrolytes}} - IR_e^{\text{membranes}} \quad (18.19)$$

where I is the current passing through the cell (A) and R is the electrical resistance of a component of the cell (Ω).

The voltage drops across the various components of a two-compartment cell with a membrane are shown in Fig. 18.4. The electrical power consumed by the cell is proportional to the sum of these potential drops; hence, to reduce the cost of operating the cell, ways must be found of reducing the larger potential drops.

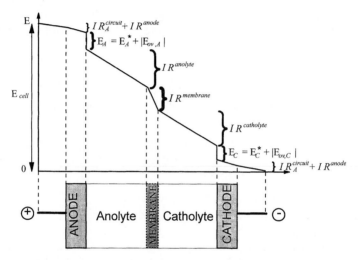

FIGURE 18.4 Schematic diagram showing typical voltage drops across each component of a two-compartment electrolysis cell (operating below the limiting current density).

Energy Yield. The energy yield Y of a cell is given by:

$$Y = \frac{\eta_p(E_A^* - E_C^*)}{E_{\text{cell}}} \quad (18.20)$$

18.4   APPLICATIONS

The applications of ion-exchange membrane reactors are divided into five categories:

1. Diffusion dialysis
2. Electrodialysis (only electrodeionization will be discussed; for other applications see Chap. 12)

3. Salt splitting using two-compartment electrolysis cells
4. Salt splitting using three-compartment electrolysis cells
5. Salt splitting using multiple-compartment bipolar membrane electrodialysis cells (see also Chap. 12)

18.4.1 Diffusion Dialysis

Diffusion dialysis is a suitable technique for treating wastewater at source and can often enable the recovery and recycling of valuable substances.

18.4.1.1 Acid Recovery. Applications of diffusion dialysis to recover acids from industrial effluent include (Jing-wen, 1987):

1. Recovery of sulfuric acid from solutions containing nickel sulfate or copper sulfate from metallurgical works
2. Recovery of sulfuric acid from solutions containing aluminum sulfate from aluminum oxide processing
3. Recovery of chromate from electroplating effluent
4. Recovery of hydrofluoric acid and fluorosilicic acid from titanium and lead processing effluent
5. Recovery of hydrochloric acid from extraction or pickling liquid
6. Recovery of sulfuric acid from iron and steel pickling effluent

Diffusion dialysis involving anion-exchange membranes is used. The waste solutions usually consist of a mixture of acids such as sulfuric, nitric, and hydrofluoric acid and aluminum, nickel, or ferric salts thereof. The acid has a higher diffusivity in the membrane phase than the salt; hence, the acid passes more rapidly from the dialysate compartments into the diffusate compartments.

Kobuchi et al. (1986) have described the application of this process to solutions containing a mixture of nitric acid, hydrofluoric acid, and ferric ions using Neosepta AFN anion-exchange membranes (Tokuyama Soda Co., Ltd.).

Sridhar and Subramaniam (1989) have investigated the application of diffusion dialysis to the recovery of sulfuric acid from cation-exchange regeneration effluent containing about 1.48% free sulfuric acid. They studied the effect of the flow rates of the dialysate and the diffusate on the quality of the recovered acid, using a model TSD-2-20 diffusion dialysis unit (Tokuyama Soda Co., Ltd.) fitted with anion-exchange membranes.

18.4.1.2 Waste Acid Neutralization with Demineralization. Neutralization of waste acid usually involves addition of calcium hydroxide, usually as slaked lime. If the waste acid is sulfuric acid, then precipitation of calcium sulfate occurs. This can be removed by sedimentation; however, it results in an increase in the hardness of the water. The technique of diffusion dialysis (using anion-exchange membranes) could provide a means of simultaneously neutralizing and desalinating the waste sulfuric acid.

In this process, the waste acid stream would pass into the diffusate compartments, and a lime slurry would be fed continuously into the dialysate compartments (see Fig. 18.5). A filter would be included in the recycle loop to remove the precipitated calcium sulfate. The difference in pH across the membrane would cause hydroxide

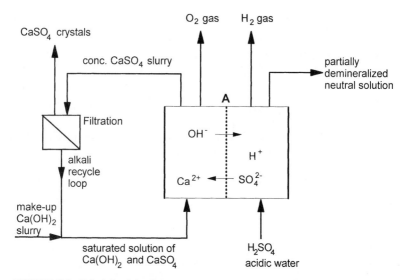

FIGURE 18.5 Principle of simultaneous neutralization and demineralization of waste sulfuric acid by diffusion dialysis. (A = anion-exchange membrane.)

ions from the lime solution to pass into the diffusate, thus neutralizing the acid stream. At the same time, sulfate ions would pass in the opposite direction into the dialysate, thus deionizing the acid stream (Simpson and Buckley, 1987b, 1988).

18.4.2 Electrodeionization

Electrodeionization (EDI) is a continuous process for producing high-purity water. The cell stack is similar to that used in electrodialysis, except that the diluate compartments are wider than usual and are filled with a mixture of cation- and anion-exchange resin (see Fig. 18.6). Equipment based on this technology has been introduced by Millipore Corporation under the trade name Ionpure™ Continuous Deionization (CDI). The equipment can be used for deionizing a wide range of solutions from concentrated brine to reverse-osmosis product water (Egozy and Dick, 1988; Ganzi et al., 1987).

The applied electrical potential difference causes the anions to migrate to the anode and the cations to the cathode. The path of the ions is restricted by the anion- and cation-exchange membranes as in electrodialysis. Hence, the water in the diluate compartments becomes deionized and the concentration of ions in the concentrate compartments increases. The ion-exchange resins in the diluate compartments maintain the conductivity in the diluate compartment at a sufficiently high level to keep the efficiency of the process high. In the lower part of the module, the concentrations of ions in the water are low, so the water is ionized at the areas of high voltage, and the protons and hydroxyl ions that are formed regenerate the cation and anion resins, respectively, causing the resins to be able to deionize the water to a greater extent.

Within reasonable limits, the quality of the product water that is obtained improves with decreasing feedwater salinity and flow rate and improves with increasing tem-

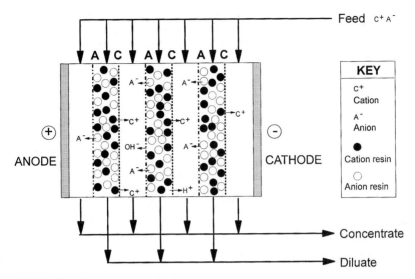

FIGURE 18.6 Principle of the operation of electrodeionization. (A = anion-exchange membrane; C = cation-exchange membrane.)

perature and applied voltage. The feedwater should be maintained at a temperature below 35°C and should contain less than 0.1 mg/L of chlorine. The feedwater should also have a very low level of sparingly soluble solids and organic matter.

18.4.3 Two-Compartment Electrolysis Cells

18.4.3.1 Generation of Chlorine/Hypochlorite. The generation of chlorine (and related products) and sodium hydroxide by the chlor-alkali process using membrane electrolysis is carried out on a large scale. For industries using relatively large quantities of chlorine as well as the other products of the chlor-alkali process (sodium hydroxide and hydrogen gas), on-site generation of chlorine is a practical alternative to the transportation and storage of large quantities of these hazardous chemicals. These systems are obtained in preassembled modularized form and may include additional equipment for processes such as brine pretreatment and sodium hypochlorite production.

The process occurring in the electrolysis cell is shown schematically in Fig. 18.3. The raw material is sodium chloride, which is dissolved in a saturator using water and depleted brine. The saturated brine is treated using clarification, filtration, and ion exchange to remove hardness minerals which may cause membrane fouling. Thereafter, it passes into the anode compartments of the electrochemical cells. The sodium ions pass through the cation-exchange membrane into the cathode compartments and the chloride ions are converted to chlorine gas at the anode. Dilute sodium hydroxide enters the cathode compartments and hydroxide ions and hydrogen gas are formed at the cathode. The sodium hydroxide solution reaches a maximum concentration of about 33% (by mass). The chlorine gas and hydrogen gas is separated from the depleted brine and concentrated sodium hydroxide solutions, respectively.

The systems are fabricated from high-quality steel and titanium. Usually DSA (dimensionally stable anodes) anodes are used. These are made from expanded titanium sheet with a mixed metal oxide coating, which promotes the evolution of chlorine gas.

18.4.3.2 Waste Acid Neutralization with Demineralization.

The technique of waste acid neutralization using diffusion dialysis (described previously) would not involve the generation of potentially explosive gases, such as hydrogen; hence, it would be suitable for underground operation. However, since the rate of diffusion is low, very large membrane surface areas would be required. Where gas-venting facilities are available, the process could be enhanced by passing a low current density $(500 \ A/m^2)$ across each pair of compartments. This would shift the equilibrium position and reduce the specific membrane area required by two orders of magnitude, compared with diffusion dialysis (Simpson and Buckley, 1988).

The application of an electrical potential difference would result in water splitting at both electrodes (the following reactions are balanced for a transfer of four electrons):

1. Water is reduced at the cathode, with the evolution of hydrogen gas

$$4H_2O + 4e^- \rightarrow 4OH^- + 2H_2\uparrow \qquad (18.21)$$

2. Water is oxidized at the anode, with the evolution of oxygen gas

$$2H_2O \rightarrow O_2\uparrow + 4H^+ + 4e^- \qquad (18.22)$$

Electroneutrality would be maintained by the transport of anions through the membrane from the catholyte (sulfuric acid stream) into the anolyte (slaked lime stream), as shown in Fig. 18.7.

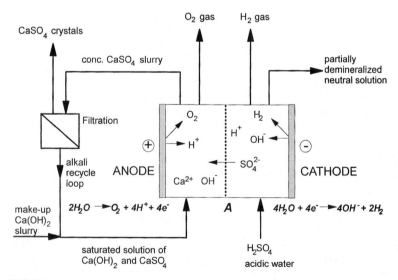

FIGURE 18.7   Principle of simultaneous neutralization and demineralization of waste sulfuric acid by electrolysis. (A = anion-exchange membrane.)

Since the cathodic reaction would produce hydroxide ions, it would aid the neutralization of the sulfuric acid. However, the anodic reaction would produce protons which would neutralize the lime; hence, extra lime would be required. This process would be suitable for applications in which it is not economically viable to recover the sulfuric acid, and the discharge of the sulfuric acid containing effluent is prohibited even when neutralized with lime in the conventional manner.

18.4.3.3   Recovery of Sodium Hydroxide.   This following process has been developed to recover sodium hydroxide from certain waste streams. It involves a pretreatment sequence in which the sodium hydroxide–containing effluent is neutralized with an acidic gas, followed by purification of the neutralized stream by cross-flow microfiltration and nanofiltration. The sodium hydroxide is then recovered using a membrane electrolysis cell. Acidic gas is simultaneously evolved and is recycled to the neutralization stage (see Fig. 18.8).

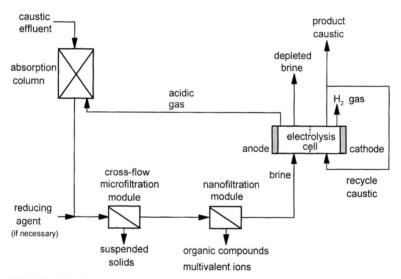

FIGURE 18.8   Schematic diagram of the sodium hydroxide recovery process involving neutralization, membrane filtration, and electrolysis.

The process was initially developed to recover and recycle sodium hydroxide from caustic effluent from the scouring of cotton fiber and has been piloted at several textile industries in South Africa (Pollution Research Group, 1990a, 1990b).

The acidic gas can be either chlorine or carbon dioxide (Simpson and Buckley, 1987a, 1987c). The carbon dioxide system is preferred because the chlorine system is hazardous and can lead to the production of oxidized organic compounds. A brief outline of the essential processes that are involved in the carbon dioxide system follows.

The carbon dioxide gas neutralizes the sodium hydroxide solution, to produce an impure sodium bicarbonate solution, and the decrease in pH causes some of the organic material to flocculate. Cross-flow microfiltration is used to remove suspended solids (see Chap. 11) and the permeate is passed into the nanofiltration stage. Nanofiltration removes further organic material and some polyvalent ions (see Chap. 9). The nanofiltration permeate is introduced into the anode compartment of a two-compartment electrolysis cell containing a cation-exchange ion-exchange membrane (see Fig. 18.9).

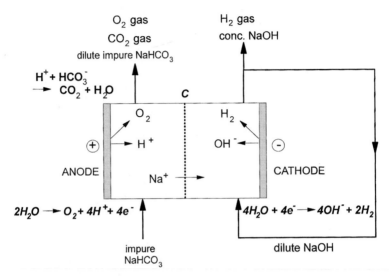

FIGURE 18.9 Principle of the electrolysis stage of the sodium hydroxide recovery process. (C = cation-exchange membrane.)

In the electrolysis cell, the main reactions are water splitting (see Eqs. [18.21] and [18.22]) as well as the following carbonate system equilibria in the anolyte:

$$2H^+ + CO_3^{2-} \leftrightarrow H^+ + HCO_3^- \leftrightarrow H_2CO_3 \leftrightarrow H_2O + CO_2\uparrow \qquad (18.23)$$

The anodic and cathodic reactions result in oxygen gas production at the anode and hydrogen gas at the cathode, while the reaction in the anolyte leads to carbon dioxide gas evolution from the anolyte.

18.4.3.4 Removal of Ammonium Nitrate from an Aqueous Effluent. The following possible application is an example of using electrolysis to convert an impure salt (ammonium nitrate) of low value into a salt of higher value (calcium nitrate) in a purified form.

The waste ammonium nitrate solution is passed into the cathode compartment of a two-compartment electrolysis cell divided by an anion-exchange membrane. A lime slurry is fed into the anode compartment via a recycle loop (see Fig. 18.10). The nitrate ions pass through the anion-exchange membrane into the anolyte and the concentration of calcium nitrate is built up by recycling this stream. The hydrogen ions produced at the anode are neutralized by the lime slurry. The hydroxide ions that are formed at the cathode combine with the ammonium ions to generate ammonium hydroxide. Ammonia gas is highly soluble in water; however, it can be stripped from the catholyte by raising the pH of the solution to about 10, heating and stripping in a conventional stripping tower (Voortman, 1992a).

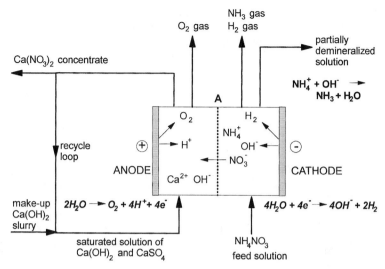

FIGURE 18.10 Principle of an electrolysis process used to remove ammonium nitrate from nitrate-rich wastewater. (A = anion-exchange membrane.)

For pure ammonium nitrate solutions, the current efficiency of this process exceeds 90 percent for the removal of over 96 percent of the nitrate. Transport of the nitrate ions through the membrane was possible when the concentration of nitrate in the anolyte was as high as 170 g/L as nitrate. The operating voltage of the cell varied from 4 to 10 V depending on the electrolyte concentration, temperature, and current density. The current density increased with decreasing nitrate concentration in the catholyte and was on average 1000 A/m^2. The specific power consumption varied from 1600 to 3500 kWh/ton, depending on current density (Voortman et al., 1992b). This application has been tested on a laboratory scale only.

18.4.3.5 Chromic Acid Regeneration with Contaminant Removal. Chromic acid is used in several industrial processes such as electroplating of chromium from chromic acid baths and etching of plastic parts prior to metallizing. During the process Cr(VI) is reduced to Cr(III) and cationic contaminants such as Cu(II) may accumulate in the liquor. A method of regenerating spent chromic acid and simultaneously removing Cu(II) ions is shown in Fig. 18.11.

The spent liquor passes into the anode compartment where anodic oxidation converts Cr^{3+} to Cr$_2$O$_7^{2-}$ with the generation of protons. The Cu^{2+} ions pass through the cation-exchange membrane into the catholyte, where they undergo electrodeposition by reduction at the copper cathode. Alternatively, if no copper is present, a nickel cathode may be used and the cathodic reaction would be the reduction of protons to hydrogen gas. The anode is usually made from a lead oxide–coated lead alloy and has a considerably larger surface area than the cathode to reduce the anodic current density. These factors favor the oxidation of Cr(III) to Cr(VI) over the oxidation of water with oxygen evolution (Pletcher and Walsh, 1990). Since chromic acid is a strong oxidizing agent, the cation-exchange membrane must be made from a highly resistant material; hence, perfluorinated membranes such as Nafion® (Du Pont) are usually used. A cylindrical cell geometry involving tubular membranes is often used.

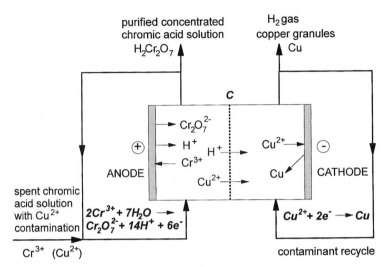

FIGURE 18.11 Principle of the process for regeneration of chromic acid baths with Cu(II) contaminant removal. (C = cation-exchange membrane.)

The efficiency of transport of copper ions from the anolyte into the catholyte is rather low since the solution is strongly acidic, so most of the current is carried by the H^+ ions. This becomes a problem if the quantity of copper contamination is high, for example, in the regeneration of chromic/sulfuric acid solutions used for etching of printed circuits. Complete removal of copper can be achieved by first passing the spent solution through the cathode compartment of the cell. The copper is deposited on the cathode and the remaining chromic acid is reduced. The chromic acid is then regenerated by passing the effluent from the cathode compartment into the anode compartment (Grot, 1982).

18.4.4 Three-Compartment Electrolysis Cells for Salt Splitting

18.4.4.1 Generation of Acids and Bases from Waste Salts. Waste streams containing high concentrations of sodium sulfate can sometimes be converted to sulfuric acid and sodium hydroxide of suitable quality for recycle. This can be carried out using a three-compartment electrolysis cell of the type illustrated in Fig. 18.12. The waste sodium sulfate stream is passed into the central compartment, which is separated from the anode and cathode by an anion-exchange membrane and a cation-exchange membrane, respectively. Sulfuric acid is produced in the anode compartment and sodium hydroxide in the cathode compartment, while the sodium sulfate in the central compartment is depleted by an equivalent amount (Millington, 1983).

Unfortunately, it is not possible to obtain a high concentration of sulfuric acid in the anolyte because it diffuses back into the central compartment. In addition, most waste sodium sulfate streams are acidic; hence, hydrogen ions migrate through the cation exchange membrane into the catholyte, neutralizing the sodium hydroxide. These effects result in a loss of current efficiency as the concentrations of sodium and sulfate in the catholyte and anolyte increase relative to that in the central compartment. Typical current efficiencies are 86 percent and 58 percent for the sodium

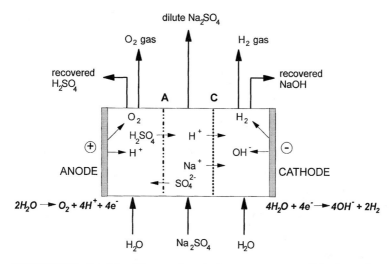

FIGURE 18.12 Principle of the operation of a three-compartment cell for the production of sulfuric acid and sodium hydroxide from sodium sulfate. (A = anion-exchange membrane; C = cation-exchange membrane.)

hydroxide and sulfuric acid streams, respectively, at concentrations of 10% by mass (Walsh, 1993).

The current efficiency can be improved by using a two-compartment cell. If the compartments are separated by an anion-exchange membrane and the sodium sulfate/sulfuric acid mixture is passed into the cathode compartment, then sulfuric acid is recovered in the anode compartment. If the separator is a cation-exchange membrane, then the sodium sulfate/sulfuric acid mixture is passed into the anode compartment and sodium hydroxide is recovered in the cathode compartment, while the anolyte becomes more acidic. Millington (1983) found that when using the two-compartment cells it was possible to recover sulfuric acid and sodium hydroxide at concentrations around 200 g/L and 100 g/L, respectively. Power consumptions of 3.85 kWh/kg (H_2SO_4) and 2.29 kWh/kg (NaOH) were obtained for the cells separated with anion- and cation-exchange membranes, respectively, when operating at a current density of 2000 A/m^2. This process is presently being commercialized by ICI Chemicals & Polymers.

A modification of the three-compartment cell is being investigated by the Electrosynthesis Company and Ormiston Mining Company (Walsh, 1993). The process involves continuous addition of ammonia to the anolyte to maintain the pH at 2; hence, sodium sulfate is converted into ammonium sulfate (used as a fertilizer) and sodium hydroxide. Since the concentration of protons at the anode is reduced, the passage of acid into the central compartment is reduced; hence, the current efficiency is increased and can be as high as 90 percent for both NaOH and $(NH_4)_2SO_4$ production. Product mass concentrations of 32% sodium hydroxide and 40% ammonium sulfate can be obtained directly in the reactor.

18.4.4.2 Recovery of Ion-Exchange Resin Regeneration Chemicals. A method of recovering sulfuric acid and sodium hydroxide from ion-exchange regeneration wastes has been described by Nott (1981). Three-compartment cells (similar to that shown in Fig. 18.12) are used; however, the process is adapted for acidic or basic regeneration waste.

Recovery of Sulfuric Acid. The principle of the process for recovering sulfuric acid from cation-exchange resin regeneration wastes is shown in Fig. 18.13.

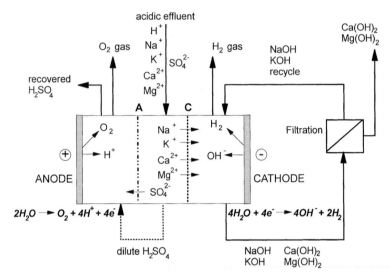

FIGURE 18.13 Principle of a process for sulfuric acid recovery from cation-exchange regeneration effluent using a three-compartment cell. (A = anion-exchange membrane; C = cation-exchange membrane.)

The sulfuric acid/sulfate salt stream enters the central compartment. The sodium, potassium, calcium, and magnesium ions pass through the cation-exchange membrane into the catholyte, and the sulfate ions pass through the anion-exchange membrane into the anolyte. A basic solution (recycled sodium and potassium hydroxide) is used as the catholyte. This leads to the precipitation of calcium and magnesium hydroxide, which can be removed by filtration. The dilute acid from the center compartment could be used as the anolyte. The sulfuric acid concentration in the anolyte increases as the process proceeds.

The extent of removal of cations, especially divalent cations, from the central compartment is important when this stream is used as the anolyte. Scaling of the cation-exchange membrane by divalent metal sulfates and hydroxides may be a problem.

Recovery of Sodium Hydroxide. The principle of the process for recovering sodium hydroxide from anion-exchange resin regeneration wastes is shown in Fig. 18.14.

The sodium hydroxide/sodium salt stream enters the central compartment. The sulfate, chloride, and bicarbonate ions pass through the anion-exchange membrane into the anolyte, and the sodium ions pass through the cation-exchange membrane into the catholyte. The anolyte is the recycled acidic solution. The dilute base from the center compartment could be used as the catholyte. The sodium hydroxide concentration in the anolyte increases as the process proceeds.

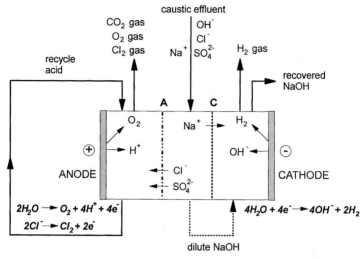

FIGURE 18.14 Principle of a process for sodium hydroxide recovery from anion-exchange regeneration effluent using a three-compartment cell. (A = anion-exchange membrane; C = cation-exchange membrane.)

The process will not remove all the silicate from the central compartment. A high silicate content in the recovered sodium hydroxide promotes silica leakage from the anion-exchange resin columns during the demineralization service cycle. Silica leakage is the most common reason for premature regeneration of anion-exchange resin columns. Organic compounds will not be removed effectively and may cause fouling of the ion-exchange membranes.

These acid and base recovery processes can significantly lower the amounts of regeneration effluents that are discharged. However, the cost of the recovery processes is likely to exceed the savings on regeneration chemicals, although the reduction in disposal costs for the regeneration effluent may compensate for this. The recovered streams will not generally be as pure as the original regenerants. When a lower-quality regenerant is used, the capacity of the ion-exchange column will be lower, and the treated water will be of lower quality. The current efficiency for cationic impurity removal for a batch acid recovery experiment was found to be around 0.15 eq/F for an average current density of 530 A/m². The current efficiency for anionic impurity removal for a batch caustic recovery experiment was found to be around 0.30 eq/F for an average current density of 650 A/m² (Nott, 1981).

18.4.5 Multiple-Compartment Bipolar Membrane Cells

18.4.5.1 Generation of Acids and Bases from Waste Salts. Salt splitting can be carried out in cells containing bipolar membranes as well as cation- and anion-exchange membranes (see Fig. 18.2*d*). The cell stacks contain as many as 200 cells between a pair of electrodes. The bipolar membranes are the sites of simultaneous hydrogen ion and hydroxyl ion production. This eliminates almost all the oxygen and hydrogen gas that would be produced at the anodes and cathodes, respectively. The cell stack (see Fig. 18.15) resembles that used in electrodialysis, except that there

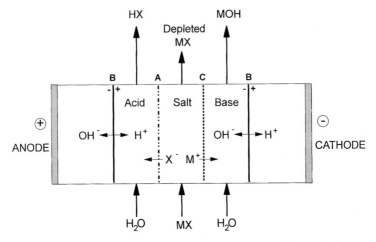

FIGURE 18.15 Principle of operation of a bipolar membrane cell for producing acid and base from a salt. (A = anion-exchange membrane; B = bipolar membrane; C = cation-exchange membrane.)

are three types of compartment (acid, salt, and base) instead of two (concentrate and diluate) (Mani, Chlanda, and Byszewski, 1988).

The salt solution (MX) passes between the cation- and anion-exchange membranes. The electrical potential applied across the cells causes the cations and anions to cross the cation- and anion-exchange membranes, respectively, and combine with the hydroxide and hydrogen ions that are formed at the bipolar membrane, to form a base and an acid.

The bipolar membrane consists of a cation-selective region, an anion-selective region, and an interface between them. The applied direct current causes the water in the membrane to dissociate. The hydrogen ions pass through the cation-selective region toward the cathode and the hydroxyl ions pass through the anion-selective region toward the anode. For efficient operation, the water permeability of the ion-exchange membrane layers must be high and the interface must be very thin and have a low resistance (Mani, 1991).

Some of the applications of bipolar membrane cells which have been commercialized by Aquatech Systems (a division of Allied-Signal Inc.) include:

1. On-site generation from salts of acids and bases for ion-exchange resin regeneration
2. Sodium alkali recycling in the pulping and bleaching industry
3. Recovery of stainless steel pickle liquor
4. Conversion of waste sodium sulfate streams into sulfuric acid and sodium hydroxide
5. Flue-gas desulfurization.

18.4.5.2 Other Cell Arrangements.
In certain applications, cell arrangements other than those shown here are more suitable (see Fig. 18.16) (Mani, 1991):

1. *Two-compartment cation cell:* Suitable for converting salts of weak acids into a mixed acid/salt stream and a relatively pure base stream.

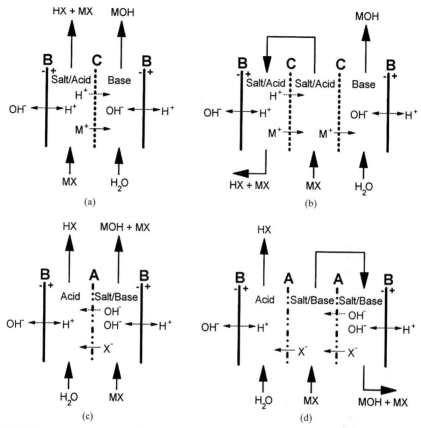

FIGURE 18.16 Other arrangements of bipolar membrane cells for producing acid and base. (A = anion-exchange membrane; B = bipolar membrane; C = cation-exchange membrane.)

2. *Two-compartment anion cell:* Suitable for converting salts of weak bases into a mixed salt/base stream and a relatively pure acid stream.

3. *Multicompartment cation cell:* Provides a mixed salt/acid stream of higher concentration than the two-compartment cation cell.

4. *Multicompartment anion cell:* Provides a mixed salt/base stream of higher concentration than the two-compartment anion cell.

18.5 CONCLUSIONS

Ion-exchange membrane reactors have a wide range of applications in wastewater treatment and the generation of wastewater treatment chemicals. They can be used to simultaneously and selectively concentrate, purify, and electrochemically convert chemical species in aqueous solution. Hence, the main applications include salt splitting (recovery of acid and alkali from a salt solution), contaminant removal, and process stream recovery.

Ion-exchange membrane reactors generally have high capital and operating costs compared with many competing technologies and, hence, should be used only when the purity requirements justify the added expense. The processes are generally only economically attractive when the recovered chemicals have a sufficiently high value or when the removal of certain impurities by the process makes the stream involved suitable for further processing or disposal.

ACKNOWLEDGMENTS

The authors would like to thank Walther G. Grot for reviewing this chapter and for his valuable comments.

SYMBOLS

a	activity of the electroactive species, mol/L^3
c	concentration of the electroactive species, mol/L^3
E	electrode potential, V
E_{cell}	cell potential, V
E_{ov}	overpotential, V
F	Faraday constant, At/mol
ΔG	Gibbs free-energy change, ML^2/t^{2}mol
I	electric current, A
i	electric current density, A/L^2
i_o	exchange current density, A/L^2
k	rate constant for the electrode process, L/t
n	number of moles of electroactive species, mol
n_e	number of electrons transferred for the reaction (as written), dimensionless
p	partial pressure of gas, M/Lt^2
q	electrical charge, At
R	gas constant, ML^2/t^{2}Tmol
R_e	electrical resistance, V/A
T	temperature, T
t	time, t
Y	energy yield, dimensionless
Y_{mat}	material yield, dimensionless

Greek

α	transfer coefficient, dimensionless
η_c	current efficiency, dimensionless

Superscripts

A   anode

C   cathode

o   standard reference state

*   equilibrium value

Subscripts

A   at the anode/anodic

C   at the cathode/cathodic

ox   oxidized species

red   reduced species

REFERENCES

Egozy, Y., and E. M. Dick, 1988, "High Purity Water by Electrodeionisation: Principle and Performance," in M. Streat (ed.), *Ion Exchange for Industry,* Ellis Horwood Ltd., Chichester, U.K. pp. 105–111.

Ganzi, G. C., Y. Egozy, A. J. Giuffrida, and A. D. Jha, April 1987, "High Purity Water by Electrodeionisation, Performance of the Ionpure™ Continuous Deionization System," *Ultrapure Water.*

Grot, W. G., 1982, "Nafion® Membrane and Its Applications," in *Electrochemistry in Industry, New Directions,* Plenum Press, New York.

Jing-wen, X., 1987, "Close-Recycling of Industrial Liquid Waste," *Desalination, 67*:299–326.

Kobuchi, Y., H. Motomura, Y. Noma, and F. Hanada, 1986, "Application of Ion Exchange Membranes to the Recovery of Acids by Diffusion Dialysis," *Journal of Membrane Science, 27*:173–179.

Mani, K. N., F. P. Chlanda, and C. H. Byszewski, 1988, "Aquatech Membrane Technology for Recovery of Acid/Base Values from Salt Streams," *Desalination, 68*:149–166.

Mani, K. N., 1991, "Electrodialysis Water Splitting Technology," *Journal of Membrane Science, 58*:117–138.

Millington, J. P., 1983, "An Electrochemical Unit for the Recovery of Sodium Hydroxide and Sulphuric Acid from Waste Streams," chap. 13 in D. S. Flett (ed.), *Ion Exchange Membranes,* Ellis Horwood Ltd., Chichester, U.K.

Nott, B. R., 1981, "Electrodialysis for Recovering Acid and Caustic from Ion-Exchange Regeneration Wastes," *Ind. Eng. Chem. Prod. Res. Dev., 20*(1):170–177.

Pletcher, D., 1992, "Electrochemical Technology for a Cleaner Environment—Fundamental Considerations," Chap. 2 in D. Genders and N. Weinberg (eds.), *Electrochemistry for a Cleaner Environment,* The Electrosynthesis Company Inc., New York.

Pletcher, D., and F. C. Walsh, 1990, *Industrial Electrochemistry,* 2d ed., Chapman and Hall, London.

Pollution Research Group, 1990a, "Water Management and Effluent Treatment in the Textile Industry: Scouring and Bleaching Effluents," Report to the Water Research Commission for Project no. 122, Pollution Research Group, University of Natal, Durban, South Africa.

Pollution Research Group, 1990b, "A Guide to the Planning, Design and Implementation of Waste-water Treatment Plants in the Textile Industry," part 3, prepared for the Water Research Commission, South Africa, by the Pollution Reseach Group, University of Natal, Durban, South Africa.

Simpson, A. E., and C. A. Buckley, 1987a, "Effluent Treatment," South African Patent No. 87/4406, Water Research Commission, Republic of South Africa.

Simpson, A. E., and C. A. Buckley, 1987b, "The Removal of Sulphuric Acid from an Aqueous Medium Containing the Acid," South African Patent No. 87/5749, Water Research Commission, Republic of South Africa.

Simpson, A. E., and C. A. Buckley, 1987c, "The Treatment of Industrial Effluents Containing Sodium Hydroxide to Enable the Reuse of Chemicals and Water," *Desalination*, **67**:409–429.

Simpson, A. E., and C. A. Buckley, 1988, "The Removal of Sulphuric Acid from Natural and Industrial Waste Waters," *Desalination*, **70**:431–442.

Sridhar, P., and G. Subramaniam, 1989, "Recovery of Acid from Cation Exchange Resin Regeneration Effluents by Diffusion Dialysis," *Journal of Membrane Science*, **45**:273–280.

Voortman, W. J., 1992a, "An Evaluation of the Technical Feasibility of Removing Ammonium Nitrate from Aqueous Effluents with Electrolysis," M.Sc.Eng. thesis, University of Natal, Durban, South Africa.

Voortman, W. J., A. E. Simpson, C. A. Kerr, and C. A. Buckley, 1992b, "Application of Electrochemical Membrane Processes to the Treatment of Aqueous Effluent Streams," *Water Science and Technology*, **25**(10):329–337.

Walsh, F. C., 1993, A First Course in Electrochemical Engineering, The Electrochemical Consultancy, Romsey, Hants, U.K.

Index

Italic numbers denote pages with relevant figures or tables. (Additional figures or tables may be included in the inclusive page numbers.)

ABOUT THE AUTHORS

The international associations who participated in the preparation of this book are the American Water Works Association, the American Water Works Association Research Foundation, Lyonnaise des Eaux, and the Water Research Commission of South Africa.